Lecture Notes in Computer Science 12240

Xingming Sun · Jinwei Wang ·
Elisa Bertino (Eds.)

Artificial Intelligence and Security

6th International Conference, ICAIS 2020
Hohhot, China, July 17–20, 2020
Proceedings, Part II

Editors
Xingming Sun
Nanjing University of Information Science
Nanjing, China

Jinwei Wang
Nanjing University of Information Science
Nanjing, China

Elisa Bertino
Purdue University
West Lafayette, IN, USA

ISSN 0302-9743 ISSN 1611-3349 (electronic)
Lecture Notes in Computer Science
ISBN 978-3-030-57880-0 ISBN 978-3-030-57881-7 (eBook)
https://doi.org/10.1007/978-3-030-57881-7

LNCS Sublibrary: SL3 – Information Systems and Applications, incl. Internet/Web, and HCI

This Springer imprint is published by the registered company Springer Nature Switzerland AG
The registered company address is: Gewerbestrasse 11, 6330 Cham, Switzerland

Preface

The 6th International Conference on Artificial Intelligence and Security (ICAIS 2020), formerly called the International Conference on Cloud Computing and Security (ICCCS), was held during July 17–20, 2020, in Hohhot, China. Over the past five years, ICAIS has become a leading conference for researchers and engineers to share their latest results of research, development, and applications in the fields of artificial intelligence and information security.

We used the Microsoft Conference Management Toolkits (CMT) system to manage the submission and review processes of ICAIS 2020. We received 1,064 submissions from 20 countries and regions, including Canada, Italy, Ireland, Japan, Russia, France, Australia, South Korea, South Africa, Iraq, Kazakhstan, Indonesia, Vietnam, Ghana, China, Taiwan, Macao, the USA, and the UK. The submissions cover the areas of artificial intelligence, big data, cloud computing and security, information hiding, IoT security, multimedia forensics, encryption, cybersecurity, and so on. We thank our Technical Program Committee (TPC) members and external reviewers for their efforts in reviewing papers and providing valuable comments to the authors. From the total of 1,064 submissions, and based on at least three reviews per submission, the program chairs decided to accept 142 papers, yielding an acceptance rate of 13%. The volume of the conference proceedings contains all the regular, poster, and workshop papers.

The conference program was enriched by a series of keynote presentations, and the keynote speakers included: Xiang-Yang Li, University of Science and Technology of China, China; Hai Jin, Huazhong University of Science and Technology (HUST), China; and Jie Tang, Tsinghua University, China. We thank them for their wonderful speeches.

There were 56 workshops organized in ICAIS 2020 which covered all the hot topics in artificial intelligence and security. We would like to take this moment to express our sincere appreciation for the contribution of all the workshop chairs and their participants. We would like to extend our sincere thanks to all authors who submitted papers to ICAIS 2020 and to all TPC members. It was a truly great experience to work with such talented and hard-working researchers. We also appreciate the external reviewers for assisting the TPC members in their particular areas of expertise. Moreover, we want to thank our sponsors: Nanjing University of Information Science and Technology, New York University, ACM China, Michigan State University, University of Central Arkansas, Université Bretagne Sud, National Natural Science Foundation of China, Tech Science Press, Nanjing Normal University, Inner Mongolia University, and Northeastern State University.

May 2020

Xingming Sun
Jinwei Wang
Elisa Bertino

Organization

General Chairs

Yun Q. Shi	New Jersey Institute of Technology, USA
Mauro Barni	University of Siena, Italy
Elisa Bertino	Purdue University, USA
Guanglai Gao	Inner Mongolia University, China
Xingming Sun	Nanjing University of Information Science and Technology, China

Technical Program Chairs

Aniello Castiglione	University of Salerno, Italy
Yunbiao Guo	China Information Technology Security Evaluation Center, China
Suzanne K. McIntosh	New York University, USA
Jinwei Wang	Nanjing University of Information Science and Technology, China
Q. M. Jonathan Wu	University of Windsor, Canada

Publication Chair

Zhaoqing Pan	Nanjing University of Information Science and Technology, China

Workshop Chair

Baowei Wang	Nanjing University of Information Science and Technology, China

Organization Chairs

Zhangjie Fu	Nanjing University of Information Science and Technology, China
Xiaorui Zhang	Nanjing University of Information Science and Technology, China
Wuyungerile Li	Inner Mongolia University, China

Technical Program Committee Members

Saeed Arif	University of Algeria, Algeria
Anthony Ayodele	University of Maryland, USA

Zhifeng Bao	Royal Melbourne Institute of Technology University, Australia
Zhiping Cai	National University of Defense Technology, China
Ning Cao	Qingdao Binhai University, China
Paolina Centonze	Iona College, USA
Chin-chen Chang	Feng Chia University, Taiwan, China
Han-Chieh Chao	Taiwan Dong Hwa University, Taiwan, China
Bing Chen	Nanjing University of Aeronautics and Astronautics, China
Hanhua Chen	Huazhong University of Science and Technology, China
Xiaofeng Chen	Xidian University, China
Jieren Cheng	Hainan University, China
Lianhua Chi	IBM Research Center, Australia
Kim-Kwang Raymond Choo	The University of Texas at San Antonio, USA
Ilyong Chung	Chosun University, South Korea
Robert H. Deng	Singapore Management University, Singapore
Jintai Ding	University of Cincinnati, USA
Xinwen Fu	University of Central Florida, USA
Zhangjie Fu	Nanjing University of Information Science and Technology, China
Moncef Gabbouj	Tampere University of Technology, Finland
Ruili Geng	Spectral MD, USA
Song Guo	Hong Kong Polytechnic University, Hong Kong, China
Mohammad Mehedi Hassan	King Saud University, Saudi Arabia
Russell Higgs	University College Dublin, Ireland
Dinh Thai Hoang	University Technology Sydney, Australia
Wien Hong	Sun Yat-sen University, China
Chih-Hsien Hsia	National Ilan University, Taiwan, China
Robert Hsu	Chung Hua University, Taiwan, China
Xinyi Huang	Fujian Normal University, China
Yongfeng Huang	Tsinghua University, China
Zhiqiu Huang	Nanjing University of Aeronautics and Astronautics, China
Patrick C. K. Hung	Ontario Tech University, Canada
Farookh Hussain	University of Technology Sydney, Australia
Genlin Ji	Nanjing Normal University, China
Hai Jin	Huazhong University of Science and Technology, China
Sam Tak Wu Kwong	City University of Hong Kong, Hong Kong, China
Chin-Feng Lai	National Cheng Kung University, Taiwan, China
Loukas Lazos	University of Arizona, USA
Sungyoung Lee	Kyung Hee University, South Korea
Chengcheng Li	University of Cincinnati, USA
Feifei Li	Utah State University, USA

Arun Kumar Sangaiah	VIT University, India
Di Shang	Long Island University, USA
Victor S. Sheng	University of Central Arkansas, USA
Zheng-guo Sheng	University of Sussex, UK
Robert Simon Sherratt	University of Reading, UK
Yun Q. Shi	New Jersey Institute of Technology, USA
Frank Y. Shih	New Jersey Institute of Technology, USA
Biao Song	King Saud University, Saudi Arabia
Guang Sun	Hunan University of Finance and Economics, China
Jianguo Sun	Harbin University of Engineering, China
Krzysztof Szczypiorski	Warsaw University of Technology, Poland
Tsuyoshi Takagi	Kyushu University, Japan
Shanyu Tang	University of West London, UK
Jing Tian	National University of Singapore, Singapore
Yoshito Tobe	Aoyang University, Japan
Cezhong Tong	Washington University in St. Louis, USA
Pengjun Wan	Illinois Institute of Technology, USA
Cai-Zhuang Wang	Ames Laboratory, USA
Ding Wang	Peking University, China
Guiling Wang	New Jersey Institute of Technology, USA
Honggang Wang	University of Massachusetts-Dartmouth, USA
Jian Wang	Nanjing University of Aeronautics and Astronautics, China
Jie Wang	University of Massachusetts Lowell, USA
Jin Wang	Changsha University of Science and Technology, China
Liangmin Wang	Jiangsu University, China
Ruili Wang	Massey University, New Zealand
Xiaojun Wang	Dublin City University, Ireland
Xiaokang Wang	St. Francis Xavier University, Canada
Zhaoxia Wang	A*STAR, Singapore
Sheng Wen	Swinburne University of Technology, Australia
Jian Weng	Jinan University, China
Edward Wong	New York University, USA
Eric Wong	The University of Texas at Dallas, USA
Shaoen Wu	Ball State University, USA
Shuangkui Xia	Beijing Institute of Electronics Technology and Application, China
Lingyun Xiang	Changsha University of Science and Technology, China
Yang Xiang	Deakin University, Australia
Yang Xiao	The University of Alabama, USA
Haoran Xie	The Education University of Hong Kong, Hong Kong, China
Naixue Xiong	Northeastern State University, USA
Wei Qi Yan	Auckland University of Technology, New Zealand

Aimin Yang	Guangdong University of Foreign Studies, China
Ching-Nung Yang	Taiwan Dong Hwa University, Taiwan, China
Chunfang Yang	Zhengzhou Science and Technology Institute, China
Fan Yang	University of Maryland, USA
Guomin Yang	University of Wollongong, Australia
Qing Yang	University of North Texas, USA
Yimin Yang	Lakehead University, Canada
Ming Yin	Purdue University, USA
Shaodi You	The Australian National University, Australia
Kun-Ming Yu	Chung Hua University, Taiwan, China
Weiming Zhang	University of Science and Technology of China, China
Xinpeng Zhang	Fudan University, China
Yan Zhang	Simula Research Laboratory, Norway
Yanchun Zhang	Victoria University, Australia
Yao Zhao	Beijing Jiaotong University, China

Organization Committee Members

Xianyi Chen	Nanjing University of Information Science and Technology, China
Yadang Chen	Nanjing University of Information Science and Technology, China
Beijing Chen	Nanjing University of Information Science and Technology, China
Baoqi Huang	Inner Mongolia University, China
Bing Jia	Inner Mongolia University, China
Jielin Jiang	Nanjing University of Information Science and Technology, China
Zilong Jin	Nanjing University of Information Science and Technology, China
Yan Kong	Nanjing University of Information Science and Technology, China
Yiwei Li	Columbia University, USA
Yuling Liu	Hunan University, China
Zhiguo Qu	Nanjing University of Information Science and Technology, China
Huiyu Sun	New York University, USA
Le Sun	Nanjing University of Information Science and Technology, China
Jian Su	Nanjing University of Information Science and Technology, China
Qing Tian	Nanjing University of Information Science and Technology, China
Yuan Tian	King Saud University, Saudi Arabia
Qi Wang	Nanjing University of Information Science and Technology, China

Lingyun Xiang	Changsha University of Science and Technology, China
Zhihua Xia	Nanjing University of Information Science and Technology, China
Lizhi Xiong	Nanjing University of Information Science and Technology, China
Leiming Yan	Nanjing University of Information Science and Technology, China
Li Yu	Nanjing University of Information Science and Technology, China
Zhili Zhou	Nanjing University of Information Science and Technology, China

Contents – Part II

Internet of Things

Green Crop Image Segmentation Based on Superpixel Blocks and Decision Tree 3
Bo Wang and Zhibin Zhang

User Behavior Credibility Evaluation Model Based on Intuitionistic Fuzzy Analysis Hierarchy Process 18
Chen Yang, Guogen Wan, Peilin He, Yuanyuan Huang, and Shibin Zhang

Research on Chain of Evidence Based on Knowledge Graph 30
Yizhuo Liu, Jin Shi, Jin Han, and Mingxin Lu

Research on the Construction of Intelligent Catalog System for New Media Information Resources 42
Yiting Li, Jin Shi, Jin Han, Mingxin Lu, and Yan Zhang

Information Security

Construction of a Class of Four-Weight Linear Codes 55
Ee Duan, Xiaoni Du, Tianxin Wang, and Jiawei Du

Designs from the Narrow-Sense Primitive BCH Codes $\mathcal{C}_{(q,q^m-1,\delta_3,1)}$ 65
Fujun Zhang, Xiaoni Du, and Jinxia Hu

A Survey on Side-Channel Attacks of Strong PUF 74
Yan Li, Jianjing Shen, Wei Liu, and Wei Zou

A Novel Visual Cryptography Scheme Shared with Edge Information Embedded QR Code 86
Fei Hu, Yuanzhi Yao, Weihai Li, and Nenghai Yu

Hiding Traces of Camera Anonymization by Poisson Blending 98
Hui Zeng, Anjie Peng, and Xiangui Kang

A Deep Learning Approach to Detection of Warping Forgery in Images 109
Tongfeng Yang, Jian Wu, Guorui Feng, Xu Chang, and Lihua Liu

The Course Design of Applied Talent Statistics Based on Fuzzy Control and Grey Measure Model 119
Li Wu and Jun Yang

Reinforcement Learning-Based Resource Allocation in Edge Computing 131
Qianyu Xie, Xutao Yang, Laixin Chi, Xuejie Zhang, and Jixian Zhang

Fuzzy Multi-objective Requirements for NRP Based on Particle Swarm Optimization 143
Yachuan Zhang, Hao Li, Rongjing Bu, Chenming Song, Tao Li, Yan Kang, and Tie Chen

A Clustering Algorithm for Wireless Sensor Networks Using Geographic Distribution Information and Genetic Algorithms 156
Yu Song, Zhigui Liu, and He Xiao

Dual Residual Global Context Attention Network for Super-Resolution 166
Jingjun Zhou, Jingbing Li, Hui Li, Jing Liu, Qianning Dai, Saqib Ali Nawaz, and Jian Shen

A Robust Zero-Watermarking Algorithm for Medical Images Using Curvelet-Dct and RSA Pseudo-random Sequences 179
Fengming Qin, Jingbing Li, Hui Li, Jing Liu, Saqib Ali Nawaz, and Yanlin Liu

Encryption Algorithm for TCP Session Hijacking 191
Minghan Chen, Fangyan Dai, Bingjie Yan, and Jieren Cheng

A Covert Information Transmission Scheme Based on High Frequency Acoustic Wave Channel 203
Huiling Li, Yayi Zou, Wenjia Yi, Ziyi Ye, and Yi Ma

Privacy Preserving Mining System of Association Rules in OpenStack-Based Cloud 215
Zhijun Zhang, Zeng Shou, Zhiyan Ning, Dan Wang, Yingjian Gao, Kai Lu, and Qi Zhang

Unsupervised Data Transmission Scheduling in Cloud Computing Environment 225
Gui Liu, Wei Zhou, Yongyong Dai, Haijiang Xu, and Li Wang

DUMPLING: Cross-Domain Data Security Sharing Based on Blockchain 236
Mingda Liu and Yijuan Shi

A Semi-quantum Group Signature Scheme Based on Bell States 246
Jinqiao Dai, Shibin Zhang, Yan Chang, Xueyang Li, and Tao Zheng

An Improved Quantum Identity Authentication Protocol for Multi-party Secure Communication 258
Peilin He, Yuanyuan Huang, Jialing Dai, and Shibin Zhang

A Mixed Mutual Authentication Scheme Supporting Fault-Detection in Industrial Internet of Things 267
Gongxin Shen

Multi-party Semi-quantum Secret Sharing Scheme Based on Bell States 280
Xue-Yang Li, Yan Chang, and Shi-Bin Zhang

A Quantum Proxy Arbitrated Signature Scheme Based on Two Three-Qubit GHZ States 289
Tao Zheng, Shi-Bin Zhang, Yan Chang, and Lili Yan

Quantum Key Distribution Protocol Based on GHZ Like State and Bell State 298
Ji-Zhong Wu and Lili Yan

Design and Implementation of Heterogeneous Identity Alliance Risk Assessment System 307
Jianchao Gan, Zhiwei Sheng, Shibin Zhang, and Yang Zhao

A Blockchain Based Distributed Storage System for Knowledge Graph Security 318
Yichuan Wang, Xinyue Yin, He Zhu, and Xinhong Hei

A Fragile Watermarking Algorithm Based on Audio Content and Its Moving Average 328
Xizi Peng, Jinquan Zhang, and Shibin Zhang

An AHP/DEA Methodology for the Public Safety Evaluation 341
Li Mao, Naqin Zhou, Tong Zhang, Wei Du, Han Peng, and Lina Zhu

Detection and Information Extraction of Similar Basic Blocks Used for Directed Greybox Fuzzing 353
Chunlai Du, Shenghui Liu, Yanhui Guo, Lei Si, and Tong Jin

Summary of Research on Information Security Protection of Smart Grid 365
Li Xu and Yanbin Sun

Framework Design of Environment Monitoring System Based on Machine Learning 380
Lingxiao Meng, Shudong Li, Xiaobo Wu, and Weihong Han

Compression Detection of Audio Waveforms Based on Stacked Autoencoders 393
Da Luo, Wenqing Cheng, Huaqiang Yuan, Weiqi Luo, and Zhenghui Liu

A Deep Reinforcement Learning Framework for Vehicle Detection and Pose Estimation in 3D Point Clouds 405
Weipeng Wang, Huan Luo, Quan Zheng, Cheng Wang, and Wenzhong Guo

A Network Security Situation Prediction Algorithm Based on BP Neural Network Optimized by SOA. 417
Ran Zhang, Min Liu, Qikun Zhang, and Zengyu Cai

Research on User Preference Film Recommendation Based on Attention Mechanism. 428
Lei Zhu, Yufeng Liu, Wei Zhang, and Kehua Yang

Traffic Anomaly Detection for Data Communication Networks. 440
Xiaoxiao Tang, Wencui Li, Jing Shen, Feng Qi, and Shaoyong Guo

Video Action Recognition Based on Hybrid Convolutional Network 451
Yanyan Song, Li Tan, Lina Zhou, Xinyue Lv, and Zihao Ma

Identification of Botany Terminology Based on Bert-BLSTM-CRF 463
Aziguli Wulamu, Ning Chen, Lijia Yang, Li Wang, and Jiaxing Shi

Correlation Analysis of Chinese Pork Concept Stocks Based on Big Data . . . 475
Yujiao Liu, Lin He, Duohui Li, Xiaozhao Luo, Guo Peng, Xiaoping Fan, and Guang Sun

Visual SLAM Location Methods Based on Complex Scenes: A Review 487
Hanxiao Zhang and Jiansheng Peng

Improvement of Co-training Based Recommender System with Machine Learning 499
Wenpan Tan, Yong He, and Bing Zhu

Privacy Security Classification (PSC) Model for the Attributes of Social Network Users 510
Yao Xiao and Junguo Liao

Clustering Analysis of Extreme Temperature Based on K-means Algorithm . . . 523
Wu Zeng, YingXiang Jiang, ZhanXiong Huo, and Kun Hu

Network Topology Boundary Routing IP Identification for IP Geolocation. . . 534
Fuxiang Yuan, Fenlin Liu, Rui Xu, Yan Liu, and Xiangyang Luo

Machine Vision Based Novel Scheme for Largely, Reducing Printing Errors in Medical Package 545
Bin Ma, Qi Li, Xiaoyu Wang, Chunpeng Wang, and Yunqing Shi

A Mutual Trust Method for Energy Internet Agent Nodes in Untrusted Environment 557
Wei She, Jiansen Chen, Xianfeng He, Xiaoyu Yang, Xuhong Lu, Zhihao Gu, Wei Liu, and Zhao Tian

Multilayer Perceptron Based on Joint Training for Predicting Popularity 570
Wei She, Li Xu, Huibo Xu, Xiaoqing Zhang, Yue Hu, and Zhao Tian

A Medical Blockchain Privacy Protection Model Based on Mimicry Defense 581
Wei Liu, Yufei Peng, Zhao Tian, Yang Li, and Wei She

Securing Data Communication of Internet of Things in 5G Using Network Steganography 593
Yixiang Fang, Kai Tu, Kai Wu, Yi Peng, Junxiang Wang, and Changlong Lu

The Use of the Multi-objective Ant Colony Optimization Algorithm in Land Consolidation Project Site Selection. 604
Hua Wang, Weiwei Li, Jiqiang Niu, and Dianfeng Liu

Image Content Location Privacy Preserving in Social Network Travel Image Sharing 617
Wang Xiang, Canji Yang, Lingling Jiao, and Qingqi Pei

A Robust Blind Watermarking Scheme for Color Images Using Quaternion Fourier Transform 629
Renjie Liang, Peijia Zheng, Yanmei Fang, and Tingting Song

A Decentralized Multi-authority ABE Scheme in Cooperative Medical Care System. 642
Jinrun Guo, Xiehua Li, and Jie Jiang

Big Data and Cloud Computing

Point-to-Point Offline Authentication Consensus Algorithm in the Internet of Things. 655
Xiao-Feng Du, Yue-Ming Lu, and Dao-Qi Han

Classification of Tourism English Talents Based on Relevant Features Mining and Information Fusion. 664
Qin Miao, Li Wu, and Jun Yang

Research on Application of Big Data Combined with Probability Statistics in Training Applied Talents 674
Li Wu and Jun Yang

Region Proposal for Line Insulator Based on the Improved Selective Search Algorithm 686
Shuqiang Guo, Baohai Yue, Qianlong Bai, Huanqiang Lin, and Xinxin Zhou

A Collaborative Filtering Algorithm Based on the User Characteristics and Time Windows 697
Dun Li, Cui Wang, Lun Li, and Zhiyun Zheng

MS-SAE: A General Model of Sentiment Analysis Based on Multimode Semantic Extraction and Sentiment Attention Enhancement Mechanism 710
Kai Yang, Zhaowei Qu, Xiaoru Wang, Fu Li, Yueli Li, and Dongbai Jia

Information Processing

Research on Psychological Counseling and Personality Analysis Algorithm Based on Speech Emotion 725
Zhaojin Hong, Chenyang Wei, Yuan Zhuang, Ying Wang, Yiting Wang, and Li Zhao

Underwater Image Enhancement Based on Color Balance and Edge Sharpening 738
Yan Zhou, Yibin Tang, Guanying Huo, and Dabing Yu

Stereo Matching Using Discriminative Feature-Oriented and Gradient-Constrained Dictionary Learning 748
Jiale Zhang, Yan Zhou, Qingwu Li, Huixing Sheng, Dabing Yu, and Xinyue Chang

Acoustic Emission Recognition Based on Spectrogram and Acoustic Features 760
Wei Wang, Weidong Liu, and Jinming Liu

Polyhedron Target Structure Information Extraction Method in Single Pixel Imaging System 769
Jin Zhang, Mengqiong Ge, Xiaoyu Shi, Zhuohao Weng, and Jian Zhang

Design and Implementation of Four-Meter Reading Sharing System Based on Blockchain 776
Baoyu Xiang, Zhuo Yu, Ke Xie, Shaoyong Guo, Meiling Dai, and Sujie Shao

Edge-Feedback ICN Cooperative Caching Strategy Based on Relative Popularity 786
Huansong Li, Zhuo Yu, Ke Xie, Xuesong Qiu, and Shaoyong Guo

Reliability Improvement Algorithm of Power Communication Network Based on Network Fault Characteristics 798
Ruide Li, Feng Wang, XinXin Zhang, Jiajun Chen, and Jie Tong

Link Prediction Based on Modified Preferential Attachment for Weighted and Temporal Networks 805
Xuehan Zhang, Xiaojuan Wang, and Lianping Zhang

Author Index 815

Contents – Part I

Artificial Intelligence

Method of Multi-feature Fusion Based on Attention Mechanism in Malicious Software Detection . . . 3
Yabo Wang and Shuning Xu

Computing Sentence Embedding by Merging Syntactic Parsing Tree and Word Embedding . . . 14
Yong Wang , Maosheng Zhong, Lan Tao, and Shuixiu Wu

An Overview of Cross-Language Information Retrieval . . . 26
Liang Zhang and Xiaobing Zhao

A Formal Method for Safety Time Series Simulation of Aerospace . . . 38
Ti Zhou, Jinying Wang, and Yi Miao

Tuple Measure Model Based on CFI-Apriori Algorithm . . . 49
Qing-Qing Wu, Xing-Shuo An, and Yan-yan Zhang

Deep Learning Video Action Recognition Method Based on Key Frame Algorithm . . . 62
Li Tan, Yanyan Song, Zihao Ma, Xinyue Lv, and Xu Dong

Optimization and Deployment of Vehicle Trajectory Prediction Scheme Based on Real-Time ANPR Traffic Big Data . . . 74
Zhe Long and Zuping Zhang

Ear Recognition Based on Gabor-SIFT . . . 86
Ying Tian, Huiwen Dong, and Libing Wang

Differentiated Services Oriented Auction Mechanism Design for NVM Based Edge Caching . . . 95
Zhenyuan Zhang, Fang Liu, Zhenhua Cai, Yihong Su, and Weijun Li

Conjunctive Keywords Searchable Encryption Scheme Against Inside Keywords Guessing Attack from Lattice . . . 107
Xiaoling Yu, Chungen Xu, and Bennian Dou

Detecting Bluetooth Attacks Against Smartphones by Device Status Recognition . . . 120
Fan Wei

Covered Face Recognition Based on Deep Convolution Generative Adversarial Networks 133
Yanru Xiao, Mingming Lu, and Zhangjie Fu

An Efficient Method for Generating Matrices of Quantum Logic Circuits. . . . 142
Zhiqiang Li, Jiajia Hu, Xi Wu, Juan Dai, Wei Zhang, and Donghan Yang

A Mobile Device Multitasking Model with Multiple Mobile Edge Computing Servers 151
Tao Deng, Yunkai Zhao, Ximing Zhang, Bin Xu, Jin Qi, and Yuntao Ma

A Localization Algorithm Based on Emulating Large Bandwidth with Passive RFID Tags 162
Die Jiang, Liangbo Xie, and Qing Jiang

An Efficient Approach for Selecting a Secure Relay Node in Fog Computing Social Ties Network 173
Mingxi Yin, Jinliang Yu, Shanshan Tu, Muhammad Waqas, Sadaqat Ur Rehman, Ghulam Abbas, and Ziaul Haq Abbas

A Research for 2-m Temperature Prediction Based on Machine Learning. . . . 184
Kaiyin Mao, Chen Xue, Changming Zhao, and Jia He

A New Scheme for Essential Proteins Identification in Dynamic Weighted Protein-Protein Interaction Networks 195
Wei Liu, Liangyu Ma, and Yuliang Tang

On Enterprises' Total Budget Management Based on Big Data Analysis 207
Hangjun Zhou, Jie Li, Jing Wang, Jiayi Ren, Peng Guo, and Jianjun Zhang

IoT Device Recognition Framework Based on Network Protocol Keyword Query 219
Zi-Xiao Xu, Qin-yun Dai, Gang Xu, He Huang, Xiu-Bo Chen, and Yi-Xian Yang

Application of ARIMA Model in Financial Time Series in Stocks. 232
Jiajia Cheng, Huiyun Deng, Guang Sun, Peng Guo, and Jianjun Zhang

A Topological Control Algorithm Based on Energy Consumption Analysis and Locomotion of Nodes in Ad Hoc Network. 244
Chaochao Li, Jiangxia Tan, Hanwen Xu, Zichuan Guo, and Qun Luo

A Simulation Platform for the Brain-Computer Interface (BCI) Based Smart Wheelchair 257
Xinru Huang, Xianwei Xue, and Zhongyun Yuan

A Learning Resource Recommendation Model Based on Fusion of Sequential Information 267
Ruofei Zhu, Zhengzhou Zhu, and Qun Guo

A Convolutional Neural Network-Based Complexity Reduction Scheme in 3D-HEVC 279
Chang Liu, Ke-bin Jia, and Peng-yu Liu

Air Quality Index Prediction Based on Deep Recurrent Neural Network 291
Zhiyu Chen, Yuzhu Zhang, Gang Liu, and Jianwei Guo

Sparse Representation for Face Recognition Based on Projective Dictionary Pair Learning 305
Yang Liu, Jianming Zhang, and Yangchun Liu

An Improved HF Channel Wideband Detection Method Based on Scattering Function 317
Lantu Guo, Yanan Liu, and Wenxin Li

Research on Quality Control of Marine Monitoring Data Based on Extreme Learning Machine 327
Yuanshu Li, Feng Liu, Kai Wang, Hai Huang, Fenglin Wei, and Hong Qi

Study on the Extraction of Target Contours of Underwater Images 339
Luyan Tong, Fenglin Wei, Yupeng Pan, and Kai Wang

Short-Term Demand Forecasting of Shared Bicycles Based on Long Short-Term Memory Neural Network Model 350
Ming Du, Danhua Cao, Xi Chen, Shurui Fan, and Zirui Li

An Effective Bacterial Foraging Optimization Based on Conjugation and Novel Step-Size Strategies 362
Ming Chen, Yikun Ou, Xiaojun Qiu, and Hong Wang

A Coding Scheme Design for the Shape Code of Standardized Yi Characters in Liangshan 375
Qiyan Hu, Xiaobing Zhao, and Liang Zhang

Police: An Effective Truth Discovery Method in Intelligent Crowd Sensing 384
Ming Zhao and Jia Jiao

Improve Data Freshness in Mobile Crowdsensing by Task Assignment 399
Ming Zhao and Shanshan Yang

Access Control Based on Proxy Re-encryption Technology for Personal Health Record Systems 411
Baolu Liu and Jianbo Xu

Generating More Effective and Imperceptible Adversarial Text Examples for Sentiment Classification 422
Xiaohu Du, Zibo Yi, Shasha Li, Jun Ma, Jie Yu, Yusong Tan, and Qinbo Wu

QR Code Detection with Faster-RCNN Based on FPN 434
Jinbo Peng, Song Yuan, and Xin Yuan

General Virtual Images Construction Using Pixel Scrambling for Face Recognition 444
Yingnan Zhao and Jie Wu

Knowledge Graph Construction of Personal Relationships 455
Yong Jin, Qiao Jin, and Xusheng Yang

Blind Spectrum Sensing Based on the Statistical Covariance Matrix and K-Median Clustering Algorithm 467
Jiawei Zhuang, Yonghua Wang, Pin Wan, Shunchao Zhang, Yongwei Zhang, and Yi Li

An Efficient Web Caching Replacement Algorithm. 479
Fan You, Tong Liu, Xiaoyu Peng, Jiahao Liang, Baili Zhang, and Yinian Zhou

Carrier-Phase Based Ranging Algorithm with Multipath Suppression in RFID System 489
Xiaohui Fu, Liangbo Xie, and Mu Zhou

Research on Vehicle Routing Problem Based on Tabu Search Algorithm 501
Jinhui Ge and Xiaoliang Liu

Unsupervised Situational Assessment for Power Grid Voltage Stability Monitoring Based on Siamese Autoencoder and k-Means Clustering 510
Xiwei Bai and Jie Tan

The Mathematical Applications of Majorization Inequalities to Quantum Mechanics 522
Mengke Xu, Zhihao Liu, Hanwu Chen, and Sihao Zheng

The Numerical Results of Binary Coherent-State Signal's Quantum Detection in the Presence of Noise 532
Wenbin Yu, Zijia Xiong, Zangqiang Dong, Yinsong Xu, Wenjie Liu, Zhiguo Qu, and Alex X. Liu

An Adaptive Parameters Density Cluster Algorithm for Data Cleaning in Big Data 543
Xiaopeng Zhang, Ruijie Lin, and Haitao Xu

Relevance and Time Based Collaborative Filtering for Recommendation 554
Aobo Zhang, Ruijie Lin, and Haitao Xu

A Displacement Data Acquisition and Transmission System Based on a Wireless Body Domain Network 564
Juan Guo, Feng Sheng, Jie Li, Chao Guo, and Dingbang Xie

Data Cleaning Algorithm Based on Body Area Network 575
Juan Guo, Yuzhuo Zong, Fang Chen, Chao Guo, and Dingbang Xie

Improving Performance of Colour-Histogram-Based CBIR Using Bin Matching for Similarity Measure 586
Martey Ezekiel Mensah, Xiaoyu Li, Hang Lei, Appiah Obed, and Ninfaakanga Christopher Bombie

A Survey on Risk Zonation of Lightning 597
Yunfei Wang and Meixuan Qu

Joint Extraction of Entity and Semantic Relation Using Encoder - Decoder Model Based on Attention Mechanism. 607
Yubo Mai, Yatian Shen, Guilin Qi, and Xiajiong Shen

Internet of Things

Image Segmentation of Manganese Nodules Based on Background Gray Value Computation 621
Ha-de Mao, Yu-liang Liu, Hong-zhe Yan, and Cheng Qian

A Portable Intelligent License Plate Recognition System Based on off-the-Shelf Mini Camera 635
Haixia Yang, Wei Zhuang, Qingfeng Zhou, Dong Dai, and Weigong Zhang

Mechanical Analysis and Dynamic Simulation of Ship Micro In-pipe Robot 646
Zhipeng Xu, Yuliang Liu, Han-de Mao, Sheng Chang, and Shuxun Li

LoRa Devices Identification Based on Differential Constellation Trace Figure. 658
Xiuting Wu, Yu Jiang, and Aiqun Hu

Improved Helmet Wearing Detection Method Based on YOLOv3 670
Peizhi Wen, Mingyang Tong, Zhenrong Deng, Qinzhi Qin, and Rushi Lan

MobiMVL: A Model-Driven Mobile Application Development Approach for End-Users 682
Zhongyi Zhai, Ke Xiang, Lingzhong Zhao, and Junyan Qian

Analysis on Buffer-Aided Energy Harvesting Device-to-Device Communications 695
Guangjun Liang, Qun Wang, Jianfang Xin, Lingling Xia, Xueli Ni, and Jie Xu

Research on Traceability of Agricultural Products Supply Chain System Based on Blockchain and Internet of Things Technology 707
Hongyi Jiang, Xing Sun, and Xiaojun Li

Design of Remote-Control Multi-view Monitor for Vehicle-Mounted Mobile Phone Based on IoT 719
Qian Ni, Meixia Ji, Ning Cao, Shouxu Du, and Wei Huo

Super-Resolution Reconstruction of Electric Power Inspection Images Based on Very Deep Network Super Resolution 729
Bowen Shi, Jilong Gao, Zhentao He, Tian Zhang, Tie Zhong, and Haipeng Chen

Single Color Image Dehazing Based on Vese-Osher Model and Dark Channel Prior Algorithm 739
Wei Weibo, Zhao Hui, and Gao Zhuzhu

Traffic-Behavioral Anomaly Detection of Endhosts Based on Community Discovery 751
Mingda Wang, Xuemeng Zhai, Hangyu Hu, and Guangmin Hu

A Lightweight Indoor Location Method Based on Image Fingerprint 763
Ran Gao and Yanchao Zhao

Performance Analysis and Optimization for Multiple Carrier NB-IoT Networks 775
Sijia Lou, En Tong, and Fei Ding

A Short Flows Fast Transmission Algorithm Based on MPTCP Congestion Control 786
Hua Zhong, PingPing Dong, WenSheng Tang, Bo Yang, and JingYun Xie

Analysis of Public Transport System Efficiency Based on Multi-layer Network 798
Lu Xu, Funing Yang, Zhezhou Yu, Qiuyang Huang, Xuan Ji, and Yuanbo Xu

Fast Video Classification with CNNs in Compressed Domain 810
Lorxayxang Kai, Yang Wu, Xiaodong Dai, and Ming Ma

Author Index 823

Internet of Things

Green Crop Image Segmentation Based on Superpixel Blocks and Decision Tree

Bo Wang[1,2] and Zhibin Zhang[1,2](✉)

[1] Inner Mon1School of Computer Science, Inner Mongolia University, Hohhot 010021, China
imucswangbo@gmail.com, cszhibin@imu.edu.cn

[2] Key Laboratory of Wireless Networks and Mobile Computing, School of Computer Science, Inner Mongolia University, Hohhot 010021, China

Abstract. When segmenting green crops, we usually use green indexes, such as Excess Green Index (ExG), Combined Indices 2 (COM2), Modified Excess Green Index (MExG) etc., which are regarded as efficient methods. However, they can't extract green crop exactly under complex environmental conditions. Particularly, they can't segment green crops from complex soil backgrounds, such as with high light or deep shadow areas in crop leaves. To address current deficiencies in green crop segmentation, this paper introduces a new crop segmentation method to extract more compete green crops. Firstly, a pre-processing procedure divides the crop images into superpixel blocks by using Simple Linear Iterative Clustering (SLIC) algorithm, then these superpixel blocks are classified into three class by using Classification And Regression Tree (CART) decision tree based on a seven-dimensional(7-D) features vector constructed in this paper: only crop blocks (OC-block), only background blocks (OB-blocks) and CBE-blocks within which crop and background both coexist. Finally, CBE-blocks are processed by making good use of the advantage of ExG to exact green crops in them. Experimental results show that the algorithm proposed in this paper can segment accurately the green crops from the soil backgrounds, even under relative complex field conditions with *Accuracy* of 98.44%, *Precision* of 91.75%, *Recall* of 91.78%, *FPR* of 0.55% and *F1_score* of 89.31%.

Keywords: Crop image segmentation · First-order gradient · SLIC · CART · Traditional green index

1 Introduction

1.1 Background

Image segmentation has always been a hot research topic. Liu et al. propose multi-objective particle swarm optimization (PSO) clustering algorithm for image segmentation [1]. Particularly, it is a key step in machine vision applications. For example, agricultural robots equipped with cameras can autonomously spray pesticides by detecting weed area and position to reduce usage of herbicides [2–6]. At present, many crop image segmentation methods have been proposed to make agricultural robots obtain

X. Sun et al. (Eds.): ICAIS 2020, LNCS 12240, pp. 3–17, 2020.
https://doi.org/10.1007/978-3-030-57881-7_1

further applications, or more accurately monitoring crop growth. Traditional crop image segmentation is to use color indices, combining with the Otsu threshold method to segment the green crops from their backgrounds [7], such as Excess Green Index (ExG) [8], Vegetative Index (VEG) [9], Excess Green minus Excess Red Index (ExGR) [10], Color Index of Vegetation Extraction (CIVE) [11], Combined Index 1 (COM1) [12] and Combined Index 2 (COM2) [13] etc. Typically, COM1 and COM2 both utilize multiple color features by fusing information of each color feature to improve segmentation quality. COM1 simultaneously applied ExG, VEG, ExGR, CIVE [8–11]. Furthermore, COM2 excludes ExGR because it classified the shadow of the maize plant as part of plant. Each selected method is controlled by a weighting factor, with the weights summing to 1. However, due to the complexity of field environments, this kind of algorithms like COM1, combining with the automatic threshold cannot adapt to all situations in field. In some results, the green crop segmented by them are not good. Besides, the regions with reflection or high light in crop image, also cannot be proceeded well by using the traditional methods above mentioned.

Concerning the learning-based approaches in crop image segmentation, most of them are based on pixels to process crop images, which need large amounts of calculation time. But all these efforts are a beneficial exploration for automatic guidance or other operations of agricultural applications. It also has many learning-based applications such as leaf diseases detection, monitoring and classifying [14–20]. Meanwhile, we don't give up to develop other effective crop image segmentation method to resolve the problems above mentioned.

Superpixel strategy is applied to divide an image into a number of closed regions so that pixels in each region have high similarly in color, brightness, texture, and even high contrast between regions. It first proposed by Ren et al. [21]. Most of them are applied in natural images, Fu et al. noise-resistant superpixel segmentation (NRSS) algorithm to hyperspectral images [22]. Superpixels have two distinct advantages: the first can reduce computational complexity; secondly, it is more feature-expressive than a single pixel. Therefore, superpixels are now widely used in machine vision. There are also many kinds of superpixel algorithms, such as mean-shift [23] and Normalized cut [24]. These two algorithms can directly generate superpixel blocks, but they cannot generate regular shapes, and the feature acquisition efficiency is also low. Later, faster and more efficient superpixel algorithms emerged: Turbopixel [25] and Simple Linear Iterative Clustering (SLIC) [26]. Turbopixel starts from the seed point and expands with color and gradient to generate superpixel blocks. The effect of the algorithm is very associated with the initial position of the seed point and cannot handles slender objects. The SLIC method uses the K-means clustering algorithm to search for pixels in a defined area to generate superpixels. Because of efficiency and simplicity, it is widely used in the image processing. However, for small targets, SLIC has to wrongly divide them into the other regions. In this paper, we propose a green crop image segmentation method based on SLIC superpixel strategy.

1.2 Motivation and the Proposed Strategy

The above methods are all pixel-based and cannot represent the local features of crop images. For example, they cannot handle the regions with deep shadow, reflection or high light in crop images that are due to complex field environments [27]. Meanwhile, the methods based-pixels consume relatively more time. Consequently, this paper proposes to use the SLIC strategy to segment the green crop images where a crop image will be divided into small image blocks, among which each holds the homogeneities of color and texture. Especially, in order to partition the image blocks, the hexagonal superpixel block in the literature [28] is used. Next, this paper adopts the decision tree classification model, combining with features of the mean and variance of each component in the CIElab space to classify the image blocks into the crops and backgrounds.

However, from the crop images processed by SLIC, we find that the results contain some inaccurate blocks, that is, there are still some certain crops and backgrounds patches coexisting within the same blocks due to weak contrasts between the crops and corresponding backgrounds. Therefore, after classifying the blocks through using decision tree, we have specially to deal with the wrong blocks. Additionally, inspired by Dalal [29] and Canny [30], we use the first-order gradient information in addition to the color features in the CIElab space to improve the accuracy of the decision tree classification model. Besides, since ExG has good ability to provide a high contrast between soil and crops, and can product near binary images [27], we use it to process the wrong blocks classified by the decision tree classification model. Experimental results show that the overall correct classification rate is also improved. The flowchart of the algorithm proposed in this paper is shown in Fig. 1.

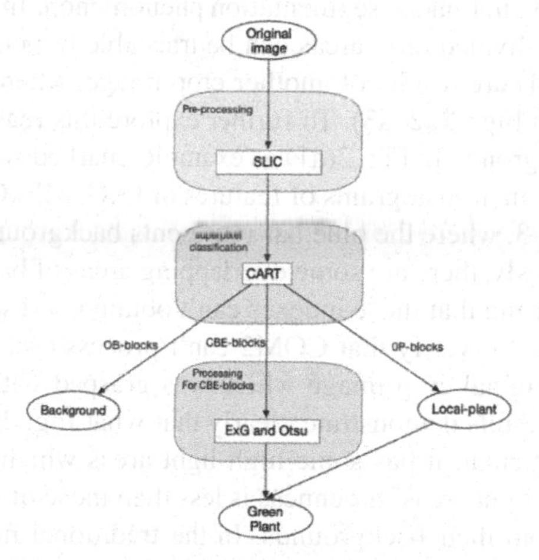

Fig. 1. Overall flowchart of the algorithm in this paper

1.3 Organization

The structure and contents of this paper are organized as follows: Sect. 2.1 explains the images acquisition in the experiments; Sect. 2.2 introduces how to divide the images into the blocks by using SLIC; Sect. 2.3 introduces how to CART classify the blocks divided based on the feature vectors we propose; Sect. 2.4 introduces post-processing for inaccurate blocks by combining ExG method. Section 3 is discussion and analysis where we compare the advantages and disadvantages with the traditional methods. In the Sect. 4, the conclusions are made about our research work.

2 Materials and Methods

2.1 Crop Image Acquisition and Training Set

In this study, most images we used were grasped in a maize trial field of Inner Mongolia University in Hohhot where we fixed the digital camera (MV-VD030SC, Shanxiwenshi Ltd, China) at a certain height and let lens face the maize, and then photographed their images between May 21 and July 11, 2011. Besides, those images were spaced by several days in order to obtain them under different illumination conditions and in different growth stages. The size of each image is 480*640. At the same time, other green crops are also taken into account, such as cabbage and broad beans. Finally, 44 images were chosen.

Figure 2 shows some original crop images and their segmentation results by traditional methods. From the Fig. 2(a2–a3), we can see that there exist under-segmentation areas in the ExG and MExG results. MExG [31] is very robust to the changing illumination conditions. The phenomenon of under-segmentation is due to some background and crop areas being similar in color. For the result of COM2 in Fig. 2(a4), there is either over-segmentation nor under-segmentation phenomenon. In addition to the above problems, the wrongly divided crop areas can be traceable to poor contrast in color or texture. Figure 2(b2–b4) are results of another crop image, where exists similar situations with the results in Fig. 2(a2–a3). To further explore this reason, we take a part of pixels of crop and background in Fig. 2(a1) as example, marked with the two red circles respectively, then make their histograms of features of ExG, MExG and COM2, respectively, as shown in Fig. 3, where the blue bar represents background and the orange bar represents crop. Obviously, there are some overlapping areas of background and crop in each feature space, causing that those indexes can't obtain good segmentation results.

Particularly, in order to verify that COM2 can't process crop images in poor contrast, we process an original crop image which was grasped with shadows, as shown in Fig. 2(c1–c4). The results demonstrate clearly that what Fig. 2(d2–d4) implicate for cabbage image. In particular, it has some high-light areas which are marked with red rectangles. Those high-light areas' greenness is less than those of the normal, which are difficult to separate from their backgrounds. In the traditional methods like ExG and COM2, though they can extract the most greenness, for the areas high-light, they can't extract well. But the results of MExG are relatively close to the ground truth images, shown in Fig. 2(a3–d3), and Fig. 2(d3) is the best segmentation result of them. Therefore, MExG is suitable for segmentation of the crop images when the backgrounds are simplex but in good illumination. Particularly, from the segmenting results, we see that MExG can process high-light or shadow areas in the images compared with COM2 and ExG. However, COM2 and ExG seem to have better ability to process the backgrounds in the images. Besides, ExG and MExG are sensitive to crop images where their backgrounds appear similar with the corresponding crops; COM2 is sensitive to high-light or shadowing areas. Summarily, these traditional greenness indexes have their respective disadvantages and advantages, and the results above indicate that only considering color feature, ones cannot obtain idea segmentation effects on green crop images. The goal of this paper is to propose a synthetic segmentation method with SLIC Superpixel and CART tree methods in order to surpass those traditional indexes.

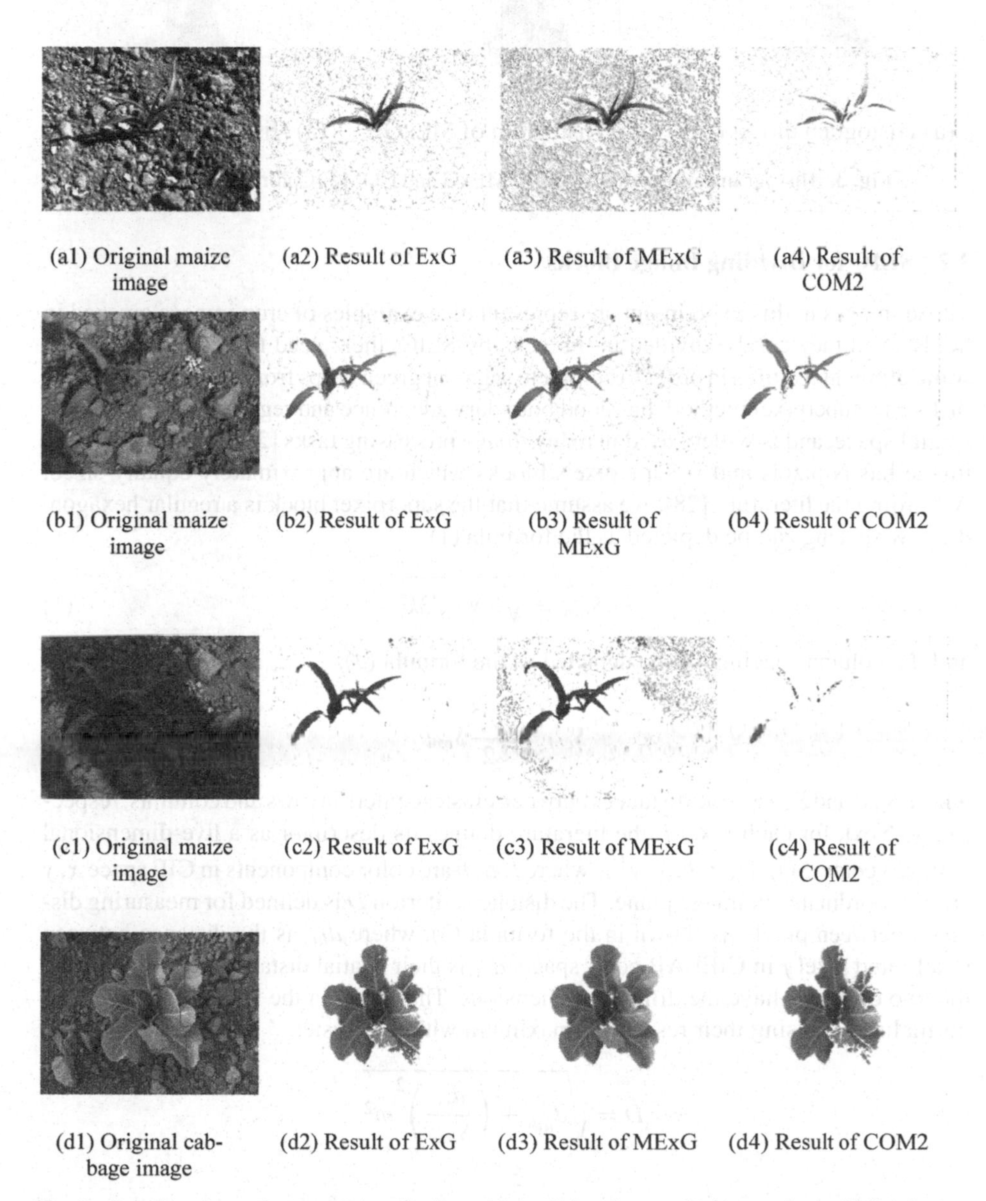

(a1) Original maize image (a2) Result of ExG (a3) Result of MExG (a4) Result of COM2

(b1) Original maize image (b2) Result of ExG (b3) Result of MExG (b4) Result of COM2

(c1) Original maize image (c2) Result of ExG (c3) Result of MExG (c4) Result of COM2

(d1) Original cabbage image (d2) Result of ExG (d3) Result of MExG (d4) Result of COM2

Fig. 2. Results of crop images segmentation with traditional green index methods (Color figure online).

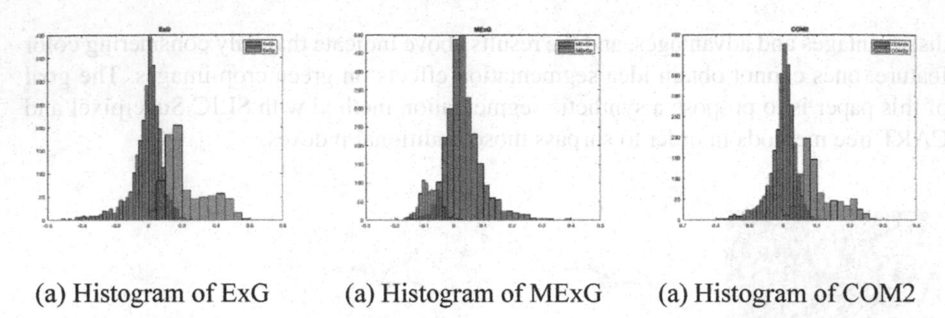

(a) Histogram of ExG (a) Histogram of MExG (a) Histogram of COM2

Fig. 3. Histograms of features of ExG, MExG and COM2 (Color figure online).

2.2 SLIC for Dividing Image Blocks

Those images in this experiment are representative examples of crop images grasped in fields. And they can be divided into blocks by SLIC, then, need to be combined with some efficient features in order to complete segment green crops from their backgrounds. SLIC as a superpixel method, has good boundary adherence and regular size in color and spatial space, and is widely used in many image processing tasks [28]. Assuming a crop image has N pixels and k Superpixels blocks which are approximately equally-sized, According the literature [28], we assume that the superpixel block is a regular hexagon, the row spacing can be depicted as the formula (1).

$$S_{row} = \sqrt{2N/\sqrt{3}k} \tag{1}$$

and the column spacing can be depicted as the formula (2).

$$S_{col} = \frac{\sqrt{3}}{2} S_{row} \tag{2}$$

where S_{row} and S_{col} are the distances between cluster centers in rows and columns, respectively. Next, for each pixel i, the literature defines its descriptor as a five-dimensional feature vector (5D), $[l, a, b, x, y]^T$, where l, a, b are color components in CIE space;x, y are its coordinates in image plane. The distance criterion D is defined for measuring distance between pixels as shown in the formula (3), where d_{lab} is the distance between pixel i and pixel j in CIELAB color space, d_{xy} is their spatial distance [26]. Obviously, the two distances have the different dimensions. Therefore, in the formula (3), they are normalized by using their respective maximum within a cluster.

$$D = \sqrt{d_{lab}^2 + \left(\frac{d_{xy}}{S_{row}}\right)^2 m^2}$$

$$d_{lab} = \sqrt{(l_j - l_i)^2 + (a_j - a_i)^2 + (b_j - b_i)^2} \tag{3}$$

$$d_{xy} = \sqrt{(x_j - x_i)^2 + (y_j - y_i)^2}$$

In the formula (3), i are seed points, j are neighborhood points; $l_j, a_j, b_j, l_i, a_i, b_i$ indicate the component of neighborhood points and seed points in CIELab space; x_j, y_j, x_i, y_i are coordinates of neighborhood points and seed points, separately. In this paper, we take the m ranging from 5 to 40 according to the literature [28].

Particularly, it is difficult to determine t he number of Superpixels k. In practice, k is larger than the preset value when the parameter m in the formula (3) is small. We empirically tried the different m values of 5, 20, 35, but taking the same k (200). The results are shown in Fig. 3. Obviously, when $m = 5$ and $m = 20$, the most divided blocks are irregular and the blocks' edges have bad adherence with crops' boundaries in images, as shown in Fig. 4(a–b), where the amplified details are marked with red rectangle, respectively. That is, the generated superpixels are different from the preset values and don't fit the crops' boundaries. When m is 35, the divided blocks are relatively regular, as shown in Fig. 4(c), but the boundary adherence is still not good, where the amplified details are also marked with red rectangle. It is adverse to the extraction of green crops.

Fig. 4. Pre-processing results by SLIC for the different m when K= 200 (Color figure online).

In order to improve boundary adherence, we increase the value of k, the results are shown in Fig. 5, where k is 200, 600, 1000, respectively, but for the same $m = 35$. As we can see that when k taking 1000, the boundary adherence between the crops' edges and the blocks' edges is relatively good, and the shapes of generated Superpixel blocks are relatively regular.

At last, in the experiment, we selected the parameter k for 1000 and m for 35. However, there still exist some inaccurate blocks in the results marked with red circle, shown as in Fig. 6(a). SLIC superpixel method expressing the features of target images is regarded as a superpixel block way, requiring that the edges of superpixel blocks fit basically the target boundaries in image. This will benefit the segmentation of green crops from backgrounds. In subsequent processing, we classify the superpixel blocks by using the CART based on a 7-D feature vector: mean and standard deviation of three components in CIELab space, as well as the first-order gradient in blocks. Then combining the traditional ExG, we can obtain the correct segmentation of the inaccurate blocks in crop images. Particularly, the first-order gradient information added improves the accuracy of the inaccurate suerpixel blocks.

Based on the analysis above, in this paper, we take some superpixel blocks as the training set of the CART where each crop image is initially divided into about 1000

(a) k=200 (b) k=600 (c) k=1000

Fig. 5. Pre-processing results by SLIC when $k = 200$, $k = 600$, $k = 1000$ respectively, for the same $m = 35$

(a) Normal maize image (b) Crop image with shadows (c) Cabbage image with high-light areas

Fig. 6. Result after pre-processing by SLIC (Color figure online).

blocks by the SLIC algorithm [26]. Especially, we define the only crop blocks denoted as OC-block, the only background blocks denoted as OB-block and the both the crop and background coexisting blocks denoted as CBE-block. Considering the number of OB-blocks is far more than that of OC-blocks or CBE-blocks, we only choose a part of OB-blocks. At last, 44 crop images are selected, in which there are 7640 image blocks involved, containing 757 OC-blocks, 6333 OB-blocks and 550 CBE-blocks.

2.3 Superpixel Blocks Classification Using CART

In this section, we utilize the CART [32] to classify the superpixel blocks divided by SLIC. Decision tree has been successfully applied in many fields [33, 34]. Besides, CART as a nonparametric technique, has sufficiently explanatory capability, unlike the other methods such as artificial neural networks (ANNs) [31], and can reveal complex associations between variables [35].

CART algorithm uses a binary-dividing procedure to split the dataset given, beginning with the parent node based on yes/no question of a single variable. After iteratively constructing the tree, all samples obtain the optimum splits with high purity [36]. Meanwhile, in order to decrease the impurity, the Gini index [37] used is defined as the formula (4), where D is original set, and D_1 and D_2 are subsets divided at the node s; Xs is the binary-dividing point with feature s at X.

$$Gini(D, X_s) = \frac{|D_1|}{D} Gini(D_1) + \frac{|D_2|}{D} Gini(D_2) \tag{4}$$

The Gini index, $Gini(D, X_s)$ is the impurity of D in binary-dividing of X_s. $|D|$ and $|D_i|$ is the number of respective samples of D and D_i. $Gini(D_i)$ is defined as the formula (5).

$$Gini(D_i) = 1 - \sum_{k=1}^{n} \left(\frac{|C_k|}{|D_i|}\right)^2 \tag{5}$$

C_k stand for the samples belonging to the kth class in D_i. The smaller the value of Gini index is, the smaller the impurity is, then the better the split results are [36]. Figure 6 shows that there exist some inaccurate blocks during iteratively clustering, which are CBE-blocks. According to the analysis above, we add gradient information to further process these CBE-blocks. A 7-D feature vector we used is defined as follows:

$$X_i = [\bar{l}_i, \bar{a}_i, \bar{b}_i, \hat{l}_i, \hat{a}_i, \hat{b}_i, \bar{g}_i] \tag{6}$$

In the formula (6), $\bar{l}_i, \bar{a}_i, \bar{b}_i$ are mean of l, a, b, respectively, in CIELab color space defined as follows:

$$\bar{l}_i = \frac{1}{m} \sum\nolimits_{j=1}^{m} l_i,\ \bar{\mathrm{a}}_{\mathrm{i}} = \frac{1}{m} \sum\nolimits_{j=1}^{m} a_i,\ \bar{\mathrm{b}}_{\mathrm{i}} = \frac{1}{m} \sum\nolimits_{j=1}^{m} b_i \tag{7}$$

Besides, $\bar{l}_i, \bar{a}_i, \bar{b}_i$ are their respective standard deviations defined as follows:

$$\hat{l}_i = \sqrt{\frac{1}{m} \sum\nolimits_{j=1}^{m} (l_j - \bar{l}_i)},\ \hat{\mathrm{a}}_i = \sqrt{\frac{1}{m} \sum\nolimits_{j=1}^{m} (a_j - \bar{a}_i)},\ \hat{\mathrm{b}}_{\mathrm{i}} = \sqrt{\frac{1}{m} \sum\nolimits_{j=1}^{m} (b_j - \bar{b}_i)} \tag{8}$$

Particularly, $\bar{g}_i$ is the mean of the first-order gradient, defined as the formula (9).

$$\bar{g}_i = \frac{1}{m} \sum\nolimits_{j=1}^{m} g_i \tag{9}$$

In the formula (9), g_i is the first-order gradient of the grayscale of each pixel, and m is the number of pixels. It can identify sharp changes about grayscale in color space. Thus, it can improve accuracy of classifying those Superpixel blocks. Inspired by Dalal [29] and Canny [30], the formula (9) can be calculated according to the formula (10).

$$g(x, y) = \sqrt{g_x^2(x, y) + g_y^2(x, y)} \tag{10}$$

In the formula (10), g_x and g_y are the horizontal and vertical gradient, respectively, and they can be calculated according to the formula (11).

$$\begin{aligned} g_x(x, y) &= H(x+1, y) - H(x-1, y) \\ g_y(x, y) &= H(x, y+1) - H(x, y-1) \end{aligned} \tag{11}$$

In the formula (11), $H(x, y)$ is grayscale value about a pixel point (x, y). In actual calculation, we first use vector $[-1, 0, 1]$, called gradient operator, to convolve with each pixel in block to get the gradient of horizontal direction, then, use $[1, 0, -1]^T$ to get its gradient of vertical direction. Doing so, we can correctly identify CBE-blocks. In the experiment, we use *accuracy*, *recall* and *discriminating power ratio* w_i [36] to explain improvement for classifying CBE-blocks. w_i is a number which changes between 0 and 1, representing the ratio of the samples i-th node in the dataset to the total samples in all terminal nodes. Table 1 is classification results of CART with and without the first-order gradient information.

Table 1. Classification results of CART with and without gradient information.

Item	*Accuracy*	*Recall*	w_i
without first-order gradient	86.77%	63.87%	0.0638
with first-order gradient	93.64%	69.47%	0.0689

Figure 7 is one typical example of the classification results by CART where Fig. 7(a) is the pre-processing result by SLIC; Fig. 7(b) shows the amplified details marked by the red rectangles in the Fig. 7(a). Obviously, there exist some CBE-blocks in Fig. 7(a) like those CBE-blocks shown in Fig. 7(b). Figure 7(c) and (d) are classification results without and with the gradient information, respectively. As we can see, in the Fig. 7(c), only one CBE-block is identified currently based on the feature vector without the gradient information, while more CBE-blocks are identified correctly in the Fig. 7 (d) based on the feature vector with the gradient information. This indicates that the CART can successfully identify the CBE-blocks through adding the first-order gradient information into the feature vector. Then, we employ ExG index to extract the corresponding parts of green crops, described in the next section.

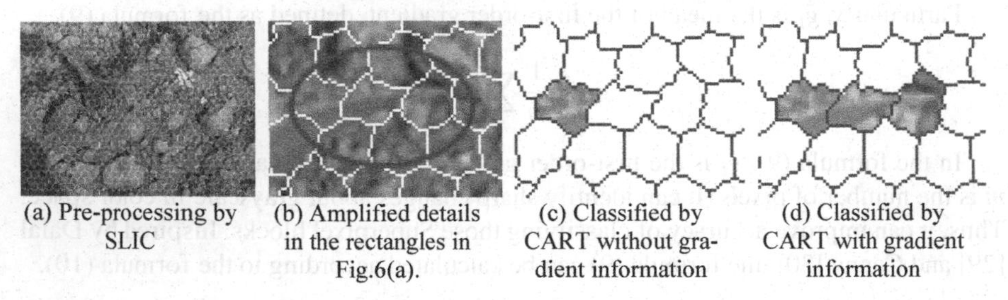

(a) Pre-processing by SLIC (b) Amplified details in the rectangles in Fig.6(a). (c) Classified by CART without gradient information (d) Classified by CART with gradient information

Fig. 7. Results of Classification by CART with and without the first-order gradient information in the blocks (Color figure online).

2.4 ExG Index for CBE-Blocks

CBE-blocks obviously exist in the results divided by SLIC according to the analysis above. Actually, those blocks which are wrongly identified generally contain end sections of crop leaves which are similar with the corresponding soil backgrounds in texture or color properties. And this is reason why SLIC cannot identify correctly them. But ExG index has strong ability to process the situations. And it can product near binary images [27]. Therefore, we try ExG to extract green crop in each CBE-block identified by CART with the first-order gradient information. For Fig. 7 (d), the corresponding result processed is shown in Fig. 8 where there are 3 CBE-blocks processed correctly. As we can see, doing so in this paper, the CBE-blocks can be segmented well like Fig. 8 shown.

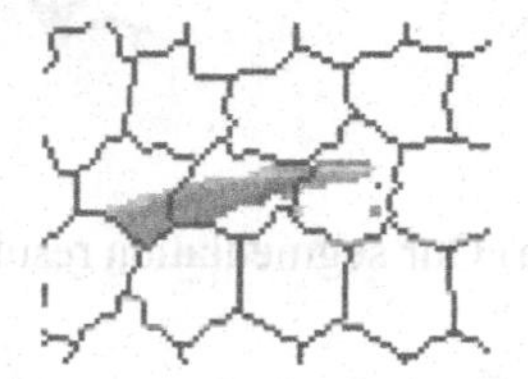

Fig. 8. Result of CBE-blocks processed by ExG.

The ExG utilized, is expressed as shown in the formulas (12-14).

$$(2g - r - b) \tag{12}$$

In the formula (12), r, g, b are regularized chromaticity coordinates:

$$r = \frac{R^*}{(R^* + G^* + B^*)}, g = \frac{G^*}{(R^* + G^* + B^*)}, b = \frac{B^*}{(R^* + G^* + B^*)} \tag{13}$$

In the formula (13), R^*, G^*, B^* are RGB values normalized 0 to 1 as follows:

$$R^* = \frac{R}{R_{max}}, G^* = \frac{R}{G_{max}}, B^* = \frac{R}{B_{max}} \tag{14}$$

In the formula (14), $R_{\max}$, $G_{\max}$, $B_{\max}$ are the maxima of red, green and blue components in color space, respectively.

3 Experiments and Discussion

We still take those typical original crops images from Fig. 2 in order to compare our segmentation results with the ground truth images. Comparison results are shown in Fig. 9. As we can see, the different illumination conditions or the soil backgrounds hardly have impact on the segmentation results processed by our algorithm. However, from the segmentation results in Fig. 2, we see that those traditional indexes such as ExG, MExG and COM2 seem to be greatly impacted by illuminant conditions or soil backgrounds. This indicates that our algorithm in this paper also has good ability to segment the green crops images like these images.

Besides, we use a public evaluation method to validate our algorithm in this paper. According to the literature [38], these evaluating indicators are defined as the formulas (15-19), where True Positive (TP) stands for amount of crop pixels classified correctly;

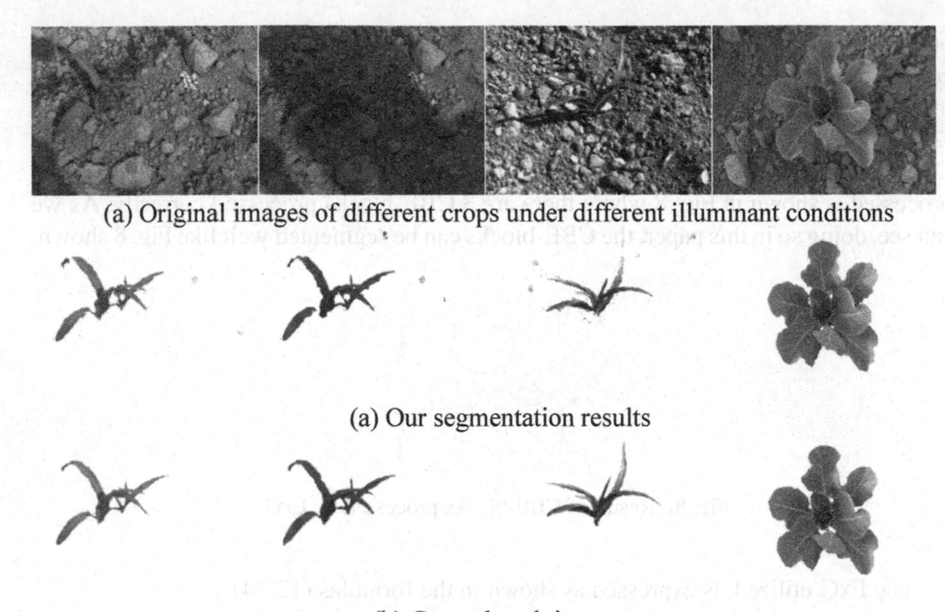

(a) Original images of different crops under different illuminant conditions

(a) Our segmentation results

(b) Ground truth images

Fig. 9. Comparisons of segmentation results of the traditional indexes, our method and ground truth

True Negative (TN) is for amount of background pixels classified correctly; False Positive (FP) represents amount of background' pixels treated wrongly as crop' pixels; False Negative (FN) is for mount of crop pixels treated wrongly as background' pixels.

Accuracy indicates proportion classified correctly in dataset, and is expressed as shown in formula (15):

$$Accuracy = \frac{TP + TN}{TP + FP + FN + TN} \tag{15}$$

Sometimes, we pay more attention to the accuracy of the partial segmentation of plants. In this case, *Precision* can be used to evaluate result of segmentation in crop image, shown in the formula (16).

$$Precision = \frac{TP}{TP + FP} \tag{16}$$

In addition, *Recall* indicates ratio of crop pixels are segmented correctly, defined as the formula (17):

$$Recall = \frac{TP}{TP + FN} \tag{17}$$

FPR indicates ratio of background pixels misclassified as crop pixels, defined as the formula (18):

$$FPR = \frac{FP}{FP + TN} \tag{18}$$

When we overall consider the *Precision* and the*Recall*, F1-score is defined as the formula (19):

$$F1\_score = 2 * \frac{Precision * Recall}{Precision + Recall} \tag{19}$$

Final comparisons of crop image segmentation in different evaluation criterions are implemented by using the method proposed in this paper, as shown in Table 2, including the proposed method in this paper surpasses the traditional method over our dataset. As we can see, *Accracy*, *Precision* and *F*1-*socre* of our algorithm are highest among corresponding results. Meanwhile, the *FPR* of our algorithm is relatively lower than those of ExG and MExG; *Fecall* is higher than those of COM2 and ExG.

Table 2. Performance of different methods

Method	Accuracy	Precision	Recall	FPR	F1-socre
ExG	96.27%	83.25%	77.44%	1.20%	78.12%
MExG	90.16%	63.75%	**91.78%**	9.62%	68.27%
COM2	96.71%	89.14%	66.24%	**0.55%**	74.25%
Proposed method	**98.44%**	**91.75%**	87.50%	0.80%	**89.31%**

4 Conclusions

This paper proposes a novel method for segmenting green crops, which integrates the advantages of SLIC superpixel algorithm, CART and ExG. In term of evaluation criteria such as *Precision*, *Recall*, *FPR* and *F1_score*, our method obtains better results than the traditional green indexes do, being insensitive to the different illumination or the soil backgrounds. This paper segments green crops from their soil backgrounds in the experiments, more than maize, cabbage and broad beans etc. It can be applied in some aspects such as three-dimensional modeling, crop growth prediction, disease detection etc. Future improvement can be considered in algorithmic acceleration to achieve higher real-time performance by using embedded FPGA architecture.

Acknowledgments. This work was supported by National Natural Science Foundation of China (No.31760254).

References

1. Liu, Z., Xiang, B., Song, Y., Lu, H., Liu, Q.: An improved unsupervised image segmentation method based on multi-objective particle swarm optimization clustering algorithm. Comput. Mat. Con. **58**(2), 451–461 (2019)

2. Davis, G., Casady, W.W., Massey, R.E.: Precision Agriculture: An introduction. Extension publications (MU) (1998)
3. Nelson, D.C., Giles, J.F.: Implications of postemergence tillage on root injury and yields of potatoes. Am. Potato J. 63(8), 445–446 (1986)
4. Eyre, M.D., Critchley, C.N.R., Leifert, C., Wilcockson, S.J.: Crop sequence, crop protection and fertility management effects on weed cover in an organic/conventional farm management trial. Eur. J. Agron. **34**(3), 153–162 (2011)
5. Liu, F., O'Connell, N.V.: Off-site movement of surface-applied simazine from a citrus orchard as affected by irrigation incorporation. Weed Sci. **50**(5), 672–676 (2002)
6. Spliid, N.H., Køppen, B.: Occurrence of pesticides in Danish shallow ground water. Chemosphere **37**(7), 1307–1316 (1998)
7. Otsu, N.: A threshold selection method from gray-level histograms. IEEE Trans. Syst. Man. Cyber. **9**(1), 62–66 (1979)
8. Woebbecke, D.M., Meyer, G.E., Von Bargen, K., Mortensen, D.A.: Color indices for weed identification under various soil, residue, and lighting conditions. Trans. ASAE **38**(1), 259–269 (1995)
9. Hague, T., Tillett, N.D., Wheeler, H.: Automated crop and weed monitoring in widely spaced cereals. Precision Agric. **7**(1), 21–32 (2006)
10. Meyer, G.E., Neto, J.C., Jones, D.D., Hindman, T.W.: Intensified fuzzy clusters for classifying plant, soil, and residue regions of interest from color images. Comput. Electron. Agric. **42**(3), 161–180 (2004)
11. Kataoka, T., Kaneko, T., Okamoto, H., Hata, S.: Crop growth estimation system using machine vision. In: Proceedings 2003 IEEE/ASME International Conference on Advanced Intelligent Mechatronics (AIM 2003), 2, b1079-b1083. IEEE, Kobe, Japan (2003)
12. Guijarro, M., Pajares, G., Riomoros, I., Herrera, P.J., Burgos-Artizzu, X.P., Ribeiro, A.: Automatic segmentation of relevant textures in agricultural images. Comput. Electron. Agric. **75**(1), 75–83 (2011)
13. Guerrero, J.M., Pajares, G., Montalvo, M., Romeo, J., Guijarro, M.: Support vector machines for crop/weeds identification in maize fields. Expert Syst. Appl. **39**(12), 11149–11155 (2012)
14. Ji-Hua, M., Bing-Fang, W., Qiang-Zi, L.: A global crop growth monitoring system based on remote sensing. In: 2006 IEEE International Symposium on Geoscience and Remote Sensing. pp. 2277–2280. IEEE (2006)
15. Anand, R., Veni, S., Aravinth, J.: An application of image processing techniques for detection of diseases on brinjal leaves using k-means clustering method. In: 2016 international conference on recent trends in information technology (ICRTIT), pp. 1–6. IEEE June 2016
16. Padol, P.B., Yadav, A.A.: SVM classifier based grape leaf disease detection. In: 2016 Conference on advances in signal processing (CASP), pp. 175–179. IEEE June 2016
17. de Luna, R. G., Dadios, E. P., Bandala, A. A.: Automated image capturing system for deep learning-based tomato plant leaf disease detection and recognition. In: TENCON 2018–2018 IEEE Region 10 Conference, pp. 1414-1419. IEEE october 2018
18. Sardogan, M., Tuncer, A., Ozen, Y.: Plant leaf disease detection and classification based on CNN with LVQ algorithm. In: 2018 3rd International Conference on Computer Science and Engineering (UBMK), pp. 382–385. IEEE September 2018
19. Tellaeche, A., Burgos-Artizzu, X.P., Pajares, G., Ribeiro, A.: A vision-based method for weeds identification through the Bayesian decision theory. Pattern Recogn. **41**(2), 521–530 (2008)
20. Kirk, K., Andersen, H.J., Thomsen, A.G., Jørgensen, J.R., Jørgensen, R.N.: Estimation of leaf area index in cereal crops using red–green images. Biosyst. Eng. **104**(3), 308–317 (2009)
21. Ren, X., Malik, J.: Learning a classification model for segmentation. ICCV, 1, 10–17 vol.1 (2003)

22. Fu, P., Xu, Q., Zhang, J., Geng, L.: A noise-resistant superpixel segmentation algorithm for hyperspectral images. Comput. Mat. Cont. **59**(2), 509–515 (2019)
23. Comaniciu, D., Meer, P.: Mean shift: a robust approach toward feature space analysis. IEEE Trans. Pattern Anal. Mach. Intell. **24**(5), 603–619 (2002)
24. Shi, J., Malik, J.: Normalized cuts and image segmentation. IEEE Trans. Pattern Anal. Mach. Intell. **22**(8), 888–905 (2000)
25. Levinshtein, A., Stere, A., Kutulakos, K.N., Fleet, D.J., Dickinson, S.J., Siddiqi, K.: Turbopixels: fast superpixels using geometric flows. IEEE Trans. Pattern Anal. Mach. Intell. **31**(12), 2290–2297 (2009)
26. Achanta, R., Shaji, A., Smith, K., Lucchi, A., Fua, P., Süsstrunk, S.: SLIC superpixels compared to state-of-the-art superpixel methods. IEEE Trans. Pattern Anal. Mach. Intell. **34**(11), 2274–2282 (2012)
27. Hamuda, E., Glavin, M., Jones, E.: A survey of image processing techniques for plant extraction and segmentation in the field. Comput. Electron. Agric. **125**, 184–199 (2016)
28. Peter, K.: Image Segmentation using SLIC SuperPixels and DBSCAN Clustering. Available on-line: peterkovesi.com/projects/segmentation, Accessed 27 Jan 2020
29. Dalal, N., Triggs, B.: Histograms of oriented gradients for human detection. In: 2005 IEEE computer society conference on computer vision and pattern recognition (CVPR), 1, pp. 886–893. IEEE (2005)
30. Canny, J.: A computational approach to edge detection. IEEE Trans. Pattern Anal. Mach. Intell. **6**, 679–698 (1986)
31. Burgos-Artizzu, X.P., Ribeiro, A., Guijarro, M., Pajares, G.: Real-time image processing for crop/weed discrimination in maize fields. Comput. Electron. Agric. **75**(2), 337–346 (2011)
32. Breiman, L.: Classification and regression trees. Routledge, London, England (2017)
33. Arifuzzaman, M., Gazder, U., Alam, M.S., Sirin, O., Mamun, A.A.: Modelling of asphalt's adhesive behaviour using classification and regression tree (CART) Analysis. Computational intelligence and neuroscience (2019)
34. Kaur, K., Kaur, K.: Failure prediction, lead time estimation and health degree assessment for hard disk drives using voting based decision trees. Comput. Mat. Cont. **60**(3), 913–946 (2019)
35. Pranali T. S., Mahesh C. A., Ragini P. M., Pawar T. S.: Decision Tree Classifier For Mining Data Stream: A Survey, International Journal of Science & Engineering Development Research, ISSN:2455–2631, 4(1), 103-106, Available: http://www.ijrti.org/papers/IJRTI1901017.pdfAccessed 27 Jan 2020
36. Vagliasindi, G., Arena, P., Murari, A.: CART data analysis to attain interpretability in a Fuzzy Logic Classifier. In: 2009 International Joint Conference on Neural Networks, pp. 3164–3171. IEEE (2009)
37. Seera, M., Lim, C.P., Tan, S.C.: A hybrid FAM-CART model for online data classification. Comput. Intell. **34**(2), 562–581 (2018)
38. Jo, K., Kweon, J., Kim, Y.H., Choi, J.: Segmentation of the main vessel of the left anterior descending artery using selective feature mapping in coronary angiography. IEEE Access **7**, 919–930 (2018)

User Behavior Credibility Evaluation Model Based on Intuitionistic Fuzzy Analysis Hierarchy Process

Chen Yang, Guogen Wan(✉), Peilin He, Yuanyuan Huang, and Shibin Zhang

Chengdu University of Information Technology, Chengdu, China
992511763@qq.com

Abstract. Based on the research of network user behavior, this paper studies and proposes an evaluation model of user behavior credibility. Firstly, the behavior evaluation index system is constructed by AHP analytic hierarchy process. Then, the information entropy is used to optimize the behavior evidence after data collection. Finally, the weight of each behavior attribute is calculated by the intuitionistic fuzzy judgment matrix, and the attribute layer and target layer of user behavior are calculated. Comprehensive credibility and based on comprehensive credibility to judge whether user behavior is credible. The simulation experiments show that the proposed application behavior credibility evaluation model has a good effect in evaluating the credibility of user behavior and can be used in increasingly complex and variable network systems.

Keywords: User behavior · Intuitionistic fuzzy theory · AHP

1 Introduction

1.1 Research Background

The behavior of network users is one of the paramount factors affects the normal operation of the network. User behavior is the key evidence that affects system security in the basic traffic data generated by network users by accessing the platform or website. One of the starting points for studying user behavior is whether that of users are credible. By combing through and analyzing the relative evidence, the behavior of users accessing the network can be grasped. Additionally, abnormal behaviors of the user with unreliable factors were detected which can evaluate the level of behavioral credibility of the user. Credibility stands for the network security-related system indicators which means judging the access behavior of users and whether a good trust relationship can be formed with the service provider in the long term as well as there are potential users of the network system risks and threats. Based on this criterion, it is necessary to study the evaluation model of the system to quantify and visualize this trusted relationship as well as give regulators a convenient and reliable management platform to control users.

X. Sun et al. (Eds.): ICAIS 2020, LNCS 12240, pp. 18–29, 2020.
https://doi.org/10.1007/978-3-030-57881-7_2

1.2 Related Work

At present, many scholars focus on the structure and algorithm of user behavior credibility evaluation. Ji Tieguo et al. [1, 2] proposed that user behavior can be classified differently by attributes in terms of AHP analytic hierarchy process, constructing the multi-dimensional behavior attribute vectors and weight size. Tian Liqin et al. [3] proposed that behavioral attributes need to be embodied, illustrated that behavior can be quantified mathematically and the credibility level of user behavior is obtained to evaluate the trust degree of user behavior. Ma Junyi et al. [4] combined fuzzy theory to study the evaluation method of user behavior. He put forward to use fuzzy set theory to describe the degree of membership which belongs to this set, so as to make a qualitative description of behavioral credibility with quantitative mathematical analysis. However, he did not fully consider the degree of the impact of the uncertainty of human factors on the evaluation value. Shen Jinchang et al. [5] combined cloud model with fuzzy theory and proposed a fuzzy comprehensive evaluation method based on cloud model. However, the model is so idealized that the evaluation subject depends on the objective mathematical model and ignores the subjective empirical evaluation.

These methods neglect subjective or objective evaluation which had a certain effect for user behavior; according to the shortcomings of previous researches, this paper proposes a user behavior credibility evaluation algorithm based on intuitionistic fuzzy analytic hierarchy to which extends the fuzzy set. The general fuzzy set can only roughly categorize the opinions of experts into a certain evaluation standard, ignoring the nuances in how experts evaluate behavior, affecting the final overall evaluation accordingly. The intuitionistic fuzzy set can accurately quantify the opinions of experts and eliminate the unicity of evaluation, so that each expert's opinions can fully influence the evaluation of user behavior.

2 Intuitionistic Fuzzy Analytic Hierarchy Process

2.1 Intuitionistic Fuzzy Theory

As the core element of the intuitionistic fuzzy set, the intuitionistic fuzzy number contains two attribute parameters, namely membership degree and non-membership degree, so that it can depict the so-called "neutral state" of "not this or the other", that is the degree of uncertainty. Experts may have such a neutral state when judging a certain factor and tend to favor a certain qualitative evaluation to varying degrees. And user behavior has the character of fuzziness exactly, a user's behavior may be either this or that and it can only be described in terms of membership to the extent to which it belongs to a possibility. Therefore, it is more accurate to use intuitionistic fuzzy theory to study whether the user behavior is credible. Atanassov [9] extended the dimension of Zadeh's fuzzy set theory and proposed the concept of intuitionistic fuzzy sets. Unlike fuzzy set theory, intuitionistic fuzzy set theory further refines the degree of uncertainty in decision-making and describes the state of "non-this or not" by means of hesitation. The theory of intuitionistic fuzzy sets is briefly introduced.

X is the domain of a given intuitionistic fuzzy set A, $A = \{<x, \mu(x), \nu(x)> | x \in U\}$, $\mu_A: X \rightarrow [0, 1]$, $x \in X \rightarrow \mu_A(x) \in [0, 1]$ and $\nu_A: X \rightarrow [0, 1]$, $x \in X \rightarrow \nu_A(x) \in$

[0, 1], $\mu_A(x)$ is the membership function of the intuitionistic fuzzy set A; $\nu_A(x)$ is the non-membership function of the intuitionistic fuzzy set A and meet the conditions $0 \leq \mu(x) + \nu(x) \leq 1$. What exists on the intuitionistic fuzzy set A is the $\pi(x)$, $\pi(x) = 1 - \mu(x) - \nu(x)$ called $\pi(x)$ as the hesitation of x. The mathematical expression of $\sigma = (\mu_\sigma, \nu_\sigma)$ is generally used to describe the intuitionistic fuzzy number.

2.2 AHP-Based Indicator System

According to the standard AHP three-tier architecture model, user behavior can be divided into target layer, attribute layer and evidence layer from top to bottom. The target layer is the user behavior of the research; the principle layer is the attribute classification that determines the behavior; the evidence layer is the basic unit of the behavior property which is the so-called behavioral evidence. The research framework based on AHP analytic hierarchy is shown in Fig. 1.

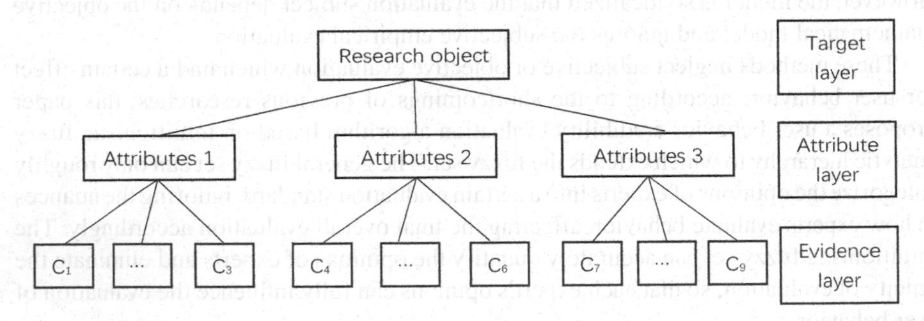

Fig. 1. AHP framework

Classifying user behavior layer by layer according to AHP principle, it not only specified the research points, but also considers the diversity of behavioral characteristics comprehensively, moreover, it is helpful to systemize the evaluation value from point to surface and from dispersion to the whole. It provides convenience and support for visual display of post-evaluation results. Specific evidence of user behavior is determined by factors that influence the user's behavior on the platform or website.

2.3 Evidence Optimization Algorithm

Normalized Pretreatment

Different evidences are not within a fixed range due to differences in categories and units and do not satisfy monotonicity. Therefore, it is necessary to preprocess the evidence with normalization in order to facilitate the following weight calculation and evaluation algorithm.

The evidence generally has three representations. The first is a number in the range of a real number, such as the number of illegal connections, the smaller the value, the more credible, the values like this are monotonic; the second is the percentage form, such as occupied storage. The resource rate is also monotonous; the third is a Boolean value of either or both, such as whether malicious information is recommended or not, represented by 0 or 1.

Let X be any real number in the interval [0, 1]. Pre-processing is the conversion of all three types of numerical evidence into values in the X interval. The value of this interval is specified to be monotonically increasing. The closer the evidence value is to the 1 evaluation result, the better the evidence value is closer to 0. The worse the evaluation result is. The evidence in the form of a real range can be mapped into the X range by the substitution formula (1).

$$f(x) = \begin{cases} \frac{x - x_{min}}{x_{max} - x_{min}} \\ \frac{x_{max} - x}{x_{max} - x_{min}} \end{cases} \tag{1}$$

Where x_{max} represents the maximum value of the evidence before preprocessing and x_{min} represents the minimum value of the evidence before processing. The values obtained after such conversion are all real numbers in the X range that satisfy the monotonically increasing.

Since the value of the percentage form and the Boolean form itself is in the range [0, 1]. It is only necessary to replace the formula (2) with uniform monotonicity.

$$f(x) = \begin{cases} x \\ \frac{x_{min} - x}{x_{max} - x_{min}} \end{cases} \tag{2}$$

Behavioral Evidence

The pre-processing stage can obtain evidence in a single data collection, but a single user behavior cannot represent the user's long-term network behavior habits, which is accidental. Therefore, periodic data collection is required for the user.

Let R be a matrix of m × n. n is representative of n kinds of behavioral evidence. m is representative of m data acquisition. ni indicates the i-th evidence of behavior, mj represents the m-th data acquisition and Rij indicates pretreatment evidence for the i-th behavior in the j-th data collection. Its expression is $C = \begin{pmatrix} C_{11} & \cdots & C_{1m} \\ \vdots & \ddots & \vdots \\ C_{n1} & \cdots & C_{nm} \end{pmatrix}$.

Entropy Weight Method Optimization Evidence

The row vector of R represents evidence of different behaviors under a certain attribute layer. Because of the uncertainty of behavior, the entropy weight method can be used to assign weights to each behavior, distinguishes the relative degree of uncertainty between indicators. The greater the degree of dispersion between the data, the greater the amount of information covered by the behavior, the greater the impact on the final evaluation. Therefore, the entropy weight method can get rid of the certain group of the collected data that has less influence on the experimental results without affecting the accuracy of the evaluation results and ensures the objectivity of the data collection. First calculate the information entropy of the i-th behavior, Calculate formula (3) as

$$e_i = -k\sum_{j=1}^{n} f_{ij} \ \ln f_{ij}(i = 1, 2, 3, \ldots, n)$$

$$f_{ij} = \frac{r_{ij}}{\sum_{j=1}^{m} r_{ij}}, k = \frac{1}{\ln m}, 0 \le e_i \le 1 \tag{3}$$

The formula for calculating the entropy weight (4) is

$$\omega_i = \frac{1 - e_i}{n - \sum_{i=1}^{n} e_i} \tag{4}$$

The optimized evidence value of an indicator can be obtained by formula (5)

$$\overline{C_n} = \sum_{i=1}^{m} \omega_i \cdot C_{ni}(n = 1, 2, \ldots, n) \tag{5}$$

An evidence optimization matrix can be obtained by obtaining optimized evidence for each behavioral indicator.

3 User Behavior Credibility Evaluation Model

3.1 Constructing an Intuitionistic Fuzzy Judgment Matrix

The evaluation itself is uncertain, different people have different views on the same thing. There is also ambiguity in evaluating a thing. In other words, it is impossible to accurately evaluate whether a certain behavior completely conforms to the evaluation under a certain scale. Intuitionistic fuzzy sets can quantify the evaluation into values that belong to a certain scale and use the degree of hesitation to reflect the ambiguity of the evaluation subject. Through the intuitionistic fuzzy judgment matrix, the weight of the same kind of behavior can be calculated and then the subjective evaluation value of the behavior is determined. The intuitionistic fuzzy evaluation table is shown in Table 1.

3.2 Consistency Test

The initial intuitionistic fuzzy judgment matrix does not necessarily have coordination consistency. The intuitionistic fuzzy judgment matrix that needs to pass the consistency test can obtain reasonable evaluation results. The test method is tested by using the distance measured of the intuitionistic fuzzy information.

$$d(\overline{R}, R) = \frac{1}{2(n-1)(n-2)} \sum_{i=1}^{n}\sum_{j=1}^{n} \left(\left|\overline{\mu}_{ij} - \mu_{ij}\right| + \left|\overline{v}_{ij} - v_{ij}\right| + \left|\overline{\pi}_{ij} - \pi_{ij}\right|\right) \tag{6}$$

Table 1. Intuitionistic fuzzy scale evaluation form

Intuitionistic fuzzy scale	Evaluation
[0.9, 0.1]	A behavior is much more important than B behavior
[0.8, 0.15]	A behavior is more important than B behavior
[0.7, 0.2]	A behavior is significantly more important than B behavior
[0.6, 0.25]	A behavior is slightly more important than B behavior
[0.5, 0.5]	A behavior is equal to B behavior
[0.4, 0.45]	A behavior is slightly less important than B behavior
[0.3, 0.6]	A behavior is significantly less important than B behavior
[0.2, 0.75]	A behavior is much less important than B behavior
[0.1, 0.9]	A behavior is much less important than B behavior

Where R is the square matrix represented by the n × n intuitionistic fuzzy set formed by the experts comparing the importance of a certain layer of indicators according to the relative level of the upper layer; $\overline{R}$ is called the intuitionistic fuzzy consistency judgment matrix. It is also a square matrix of n × n intuitionistic fuzzy sets, but it is obtained by matrix transformation of the intuitionistic fuzzy judgment matrix R. Its construction method is as follows

When j > i + 1, let $\overline{r}_{ij} = (\overline{\mu}_{ij}, \overline{\nu}_{ij})$, among:

$$\overline{\mu}_{ij} = \frac{\sqrt[j-i-1]{\prod_{t=i+1}^{j-1} \mu_{it}\mu_{tj}}}{\sqrt[j-i-1]{\prod_{t=i+1}^{j-1} \mu_{it}\mu_{tj}} + \sqrt[j-i-1]{\prod_{t=i+1}^{j-1} (1-\mu_{it})(1-\mu_{tj})}}, j > i+1 \quad (7)$$

$$\overline{\nu}_{ij} = \frac{\sqrt[j-i-1]{\prod_{t=i+1}^{j-1} \nu_{it}\nu_{tj}}}{\sqrt[j-i-1]{\prod_{t=i+1}^{j-1} \nu_{it}\nu_{tj}} + \sqrt[j-i-1]{\prod_{t=i+1}^{j-1} (1-\nu_{it})(1-\nu_{tj})}}, j > i+1 \quad (8)$$

1. When j = i + 1, let $\overline{r}_{ij} = (\mu_{ij}, \nu_{ij})$
2. When j < I, let $\overline{r}_{ij} = (\overline{\nu}_{ji}, \overline{\mu}_{ji})$

Intuitionistic fuzzy consistency judgment matrix can be obtained by the above construction method $\overline{R} = \left(\overline{r}_{ij}\right)_{n\times n}$. Test the consistency of R with the formula (1). If d($\overline{R}$, R) < 0.1, then $\overline{R}$ Pass the consistency test. Otherwise, iterative parameters σ need to be introduced. The new intuitionistic fuzzy consistency judgment matrix is reconstructed by adjusting the value parameter of σ and the measurement distance is calculated once for each parameter adjustment until it finally passes the consistency test. In this paper, the parameter σ has a value range of $\sigma \in [0, 1]$ and iteratively tests from the upper limit with a decreasing value of 0.01. The specific construction method is:

$$\widetilde{\mu}_{ij} = \frac{(\mu_{ij})^{1-\sigma}(\overline{\mu}_{ij})^{\sigma}}{(\mu_{ij})^{1-\sigma}(\overline{\mu}_{ij})^{\sigma} + (1-\mu_{ij})^{1-\sigma}(1-\overline{\mu}_{ij})^{\sigma}}, i, j = 1, 2, \ldots n \tag{9}$$

$$\widetilde{\nu}_{ij} = \frac{(\nu_{ij})^{1-\sigma}(\overline{\nu}_{ij})^{\sigma}}{(\nu_{ij})^{1-\sigma}(\overline{\nu}_{ij})^{\sigma} + (1-\nu_{ij})^{1-\sigma}(1-\overline{\nu}_{ij})^{\sigma}}, i, j = 1, 2, \ldots n \tag{10}$$

Through iterative calculation, we finally find a σ that satisfies the requirements, so that the constructed intuitionistic fuzzy judgment consistency judgment matrix $\widetilde{R} = (\widetilde{r}_{ij})_{n\times n}$ can pass the consistency test of formula (6) among $\widetilde{r}_{ij} = (\widetilde{\mu}_{ij}, \widetilde{\nu}_{ij})$.

3.3 Intuitionistic Fuzzy Weight

The intuitionistic fuzzy weight indicates the importance of the expert's evaluation of a certain behavioral index. Generally, the higher the degree of membership and the lower the degree of hesitation, the greater the reference value of the expert for a certain evaluation. The intuitionistic fuzzy judgment matrix passed the consistency test can calculate the weight of each index of each layer from Eq. (11)

$$\omega_i = \left(\frac{\sum_{j=1}^{n} \overline{\mu}_{ij}}{\sum_{i=1}^{n}\sum_{j=1}^{n} (1-\overline{\nu}_{ij})}, 1 - \frac{\sum_{j=1}^{n} (1-\overline{\nu}_{ij})}{\sum_{i=1}^{n}\sum_{j=1}^{n} \overline{\mu}_{ij}} \right), i = 1, 2, 3, \ldots\ldots, n \tag{11}$$

The intuitionistic fuzzy weights of each layer of similar indicators are sorted according to the intuitionistic fuzzy ranking function. The larger the number ω, the greater the importance. The formula for the intuitionistic fuzzy sorting function is

$$\rho(\omega_i) = 0.5(1+\pi)(1-\mu), i = 1, 2, 3, \ldots, n \tag{12}$$

Then empower according to the proportion of $\rho(\omega_i)$ score in n evidence

$$\omega_i' = \frac{\omega_i}{\sum_{i=1}^{n} \omega_i}, (i = 1, 2, 3, \ldots, n) \tag{13}$$

3.4 Fuzzy Comprehensive Evaluation

The evaluation value of the corresponding attribute layer can be obtained by weighting the weight of the secondary index corresponding to the optimization evidence. The weighted calculation of the attribute evaluation value of each attribute layer and the corresponding first-level index weight can obtain the comprehensive evaluation value of the target layer. Its weighting formula is:

$$E = \sum_{i=1}^{n} \omega_i' C_i \tag{14}$$

4 Simulation Experiment and Analysis

Now take a precision instrument and equipment sharing platform as the experimental basis, simulate the user behavior of the operator on this platform within one month and calculate the credit of the user on the platform based on the experimental data. This experiment assumes that the user is a legitimate operator who has been authenticated without external interference, to ensure the effectiveness of the experiment and the user has no obvious malicious attack network behavior.

After the administrator logs in to the large-scale instrument and equipment online control system, the administrator can operate and control the platform within its authority. Commonly used functions include monitoring instrument status; device information extraction; device addition and deletion check and change; unit management; file uploading. The experiment will collect the log message of the administrator for one month or so in the login status of the platform and organize the relevant data into preparation for preprocessing.

4.1 Constructing Behavior Evaluation Index System

Establish user behavior model index system based on actual situation of experimental platform (As shown in Fig. 2).

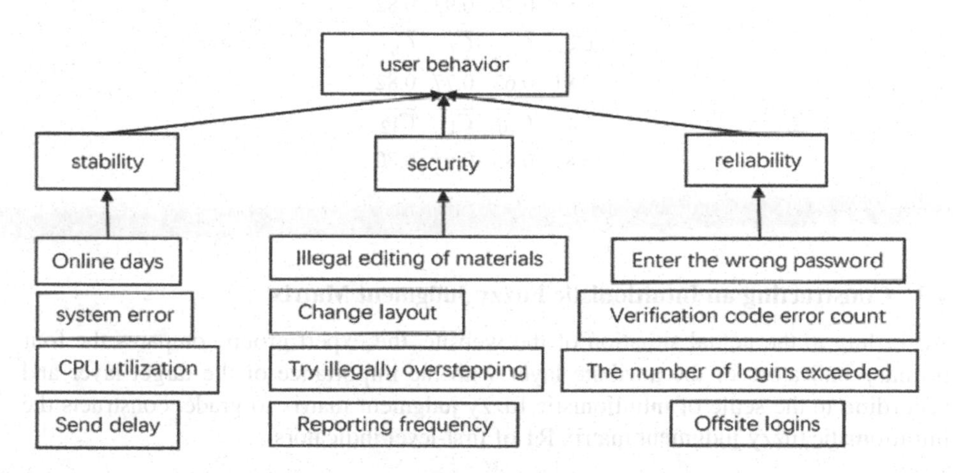

Fig. 2. User behavior indicator system

The closer the value is to 1, the less the user's mis-operation on the platform and the higher the reliability.

4.2 Pretreatment Optimization Evidence

According to the index system, the user behavior evidence is collected and the behavior evidence is preprocessed according to formulas (1) and (2) and the behavior evidence

Table 2. Stability evidence pretreatment table

	1	2	3	4	5
C1	0.95	0.72	0.89	0.93	0.87
C2	0.77	0.41	0.92	0.37	0.56
C3	0.90	0.87	0.92	1	0.85
C4	0.56	0.67	0.87	0.97	0.91

preprocessing table is constructed, as shown in Table 2. Safety evidence and Reliability evidence table is same as the Table 2.

Calculate the information entropy and entropy weight of each data acquisition under the same attribute layer according to formulas (3) and (4), determine the influence factor of the data collection sample and calculate the optimization value of each evidence according to formula (5). The optimized results are shown in Table 3.

Table 3. Evidence optimization table

$\overline{C_1}$	$\overline{C_2}$	$\overline{C_3}$	$\overline{C_4}$
0.87	0.46	0.93	0.82
$\overline{C_5}$	$\overline{C_6}$	$\overline{C_7}$	$\overline{C_8}$
0.81	0.62	0.77	0.82
$\overline{C_9}$	$\overline{C_{10}}$	$\overline{C_{11}}$	$\overline{C_{12}}$
0.81	0.85	0.65	0.80

4.3 Constructing an Intuitionistic Fuzzy Judgment Matrix

According to the actual situation of the website, the expert group compares the four primary indicators of the attribute layer with the importance of the target layer and according to the scale of intuitionistic fuzzy judgment matrix to grade, constructs the intuitionistic fuzzy judgment matrix R1 of first-level indicators:

$$R_1 = \begin{bmatrix} (0.5, 0.5) & (0.451, 0.532) & (0.525, 0.412) \\ (0.549, 0.434) & (0.5, 0.5) & (0.615, 0.36) \\ (0.475, 0.462) & (0.385, 0.59) & (0.5, 0.5) \end{bmatrix}$$

The initial intuitionistic fuzzy judgment matrix R for expert scoring needs to pass the consistency test. First, the intuitionistic fuzzy judgment matrix is transformed into the intuitionistic fuzzy consistency judgment matrix by formulas (4) and (5) $\overline{R_1}$:

$$\overline{R_1} = \begin{bmatrix} (0.5, 0.5) & (0.451, 0.532) & (0.568, 0.39) \\ (0.532, 0.451) & (0.5, 0.5) & (0.615, 0.36) \\ (0.39, 0.568) & (0.36, 0.615) & (0.5, 0.5) \end{bmatrix}$$

Then the rationality of the intuitionistic fuzzy consistency judgment matrix is tested by the distance measure formula (3) of the intuitionistic fuzzy information. After the formula calculates d = 0.095022, and d < 0.1, the consistency judgment matrix passes the consistency detection. If the consistency check is not passed, the intuitionistic fuzzy judgment consistency matrix needs to be reconstructed by Eqs. (7) and (8). In the same way, the expert group scores the importance of each evidence and the attribute layer corresponding to the evidence according to the scale, constructs the intuitionistic fuzzy secondary index intuitionistic fuzzy judgment matrix and the intuitionistic fuzzy consistency judgment matrix of R_{2a}, R_{2b} and R_{2c}. It can be calculated in the same way. After calculation, the first constructed intuitionistic fuzzy consistency judgment matrix did not pass the consistency test and after repeated iterations of Eqs. (9) and (10). Finally, when the σ is 0.86, 0.93 and 0.79. Respectively, the consistency test is passed. Constructed $\widetilde{R}_{2a}$, $\widetilde{R}_{2b}$, $\widetilde{R}_{2c}$. Layout reasons only show $\widetilde{R}_{2a}$

$$\widetilde{R}_{2a} = \begin{bmatrix} (0.5, 0.5) & (0.69, 0.27) & (0.74, 0.23) & (0.91, 0.05) \\ (0.31, 0.65) & (0.5, 0.5) & (0.72, 0.24) & (0.852, 0.13) \\ (0.26, 0.71) & (0.28, 0.68) & (0.5, 0.5) & (0.79, 0.19) \\ (0.09, 0.87) & (0.148, 0.834) & (0.21, 0.77) & (0.5, 0.5) \end{bmatrix}$$

4.4 Calculating Intuitionistic Fuzzy Weights

The subjective weight of the intuitionistic fuzzy set scored by each evidence expert is calculated by Eq. (11). Then, the subjective weights of the intuitionistic fuzzy sets are sorted according to the formula (12) and the sorting results are weighted by the formula (13). The weight information of each level of indicators is shown in Table 4.

Table 4. Index weights at all levels

	Secondary indicator intuitionistic fuzzy set	Secondary indicator weight	Primary indicator intuitionistic fuzzy set	First-level indicator weight
C1	(0.362, 0.607)	0.21	(0.331, 0.643)	0.33
C2	(0.293, 0.68)	0.23		
C3	(0.2, 0.777)	0.26		
C4	(0.1, 0.887)	0.3		
C5	(0.275, 0.697)	0.24	(0.359, 0.617)	0.314
C6	(0.198, 0.782)	0.26		
C7	(0.243, 0.735)	0.25		
C 8	(0.24, 0.739)	0.25		
C 9	(0.341, 0.626)	0.22	(0.273, 0.702)	0.357
C 10	(0.322, 0.651)	0.22		
C 11	(0.204, 0.776)	0.26		
C 12	(0.089, 0.9)	0.3		

4.5 Calculate User Behavior Credibility

The obtained index weight and the optimization matrix are respectively calculated by formula (14) to calculate the evaluation value of each attribute in the attribute layer and the comprehensive evaluation value of the target layer. After the second-level index weight calculation, the evaluation value of user behavior stability is 0.779, the safety evaluation value is 0.753 and the reliability evaluation value is 0.775. Finally, based on the weight of the first-level indicators, the comprehensive evaluation value of user behavior is 0.769. The final evaluation results can be graded by relevant rating criteria and each level attribute can be assessed. It can be seen from the evaluation results that the user's overall behavior credibility level belongs to the upper middle level and the credibility value of each horizontal attribute tends to be stable. The trust management platform can control the user accordingly.

5 Conclusion

The fuzzy comprehensive evaluation matrix based on the intuitionistic fuzzy theory is better than the ordinary fuzzy matrix in quantifying the subjective opinions of experts on the degree of importance of relevant indicators, which makes the evaluation system flexible and the data remains undistorted. The weighted average selection of the evidence optimization algorithm can make every data collection play a certain impact, instead of taking large and small operators to highlight the decisive influence of one collection while ignoring the influence degree of other collection times. In the end, the evaluation value can achieve accurate quantitative analysis and qualitative rating, which makes it convenient and intuitive to obtain the evaluation results. The advantage of this method is not only the accuracy of the evaluation description, but also the ease of programming. The computational complexity of the algorithm is low which can be used for larger-scale indicator data as well as it can make scientific user behavior evaluation for network platform in different environment. However, since the intuitionistic fuzzy judgment matrix is obtained from expert experience, the construction time of this part of matrix is affected by the actual environment.

Acknowledgments. The authors would like to thank the reviewers for their detailed reviews and constructive comments, which have helped improve the quality of this paper. This work was supported in part by the National Key Research and Development Project of China (No. 2017YFB0802302).

References

1. Ji, T., Tian, L., Hu, Z., Sun, J.: A user behavior evaluation method based on AHP in trusted network. Comput. Eng. Appl. **42**(19), 123–126 (2007)
2. Jiang, Z., Li, S., Yin, C.: Evaluation of network user behavior credibility based on multi-dimensional decision attributes. Appl. Res. Comput. **28**(06), 2289–2293+2320 (2011)
3. Lin, C., Tian, L., Wang, Y.: Research on user behavior credibility in trusted network. Comput. Res. Dev. **45**(12), 2033–2043 (2008)

4. Ma, J., Zhao, Z., Ye, X.: Evaluation of trusted network user behavior based on fuzzy decision analysis. Comput. Eng. **37**(13), 125–127+131 (2011)
5. Shen, J., Du, S., Luo, W., Luo, J., Yang, Q., Chen, Z.: Fuzzy comprehensive evaluation method based on cloud model and its application. Fuzzy Syst. Math. **26**(06), 115–123 (2012)
6. Zhang, J., Zhang, X., Zhang, G.: User behavior evaluation method based on fuzzy comprehensive strategy. Comput. Technol. Dev. **27**(05), 138–143 (2017)
7. Lü, Q.: Trustworthy Evaluation Model Based on Cloud User Behavior. Hebei University (2015)
8. Wu, Q., Bao, B.: Comprehensive framework for user behavior reliability evaluation. Digital Library Forum (05), 54–60 (2017)
9. Atanassov, K.T.: Intuitionistic fuzzy sets. Fuzzy Sets Syst. **20**(1), 87–96 (1986)
10. Ning, M., Guan, J., Liu, P., Zhang, Z., O'Hare, G.M.P.: GA-BP air quality evaluation method based on fuzzy theory. Comput. Mater. Continua **58**(1), 215–227 (2019)
11. Li, D., Wu, H., Gao, J., Liu, Z., Li, L., Zheng, Z.: Uncertain knowledge reasoning based on the fuzzy multi entity Bayesian networks. Comput. Mater. Continua **61**(1), 301–321 (2019)
12. Tan, T., Wang, B., Tang, Y., Zhou, X., Han, J.: A method for vulnerability database quantitative evaluation. Comput. Mater. Continua **61**(3), 1129–1144 (2019)

Research on Chain of Evidence Based on Knowledge Graph

Yizhuo Liu[1], Jin Shi[1(✉)], Jin Han[2], and Mingxin Lu[1]

[1] School of Information Management, Nanjing University, Nanjing 210093, China
shijin@nju.edu.cn

[2] School of Computer and Software, Nanjing University of Information Science and Technology, Nanjing 210044, China

Abstract. Evidence plays an extremely important role in legal proceedings, historical research, and diplomatic disputes. In recent years, more and more evidence data has been presented in the form of electronic data. Therefore, the extraction, organization, and validation of knowledge in electronic evidence also seem more and more important. In order to effectively organize the evidence data and construct the Chain of Evidence that can meet the practical needs, this paper uses Knowledge Graph based method to conduct relevance of evidence research. Firstly, through Knowledge Extraction, Knowledge Fusion, Knowledge Reasoning, the evidence data Knowledge Graph is constructed. After that, the Chain of Evidence will be built through evidence correlation and evidence influence evaluation, and finally different forms of knowledge presentation are carried out for different users. In this paper, the work and process of evidence chain association based on knowledge map are introduced, and the possible research direction in the future is prospected.

Keywords: Knowledge graph · Chain of evidence · Knowledge extraction

1 Introduction

Evidence refers to the basis for determining the facts of a case according to the rules of lawsuit. Evidence is of great significance for the parties to conduct litigation activities, safeguard their legitimate rights and interests, for the court to find out the facts of the case, and for the correct judgment according to law. The issue of evidence is the core issue of litigation. In the trial process of any case, it is necessary to reproduce the original appearance of the event through the evidence chain formed by evidence and evidence, and the judgment based on sufficient evidence can be just.

According to the evidence science, whether the evidence has the evidentiary power for the facts of a case, and the magnitude of the evidentiary power, depends on whether the evidence is related to the facts of a case, as well as the close and strong degree of the connection. In order to improve the proof power of evidence, the concept of Chain of Evidence is introduced. Chain of Evidence is a legal term, which refers to the chain formed by a series of objective facts and objects. Within the scope of law, it is the sequential evidence of documents or records in chronological order, record

X. Sun et al. (Eds.): ICAIS 2020, LNCS 12240, pp. 30–41, 2020.
https://doi.org/10.1007/978-3-030-57881-7_3

keeping, control, transmission, analysis and physical or electronic configuration. The public security criminal investigators need to collect evidence extensively in the process of solving the case. When the collected witness testimony and trace evidence are orderly connected to form the main link of the criminal suspect's committing the case, and can completely prove the criminal process, then they can judge the suspect and take necessary criminal investigation measures. During the trial, the court must determine the criminal suspect's guilt or innocence according to the evidence chain provided by the public security organ. The Chain of Evidence should be able to prove the object of proof of the case. As the initial link of proof, the object of proof refers to the facts of the case that need to be proved by evidence in the proving activities. The object of proof in criminal procedure mainly refers to the facts about the elements of criminal act and the circumstances of sentencing. The evidence in the Chain of Evidence must be able to corroborate each other and eliminate reasonable doubt. Mutual Corroboration of evidence is the activity of comparing and testing a certain evidence with other evidence in order to judge the authenticity of evidence and the strength of proof, and then proving the facts of a case.

At present, the existing evidence data show the characteristics of subject diversification, content subject diversification, technical method diversification and data language diversification. Moreover, in the fields of criminal procedure and historical research, the evidence data gradually make the transformation from text to data. With the popularization of the Internet and the progress of science and technology, the sources of information are becoming more and more diverse. The literature sorting tools used in the past have been difficult to meet the needs of multi-channel knowledge discovery. The limitations of evidence association based on text processing are becoming more and more obvious, and more attention is paid to the processing of electronic evidence.

2 Related Works

2.1 Knowledge Graph

Knowledge Graph was first published by Google. Its main function is to improve the quality of returned answers of search engines and the efficiency of user queries. Because Knowledge Graph contains a lot of structured knowledge and special storage structure, it plays an important role in many Natural Language Processing (NLP) applications, such as Question Answering System [1], Entity Linking [2] and so on. Knowledge Bases Completion (KBC) [3, 4] has become a hot research topic. It mainly refers to adding new entities, relationships, entity attributes and attribute values to the Knowledge Graph.

Knowledge Graph is a kind of structured data set used to describe the entities and relations of the objective world. It was first proposed by Google. Based on wikidata and freebase databases, as well as other public databases, it uses semantic search information from multiple sources to enhance search engine search results. Microsoft has developed a Bing search engine based on satori knowledge base [5], which can provide a variety of search services. Microsoft's Person Cubic Meter is a new type of social search engine, which can automatically extract names, location names, organization names and other information. It has an algorithm that can automatically calculate the possibility of the relationship between inputs. Facebook Graph Search [6] is a semantic search engine designed to provide users with answers through natural language queries rather than link lists.

Wang [7] sets up a mixed corpus that connects general knowledge and special knowledge, use CRF model to segment words, and use a new statistical method to analyze semantic association to build a knowledge graph. Rospocher [8] uses NLP and Semantic Web technology to automatically generate knowledge graph from news articles. Borders and Gabriovich [9] use highly multi-source heterogeneous data, and carry out Named Entity Recognition and Entity Disambiguation according to the data specification of Freebase, and then build a web scale knowledge graph. Niu and Huang [10] constructs a binary relation knowledge base, embeds the elements into the knowledge graph through transg, and proposes an algorithm to distinguish different word vectors.

2.2 Chain of Evidence

Panesar Walawege [11] proposed a conceptual model to construct a seamless evidence chain by capturing information and traceability links that meet the requirements of IEC 61508, in order to represent the evidence of software security disputes. Ahmad [12] proposes a conceptual method of collecting evidence in Intranet Based on linked audit log model, which improves the efficiency of event configuration by evidence chain. Hayes [13] designed a logic chain based on evidence, taking the potential impact of a metal concentration change on the environment as the logic chain to prove the impact of trace metal pollution on the ecosystem. Jasmin [14] used Ontology technology to realize the evidence chain of electronic evidence in order to retain effective evidence.

Although the construction and application of Chain of Evidence has been studied at home and abroad, there are many problems in the constructed chain, such as the data format is not standardized, the proving power is insufficient, and the query is difficult. Therefore, this paper constructs a knowledge extraction framework for multi-source heterogeneous data to build Chain of Evidence based on Knowledge Graph. In this study, natural language process (NLP) technology is used to process evidence data, and multi-source heterogeneous data are deeply associated and mined through knowledge association. On the basis of metadata extraction and Ontology construction of evidence data, knowledge association of evidence data is achieved, and a dynamic Knowledge Graph is constructed. It also reveals and organizes the relationship between evidences. The Knowledge Graph oriented to evidence data can show the patterns and characteristics of a large number of evidences, and the constructed Knowledge Graph can help the relevant management departments and the public to organize, retrieve and understand the massive information. In the process of evidence retrieval, knowledge in the Knowledge Graph is presented in the form of ordered Chain of Evidence, and intelligent recommendation, question answering and other functions are fused. The whole process of evidence association includes identification and collection, analysis, storage, preservation, transportation and use. This paper mainly deals with the analysis, organization and storage of evidence.

3 Research Methodologies

3.1 Overall Process

The purpose of constructing Knowledge Graph is to express all kinds of concepts, entities and their related relationships in a large number of evidence data vividly and accurately. These nodes, which are composed of entities and concepts, are connected as a whole by the relationship and constitute the Knowledge Graph of evidence, that is, the network map representing the relationship between knowledge. The construction of Knowledge Graph of evidence data helps to aggregate a large number of concept topics, so as to achieve rapid response and reasoning of knowledge.

The main research process includes data collection and preprocessing, model building, Knowledge Extraction, Knowledge Fusion, Knowledge Reasoning and Knowledge Verification. Figure 1 is the construction process of Knowledge Graph based on multi-source heterogeneous data. Firstly, in order to obtain the available data after preprocessing, the evidence data is divided into text data and non text data that can be processed directly, and data cleaning and data annotation are carried out respectively. Then, model construction and Knowledge Extraction are carried out. Entity type and relationship type to be extracted are defined through Ontology model, and named based on this construction Entity Recognition model and Entity Relationship Extraction model. Then through Knowledge Fusion and Knowledge Verification, reduce the error and redundant information of extracted knowledge, and determine the credibility of knowledge; finally, use the evaluated knowledge to build Knowledge Graph and carry out Knowledge Reasoning, and use graphic database to visualize Knowledge Graph.

3.2 Ontology Design

Domain Ontology describes the relationship between concepts and concepts in specific fields (such as medicine and Geography) in a formal way, and defines the data patterns in the Knowledge Graph. Ontology knowledge base includes concepts, classes, relationships and instances. Domain Ontology is an important tool to help domain experts to standardize and annotate domain knowledge. Ontology is a kind of philosophy theory, which defines a set of representation primitives to model domain knowledge in the context of computer and information science. Ontology is the standard of conceptual modeling, an abstract model to describe the objective world, and a formal definition of concepts and their relations. A group of individual instances of ontology and classes constitute a knowledge base. Alobaidi and others [15] integrated semantic extension, syntactic pattern, Knowledge Graph and other technologies, and put forward an Ontology automatic generation framework based on the way of linking medical ontology, which can better solve the problem of automatic generation process of medical ontology. Bravo [16] defines an Ontology description model for the academic background description of public universities, and evaluated the model with Ontology description ability and domain coverage, and achieved good results. Jasmin [14] uses Ontology technology to achieve the evidence chain of electronic evidence inorder to retain effective evidence.

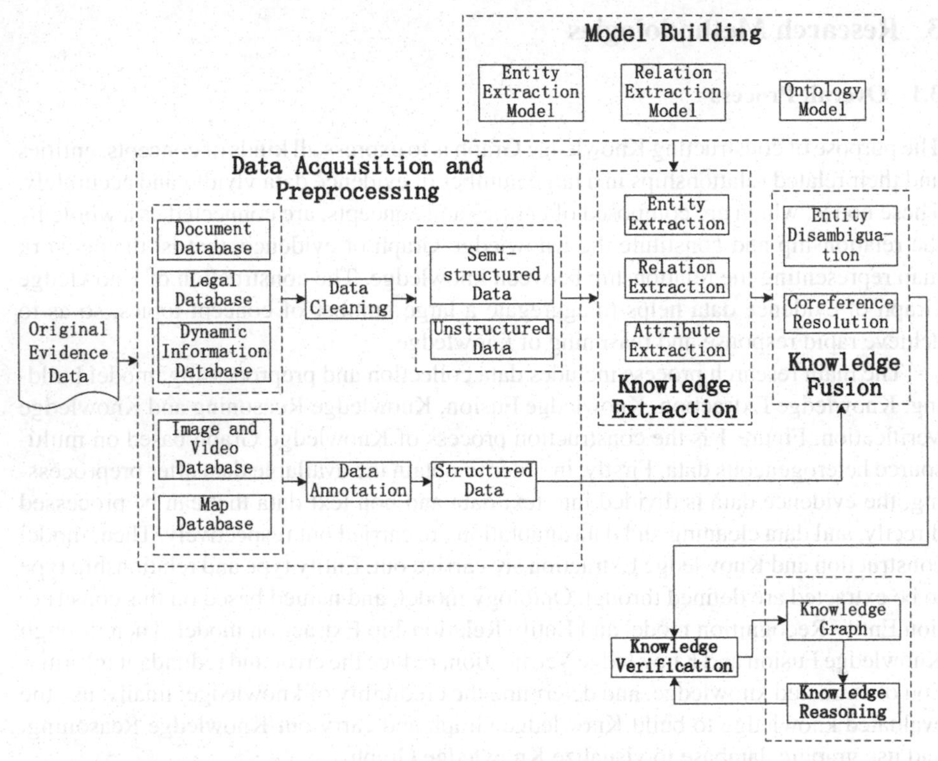

Fig. 1. Construction process of evidence knowledge graph

3.3 Construction of Knowledge Graph

Data Acquisition and Preprocessing. The data used in this paper includes five kinds of databases: Document Database, Legal Database, Dynamic Information Database, Image and Video Database and Map Database. The databases that can be processed directly by computer include Document Database, Legal Database and Dynamic Information database. The data in these three kinds of databases can be cleaned, including missing value processing, completion, noise removal and desensitization. The Image and Video Database and Map Database that cannot be processed directly will be premarked, including the subject, author, organization, data source, publication time and other metadata for subsequent processing.

The data after data cleaning and data annotation can be divided into structured data set, semi-structured data set and unstructured data set. For semi-structured and unstructured data set after data integration and transformation, knowledge extraction is required, recognizer is trained through Deep Learning, and then available knowledge is acquired; for structured data set after data annotation, knowledge extraction is required by establishing the mapping relationship (Entity Linking) between the concept in the database and the ontology in the Knowledge Graph, the corresponding attributes and relationships are automatically extracted. Knowledge fusion is directly carried out without knowledge extraction module.

Knowledge Extraction. The basic unit of Knowledge Graph is composed of "entity relationship entity" (e-r-e) triplet, which can be expressed as follows:

$$G = (E, R, S)$$

Where E represents the set of entities in the knowledge base, R represents the set of relationships among entities in the knowledge base, and S represents the set of triples in the knowledge base.

The task of Knowledge Extraction is to automatically extract entities, relationships and other implicit semantic knowledge from semi-structured data sets and unstructured data sets through computer models. Knowledge Extraction is the foundation of Knowledge Graph construction, including Entity Extraction, Relationship Extraction and Attribute Extraction. The main purpose of this study is to extract and organize knowledge from semi-structured and unstructured evidence.

Entity Extraction, also known as Named Entity Recognition (NER), refers to the automatic recognition of named entities from text data sets. The quality of Entity Extraction has a great impact on the efficiency and quality of subsequent knowledge acquisition, so it is the most critical part of Knowledge Extraction. Named Entity Recognition technology is mainly divided into rule based Entity Recognition, Machine Learning based Entity Recognition and Deep Learning based Entity Recognition. In the early days, the work of Named Entity Recognition was mostly to build limited rules manually, and to find strings matching these rules from the text. This method costs too much labor cost and time cost, and single domain rules are difficult to transplant to other knowledge fields, so more and more scholars turn their attention to entity recognition technology based on Machine Learning and Deep Learning.

Machine Learning is a popular method in entity recognition at present. The common Machine Learning methods include Maximum Entropy (ME), Support Vector Machine (SVM), Hidden Markov, Conditional Random Field (CRF) and Deep Learning model. In common Named Entity Recognition tasks, CRF and SVM have very similar effects. In the study of English medical entity recognition, Jiang [17] found that the recognition effect of CRF is better than that of SVM. Lei [18] compares CRF, SVM, SSVM and ME, then found that SSVM is slightly better than other methods in Chinese medical entity recognition. In the field of text mining, Deep Learning can better represent the semantic features of large-scale text data through learning and training, and has a good performance in entity recognition. Lample [19] proposed the structure of the combination of BiLSTM and CRF model for label prediction. The combination model achieved the best results in the task of entity recognition in four languages. Ma and Hovy [20] added character level embedding to the BiLSTM-CRF model, and the F value of entity recognition on the corpus of CoNLL 2003 reached 91.21%.

Relation extraction is one of the basic components of information extraction, which refers to the task of automatically extracting the implicit relations between entities in text in the process of natural language processing. In terms of implementation, according to the dependence of manual annotation data, it can be subdivided into supervision based, semi supervision based, unsupervised and open domain oriented extraction. In recent years, Deep Learning has also been applied to relationship extraction, and achieved better results than traditional Machine Learning methods.

Knowledge Fusion. Knowledge Fusion includes Entity Disambiguation and Reference Resolution. Although structured entities and relationships are extracted from the model, there are still quite a lot of redundant or error information in these knowledge. Knowledge Fusion is to solve this problem. Because there are many types of data in evidence data, multiple Knowledge Graphs can be constructed, so it is necessary to realize the integration of various types of evidence data resources based on association data, so as to build a complete Chain of Evidence based on Knowledge Graph and semantic web. The Knowledge Fusion of evidence data is a process that takes the knowledge element with consistent structure description as input, merges and unites the input knowledge elements according to corresponding rules. Then generate new available knowledge objects. The module of knowledge extraction and fusion firstly extracts knowledge from the distributed and heterogeneous evidence data sources at the structural level. Then to provide a consistent data structure, it eliminates the syntactic and semantic differences of the distributed and heterogeneous data through matching and transformation. Secondly, according to the classification rules of knowledge use, type and application, generates data knowledge through ontology identification, knowledge construction and other technologies. Finally, set the corresponding fusion rules and algorithms according to the problem solving, and fuse the knowledge elements with the support of knowledge reasoning and fusion technology. Then we can generate the associated knowledge that can be applied to the evidence chain, and dynamically update it into the constructed Knowledge Graph.

Entity Ambiguity refers to the fact that a named entity may be related to multiple entity concepts. In this study, Entity linking is chosen to solve entity ambiguity. Entity linking is one of the sub tasks of natural language processing, which aims to identify the concept entities in the text that are linked to the specific entities in the related knowledge base. For an entity to be linked, the entity linking module in this paper mainly includes three steps:

a. Generate candidate entity set from the entity database by directly retrieving the entity in the knowledge base.
b. Sort candidate entities, the main methods can be divided into three categories: based on popularity, based on semantic relevance, based on supervised learning.
c. When the candidate entity set cannot be generated in the knowledge base, the approximate entity is given by calculating the semantic similarity of the text.

Blanco [2] proposes a Named Entity Linking algorithm using Graph Distance, which belongs to the entity sorting based on semantic similarity in the second step above. The candidate selection is formalized as the optimization problem of the underlying concept graph, and the value to be optimized is the average distance between the nodes in the graph. Cifariello [21] combined the language modeling technology based on text evidence with Wikipedia through entity linking. Huang [22] compares a variety of Entity Linking methods, and finally chose a method of using popularity. Combined with Baidu Encyclopedia data and Word2vec technology, they proposed a Chinese Micro blog Entity Linking method, and achieved good results.

In the extracted knowledge, there may also be a problem that an entity corresponds to multiple concept terms. It refers to the process of digestion according to the definition of Automatic Content Extraction (ACE) evaluation by National Institute of Standards and Technology (NIST) of the United States to determine the entity it points to in the standard knowledge base for the expression in the text.

The Hobbs algorithm [23] is one of the earliest pronoun resolution algorithms, which is mainly based on the syntactic analysis tree for correlation search. It includes two algorithms: one is completely based on syntactic knowledge, also known as simple Hobbs algorithm; the other is both considering syntactic knowledge and semantic knowledge; Denis and Baldridge [24] use integer linear programming to infer the referential detection of noun phrases (expression detection) and coreference resolution increase the F value of ACE data by 3.7%–5.3%. Lu [25] Proposed a WSD method based on knowledge graph, using graph scoring algorithm to evaluate the importance of word node and determine the correct meaning of ambiguous words.

Knowledge Description. The purpose of evidence association based on Knowledge Graph is to reduce the workload of scholars or other users, save their time cost, and help to achieve the functions of reading guide, QA system and recommendation. Therefore, the description of entity nodes in Knowledge Graph is also the focus of this study. The description of evidence knowledge node mainly refers to the description of evidence data Ontology, including data author, publication time, data subject and data source; in addition, the description of data related nodes also includes the description of various related evidence quantity and types.

By describing the nodes in the Knowledge Graph, we can realize the function of guide reading, that is, users do not need to browse the Knowledge Graph or retrieve the relevant nodes on a large scale, only need to browse the interested evidence nodes, then they can understand the various attributes of the data and the overview of the data. At the same time, the nodes that record users' feedback or use records can be analyzed through the description of the nodes, which can make user portraits, make relevant recommendations, and return knowledge nodes that may be of interest to users in the user retrieval results.

3.4 Chain of Evidence Based on Knowledge Graph

Evidence Association Model Based on Deep Relation Reasoning. Knowledge Reasoning in evidence Knowledge Graph refers to the establishment of a new relationship between geographic entities based on the entity relationship data in the knowledge base and computer reasoning, so as to expand and enrich the knowledge network.

Knowledge reasoning can be roughly divided into symbol based reasoning and statistical based reasoning. Symbol based reasoning mainly uses related rules to infer new entity relationship from existing entity relationship, and detects logical conflict of Knowledge Graph. Based on statistical reasoning is to use machine learning method to learn new entity relationship from Knowledge Graph through statistical laws, mainly including entity relationship learning method, type reasoning method and pattern induction method. It can also be divided into: description based reasoning, rule mining based reasoning, probability based reasoning and representation based learning and neural network based reasoning. Because the knowledge base is large enough and the knowledge network is rich enough, the implicit relationship and knowledge can be inferred from the knowledge base.

Dkrl algorithm [26] based on convolutional neural network takes the semantic information of entity description into account better. It can solve the problem of data sparsity and improve the ability of model to distinguish relations by reducing the dimension of data features through convolutional neural network. After getting the potential relationship between the evidence entities, it can provide auxiliary reasoning function according to the user query, help the user to find the potential relationship between the evidences, and build the evidence chain based on the Knowledge Graph.

Recommendation Sorting Algorithm Based on Evidence Correlation and Evidence Influence. In the process of building the Knowledge Graph of evidence data, knowledge evaluation, including the evaluation of team evidence, is also an important part, that is, to judge whether the evidence data has the value and value of proving the facts to be proved. In this paper, the theoretical area of influence graph is used as a reference standard for knowledge proof evaluation. Through the methods of likelihood ratio, Bayes decision and D-S (Dempster/Shafer) mathematical model, we can provide a big premise to judge the relevance of evidence, that is, the posterior probability p, and then we can scientifically standardize and quantify each link of the evidence chain formation, and construct its evidence ability review rules to present the process and results of the proof standard in a relatively intuitive form Finally, we can build a complete evidence Knowledge Graph through deep learning. Yang [27] proposes a spatial clustering algorithm based on unsupervised method, which uses different discount factors to modify the evidence set, and then uses D-S rules for information fusion.

Scholars generally divide evidence into documentary evidence, oral evidence record and written testimony from the theoretical level. Different types of evidence also have different "Evidence Influence" in the judicial scene. How to arrange the data stored in the Knowledge Graph based on the query of users is one of the research focuses of this paper. At present, the evidence superiority standard commonly used in solving territorial disputes will be included in the recommended algorithm system as an important reference for feature weight adjustment.

Knowledge Graph representation learning can map the entity of Knowledge Graph to a low dimensional space, vectorize the entity of Knowledge Graph, represent the structural information of Knowledge Graph through vectors, improve the efficiency of inter entity relationship calculation, and apply to the entity relationship calculation of large-scale Knowledge Graph. Node2vec [28] is an algorithm framework proposed by Aditya Grover and jure leskovec in 2016. It uses Word2vec's ideas for reference, defines the objective function as shown in Formula (1), maximizes the preservation of local neighbor structure of nodes, and generates node communities by second-order random walk. Node2vec algorithm can combine the local and macro information in the network to extract network features, which has good adaptability.

$$max_f \sum_{u \in V} \log P(N_S(u)|f(u)) \tag{1}$$

In this paper, the Node2vec network representation algorithm based on deep learning is proposed to embed Knowledge Graph entities into k-dimensional space, so as to better find the correlation between evidences. Combined with sorting learning technology and deep learning technology, a recommended sorting algorithm for evidence correlation and evidence influence is implemented. The algorithm flow is shown in Fig. 2.

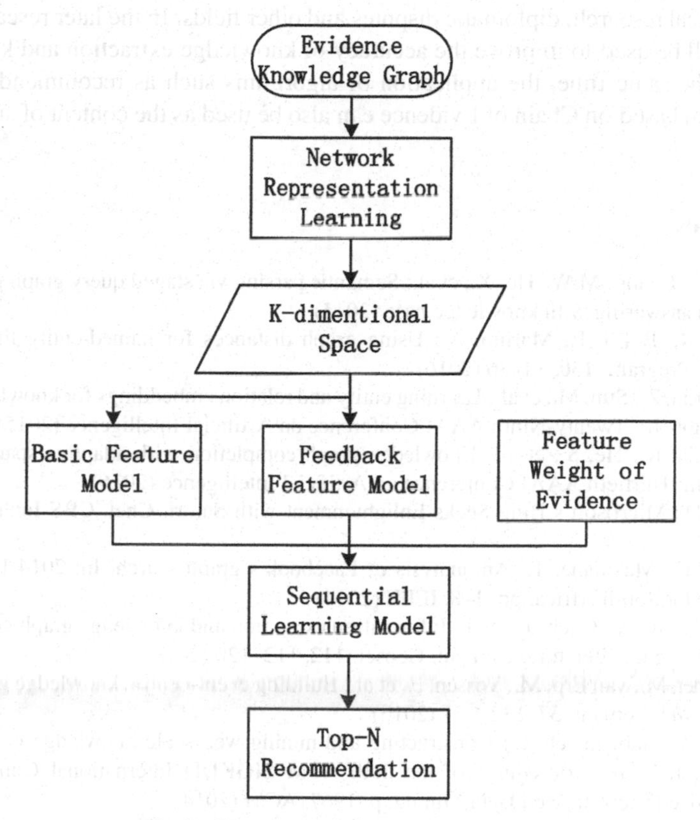

Fig. 2. Recommendation sorting algorithm

4 Conclusions

The analysis, processing and organization of large-scale, multi-source and heterogeneous evidences are expanding rapidly in the current data, and the research on the construction of evidence chain based on electronic data is becoming more and more important. Through the analysis of evidence data, this paper carries out evidence Ontology design, evidence data cleaning and data pre-processing, pre-labeling; Knowledge Extraction through Machine Learning, including Entity Extraction and Relationship Extraction; Entity Disambiguation by Entity Link, and then achieves Knowledge Fusion. Finally,

Knowledge Verification and Knowledge Reasoning are proposed, then construct Knowledge Graph based on extracted effective knowledge. After that, through the recommendation sorting algorithm based on evidence correlation and evidence influence, the Chain of Evidence is completed. The purpose of the evidence chain framework based on Knowledge Graph constructed in this study is to shorten the distance between knowledge, not limited to a certain field, which can be used as an intelligence tool in legal proceedings, historical research, diplomatic disputes and other fields. In the later research, deep learning will be used to improve the accuracy of knowledge extraction and knowledge fusion; at the same time, the application of algorithms such as recommendation and Q&A system based on Chain of Evidence can also be used as the content of subsequent research.

References

1. Yih, S.W., Chang, M.W., He, X., et al.: Semantic parsing via staged query graph generation: question answering with knowledge base (2015)
2. Blanco, R., Boldi, P., Marino, A.: Using graph distances for named-entity linking. Sci. Comput. Program. **130**, 24–36 (2016)
3. Lin, Y., Liu, Z., Sun, M., et al.: Learning entity and relation embeddings for knowledge graph completion. In: Twenty-Ninth AAAI Conference on Artificial Intelligence (2015)
4. Ji, G., Liu, K., He, S., et al.: Knowledge graph completion with adaptive sparse transfer matrix. In: Thirtieth AAAI Conference on Artificial Intelligence (2016)
5. Farber, D.: Microsoft's Bing Seeks Enlightenment with Satori. Cnet. CBS Interactive Inc. (2013)
6. Khan, Z.C., Mashiane, T.: An analysis of Facebook's graph search. In: 2014 Information Security for South Africa, pp. 1–8. IEEE (2014)
7. Wang, C., Ma, X., Chen, J., et al.: Information extraction and knowledge graph construction from geoscience literature. Comput. Geosci. **112**, 112–120 (2018)
8. Rospocher, M., van Erp, M., Vossen, P., et al.: Building event-centric knowledge graphs from news. J. Web Semant. **37**, 132–151 (2016)
9. Bordes, A., Gabrilovich, E.: Constructing and mining web-scale knowledge graphs: KDD 2014 tutorial. In: Proceedings of the 20th ACM SIGKDD International Conference on Knowledge Discovery and Data Mining, p. 1967. ACM (2014)
10. Niu, B., Huang, Y.: An improved method for web text affective cognition computing based on knowledge graph. CMC-Comput. Mater. Continua **59**(1), 1–14 (2019)
11. Panesar-Walawege, R.K., Sabetzadeh, M., Briand, L., et al.: Characterizing the chain of evidence for software safety cases: a conceptual model based on the IEC 61508 standard. In: 2010 Third International Conference on Software Testing, Verification and Validation, pp. 335–344. IEEE (2010)
12. Ahmad, A.: The forensic chain of evidence model: improving the process of evidence collection in incident handling procedures. In: The 6th Pacific Asia Conference on Information Systems (2002)
13. Hayes, F., Spurgeon, D.J., Lofts, S., et al.: Evidence-based logic chains demonstrate multiple impacts of trace metals on ecosystem services. J. Environ. Manag. **223**, 150–164 (2018)
14. Cosic, J., Cosic, G., Ćosić, J., et al.: Chain of custody and life cycle of digital evidence. Comput. Technol. Appl. **3**, 126–129 (2012)
15. Alobaidi, M., Malik, K.M., Hussain, M.: Automated ontology generation framework powered by linked biomedical ontologies for disease-drug domain. Comput. Methods Programs Biomed. **165**, 117–128 (2018)

16. Bravo, M., Reyes-Ortiz, J.A., Cruz-Ruiz, I., et al.: Ontology for academic context reasoning. Procedia Comput. Sci. **141**, 175–182 (2018)
17. Jiang, M., Chen, Y., Liu, M., et al.: A study of machine-learning-based approaches to extract clinical entities and their assertions from discharge summaries. J. Am. Med. Inform. Assoc. **18**(5), 601–606 (2011)
18. Lei, J., Tang, B., Lu, X., et al.: A comprehensive study of named entity recognition in Chinese clinical text. J. Am. Med. Inform. Assoc. **21**(5), 808–814 (2013)
19. Lample, G., Ballesteros, M., Subramanian, S., et al.: Neural architectures for named entity recognition. In: Proceedings of NAACL-HLT, pp. 260–270 (2016)
20. Ma, X., Hovy, E.: End-to-end sequence labeling via bi-directional LSTM-CNNs-CRF. arXiv preprint arXiv:1603.01354 (2016)
21. Cifariello, P., Ferragina, P., Ponza, M.: Wiser: a semantic approach for expert finding in academia based on entity linking. Inf. Syst. **82**, 1–16 (2019)
22. Huang, D., Wang, J.: An approach on Chinese microblog entity linking combining baidu encyclopaedia and word2vec. Procedia Comput. Sci. **111**, 37–45 (2017)
23. Hobbs, J.R.: Resolving pronoun references. Lingua **44**(4), 311–338 (1978)
24. Denis, P., Baldridge, J.: Joint determination of anaphoricity and coreference resolution using integer programming. In: Human Language Technologies 2007: The Conference of the North American Chapter of the Association for Computational Linguistics; Proceedings of the Main Conference, pp. 236–243 (2007)
25. Lu, W., Meng, F., Wang, S., et al.: Graph-based chinese word sense disambiguation with multi-knowledge integration. Comput. Mater. Continua **61**(1), 197–212 (2019)
26. Xie, R., Liu, Z., Jia, J., et al.: Representation learning of knowledge graphs with entity descriptions. In: Thirtieth AAAI Conference on Artificial Intelligence (2016)
27. Yang, K., Tan, T., Zhang, W.: An evidence combination method based on DBSCAN clustering. Comput. Mater. Continua **57**(2), 269–281 (2018)
28. Grover, A., Leskovec, J.: node2vec: scalable feature learning for networks. In: Proceedings of the 22nd ACM SIGKDD International Conference on Knowledge Discovery and Data Mining, pp. 855–864. ACM (2016)

Research on the Construction of Intelligent Catalog System for New Media Information Resources

Yiting Li[1], Jin Shi[1(✉)], Jin Han[2], Mingxin Lu[1], and Yan Zhang[2]

[1] School of Information Management, Nanjing University, Nanjing 210093, China
shijin@nju.edu.cn

[2] School of Computer and Software, Nanjing University of Information Science and Technology, Nanjing 210044, China

Abstract. With the electronic transformation of traditional paper media, the integration of government new media platforms, the establishment of various institutions on social, music, video, shopping and other websites, and the rise of We-Media, massive new media information resources come out. In order to solve the problem of organizing and managing such information resources, and meet the needs of people's deep use, propose an intelligent catalog system based on the development status of new media information resources. The intelligent catalog system for new media information resources combines massive data, bibliography theories, various catalogs, emerging intelligent technologies, catalog applications, etc., consisting of catalog data, catalog set, catalog application and support and guarantee. Metadata, classification and coding, and knowledge graph are key technologies for the construction of intelligent catalog system. Through the combination of bibliographic theory and information technologies, new media information resources can be better managed and utilized, and it can be fulfilled to provide personalized, diversified, integrated, accurate knowledge services for the users.

Keywords: Intelligent catalog system · New media information resources · Catalog application

1 Introduction

As a tool to reveal the external and content characteristics of literature and books, the catalog is widely used in the organization and management of information resources. The establishment of information resource catalog system has also become a key part of information resource construction.

However, people's understanding of the connotation and extension of the catalog are always limited by the technology and thinking of the era. The working objects and application forms of the traditional catalog do not take the growth trend of network information resources into account, which makes the existing catalog system difficult to cope with some problems. That is the organization and management of massive new media information resources brought about by the outbreak of five-year short texts and

X. Sun et al. (Eds.): ICAIS 2020, LNCS 12240, pp. 42–52, 2020.
https://doi.org/10.1007/978-3-030-57881-7_4

short videos. The optimization and transformation of the catalog system have become the trend of the times.

This paper analyzes the development status of new media information resources. Then based on the characteristics of information resources and people's needs, this paper proposes the intelligent catalog system for new media information resources which integrates various catalogs, theories, methods and information technologies. It is expected to provide reference for the construction and the management of new media information resources catalog, and help to optimize and upgrade catalog services.

2 Analysis of the Current Situation of the Use of New Media Information Resources

With the electronic transformation of traditional paper media, the integration of government new media platforms, the establishment of various institutions on social, music, video, shopping and other websites, and the rise of We-Media, the development of online media has ushered in a new wave. The threshold for publishing and interacting with various types of information has been greatly reduced, resulting in a large amount of content data collected, displayed and pushed by the website platform, and operational data of interaction between the users and the information providers.

This type of information resources, which rely on the Internet and have a huge upgrade and breakthrough in terms of quantity, forms, distribution channels, and communication influence, are the new media information resource studied in this paper. It is characterized by a large number, a wide variety, rapid growth and a complex relationship between resources.

The convenience and high speed of the Internet allows people to freely access and utilize new media information resources anytime, anywhere to meet their learning, entertainment, social, work and other needs. Nowadays, people's requirements for the use of new media information resources are mainly convenience, rationality, depth and interaction. However, due to new media information resources in the early stage of development, there are still some shortcomings in the organization and management of this emerging resources.

Websites and platforms conduct inaccurate information indexing, and the items settings of catalog have problems such as crossover and incomplete coverage [1], which affects the user's resource acquisition and positioning. In terms of resource utilization, the use of functions such as search and recommendation are most common. However, due to the complexity of the resources themselves and the individual users, and the granularity of the information resources organization, the functions of searching thoroughly and precisely and recommend accurately still have a lot of room for improvement. It is still unable to provide users with high-quality information services and knowledge service.

3 Overview of Intelligent Catalog System for New Media Information Resources

3.1 Concept of Intelligent Catalog System for New Media Information Resources

Library and Information Science defines the catalog system as the established catalogs and their complementary and interconnected organic whole. At the same time, the catalog system clarifies the types of catalogs and their quantity, the functions of each catalog and relationships between various catalogs. In recent years, in the fields of government affairs, education, industry and commerce, medical care, and urban construction, in order to promote the unified management and utilization of distributed information resources, many specialized information resource catalog systems have been designed and constructed, and relevant work processes and institutional standards have been formulated[2–5].

Different from the form and content defined by the above catalog systems, the intelligent catalog system for new media information resources not only includes various catalogs that reveal the shape and content characteristics of new media information resources, but also has to undertake the educational mission of the catalog. It will combine massive data, bibliography theories, various catalogs, emerging intelligent technologies, catalog applications, etc., to establish a integrated system. The intelligent catalog system for new media information resources can meet demands for deep utilization of network information resources, to achieve personalized, diversified, integrated, accurate knowledge services.

3.2 Functions of Intelligent Catalog System for New Media Information Resources

Resource Organization and Management. In view of the multi-source heterogeneity of new media information resources, the basic function of the intelligent catalog system is to organize and integrate massive data according to unified standards, and then formulate corresponding storage and management rules to facilitate the platform itself and users' subsequent exploitation. For image, audio, video and other types of data resources, not only in simple labeling and classification, but also rely on the support of new technologies, updating the methods and means of upgrading data organization. According to the external and content characteristics, and different needs of data, appropriate description and cataloging formats will be developed for more detailed processing and storage. The intelligent catalog system emphasizes intelligence of the concepts and actions of organizing data resources. The ultimate goal is to comprehensively consider users' needs and organized content, and improve the speed and accuracy of identifying and organizing data.

Resource Location and Access. Various website applications provide a platform for information release and interaction between organizations and individuals. However, when users have reading requirements for specific resource, they often encounter obstacles due to the complexity. Therefore, the intelligent catalog system provides accurate location and access functions to users. Users use search engines to search for certain resource, or use a catalog tool to locate according to the hierarchical structure. This function relies on detailed and accurate catalog establishment specifications and proper and comprehensive catalog application development.

Resource Utilization and Mining. The intelligent catalog system also starts from the practicity of bibliography, taking user's deep requirements for using and mining data into account, and incorporating the functions of guiding, recommendation, etc. into the scope of establishment. In addition to the rational organization, accurate position and queries of resources, users are given the means to explore the relationship between resources and the hidden knowledge. By guiding users to read, query, utilize, and recommending new media information resources that meet user needs and preferences, catalog's roles of studying and researching are truly implemented.

4 Architecture of Intelligent Catalog System for New Media Information Resources

According to the definition of intelligent catalog system for new media information resources and its functions, the intelligent catalog system consists of catalog data, catalog set, catalog application and support and guarantee. The overall framework is shown in Fig. 1.

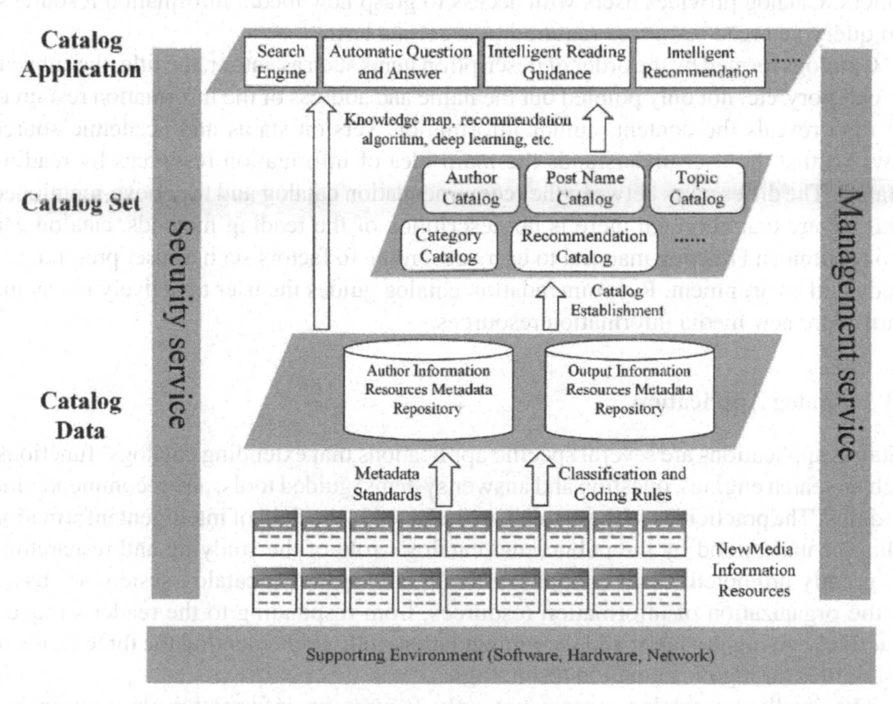

Fig. 1. Architecture of the new media information resource intelligent catalog system

4.1 Catalog Data

New media information resources from social networking, music, video, shopping and other platforms follow unified metadata standards, classification and coding rules to form metadata about the authors and their post, which are stored in the author information resources metadata repository and output information resource metadata repository. The establishment of metadata and the design of classification and coding are all aimed at uniqueness, integrity, systemicity and scalability, so as to avoid problems such as indexing errors and unclear references.

The output information resource metadata repository includes both metadata about post, as well as metadata about text, pictures, audio, and video, etc. in the post. Output information resource metadata will establish corresponding affiliation with its authors, to facilitate the development and use of subsequent retrieval, reading, recommendation and other functions.

4.2 Catalog Set

Catalog set is a collection of author catalog, post name catalog, topic catalog, category catalog, recommendation catalog and so on, based on the metadata describing the author and its output. Catalog guides the user's reading mainly consists of four points: specifying reading material, giving reading orders, explaining reading methods and indicating readers. Catalog provides users with access to grasp new media information resources, and query and read resources required by a certain term.

Catalogs formed by the order of description items such as author, the title, the subject, the category, etc. not only pointed out the name and address of the information resource, but also reveals the content, author information, version status and academic source flow. So that the user understands the main idea of information resources by reading catalog. The differences between the recommendation catalog and the above-mentioned catalogs are that although there is no description of the reading methods, catalog can also recommend reading material to users according to factors such as user preferences, needs and environment. Recommendation catalog guides the user to actively reach and learn more new media information resources.

4.3 Catalog Application

Catalog applications are several specific applications that extending catalogs' functions, such as search engines, question and answer systems, guided tools, and recommendation modules. The practicities of the catalog are reflected in the field of intelligent information behavior understood by the public, and catalog's role of the studying and researching are greatly promoted. Catalog applications in the intelligent catalog system are based on the organization of information resources, from responding to the reader's request to actively giving guidance and recommendation, fully implementing the three levels of practicities through information technologies.

The intelligent catalog system not only focuses on information description and knowledge organization, but uses this part as the basis seeking to expand the role of reading and studying. In the past, catalog is to help people find the address and then get the original text. Nowadays, it is necessary to improve the retrieval efficiency. People read the article for the sake of understanding, and now they need to study how to quickly and accurately answer the user's questions after organizing the knowledge unit. The original author catalog, topic catalog, recommendation catalog teach people to read, and now we need to explore more ways to guide and recommend.

Intelligent Retrieval. In traditional bibliography, the first step of retrieval is to find the clue or address from catalog, and then get the original text. With the increase of data, people's requirements for the convenience of information acquisition and the enhancement of search awareness, two steps merging becomes the norm. Search engines have become the most commonly used search tool. As of June 2019, the number of search engine users in China reached 695 million, accounting for 81.3% of the total netizens [6].

Intelligent retrieval is continue to seek breakthroughs and developments at a superior level today, further improving search speed, precision and recall. Intelligent retrieval reduces the use threshold of search tools through concept level search, promoting the universalization of information search and use. More importantly, intelligent retrieval must always focus on the user and accurately identify user characteristics and needs, so as to provide users with the most satisfactory personalized search results.

Intelligent retrieval is mainly composed of organization of data, process of retrieval conditions, hit and presentation of retrieval results. The organization of data relies on the establishment of standard metadata repository, which lays a good foundation for intelligent retrieval. The data only can be returned to the user when its accurately described features matching with the retrieval conditions [7]. The processing of the retrieval condition and the hit of the retrieval result are the progress of understanding and conversing user's retrieval expression and matching with the data set. When the query wrongly spelled, it is necessary to make a reminder and give a possibly correct expression, which is the level of intelligence that many search engines have achieved. The presentation of the search results involves the ranking of hit results and the recommendation of other related content. In most scenarios, it depends on the similarity between retrieval condition and the result, and then sorting in descending order. The most ideal presentation method is that the user can make different choices.

Intelligent Question and Answer. Intelligent retrieval is based on controlled language or more standardized retrieval language, while intelligent question and answer can respond to the question of the natural language expression and return answers. It is a further expression that responding to the user request, expanding the scope of catalog application. In the past, when people were confused, they could read books to find answers, while the catalog provided a way for positioning and played a guiding role. By directly giving the answers to satisfy user's desire, the user is given the interactive experience, which meets the requirements of intelligence.

Based on the knowledge base, the intelligent question and answer accurately understands the user's problem through natural language processing, and then matches the standardly expressed problem with the knowledge in the knowledge base. At last, the answer that best meets the user's request can be returned. Because the matching of questions and knowledge in the intelligent question and answer and the hit of the answer are similar to intelligent retrieval, there are some similarities in the key technologies involved, such as knowledge graph, word vector, blur match, etc. Considering the user's direct demand for question and answer, intelligent question and answer needs to complete the screening of hit answers, mainly relying on sorting learning technology. That is selecting different features from the perspectives of word frequency, grammar, semantics and so on, and setting weights to score and sort the hit answers. The final number of answers is determined by the service provider and user [8].

Intelligent Reading Guidance. Guide reading in intelligent catalog system is different from presenting organized documents and guiding readers to find what they need through category catalog, author catalog, article name catalog, topic catalog, library collection catalog and so on. Technologies of wireless communication, Internet of Things, data mining, and visualization support the disclosure and utilization of data, information, and literature in multiple scenarios.

Intelligent reading guidance for new media information resources is to judge the user's reading needs and preferences through data mining. Develop good reading habits and interests in the process of teaching reading online news, posts in Twitter, Weibo and WeChat, or watching videos. Users can learn professional or non-professional knowledge and skills through reading, which is also part of the knowledge service.

Intelligent Recommendation. The forms of catalog's recommendation function previously have been recommendation catalog, book reviews, and abstracts, so as to popularize knowledge of a particular topic or to guide self-study. Nowadays, faced with the complexity of information resources, it is required to improve the understanding of users and the accuracy of matching users and content to achieve a higher level of recommendation. So that catalog's leading role can be full played, and users' enthusiasm of using the catalog tools and learning knowledge can be mobilized.

There are two main ways to implement intelligent recommendation for new media information resources. One is to use content recommendation by search engine, that is, to recommend information content or other search terms in certain areas of the homepage or pop-up windows, thereby giving users channels of expanding search and reading. At present, some mainstream search engines have used the platform entry advantages to connect news, short videos and other content to launch information flow products to develop recommended services.

Another implementation method is a personalized customized dynamic recommendation catalog. Dynamic is the variability of catalog. Forms of catalog and recommended content will be changed as the user's interests and habits change. For example, website sets up a dynamic catalog for different users on a dedicated page or section. First, it accurately indexes the content displayed by the website in the work of intelligent organization, and then tag the user according to his/her operations of browsing, collecting, loving or sharing. The tags will be periodically modified and updated. At last, personalized recommendation directory that meets the user's recent preferences and needs can be formed through matching user tags with website content tags.

4.4 Support and Guarantee

Support and guarantee are the necessary conditions for maintaining the organization, management and utilization of information resources in the intelligent catalog system.

They are mainly composed of supporting environment, standard specification, security and management. The supporting environment includes the software, hardware, and network of the system. The standard specification is used to make standards of catalog system, such as metadata creation standards, classification and coding rules, and catalog establishment rules. Security and management are security services and management services to maintain the operation of the catalog system, so that the catalog practice can be carried out in accordance with the relevant standards and management specifications of the state regarding information security.

5 Key Technologies for the Construction of Intelligent Catalog System for New Media Information Resources

5.1 Metadata

Metadata is a method of organizing data that is widely recognized and applied. Metadata is data that describes data and can be thought of as an electronic catalog. The items describing the data content are automatically extracted from the massive data of various sources and forms, and then manually filtered and supplemented to complete the standardized, orderly and efficient data organization. The efficiency of cataloging based on metadata technology is about three times that of manual cataloging and has high reusability. After building the catalog metadata of a certain subject area, it can be applied to other fields with only minor adjustments. It has broad application prospects in the data organization of shopping, music, video and other website platforms that collect various thematic data. Refer to the Chinese national standard *Government information resource catalog system - Part 3: Core metadata*. The core metadata of the author information resource and output information resource are described in Table 1 and Table 2.

5.2 Classification and Coding

The classification of new media information resources is based on scientificity, systemicity, extendability, compatiblity and practicity [9]. Information about author and output from various sources are distinguished and classified according to certain methods. Referring to the Chinese national standard *Government information resource catalogue system - Part 4: Government information resource classification*, the classification of output information resources is still based on topic, which can be divided into 21 first-class categories. Among them, “integrated government affairs”, “information industry”, “tourism and service industry”, “culture, health and sports”, “civil affairs and community” are several first-class categories of big amount of output information resources. Secondary classes have changed due to main sources of new media information resources and future developments. The classification of author information resources is relatively simple, divided into government affairs, finance, health, culture, sports, entertainment, food, beauty, pets and games.

The new media information resource coding is to give the author, the output and specific content in it symbols with certain rules and easy to be processed by the computer and the person, and then form a collection of code elements, which often reflect the way of

Table 1. Author information resource core metadata.

Metadata item	Name	Definition	Data type	Range
Name of author	authorName	The name author used to introduce himself when publishing information resources	string	No limits
Brief introduction of author	authorIntroduction	Text that briefly introduce the author	string	No limits
Identifier of metadata	metadataIdentifier	Unique identifier of the metadata	string	No limits
Name of organization	organizationName	Name of organization in charge	string	No limits
Date of update	metadatadateUpdate	Date when updating metadata	Date type	Executed according to GB/T 7408-2005, the format is CCYY-MM-DD

classification. According to the Chinese National Standard *Basic principles and methods for information classifying and coding*, coding needs to conform to the principles of uniqueness, rationality, expandability, conciseness, applicability and normativeness.

In this study, the combination code is used to digitally identify the author's information resources and output information resources. The combination code is a conforming code composed of some code segments, showing different characteristics of the coding object. The Chinese citizenship number is a typical combination code. The author information resources can be encoded in classification and the order of storage into the database. Output information resources coding needs to take classification, type, release time, order of storage into the database, and other factors into consideration.

5.3 Knowledge Graph

Knowledge map is an important technology that can realize the functions of searching, question answering and recommendation in catalog applications. Intelligent retrieval needs to understand and transform user's search expression, and then match the data set. Intelligent retrieval can use the knowledge map technology to transform the query as a triple of "entity-relationship-entity" or "entity-attribute-attribute value", and then conduct similarity calculation and matching with knowledge in the knowledge base. The search engine based on knowledge map technology can also directly give answers or return integrated forms of search results such as text, pictures, videos, etc., eliminating the time for users to filter and compare.

Table 2. Output information resource core metadata.

Metadata item	Name	Definition	Data type	Range
Name of output	outputTitle	Brief title of the output information resource content	string	No limits
Date of publishing output	outputdateOfPublication	Date when author published the information resource	Date type	Executed according to GB/T 7408-2005, the format is CCYY-MM-DD
Time of publishing output	outputtimeOfPublication	Time when author published the information resource	Date type	Executed according to GB/T 7408-2005, the format is XX:YY
Type of output	outputType	Type of published information resources	string	No limits
Abstract of output	outputAbstract	Text that briefly introduce the output.	string	No limits
Name of author of output	outputauthorName	Name of author who published the information resource	string	No limits
Keyword	keyword	A generic word, formal word or phrase used to summarize the main content of a new media information resource	string	No limits
Name of category	categoryname	Name of category that the output information resource is classified to	string	No limits
Code of category	categorycode	Code corresponding to the category	string	No limits
Identifier of metadata	metadataIdentifier	Unique identifier of the metadata	string	No limits
Name of organization	organizationName	Name of organization in charge	string	No limits
Date of update	metadatadateUpdate	Date when updating metadata	Date type	Executed according to GB/T 7408-2005, the format is CCYY-MM-DD

The matching of questions and knowledge in the intelligent question and answer and the hit of the answer are similar to intelligent retrieval. Therefore, the intelligent question and answer can also accurately understand the user's problem through the knowledge map technology on the basis of building the knowledge base [10]. Then match the problem after the specification expression with the knowledge in the knowledge base, so that can find the answer that best meets the user's request as the answer.

Knowledge map describes the entities of the physical world, the attributes of entities, and the relationships between entities in symbolic form, and can construct the architecture of knowledge points. Therefore, all entities can be organized according to the method of attribute graph clustering [11]. User preference habits can be discovered according to entities that user usually search and browse, using the attribute map clustering, thereby realizing specific content for different users.

6 Conclusion

In order to solve the contradiction between the continuous growth of new media information resources and the need for people to deeply utilize information resources, the optimization and transformation of the catalog system is imperative. This paper researches the development status of new media information resources and people's deep use needs, proposing the intelligent catalog system. Massive data, bibliography theories, various catalogs, emerging intelligent technologies, catalog applications, and other necessary conditions for support and guarantee contribute to the management and use of new media information resources collaboratively. It is expected that the intelligent catalog system can provide reference for the construction and management of new media information resources catalog, and help to optimize and upgrade catalog services.

References

1. Jingang, L., Shibin, L., Wei, L.: The design and implementation of the IDS catalogue data archive system. In: Zhang, T. (ed.) International Conference on Instrumentation, Measurement, Circuits and Systems 2011, LNCS, vol. 127, pp. 837-844. Springer, Heidelberg (2012)
2. Stillerman, J., Fredian, T., Greenwald, M., et al.: Data catalog project—a browsable, searchable, metadata system. Fusion Eng. Des. **112**, 995–998 (2016)
3. Siabato, W., Claramunt, C., Manso-Callejo, M.Á., et al.: TimeBliography: a dynamic and online bibliography on temporal GIS. Trans. GIS **18**(6), 799–816 (2015)
4. Truitt, A., Young, P.A., Spacek, A., et al.: A catalog of stellar evolution profiles and the effects of variable composition on habitable systems. Astrophys. J. **804**(2), 145–160 (2015)
5. Zhong, S.: Heterogeneous memristive models design and its application in information security. Comput. Mat. Continua **60**(2), 465–479 (2019)
6. China Internet Network Information Center. http://www.cac.gov.cn/2019-08/30/c_1124938750.htm. Accessed 28 Oct 2019
7. Lacasta, J., Lopez-Pellicer, F.J., Espejo-García, B., et al.: Aggregation-based information retrieval system for geospatial data catalogs. Int. J. Geogr. Inf. Sci. **31**(8), 1583–1605 (2017)
8. Weijin, J., Jiahui, C., Yirong, J., et al.: A new time-aware collaborative filtering intelligent recommendation system. Comput. Mater. Continua **61**(2), 849–859 (2019)
9. BoberićKrstićev, D., Tešendić, D.: Mixed approach in creating a university union catalogue. Electronic Library **33**(6), 970–989 (2015)
10. Oramas, S., Ostuni, V.C., Noia, T.D., et al.: Sound and music recommendation with knowledge graphs. Acm Trans. Intell. Syst. Technol. **8**(2), 1–21 (2016)
11. Bohan, N., Yongfeng, H.: An improved method for web text affective cognition computing based on knowledge graph. Comput. Mater. Continua **59**(1), 1–14 (2019)

Information Security

Construction of a Class of Four-Weight Linear Codes

Ee Duan(✉), Xiaoni Du, Tianxin Wang, and Jiawei Du

College of Mathematics and Statistics, Northwest Normal University, Lanzhou 730070, Gansu, People's Republic of China

eeduan@126.com, ymldxn@126.com, wangtianxin1019@163.com, 3230096738@qq.com

Abstract. Linear codes with a few weights are of importance in consumer electronics, communications and data storage systems, and they have many applications in secret sharing schemes, authentication codes, association schemes and strongly regular graphs. In this paper, by applying the defining set theory, a class of four-weight linear codes over $\mathbb{F}_p$ are constructed. Then, we use exponential sums to determine their weight distributions explicitly. Furthermore, An example is given to show the correctness of the results by Magma program.

Keywords: Linear codes · Weight distributions · Gauss sums

1 Introduction

Throughout this paper, let p be an odd prime and m a positive integer. Let $\mathbb{F}_q$ denote the finite field with $q = p^m$ elements and $\mathbb{F}_q^* = \mathbb{F}_q \backslash \{0\}$. Let Tr and Tr_s denote the absolute *trace function* from $\mathbb{F}_q$ onto $\mathbb{F}_p$ and $\mathbb{F}_{p^s}$ onto $\mathbb{F}_p$, respectively.

An $[n, k, d]$ linear code $\mathcal{C}$ over $\mathbb{F}_q$ is a k-dimensional subspace of $\mathbb{F}_q^n$ with minimum Hamming distance d. Let A_i be the number of codewords with Hamming weight i in a code $\mathcal{C}$. The *weight enumerator* of $\mathcal{C}$ is defined by

$$1 + A_1 z + A_2 z^2 + \ldots + A_n z^n,$$

and the sequence $(1, A_1, \ldots, A_n)$ is called the *weight distribution* of $\mathcal{C}$. If $|\{1 \leq i \leq n : A_i \neq 0\}| = t$, then we call $\mathcal{C}$ a *t-weight code.*

Linear codes with a few weights can be applied in association schemes [2], authentication codes [6], combination designs [3], secret sharing schemes [15] and strongly regular graphs [17]. Since the weight distribution not only gives the minimum weight of the code, which determines the error correction ability of the one, but also allows the computation of the error probability of error detection and correction with respect to some error detection and error correction algorithms. Hence, the weight distribution attracts much attention in coding theory and much work focus on the determination of the weight distributions of linear codes. Therefore, it is an important research topic in coding theory.

Supported by National Natural Science Foundation of China (61772022).

X. Sun et al. (Eds.): ICAIS 2020, LNCS 12240, pp. 55–64, 2020.
https://doi.org/10.1007/978-3-030-57881-7_5

L. D. Baumert and R. J. Mceliece firstly propose a new method to construct linear codes in 1972 [1]. Later, in 2007, with this method C. Ding and H. Niederreiter proposed that linear codes with a few weights can be constructed by selecting a properly defining set, which is a subset of finite fields [7]. That is, for any nonempty set $D = \{d_1, d_2, \ldots, d_n\} \subseteq F_q$, a *linear code* of length n over $\mathbb{F}_p$ is defined as

$$\mathcal{C}_D = \{(Tr(xd_1), Tr(xd_2), \ldots, Tr(xd_n)) : x \in \mathbb{F}_q\}, \tag{1}$$

the set D is called the *defining set* of linear code $\mathcal{C}_D$. After that, How to choose defining sets for good linear codes interesting and important. many classes of linear codes with a few weights were obtained by the method (See Refs. [1,5,7,9–11,13,14,16,18] for example). From the proper selection of D, many linear codes can be obtained. There is a recent survey on three-weight cyclic codes [19]. Some interesting two-weight and three-weight codes were presented in Refs. [2,20–28]. In particular, some optimal codes and almost optimal codes were given in [5,10,11,13,14,16].

In this paper, based on the above construction method, we will construct a class of linear codes with length p^{m-2} and determine the weight distribution of the codes explicitly by using character sums over finite fields.

2 Preliminaries

In this section, we introduce some basic conception on character sum.

An additive character over $\mathbb{F}_q$ is a non-zero function χ from $\mathbb{F}_q$ to the set of complex numbers of absolute value 1 such that $\chi(x+y) = \chi(x)\chi(y)$ for any pair $(x, y) \in \mathbb{F}_q^2$. For each $u \in \mathbb{F}_q$, the function

$$\chi_u(v) = \zeta^{Tr(uv)}, \ v \in \mathbb{F}_q,$$

denotes an *additive character* over $\mathbb{F}_q$, where $\zeta = e^{2\pi\sqrt{-1}/p}$ is a primitive p-th root of unity. Since $\chi_0(v) = 1$ for all $v \in \mathbb{F}_q$, which is the trivial additive character over $\mathbb{F}_q$, we call χ_1 the *canonical additive character* over $\mathbb{F}_q$ and we have $\chi_u(x) = \chi_1(ux)$ for all $u \in \mathbb{F}_q$. The additive character satisfies the orthogonal property [12], that is

$$\sum_{v \in \mathbb{F}_q} \chi_u(v) = \begin{cases} q \text{ if } u = 0, \\ 0 \text{ if } u \neq 0. \end{cases} \tag{2}$$

Let g be a fixed primitive element of $\mathbb{F}_q$. For each $j = 0, 1, \ldots, q-2$, the function $\lambda_j(g^k) = e^{2\pi\sqrt{-1}jk/(q-1)}$ for $k = 0, 1, \ldots, q-2$ defines a *multiplicative character* over $\mathbb{F}_q$. We call $\eta' = \lambda_{(p-1)/2}$ the *quadratic characters* over $\mathbb{F}_p$. The *Gauss sums* over $\mathbb{F}_p$ is defined by

$$G'(\eta') = \sum_{v \in \mathbb{F}_p} \eta'(v)\chi_1'(v), \tag{3}$$

where χ_1' are the canonical multiplicative and additive character of $\mathbb{F}_p$, respectively.

We define *quadratic residual* and *non-quadratic residual module* p as follows

$$S_p = \{x^2 : x \in \mathbb{F}_p^*\}, NS_p = \mathbb{F}_p^* \backslash S_p. \quad (4)$$

We will use Eq. (2)–(4) frequently to construct the codes and determine their parameters.

3 Main Results

In this section, we present the main results, including the construction, length, dimension and weight distribution of the linear code $\mathcal{C}_D$. The proofs will be given in Sect. 4.

We begin this section by selecting defining set

$$D = \{x \in \mathbb{F}_q : Tr(x) = 1, Tr(x^{p^s+1}) \in S_p\}, \quad (5)$$

to construct linear code

$$\mathcal{C}_D = \{(Tr(ax)_{x \in D}) : a \in \mathbb{F}_q\}, \quad (6)$$

where $m = 2s$, $p \mid m$, $p \equiv 3$ $(mod\ 4)$ and s is a positive integer.

For the code $\mathcal{C}_D$ given in Eq. (6), we have the following Theorem is

Theorem 1. *If* $p \mid m$ *and* $p \equiv 3$ $(mod\ 4)$, *Then the weight distribution of the code* $\mathcal{C}_D$ *with the parameter* $[\frac{p-1}{2}p^{m-2}, m]$ *is shown in Table 1.*

Table 1. The weight distribution of $\mathcal{C}_D$

Weight	Multiplicity
0	1
$p^{m-2}(p-1)/2$	$p-1$
$p^{m-3}(p-1)^2/2$	$(p^m + p^{m-1} - p^2 - p)/(p+1)$
$(p^{m-2} - p^{m-3} - p^{s-2})(p-1)/2$	$(p^{m+1} + p^m - p^{m-1} - p^{m-2})/(2(p+1))$
$(p^{m-1} - 2p^{m-2} + p^{m-3} + p^{s-2} + p^{s-1})/2$	$(p^{m+1} - p^m - p^{m-1} + p^{m-2})/(2(p+1))$

Example 1. Let $(p, m) = (3, 6)$, then the code $\mathcal{C}_D$ has parameters $[81, 6, 51]$ and weight enumerator $1 + 324z^{51} + 240z^{54} + 162z^{60} + 2z^{81}$, which drived from Magma program confirmed the results given in Theorem 1.

4 Auxiliary Lemmas and Proofs of the Main Results

We begin this section by introducing some known facts on exponential sums. Lemma 1 on Weil sums was given by R. S. Coulter [4] which we will convenient to calculate the length and the weight distribution of $\mathcal{C}_D$.

Lemma 1 [4]. *Let $m = 2s$, where s is a positive integer. For any $a \in \mathbb{F}_{p^s}^*$ and $b \in \mathbb{F}_{p^m}$, one has*

$$\sum_{x \in \mathbb{F}_q} \zeta^{Tr_s(ax^{p^s+1})+Tr(bx)} = -p^s \zeta^{-Tr_s(\frac{bp^s+1}{a})}.$$

Lemmas 2 and 3 given by R. Lidl and H. Niederreiter in [12] are well known results on Exponential sums, which will be used later to prove the main results of this paper.

Lemma 2 [12]. *If $f(x) = a_2x^2 + a_1x + a_0 \in \mathbb{F}_p[x]$ with $a_2 \neq 0$, then*

$$\sum_{x \in \mathbb{F}_p} \zeta^{Tr(f(x))} = \zeta^{Tr(a_0 - a_1^2(4a_2)^{-1})} \eta(a_2) G'(\eta').$$

Lemma 3 [12]. *With the notation above, one has*

$$G'(\eta') = \sqrt{p^*},$$

where $p^* = (-1)^{\frac{p-1}{2}} p$.

The following lemmas are essential to determine the length and the weight distribution of $\mathcal{C}_D$.

Lemma 4. *Put $N(1, l) = |\{x \in \mathbb{F}_q : Tr(x) = 1, Tr(x^{p^s+1}) = l\}|$, for any $l \in S_p$, $p \mid m$, we have*

$$N(1, l) = p^{m-2}.$$

Proof: From the orthogonal property of additive characters in Eq. (2), we have

$$\begin{aligned} N(1, l) &= \sum_{x \in \mathbb{F}_q} \Big(\frac{1}{p} \sum_{y \in \mathbb{F}_p} \zeta^{y(Tr(x)-1)}\Big)\Big(\frac{1}{p} \sum_{z \in \mathbb{F}_p} \zeta^{z(Tr(x^{p^s+1})-l)}\Big) \\ &= p^{m-2} + \frac{1}{p^2}(\Omega_1 + \Omega_2 + \Omega_3), \end{aligned}$$

where

$$\Omega_1 = \sum_{x\in\mathbb{F}_q}\sum_{y\in\mathbb{F}_p^*}\zeta^{y(Tr(x)-1)} = \sum_{y\in\mathbb{F}_p^*}\zeta^{-y}\sum_{x\in\mathbb{F}_q}\zeta^{yTr(x)} = 0,$$

$$\Omega_2 = \sum_{x\in\mathbb{F}_q}\sum_{z\in\mathbb{F}_p^*}\zeta^{z(Tr(x^{p^s+1})-l)} = \sum_{z\in\mathbb{F}_p^*}\zeta^{-lz}\sum_{x\in\mathbb{F}_q}\zeta^{zTr(x^{p^s+1})}$$

$$= -p^s\sum_{z\in\mathbb{F}_p^*}\zeta^{-lz} = p^s,$$

$$\Omega_3 = \sum_{x\in\mathbb{F}_q}\sum_{y\in\mathbb{F}_p^*}\zeta^{y(Tr(x)-1)}\sum_{z\in\mathbb{F}_p^*}\zeta^{z(Tr(x^{p^s+1})-l)}$$

$$= -p^s\sum_{z\in\mathbb{F}_p^*}\zeta^{-lz}\sum_{y\in\mathbb{F}_p^*}\zeta^{-\frac{m}{4z}y^2-y}$$

$$= -p^s\sum_{z\in\mathbb{F}_p^*}\zeta^{-lz}\sum_{y\in\mathbb{F}_p^*}\zeta^{-y} = -p^s.$$

Thus, we complete the proof of the lemma. □

Obviously, the length of the code $\mathcal{C}_D$ is $n = \frac{p-1}{2}N(1,l) = \frac{p-1}{2}p^{m-2}$. For any $a \in \mathbb{F}_q$, $l \in S_p$ and $\mathbf{c}(a) \in \mathcal{C}_D$, let

$$T_{a,l} = |\{x \in \mathbb{F}_q : Tr(x) = 1, Tr(x^{p^s+1}) = l, Tr(ax) = 0\}|,$$

then the weight of $\mathbf{c}(a)$ is

$$w(\mathbf{c}(a)) = n - \sum_{l\in S_p} T_{a,l}.$$

Then, use Eq. (2) and (3) again, we have

$$\begin{aligned}
T_{a,l} &= |\{x \in \mathbb{F}_q : Tr(x) = 1, Tr(x^{p^s+1}) = l, Tr(ax) = 0\}| \\
&= \sum_{x\in\mathbb{F}_q}\Big(\frac{1}{p}\sum_{y\in\mathbb{F}_p}\zeta^{y(Tr(x)-1)}\Big)\Big(\frac{1}{p}\sum_{z\in\mathbb{F}_p}\zeta^{z(Tr(x^{p^s+1})-l)}\Big)\Big(\frac{1}{p}\sum_{w\in\mathbb{F}_p}\zeta^{wTr(ax)}\Big) \\
&= p^{m-3} + \frac{1}{p^3}(\Omega_1 + \Omega_2 + \Omega_3 + \psi_1 + \psi_2 + \psi_3 + \psi_4) \\
&= p^{m-3} + \frac{1}{p^3}(\psi_1 + \psi_2 + \psi_3 + \psi_4), \qquad (7)
\end{aligned}$$

where $\psi_i,\ 1 \le i \le 4$ are showed as follows

$$\psi_1 = \sum_{x\in\mathbb{F}_q}\sum_{w\in\mathbb{F}_p^*}\zeta^{wTr(ax)} = \sum_{w\in\mathbb{F}_p^*}\sum_{x\in\mathbb{F}_q}\zeta^{wTr(ax)} = \begin{cases} p^m(p-1) & a=0,\\ 0 & a\neq 0,\end{cases}$$

$$\psi_2 = \sum_{x\in\mathbb{F}_q}\sum_{y\in\mathbb{F}_p^*}\zeta^{y(Tr(x)-1)}\sum_{w\in\mathbb{F}_p^*}\zeta^{wTr(ax)} = \sum_{y\in\mathbb{F}_p^*}\zeta^{-y}\sum_{w\in\mathbb{F}_p^*}\sum_{x\in\mathbb{F}_q}\zeta^{Tr((y+wa)x)}$$

$$= \begin{cases} \sum_{y\in\mathbb{F}_p^*}\zeta^{-y}(p^m + \sum_{w\neq -a^{-1}y}\sum_{x\in\mathbb{F}_q}\zeta^{Tr((y+wa)x)}) & a\in\mathbb{F}_p^*,\\ 0 & a\notin\mathbb{F}_p^*,\end{cases}$$

$$= \begin{cases} -p^m & a\in\mathbb{F}_p^*,\\ 0 & a\notin\mathbb{F}_p^*,\end{cases}$$

$$\psi_3 = \sum_{x\in\mathbb{F}_q}\sum_{z\in\mathbb{F}_p^*}\zeta^{z(Tr(x^{p^s+1})-l)}\sum_{w\in\mathbb{F}_p^*}\zeta^{wTr(ax)} = \sum_{z\in\mathbb{F}_p^*}\zeta^{-zl}\sum_{w\in\mathbb{F}_p^*}(-p^s)\zeta^{-Trs(\frac{(wa)^{p^s+1}}{2z})}$$

$$= \begin{cases} p^s(p-1) & Tr(a^{p^s+1})=0\\ -p^s\sum_{z\in\mathbb{F}_p^*}\zeta^{-zl}(\eta'(-\frac{Tr(a^{p^s+1})}{4z})G'(\eta')-1) & Tr(a^{p^s+1})\neq 0\end{cases}$$

$$= \begin{cases} p^s(p-1) & Tr(a^{p^s+1})\in S_p\cup\{0\},\\ -p^s(p+1) & Tr(a^{p^s+1})\in NS_p,\end{cases}$$

$$\psi_4 = \sum_{x\in\mathbb{F}_q}\sum_{y\in\mathbb{F}_p^*}\zeta^{y(Tr(x)-1)}\sum_{z\in\mathbb{F}_p^*}\zeta^{z(Tr(x^{p^s+1})-l)}\sum_{w\in\mathbb{F}_p^*}\zeta^{wTr(ax)}.$$

For the value of ψ_4, we have the Lemma 5.

Lemma 5. *With the notation above, we have*

(1) for $Tr(a^{p^s+1})=0$,

$$\psi_4 = \begin{cases} -p^s(p-1) & Tr(a)=0,\\ p^s & Tr(a)\neq 0.\end{cases}$$

(2) for $Tr(a^{p^s+1})\neq 0$,

if $Tr(a)=0$*, then*

$$\psi_4 = \begin{cases} -p^s(p-1) & Tr(a^{p^s+1})\in S_p,\\ p^s(p+1) & Tr(a^{p^s+1})\in NS_p,\end{cases}$$

if $Tr(a)\neq 0$*, then*

$$\psi_4 = \begin{cases} p^s & Tr(a^{p^s+1})\in S_p,\\ p^s+2p^{s+1}-p^{s+2} & Tr(a^{p^s+1})\in NS_p,\ l=-\frac{Tr(a^{p^s+1})}{Tr^2(a)},\\ p^s+2p^{s+1} & Tr(a^{p^s+1})\in NS_p,\ l\neq -\frac{Tr(a^{p^s+1})}{Tr^2(a)}.\end{cases}$$

Proof: From the Lemma 1, we have

$$\begin{aligned}
\psi_4 &= \sum_{x\in\mathbb{F}_q}\sum_{y\in\mathbb{F}_p^*}\zeta^{y(Tr(x)-1)}\sum_{z\in\mathbb{F}_p^*}\zeta^{z(Tr(x^{p^s+1}-l))}\sum_{w\in\mathbb{F}_p^*}\zeta^{wTr(ax)}\\
&= \sum_{y\in\mathbb{F}_p^*}\zeta^{-y}\sum_{z\in\mathbb{F}_p^*}\zeta^{-zl}\sum_{w\in\mathbb{F}_p^*}\sum_{x\in\mathbb{F}_q}\zeta^{Trs(2zx^{p^s+1})+Tr((y+wa)x)}\\
&= \sum_{y\in\mathbb{F}_p^*}\zeta^{-y}\sum_{z\in\mathbb{F}_p^*}\zeta^{-zl}\sum_{w\in\mathbb{F}_p^*}(-p^s)\zeta^{-Trs(\frac{(y+wa)^{p^s+1}}{2z})}\\
&= -p^s\sum_{y\in\mathbb{F}_p^*}\zeta^{-y}\sum_{z\in\mathbb{F}_p^*}\zeta^{-zl}\sum_{w\in\mathbb{F}_p^*}\zeta^{-\frac{w^2Tr(a^{p^s+1})+2ywTr(a)+my^2}{4z}}\\
&= -p^s\sum_{y\in\mathbb{F}_p^*}\zeta^{-y}\sum_{z\in\mathbb{F}_p^*}\zeta^{-zl}\sum_{w\in\mathbb{F}_p^*}\zeta^{-\frac{Tr(a^{p^s+1})}{4z}w^2-\frac{yTr(a)}{2z}w}.
\end{aligned}$$

We prove this lemma in the following cases,

(1) for $Tr(a^{p^s+1})=0$,

$$\begin{aligned}
\psi_4 &= -p^s\sum_{y\in\mathbb{F}_p^*}\zeta^{-y}\sum_{z\in\mathbb{F}_p^*}\zeta^{-zl}\sum_{w\in\mathbb{F}_p^*}\zeta^{-\frac{yTr(a)}{2z}w}\\
&= \begin{cases}-p^s(p-1) & Tr(a)=0,\\ p^s & Tr(a)\neq 0.\end{cases}
\end{aligned}$$

(2) for $Tr(a^{p^s+1})\neq 0$, from Lemmas 2 and 3, if $Tr(a)=0$, then

$$\begin{aligned}
\psi_4 &= -p^s\sum_{y\in\mathbb{F}_p^*}\zeta^{-y}\sum_{z\in\mathbb{F}_p^*}\zeta^{-zl}\sum_{w\in\mathbb{F}_p^*}\zeta^{-\frac{Tr(a^{p^s+1})}{4z}w^2}\\
&= -p^s\sum_{y\in\mathbb{F}_p^*}\zeta^{-y}\sum_{z\in\mathbb{F}_p^*}\zeta^{-zl}(\ zeta^{0-0}\eta'(-\frac{Tr(a^{p^s+1})}{4z})G'(\eta')-1)\\
&= -p^s\sum_{y\in\mathbb{F}_p^*}\zeta^{-y}(G'(\eta')\sum_{z\in\mathbb{F}_p^*}\zeta^{-zl}\eta'(-z)\eta'(Tr(a^{p^s+1}))+1)\\
&= \begin{cases}-p^s(p-1) & Tr(a^{p^s+1})\in S_p,\\ p^s(p+1) & Tr(a^{p^s+1})\in NS_p,\end{cases}
\end{aligned}$$

and if $Tr(a) \neq 0$, then

$$\psi_4 = -p^s \sum_{y\in\mathbb{F}_p^*} \zeta^{-y} \sum_{z\in\mathbb{F}_p^*} \zeta^{-zl}(\zeta^{-\frac{4z^3}{y^2Tr^2(a)Tr(a^{p^s+1})}} \eta'(-\frac{Tr(a^{p^s+1})}{4z})G'(\eta') - 1)$$

$$= -p^s \sum_{y\in\mathbb{F}_p^*} \zeta^{-y} \sum_{z\in\mathbb{F}_p^*} \zeta^{-zl}[\zeta^{\frac{Tr^2(a)}{4zTr(a^{p^s+1})}y^2} \eta(-\frac{Tr(a^{p^s+1})}{4z})G'(\eta') - 1]$$

$$= -p^s \sum_{z\in\mathbb{F}_p^*} \zeta^{-zl}(\eta'(-\frac{Tr(a^{p^s+1})}{4z})G'(\eta') \sum_{y\in\mathbb{F}_p^*} \zeta^{\frac{Tr^2(a)}{4zTr(a^{p^s+1})}y^2-y} + 1)$$

$$= p^s - p^sG'(\eta') \sum_{z\in\mathbb{F}_p^*} \zeta^{-zl}\eta'(-\frac{Tr(a^{p^s+1})}{4z}) \sum_{y\in\mathbb{F}_p^*} \zeta^{\frac{Tr^2(a)}{4zTr(a^{p^s+1})}y^2-y}$$

$$= p^s - p^sG'(\eta') \sum_{z\in\mathbb{F}_p^*} \zeta^{-zl}\eta'(-\frac{Tr(a^{p^s+1})}{4z})(\zeta^{-\frac{Tr(a^{p^s+1})}{Tr^2(a)}z} \eta'(\frac{Tr^2(a)}{4zTr(a^{p^s+1})})G'(\eta') - 1)$$

$$= p^s + p^sG'(\eta') \sum_{z\in\mathbb{F}_p^*} \zeta^{-zl}\eta'(-z)\eta'(Tr(a^{p^s+1})) - p^sG'^2(\eta') \sum_{z\in\mathbb{F}_p^*} \zeta^{-z(l+\frac{Tr(a^{p^s+1})}{Tr^2(a)})}\eta'(-1)$$

$$= \begin{cases} p^s & Tr(a^{p^s+1}) \in S_p, \\ p^s + 2p^{s+1} - p^{s+2} & Tr(a^{p^s+1}) \in NS_p,\ l = -\frac{Tr(a^{p^s+1})}{Tr^2(a)}, \\ p^s + 2p^{s+1} & Tr(a^{p^s+1}) \in NS_p,\ l \neq -\frac{Tr(a^{p^s+1})}{Tr^2(a)}. \end{cases}$$

Thus, we prove of the lemma. □

From all the discussions above, we get the following lemma.

Lemma 6. *With the notation above, we have*

(1) if $a = 0$, then $T_{a,l} = p^{m-2}$.
(2) if $a \in \mathbb{F}_p^*$, then $T_{a,l} = 0$.
(3) if $a \in \mathbb{F}_q^* \backslash \mathbb{F}_p^*$,
(i) $Tr(a) = 0$, $Tr(a^{p^s+1}) \in S_p$, then $T_{a,l} = p^{m-3}$.
(ii) $Tr(a) \neq 0$, $Tr(a^{p^s+1}) \in S_p \cup \{0\}$, then $T_{a,l} = p^{m-3} + p^{s-2}$.
(iii) $Tr(a) = 0$, $Tr(a^{p^s+1}) \in NS_p$, then

$$T_{a,l} = \begin{cases} p^{m-3} + p^{s-2} & l \neq -\frac{Tr(a^{p^s+1})}{Tr^2(a)}, \\ p^{m-3} + p^{s-2} + p^{s-1} & l = -\frac{Tr(a^{p^s+1})}{Tr^2(a)}. \end{cases}$$

Proof: From the value of $\psi_1 - \psi_4$, and Eq. (7), we calculate the value of T_a, Thus we complete the proof of the lemma. □

Now we are ready to prove Theorem 1.

Proof of Theorem 1: The non-zero weight of the lines Table 1 is denoted by w_i, and the corresponding multiplicity is A_{w_i}, $1 \leq i \leq 4$. It follows from

Lemmas 4, 6, and by the Pless Power moments [8] that

$$\sum_{i=1}^{4} A_{w_i} = p^m - 1,$$

$$\sum_{i=1}^{4} w_i A_{w_i} = \frac{1}{2}p^{m-1}(p-1)(p^{m-1} - p^{m-2}),$$

$$\sum_{i=1}^{4} w_i^2 A_{w_i} = \frac{1}{2}(p^{m-2})^2(p-1)^2(\frac{p-1}{2}p^{m-1} - \frac{p-1}{2}p^{m-2} + 1).$$

Solved the equations , we can get the A_{w_i}, $1 \le i \le 4$. Therefore, we complete the proof. □

5 Concluding Remarks

In this paper, we constructed a class of linear codes with four-weight by a properly selection of the defining set, then we determined their length, dimension and weight distribution using character sums over finite fields.

Acknowledgements. The authors are grateful to the reviewers and editors for their comments and suggestions that improved the quality of this paper.

References

1. Baumort, L.D., Mcelicce, R.J.: Weights of irreducible cyclic codes. Inf. Control **20**(2), 158–175 (1972)
2. Calderbank, A.R., Goethals, J.M.: Three-weight codes and association schemes. Philips J. Res. **39**(4), 143–152 (1984)
3. Calderbank, A.R., Kantor, W.M.: The geometry of two-weight codes. Bull. Lond. Math. Soc. **18**(2), 97–122 (1986)
4. Coulter, R.S.: Further evaluations of Weil sums. Acta Arithmetica **86**(3), 217–226 (1998)
5. Du, X., Li, X., Wan, Y.: A class of linear codes with three and five weights. Acta Electronica Sinica (to appear)
6. Ding, C., Helleseth, T., Klove, T.: A generic construction of Cartesian authentication codes. IEEE Trans. Inf. Theor. **53**(6), 2229–2235 (2007)
7. Ding, C., Niederreiter, H.: Cyclotomic linear codes of order 3. IEEE Trans. Inf. Theor. **53**(6), 2274–2277 (2007)
8. Huffman, W.C., Pless, V.: Fundamentals of Error-Correcting Codes. Cambridge University Press, Cambridge (2003)
9. Heng, Z., Yue, Q.: Complete weight distributions of two classes of cyclic codes. Crypt. Commun. **9**(3), 323–343 (2017)
10. Li, C., Feng, K.: The construction of a class of linear codes with good parameters. Acta Electronica Sinica **31**(1), 51–53 (2003)

11. Luo, J., Feng, K.: On the weight distribution of two classes of cyclic codes. IEEE Trans. Inf. Theor. **54**(12), 5332–5344 (2008)
12. Lidl, R., Niederreiter, H.: Finite Fields. Cambridge University Press, Cambridge (1997)
13. Tan, P., Zhou, Z., Tang, D.: The weight distribution of a class of two-weight linear codes derived from Kloosterman sums. Crypt. Commun. **10**(2), 291–299 (2018)
14. Tang, C., Qi, Y., Huang, D.: Two-Weight and Three-Weight linear codes from square functions. IEEE Commun. Lett. **20**(1), 29–32 (2016)
15. Yuan, J., Ding, C.: Secret sharing schemes from three classes of linear codes. IEEE Trans. Inf. Theor. **52**(1), 206–212 (2006)
16. Yang, S., Yao, Z.: Complete weight enumerators of a family of three-weight linear codes. Des. Codes Crypt. **82**(3), 663–674 (2017)
17. Calderbank, A.R., Goethals, J.M.: Three-weight codes and association schemes. Res **18**, 97–122 (1986)
18. Ding, C., Wang, X.: A coding theory construction of new systematic authentication codes. Theoret. Comput. Sci. **330**, 81–99 (2005)
19. Ding, C., Li, C., Li, N., Zhou, Z.: Three-weight cyclic codes and their weight distributions. Preprint (2014)
20. Courteau, B., Wolfmann, J.: On triple-sum-sets and two or three weight codes. Discrete Math. **50**, 179–191 (1984)
21. Choi, S.-T., Kim, J.-Y., No, J.-S., Chung, H.: Weight distribution of some cyclic codes. In: International Symposium on Information Theory, pp. 2911–2913. IEEE Press (2012)
22. Feng, K., Luo, J.: Value distribution of exponential sums from perfect nonlinear functions and their applications. IEEE Trans. Inform. Theor. **53**(9), 3035–3041 (2007)
23. Li, C., Yue, Q., Li, F.: Weight distributions of cyclic codes with respect to pairwise coprime order elements. Finite Fields Appl. **28**, 94–114 (2014)
24. Li, C., Yue, Q., Li, F.: Hamming weights of the duals of cyclic codes with two zeros. IEEE Trans. Inform. Theor. **60**(7), 3895–3902 (2014)
25. Rao, A., Pinnawala, N.: A family of two-weight irreducible cyclic codes. IEEE Trans. Inform. Theor. **56**(6), 2568–2570 (2010)
26. Li, C., Xu, G., Chen, Y., Ahmad, H., Li, J.: A new anti-quantum proxy blind signature for blockchain-enabled Internet of Things. CMC: Comput. Mater. Contin. **61**(2), 711–726 (2019)
27. He, Q., et al.: A weighted threshold secret sharing scheme for remote sensing images based on Chinese remainder theorem. CMC: Comput. Mater. Contin. **58**(2), 349–361 (2019)
28. Zhao, Y., Yang, X., Li, R.: Design of feedback shift register of against power analysis attack. CMC: Comput. Mater. Contin. **58**(2), 517–527 (2019)

Designs from the Narrow-Sense Primitive BCH Codes $\mathcal{C}_{(q,q^m-1,\delta_3,1)}$

Fujun Zhang(✉), Xiaoni Du, and Jinxia Hu

College of Mathematics and Statistics, Northwest Normal University, Lanzhou 730070, Gansu, People's Republic of China
zfj1995n@126.com, ymldxn@126.com, jinxiahu@126.com

Abstract. The interplay between coding theory and t-designs has been a very interesting topic for researchers. It is well known that the supports of all codewords with a fixed weight in a code may form a t-design under certain conditions. In this paper, we will determine explicitly the parameters of many infinite families of 2-designs from the extended of the primitive BCH codes $\mathcal{C}_{(q,q^m-1,\delta_3,1)}$ with $\delta_3 = (q-1)q^{m-1} - 1 - q^{\lfloor (m+1)/2 \rfloor}$ and odd prime q given by Li et al. [18] and point out that their duals hold 2-designs.

Keywords: 2-Designs · BCH codes · Affine-invariant codes · Weight distributions

1 Introduction

We begin with a brief recall of t-designs. Let $\mathcal{P}$ be a set with $v > 1$ elements and $\mathcal{B}$ a set of k-subsets of $\mathcal{P}$, where k is a positive integer with $1 \leq k \leq v$. Let t be a positive integer with $t \leq k$. If every t-subset of $\mathcal{P}$ is contained in exactly λ elements of $\mathcal{B}$, then the pair $\mathbb{D} = (\mathcal{P}, \mathcal{B})$ is called a t-(v, k, λ) design, or simply t-design. The elements of $\mathcal{P}$ are called points, and those of $\mathcal{B}$ are referred to as blocks. We often use b to denote the number of blocks in $\mathcal{B}$. Such t-design is called simple since there is no repeated blocks in $\mathcal{B}$. A t-design is called symmetric if $v = b$ and trivial if $k = t$ or v. In this paper, we only study simple t-design with $t < k < v$. Moreover, in a t-design, the number of blocks in a t-design is given by

$$b\binom{k}{t} = \lambda\binom{v}{t}. \tag{1}$$

t-Designs are probably the most-learned type of design. As an significant branch of modern combinatorial theory, t-Designs have very significant applications in coding theory, cryptography, communications and statistics. It is well known that the interplay between t-designs and codes goes in the two aspects. On one hand, a linear code over any finite field can be derived from the incidence matrix of a t-design and much progress has been made and documented in

Supported by National Natural Science Foundation of China (61772022).

X. Sun et al. (Eds.): ICAIS 2020, LNCS 12240, pp. 65–73, 2020.
https://doi.org/10.1007/978-3-030-57881-7_6

[1,5,20,21]. On the other hand, linear and nonlinear codes could be employed to construct t-designs with $t = 2, 3$. Sporadic 4-designs and 5-designs with fixed parameters were derived from binary and ternary Golay codes. There is no known general result that characterizes all codes whose codewords of a given weight form t-designs. There are, however, result that show that codes satisfying certain conditions have weights where codewords of that weight hold a t-design. In general, there are two standard approaches to obtain t-designs from linear codes. One is to apply the Assmus-Mattson Theorem [15] which is the sufficient condition and we refer the readers to [8,10] for details. The other is to study the automorphism group of a linear code $\mathcal{C}$. If the permutation part of the automorphism group acts t-transitively on the code $\mathcal{C}$, then $\mathcal{C}$ holds t-designs [15]. Only a small amount of work in this direction has been done. For example, Ding et al. [9] and Du et al. [12–14] have derived infinite families of 3-designs and 2-designs from some different classes of affine-invariant codes, respectively. Both of them are based on the support designs. The main idea is that indexed the coordinates of a codeword in $\mathcal{C}$ by $(0, 1, \ldots, n-1)$, for each i with $A_i \neq 0$, let $\mathcal{B}_i$ denote the set of the supports of all codewords with weight i and $\mathcal{P} = \{0, 1, \ldots, n-1\}$. The pair $(\mathcal{P}, \mathcal{B}_i)$ may be a t-(n, i, λ) design for some positive λ which can be explicitly determined by Eq. (1) [8] (we will explain the definition later). Many other constructions of t designs can be founded in [3,19,22–24].

In this paper, our mainly contributions are to determine explicitly the parameters of several infinite families of 2-designs from the extended of the primitive BCH codes $\mathcal{C}_{(q,q^m-1,\delta_3,1)}$ for odd prime q and meanwhile point out that their duals hold 2-designs. Furthermore, we use Magma program to give some examples.

2 Preliminaries

In this section, we mainly present some basic facts on coding theory and 2-designs.

We start this section by introducing some basic concepts of codes. Throughout this paper, we assume that q is an odd prime. Let $\mathbb{F}_q$ be the finite field with q elements and $\mathbb{F}_q^* = \mathbb{F}_q \backslash \{0\}$. An $[n, k, d]$ linear code $\mathcal{C}$ over $\mathbb{F}_q$ is called cyclic provided that for each codeword

$$\mathbf{c} = (c_0, c_1, \ldots, c_{n-1}) \in \mathcal{C},$$

implies

$$(c_{n-1}, c_0, \ldots, c_{n-2}) \in \mathcal{C}.$$

Any cyclic code $\mathcal{C}$ can be expressed as $\mathcal{C} =< g(x) >$, where $g(x)$ is monic and has the least degree. The polynomial $g(x)$ is called the generator polynomial and

$$h(x) = (x^n - 1)/g(x)$$

is referred to as the parity check polynomial of $\mathcal{C}$. The dual of a cyclic code $\mathcal{C}$, denoted $\mathcal{C}^{\perp}$, is a cyclic code with length n, dimension $n - k$ and generator polynomial

$$x^k h(x^{-1})/h(0).$$

The extended code $\overline{\mathcal{C}}$ to be the code

$$\overline{\mathcal{C}} = \{(c_0, c_1, \ldots, c_{n-1}, c_n) : (c_0, c_1, \ldots, c_{n-1}) \in \mathcal{C} \text{ with } \sum_{i=0}^{n} c_i = 0\}.$$

Let A_i be the number of codewords with Hamming weight i in $\mathcal{C}$. The weight enumerator of $\mathcal{C}$ is defined by $\sum_{i=0}^{n} A_i z^i$, and the sequence

$$(1, A_1, \ldots, A_n)$$

is called the weight distribution of the code $\mathcal{C}$. If

$$|\{1 \leqslant i \leqslant n : A_i \neq 0\}| = \omega,$$

then we call $\mathcal{C}$ a ω-weight code. For any codeword $\mathbf{c} \in \mathcal{C}$, we define the number of non-zero components in $\mathcal{C}$ as support of $\mathcal{C}$ by

$$Suppt(\mathbf{c}) = \{0 \leq i \leq n-1 : c_i \neq 0\}.$$

Denote the weight enumerator of $\mathcal{C}^{\perp}$ by $A^{\perp}(z)$. The following result, called MacWilliams Identity [15], shows that $A(z)$ and $A^{\perp}(z)$ can be derived from each other.

Lemma 1 [15]. *Let $\mathcal{C}$ be an $[n, k, d]$ code over $\mathbb{F}_q$ with weight enumerator $A(z) = \Sigma_{i=0}^{n} A_i z^i$. Then*

$$A^{\perp}(z) = q^{-k}(1 + (q-1)z)^n A(\frac{1-z}{1+(q-1)z}).$$

Now we introduce relevant knowledge on t-designs from the affine-invariant codes. The set of coordinate permutations that map a code $\mathcal{C}$ to itself forms a group, which is called the *permutation automorphism group* of $\mathcal{C}$ and denoted by $PAut(\mathcal{C})$. The *affine group* $GA_1(\mathbb{F}_q)$ is defined by the set of all permutations

$$\sigma_{a,b} : x \mapsto ax + b$$

of $\mathbb{F}_q$, where $a \in \mathbb{F}_q^*$ and $b \in \mathbb{F}_q$. An affine-invariant code is an extended cyclic code $\overline{\mathcal{C}}$ over $\mathbb{F}_q$ such that

$$GA_1(\mathbb{F}_q) \subseteq PAut(\overline{\mathcal{C}}) \text{ [15]}.$$

The next lemma tells us that we can get new affine-invariant codes from the old ones.

Lemma 2 [6]. *The dual of an affine-invariant code $\overline{\mathcal{C}}$ over $\mathbb{F}_q$ of length $n+1$ is also affine-invariant.*

Lemma 2 and the following Theorem 1 given by Ding in [6] are very powerful tools in constructing 2-designs from affine-invariant code. We will employ them to present our main result in this paper.

Theorem 1 [6]. *Let $\overline{A_i}$ denote the weight distribution of $\overline{\mathcal{C}}$. For each i with $\overline{A_i} \neq 0$ in an affine-invariant code $\overline{\mathcal{C}}$, the supports of the all codewords of weight i form a 2-design.*

Theorem 1 is very attractive in the sense that they determine the existence of 2-design. The following lemma mainly tells us the relation of the codewords with the same support and meeting certain conditions in a linear code $\mathcal{C}$(that is, they must be scalar multiples of each other), which will be used together with Eq. (1) to calculate the parameters of 2-designs later.

Lemma 3 [15]. *Let $\mathcal{C}$ be a linear code over $\mathbb{F}_q$ with minimum weight d and length n. Let w be the largest integer with $w \leq n$ satisfying*

$$w - \lfloor \frac{w+q-2}{q-1} \rfloor < d.$$

Let $\mathbf{c_1}$ and $\mathbf{c_2}$ be two codewords of weight i with $d \leq i \leq w$ and $Suppt(\mathbf{c_1}) = Suppt(\mathbf{c_2})$. Then

$$\mathbf{c_1} = a\mathbf{c_2}$$

for some $a \in \mathbb{F}_q^*$.

3 Narrow-sense Primitive BCH Codes $\mathcal{C}_{(q,q^m-1,\delta_3,1)}$

In this section, we introduce some facts and known results on Narrow-sense primitive BCH codes $\mathcal{C}_{(q,q^m-1,\delta_3,1)}$, where

$$\delta_3 = (q-1)q^{m-1} - 1 - q^{\lfloor (m+1)/2 \rfloor}. \tag{2}$$

Now we introduce the definition of $\mathcal{C}_{(q,q^m-1,\delta_3,1)}$. Let $n = q^m - 1$ for a positive integer m and let α be a generator of $\mathbb{F}_{q^m}^*$, which is the multiplicative group of $\mathbb{F}_{q^m}$. For any integer i with $0 \leq i \leq n-1$, let $m_i(x)$ denote the minimal polynomial of α^i over $\mathbb{F}_q$. For δ_3, we define

$$g_{(q,q^m-1,\delta_3,1)} = lcm(m_1(x), m_2(x), \ldots, m_{\delta_3-1}(x)),$$

where lcm denotes the least common multiple of these minimal polynomials. Denote the cyclic code of length n over $\mathbb{F}_q$ with generator polynomial $g_{(q,q^m-1,\delta_3,1)}$ by $\mathcal{C}_{(q,q^m-1,\delta_3,1)}$, then $\mathcal{C}_{(q,q^{m-1},\delta_3,1)}$ is narrow-sense primitive BCH code with designed distance δ_3. It is known that the code $\mathcal{C}_{(q,q^m-1,\delta_3,1)}$ is equivalent to the following code [11]

$$\mathcal{C}_{\delta_3} = \{(Tr(ax^{1+q^h} + bx^{1+q^{h+1}} + cx) + e)_{x \in \mathbb{F}_{q^m}^*}$$

$$: a, b, c \in \mathbb{F}_{q^m}, e \in \mathbb{F}_q\},$$

where Tr is the trace function from $\mathbb{F}_{q^m}$ onto $\mathbb{F}_q$ and $h = \lfloor (m-1)/2 \rfloor + 1$. Charpin pointed out [4] that it is a well-known hard problem to determine the parameters of BCH codes. So far, we have very limited knowledge of BCH codes, because of the dimension and minimum distance of BCH codes are in general open problems. Only a few know results on BCH codes can be determined in [2,7,16,17] .

By above equation, we can get

$$\overline{\mathcal{C}_{(q,q^m-1,\delta_3,1)}} = \{(Tr(ax^{1+q^h} + bx^{1+q^{h+1}} + cx) + e)_{x\in\mathbb{F}_{q^m}}$$

$$: a, b, c \in \mathbb{F}_{q^m}, e \in \mathbb{F}_q\}.$$

From above equation, we have

$$\overline{\mathcal{C}_{(q,q^m-1,\delta_3,1)}} = \{(Tr(ax^{1+q^h} + bx^{1+q^{3h}} + cx) + e)_{x\in\mathbb{F}_{q^m}}$$

$$: a, b, c \in \mathbb{F}_{q^m}, e \in \mathbb{F}_q\},$$

when $m \geq 5$ is odd and

$$\overline{\mathcal{C}_{(q,q^m-1,\delta_3,1)}} = \{(Tr(ax^{1+q^h} + bx^{1+q^{h+1}} + cx) + e)_{x\in\mathbb{F}_{q^m}}$$

$$: a \in \mathbb{F}_{q^{\frac{m}{2}}}, b, c \in \mathbb{F}_{q^m}, e \in \mathbb{F}_q\},$$

when $m \geq 4$ is even.

Then, the following Lemma 4 gives the weight distribution of $\overline{\mathcal{C}_{(q,q^m-1,\delta_3,1)}}$ given by Li et al. [18].

Lemma 4 [18]. *Let q be an odd prime and $m \geq 4$ an integer. Then, the code $\overline{\mathcal{C}_{(q,q^m-1,\delta_3,1)}}$ is a $[q^m, 3m+1, q^m - q^{m-1} - q^{\frac{m+1}{2}}]$ code with the weight distribution $\overline{A_i}$ given in Table 1 when m is odd and is a $[q^m, \frac{5m}{2}+1, q^m - q^{m-1} - q^{\frac{m}{2}}]$ code with the weight distribution $\overline{A_i}$ given in Table 2 when m is even.*

Table 1. Weight distribution of $\overline{\mathcal{C}_{(q,q^m-1,\delta_3,1)}}$ for odd m.

Weight	Multiplicity
0	1
$(q-1)q^{m-1} \mp q^{\frac{m+1}{2}}$	$\frac{1}{2}q^{m-2}(q^{m-1}-1)(q^m-1)/(q+1)$
$(q-1)(q^{m-1} \mp q^{\frac{m-1}{2}})$	$\frac{1}{2}q^{m-1}(q^{m-1} \pm q^{(m-1)/2})(q^m-1)$
$(q-1)q^{m-1} \mp q^{\frac{m-1}{2}}$	$\frac{1}{2}q^{m-1}(q^{m+3} - q^{m+2} - q^{m-1} \mp q^{\frac{m+3}{2}} \pm q^{\frac{m-1}{2}} + q^3)(q^m-1)/(q+1)$
$(q-1)q^{m-1}$	$q((q-1)(2q^{2m-2} + q^{2m-4} + q^{m-2}) + q^{m-3} + 1)(q^m-1)$
q^m	$q-1$

One can see that the code $\overline{\mathcal{C}_{(q,q^m-1,\delta_3,1)}}$ is eight-weight if m is odd and nine-weight if m is even.

Table 2. Weight distribution of $\overline{\mathcal{C}(q,q^m-1,\delta_3,1)}$ for even m.

Weight	Multiplicity
0	1
$(q-1)q^{m-1}-q^{\frac{m}{2}}$	$\frac{1}{2}q^{\frac{m}{2}-2}(q^{m+2}-q^m+2q^{m-1}-2q^{\frac{m}{2}})(q^m-1)/(q+1)$
$(q-1)(q^{m-1}-q^{\frac{m}{2}-1})$	$\frac{1}{2}q^{m+1}(q^{\frac{m}{2}}+1)(q^m-1)/(q+1)$
$(q-1)q^{m-1}-q^{\frac{m}{2}-1}$	$\frac{1}{2}q^m(q^{\frac{m}{2}}-1)(q^{m+1}-2q^m+q)$
$(q-1)q^{m-1}$	$q(1+q^{\frac{3m}{2}-1}-q^{\frac{3m}{2}-2}+2q^{\frac{3m}{2}-3}-q^{m-2})(q^m-1)$
$(q-1)q^{m-1}+q^{\frac{m}{2}-1}$	$\frac{1}{2}q^{m+1}(q-1)(q^{\frac{m}{2}}+1)(q^m-1)/(q+1)$
$(q-1)(q^{m-1}+q^{\frac{m}{2}-1})$	$\frac{1}{2}q^m(q^{\frac{m}{2}}-1)(q^{m+1}-2q^m+q)/(q-1)$
$(q-1)q^{m-1}+q^{\frac{m}{2}}$	$\frac{1}{2}q^{\frac{3m}{2}-2}(q-1)(q^m-1)$
$(q-1)(q^{m-1}+q^{\frac{m}{2}})$	$q^{m-2}(q^{\frac{m}{2}-1}-1)(q^m-1)/(q^2-1)$
q^m	$q-1$

4 2-Designs from the Narrow-sense Primitive BCH Codes $\mathcal{C}_{(q,q^m-1,\delta_3,1)}$

In this section, we will determine explicitly the parameters of 2-designs from the code $\overline{\mathcal{C}_{(q,q^m-1,\delta_3,1)}}$ and point out that $\overline{\mathcal{C}_{(q,q^m-1,\delta_3,1)}}^{\perp}$ holds 2-designs.

We first present a known result on $\overline{\mathcal{C}_{(q,q^m-1,\delta_3,1)}}$ given by Ding in [6], which can be used to obtain the next theorem.

Lemma 5 [6]. *For δ_3, $\overline{\mathcal{C}_{(q,q^m-1,\delta_3,1)}}$ is an affine-invariant. And thus, the supports of the codewords of weight $i>0$ in $\overline{\mathcal{C}_{(q,q^m-1,\delta_3,1)}}$ form a 2-design, provided that $\overline{A}_i \neq 0$.*

With the help of Lemma 5, we now describe several 2-designs from the narrow-sense primitive BCH codes $\mathcal{C}_{(q,q^m-1,\delta_3,1)}$ in the following theorem.

Theorem 2. *Let q be an odd prime and $m \geq 4$ an integer. Then for odd m, $\overline{\mathcal{C}_{(q,q^m-1,\delta_3,1)}}$ holds 2-(q^m,i,λ) designs with the following pairs of (i,λ):*

- $(i,\lambda)=\Big((q-1)q^{m-1}\mp q^{\frac{m+1}{2}}, \frac{1}{2}(q^{m-1}-1)((q-1)q^{m-3}\mp q^{\frac{m-3}{2}})((q-1)q^{m-1}\mp q^{\frac{m+1}{2}}-1)/(q^2-1)\Big)$.
- $(i,\lambda)=\Big((q-1)(q^{m-1}\mp q^{\frac{m-1}{2}}), \frac{1}{2}(q^{2m-3}-q^{m-2})((q-1)(q^{m-1}\mp q^{\frac{m+1}{2}})-1)\Big)$.
- $(i,\lambda)=\Big((q-1)q^{m-1}\mp q^{\frac{m-1}{2}}, \frac{1}{2}(q^{m+2}-q^{m+1}-q^{m-2}\mp q^{\frac{m+1}{2}}\pm q^{\frac{m-3}{2}}+q^2)((q-1)q^{m-1}\mp q^{\frac{m-1}{2}})((q-1)q^{m-1}\mp q^{\frac{m-1}{2}}-1)/(q^2-1)\Big)$.
- $(i,\lambda)=\Big((q-1)q^{m-1}, ((q-1)(2q^{2m-2}+q^{2m-4}+q^{m-2})+q^{m-3}+1)((q-1)q^{m-1}-1)\Big)$.

For even m, it holds 2-(q^m, i, λ) designs with the following pairs of (i, λ):

- $(i,\lambda)=\left((q-1)q^{m-1}-q^{\frac{m}{2}}, \frac{1}{2}(q^{\frac{m}{2}+2}-q^{\frac{m}{2}}+2q^{\frac{m}{2}-1}-2)((q-1)q^{m-3}-q^{\frac{m}{2}-2})((q-1)q^{m-1}-q^{\frac{m}{2}}-1)/(q^2-1)\right)$.
- $(i,\lambda)=\left((q-1)(q^{m-1}-q^{\frac{m}{2}-1}), \frac{1}{2}q(q^{\frac{m}{2}}+1)(q^{m-1}-q^{\frac{m}{2}-1})((q-1)(q^{m-1}-q^{\frac{m}{2}-1})-1)/(q+1)\right)$.
- $(i,\lambda)=\left((q-1)q^{m-1}-q^{\frac{m}{2}-1}, \frac{1}{2}(q^{m+1}-2q^m+q)(q^{\frac{m}{2}}-1)((q-1)q^{m-1}-q^{\frac{m}{2}-1})((q-1)q^{m-1}-q^{\frac{m}{2}-1}-1)/(q-1)(q^m-1)\right)$.
- $(i,\lambda)=\left((q-1)q^{m-1}, (1+q^{\frac{3m}{2}-1}-q^{\frac{3m}{2}-2}+2q^{\frac{3m}{2}-3}-q^{m-2})((q-1)q^{m-1}-1)\right)$.
- $(i,\lambda)=\left((q-1)q^{m-1}+q^{\frac{m}{2}-1}, \frac{1}{2}q(q^{\frac{m}{2}}+1)((q-1)q^{m-1}+q^{\frac{m}{2}-1})((q-1)q^{m-1}+q^{\frac{m}{2}-1}-1)/(q+1)\right)$.
- $(i,\lambda)=\left((q-1)(q^{m-1}+q^{\frac{m}{2}-1}), \frac{1}{2}(q^{m+1}-2q^m+q)(q^{\frac{m}{2}}-1)(q^{m-1}+q^{\frac{m}{2}-1})((q-1)(q^{m-1}+q^{\frac{m}{2}-1})-1)/(q-1)(q^m-1)\right)$.
- $(i,\lambda)=\left((q-1)q^{m-1}+q^{\frac{m}{2}}, \frac{1}{2}q^{\frac{m}{2}-2}((q-1)q^{m-1}+q^{\frac{m}{2}})((q-1)q^{m-1}+q^{\frac{m}{2}}-1)\right)$.
- $(i,\lambda)=\left((q-1)(q^{m-1}+q^{\frac{m}{2}}), (q^{\frac{m}{2}-1}-1)(q^{m-3}+q^{\frac{m}{2}-2})((q-1)(q^{m-1}+q^{\frac{m}{2}})-1)/(q^2-1)\right)$.

Proof. According to Lemma 5, $\overline{\mathcal{C}_{(q,q^m-1,\delta_3,1)}}$ holds 2-designs. By Lemma 3, one can prove that the number of supports of all codewords with weight $i \neq 0$ in the code $\overline{\mathcal{C}_{(q,q^m-1,\delta_3,1)}}$ is equal to $\overline{A_i}/(q-1)$ for each i, where $\overline{A_i}$ are given in Tables 1–2. Then the desired conclusions follow from Lemma 5 and Eq. (1). After a simple calculation, we completed the proof of the theorem. □

The next theorem tells us that $\overline{\mathcal{C}_{(q,q^m-1,\delta_3,1)}}^{\perp}$ holds 2-designs.

Theorem 3. *For $\delta_3 = (q-1)q^{m-1} - 1 - q^{\lfloor (m+1)/2 \rfloor}$, then $\overline{\mathcal{C}_{(q,q^m-1,\delta_3,1)}}^{\perp}$ is an affine-invariant. And thus the supports of the codewords of weight $i > 0$ in $\overline{\mathcal{C}_{(q,q^m-1,\delta_3,1)}}^{\perp}$ form a 2-design, provided that $\overline{A_i}^{\perp} \neq 0$.*

Proof. The desired conclusion follows from Lemma 2 and Theorem 1. Thus, the proof of is completed. □

Remark 1. With the help of Theorem 3, we can obtain some infinite families of 2-designs from $\overline{\mathcal{C}_{(q,q^m-1,\delta_3,1)}}^{\perp}$. However, it is a tedious thing to calculate the weight enumerator of $\overline{\mathcal{C}_{(q,q^m-1,\delta_3,1)}}^{\perp}$ by using Tables 1–2 and Lemma 1. The reader is cordially invited to settle it.

The following examples from Magma program confirm the main results in Theorem 2.

Example 1. If $(q,m)=(3,4)$, then $\delta_3=44$ and the code $\overline{\mathcal{C}_{(q,q^m-1,\delta_3,1)}}$ has parameters $[81,11,45]$ and weight enumerator

$$1+6840z^{45}+24300z^{48}+27216z^{51}+49920z^{54}+48600z^{57}+$$

$$13608z^{60}+6480z^{63}+180^{72}+2z^{81}.$$

It gives 2-$(81,i,\lambda)$ designs with the following pairs (i,λ) :

$$(45,1045),(48,4230),(51,5355),(54,11024),$$

$$(57,11970),(60,3717),(63,1953),(72,71),$$

which confirmed the results given in Theorem 2 for even m.

Example 2. If $(q,m)=(3,5)$, then $\delta_3=134$ and the code $\overline{\mathcal{C}_{(q,q^m-1,\delta_3,1)}}$ has parameters $[243,16,135]$ and weight enumerator

$$1+65340z^{135}+882090z^{144}+10408662z^{153}+20158116z^{162}+48600z^{57}+$$

$$10761498z^{171}+705672^{180}+65340^{189}+z^{242}.$$

When i is the minimum nonzero weight, that is $i=135$, it gives

$$2-(243,135,10050).$$

which confirmed the result given in Theorem 2 for even m.

5 Concluding Remarks

In this paper, with the properties of affine-invariant codes, we determined explicitly the parameters of many infinite families of 2-designs from the extended of the primitive BCH codes $\mathcal{C}_{(q,q^m-1,\delta_3,1)}$ and pointed out that the $\overline{\mathcal{C}_{(q,q^m-1,\delta_3,1)}}^{\perp}$ holds 2-design. We also illustrated the validity of our results via some examples given by Magma program. Meanwhile, we left an open problem for interested readers.

Acknowledgements. The authors would like to thank the anonymous reviewers and the editor for their valuable comments and suggestions, which improved enormously the presentations and quality of the paper. This research was supported by *NSFC* (Grant Nos. 6177022).

References

1. Assmus Jr., E.F., Key, J.D.: Designs and Their Codes. Cambridge University Press, Cambridge (1992)
2. Augot, D., Sendrier, N.: Idempotents and BCH bound. IEEE Trans. Inf. Theor. **40**(15), 204–207 (1994)

3. Beth, T., Jungnickel, D., Lenz, H.: Design Theory. Cambridge University Press, Cambridge (1999)
4. Charpin, P.: Open problems on cyclic codes. In: Pless, V.S., Huffman, W.C. (eds.) Handbook of Coding Theory, vol. 1, pp. 963–1063. Elsevier, Amsterdam, The Netherlands (1998)
5. Ding, K., Ding, C.: A class of two-weight and three-weight codes and their applications in secret sharing. IEEE Trans. Inf. Theor. **61**(11), 5835–5842 (2015)
6. Ding, C.: Design from Linear Code. World Scientific, Singapore (2018)
7. Ding, C.: Parameters of several classes of BCH codes. IEEE Trans. Inf. Theor. **61**(8), 5322–5330 (2015)
8. Ding, C., Li, C.: Infinite families of 2-designs and 3-designs from linear codes. Discret Math. **340**(10), 2415–2431 (2017)
9. Ding, C., Tang, C., Vladimir, D.T.: Linear codes of 2-designs associated with subcodes of the ternary generalized Reed-Muller codes. arXiv http://arxiv.org/abs/1907.13032. [math.CO] (2019)
10. Ding, C.: Infinite families of 3-designs from a type of five weight codes. Crypt. Commun. **86**(3), 703–719 (2018)
11. Ding, C., Fan, C., Zhou, Z.: The dimension and minimum distance of two classes of primitive BCH codes. Finite Fields Appl. **36**(2), 237–263 (2017)
12. Du, X., Wang, R., Fan, C.: Infinite families of 2-designs from a class of cyclic codes. J. Comb. Des. **28**(3), 157–170 (2019)
13. Du, X., Wang, R., Tang, C., Wang, Q.: Infinite families of 2-designs from linear codes. Appl. Algebra Eng. Commun. Comput., 1–19 (2020). https://doi.org/10.1007/s00200-020-00438-8
14. Du, X., Wang, R., Tang, C., Wang, Q.: Infinite families of 2-designs from two classes of binary cyclic codes with three nonzeros. Finite Fields Appl. (2019, to appear)
15. Huffman, W.C., Pless, V.: Fundamentals of Error Correcting Codes. Cambridge University Press, Cambridge (2003)
16. Kasami, T., Lin, S.: Some results on the minimum weight of primitive BCH codes. IEEE Trans. Inf. Theor. **18**(10), 824–825 (1972)
17. Li, C., Ding, C., Li, S.: LCD cyclic codes over finite fields. IEEE Trans. Inf. Theor. **18**(3), 4344–4356 (2018)
18. Li, C., Wu, P., Liu, F.: On two classes of primitive BCH codes and some related codes. IEEE Trans. Inf. Theor. (2018). https://doi.org/10.1109/TTT.2018.2883615
19. MacWlliams, F.J., Sloane, N.J.A.: The Theory of Error-Collecting Codes. Elsevier, Amsterdam (1977)
20. Tonchev, V.D.: Codes and design. In: Pless, V.S., Huffman, W.C. (eds.) Handbook of Coding Theory, vol. II, pp. 1229–1268. Elsevier, Amsterdam (1998)
21. Tonchev, V.D.: Handbook of Combinatorial Designs, 2nd edn., pp. 677–701. CRC Press, New York (2007)
22. Xie, J., Sun, D., Cai, J., Cai, F.: Waveband selection with equivalent prediction performance for FTIR/ATR spectroscopic analysis of COD in suger refinery waste water. CMC: Comput. Mater. Contin. **59**(2), 687–695 (2019)
23. Zhao, W., Li, P., Zhu, C., Liu, D., Liu, X.: Defense against poisoning attack via evaluating training samples using multiple spectral clustering aggregation method. CMC: Comput. Mater. Contin. **59**(3), 817–832 (2019)
24. Zhang, Y., Deng, Y., Liu, Y., Wang, L.: Dynamic modeling and stability analysis of tilt wing unmanned aerial vehicle during translation. CMC: Comput. Mater. Contin. **59**(3), 833–851 (2019)

A Survey on Side-Channel Attacks of Strong PUF

Yan Li(✉), Jianjing Shen, Wei Liu, and Wei Zou

Information Engineering, University of Strategic Support Force, Zhengzhou 450002, Henan, China
15136288109@139.com

Abstract. With the application domain extension and the in-depth research on physical unclonable function, the security of PUF has attracted more and more attention. Various attack methods have been emerged, among which side channel attacks have advantages in modeling PUF with non-linear structures. Based on the research of strong PUF attacks, this paper classifies the existing side channel analysis methods. According to the unified symbol rules, the principles of PUF error injection, reliability attack and power analysis are analyzed. Finally, the future development prospects of PUF side channel attacks are discussed.

Keywords: Strong physical unclonable function · Side-channel attack · Machine learning

1 Introduction

The physical unclonable function (PUF) is a pseudo-random function implemented using a physical structure and used to extract the differences in the physical characteristics of the chip. PUF is tamper-resistant and can be implemented on very few hardware resources [1,2].

With the upsurge of artificial intelligence research, the application of machine learning in the field of strong PUF attack has been widely studied. Logic regression, support vector machine, evolutionary strategy, in-depth learning and ensemble learning can attack one or more PUFs successfully. Machine learning has become a common method to attack PUFs [3–5]. However, when the number of non-linear components such as XOR logic is large, or the size of PUF is large, the attack difficulty increases exponentially.

Side channel attacks are not the same as traditional attack methods. The former mainly analyzes the vulnerability brought about by the physical implementation of the cryptosystem, while the latter mainly analyzes the vulnerability of the search algorithm [6]. Common side channel attacks include timing, power consumption, electromagnetic leakage, fault injection, etc. For the strong PUF, it is difficult to attack the PUF directly and successfully by relying on the information obtained from the traditional side channel attack, and establish the data model of the target. However, when combined with machine learning, side channel attacks can effectively reduce the computational complexity and attack time

X. Sun et al. (Eds.): ICAIS 2020, LNCS 12240, pp. 74–85, 2020.
https://doi.org/10.1007/978-3-030-57881-7_7

of machine learning in PUF modeling by providing additional information, thus becoming the mainstream PUF attack method at present [7,8].

2 Analysis of the Current Research Status of Strong PUF Attack

2.1 Strong PUF and its Mathematical Model

PUF generally refers to a circuit topology structure that extracts mismatch parameters from manufacturing variations, which is formally defined as based on the inherent random characteristics k introduced in the device manufacturing process.

PUF generally refers to a circuit topology that extracts mismatch parameters from manufacturing variations, which is formally defined as mapping the excitation c selected in finite space c to the response in finite space r based on the inherent randomness k introduced in the device manufacturing process. Physical system, expressed as: $PUF_k : C \rightarrow R.$

Researchers have used the entropy source generated by the difference in device manufacturing process to design a variety of PUF circuits. Common weak PUFs include Optical PUF, Memristor Crossbar PUF [9] and Arbiter-based PUF.

Arbiter PUF using n-tier multi-channel selector circuit structure of the cascade, the same signal along a different path input level 1 multiplexer, excitation vector of each corresponding level multiplexer, parallel path or cross paths is used to select, two way along a different path transmission signal, extraction and transmission delay, by arbitration form response output (Fig. 1).

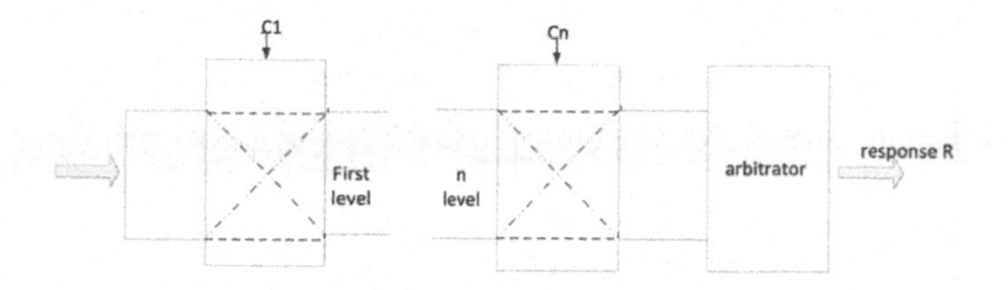

Fig. 1. Arbiter PUF circuit structure

Assume that the delay difference of the i-th level is ΔD_i, and when the input is c_i, $\delta_{c_i,i}$ represents the delay difference of the i-th stage,so $\Delta D_i = \Delta D_{i-1} * (1 - 2 * c_i) + \delta_{c_i,i}$. Thus, the calculation formula $\Delta D_n = \overrightarrow{w}^T \overrightarrow{\Phi}$ of the delay difference ΔD_n at the end of the arbiter is obtained, which is the scalar product of the delay vector $vecw$ and the feature vector $\overrightarrow{\Phi}$. Among them, the feature vector by the vector $\overrightarrow{C}$: $\overrightarrow{\Phi} = \prod_{l=i}^{n} (1 - 2c_l)$. The delay vector $\overrightarrow{w} = (w_1, \ldots, w_{n+1})$ can be used to uniquely describe the arbiter PUF. If you know the delay parameters of each stage multiplexer, it is easy to get the model of the arbiter PUF. The delay parameter cannot be directly obtained by measurement, but after obtaining a sufficient number of excitation response pairs, the delay vector can be solved by the least squares method to obtain the mathematical model of the arbiter PUF.

2.2 Research on PUF Structure Against Machine Learning Attacks

Lightweight Secure PUF and XOR Arbiter PUF. A more common security enhancement method is to add nonlinear elements, such as adding XOR logic to the standard arbiter PUF, to form XOR arbiter PUF and lightweight security PUF. The difficulty of using the machine learning method to attack the XOR arbiter is exponentially increasing with the number of switching switches. The lightweight security PUF is composed of a multi-channel arbiter PUF. The excitation forms a different sequence through the input network and is sent to each channel arbiter PUF. Each channel responds through the output network to generate a response (Figs. 2 and 3).

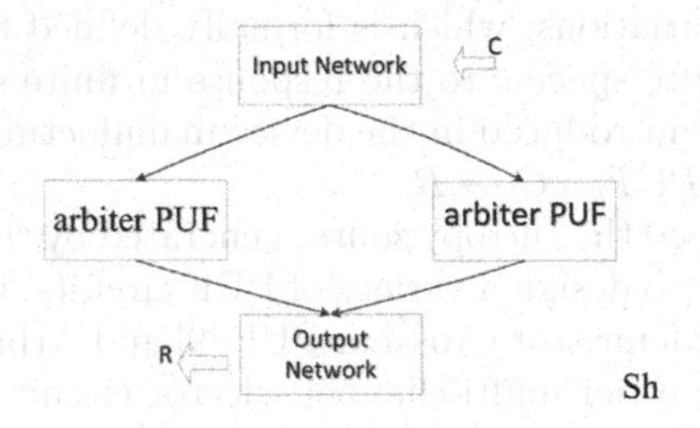

Fig. 2. Circuit structure of lightweight safety PUF

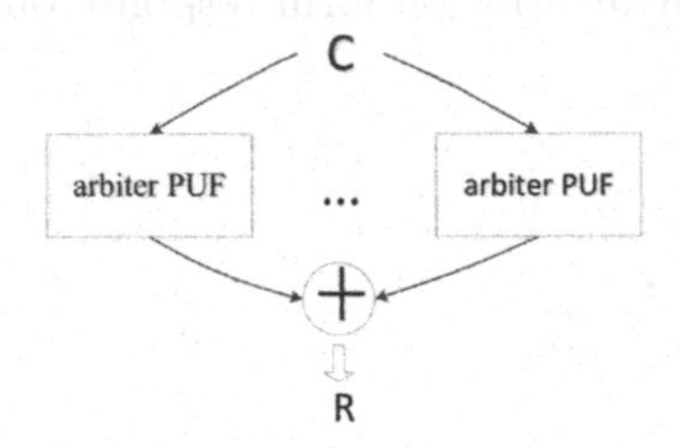

Fig. 3. Circuit structure of XOR PUF

Sahoo proposed two different cryptanalysis attack methods, successfully attacking combined PUF and lightweight security PUF. In order to increase the resistance to machine learning, nonlinear elements are added when designing PUF. The most common method is to add XOR logic. Multiple arbiter PUFs are used in the XOR PUF. These arbiter PUFs using the same excitation, get a response after Exclusive OR as xor PUF response output. The greater the number of arbiters included, the stronger the ability of XOR PUFs to combat machine learning attacks. When the machine learning method encounters a PUF with a complicated structure and a large scale, its operation time often exceeds the practical range [18].

SLPUF and Controlled PUF. One of the ideas against machine learning attacks is to hide the true stimulus or response of the PUF, using only auxiliary data or response part of the sequence in the authentication application. Such as Reverse Fuzzy Extractor Protocol [11], SLPUF [12], etc. However, the auxiliary data and the excitation generated by the LFSR excitation generator also leak the corresponding information. Becker uses the evolutionary strategy to obtain an accurate PUF model from the leaked unreliable information, thus cracking RFEP and SLPUF. Controlled PUF proposed by Gassend, by increasing the input circuit prevents attackers random excitation, while hiding the PUF response and outputting the response after shift processing [13]. But controlled PUF depend on error-correcting codes, error-correcting codes can be used as the data provided to ML.

3 Side Channel Attack Against Strong PUF

3.1 Side Channel Attack

The method of cryptanalysis using side channel leakage information is called bypass analysis or side channel attack [6]. Since the introduction of side-channel attacks in 1996, cryptanalysts have carried out a lot of research work. Side-channel attacks have constituted a huge threat to the security of various cryptographic algorithms on the device.

In 2013, Merli conducted electromagnetic side channel attack on RO PUF for the first time [15]. In the same year, Delvaux used the reproducibility of the response as the side channel information to model the arbitrator PUF [16], realizing the ideas put forward by Ruhrmair in the literature [3]. Although the performance is not as good as the machine learning attack, this method only uses the reliability information to model the arbiter PUF, which opens up a precedent for strong PUF side channel attacks. Subsequently, Delvaux proposed a fault injection attack that affects the repeatability of the PUF response by changing the environmental conditions such as the supply voltage and operating temperature [17], thereby improving the attack efficiency. In 2014, Ruhrmair proposed a power consumption and time-side channel attack method for XOR PUF and lightweight security PUF [18]. Using simple power analysis and delay measurement circuit, the side channel attack was combined with the existing machine learning modeling attack. Improved performance of modeling attacks. Becker studied two methods of passive power side channel and active fault injection for arbiter-based PUF, and verified it on controlled PUF [19].The power-side channel attack uses the correlation power analysis, and the correlation coefficient of the power trajectory is used as the fitness function, combined with the covariance matrix adaptive evolution strategy (CMA-ES) to form the side channel CMA-ES attack method. Active fault injection uses a change in the supply voltage to cause a response bit to reverse fault, and combines the unreliability of the response with machine learning to form a hybrid attack method. Becker further changed the active attack to passive attack, using the unreliable correlation coefficient as the fitness function, and successfully attacked the XOR PUF [20] based on CMA-ES.

3.2 Side Channel Attack Based on PUF Reliability

The reliability of the PUF is usually measured by the bit error rate, and the bit error rate Rel is the average of the number of flipped bits in the response of the same stimulus multiple test. l is the digits of response, m is the number of tests, and $HD(R_i, R_{i,t})$ is the Hamming distance between the response of the t th test and the original response. $Rel = 1 - BER = 1 - \frac{1}{m}\sum_{t=1}^{m}\frac{HD(R_i,R_{i,t})}{l}$.

The ideal value of reliability is 1, that is, ideally, no matter how many experiments are repeated in any environment, the response of the same stimulus should be consistent; However, in the actual environment, PUF is inevitably affected by many factors such as temperature, voltage and aging, resulting in a change in response. For the attacker, not only does the PUF's stimulus response contain useful information, but the reliability of the response, the fluctuations in the response, can also provide valuable information [20]. Instead of using an excitation response pair, the PUF is modeled using only the reliability of the response, known as a reliability-based side channel attack.

Repeatability refers to the short-term reliability [16] of PUF under the influence of noise (excluding the long-term effects of device aging). Make full use of the effect of random noise in repeated measurements, and use the repeatability of response as side channel information for arbiter PUF modeling.
ΔD_{PUF} :PUF delay difference caused by incentive, ΔD_{noise} : Delayed effects of accumulated noise
Δt_{Arb}:Delay introduced by the arbiter bias, ΔD_{PUF} and ΔD_{noise} meet the Gaussian distribution with a mean of 0, standard deviations are σ_V and σ_N respectively. After considering the delay introduced by the arbiter, the output of the PUF is determined by ΔD_{PUF}-Δt_{Arb}. Combined with the probability density function of noise, the formula for repeatability $R \in [0,1]$ is $R(\Delta D_{PUF}) = \frac{1}{2}erfc(\frac{\Delta t_{Arb}-\Delta D_{PUF}}{\sqrt{2}\sigma_N})$, where $erfc$(t)$= (2/\sqrt{\pi})\int_t^{\infty} e^{-z^2}dz$, $\Delta D_{PUF}(R) = \Delta t_{Arb} - \sqrt{2}\sigma_N erfc^{-1}(2R)$.

The value of the repeatability $R \in \{0, (1/M), \ldots, (M-1/M), 1\}$ indicates the number of times the response value obtained by using the same stimulus for times is
1. When ΔD_{PUF}=Δt_{Arb}, R=1/2. The farther R is from 1/2, the higher the repeatability of the response (Fig. 4).

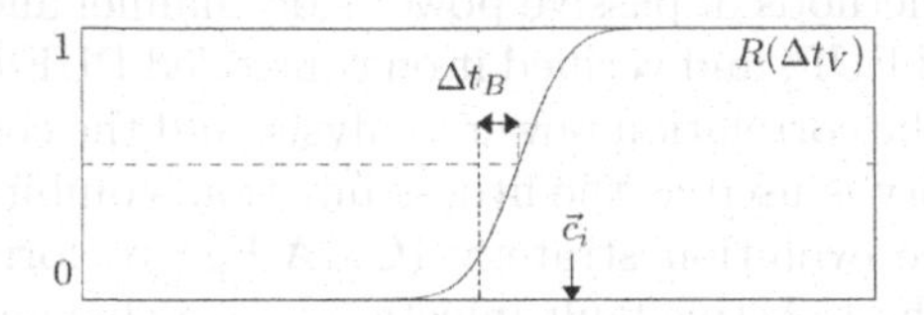

Fig. 4. Repeatability and excitation curve

As can be seen from Fig. 4, R is linear with the excitation $\overrightarrow{c}_i$ in the range of [0.1, 0.9]. After collecting N $\{\overrightarrow{c}_i, R_i\}$ combinations in the linear region, the delay model is established by least squares method. The key to implementing the attack is to quickly find the available $\{\overrightarrow{c}_i, R_i\}$ combination, but usually the response of the arbiter PUF is relatively stable, only about 10% of the response repeatability will be in the linear range of [0.1, 0.9]. This method is less efficient than machine learning methods and cannot be used for XOR PUF. If only $D_{noise}$is considered, Δt_{Arb}v is ignored and the PUF model becomes $\Delta D = \Delta D_{PUF}+D_{noise}= \overrightarrow{w}^T\overrightarrow{\phi}+D_{noise}$. If the delay difference ΔD_{PUF} is large, the noise has no effect on the response, but if ΔD_{PUF} is small, the response is largely dependent on D_{noise}. The PUF response is measured repeatedly using the same stimulus. If the response is unstable for a given stimulus, the delay ΔD_{PUF} corresponding to the stimulus is judged to be close to 0: $|\Delta D_{PUF}| < \varepsilon$. The goal of the fitness function is to select the most appropriate $\overrightarrow{w}$ from a given set of PUF models. The operation process of the reliability-based CMA-ES algorithm is similar to the traditional CMA-ES algorithm except that the parameter ε is added, and the other steps are the same. Proceed as follows:

1. Select the excitation $\overrightarrow{\phi}_i$ to repeat the test to the PUF l times, get l responses $r_{i,1}, r_{i,2},\cdots, r_{i,l}$, calculate the reliability $h_i = \left|\frac{l}{2} - \sum_{j=1}^{l} r_{i,j}\right|$.
2. In order to test the fitness of the guessing model, the guessing reliability $\tilde{h}_i$ is calculated: if $\left|\overrightarrow{w}^T\overrightarrow{\phi}\right| > \varepsilon$, then $\tilde{h}_i$=1; otherwise $\tilde{h}_i$= 0.
3. Select n excitations $\overrightarrow{\phi}_1, \overrightarrow{\phi}_2, ..., \overrightarrow{\phi}_n$ and repeat the above process to obtain the reliability vector $h = \{h_1, ..., h_n\}$ and the guess reliability vector $\tilde{h} = \left\{\tilde{h}_1, ..., \tilde{h}_n\right\}$. Calculate the Pearson correlation coefficient between them, the correlation coefficient is higher, the model is more suitable.

From the perspective of reliability, the more the arbiter PUFs included in the XOR PUF, the lower the reliability of the XOR PUF. Becker designed an attack method for the XOR arbiter based on the divide-and-conquer. The reason why the division strategy can be adopted is because the reliability of the XOR arbiter is determined by the n-channel arbiter PUF, the weight of each channel is the same, and the flip of any one PUF response will affect the result of XOR. When modeling a certain path in the n-channel arbiter PUF, other n−1 way PUFs can be regarded as noise from the perspective of machine learning. Since the noise has little effect on the performance of the algorithm, the CMA-ES algorithm with the correlation coefficient as the fitness function is used to model each PUF in turn, and the model of the XOR arbiter is obtained after repeating n times. However, CMA-ES is a non-deterministic machine learning algorithm. When it cannot converge to the optimal solution, it needs to restart the learning process. The n-channel PUF in the XOR arbiter shares the same stimulus. If the machine learning method converges to the same PUF every time, modeling cannot be implemented. The probability characteristics of the CMA-ES algorithm allows for the same stimulus to converge to different PUFs per run [19]. In order to improve efficiency, the attack adopts a two-step approach: firstly, most of the PUF models are trained using the side channel CMA-ES, and then the remaining models are trained using classical machine learning methods such as logistic regression. The complexity of the XOR PUF has been successfully reduced from exponential to polynomial.

3.3 Fault Attack

Fault attack is one of the common side channel attack methods. In special cases, if there is human interference or in a harsh environment, the PUF may have a response error. The method of modeling the PUF using these error messages and fault behavior is called a PUF fault attack. Due to the inherent tamper resistance of the PUF, intrusive and semi-intrusive fault injections are difficult to function, and non-intrusive fault injection is often used to attack the PUF. The arbiter PUF has a problem of unstable response in practical applications. The cause of this phenomenon is thermal noise and environmental changes. When the path delay difference is close to zero, the response result is largely determined by noise. If the PUF can be physically touched, the attacker can attack by adjusting the supply voltage and changing the ambient temperature.

Delvaux by using 65 nm CMOS chip technology, respectively for RO PUF and arbitrator PUF was tested, and put forward can response error induced by changing the environment, increase the number of unstable incentives for corresponding targeted produce contains lateral excitation response of channel information, thus improve the attack efficiency [16]. Becker also used fault injection when attacking Controlled PUF [20]. The basic idea is: observe the response after gradually adjusting the power supply voltage under the condition of constant excitation. If the response is changed, the bit flip fault is induced as F=1, otherwise it is recorded as F= 0. The situation in which F=1 occurs is most likely because the delay difference corresponding to the PUF under a particular excitation is close to zero, that's $|\Delta D_{PUF}| < \tau$. To test the PUF model, define the delay difference of the stimulus i as $\Delta\hat{D}_i$ and define the hypothesis vector $\hat{F_\text{i}}$, if $\left|\Delta\hat{D}_i\right| < \hat{\tau}$, otherwise $\hat{F_\text{i}}$=1. The correlation coefficient between the assumed $\hat{F_\text{i}}$ and the actually measured F_i is used as a fitness function to determine whether the CMA-ES algorithm is suitable for each round. The above fault injection is used as an auxiliary means for reliability side channel attacks, which is used to accelerate the collection of reliability information and improve efficiency.

3.4 Power Attack

A large number of researchers have proposed different types of power analysis methods, such as simple power analysis (SPA) and differential power analysis (DPA) [21] proposed by Kocher, and template attacks proposed [22] by Chari et al. Based on the simulation test of the internal circuit of the arbiter PUF, Ruhrmair designed a side channel attack method for extracting PUF response information by simple power analysis [23]. Measure the voltage/current value on power supply pins and calculate the power trace. The response of the arbiter PUF can be determined by the power value. This response value obtained by simple power analysis can be used to attack a more complex PUF such as a combined PUF, XOR PUF. This type of PUF contains a multi-channel arbiter PUF, and the response of each arbiter PUF as an intermediate result cannot be directly read from the outside.

The excitation of each arbitrator PUF in the lightweight security PUF is different, and the corresponding power side channel attack can adopt the divide-and-conquer strategy. By continuously designing and adjusting the excitation combination and combining the power analysis to obtain the total number of PUFs with a response of "1", the excitation response of each arbiter PUF can be distinguished. This splits the combined PUF into a number of separate arbiter PUFs, and the model parameters are obtained by simultaneous equations. The simple power analysis method can successfully attack the lightweight security PUF, but XOR PUF invalid. Because the arbiter PUF in the XOR PUF shares the same excitation, the total number of responses obtained by simple power analysis cannot distinguish the single arbiter PUF. Coupled with the interference of various types of noise, it is more and more difficult to obtain ideal power information.

In order to solve this problem, Ruhrmair proposed a side channel attack method using differential power analysis, called excitation-based response prediction [18]. Differentiate subtle changes by comparing the power trajectories before and after the response, extract the power of exclusive-or gate and convert it into an identifiable form related to the response. Let $r_i\left(C\right) \in \{0, 1\}$ denote that when the excitation is C, the output of the i th of the k arbiter PUFs in the XOR PUF, the side channel information is $n = \sum_i r_i(C)$. A gradient optimization algorithm similar to logistic regression is used to minimize the mean square error between the side channel model $f\left(w, C\right) = \sum\limits_i \Theta(w_i^T \overrightarrow{\phi}_i)$ and the actual output n.

$l\left(M, w\right) = \sum\limits_{(C,t)\in M} \left(f\left(w, C\right) - n\right)^2$The corresponding gradient is $\nabla l\left(M, w\right) = \sum\limits_{(C,t)\in M} 2(f\left(w, C\right) - n)\nabla f\left(w\right)$, $\nabla f\left(w_j\right) = \sigma(w_j^T \overrightarrow{\phi}_j)(1 - \sigma(w_j^T \overrightarrow{\phi}_{\mathrm{j}}))\overrightarrow{\phi}_{\mathrm{j}}$.

Use the RProp algorithm to find the solution $\hat{w}$ that minimizes l. This model depends only on the direction, independent of the length $\|w_i\|$ of the weight vector. Using the side channel information, the gradient of the arbiter PUF weight vector w_j depends only on the weight vector itself. Using a two-step optimization method, first, the PUF model is optimized based on the gradient using the side channel information until more than 95% of the XOR arbiter outputs can be replicated. The model is then continuously optimized using standard logistic regression algorithms [19].

For lightweight safety PUF, Becker uses a method of correlation power analysis [20]. The power curve of the arbiter PUF is analyzed by simulation, and the Pearson correlation coefficient between the real PUF power curve and the prediction model power curve with different accuracy is calculated in turn, and the conclusion is drawn: the higher the accuracy of the model, the greater the value of the correlation coefficient between powers. The results of the model-related power analysis can be used to predict the accuracy of the model and determine the pros and cons of the model.

CMA-ES is combined with related power analysis. The original prediction model is randomly generated by using evolutionary strategy. The accuracy of the prediction model is judged based on the correlation power analysis. The power correlation coefficient is used as the fitness function, and the model with high accuracy is reserved as the parent. Convergence to an approximate optimal solution. In the solution process, in order to further improve the speed and accuracy, the matching number between the response strings can be used as the fitness function after the model accuracy reaches a certain threshold. Based on the idea of divide and conquer, when using the side channel CMA-ES algorithm attack, the response of the other arbiter PUF module is regarded as noise for each arbiter PUF module.

These methods can be individually applied to attack arbiter, but for a more complex PUF, modeling using only side channel attacks cannot be achieved. A more common application is the combination of side channel attacks and machine learning methods to form a hybrid attack. The analysis shows that the hybrid side performance of the power side channel plus logistic regression is the best. However, CMA-ES also has its unique advantage. When the circuit structure of the PUF is unknown, the structural model and internal parameters can still be inferred as long as there is sufficient excitation response pair.

4 The Dilemma and Prospect of PUF Side Channel Attack

4.1 The Dilemma of Side Channel Attack

Although the side channel approach opens up new avenues for attacking PUFs with complex structures, it reduces the difficulty of attack. However, side channel attacks also face many difficulties in practical applications. Firstly, side channel information such as power and electromagnetic are usually measured. But in practical applications, environmental noise and measurement noise will have a large impact on the measurement results, which will affect the correctness and availability of the extracted side channel information. Secondly, certain information measurement methods or semi-invasive attack methods require special equipment or conditions, and they do not have universal applicability. The delay measurement circuit in the time-side channel attack proposed by Ruhrmair is implemented by the internal logic of the FPGA and cannot be applied to most PUFs.

4.2 Research Prospects in the Field of PUF Side Channel Security

In recent years, new PUF side channel attack methods have emerged in large numbers. The attack path has added special methods such as photon side channel and Remanence leakage [27]. The attack range covers a variety of strong PUF and weak PUF [28–30], and the attack object is extended from the classic structure PUF to the PUF proposed by the new design.

In the face of increasingly serious security threats, side channel defense has gradually become a hot issue in PUF research, and some methods to prevent side channel information leakage have been applied to PUF design. For example,increasing the symmetry of the circuit to reduce information leakage. In the arbitrator part, two triggers with the same structure but opposite outputs are used. In theory, the power consumption is the same whether the output is 0 or 1. In addition, artificially adding noise to cover subtle changes in measured values, using special circuit structures to reduce side channel leakage can also resist side channel attacks to some extent. For example, Yuan Cao proposed a RO PUF based on a current-limited delay unit to protect against electromagnetic side channel attacks by reducing electromagnetic radiation.

Side channel attacks also have far-reaching effects in the PUF design field. In recent years, the new PUF structure proposed will improve the anti-side channel attack capability as the main target. The FSM-PUF proposed by Yansong Gao limits the possibility of extracting response reliability information by prohibiting the repetition of the same excitation, thus avoiding reliability-based side channel attacks. Side channel analysis is also used as one of the evaluation methods for PUF design. The combined PUF designed by Sahoo with the arbiter PUF as the basic component is analyzed and verified by Becker's reliability side channel plus CMA-ES hybrid attack model, which proves that the combined PUF has the ability to defend against the hybrid attack.

With the increasing emphasis on the side channel attacks by PUF designers and the use of multiple defense methods, the traditional attack methods are facing the danger of gradual failure. Studying the new PUF side channel analysis method has become the focus of researchers. The proposed defense countermeasures can only deal with one or several types of side channel attacks. Combining different side channel analysis strategies can provide more target information and improve the efficiency and accuracy of machine learning methods.

5 Conclusion

Based on the introduction of the strong PUF mathematical model, the paper analyzes the existing strong PUF side channel attacks by using uniform symbol rules according to the types of reliability, power analysis and fault injection and compare the performance, attack objects and applicable occasions of different types of hybrid attacks. Finally, the paper looks forward to the application prospect of side channel attack in PUF modeling, and emphasizes the importance of side channel analysis in the field of PUF design: in the PUF design or security evaluation, it is necessary to fully consider its ability to resist side channel attacks.

References

1. Ravikanth, P.S.: Physical one-way functions. Science **297**(5589), 2026–2030 (2002)
2. Herder, C., Yu, M.D., Koushanfar, F., et al.: Physical unclonable functions and applications a tutorial. Proc. IEEE **102**(8), 1126–1141 (2014)
3. Rührmair, U., Holcomb, D.E.: PUFs at a glance. In: Design, Automation & Test in Europe Conference & Exhibition. IEEE, pp. 1–6 (2014)
4. Ganji, F., Tajik, S., Fäßler, F., Seifert, J.-P.: Strong machine learning attack against PUFs with no mathematical model. In: Gierlichs, B., Poschmann, A.Y. (eds.) CHES 2016. LNCS, vol. 9813, pp. 391–411. Springer, Heidelberg (2016). https://doi.org/10.1007/978-3-662-53140-2_19
5. Guo, Q., Ye, J., Gong, Y., et al.: Efficient attack on non-linear current mirror PUF with genetic algorithm. In: Asian Test Symposium, pp. 49–54. IEEE (2016)
6. Koeune, F., Standaert, F.-X.: A tutorial on physical security and side-channel attacks. In: Aldini, A., Gorrieri, R., Martinelli, F. (eds.) FOSAD 2004–2005. LNCS, vol. 3655, pp. 78–108. Springer, Heidelberg (2005). https://doi.org/10.1007/11554578_3
7. Xu, X., Burleson, W.: Hybrid side-channel/machine-learning attacks on PUFs: a new threat? In: Design, Automation and Test in Europe Conference and Exhibition, pp. 1–6. IEEE (2014)
8. Fukushima, S., et al.: Delay PUF assessment method based on side-channel and modeling analyzes: the final piece of all-in-one assessment methodology. In: IEEE Trustcom BigDataSE ISPA, pp. 201–207. Institute of Electrical and Electronics Engineers (2016)
9. Liu, Y., Xie, Y., Bao, C., et al.: A combined optimization-theoretic and side-channel approach for attacking strong physical unclonable functions. IEEE Trans. Very Large Scale Integ. Syst. **26**(1), 73–81 (2017)
10. Vijayakumar, A., Patil, V.C., Prado, C.B., et al.: Machine learning resistant strong PUF: possible or a pipe dream? In: IEEE International Symposium on Hardware Oriented Security and Trust, pp. 19–24. IEEE (2016)
11. Herrewege, A.V., Katzenbeisser, S., Maes, R., et al.: Reverse fuzzy extractors: enabling lightweight mutual authentication for PUF-enabled RFIDs. In: International Conference on Financial Cryptography and Data Security, FC 2012, pp. 374–389 (2012)
12. Majzoobi, M., Rostami, M., Koushanfar, F., et al.: Slender PUF protocol: a lightweight, robust, and secure authentication by substring matching. In: IEEE Symposium on Security and Privacy Workshops, pp. 33–44. IEEE Computer Society (2012)
13. Gassend, B., Clarke, D., Dijk, M.V., et al.: Controlled Physical random functions. In: 2002 Proceedings of the Computer Security Applications Conference, pp. 149–160. IEEE (2002)
14. Merli, D., Schuster, D., Stumpf, F., Sigl, G.: Side-channel analysis of PUFs and Fuzzy extractors. In: McCune, J.M., Balacheff, B., Perrig, A., Sadeghi, A.-R., Sasse, A., Beres, Y. (eds.) Trust 2011. LNCS, vol. 6740, pp. 33–47. Springer, Heidelberg (2011). https://doi.org/10.1007/978-3-642-21599-5_3
15. Merli, D., Heyszl, J., Heinz, B., et al.: Localized electromagnetic analysis of RO PUFs. In: IEEE International Symposium on Hardware-Oriented Security and Trust, pp. 19–24. IEEE (2013)
16. Delvaux, J., Verbauwhede, I.: Side channel modeling attacks on 65 nm arbiter PUFs exploiting CMOS device noise. In: IEEE International Symposium on Hardware-Oriented Security and Trust, pp. 137–142. IEEE (2013)

17. Delvaux, J., Verbauwhede, I.: Fault injection modeling attacks on 65 nm arbiter and RO sum PUFs via environmental changes. IEEE Trans. Circ. Syst. I Regul. Pap. **61**(6), 1701–1713 (2014)
18. Rührmair, U., et al.: Efficient power and timing side channels for physical unclonable functions. In: Batina, L., Robshaw, M. (eds.) CHES 2014. LNCS, vol. 8731, pp. 476–492. Springer, Heidelberg (2014). https://doi.org/10.1007/978-3-662-44709-3_26
19. Becker, G.T.: The gap between promise and reality: on the insecurity of XOR arbiter PUFs. In: Güneysu, T., Handschuh, H. (eds.) CHES 2015. LNCS, vol. 9293, pp. 535–555. Springer, Heidelberg (2015). https://doi.org/10.1007/978-3-662-48324-4_27
20. Becker, G.T.: On the Pitfalls of using arbiter-PUFs as building blocks. IEEE Trans. Comput. Aided Des. Integr. Circ. Syst. **34**(8), 1295–1307 (2015)
21. Messerges, T.S., Dabbish, E.A., Sloan, R.H.: Investigations of power analysis attacks on smartcards. In: Usenix Workshop on Smartcard Technology on Usenix Workshop on Smartcard Technology, p. 17. USENIX Association (1999)
22. Chari, S., Rao, J.R., Rohatgi, P.: Template attacks. In: Kaliski, B.S., Koç, K., Paar, C. (eds.) CHES 2002. LNCS, vol. 2523, pp. 13–28. Springer, Heidelberg (2003). https://doi.org/10.1007/3-540-36400-5_3
23. Mahmoud, A., RÄuhrmair, U., Majzoobi, M., Koushanfar, F.: Combined modeling and side channel attacks on strong PUFs. IACR Cryptol. ePrint Arch. (2013)
24. Tajik, S., Lohrke, H., Ganji, F., et al.: Laser fault attack on physically unclonable functions. In: The Workshop on Fault Diagnosis & Tolerance in Cryptography, pp. 85–96. IEEE Computer Society (2015)
25. Skorobogatov, S.: Semi-invasive attacks - a new approach to hardware security analysis. Technical report UCAM-CL-TR-630. University of Cambridge, Computer Laboratory (2005)
26. Kumar, R., Burleson, W.: Side-Channel assisted modeling attacks on feed-forward arbiter PUFs using silicon data. In: Mangard, S., Schaumont, P. (eds.) RFIDSec 2015. LNCS, vol. 9440, pp. 53–67. Springer, Cham (2015). https://doi.org/10.1007/978-3-319-24837-0_4
27. Zeitouni, S., Oren, Y., Wachsmann, C., et al.: Remanence decay side-channel: the PUF case. IEEE Trans. Inf. Forensics Secur. **11**(6), 1106–1116 (2016)
28. Ruhrmair, U., Schlichtmann, U., Burleson, W.: Special session: how secure are PUFs really? On the reach and limits of recent PUF attacks. In: Design, Automation and Test in Europe Conference and Exhibition, p. 346. IEEE (2014)
29. Kumar, R., Burleson, W.: Hybrid modeling attacks on current-based PUFs. In: IEEE International Conference on Computer Design, pp. 493–496. IEEE (2014)
30. Merli, D., Schuster, D., Stumpf, F., et al.: Semi-invasive EM attack on FPGA RO PUFs and countermeasures. In: The Workshop on Embedded Systems Security, pp. 1–9. ACM (2011)
31. Jiang, X., Liu, M., Yang, C., Liu, Y., Wang, R.: A blockchain-based authentication protocol for WLAN mesh security access. Comput. Mater. Continua **58**(1), 45–59 (2019)
32. Cui, J., Zhang, Y., Cai, Z., Liu, A., Li, Y.: Securing display path for security-sensitive applications on mobile devices. Comput. Mater. Continua **55**(1), 017–035 (2018)
33. Zhang, H., Yi, Y., Wang, J., Cao, N., Duan, Q.: Network security situation awareness framework based on threat intelligence. Comput. Mater. Continua **56**(3), 381–399 (2018)

A Novel Visual Cryptography Scheme Shared with Edge Information Embedded QR Code

Fei Hu, Yuanzhi Yao(✉), Weihai Li(✉), and Nenghai Yu

CAS Key Laboratory of Electromagnetic Space Information,
University of Science and Technology of China, Hefei 230027, Anhui, China
feih@mail.ustc.edu.cn, {yaoyz,whli,ynh}@ustc.edu.cn

Abstract. QR code has been widely used in our daily life due to its convenience, and the security of QR code becomes a rising problem. In this paper, we introduce a new algorithm that adds the edge of the gray image to the QR code, and the edge information is captured by the Canny detector. And then we use the random grid based visual cryptography scheme (RG-VCS) to generate shares. Each share of the VCS is an individual QR code that can be scanned by a QR code reader, and the secret information can be decoded by purely stacking without computation. Meanwhile, we can also recover the secret image with better visual quality with the help of computation.

Keywords: Visual secret sharing scheme · QR code · Data hiding

1 Introduction

The visual cryptography scheme (VCS) is an interesting research topic, which was firstly introduced by Naor and Shamir in 1994 [1]. VCS is a secret sharing method using a noise-like image to hide and share information by divided a binary image into several shares that can recover the image almost the same by stacking a certain number of shares together. The advantage of the scheme is that the decoding process is based on the human visual system. Due to the decryption phase does not need too much computation, it has drawn many researchers' attention, and more research results can be found in publication [2, 18–20].

Since the decryption is in the manner of overlapping required number of shares together to reveal the secret image, which is known as the "stack model", there are two kinds of construction models, XOR-based VCS [3] and RG-based VCS [4].

Although some researchers applied VCS in authentication, watermark and some other domain, the shares were usually meaningless and can hardly save information. In this paper, we propose a scheme that combines QR code with VCS to overcome the problem.

Nowadays, QR code is everywhere in our daily life. Denso Wave Incorporated [7] invented a new two-dimensional code, named quick response code. Since QR codes have a wide range of applications, the security of the code becomes an emergency problem.

X. Sun et al. (Eds.): ICAIS 2020, LNCS 12240, pp. 86–97, 2020.
https://doi.org/10.1007/978-3-030-57881-7_8

Many researches [9, 12–15] have done to solve the problem. But they are all based on the encryption or stenography techniques, which will be computation consumptive.

Considering the advantage of VCS and QR code, Liu et al. [9] proposed two-level QR codes that utilize the error correction capacity and recognition pattern of QR code. However, it still does not have any visual signal.

In this paper, we propose a new QR code by combining visual cryptography. And we can get visual shares that are readable by QR code scanners and contain the secret information. Furthermore, we can decrypt by stacking them together without computation with low contrast, while a better result with the help of computation.

The remainder of the paper is organized as follows. Section 2 introduces the VCS, QR code, and Canny edge detection, and Sect. 3 illustrates our propose VCS with modified QR code, and Sect. 4 shows the experimental result of our scheme and compared with some related works, finally conclude this paper in Sect. 5.

2 Related Work

2.1 Visual Cryptography Scheme

Visual cryptography scheme (VCS), also called image secret sharing scheme, was proposed in 1994 by Naor and Shamir, and it is a special secret sharing algorithm in which the object shares are noise-like images. In the conventional VCS, a secret image I is encrypted into n random-looking shares, and we can almost recover the secret image I by superimposing any group of k shares in the n shares, while less than k shares cannot get any message about the original image. The proposed scheme is denoted as *(k, n)-VCS*. Due to the simplicity of encryption and security of "one-time-pad", researchers drew much attention to it, and got many extraordinary results.

Generally, binary images are used in VCS, and the shares are binary images, too. We denote black pixels with 1 while white 0 in this paper, and Table 1 illustrates a (2, 2) VCS algorithm. Pixels from the secret image are separated into a pair of sub-pixels in two shares, the sub-pixels are randomly chosen from the following two columns with 50-50 percent, and we can get a randomly arranged image with the same amount of black and white pixels. Furthermore, a single share has no information of the secret image, when two shares are stacked, the original black pixel will be fully recovered while white pixel will be recovered as one black pixel and a white pixel. Figure 1 displays a demo of (2, 2) VCS, we can see that the stacked image can almost show the original image message with low contrast and pixel extension.

Revenkar and et al. [5] use VCS to achieve the iris authentication by respectively store standard iris information in the database and user's ID card. Rosking and et al. [6] proposed a method combining the VCS with the vote system. Liu et al. [9] proposed two-level QR codes that utilize the error correction capacity and recognition pattern of QR code. Wan and et al. [12] illustrated a scheme that QR codes carry the shared information and the secret information can be revealed by stacking or calculation.

Random Grids Based VCS. Random grid based VSC (RG-VCS) [4] can avoid the pixel extension, and does not need to consider how to design the share matrix. In light of the above advantages, many researchers show more interest in RG-VCS. It utilizes

Table 1. The overview of (2, 2) VCS

Secret pixel				
Share 1				
Share 2				
Probability	50%	50%	50%	50%
Stacked result				

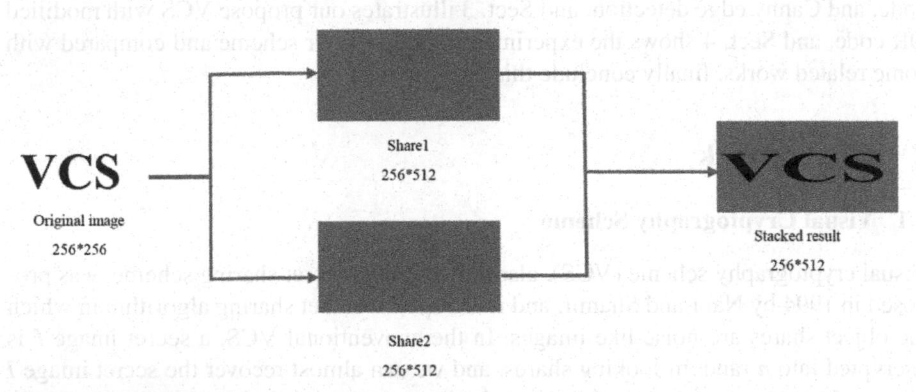

Fig. 1. An application of (2, 2) VCS. In the above picture, a binary picture is encrypted into two meaningless images which have no message about the original image, but the size of the image doubled, and the stacked result can reveal the original image with contract loss.

the random grid theorem that two noise-looking random grids generated to hide the secret image, and all shares are the same size with the secret image. And the decryption progress is identical with traditional VCS by stacking. The algorithm of RG-based VCS is in Table 2.

Table 2. Algorithm of RG-based VCS

Algorithm 1.(k, n) RG-based VCS	
Input:	A M*N binary secret image I, global parameter k and n
Output:	n random grid share images $S_1, S_2, \ldots, S_n$.
Step 1:	For every pixel (i, j) in secret image repeat steps 2-6.
Step 2:	Randomly select $b_1, b_2, \ldots, b_k \in \{0,1\}$.
Step 3:	If $I(i,j) = b_1 \oplus b_2 \oplus \ldots \oplus b_k$, go to step 5; else go to step 4.
Step 4:	Randomly choose $p \in \{1, 2, \ldots, k\}$, let $b_p = \overline{b_p}$.
Step 5:	Randomly select $b_{k+1}, b_{k+2}, \ldots, b_n \in \{0, 1\}$.
Step 6:	Randomly replace $b_1, b_2, \ldots, b_n$ to $S_1(i, j), S_2(i, j), \ldots, Sn(i, j)$.
Step 7:	Output n share images $S_1, S_2, \ldots, S_n$.

For example, in a (2, 2) random grid VCS, we first generate two random grid S_1 and S_2. Then we adjust some pixels use Eq. (1). And we use Eq. (2) to decrypt. If the pixel in secret image I is black (denoted as 1), in the recovery image, it will be always black. Otherwise, if white, the recover result will be randomly white or black with a probability of 50%,50%.

$$S_2(i,j) = \begin{cases} S_1(i,j) \text{ if } I(i,j) = S_1(i,j) \oplus S_2(i,j) \\ \overline{S_1(i,j)} \text{ if } I(i,j) \neq S_1(i,j) \oplus S_2(i,j) \end{cases} \tag{1}$$

$$I'(i,j) = S_1(i,j) \otimes S_2(i,j) \tag{2}$$

Figure 2 displays a demo of (2, 2) RG-VCS.

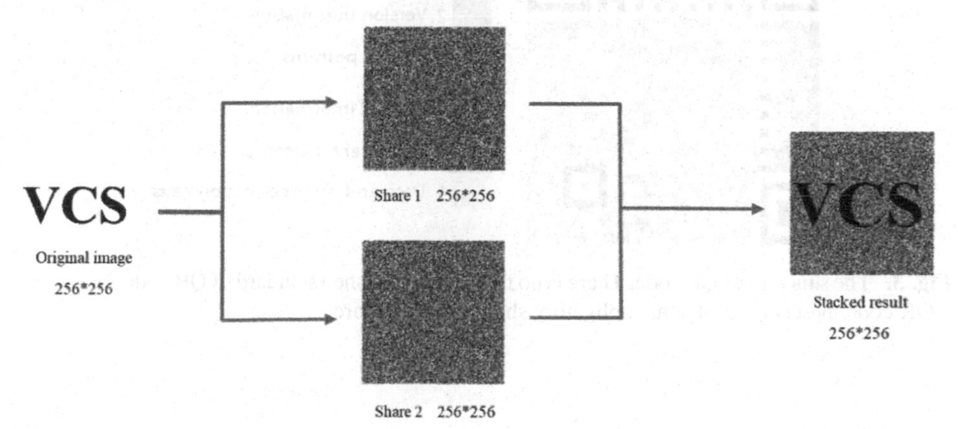

Fig. 2. An application of (2, 2) RG-VCS. As is shown in the above picture, a binary picture is encrypted into two meaningless images. Contract with traditional VCS, the shares have no pixel extension, while the stacked result still has contract loss.

2.2 Quick Response Code

Quick response code (QR code) is a kind of two-dimensional code which was invented by Denso Wave Incorporate in 1994. QR code consists of many black and white square, there are several obligatory elements in a QR code: position patterns, version information, timing patterns, formation information, alignment patterns, data and error correction keys, as is illustrated in Fig. 3 [8].

The position patterns, alignment patterns and timing patterns are used to locate the square and verify the structure of code. And the version information pattern contains the encoded information about the version of the code. The versions of it are directly connected with their dimensions and the amount of data that should be encoded.

According to Cheng and et al. [9], a QR code can be decoded successfully if we replace a square with a size of $n \times n$ cells with the central keeping unchanged. And this characteristic has a great contribution to this paper.

The QR code has error correction capability to restore data guarding against the code is damaged. The QR code error correction feature is implemented by adding Reed-Solomon codes [16] to the original data. Four error correction levels are available for users to choose, L, M, Q and H, each of them with the error correction capacity 7%, 15%, 25% and 30%. Raising the level improves error correction capability but also increases the amount of data QR code size.

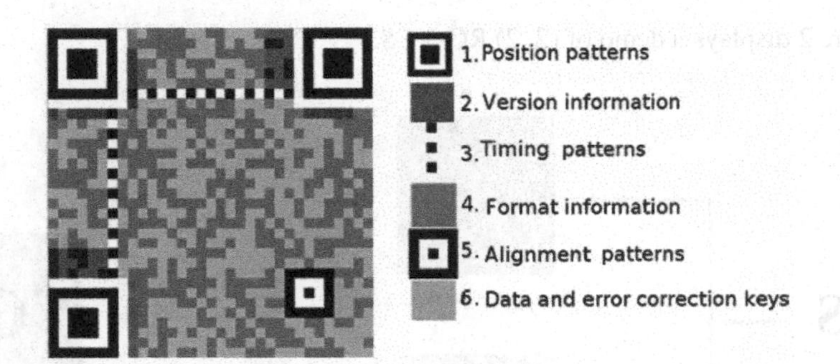

Fig. 3. The structure of QR code. There is no formally established standard of QR code. However, a QR code must consist of some obligatory shown in the picture.

2.3 Edge Detection

Edge information is the most important structure property in an image, and we can remove lots of irrelevant information to reduce the data size. There are many edge detection algorithm, here we use Canny detection [16] to accomplish our scheme.

For a grayscale image, a Canny edge detector first uses a Gaussian filter to remove the noise which might influence the accuracy of edge location. And the Gaussian filter is defined in Eq. 3.

$$G(x, y) = \frac{1}{2\pi\sigma^2}\exp\left(-\frac{x^2 + y^2}{2\pi\sigma^2}\right) \tag{3}$$

Then the detector calculates the edge gradient and direction for each pixel. After that, the non-maxima suppression algorithm is applied to get a more accurate edge candidate. Finally, by double threshold method and suppression of isolated low threshold candidates, we get an edge image (Figs. 4 and 5).

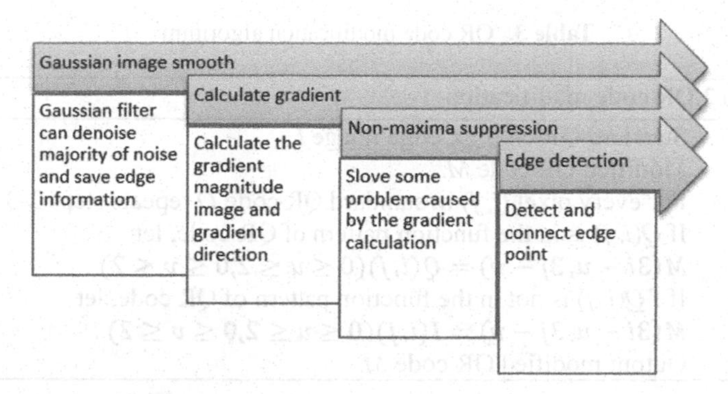

Fig. 4. The chart above shows the main steps of Canny Edge detection.

Fig. 5. The images above show the result of the Canny edge detector. As is illustrated in the right picture, the details of the image edge are well captured.

3 Image Sharing Scheme with Modified QR Code

In this section, we will introduce a new scheme that combines the QR code with VCS. According to Cheng et al. [9], we can modify the pixel's structure of QR code and let the code readable by scanner. Based on this property, we first insert an image into an original QR code and then applying VCS algorithm to generate n QR shares that contain the secret information.

Based on the previous work, we change every pixel of QR code into a 3×3 matrix, which is composed of the value of QR code's pixel and the edge of visual image. And as is introduce in Sect. 2.3, the edge information is captured by the Canny edge detector, in which we denote '1' as edge while others '0'.

For a standard QR code, according to Fu [15], in all 9 pixels of the sub-modules, we put the edges of the visual mask into the outer 8 pixels and the central pixel keeps the value of original QR code. The QR code modification algorithm is illustrated in Table 3.

In the algorithm below, a QR code can be modified and embed the edge image into it. To make the new QR code Q more quickly read by the scanner, we adjust the distribution of outer 8 pixels. If the central pixel of the sub-module is black, denoted as '1', we randomly choose a threshold $t \in [5, 8]$, for example, 6, and randomly make t pixels into black. Similarity, if the central pixel of the sub-module is white, denoted as

Table 3. QR code modification algorithm

Algorithm 2. QR code modification	
Input:	Standard QR code *Q*, edge image *I*.
Output:	Modified QR code *M*.
Step 1:	For every pixel (i, j) in standard QR code *Q* repeats steps 2-3
Step 2:	If $Q(i, j)$ is in the function pattern of QR code, let $M(3i-u, 3j-v) = Q(i,j)(0 \le u \le 2, 0 \le v \le 2)$.
Step 3:	If f $Q(i, j)$ is not in the function pattern of QR code, let $M(3i-u, 3j-v) = I(i,j)(0 \le u \le 2, 0 \le v \le 2)$.
Step 4:	Output modified QR code *M*.

'0', we randomly choose a threshold $t \in [1, 4]$, for example, 2, and randomly make t pixels into white.

Furthermore, the integration of edge information with QR code keeps the error correction mechanism unchanged. Therefore, we propose a new VCS combining with the frame of QR code that uses the error correction capacity.

The main idea of VCS for QR code is to generate pixel responding shares based on QR code, and we accomplish the encryption procedure by RG-based VCS, which is introduced in Sect. 2.1. The pixels' values of QR code are changed in the range of error correction capacity so that each share of the VCS can be read by a scanner. And according to the property of VCS, the secret information can be revealed by human eyes by stacking required shares together without computation. The algorithm of shares generation progress is shown below (Table 4).

Table 4. The visual cryptography scheme based on QR code

Algorithm 3. The visual cryptography scheme based on QR code	
Input	A binary secret image *S*, modified QR code $M_1, \ldots, M_n$, and VCS common parameters *k* and *n*.
Output	QR code shares $QS_1, \ldots, QS_n$
Step 1	For every pixel (i, j) in secret image S repeat steps 2-3
Step 2	Randomly select *k* integers in the range of *n*
Step 3	If $S(i,j) \neq QS_1(i,j) \oplus, \ldots, \oplus QS_k(i,j)$, randomly select $p \in [1, k]$, and let $QS_p(i,j) = \overline{QS_p(i,j)}$.
Step 4	Output *n* QR code shares $QS_1, \ldots, QS_n$

Though the QR code has a certain error correction capacity that allows us to change the pixel for our scheme, we should note that the modified region should be in the data pattern, and the function pattern will be kept stable.

The QR code shares generation architecture of the scheme is illustrated in Fig. 6.

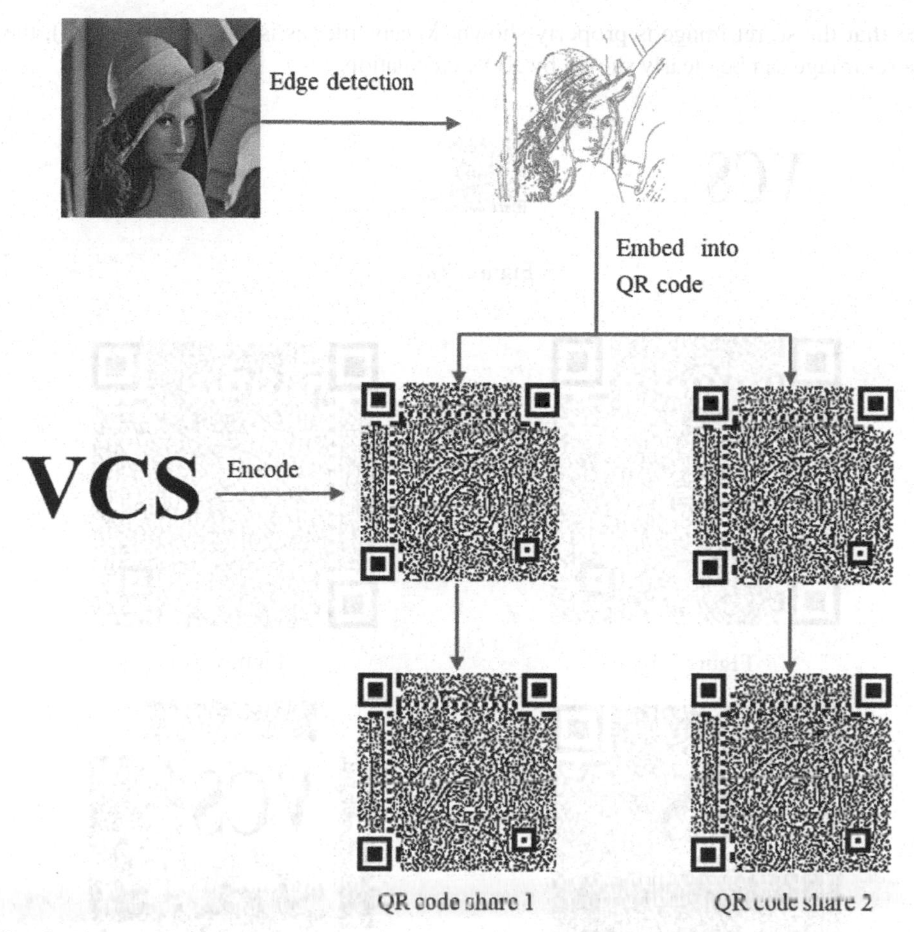

Fig. 6. The above image shows the throughout process of QR code sharing scheme. We first apply the algorithm 2 to embed the edge image into QR code, and the apply algorithm 3 to generate final QR code shares. All QR codes can be read by scanner, and we can decrypt by simply stacking them together or calculate by XOR operation.

4 Experimental Results and Analysis

4.1 Experimental Results

In this section, we utilize our scheme in a (2, 2) QR code based VCS and a (3, 4) QR code based VCS. The QR code is of version 6 and with error correction level H, and the size of secret image of character 'VCS' is of 45 × 60 within the error correction level. The QR code is generate by [7].

Figure 7 shows a (2, 2) QR code scheme that generates two shares, the original QR codes are encrypted with 'a1' and 'ab2' separately. Figure 7(b) and Fig. 7(c) show the two shares, due to the edge information embedded, the size of the shares are 3 times of the original QR codes. And Fig. 7(d) shows the stacked result of two shares, and we can see that the secret image is properly shown. Meanwhile, as is shown in Fig. 7(e), the secret image can be clearly shown by XOR calculation.

Fig. 7. Result of QR code version 6 with error correction level H, and the two shares (b) and (c) contain edge information and can be read. The stacking result can show the secret information properly, while we can calculate the secret image clearly.

The Fig. 8 shows that a (3, 4) QR based VCS, and the original QR codes information are 'a1', 'ab2', 'abc3' and 'abcd4' separated. Figure 8(b)–(e) are the 4 shares, and when we stack 2 pictures together (Fig. 8(f)), we can get nothing from the result, and when we put 3 pictures (Fig. 8(g)), the secret image is revealed with some visual loss, similarity with (2, 2) scheme, we can also get the original image by calculation (Fig. 8(h)).

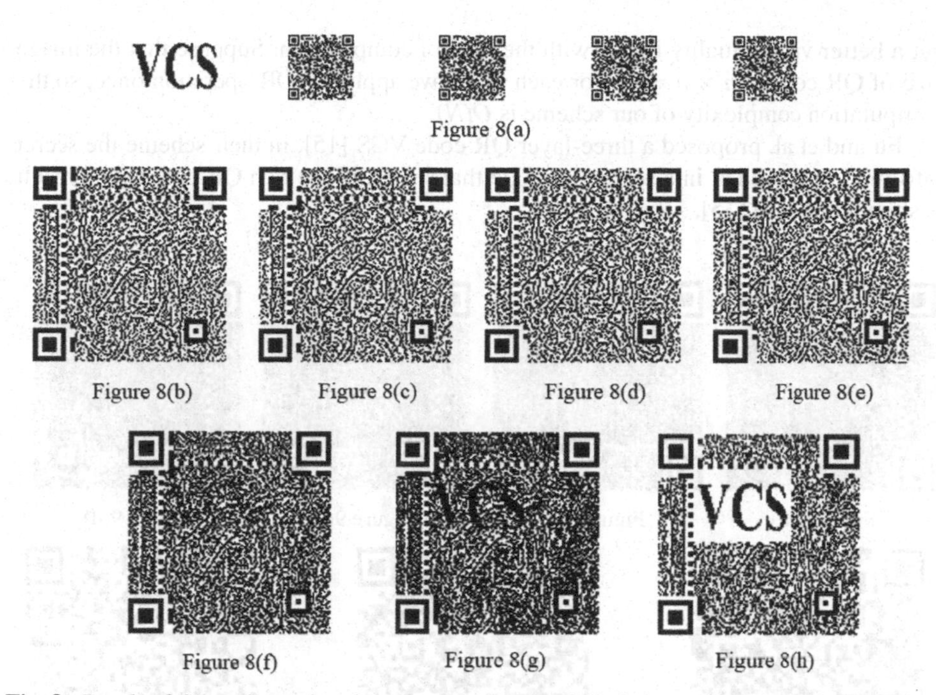

Fig. 8. Result of QR code version 6, error correction level H, 4 shares (b)–(e) are the modified QR images contain edge information and secret shares, each share's size is 3 times of the original QR code. (f) is the stack of 2 pictures, as we can see that, the result cannot give any information of the secret image, while we stack 3 pictures, which is shown in (g), we can see the characters in the picture. And in (h) we get the secret image by calculation.

4.2 Compared with Related Schemes

In this section, we compare our scheme with other VCS and the result is in Table 5.

Table 5. Comparisons of this paper and related works

Paper	Visual readability	Machine readability	Decryption complexity
[9]	No	Yes	*O(N)*
[11]	Yes	No	*O(N)*
[12]	No	Yes	*O(N)*
[13]	No	No	*O(N)*
[14]	No	Yes	*O(N)*
[15]	Yes	No	*O(N)*
Our paper	Yes	Yes	*O(N)*

It is clear that the secret image can be seen by human eyes when stacking two shares directly with some contract loss. On the other side, we can also use XOR operation to get a better visual quality image with the help of computation. Suppose that the image size of QR code is $n \times n = N$, for each pixel, we apply a XOR operation once, so the computation complexity of our scheme is $O(N)$.

Fu and et al. proposed a three-layer QR code VCS [15], in their scheme the secret information is hidden in a secret QR code that shared in the rich QR codes. The result is shown in Fig. 9 [15].

Fig. 9. Fig. 9 is the result of Fu and et al. [15], the secret image is shared in (a)–(d), and each of them can be decoded by QR readers. And the (e)–(f) are the XOR results of QR codes. The result is calculated by a decoding algorithm and then scanned by QR reader.

In comparison of [15], the QR code of our scheme is of less mask image information, and only keeps the edge image that can be used as authentication and so on. And in the decoding phrase, our scheme can directly get the secret information by stacking shares together without computation, meanwhile, with the help of image processing equipment, we can also get the secret information by XOR operation and do not need additional operation.

5 Conclusion

VCS is a convenient algorithm that shares secret information and is computation free for binary image. Based on this property, we apply QR code as the carrier of shares and make all shares to be QR codes. Due to the error correction capacity, the modified QR code can also be read. Furthermore, before injecting secret information into a QR code, we embed edge image into the code, which can help us in QR code authentication. Experimental results show that our scheme can not only reveal the secret image by stacking without computation but also recover the secret image by computation with great visual quality.

Acknowledgment. This work was supported by the National Key Research and Development Program of China under Grant 2018YFB0804101 and the National Natural Science Foundation of China under Grant 61802357.

References

1. Naor, M., Shamir, A.: Visual cryptography. In: Advances in Cryptology-Eurocrypt. Workshop on the Theory and Application of Cryptographic Techniques (1995)
2. Liu, F., Yan, W.Q.: Visual Cryptography for Image Processing and Security, vol. 2. Springer, Cham (2014). https://doi.org/10.1007/978-3-319-09644-5
3. Wang, D.S., et al.: Two secret sharing schemes based on Boolean operations. Pattern Recogn. **40**(10), 2776–2785 (2007)
4. Chen, T.H., Li, K.C.: Multi-image encryption by circular random grids. Inf. Sci. **189**, 255–265 (2012)
5. Revenkar, P.S., Gandhare, W.Z.: Secure iris authentication using visual cryptography. Int. J. Comput. Sci. Inf. Secur. **7**(3), 217–221 (2010)
6. Rosking, J.A., Emign, A.T.: Visual cryptography and voting technology using a pair of enhanced contrast glyphs in overlay, USA, US, no. 7,667,871 (2010)
7. Standard Test Images (Online) (2004). https://www.qrcode.com/zh/. Accessed 6 Nov 2019
8. The Structure of QR Code (Online). http://qr.biz/articles/the_structure_of_qr_code. Accessed 6 Nov 2019
9. Cheng, Y., Fu, Z., Yu, B., Shen, G.: A new two-level QR code with visual cryptography scheme. Multimedia Tools Appl. **77**(16), 20629–20649 (2017). https://doi.org/10.1007/s11042-017-5465-4
10. Yang, C.N., et al.: k out of n region-based progressive visual cryptography. IEEE Trans. Circuits Syst. Video Technol. **29**(1), 252–262 (2019)
11. Yang, C.N., Sun, L.Z., Cai, S.R.: Extended color visual cryptography for black and white secret image. Theory Comput. Sci. **609**(P1), 143–161 (2016)
12. Wan, S., Lu, Y., Yan, X., et al.: Visual secret sharing scheme for (k, n) threshold based on QR code with multiple decryptions. J. Real-Time Image Proc. **9**, 1–16 (2017)
13. Wang, G., Liu, F., Yan, W.Q.: 2D barcodes for visual cryptography. Multimedia Tools Appl. **75**(2), 1223–1241 (2016)
14. Liu, Y., Fu, Z.X., Wang, Y.W.: Two-level information management scheme based on visual cryptography and QR code. Appl. Res. Comput. **33**, 3460–3463 (2016)
15. Fu, Z.X., Cheng, Y., Yu, B.: Rich QR code with three-layer information using visual secret sharing scheme. Multimedia Tools Appl. **78**(14), 19861–19875 (2019). https://doi.org/10.1007/s11042-019-7333-x
16. Wicker, S.B., Bhargava, V.K.: Reed-Solomon codes and their applications. Commun. ACM **24**(9), 583–584 (1994)
17. Canny, J.: A computational approach to edge-detection. IEEE Trans. Pattern Anal. Mach. Intell. **8**(6), 679–698 (1986)
18. Yang, Y.W., et al.: Watermark embedding for direct binary searched halftone images by adopting visual cryptography. Comput. Mater. Continua **55**(2), 255–265 (2018)
19. Yi, C., Zhi, L.Z., et al.: Coverless information hiding based on the molecular structure images of material. Comput. Mater. Continua **54**(2), 197–207 (2018)
20. Di, X., et al.: High capacity data hiding in encrypted image based on compressive sensing for nonequivalent resources. Comput. Mater. Continua **58**(1), 1–13 (2019)

Hiding Traces of Camera Anonymization by Poisson Blending

Hui Zeng[1], Anjie Peng[1](✉), and Xiangui Kang[2]

[1] School of Computer Science and Technology, Southwest University of Science and Technology, Mianyang, China
penganjie200012@163.com

[2] School of Data and Computer Science, Sun Yat-sen University, Guangzhou, China

Abstract. Sensor noise caused by photo response non-uniformity (PRNU) has been widely accepted as a reliable fingerprint for source camera identification (SCI). An interesting research topic in this area concerns the repudiability of PRNU-based SCI, which includes methods of removing or synthesizing the fingerprint on which forensic methods rely. Removing the PRNU fingerprint from a given image, also known as camera anonymization, is important for privacy protection and anti-tracking. However, camera anonymization sometimes introduces annoying visual artifacts in the resultant image. In this work, Poisson blending is used to hide the traces left by camera anonymization. Theoretical analysis and experimental results show that the proposed method can suppress the visual artifacts caused by camera anonymization effectively while maintaining the anti-forensic effectiveness.

Keywords: Photo response non-uniformity · Source camera identification · Camera anonymization · Poisson blending

1 Introduction

Tracking the source device from a questioned image or video is an important topic in multimedia forensics. Many methods of doing so have been proposed in the literature, such as by analyzing the pixel defects [1], dark current noise [2], sensor dust [3], noise model [4], and so on. Among these methods, the photo response non-uniformity (PRNU)-based ones are most promising since they can identify individual cameras, even of the same model, and are robust against common post-processing methods [5–11]. However, recent studies show that the signature on which these source camera identification (SCI) methods are based can be removed or substituted [12–18], which motivates researchers to reevaluate the security of the PRNU-based SCI methods [19–21].

In this paper, we take the role of a forger who wants to hide the origin information of a questioned image or video by removing the camera fingerprint from it. This technique is called *camera anonymization* and can be used for privacy protection [22] and anti-tracking. Existing methods of camera anonymization can be classified into two categories. The first category assumes that the camera fingerprint is available or can

X. Sun et al. (Eds.): ICAIS 2020, LNCS 12240, pp. 98–108, 2020.
https://doi.org/10.1007/978-3-030-57881-7_9

be accurately estimated from the image [13, 14, and 15], e.g., the camera is owned by the forger himself. In this case, camera anonymization can be achieved by first estimating the magnitude of the camera fingerprint of a questioned image or video, and then subtracting it [14]. Methods in the second category are more general since they do not rely on the knowledge of the camera fingerprint, but typically work by modifying pixel values and scrambling their positions to destroy the underlying camera fingerprint. For instance, seam-carving [23] and PatchMatch [24] are utilized to remove the camera fingerprint in [16] and [17], respectively. More recently, [18] proposed an anonymization method by first deleting blocks from an image and then reconstructing these blocks as an image inpainting problem [25]. To address the blurring problem around edges, the pixels around the edges are substituted with those from a denoised image. However, due to the robustness nature of PRNU against common post-processing, destroying the underlying camera fingerprint results in an unavoidable loss of image quality. In [18], some unexpected visual artifacts exist along the edges in the anonymized image, which would put the camera anonymization in an awkward position where a forensic investigator can easily point out these suspicious traces and refuse to draw a conclusion from the questioned image. As a matter of fact, for image anti-forensics, retaining the quality of the image is even more challenging and more important than making it forensically undetectable. Otherwise, we can simply destroy the image to make it undetectable. In this respect, anti-forensic is similar to another main track in multimedia security area, steganography [26, 27], which try her best to circumvent steganalysis [28, 29] while assuring the media is decodable in the recipient. How to maintain the image quality as far as possible while successfully removing the camera fingerprint is the motivation for this study.

In this paper, Poisson blending [30] is used to combine the image reconstructed by inpainting with the denoised image, which can successfully hide the unexpected artifacts around the edges as that in [18]. Specifically, a two-step blending scheme is proposed to assure consistency of the pixel values around the edges. In the first step, the denoised image is taken as the source image and the reconstructed image is taken as the target image. In the second step, the image resulting from the first step is taken as the source image and the denoised image is taken as the target image.

The contributions of this work are summarized as following:

1) One type of visual artifacts existed in the state-of-the-art camera anonymization method, which we called *wiping traces*, is pointed out and analyzed.
2) To hide the *wiping traces*, a modified Poisson blending method is proposed to combine the inpainting image with the denoising image seamlessly.
3) Experimental results on different datasets show that the proposed method can achieve better visual quality compared to [18] while maintaining forensic undetectability.

The rest of this paper is organized as follows. In the next section, we briefly review of the process of PRNU-based SCI and the camera anonymization method. The details of the proposed method are provided in the third section. Experimentally study is presented in the fourth section, and the conclusion is drawn in Sect. 5. In this paper, symbols in **boldface** represent either vectors or matrices.

2 PRNU-Based Camera Identification and Anonymization

In this section, we firstly briefly review the PRNU-based camera identification scheme and then introduce a newly emerging camera anonymization method [18].

2.1 Camera Identification

1. The sensor pattern noise (SPN) $\boldsymbol{W_I}$ is extracted from the image $\boldsymbol{I}$ in question by subtracting a denoised version of $\boldsymbol{I}$ from itself.

$$\boldsymbol{W_I} = \boldsymbol{I} - F(\boldsymbol{I}) \tag{1}$$

where F is a denoising filter. In this study, we follow [18] to use the wavelet-based denoising filter described in Appendix A of [5].
2. The fingerprint of a reference camera, $\boldsymbol{K}$, is estimated by averaging (or weighted averaging [6]) of the SPN of a number of photographs taken with this camera.
3. The correlation $\rho(\boldsymbol{W_I}, \boldsymbol{IK})$ is used to measure the similarity between $\boldsymbol{W_I}$ and $\boldsymbol{K}$, such as the normalized cross correlation (NCC), which is defined as:

$$\rho(\boldsymbol{W_I}, \boldsymbol{IK}) = \frac{(\boldsymbol{W_I} - \overline{\boldsymbol{W_I}}) \cdot (\boldsymbol{IK} - \overline{\boldsymbol{IK}})}{\|\boldsymbol{W_I} - \overline{\boldsymbol{W_I}}\| \|\boldsymbol{IK} - \overline{\boldsymbol{IK}}\|} \tag{2}$$

where $\cdot$ denotes the inner product, and $\overline{\boldsymbol{W_I}}, \overline{\boldsymbol{IK}}$ are the means of $\boldsymbol{W_I}$ and $\boldsymbol{IK}$, respectively. If $\boldsymbol{I}$ is not taken with the reference camera, $\boldsymbol{W_I}$ will be uncorrelated with $\boldsymbol{IK}$, that is, $\rho(\boldsymbol{W_I}, \boldsymbol{IK}) \approx 0$. If $\boldsymbol{I}$ is taken with the reference camera, $\boldsymbol{W_I}$ will share a common component with $\boldsymbol{IK}$, and thus $\rho(\boldsymbol{W_I}, \boldsymbol{IK})$ usually show a positive value. The greater the value of ρ is, the higher the probability that $\boldsymbol{I}$ is taken by the reference camera. In practice, to determine whether a test image is from a given camera, Neyman-Pearson approach is used to choose a threshold τ according to a given false positive rate (FPR).

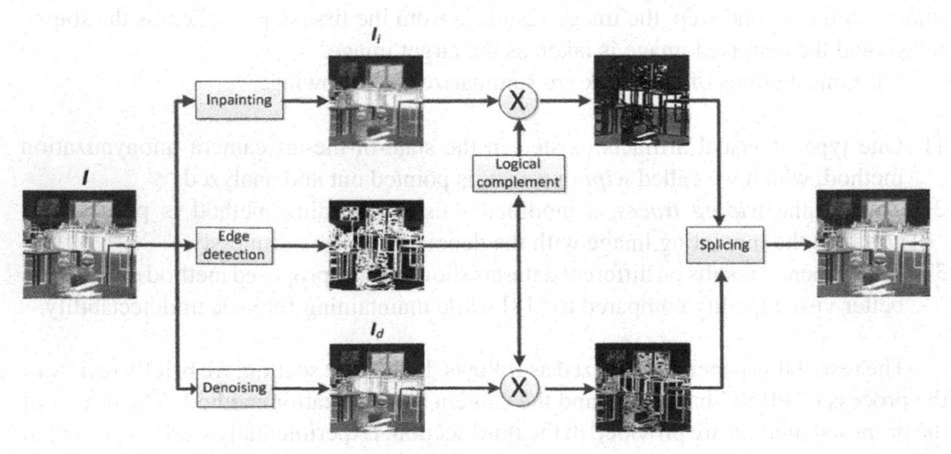

Fig. 1. Diagram of the camera anonymization [18].

2.2 Camera Anonymization Identification

The objective of camera anonymization is to remove or attenuate the PRNU traces of a given image $\boldsymbol{I}$ so that the source camera of $\boldsymbol{I}$ cannot be tracked, that is, $\rho(\boldsymbol{W}_I, \boldsymbol{IK}) < \tau$. If the camera is available (which implies that an estimate of $\boldsymbol{K}$ is available), camera anonymization can be achieved by first estimating the magnitude of the PRNU of $\boldsymbol{I}$, and then subtracting it from $\boldsymbol{I}$ [14]. If $\boldsymbol{K}$ is unavailable, the anonymization task is more complex. Recently, [18] proposed an anonymization method entailing first deleting blocks of $\boldsymbol{I}$ and then reconstructing these blocks as an image inpainting problem. However, the image reconstructed by inpainting ($\boldsymbol{I}_i$) is rather blurry around the edges (see Fig. 2(b)). Hence, the pixels of edges are further substituted with those from a denoised version of $\boldsymbol{I}$ ($\boldsymbol{I}_d$) [31]. The complete diagram of the method in [18] is illustrated in Fig. 1.

As stated in [18] and also verified by our experiments, this method [18] can successfully achieve the objective of camera anonymization (see Fig. 5(a)). However, some unexpected visual artifacts exist in the anonymized image. Figure 2 shows an example of a camera anonymized image. Take Fig. 2(c) and its magnified version, Fig. 2(d), as an example, traces like pencil wiping can be observed along strong edges (indicated by yellow arrows; we call such artifact as *wiping traces* hereafter). The reason for the wiping traces is the inconsistency between $\boldsymbol{I}_i$ and $\boldsymbol{I}_d$ along the splicing boundary. Note that the final step in Fig. 1 is indeed splicing of $\boldsymbol{I}_i$ with $\boldsymbol{I}_d$ according to the edge mask. Looking for an alternative strategy to combine $\boldsymbol{I}_i$ with $\boldsymbol{I}_d$ rather than splicing directly is the motivation for the proposed method.

Fig. 2. The *wiping traces* left by the camera anonymization method [18]: (a) original image, (b) the image reconstructed by inpainting, (c) anonymized image, (d) magnified region of (c).

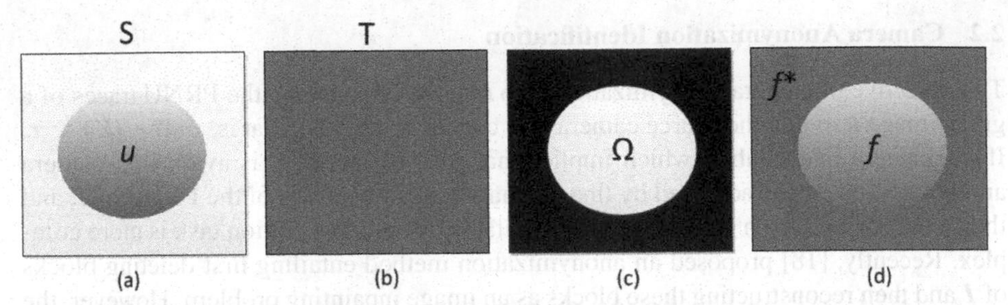

Fig. 3. Notation in Poisson blending, a source image S is inserted in the target image T according to a mask Ω. (a) Source image. (b) Target image. (c) Binary mask. (d) Result of Poisson blending.

3 Proposed Method

In light of the analysis concerning the wiping traces of the anonymized image in Sect. 2, we propose an alternative anonymization method to hide these traces in this section. Specifically, we substitute the splicing operation in [18] with a Poisson blending operation, which can result in more realistic boundaries.

If we want insert a source image $\boldsymbol{S}$ (foreground) into a target image $\boldsymbol{T}$ (background) with a mask $\boldsymbol{\Omega}$ (see Fig. 3), the Poisson blending [30] can be expressed as

$$\min_f \int_{\Omega} |\nabla f - \nabla u| \mathrm{d}\Omega \; with \; f|_{\partial\Omega} = f^*|_{\partial\Omega} \tag{3}$$

where u is the pixel values of $\boldsymbol{S}$ in $\boldsymbol{\Omega}$, f is the pixel values in $\boldsymbol{\Omega}$ after blending, f^* is the pixel values outside $\boldsymbol{\Omega}$ after blending, $\partial\Omega$ is the boundary of $\boldsymbol{\Omega}$, and ∇ is the gradient operator. The solution to this minimization problem implies that the textures of $\boldsymbol{S}$ are kept in $\boldsymbol{\Omega}$, while making the transition area along $\partial\Omega$ as smooth as possible. In our preliminary study, we found the Poisson blending can attenuate the wiping traces of anonymization to a certain extent; however, these artifacts are still visible if observed

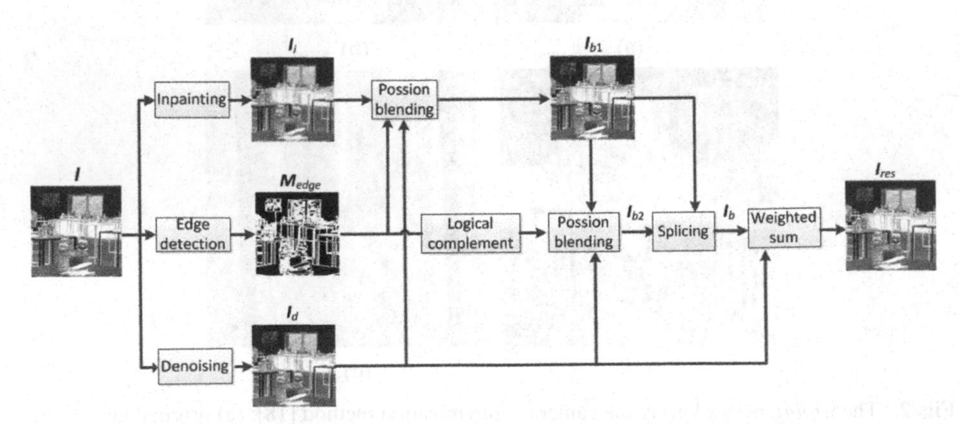

Fig. 4. The diagram of the proposed camera anonymization method.

carefully. Motivated by [32], we first take $\boldsymbol{I}_i$ as the target image and, $\boldsymbol{I}_d$ as the source image and blend them together according to the edgy mask $\boldsymbol{M}_{edge}$. That is, in the image $\boldsymbol{I}_{b1}$ resulting from blending, the smooth area is taken directly from $\boldsymbol{I}_i$ and the textured area is transformed from $\boldsymbol{I}_d$. In the next step, we take $\boldsymbol{I}_d$ as the target image, $\boldsymbol{I}_{b1}$ as the source image and blend them together according to the logical complement of $\boldsymbol{M}_{edge}$. That is, in the image $\boldsymbol{I}_{b2}$ resulting from blending, the smooth area is transformed from $\boldsymbol{I}_{b1}$ and the textured area is taken directly from $\boldsymbol{I}_d$. In the third step, the textured area of $\boldsymbol{I}_{b1}$ and the smoothed area of $\boldsymbol{I}_{b2}$ are combined to form $\boldsymbol{I}_b$. Note that there are two advantages of $\boldsymbol{I}_b$: firstly, the textured area of $\boldsymbol{I}_b$ is as distinct as that of $\boldsymbol{I}_d$ due to the gradient-preserving nature of Poisson blending; secondly and the most importantly, since both the textured area and the smoothed area have been transformed, the inconsistency between $\boldsymbol{I}_i$ and $\boldsymbol{I}_d$ along the splicing boundary analyzed in Sect. 2 is further suppressed. The final resulting image is weighted by $\boldsymbol{I}_b$ and $\boldsymbol{I}_d$ to further enhance the visual quality.

$$\boldsymbol{I}_{res} = w\boldsymbol{I}_d + (1 - w)\boldsymbol{I}_b \tag{4}$$

where the weight w can balance a tradeoff between the undetectability and visual quality of $\boldsymbol{I}_{res}$. Larger w results an image of high visual quality at the expense the undetectability. We experimentally set $w = 0.2$ in this study. A complete diagram of the proposed method is shown in Fig. 4.[1]

Table 1. Images used in the experiments.

Database	Camera	No. of nature images for test	No. of flat images for estimating camera fingerprint	Resolution
DID	Ricoh_GX100_C0	189	25	2736 × 3648
	Ricoh_GX100_C1	164	25	2736 × 3648
	Ricoh_GX100_C2	167	25	2736 × 3648
	Rollei_RCP_C0	178	50	2304 × 3072
	Rollei_RCP_C1	160	50	2304 × 3072
VISION	D01_Samsung_GalaxyS3Mini	200	25	1920 × 2560
	D02_Apple_iPhone4s	200	50	2448 × 3264
	D09_Apple_iPhone4	200	50	1936 × 2592
	D10_Apple_iPhone4s	178	25	2448 × 3264
	D11_Samsung_GalaxyS3	200	50	2448 × 3264

[1] The code is available in www.escience.cn/people/Zenghui.

4 Experimental Results

To evaluate the performance of the proposed method, experiments were conducted on two widely used databases: Dresden Image Database (DID) [33] and VISION [34]. Each database responsible for five cameras, results a total of 10 cameras and about 2000 test images of JPEG format. Table 1 shows the camera model, number of test images, number of images for calculating camera fingerprints (Only flat field images are used for better quality of fingerprints) and original resolution. The cameras used in DID are traditional cameras and those used in VISION are smartphone cameras. Once one camera is fixed as the reference camera, the images taken with this camera are used as positive samples and the images from all the other cameras are used as negative samples. In total, there are about 2000 positive samples, 2000 anonymized samples obtained with the method in [18], 2000 anonymized samples obtained with the proposed method, and about 200 $\times$ 9 $\times$ 10 = 18000 negative samples. All the images are cropped from the center to a size of 512 $\times$ 512 as was done in [18].

4.1 PRNU Undetectability

In this subsection, the distribution of the *NCC* values of the positive class and negative class are obtained with the positive and negative samples and then used to evaluate the undetectability of the anonymized images. For each of the test images, two versions of the anonymized image are created with the method presented in [18] and the proposed method. The histograms of the *NCC* values of the positive samples, the negative samples, and the anonymized samples are shown in Fig. 5. Figures 5(a) and (b) illustrate the camera anonymization effect of the method presented in [18] and the proposed method, respectively. It is observed from both the Fig. 5(a) and Fig. 5(b) that the distribution center of the anonymized samples (yellow bars) is very close to that of the negative

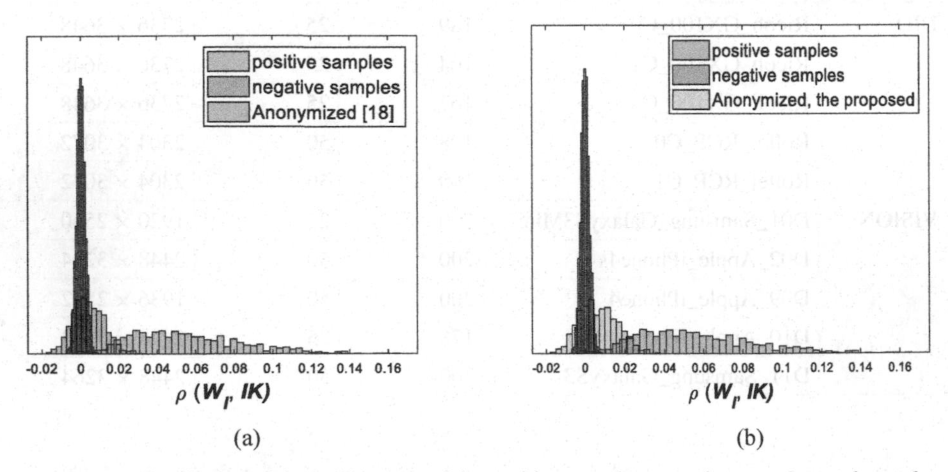

Fig. 5. The distribution of the NCC values of the positive samples, negative samples, and samples anonymized by (a) the method in [18], (b) the proposed method. Both camera anonymization schemes can successfully attenuate the PRNU factor in the test images. (Color figure online)

samples (red bars) and far away from that of the positive samples (blue bars), which implies a high probability of successful anonymization.

4.2 Visual Imperceptibility

Like other anti-forensic techniques, camera anonymization is constrained not only by the forensic undetectability, but also by the perceptual quality [12], and the latter is typically more challenging to enforce. In this subsection, both quantitative and qualitative comparisons of the perceptual quality are made between the anonymized images created by the proposed method and the method presented in [18]. For quantitative comparison, we use the structural similarity index (SSIM) [35] and multi-scale structural similarity index (MS-SSIM) [36] reference to the original image to evaluate the visual quality of the anonymized images. Table 2 lists the averaged SSIM metric and MSSSIM metric of the anonymized images by [18] and the proposed method. It can be concluded from both metrics that the resultant image quality of the proposed method is better than that of [18]. For example, the averaged SSIM metric with the proposed method is higher than that with the method in [18] by 0.007. The standard deviation of the metrics are also provided in Table 2. It is observed that the quality of the resultant image with the proposed method is more consistent than that with the method in [18].

One may argue that the improvement of the proposed method over [18] is marginal. This is because the difference between the proposed method and [18] is mainly in the area along strong edges. Although such area only takes over a small part of the whole image, it is critical for human perception due to the fact that the human visual system

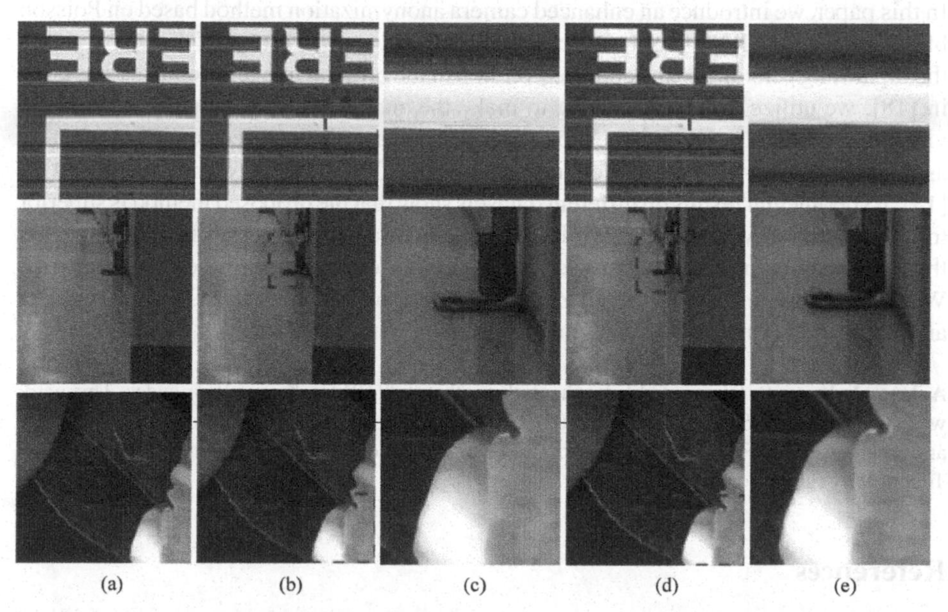

Fig. 6. The visual quality comparison: (a) the original image, (b) image anonymized by the method proposed in [18], (c) magnified region of (b), (d) image anonymized by the proposed method, (e) magnified region of (d).

Table 2. Quantitative comparison of the image quality of camera anonymization methods, the best results are in **bold**.

	SSIM		MSSSIM	
	Mean	Standard deviation	Mean	Standard deviation
[18]	0.947	0.036	0.973	0.013
Proposed	**0.954**	0.032	**0.974**	0.011

(HVS) is highly sensitive to the structural distortions [37, 38]. Figure 6 illustrates three examples of camera anonymization. In the left column are the original images from VISION, various indoor and outdoor scenes. The anonymized images with the method presented in [18] and the proposed method are shown in the second and fourth columns, respectively. The comparative observation of Figs. 6(b) and (d) show that the wiping traces of [18] have been successfully attenuated by the proposed method. Taking the second image (the second row) as an example, distinct wiping traces exist around the extinguisher in the image produced by the method presented in [18] (the third column), whereas these traces are invisible in the image resulting from the proposed method (the fifth column). In practice, the wiping traces around the extinguisher are likely to arouse suspicious about the integrity of the image and make camera anonymization in vain.

5 Conclusion

In this paper, we introduce an enhanced camera anonymization method based on Poisson blending against PRNU-based source camera identification. Instead of directly splicing the inpainting version and the denoised version of the original image together as was done in [18], we utilize Poisson blending to make the anonymized image more convincing visually. Specifically, we take the inpainting version and the denoised version alternately as the background in Poisson blending, and successfully hide the wiping traces left by [18]. Experimental results on different datasets show that the proposed method is superior to [18] in terms of visual quality while maintaining forensic undetectability. We hope that this work can help researchers revisit the reliability of the sensor-based SCI methods. We note that some researchers utilize convolutional neural network [39, 40] for camera anonymization purpose [41], and plan to compare with it in future.

Acknowledgment. We would like to thank the authors of [18] for sharing their codes. This work was supported by NSFC (grant no. 61702429), China Scholarship Council (no. 201908515095), and the Research Fund for the Doctoral Program of Southwest University of Science and Technology (grant no. 18zx7163).

References

1. Geradts, Z., Bijhold, J., Kieft, M., Kurosawa, K., Kuroki, K., Saitoh, N.: Methods for identification of images acquired with digital cameras. In: Proceedings of SPIE, Enabling Technologies for Law Enforcement Security, vol. 4232, pp. 505–512x (2001)

2. Kurosawa, K., Kuroki, K., Saitoh, N.: CCD fingerprint method identification of a video camera from videotaped images. In: Proceedings of ICIP, pp. 537–540 (2002)
3. Dirik, A.E., Sencar H.T., Memon, N.: Source camera identification based on sensor dust characteristics. In: IEEE Workshop on Signal Processing Applications for Public Security & Forensics (2007)
4. Thai, T.H., Cogranne, R., Retraint, F.: Camera model identification based on the heteroscedastic noise model. IEEE Trans. Image Process. **23**(1), 250–263 (2014)
5. Lukas, J., Fridrich, J., Goljan, M.: Digital camera identification from sensor pattern noise. IEEE Trans. Inf. Forensics Secur. **1**, 205–214 (2006)
6. Chen, M., Fridrich, J., Goljan, M., Lukas, J.: Determining image origin and integrity using sensor noise. IEEE Trans. Inf. Forensics Secur. **3**(1), 74–90 (2008)
7. Li, C.T.: Source camera identification using enhanced sensor pattern noise. IEEE Trans. Inf. Forensics Secur. **5**(2), 280–287 (2010)
8. Kang, X., Li, Y., Qu, Z., Huang, J.: Enhancing source camera identification performance with a camera reference phase sensor pattern noise. IEEE Trans. Inf. Forensics Secur. **7**(2), 393–402 (2012)
9. Zeng, H., Kang, X.: Fast source camera identification using content adaptive guided image filter. J. Forensic Sci. **61**(2), 520–526 (2016)
10. Al-Ani, M., Khelifi, F.: On the SPN estimation in image forensics: a systematic empirical evaluation. IEEE Trans. Inf. Forensics Secur. **12**(5), 1067–1081 (2017)
11. Rosenfeld, K., Sencar, H.T.: A study of the robustness of PRNU-based camera identification. In: IS&T/SPIE Electronic Imaging (EI). International Society for Optics and Photonics (2009)
12. Gloe, T., Kirchner, M., Winkler, A., Bohme, R.: Can we trust digital image forensics? In: 15th International Conference on Multimedia, pp. 78–86 (2007)
13. Li, C.-T., Chang, C.-Y., Li, Y.: On the repudiability of device identification and image integrity verification using sensor pattern noise. In: Weerasinghe, D. (ed.) ISDF 2009. LNICST, vol. 41, pp. 19–25. Springer, Heidelberg (2010). https://doi.org/10.1007/978-3-642-11530-1_3
14. Zeng, H., Chen, J., Kang, X., Zeng, W.: Removing camera fingerprint to disguise photograph source. In: Proceedings of ICIP, pp. 1687–1691 (2015)
15. Bonettini, N., Bondi, L., Güera, D., et al.: Fooling PRNU-based detectors through convolutional neural networks. In: The 26th European Signal Processing Conference (EUSIPCO), pp. 957–961 (2018)
16. Dirik, A.E., Sencar, H.T., Memon, N.: Analysis of seam-carving-based anonymization of images against PRNU noise pattern-based source attribution. IEEE Trans. Inf. Forensics Secur. **9**(12), 2277–2290 (2014)
17. Entrieri, J., Kirchner, M.: Patch-based desynchronization of digital camera sensor fingerprints. IS&T Electron. Imaging (EI) **87**, 1–9 (2016)
18. Mandelli, S., Bondi, L., Lameri, S., et al.: Inpainting-based camera anonymization. In: Proceedings of ICIP, pp. 1522–1526 (2007)
19. Goljan, M., Fridrich, J., Chen, M.: Defending against fingerprint-copy attack in sensor-based camera identification. IEEE Trans. Info. Forensics Secur. **6**(1), 227–236 (2011)
20. Zeng, H.: Rebuilding the credibility of sensor-based camera source identification. Multimed. Tools Appl. **75**(21), 13871–13882 (2016)
21. Zeng, H., Liu, J., Yu, J., et al.: A framework of camera source identification bayesian game. IEEE Trans. Cybern. **47**(7), 1757–1768 (2017)
22. Wang, P., Wang, Z., Chen, T., Ma, Q.: Personalized privacy protecting model in mobile social network. Comput. Mater. Continua **59**(2), 533–546 (2019)
23. Avidan, S., Shamir, A.: Seam carving for content-aware image resizing. ACM Trans. Graph. (Proc. SIGGRAPH) **26**(3) (2007)

24. Barnes, C., Shechtman, E., Finkelstein, A., Goldman, D.B.: Patch-Match: a randomized correspondence algorithm for structural image editing. ACM Trans. Graph. (Proc. SIGGRAPH) **28**(3) (2009)
25. Papafitsoros, K., Schoenlieb, C.B., Sengul, B.: Combined first and second order total variation inpainting using split Bregman. Image Process. Line (IPOL) **3**, 112–136 (2013)
26. Bas, P., Filler, T., Pevný, T.: "Break Our steganographic system": the ins and outs of organizing BOSS. In: Filler, T., Pevný, T., Craver, S., Ker, A. (eds.) IH 2011. LNCS, vol. 6958, pp. 59–70. Springer, Heidelberg (2011). https://doi.org/10.1007/978-3-642-24178-9_5
27. Meng, R., Rice, S.G., Wang, J., Sun, X.: A fusion steganographic algorithm based on faster R-CNN. Comput. Mater. Continua **55**(1), 001–016 (2018)
28. Fridrich, J., Kodovský, J.: Rich models for steganalysis of digital images. IEEE Trans. Info. Forensics Secur. **7**(3), 868–882 (2012)
29. Kang, Y., Liu, F., Yang, C., et al.: Color image steganalysis based on residuals of channel differences. Comput. Mater. Continua **59**(1), 315–329 (2019)
30. Perez, P., Gangnet, M., Blake, A.: Poisson image editing. ACM Trans. Graph. **22**, 313–318 (2003)
31. Dabov, K., Foi, A., Katkovnik, V., Egiazarian, K.: Image denoising by sparse 3D transform-domain collaborative filtering. IEEE Trans. Image Process. **16**(8), 2080–2095 (2007)
32. Afifi, M., Hussain, K.F.: MPB: a modified poisson blending technique. Comput. Vis. Media **1**(4), 331–341 (2015)
33. Gloe, T., Bohme, R.: The Dresden image database for benchmarking digital image forensics. J. Digit. Forensic Pract. **3**(2–4), 150–159 (2010)
34. Shullani, D., Fontani, M., Iuliani, M., Al Shaya, O., Piva, A.: VISION: a video and image dataset for source identification. EURASIP J. Inf. Secur. **2017**, 15 (2017). https://doi.org/10.1186/s13635-017-0067-2
35. Wang, Z., Bovik, A.C., Sheikh, H.R., Simoncelli, E.P.: Image quality assessment: from error visibility to structural similarity. IEEE Trans. Image Process. **13**(4), 600–612 (2004)
36. Wang, Z., Simoncelli, E.P., Bovik, A.C.: Multi-scale structural similarity for image quality assessment. In: Proceedings of IEEE Asilomar Conference on Signals, Systems and Computers, pp. 1398–1402 (2003)
37. Teo, P.C., Heeger, D.J.: Perceptual image distortion. In: Proceedings of IEEE International Conference on Image Processing, pp. 982–986 (1994)
38. Zhou, W., Bovik, A.C.: Mean squared error: love it or leave it? A new look at Signal Fidelity Measures. IEEE Signal Process. Mag. **26**(1), 98–117 (2009)
39. LeCun, Y., Bottou, L., Bengio, Y., Haffner, P.: Gradientbased learning applied to document recognition. Proc. IEEE **86**(11), 2278–2324 (1998)
40. Chen, J., Kang, X., Liu, Y., Wang, Z.J.: Median filtering forensics based on convolutional neural networks. IEEE Signal Process. Lett. **22**(11), 1849–1853 (2015)
41. Bonettini, N., Bondi, L., Guera, D., et al.: Fooling PRNU-based detectors through convolutional neural networks. In: European Signal Processing Conference (EUSIPCO), pp. 957–961 (2018)

A Deep Learning Approach to Detection of Warping Forgery in Images

Tongfeng Yang(✉), Jian Wu, Guorui Feng, Xu Chang, and Lihua Liu

Shandong University of Political Science and Law, Jinan, Shandong, China
yangtf2014@126.com

Abstract. In recent years, image forensics has received full attention from researchers. A large number of algorithms for image smoothing, JPEG compression, copy-move, and shear tampering were published. However, there are still many image tampering algorithms that are not involved. In this paper, we publish a dataset of image warping, which contains more than 10000 images, and propose a novel convolutional neural network called **DWF-CNN** to identify warped images. In experiments, we compared the performance with 4 alternative networks. The proposed network with the preprocessing layer of the SRM layer and Bayar convolutional layer got the best result, which reached to the accuracy of 99.36%. The experiments also showed that the network with the regular convolutional layer performed even worse than a random guess. It illustrates the importance of the well-designed preprocessing layer in this research area again.

Keywords: Image forensics · Convolutional neural networks · Image warping

1 Introduction

With the popularity of digital cameras and mobile phones, more and more digital images and videos have been captured and published, and the number of pictures on the Internet has increased dramatically. Users can easily prettify, modify, and tamper the content of images with common image processing software, e.g. Photoshop. These software are designed to be easy to use and allows image tampering without expertise. Moreover, some jobs specializing in image processing, such as advertising design and graphic designers, have been derived. Related courses are offered in most computer-related majors too.

Consequently, image forgery is becoming a rampant problem. Some approaches were proposed to protect the authenticity of image content by adding watermarks [4, 8], hashes and etc., which called active image forensic. Images need to be preprocessed before they

Supported by Program for Young Innovative Research Team in Shandong University of Political Science and Law(Network Information Security and Forensics, Intelligent Information Processing);Big data and Artificial Intelligence Legal Research Collaborative Innovation Center of Shandong University of Political Science and Law; Key Laboratory of Evidence-Identifying in Universities of Shandong (SDUPSL); Projects of Shandong Province Higher Educational Science and Technology Program under Grant No.J16LN19, J18KA357, J18KA383; Projects of Shandong University of Political Science and Law under Grant No. 2016Z03B, 2015Z03B,2019KQ13Z.

X. Sun et al. (Eds.): ICAIS 2020, LNCS 12240, pp. 109–118, 2020.
https://doi.org/10.1007/978-3-030-57881-7_10

are published. In contract, passive techniques that need no image preprocessing are more useful but more challenging.

In recent years, a variety of image tampering detection techniques have been proposed, each algorithm targets one or several tampering methods. Image smoothing [16, 23], splicing [6, 11, 21], JPEG compression [1, 3, 12, 14, 18], and copy-move tampering [5, 15, 19, 20, 22, 27] are the most concerned tampering methods.

However, many tampering methods are not involved, such as image warping, image overlay, and content recognition in recent years. These algorithms are also used frequently in image processing.

Most proposed methods based on deep learning use a preprocessing layer to reduce the interference from image contents on image forensics. It is because, with an image processing software, the forgery images tend to close to the authentic ones not only visually but also statistically.

Fridrich [9] proposed 15 kernels for steganalysis. Zhou [25] selected 3 of those well-designed kernels as the first layer of their network for image tamping detection and achieved state-of-the-art performance compared to alternative methods. This layer often called SRM or SRMConv2d and is proved to be very effective in several follow-up works [15, 24].

Bayar [2] proposed a novel constrained convolutional layer with kernels having fixed weight -1 in their center. The network can handle multiple types of image forensics tasks.

Image warping is a prevalent method of image tampering. In photoshop, it called "liquefy". With this method, the face and shape of a person can be adjusted. This challenges the authenticity of images on the Internet (Fig. 1).

Fig. 1. Sample images of image warping. The left is authentic and right is warped.

In this paper, we propose a novel CNN network for image warping forgery. The network consists of two blocks: preprocessing block and regular CNN. We test the first block of 5 forms, and compared their performances and analyzed the results.

Section 2 describes the method of building the public image warping dataset. Section 3 gives the architecture of our proposed CNN network and Sect. 4 demonstrates the result of our experiments. Section 5 gives the conclusion.

2 Image Warping Dataset

As far as we know, there is no image warping dataset yet. Manually building a dataset of sufficient size is very labor-intensive. Fortunately, it is possible to construct the dataset using algorithms. The algorithm proposed by [10] is employed in this paper as it has many advantages: only pixels in the circular selection will be distorted, the father to the center of the circle, the smaller the distortion of pixels, and no changes on the edge of the circle, the image changes are uniform and natural.

We cut the images in the authentic set(AU) of CASIA v2.0 [7] to 256 × 256 as negative samples and the warped images as positive ones.

The images bigger than 256 × 256 were cut to 256 × 256, and smaller ones were discarded. We randomly selected parameters of image warping and warped each image.

Equation 1 gives the warping algorithm with is called liquefy in Photoshop, while x is coordinates of the pixels in the warped image, and u is coordinates of the pixels in the source image. c and r donate the center coordinate and radius of the circle, and m donates the coordinate of user's mouse.

$$u = x - \left(\frac{r^2 - |x - c|^2}{r^2 - |x - c|^2 + |m - c|^2} \right)(m - c) \tag{1}$$

The warped image on the Internet is visually different from the original image. We set r to 100. Even so, it can be seen from the Fig. 2 that the visual difference between the distorted image and the original image is still tiny. If there is no original image for comparison, it is almost difficult to judge whether it has been warped.

Fig. 2. Sample images of the dataset. Images in the top line are authentic and the bottom ones are warped.

We then randomly select the center of circle c and make sure the whole circle is in the image. We build vector d with the length of r/2 and random selected direction θ (Eq. 2), then let m $= d + c$.

$$d = \left\{ \frac{r\cos(\theta)}{2}, \frac{r\sin(\theta)}{2} \right\}, \theta \in [0, 2\pi] \tag{2}$$

Finally, we got 10066 images, while 5033 images are warped, and the others are original. All images have a size of 256 x 256. The dataset now is available on Github[1].

[1] https://github.com/taiji1985/ImageWarpingDataset.

3 Proposed Convolutional Neural Network

3.1 The Structure of Typical CNN Network

A typical CNN [26, 28, 29] network consists of several convolutional layers and fully connected layers. The input of the first convolutional layer is the image from the dataset, while the input of other layer is the output of the previous layer which called feature map. Let us donate the feature map (output) of the layer n by F^n, the kernel and bias of convolution by W^n and B^n, convolution operator by $*$, activation function by ϕ. We have:

$$F^n = \phi(F^{n-1} * W^n + B^n) \tag{3}$$

where F^0 is the input image and F^{n-1} donate the output of the above layer while $n > 0$.

In the pooling layer, image (or feature map) is downsampled to reduce the number of parameters and the computation load. Max-pooling and average-pooling are the most commonly used.

The fully connected layer is a linear layer with an activation function. The number of its parameters is product of the size of the input vector and output vector. As it has so many parameters, it is very easy to over-fitting. A technique called dropout is adopted. It randomly drops out some dimensions of the input vector in each training step, which makes the rest dimensions more discriminate.

3.2 The Overview of Proposed CNN Network

Proposed **DWF-CNN** network is shown in Fig. 3 which consists of two blocks: the preprocessing block and classification block. We use SRMConv2d [25] layers with 9 filters and BayarConv2d [2] layers with 3 filters and concatenate their output in channel dimension as the preprocessing block.

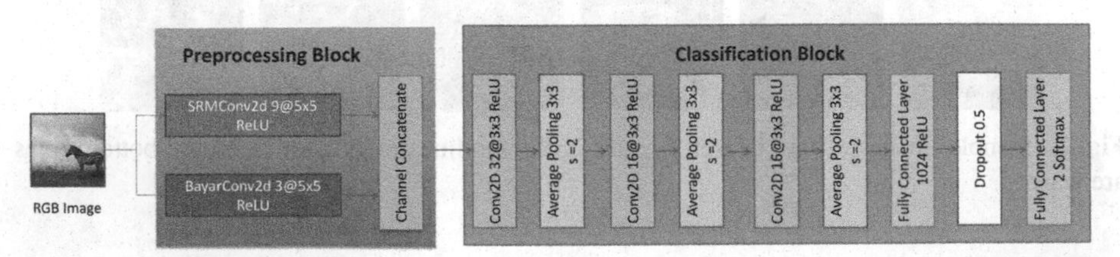

Fig. 3. The architecture of CNN network proposed in this paper. The network contains two block. The preprocessing block contains preprocessing layer to improve the classification performance and classification block is a well-designed CNN network for classification.

In classification block, we use 3 convolutional layers with a kernel size of 3×3 and ReLU activation function. Each of those layers is followed by one average pooling layer with a size of 3×3 and a step of 2. Two fully connected layers are employed at the end of the network with 1024 and 2 neurons respectively. The first fully connected layer use ReLU activation function and the last one uses softmax for output.

3.3 The Preprocessing Block

We build the SRMConv2d layer with 9 filters by the 3 kernels given in Eq. 4. The input of this layer is an image with 3 color channels: red, green and blue, so the shape of the filter is (3, 5, 5).

$$K_1 = \frac{1}{4}\begin{bmatrix} 0 & 0 & 0 & 0 & 0 \\ 0 & -1 & 2 & -1 & 0 \\ 0 & 2 & -4 & 2 & 0 \\ 0 & -1 & 2 & -1 & 0 \\ 0 & 0 & 0 & 0 & 0 \end{bmatrix}, K_2 = \begin{bmatrix} -1 & 2 & -2 & 2 & -1 \\ 2 & -6 & 8 & -6 & 2 \\ -2 & 8 & -12 & 8 & -2 \\ 2 & -6 & 8 & -6 & 2 \\ -1 & 2 & -2 & 2 & -1 \end{bmatrix}, K_3 = \begin{bmatrix} 0 & 0 & 0 & 0 & 0 \\ 0 & 0 & 0 & 0 & 0 \\ 0 & 1 & -2 & 1 & 0 \\ 0 & 0 & 0 & 0 & 0 \\ 0 & 0 & 0 & 0 & 0 \end{bmatrix} \quad (4)$$

Let $\mathbf{0}$ donate the 5×5 zero matrix, the 9 filters can be represented by Eq. 5. In other words, we let the kernel K_i given above as the weights for one input channel, and the weights for the other two channels are set to zero matrix. It can separately process the three channels of the image to extract features with higher discrimination.

$$\begin{bmatrix} K_1 & \mathbf{0} & \mathbf{0} \\ \mathbf{0} & K_1 & \mathbf{0} \\ \mathbf{0} & \mathbf{0} & K_1 \\ K_2 & \mathbf{0} & \mathbf{0} \\ \mathbf{0} & K_2 & \mathbf{0} \\ \mathbf{0} & \mathbf{0} & K_2 \\ K_3 & \mathbf{0} & \mathbf{0} \\ \mathbf{0} & K_3 & \mathbf{0} \\ \mathbf{0} & \mathbf{0} & K_3 \end{bmatrix} \quad (5)$$

BayarConv2d is a constrained convolutional layer while the center value of the filters fixes to -1, and the sum of other values in one kernel fix to 1. The amount of all weights normalize to zero. We use 3 such filters in our network.

Let w_{ij} donates the weight of 1 filter of BayarConv2d layer, we have:

$$\begin{cases} \sum_i \sum_j w_{ij} = 0 \\ w_{0,0} = -1 \end{cases} \quad (6)$$

We merge the output of the SRMConv2d layer and BayarConv2d layer in the channel dimension and get 12 channels at all. We also tested other forms of preprocessing block (Fig. 4) in our experiments. The above form got better accuracy.

The Preprocessing Block (b) combines SRMConv2d, BayarConv2d, and regular convolutional layers and gets 32 channels for output. This structure tried to get a better result but failed. Block (c–e) contains a single type of preprocessing layer to test the performance of each one. Block (c) and (d) got results lower than the proposed effect.

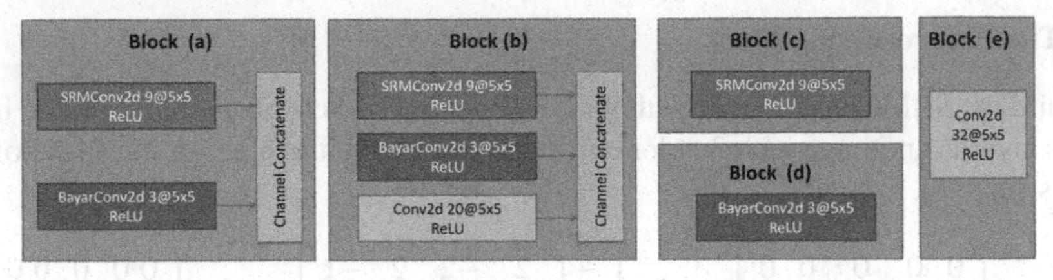

Fig. 4. The Preprocessing Block (b) combines SRMConv2d, BayarConv2d, and regular convolutional layers and gets 32 channels for output. This structure tried to get a better result but failed. Block (c-e) contains a single type of preprocessing layer to test the performance of each one. Block (c) and (d) got results lower than the proposed effect.

3.4 The Classification Block

The first convolutional layer in classification block has 32 filters of 3×3, and the second and third convolutional layer has 16 filters with the same size. They all use the ReLU activation function. We use symmetric padding in each of those layers.

In VGG [17] network, the number of filters grows twice after each two-layer, while the number of our network decreases. We had tested the architecture like VGG, but it performed pool as it had too many parameters to converge.

Those convolutional layers are followed by three 3×3 average pooling layer instead of 2×2 pooling used in classic CNN network like Alexnet [13] or VGG [17]. The bigger size increases the receptive field.

We employ two fully connected layers and use a dropout layer between them with a drop rate of 0.5.

4 Experiments

We split our image warping dataset randomly into two parts by 9:1, the first part for training, and the second part for evaluation.

We tested five forms of preprocessing block demonstrated by Fig. 4: (a) the proposed block with SRMConv and BayarConv2d, (b) block with SRMConv2d, BayarConv2d and normal Conv2d which has 32 channels at all. (c) block with only the SRMConv2d layer with 9 filters. (d) block with only BayarConv2d layer with 3 filters. (e) block with a normal Conv2d layer.

We used Google Colaboratory[2] in our experiments, which is free. It has 24G GDDR5 memory and a speed of 2.91 tflops for each runtime. The optimizer Adam with learning rate 0.001 and loss function of sparse categorical cross-entropy were employed in our networks.

The result of our experiments is impressive. It shows that the CNN network with regular convolutional layer performed worse than random guess during continuous training, but our model achieved an accuracy of up to 99.36%, the best results of each model shown in Table 1.

[2] https://colab.research.google.com.

Table 1. Classification accuracies of several CNN networks with different pre-processing block.

Form of Processing Block	Accuracy
(a) SRMConv2d and BayarConv2d	**99.36%**
(b) SRMConv2d,BayarConv2d and Conv2d	98.20%
(c) SRMConv2d	98.94%
(d) BayarConv2d	92.38%
(e) Conv2d	53.54%

Firstly, we tested the block with the regular convolutional layer (preprocessing block(e)) and showed the result in Fig. 5. The experiment showed that accuracy continuously declined to 34.92% while we trained 100 epochs. The validation accuracy shows that this model misclassified the images more than the correct classification and worse than the random guess dues to that the CNN network always ignores the deformation of images. CNNs are better at identifying the content of images, including deformed content. It is an advantage in content recognition, but a disadvantage in distortion detection.

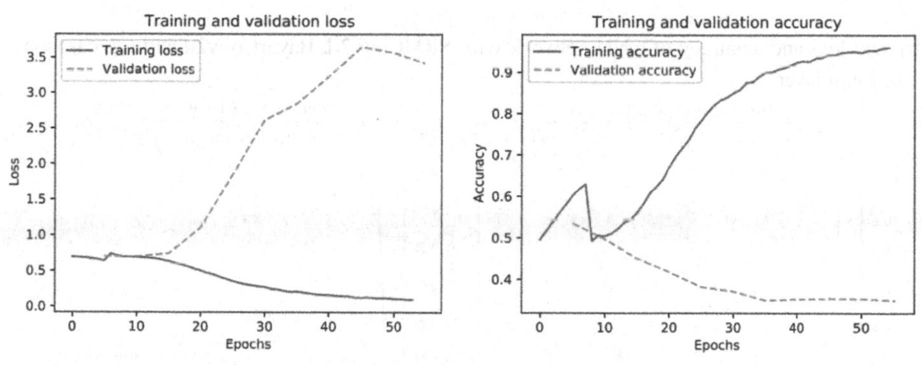

Fig. 5. The loss and accuracy of CNN network with regular convolutional layer in preprocessing block.

We also tested the CNN network with the BayarConv2d layer with 3 filters, which got an accuracy of 92.38% shown in Fig. 5 (d). The training accuracy reaches 97% after 13 epochs, and increase slowly after that, while the validation accuracy grows slowly and unhesitatingly.

The CNN network with the SRMConv2d layer got an accuracy of 98.94% quickly, after 10 epochs. The network trained faster than the others while the SRMConv2d layer is not trainable, and the total number of parameters in the network was smaller.

The results of the proposed network with preprocessing block(a) show in Fig. 6 (a), which got an accuracy of 99.36%. Note that the preprocessing block of this network contains SRMConv2d layer and BayarConv2d layer but no regular convolutional layer.

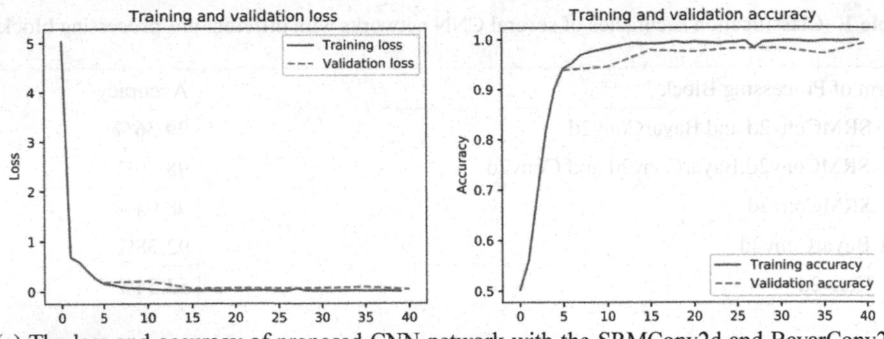

(a) The loss and accuracy of proposed CNN network with the SRMConv2d and BayarConv2d layers.

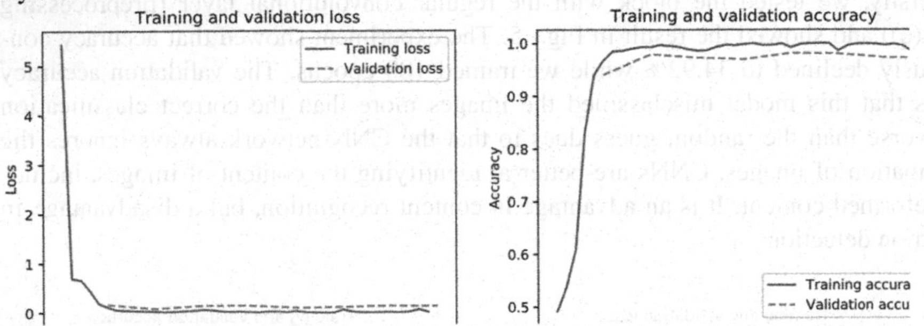

(b) The loss and accuracy of CNN network with SRMConv2d, BayarConv2d and regular Convolutional layer.

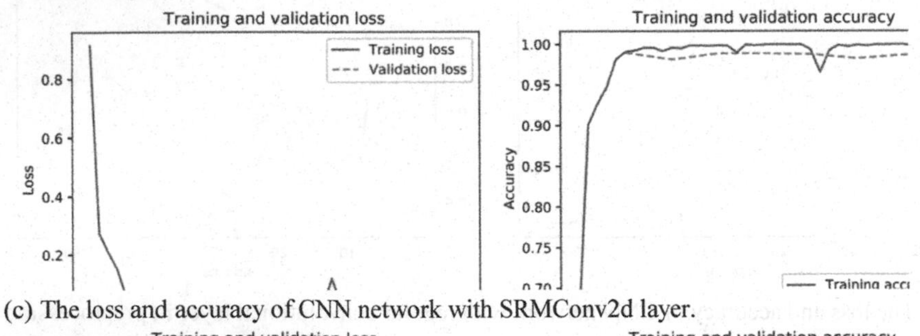

(c) The loss and accuracy of CNN network with SRMConv2d layer.

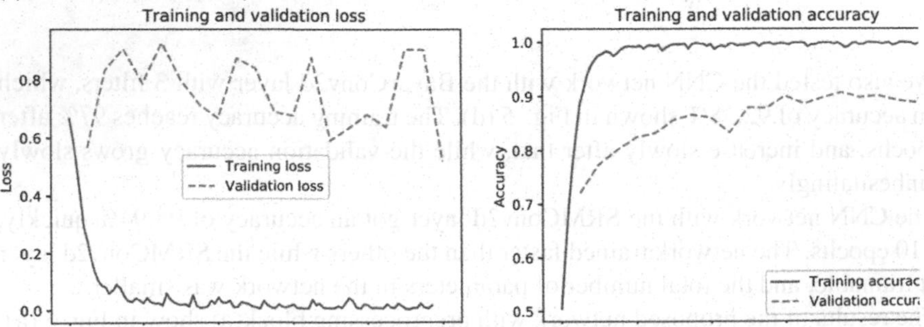

(d) The loss and accuracy of CNN network with BayarConv2d layer.

Fig. 6. The loss and accuracies of CNN network with 4 forms of the preprocessing block.

In contrast, we used all of the layers included SRMConv2d, BayarConv2d, and regular convolutional layer in preprocessing block but got worse accuracy of 98.20% shown in Fig. 6 (b). Combined with the above experiments, it can be concluded that the regular convolutional layer has a negative impact on accuracy.

5 Conclusion

Image warping is a common method of beautifying and forging in image processing. We proposed a dataset with more than 10000 images and proposed an effective identification method. We intensely studied the influence of the preprocessing layer on the classification results through experiments. The effect of traditional CNN with regular convolutional layers is deplorable because of the visual similarity between the distorted image and the original image, even lower than randomly guess. The well-designed preprocessing layer proposed can reduce the impact of content on the image and focus on image tampering forensics.

References

1. Barni, M., et al.: Aligned and non-aligned double JPEG detection using convolutional neural networks. CoRR abs/1708.00930 (2017) http://arxiv.org/abs/1708.00930
2. Bayar, B., Stamm, M.C.: Constrained convolutional neural networks: a new approach towards general purpose image manipulation detection. IEEE Trans. Inf. Forensics. Secur. **13**(11), 2691–2706 (2018)
3. Bin, L., Shi, Y.Q., Jiwu, H.: Detecting doubly compressed JPEG images by using mode based first digit features. In: 2008 IEEE 10th Workshop on Multimedia Signal Processing. pp. 730–735 (2008). https://doi.org/10.1109/MMSP.2008.4665171
4. Chan, K.C., Moon, Y.S., Cheng, P.S.: Fast fingerprint verification using subregions of fingerprint images. IEEE Trans. Circ. Syst. Video Technol. **14**(1), 95–101 (2004). https://doi.org/10.1109/TCSVT.2003.818358
5. Cozzolino, D., Poggi, G., Verdoliva, L.: Efficient dense-field copy–move forgery detection. IEEE Trans. Inf. Forensics Secur. **10**(11), 2284–2297 (2015). https://doi.org/10.1109/TIFS.2015.2455334
6. Cozzolino, D., Poggi, G., Verdoliva, L.: Splicebuster: a new blind image splicing detector. In: 2015 IEEE International Workshop on Information Forensics and Security (WIFS), pp. 1–6, November 2015 https://doi.org/10.1109/WIFS.2015.7368565
7. Dong, J., Wang, W., Tan, T.: Casia image tampering detection evaluation database, pp. 422–426, July 2013 https://doi.org/10.1109/ChinaSIP.2013.6625374
8. Farid, H., Lyu, S.: Higher-order wavelet statistics and their application to digital forensics. In: 2003 Conference on Computer Vision and Pattern Recognition Workshop. vol. 8, pp. 94–94 (June 2003). https://doi.org/10.1109/CVPRW.2003.10093
9. Fridrich, J., Kodovsky, J.: Rich models for steganalysis of digital images. IEEE Trans Inf. Forensics Secur. **7**(3), 868–882 (2012). https://doi.org/10.1109/TIFS.2012.2190402
10. Gustafsson, A.: Interactive image warping. https://www.gson.org/thesis/warping- thesis.pdf (1993). Accessed 19 July 2008
11. Huh, M., Liu, A., Owens, A., Efros, A.A.: Fighting fake news: Image splice detection via learned self-consistency. CoRR abs/1805.04096 (2018). http://arxiv.org/abs/1805.04096
12. Korus, P., Huang, J.: Multi-scale fusion for improved localization of malicious tampering in digital images. IEEE Trans. Image Process. **25**(3), 1312–1326 (2016)

13. Krizhevsky, A., Sutskever, I., Hinton, G.E.: Imagenet classfication with deep convolutional neural networks. Commun. ACM **60**(6), 84–90 (2017)
14. Pevny, T., Fridrich, J.: Detection of double-compression in jpeg images for applications in steganography. IEEE Trans. Inf. Forensics Secur. **3**(2), 247–258 (2008). https://doi.org/10.1109/TIFS.2008.922456
15. Rao, Y., Ni, J.: A deep learning approach to detection of splicing and copy-move forgeries in images. In: 2016 IEEE International Workshop on Information Forensics and Security (WIFS), pp. 1–6 (Dec 2016). https://doi.org/10.1109/WIFS.2016.7823911
16. Rhee, K.H.: Median filtering detection using variation of neighboring line pairs for image forensics. J. Electron. Imag. **25**(5), 1–13 (2016)
17. Simonyan, K., Zisserman, A.: Very deep convolutional networks for large-scale image recognition. arXiv 1409.1556 pp. 1–7 (2014)
18. Taimori, A., Razzazi, F., Behrad, A., Ahmadi, A., Babaie-Zadeh, M.: Quantization-unaware double JPEG compression detection. J. Math. Imaging Vis. **54**(3), 269–286 (2016). https://doi.org/10.1007/s10851-015-0602-z
19. Wen, B., Zhu, Y., Subramanian, R., Ng, T., Shen, X., Winkler, S.: Coverage—a novel database for copy-move forgery detection. In: 2016 IEEE International Conference on Image Processing (ICIP), pp. 161–165 Sep 2016
20. Wu, Y., Abd-Almageed, W., Natarajan, P.: Image copy-move forgery detection via an end-to-end deep neural network. In: 2018 IEEE Winter Conference on Applications of Computer Vision (WACV), pp. 1907–1915 March 2018
21. Wu, Y., Abd-Almageed, W., Natarajan, P.: Deep matching and validation network an end-to-end solution to constrained image splicing localization and detection. CoRR abs/1705.09765 (2017), http://arxiv.org/abs/1705.09765
22. Wu, Y., Abd-Almageed, W., Natarajan, P.: BusterNet: detecting copy-move image forgery with source/target localization: 15th European Conference, Munich, Germany, September 8–14, 2018, Proceedings, Part VI, pp. 170–186 (2018)
23. Yang, J., Ren, H., Zhu, G., Huang, J., Shi, Y.-Q.: Detecting median filtering via two-dimensional AR models of multiple filtered residuals. Multimed. Tools. Appl. **77**(7), 7931–7953 (2017). https://doi.org/10.1007/s11042-017-4691-0
24. Yue Wu, W.A., Natarajan, P.: Mantra-net: Manipulation tracing network for detection and localization of image forgerieswith anomalous features (2019)
25. Zhou, P., Han, X., Morariu, V., Davis, L.: Learning rich features for image manipulation detection, pp. 1053–1061 (2018). https://doi.org/10.1109/CVPR.2018.00116
26. Zhang, J., Li, Y., Niu, S., Cao, Z., Wang, X.: Improved fully convolutional network for digital image region forgery detection. Comput. Mater. Con. **60**(1), 287–303 (2019)
27. Cui, Q., McIntosh, S., Sun, H.: Identifying materials of photographic images and photorealistic computer generated graphics based on deep CNNs. Comput. Mater. Con. **055**(2), 229–241 (2018)
28. Lecun, Y., Bottou, L., Bengio, Y., Haffner, P.: Gradient-Based learning applied to document recognition. Proc. IEEE **86**, 2278–2324 (1998)
29. Fang, W., Zhang, F., Sheng, V.S., Ding, Y.: A method for improving CNN-based image recognition using DCGAN. Comput. Mater. Con. **57**(1), 167–178 (2018)

The Course Design of Applied Talent Statistics Based on Fuzzy Control and Grey Measure Model

Li Wu[1](✉) and Jun Yang[2]

[1] School of Mathematics and Computer Science, Aba Teachers University, Wenchuan 623002, Sichuan, China
sansheng8210nll@163.com

[2] Chongqing Vocational College of Transportation, Chongqing 402247, China

Abstract. In order to improve the teaching quality of probability theory and statistics course under the goal of cultivating applied talents, a design model of probability theory and statistics course for applied talents based on SPSS software is proposed. The distributed structure model of probability theory and statistics course is constructed under the goal of cultivating applied talents, and the constraint parameters of teaching objectives of probability theory and statistics course are designed by using the method of quantitative characteristic analysis. The explanatory variable model and control variable model of probability theory and statistics course teaching are constructed, and the teaching planning design of probability theory and statistics course is carried out by using descriptive statistical analysis method and regression analysis method. The quantitative regression analysis model of course design and teaching evaluation of probability theory and statistics course is established, and the optimization design of applied talent probability theory and statistics course is carried out by combining fuzzy control and grey measure method. The effectiveness of probability theory and statistics course of applied talents is analyzed in SPSS software environment. The result of statistical analysis shows that this model can improve the positive effect of course teaching. The robustness of teaching results promotes the innovation and cultivation of probability theory and statistical applied talents.

Keywords: SPSS software · Applied talents · Probability theory and statistics · Course design

1 Introduction

Probability theory is a branch of mathematics that studies the quantitative law of random phenomena. The stochastic phenomenon is relative to the decisive phenomenon. A phenomenon that is bound to happen under certain conditions is called a decisive phenomenon. For example, at standard atmospheric pressure, when the pure water is heated to 100°, the water will boil. Random phenomenon is that when the basic conditions remain unchanged, every test or observation can not be sure what kind of results will

X. Sun et al. (Eds.): ICAIS 2020, LNCS 12240, pp. 119–130, 2020.
https://doi.org/10.1007/978-3-030-57881-7_11

happen. For example, tossing a coin may lead to a positive or negative side. The realization of random phenomena and its observation are called random tests. Every possible outcome of a random test is called a basic event. A group or a group of basic events is collectively called random events or events. Typical random experiments include throwing dice, throwing coins, playing cards and Roulette Games. The probability of an event is a measure of the likelihood of an event occurring. Although the occurrence of an event in a random trial is accidental, the random trials that can be repeated in large quantities under the same conditions often show obvious quantitative rules.

The nature of the subject determines that the characteristic of the course of probability Theory and Mathematical Statistics is that it is mainly applied, and the training of applied talents becomes the important goal of the course of probability Theory and Mathematical Statistics. At present, the course of probability Theory and Mathematical Statistics is regarded as a compulsory course for the major of science and engineering, which is closely related to the characteristics of the subject [1]. Because the in-depth study of related majors needs to seek relevant research tools and research methods from probability and statistics, it needs probability and statistics as theoretical support. Because of this, in the traditional teaching mode, the probability and statistics course is generally regarded as the same nature as other basic mathematics courses such as Advanced Mathematics, teaching with the same teaching mode, paying attention to the proof and derivation of theorems and formulas. Paying attention to the complicated calculation and neglecting the practical utility of the course causes the disconnection between the theory and the application, the course teaching is boring, and the students do not know where the use of the course is, which results in the students' weariness of learning and the decline of the teaching quality [2]. In order to improve the teaching effect of probability theory and statistics course, it is necessary to study the design method of probability theory and statistics course under the goal of cultivating applied talents [3].

2 Teaching Design of Probability Theory and Statistics Course Under the Training Target of Applied Talents

2.1 An Overview of Curriculum Teaching

Probability theory and mathematical statistics are the mathematical disciplines to study the statistical regularity of random phenomena. It is a backbone theory course in the teaching plan of science and engineering major in Colleges and universities. Through the study of this course, the students can grasp the basic concepts and methods of probability theory and mathematical statistics, cultivate the students' ability to solve the related random practical problems by using the probability characteristics of random variables, understand the ideas and methods of mathematical statistics, master the common statistical methods and cultivate the ability of the statistical problems related to the statistics [4]. The basic theory of probability and statistics is laid down in the course of study and in engineering and scientific research.

2.2 Teaching Goal

Probability theory and mathematical statistics are the mathematical disciplines to study the statistical regularity of random phenomena. It is a backbone theory course in the teaching plan of science and engineering major in Colleges and universities. Through the study of this course, the students can master the basic concepts and methods of probability theory and mathematical statistics, cultivate the students' ability to solve practical problems, and lay the basis for the theoretical basis of probability and statistics for the study of the follow-up courses and the work of engineering and scientific research.

2.3 Teaching Content

When the teaching course of mathematical statistics is squeezed greatly, the present situation of emphasizing probability and neglecting statistics and emphasizing theory and application is caused. Many practical methods of mathematical statistics, such as ANOVA and regression analysis, have not come to an end. In response to this problem, what we have to do is to revise the syllabus, rewrite the textbooks, and compress the chapters and hours of the probability theory that we have learned in high school. Appropriate addition of relevant chapters and hours of mathematical statistics, which are closely related to the major of the course, to achieve the purpose of learning for practical use. In addition, in line with the goal of cultivating applied talents, we should properly strengthen the teaching of statistical software in the teaching link, introduce to students the application of statistical software such as SAS, SPASS, S-Plus, Minitab, and find out the corresponding points between these software and relevant professional cases [5].

Table 1. The content structure of the course

Knowledge teaching unit	Class hour	Remarks
Event and probability	18	Probability theory
Discrete random variable	4	
Continuous random variable	6	
The digital characteristics of random variables	20	
The law of large numbers and the central limit theorem	8	
Basic concepts of mathematical statistics	10	
Parameter estimation	8	Mathematical statistics
Hypothesis test	3	
Summary	1	
Total	78	

The so-called "follow the Frontier" is to keep pace with the times in the arrangement of teaching content and to keep up with the domestic and international related research results. Today's society is a society of information, which changes rapidly and is influenced by objective conditions. The content of teaching materials is inevitably lagging behind to a certain extent. From the angle of cultivating practical talents and optimizing teaching contents and reforming the teaching system, Permeate the latest academic ideas into the daily teaching, good at classical theories such as Bayesian decision theory, reliability theory, information theory and so on to give the atmosphere and characteristics of the times, In preparing lessons, we should look for popular cases to explain the practical application of these theories, so that students can use perceptual learning in class and perceive the charm of the curriculum [6–8]. Table 1 is the content structure of the course

2.4 Teaching Idea

In view of the existing problems in the teaching of statistics, we can not blindly copy the foreign teaching mode, realize the form of "international integration", and can not avoid the problem of existence because of personal gain and loss. Colleges and universities should be allowed to carry out "diversification" and "characteristic" independent running experiments, so as to find a new teaching mode suitable for China's national conditions.

2.5 Teaching Methods

Teaching methods, curriculum design should be more interactive, and it should be good at examples. The traditional teaching process of probability and statistics often adopts the mode of "full hall irrigation", in which students lose their motivation and interest in learning quickly. In terms of teaching methods, we need to interact with students in the classroom as necessary, for example, when we talk about birthday models in classical generalizations, we can first ask students to talk about existing teaching classes. Are at least two students more likely to have birthdays on the same day or less? The formula of event probability in classical probability type is used to prove that the learning enthusiasm of students is aroused effectively. In the course of preparing classes, we can compile the application cases of probability statistics in different fields according to the professional background of different teaching classes [9]. For example, for the students majoring in engineering management, we can construct cases related to the data processing of engineering cost. Let students know that they can use their theoretical knowledge of probability and statistics to predict the risks and profitability of engineering projects; or, for students majoring in finance and economics, Usable cases: "A life insurance company designed a kind of personal accident insurance in the face of a certain group of people, and the insured pays a premium of 120 yuan to the insurance company, if the insured dies accidentally within one year", The family can get 20000 yuan from the insurance company. If the accident mortality rate for this group of people in one year is 0.002, the cost of selling an insurance policy is $20 and the cost of settlement is $500, how many such customers will the insurance company have to develop in order to have a 99.9% probability of not losing money? "Let students participate in solving some mathematical models driven by such problems, deepen students' understanding of probability and statistics knowledge, and cultivate students' ability to use theoretical knowledge to solve practical problems, which is consistent with the goal of cultivating application-oriented talents [10].

3 Application and Analysis of SPSS Statistical Software in the Course Design of Probability Theory and Mathematical Statistics

3.1 Constraint Parameter Design and Mathematical Model Construction of Teaching Objectives in Probability and Statistics Courses

The distributed structure model of probability theory and statistics course is constructed under the goal of cultivating applied talents, and the constraint parameters of teaching objectives of probability theory and statistics course are designed by using the method of quantitative characteristic analysis. Using related resources, such as laboratory, to cultivate students' ability to process and process data by using relevant statistical software, such as SPASS, SAS, Minitab, and to cultivate innovative application talents from the results of analysis [11]. The stochastic functional differential equation is used to construct the robust convergence model of course teaching. In the perturbed characteristic functional convex combination model, the robustness degree S is divided into $\mathbf{X}^{(0)}$ part S^N, that is, $\mathbf{X}^{(1)}, \mathbf{X}^{(2)}, \cdots, \mathbf{X}^{(N)}$. By using the method of full sample analysis, the symmetric solution of the LMEs of the composite model is solved [12], and the following results are obtained:

$$W_N = \max_{1 \le i \le N} \left\| W(\mathbf{X}^{(i)}) \right\|_\infty \tag{1}$$

Assuming that the teaching objective parameter $\{q_N\}$ monotonously increments $q_N \ge 1$ of probability theory and statistics course, and when $N \to \infty$, obtains over-confidence positive correlation characteristic quantity $q_N \to \infty$, the innovation teaching characteristic functional of probability theory and statistics course after finite step calculation is obtained:

$$\alpha_N = \min_{1 \le i \le N} \{\inf F(\mathbf{X}^{(i)}, q_N)\} \tag{2}$$

Under the constraints of the ability to acquire knowledge, we obtain:

$$\dot{x}(t) = Ax(t) + Bx(t - d_1(t) - d_2(t))\sqrt{2}, x(t) = \phi(t) \quad t \in [-h, 0] \tag{3}$$

Under the similar empirical distribution, the input characteristic vectors of innovative talents training in probability theory and statistics course are:

$$x(t) = (x_0(t), x_1(t), \cdots, x_{k-1}(t))^T \tag{4}$$

For the Euclidean distance of the weighted vector $x(t)$, the talent training among individuals is expressed as follows:

$$d_j = \sum_{i=0}^{k-1} (x_i(t) - \omega_{ij}(t))^2, \; j = 0, 1, \cdots, N-1 \tag{5}$$

Where $\omega_j = (\omega_{0j}, \omega_{1j}, \cdots, \omega_{k-1,j})^T$, $\forall \varepsilon > 0, \exists \hat{N} > 0$, while $N > \hat{N}$, $\left| \min_{x \in X^{(0)}} e = (f(\mathbf{x}) - \alpha_N) \right| < \varepsilon$, the descriptive statistical analysis method and regression analysis method are used to design the teaching planning of probability theory and statistics course, the quantitative regression analysis model of probability theory and statistical course design and teaching evaluation is established, which is expressed as follows:

$$Q^{-}.R^{-} = Q^{-}.R_{tt} - \frac{Q^{-}}{R_{tt}}(R_{*t}.R_{*t})_{\backslash tt} = Q^{-}.R_{tt} + (R_{*t}.R_{*t})_{\backslash tt} = I \quad (6)$$

The upper expression shows that the set S_s has a negative effect on the innovation performance of the cluster with the set of samples, and the boundary value solution vector model for the innovation evaluation of probability theory and statistical courses is convergent. Under the technology of forward regulation [13], the teaching planning model of probability theory and statistics course is obtained as follows:

$$\begin{aligned} 0 &\leq \left| \min_{x \in X^{(0)}} (f(\mathbf{x}, q_k) - \alpha_N) \right| \\ &= \left| \min_{x \in X^{(0)}} (f(\mathbf{x}, q_k) - \inf F(\mathbf{X}^{(k)}, q_k) \right| \\ &= \left| \min_{x \in X^{(k)}} (f(\mathbf{x}, q_k) - \inf F(\mathbf{X}^{(k)}, q_k) \right| \\ &\leq K \left\| W(\mathbf{X}^{(k)}) \right\|_{\infty} \end{aligned} \quad (7)$$

While $N \to \infty$, $\left|\mathbf{X}^{(k)}\right| \to 0$, the semi-simple eigenvalues are obtained to characterize the asymptotic stability of polynomials [14]. In this case, $\left\| W(\mathbf{X}^{(k)}) \right\|_{\infty} \to 0$, the perturbation eigenvalues of the teaching programming of probability and statistics courses are obtained by the perturbation clipping theorem of stochastic functional differential equations:

$$\left| \min_{x \in X^{(0)}} (f(\mathbf{x}, q_k) - \alpha_N) \right| \to 0 \quad (8)$$

Then $\forall \varepsilon > 0, \exists N_1 > 0$, while $N > N_1$, in the invariant convex combination model subspace, the technique locking effect is introduced, and the boundary value solution is obtained:

$$\left| \min_{x \in X^{(0)}} (f(\mathbf{x}, q_k) - \alpha_N) \right| < \frac{\varepsilon}{2} \quad (9)$$

According to the convex combinatorial optimization theorem of complex value function, the explanatory variable model and control variable model of probability theory and statistics course teaching are constructed, and the descriptive statistical analysis method and regression analysis method are used to carry out the probability and statistics course teaching planning and design [15].

3.2 Regression Analysis Model of Probability Theory and Statistical Course Teaching

The quantitative regression analysis model of curriculum design and teaching evaluation of probability theory and statistics course is established, while $f(\mathbf{x}, q_k) \to f(\mathbf{x})$, then

$\exists N_2 > 0$, while:

$$\left| \min_{x \in X^{(0)}} (f(\mathbf{x}, q_k) - f(\mathbf{x})) \right| < \frac{\varepsilon}{2} \tag{10}$$

The measure of two variables is used to quantify the degree of proximity of the system and get $\hat{N} = \max\{N_1, N_2\}$, when $N > \hat{N}$, the statistical values of the main samples satisfy:

$$\begin{aligned} &\left| \min_{x \in X^{(0)}} e = (f(\mathbf{x}) - \alpha_N) \right| \\ &= \left| \min_{x \in X^{(0)}} [(f(\mathbf{x}, q_k) - \alpha_N) - (f(\mathbf{x}, q_k) - f(\mathbf{x}))] \right| \\ &< \frac{\varepsilon}{2} + \frac{\varepsilon}{2} = \varepsilon \end{aligned} \tag{11}$$

It is known that the boundary value convex combinatorial model of stochastic functional differential equation is asymptotically convergent based on Cauchy convergence condition. The descriptive statistical analysis method and regression analysis method are used to design the teaching planning of probability and statistical courses, and the technical proximity degree and the system proximity degree between teaching organizations in the course of curriculum innovation design are defined as follows [16]:

$$u = \sum_{j=0}^{n} a_j s n^j \xi \tag{12}$$

The estimated coefficient of curriculum teaching evaluation is:

$$O(u(\xi)) = n \tag{13}$$

According to the theory of translation and extension of equilibrium point, the tight degree of connection between nodes is obtained, and the complex differential equation model is obtained as [17]:

$$u_{tx} + qu_{yy} + 6(uu_x)_x + u_{xxx} = 0 \tag{14}$$

The effectiveness of curriculum design is evaluated by means of piecewise regression analysis, and its mathematical model is obtained as follows [18]:

$$\begin{aligned} \min \quad & F(x) = (f_1(x), f_2(x), \ldots, f_m(x))^T \\ s.t. \quad & g_i \leq 0, \qquad i = 1, 2, \ldots, q \\ & h_j = 0, \qquad j = 1, 2, \ldots, p \end{aligned} \tag{15}$$

The cooperative innovation mechanism is used for quantitative regression analysis [16, 17], the non-redundant relationship between the two links under the curriculum innovation system is obtained [19]. The quantitative characteristics of probability and statistics courses for applied talents are analyzed under the environment of

SPSS software. In the explanatory variables, the evaluation values are expressed as follows [20]:

$$\inf F(\mathbf{X}, q) - \inf f(\mathbf{X}, q) \leq \frac{K}{2}\|W(\mathbf{X})\|_{\infty}^{2} \tag{16}$$

Based on the above analysis, combined with fuzzy control and grey measurement method, the optimization design of probability theory and statistics course for applied talents is carried out.

4 Empirical Analysis

In order to test the application performance of this design model in the optimization design of probability theory and statistics course under the goal of cultivating applied talents, the experiment is carried out with SPSS software, the descriptive statistical results of the effectiveness analysis of probability theory and statistical curriculum design under the goal of cultivating applied talents are shown in Table 2. The minimum value, maximum value, mean value and variance of the constraint parameter are 1, 85, 2.093 and 3.473 respectively, and the mean value is less than the variance.

Table 2. Descriptive statistical analysis

	Correlation coefficient	Contribution weight	Decision statistics
SC	7.543	3.554	5.675
PC	3.134	1.321	3.654
Size	6.564	3.242	5.323
MB	8.213	3.754	8.754
ALR	7.653	2.543	5.276
Profitability	6.764	3.434	6.898

According to the results of descriptive statistical analysis and correlation analysis, the effectiveness of probability theory and statistics course of applied talents is analyzed under the environment of SPSS software, and the results of empirical analysis are shown in Fig. 1.

The analysis shows that using this model to design the probability theory and statistics course of applied talents has good effectiveness. In order to verify the reliability of the research conclusion, the robustness test is carried out, the zero expansion negative binomial regression is used as the basic set of the model, and the results of robust analysis are shown in Fig. 2.

The analysis shows that this model can improve the positive effect of curriculum teaching and promote the innovation and cultivation of probability theory and statistical applied talents.

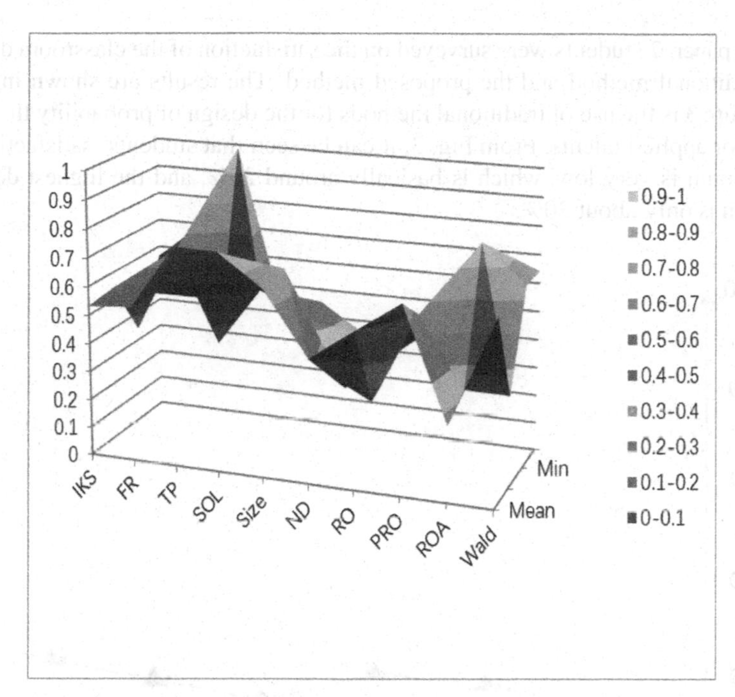

Fig. 1. Results of statistical analysis on the effectiveness of probability and statistics courses for applied talents under the environment of SPSS software

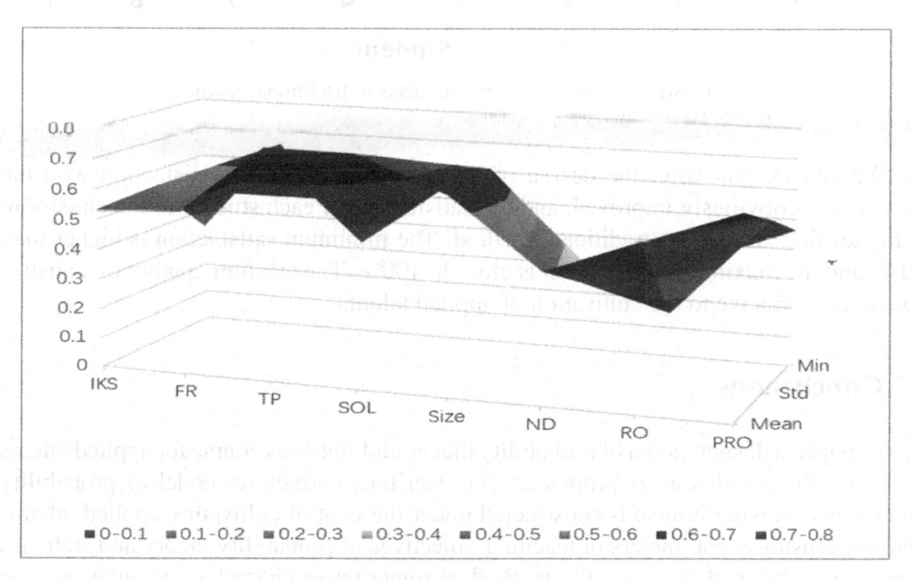

Fig. 2. Results of robustness analysis

In this paper, 7 students were surveyed on the satisfaction of the classroom designed by the traditional method and the proposed method. The results are shown in Figs. 3 and 4. Figure 3 is the use of traditional methods for the design of probability theory and statistics for applied talents. From Fig. 3, it can be seen that students' satisfaction with the curriculum is very low, which is basically around 20%, and the highest degree of satisfaction is only about 30%.

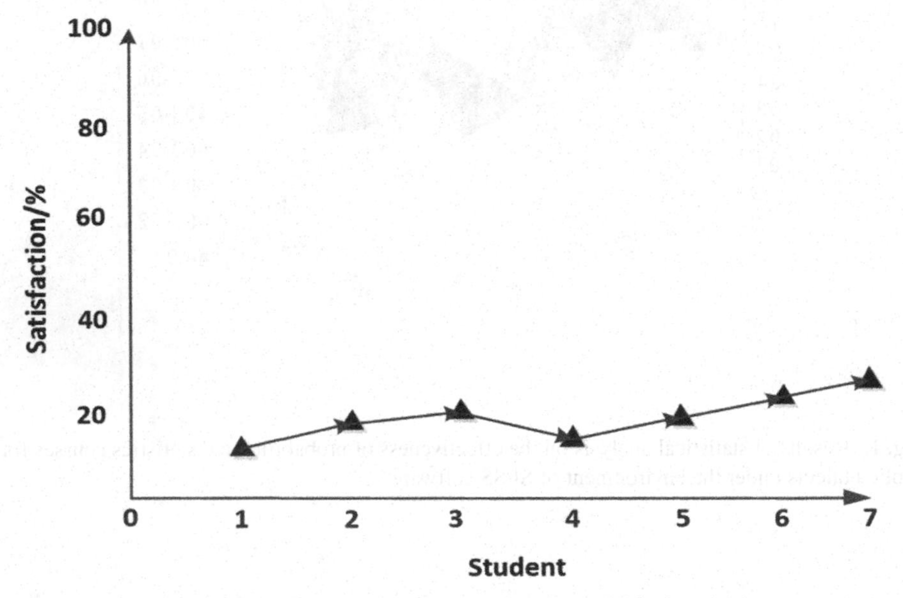

Fig. 3. Students' satisfaction with class in traditional methods

We can see that after the design of the class, the students' satisfaction with the classroom is obviously improved, and the satisfaction of each student to the classroom is higher than that of the traditional method. The minimum satisfaction is higher than 40%, and the maximum satisfaction is close to 100%. The teaching quality of statistics course is conducive to the cultivation of applied talents.

5 Conclusions

In this paper, a design model of probability theory and statistics course for applied talents based on SPSS software is proposed. The distributed structure model of probability theory and statistics course is constructed under the goal of cultivating applied talents, and the constraint parameters of teaching objectives of probability theory and statistics course are designed by using the method of quantitative characteristic analysis. The explanatory variable model and control variable model of probability theory and statistics course teaching are constructed, and the teaching planning design of probability theory and statistics course is carried out by using descriptive statistical analysis method and regression analysis method. The effectiveness of probability theory and statistics course of applied talents is analyzed in SPSS software environment. The result of statistical analysis shows that this model can improve the positive effect of course teaching. The robustness of teaching results promotes the innovation and cultivation of probability theory and statistical applied talents. This model has good application value in probability theory and statistical curriculum design of applied talents.

Research Programs:

1. The Research Program of National Natural Science Foundation—Research on Construction Theory and Related Information Calculation of High Efficiency High Rank Space Time Code, No: 11861001.
2. The general program of Aba Teachers University—Applied Research on Factor Analysis in Multivariate Statistics, No: ASB18-05.

References

1. Cheng, G., Wang, Y.: Multi-mobile Agent collaborative control data separation method in wireless sensor network. J. Comput. Appl. **35**(4), 910–915 (2015)
2. Zuxiong, L.I.: Periodic solution for a modified Leslie-Gower model with feedback control. Acta Math. Appl. Sinica **38**(1), 37–52 (2015)
3. Tan, Z.: The asymptotic relation between the maxima and sums of discrete and continuous time strongly dependent gaussian processes. Acta Math. Appl. Sinica **38**(1), 27–36 (2015)
4. Ding, Y., Dai, H., Wang, S.: Image quality assessment scheme with topographic independent components analysis for sparse feature extraction. Electron. Lett. **50**(7), 509–510 (2014)
5. Manikandan, L., Selvakumar, R.K.: A new survey on block matching algorithms in video coding. Int. J. Eng. Res. **3**(2), 121–125 (2014)
6. Bdi, H., Williams, L.J.: Principal component analysis. Wiley Interdisc. Rev.: Comput. Stat. **2**(4), 433–459 (2010)
7. Dai, S., Lü, K., Zhai, R., Dong, J.: Lung segmentation method based on 3D region growing method and improved convex hull algorithm. JEIT **38**(9), 2358–2364 (2016)
8. Ma, G., Jian, J., Jiang, X.: An improved fletcher-reeves conjugate gradient method with descent property. Acta Math. Appl. Sinica **38**(1), 89–97 (2015)
9. Kumar, A., Pooja, R., Singh, G.K.: Design and performance of closed form method for cosine modulated filter bank using different windows functions. Int. J. Speech Technol. **17**(4), 427–441 (2014). https://doi.org/10.1007/s10772-014-9242-8
10. Rajapaksha, N., Madanayake, A., Bruton, L.T.: 2D space- time wave-digital multi-fan filter banks for signals consisting of multiple plane waves. Multidimension. Syst. Signal Process. **25**(1), 17–39 (2014)
11. Huang, H., Ge, X., Chen, X.: Density clustering method based on complex learning classification system. J. Comput. Appl. **37**(11), 3207–3211 (2017)
12. Chen, H., Chen, P., Yang, H., et al.: Empirical analysis of offshore and onshore RMB interest rate pricing: Based on the spillover index and its dynamic path. Int. Financ. Res. **350**(6), 86–96 (2016)
13. Yao, L., Yao, W.X.: Research on the policy effect of incremental expansion of margin and securities lending: Based on the multi period DID model and Hausman's test. Int. Financ. Res. **349**(5), 85–96 (2016)
14. Bai, M., Qin, Y.: Short-sales constraints and liquidity change: cross-sectional evidence from the Hong Kong market. Pac.-Basin Financ. J. **26**, 98–122 (2014)

15. Jin, Y., Jia, S.: Study on the influence of the introduction of leverage ratio on the asset structure of commercial banks. Int. Financ. Res. **350**(6), 52–60 (2016)
16. Zhou, J., Zhou, Y., Cao, S., et al.: Multivariate logistic regression analysis of postoperative complications and risk model establishment of gastrectomy for gastric cancer: a single-center cohort report. Scand. J. Gastroenterol. **51**(1), 8–15 (2016)
17. Kerns, L., Chen, J.: A note on range regression analysis. J. Appl. Probabil. Stat. **11**(2), 19–27 (2016)
18. Chen, W., Feng, G., Zhang, C., Liu, P., Ren, W., Cao, N., Ding, J.: Development and application of big data platform for garlic industry chain. Comput. Mater. Continua **58**(1), 229–248 (2019)
19. Wang, X., Jiang, J., Zhao, S., Bai, L.: A fair blind signature scheme to revoke malicious vehicles in VANETs. Comput. Mater. Continua **58**(1), 249–262 (2019)
20. Jayashree, N., Bhuvaneswaran, R.S.: A robust image watermarking scheme using Z-transform, discrete wavelet transform and bidiagonal singular value decomposition. Comput. Mater. Continua **58**(1), 263–285 (2019)

Reinforcement Learning-Based Resource Allocation in Edge Computing

Qianyu Xie, Xutao Yang, Laixin Chi, Xuejie Zhang, and Jixian Zhang(✉)

School of Information Science and Engineering, Yunnan University, Kunming 650504, Yunnan, People's Republic of China
denonji@163.com

Abstract. The problem of online resource allocation in edge computing has become a research hotspot. Meanwhile, reinforcement learning (RL) is suitable for solving online problems. In this paper, we combine edge computing online resource allocation with RL. This combination enables edge computing resource providers to obtain more social welfare and improve resource utilization. Specifically, we define a dynamic resource allocation problem for edge computing: edge equipment users request resources from a nearby edge computing server (ECS), the amount of resources required varies among the users, and there are time limits for the completion of the requested tasks. Since this resource allocation problem is NP-hard, it cannot be solved in polynomial time. Therefore, we propose an algorithm based on the policy-gradient algorithm in RL to solve the problem. Our approach is experimentally compared with existing research in terms of social welfare and resource utilization, for which it achieves good results.

Keywords: Edge computing · Resource allocation · Reinforcement learning · Policy-gradient

1 Introduction

In recent years, due to the rapid development of the Internet of Things (IoT), a large amount of edge equipment has been needed to interact with a server and must support intensive operations at low latency. We can use the mobile cloud to connect to edge equipment [1], and the cloud has abundant resources. However, the distance between the edge equipment and the mobile cloud computing center is substantial, and the data transfer network burden will increase with frequent interactions, thereby reducing the real-time response of the service. We can solve this problem by deploying an ECS with rich resources near edge equipment instead of connecting through mobile cloud computing centers.

This work is supported in part by the National Natural Science Foundation of China (Nos. 61762091, 61662088 and 11663007), the Project of the Natural Science Foundation of Yunnan Province of China (2019FB142), and the Program for Excellent Young Talents, Yunnan University, China.

X. Sun et al. (Eds.): ICAIS 2020, LNCS 12240, pp. 131–142, 2020.
https://doi.org/10.1007/978-3-030-57881-7_12

A schematic diagram of EC is shown in Fig. 1. EC has become a hot topic because it migrates the original mobile cloud computing model to the edge [2] (edge equipment: mobile phones, cameras, fax machines, etc.), thus reducing the computational load on the computing center.

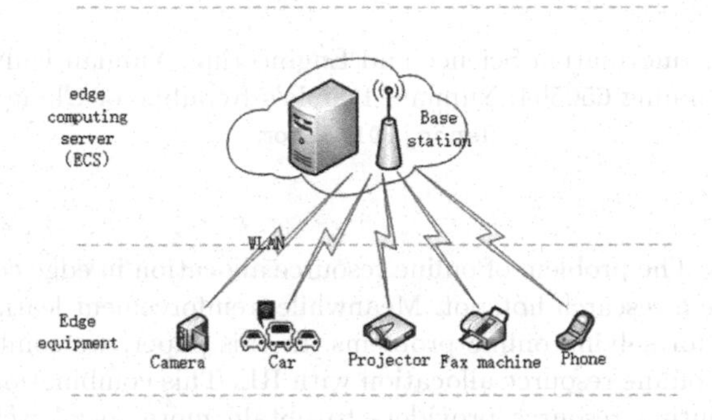

Fig. 1. Edge computing model.

Resource allocation is the main problem in EC. When an edge equipment user makes a resource request to a nearby ECS to complete a calculation, the ECS schedules resources according to a certain mechanism. This allocation can be equivalent to the multidimensional knapsack problem (MKP), which is NP-hard. We need to design a mechanism to improve the utility of the base system. In the conventional resource allocation algorithm, although an optimal solution is obtainable, the time complexity of the calculation is relatively high; on the other hand, the approximate algorithm or the heuristic algorithm results in low computational efficiency but low allocation accuracy compared with the optimal solution.

Fortunately, here, we can use the RL algorithm. An RL Agent can perform an Action based on the existing Environment and Rewards, and the Action can change the Environment; this process loops until a satisfactory solution is found. RL can handle dynamic problems, and because of the adaptability of RL, the algorithm can obtain good results based on past experience. Thus, to improve the utility and reduce the time complexity, this paper considers using reinforcement learning to solve the problem of edge computing resource allocation (ECRA). The resource allocation problem is continuous, so we use the policy-gradient algorithm in RL.

The sections of this paper are organized as follows: Sect. 2 briefly describes the research status of the field of ECRA and RL methods, Sect. 3 establishes a reasonable dynamic ECRA model, and Sect. 4 describes the RL method in ECRA, as well as the reasonable use of the distribution field and the design algorithm implementation. Section 5 includes comparative analysis of experimental results, and Sect. 6 provides a summary.

2 Related Work

In recent years, many studies have achieved encouraging results with respect to ESC's provision of resource services for edge equipment. Zeng et al. [3] considered introducing cloud computing technology into the edge network to achieve optimal task scheduling and optimal virtual machine allocation. Due to the wide variety of edge equipment, Gu [4] and Cai et al. [5] proposed a mixed integer nonlinear programming model to solve resource optimization configuration problems. In the resource allocation algorithm, Marinescu D et al. used integer programming to find the optimal solution [6]. Kiani et al. [7] proposed a model for the user to configure the edge servers from near to far. Nejad M M et al. proposed a PTAS algorithm for multitasking scheduling [8,9]. Shi W and Zhang L transformed the online auction resource allocation problem into a continuous static multiround static resource allocation problem [10], and Liu X, Li W, Zhang X et al. designed a resource management problem in a heterogeneous physical machine environment. The approximate solution is obtained based on the multidimensional multimapping mechanism of combined auction and the corresponding efficient algorithm [11]. Zhang J et al. propose three resource allocation predict algorithms based on linear regression, logistic regression and Support Vector Machine(SVM) for the different scale of problem [12]. Although the heuristic algorithm [13,14] can satisfy the monotonicity, greedy algorithms on resource allocation [15] is fast but the effect of solving the multiresource allocation problem is not good.

Therefore, research on resource allocation can be based on machine learning theory [16]. To dynamically allocate ECS resources, the problem is extended to online mechanism design [17], including several model-free methods [18–20]. However, these algorithms cannot make predictions, and resources are currently over-allocated, resulting in high-value tasks or urgent tasks that cannot be assigned in the next step. The RL-based method is used in other fields and can make good predictions [21,22]; furthermore, it can automatically optimize the mechanism [23].

Currently, RL is applied in manufacturing process control, task scheduling, robot game design, etc. For example, Christopher proposed an algorithm for controlling robot arm motion, and Peter Stone proposed a soccer algorithm [24]. In some scheduling algorithms, reinforcement learning has also been used. Moore et al. studied how RL controls the manufacturing process in practice [25]. Jiang Guofei et al. used the Q-learning algorithm to implement inverted control [26]. In some scheduling tasks, for example, Robert Crites et al. studied using reinforcement learning in an elevator scheduling algorithm in early high-rise buildings [27] and achieved results superior to several existing elevator scheduling algorithms. Studies have shown that RL is more accurate than manual and conventional controllers. Therefore, RL can be used to solve the problems existing in the conventional resource allocation problem algorithm, and the results are stable [28].

3 The Problem of Edge Computing Resource Allocation

We consider the resources required for edge equipment as tasks and temporarily do not consider bandwidth limitations. It is assumed that there are a total of n types of resources in the ECS, a set of resource types $R = \{1, 2, \ldots, n\}$, and a single type $r \in R$; the unit cost of each resource is $\boldsymbol{P} = (p_1, p_2, \cdots, p_n)$, and the total resource capacity is $\boldsymbol{C} = (c_1, c_2, \ldots, c_n)$. The running time slots of the system is $t \in [1, T]$. The user is $i \in U$, $U = \{1, 2, \ldots, m\}$. The user's task arrival time slot is $a_i \in [1, T]$, and the deadline time slot is $d_i \in [1, T]$. The resource requirement number is $\boldsymbol{k}_i$, $\boldsymbol{k}_i = (k_{i1}, k_{i2}, \ldots, k_{in})$. The task execution time is e_i time slots, and the user's bid is b_i. Therefore, the task of user i can be described as $\theta_i = (\boldsymbol{k}_i, a_i, d_i, e_i, b_i)$. In the process of task dynamic arrival, task θ_i must be assigned before the latest start time $l_i (l_i = d_i - e_i + 1)$; otherwise, it will be abandoned. Once the task execution begins, it cannot be interrupted until it is complete (the task is non-preemptive). We need to judge whether each task is allocated for maximizing social welfare. The integer programming model for maximizing social welfare can be described by the bid b_i of the winning task as:

$$Maximize: \quad V = \sum_{i \in U} \sum_{t=a_i}^{d_i - e_i + 1} b_i \cdot x_{it} \tag{1}$$

$$s.t. \sum_{i \in U} \sum_{\omega = t - e_i + 1}^{t} k_{ir} \cdot x_{iw} \leq c_r \forall r = 1, 2, \cdots, n, \forall t \in [1, T] \tag{1a}$$

$$\sum_{t=a_i}^{d_i - e_i + 1} x_{it} \leq 1, \forall i \in U \tag{1b}$$

$$x_{it} \in \{0, 1\}, \forall i \in U, t \in [1, T] \tag{1c}$$

where V indicates the total social welfare. x_{it} denotes whether the task of user i is allocated at time t, where $x_{it} = 1$ denotes that allocation is successful at time slot t; otherwise, the allocation fails. (1) means that the objective function is maximizing social welfare, and (1a) represents that the total amount of each resource allocated to the task does not exceed the resource capacity at any time slot. (1b) denotes that the same task can only be allowed once. (1c) represents that θ_i has two allocation results.

In the literature [29], the author's model does not consider the time limit. In this paper, we constrain the submission time, and deadlines are added to make the model more closely approximate the actual edge computing resource allocation problem.

4 Algorithm Design

4.1 Reinforced Learning Framework Design

In the general operation process of RL, assume that the Markov process is in the $t-th$ step; then, the main elements of RL include Agent, Environment,

State (s_t), Reward (r_t), and Action (the meaning of Action is the same as x_{it}, so we use x_{it} to describe Action).

Each element is defined as follows: Agent is a strategic decision system (which must be trained). Environment means the information that the Agent cannot change in the Markov process of RL. State is the quantification of the Environment. Action is the action space of the Agent. Reward indicates the degree of Reward and punishment for the Action of the Agent. The relationship of each element is shown in Fig. 2.

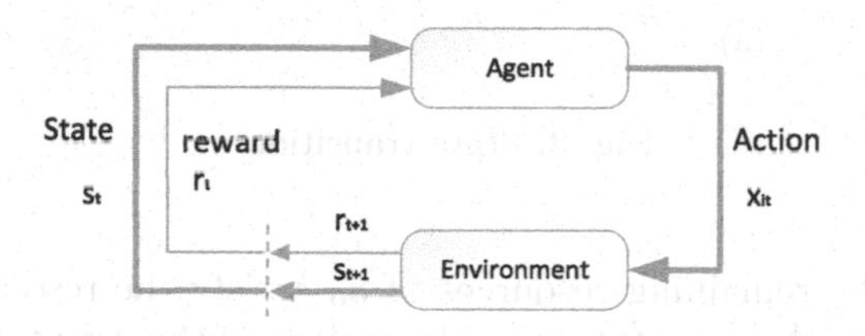

Fig. 2. RL process of operation.

If we want to combine RL with ECRA, each element in RL is defined as follows:

- **Agent**: Its specific meaning is an allocation strategy decision system (to determine whether the task is allocated resources).
- **Environment**: Regarded as the waiting tasks to be allocated and the number of resources that have been occupied.
- **State**: Quantification of the waiting tasks to be allocated and the number of resources that have been occupied.
- **Action**: There are two Actions: 0 or 1; 0 means refusal to allocate resources, while 1 indicates allocation.
- **Reward**: A favorable Reward setting can make convergence of the training process faster. If we want to maximize social welfare, then the Reward can be equal to the user's bid for each winning task. The Reward settings are shown in (2):

$$r_t = b_i x_{it} \tag{2}$$

We use an example to explain the State transition of ECRA in RL.

Assume that resources are occupied as shown in Fig. 3; there are two types of resources: CPU and memory. The resource capacities of both are 5, and the resources occupied from now through the next 8 time steps are recorded. Thus, the occupied resources of the CPU can thus be described as follows: $\boldsymbol{o}_1 = (5,4,4,3,3,3,0,0)$; memory: $\boldsymbol{o}_2 = (4,4,4,4,4,3,0,0)$. If the task currently waiting to be assigned is θ_i, then we can express the State as $\boldsymbol{s}_0 = (\theta_i, \boldsymbol{o}_1, \boldsymbol{o}_2)$. If the task is $\theta_i{}'$, then the State can be described as $\boldsymbol{s}_3 = (\theta_i{}', \boldsymbol{o}_1, \boldsymbol{o}_2)$.

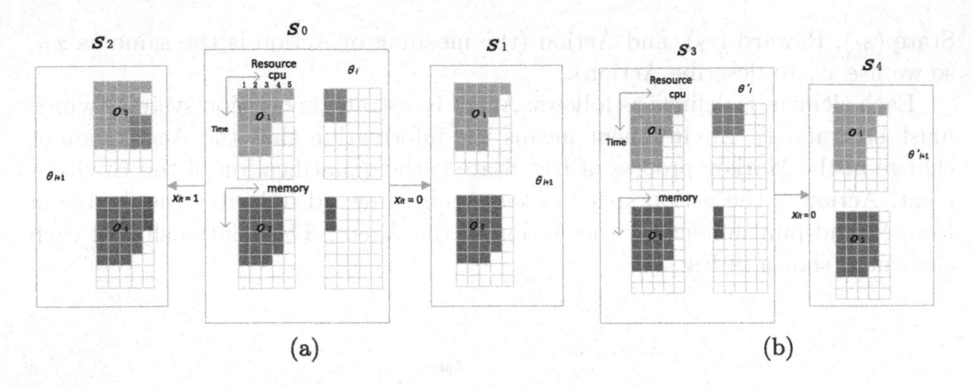

(a) (b)

Fig. 3. State transition.

In Fig. 3(a), the remaining resources of $\boldsymbol{s}_0$ satisfy the resources required by task θ_i. In this case, there are two possible Actions of the Agent. If the Agent performs action x_{it}=0, then the State transition is $\boldsymbol{s}_0 \rightarrow \boldsymbol{s}_1$. The task in $\boldsymbol{s}_1$ becomes θ_{i+1}, and the occupied resources have not changed; then, $\boldsymbol{s}_1$= $(\theta_{i+1}, \boldsymbol{o}_1, \boldsymbol{o}_2)$. If x_{it}=1, then the State transition is $\boldsymbol{s}_0 \rightarrow \boldsymbol{s}_2$, and the occupied CPU in $\boldsymbol{s}_2$ becomes as follows: $\boldsymbol{o}_3 = (5, 4, 4, 5, 5, 5, 0, 0)$; memory: $\boldsymbol{o}_4$= $(5, 5, 5, 4, 4, 3, 0, 0)$. If the task in $\boldsymbol{s}_2$ is θ_{i+1}, then $\boldsymbol{s}_2$= $(\theta_{i+1}, \boldsymbol{o}_3, \boldsymbol{o}_4)$.

In Fig. 3(b), we consider the allocation of $\theta_i{}'$ in State $\boldsymbol{s}_3$, the number of required CPUs is 3, the execution time is 3, and the remaining resources of $\boldsymbol{s}_3$ cannot satisfy $\theta_i{}'$. The Action If and only if is x_{it}=0. The State transition is thus $\boldsymbol{s}_3 \rightarrow \boldsymbol{s}_4$, the occupied resources have not changed, and the task in $\boldsymbol{s}_4$ becomes θ'_{i+1}; thus, $\boldsymbol{s}_4$= $(\theta'_{i+1}, \boldsymbol{o}_1, \boldsymbol{o}_2)$.

According to the definition of Reward, the expected Reward of the system can be defined as $Q = \sum_{t=0}^{\infty} \gamma^t r_t$, where γ is the discount rate, $\gamma \in (0, 1]$, which means the reduction of Rewards during the accumulation. In resource allocation, Q_t expresses the predicted value of the Reward in the $t-th$ step. We use Q_{t_target} to describe the true expected Reward in the $t - th$ step as follows:

$$Q_{t\_target} = r_t + \gamma Q_{t+1} \tag{3}$$

Q_{t+1} is the predicted Reward in the $(t + 1) - th$ step. The updated function of Q_t is then:

$$Q_t \leftarrow Q_t + \alpha(Q_{t\_target} - Q_t) \tag{4}$$

where α is the learning rate.

Algorithm 1. RL_ECRA

Require: Current task $\theta_i = (\boldsymbol{k}_i, a_i, d_i, e_i, b_i)$; current time t; currently waiting tasks set θ^t; allocation strategy Δ_g; resource occupation $\boldsymbol{o}_1, \ldots, \boldsymbol{o}_n$; $\boldsymbol{s}_{last} \leftarrow 0$; $x_{last} \leftarrow 0$; $v_{last} \leftarrow 0$;

Ensure: State $\boldsymbol{s}_{last}$; Action x_{last}; Reward value v_{last}; Allocation strategy table Δ_g;

1: **for all** $\theta_i \in \theta^t$ **do**
2: $\quad \delta \leftarrow \Delta_g(\theta_i)$
3: $\quad$ **if** $\delta = 0$ **then**
4: $\quad\quad$ **return** $\boldsymbol{s}_{last}, x_{last}, v_{last}, \Delta_g$ /* Remaining resources are insufficient for θ_i */
5: $\quad$ **end if**
6: $\quad \boldsymbol{s} \leftarrow [1, d_i - t - (d_i - e_i), \theta_i, b_i, b_i/\rho_i, \boldsymbol{o}_1, \ldots, \boldsymbol{o}_n]$ /* State definition */
7: $\quad \boldsymbol{w} \leftarrow POL\_GRA(\boldsymbol{s}_{last}, x_{last}, v_{last}, \boldsymbol{s})$ /* Update weight */
8: $\quad q \leftarrow NORMALIZE(\boldsymbol{s}) \cdot \boldsymbol{w}$
9: $\quad \boldsymbol{s}_{last} \leftarrow \boldsymbol{s}$
10: $\quad x_{last} \leftarrow ACT\_JUD(q)$ /* here x_{last} is the Action in the $t-th$ step */
11: $\quad v_{last} \leftarrow b_i \cdot x_{last}$
12: $\quad$ **if** $x_{last} = 1$ **then**
13: $\quad\quad \Delta_g \leftarrow \Delta_g + \theta_i$ /* Update strategy table with θ_i occupied resources */
14: $\quad$ **else**
15: $\quad\quad \Delta_g$ unchanged
16: $\quad$ **end if**
17: $\quad$ **return** $\boldsymbol{s}_{last}, x_{last}, v_{last}, \Delta_g$
18: **end for**

4.2 Edge Computing Resource Allocation Algorithm Based on Reinforcement Learning (RL_ECRA)

In this paper, we propose a reasonable algorithm to solve the ECRA. First, we define a sorting rule ρ to sort waiting tasks. The rule can be considered as earliest deadline first (EDF), etc., and we define ρ as:

$$\rho_i = \frac{b_i}{\boldsymbol{k}_i \cdot Normalization(\boldsymbol{P})} \tag{5}$$

In Algorithm 1, in the $t-th$ step, State is defined as $\boldsymbol{s}$, while the task waiting for a decision is θ_i. Agent can take Action x_{it} of 0 or 1. $\boldsymbol{w}$ represents the weight vector (line 7). $\boldsymbol{w}$ and the normalized $\boldsymbol{s}$ dot product produce the q value (line 8), which serves as input in the ACT-JUD function, and the ACT-JUD function returns an Action. Lines 9–11 save the current State, Action and Reward as last step data, which are used in the next step to update the weight.

The weight $\boldsymbol{w}$ update algorithm is shown in Algorithm 2. Algorithm 2 uses the gradient descent strategy to fit the function about $\boldsymbol{w}$. The main process of Algorithm 2 is that we store the m group history data (lines 2–4) as a batch: if the history data number does not reach the m group, then $\boldsymbol{w}$ (line 5) is returned, and when the m group is reached, $\boldsymbol{w}$ will be updated (lines 7–13). We update the weights (line 13) using the gradient strategy for linear fitting to obtain a stable value of $\boldsymbol{w}$ so that the q value is close to the actual expected Reward Q,

Algorithm 2. POL_GRA

Require: State of two steps s_{last} and s; Action x_{last}; Reward v_{last}; counter initial value of 0; gradient decrease difference ∂
Ensure: weight vector w
1: $s_{last}, s \leftarrow NORMALIZE(s_{last}, s)$ /* normalize s_{last} and s */
2: $batch[counter] \leftarrow (s_{last}, x_{last}, v_{last}, s)$
3: $counter \leftarrow counter + 1$
4: **if** $counter < m$ **then**
5: **return** w
6: **end if**
7: $Q^m \leftarrow s_{last}^m \cdot w$ /* obtain m predicted values Q in the $t - th$ step */
8: $Q_-^m \leftarrow s^m \cdot w$ /* obtain m predicted values Q_- in the $(t+1) - th$ step */
9: $Q^m \leftarrow Q^m + \alpha(v_{last}^m \cdot a_{last}^m + \gamma \cdot Q_-^k - Q^m)$
10: **while** $\nabla J(w) \geq \partial$ **do**
11: $J(w) \leftarrow \frac{1}{2m}(s_{last} \cdot w - Q)^T (s_{last} \cdot w - Q)$
12: $\nabla J(w) \leftarrow \frac{1}{m} s_{last}^T (s_{last} \cdot w - Q)$
13: $w \leftarrow w - \alpha \nabla J(w)$
14: **end while**
15: $batch \leftarrow \emptyset$
16: **return** w

and the update method of Q is as shown in (4). In the linear fitting process, the loss function $J(w)$ is (6):

$$J(w) = \frac{1}{2m}(s_{last} \cdot w - Q)^2 \tag{6}$$

To find the minimum value of the loss function, we write $J(w)$ as a form of matrix multiplication:

$$J(w) \leftarrow \frac{1}{2m}(s_{last} \cdot w - Q)^T (s_{last} \cdot w - Q) \tag{7}$$

Finally, we use the $\varepsilon - greedy$ algorithm to select the Action. As shown in Algorithm 3, the main idea of the algorithm is as follows: We obtain a random number $\varepsilon \in (0, 1)$. Assuming that the selected Action is 1, we first use q to compute the predicted Reward Q_i and true expected Reward Q_{i_target} (lines 1–2). If $\xi \leq \varepsilon$, an Action value of 0 or 1 is randomly returned. If $\xi < \varepsilon$ and $Q_{i\_target} \geq Q_i$, we return 1; otherwise, 0 is returned (line 3). For example, $\varepsilon = 0.95$ means that 95% of ACT-JUD's return of Action is based on the values of Q_{i_target} and Q_i, and 5% of the return is random; the random return of Action ensures the exploratory nature of RL.

5 Experimental Results

5.1 Experimental Settings

The DAS-2 experimental data of the Grid Workloads are used as a test dataset to simulate user requirements. The data are provided by ASCI

Algorithm 3. ACT_JUD

Require: State and weight dot product results q; Currently waiting task θ_i; random values ε, $\varepsilon \in (0,1)$

Ensure: chosen; chosen value is 0 or 1*/

1: $Q_t \leftarrow q$

2: $Q_{t\_target} \leftarrow b_i + \gamma Q_{t+1}$

3: $\prod(q,\xi) \leftarrow \begin{cases} \text{Random return } 0,1 & if\, \xi \leq \varepsilon \\ if\ Q_{t\_target} \geq Q_t, return\ 1 & \text{others} \\ if\ Q_{t\_target} < Q_t, return\ 0 & \end{cases}$

4: $\xi \sim U(0,1)$

5: $chosen \leftarrow \prod(q_0, q_1, \xi)$

6: **return** *chosen*

(Advanced School for Computing and Imaging). The file contains the user submitted task ID and resource requirements. The experimental platform configuration includes an i5-8400 CPU, 8 GB memory, and 1 TB of storage.

We set the number of samples to 10000 and the number of iterations per sample to 5000. The resource capacities of the CPU, memory, and storage size are 100, 100, and 2000. The number of sample data for updating weight m is 20. We use Python to implement the RL_ECRA algorithm and the VCG_VMAP algorithm from the literature [30], while using CPLEX to implement the optimal algorithm.

5.2 Analysis of Results

In Fig. 4, we compare the social welfare of three algorithms for different sample scales. The sample scale is set to 50, 100, 200, and 500; the sample scale of 500 is the maximum size that the optimal algorithm can solve.

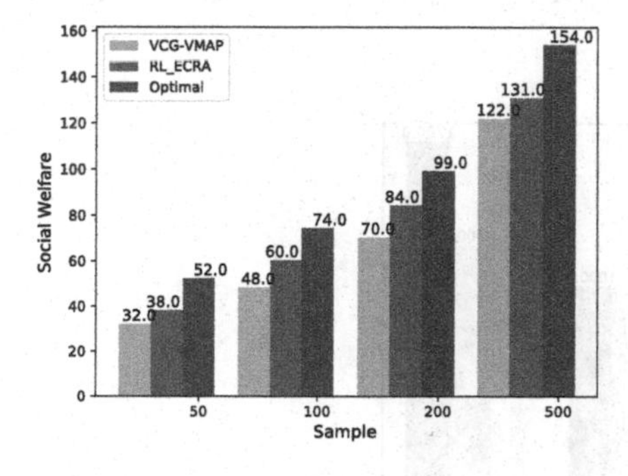

Fig. 4. Social welfare comparison.

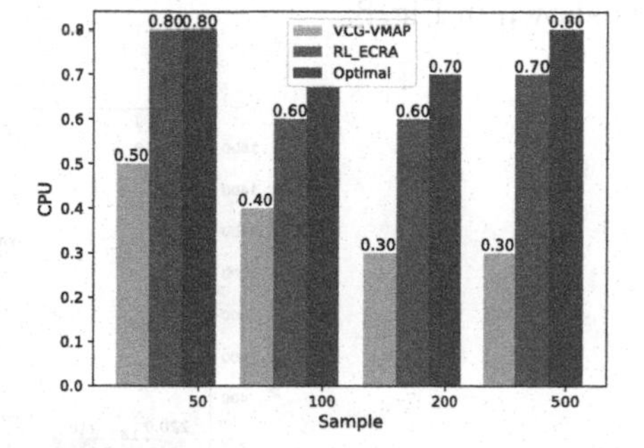

Fig. 5. Utilization of CPU.

For the four different sample scales, it can be observed that the social welfare results obtained by the RL_ECRA algorithm proposed in this paper are better than those of the VCG_VMAP algorithm. However, they are slightly worse than those of the optimal solution. This is primarily because the reinforcement learning algorithm will dynamically adjust the allocation scheme according to different resource requirements.

We compare the utilization of the CPU, storage, and memory in Fig. 5, Fig. 6, and Fig. 7.

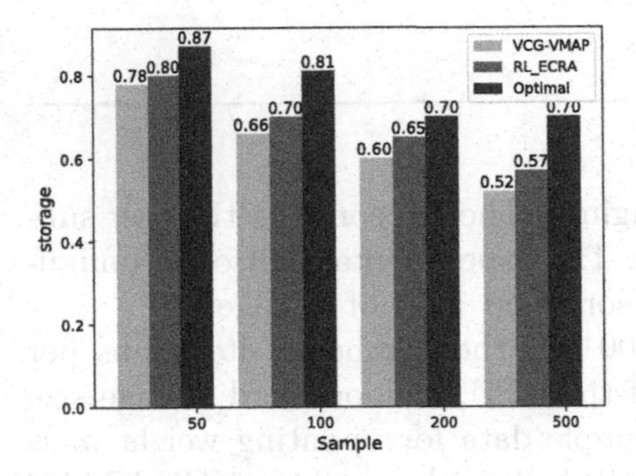

Fig. 6. Utilization of storage.

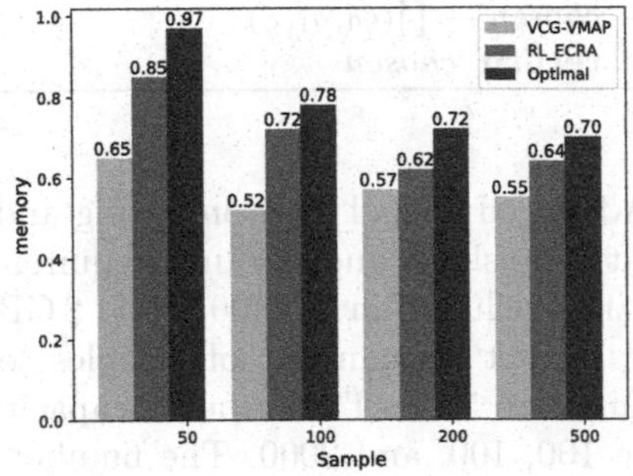

Fig. 7. Utilization of memory.

From Fig. 5, Fig. 6, and Fig. 7, it can be observed in the results of the RL_ECRA algorithm that the utilization values of the CPU, memory, and storage are higher than those of the VCG_VMAP algorithm.

In the RL_ECRA algorithm, there are still some parameters that will affect the experimental results. For example, the random probability ε for the action in the algorithm will affect the speed of convergence. For three different values of ε of 0.7, 0.8, and 0.9 on three different sample scales, the convergence speed is shown in Fig. 8:

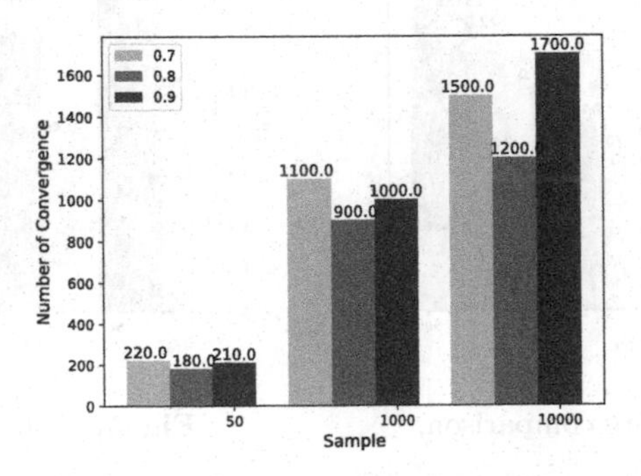

Fig. 8. Different convergence speeds for different ε on different sample scales.

6 Conclusion

In this paper, the proposed RL_ECRA algorithm uses the reinforcement learning algorithm to design a mechanism to solve the online resource allocation problem of edge computing. The algorithm result is better than those of some conventional algorithms, and reinforcement learning is adaptive and applicable to different data, which proves that reinforcement learning can be used to optimize online resource allocation. However, this paper ignores the impact of bandwidth on resource allocation in edge computing, so it is worthwhile to continue this research to make the model more closely approximate the real situation.

References

1. Satyanarayanan, M., Bahl, P., Cceres, R., Davies, N.: The case for vm-based cloudlets in mobile computing. IEEE Pervasive Comput. **8**(4), 14–23 (2009)
2. Huang, X., Rong, Y., Kang, J., He, Y., Yan, Z.: Exploring mobile edge computing for 5G-enabled software defined vehicular networks. IEEE Wirel. Commun. **24**(6), 55–63 (2018)
3. Zeng, D., Lin, G., Song, G., Cheng, Z., Shui, Y.: Joint optimization of task scheduling and image placement in fog computing supported software-defined embedded system. IEEE Trans. Comput. **65**(12), 3702–3712 (2016)
4. Gu, L., Zeng, D., Guo, S., Barnawi, A., Xiang, Y.: Cost-efficient resource management in fog computing supported medical CPS. IEEE Trans. Emerg. Top. Comput. **99**, 113–123 (2017)
5. Chi, K., Zhu, Y.H., Li, Y., Zhang, D., Leung, V.: Coding schemes to minimize energy consumption of communication links in wireless nanosensor networks. IEEE IoT J. **3**(4), 480–493 (2016)
6. Marinescu, D.C.: Cloud Computing: Theory and Practice. Elsevier, Amsterdam (2013)
7. Kiani, A., Ansari, N.: Toward hierarchical mobile edge computing: an auction-based profit maximization approach. IEEE IoT J. **99**, 2082–2091 (2016)
8. Nejad, M.M., Mashayekhy, L., Grosu, D.: A family of truthful greedy mechanisms for dynamic virtual machine provisioning and allocation in clouds. In: IEEE Sixth International Conference on Cloud Computing, pp. 594–603 (2013)
9. Mashayekhy, L., Nejad, M.M., Grosu, D.: A ptas mechanism for provisioning and allocation of heterogeneous cloud resources. IEEE Trans. Parallel Distrib. Syst. **26**(9), 2386–2399 (2014)
10. Shi, W., Zhang, L., Wu, C., Li, Z., Lau, F., An online auction framework for dynamic resource provisioning in cloud computing. In: ACM SIGMETRICS Performance Evaluation Review, vol. 42, no. 1, pp. 71–83. ACM (2014)
11. Liu, X., Li, W., Zhang, X.: Strategy-proof mechanism for provisioning and allocation virtual machines in heterogeneous clouds. IEEE Trans. Parallel Distrib. Syst. **29**(7), 1650–1663 (2017)
12. Zhang, J., Xie, N., Zhang, X., Li, W.: Supervised learning based truthful auction mechanism design in cloud computing. J. Electron. Inf. Technol. **41**(5), 1243 (2019)
13. Zhang, J., Xie, N., Zhang, X., Li, W.: An online auction mechanism for cloud computing resource allocation and pricing based on user evaluation and cost. Future Gener. Comput. Syst. **89**, 286–299 (2018)

14. Zhang, J., Xie, N., Li, W., Yue, K., Zhang, X.: Truthful multi requirements auction mechanism for virtual resource allocation of cloud computing. J. Electron. Inf. Technol. **40**(1), 25–34 (2018)
15. Guo, Y., Liu, F., Xiao, N., Chen, Z.: Task-based resource allocation bid in edge computing micro datacenter. Comput. Mater. Continua **58**, 777–792 (2019)
16. Zhang, J., Xie, N., Zhang, X.: Machine learning based resource allocation of cloud computing in auction. Comput. Mater. Continua **56**, 123–135 (2018)
17. Nisan, E.B.N., Roughgarden, T., Tardos, E., Vazirani, V.V.: Algorithmic game theory. Commun. ACM **53**(7), 78–86 (2009)
18. Mashayekhy, L., Fisher, N., Grosu, D.: Truthful mechanisms for competitive reward-based scheduling. IEEE Trans. Comput. **65**(7), 2299–2312 (2016)
19. Azar, Y., Kalp-shaltiel, I., Lucier, B., Menache, I., Naor, J., Yaniv, J.: Truthful online scheduling with commitments. In: Sixteenth ACM Conference on Economics and Computation, pp. 715–732 (2015)
20. Porter, R.: Mechanism design for online real-time scheduling. In: Proceedings of the 5th ACM Conference on Electronic Commerce, pp. 61–70. ACM (2004)
21. Parkes, D.C., Singh, S.: An MDP-based approach to online mechanism design. In: International Conference on Neural Information Processing Systems, pp. 791–798 (2003)
22. Stein, S., Gerding, E., Robu, V., Jennings, N.R.: A model-based online mechanism with pre-commitment and its application to electric vehicle charging. In: International Conference on Autonomous Agents and Multiagent Systems (2012)
23. Sandholm, T.: Automated mechanism design: a new application area for search algorithms. In: International Conference on Principles and Practice of Constraint Programming, pp. 19–36 (2003)
24. Yamada, K.: Layered learning in multiagent systems: a winning approach to soccer. Syst. Control Inf. **45**, 285 (2001)
25. Moore, A.W.: Variable resolution dynamic programming: efficiently learning action maps in multivariate real-valued state-spaces. In: Machine Learning Proceedings, vol. 1991, pp. 333–337. Elsevier (1991)
26. Jiang, G., Wu, C.: Learning to control an inverted pendulum using q-learning and neural networks. Acta Automatica Sinica **24**, 666–669 (1998)
27. Crites, R.H., Barto, A.G.: Elevator group control using multiple reinforcement learning agents. Mach. Learn. **33**(2–3), 235–262 (1998)
28. Hernandez-Leal, P., Kaisers, M., Baarslag, T., de Cote, E.M.: A survey of learning in multiagent environments: dealing with non-stationarity. arXiv preprint arXiv:1707.09183 (2017)
29. Mao, H., Alizadeh, M., Menache, I., Kandula, S.: Resource management with deep reinforcement learning. In: Proceedings of the 15th ACM Workshop on Hot Topics in Networks, pp. 50–56. ACM (2016)
30. Mashayekhy, L., Nejad, M.M., Grosu, D., Vasilakos, A.V.: An online mechanism for resource allocation and pricing in clouds. IEEE Trans. Comput. **65**(4), 1172–1184 (2015)

Fuzzy Multi-objective Requirements for NRP Based on Particle Swarm Optimization

Yachuan Zhang, Hao Li, Rongjing Bu, Chenming Song, Tao Li, Yan Kang(✉), and Tie Chen

YunNan University, Kunming 650500, China
1466463308@qq.com

Abstract. In software engineering, the development of software products raises a new set of development requirements each time. Considering the interaction between requirements, how to select an optimal subset of requirements becomes an important problem. In this paper, a fast method of requirements optimization is proposed, which can select an optimal subset from the next release of product development requirements under the limitation of user satisfactions and cost. The multiple requirements in this paper are limited by user satisfaction and cost. We mainly make the following contributions: (1) We define this problem as multi-objective problem for optimization. (2) Then particle swarm optimization (PSO) algorithm is used to adjust the convergence parameters of multiple object to search the optimal solution quickly. (3) Finally, the results of the algorithm are evaluated by using NDS number and time of multi-objective problem through fuzzy simulation data. Experimental results show that the algorithm is efficient and reliable, and can help developers make reasonable decisions.

Keywords: Software engineering · Software requirement · PSO · Multi-objective optimization

1 Introduction

Now days, software engineering is a critical job in software development with help of which person can clearly get and convey the information about software works. Boehm [1] defined software engineering is the practical application of scientific knowledge in the design and construction of computer programs and the associated documentation required to develop, operate, and maintain them.

Software requirement is a most importance phase during software life cycle. The process to gather the software requirements from client, analyze and document them is known as requirement engineering. The effective search results of requirements from the client is a big challenge for requirement engineer due to poor communication, which leads toward a low satisfaction rate and a high cost for a project. So, the requirement gathering stage is the most important and challenging stage of the Software Development Life Cycle (SDLC) [2].

X. Sun et al. (Eds.): ICAIS 2020, LNCS 12240, pp. 143–155, 2020.
https://doi.org/10.1007/978-3-030-57881-7_13

The problem of selecting an optimal subset of requirements that is known as the next release problem (NRP) was first introduced by Bagnall et al. [3]. In NRP, the target may use some optimization to select an optimal subset of requirement that can include highest satisfactions to client and the lowest cost. It is required to satisfy the requirements of some elements of the soft and hard constraints in a certain time and space. Meanwhile, the NRP problem is considered an NP-hard problem, but it cannot be solved efficiently by using exact optimization algorithms for big data problem.

The most important problem for a software company is to determine which features should be included in the next release of their products, in such a way that the highest possible number of customers get satisfied while entailing the minimum cost for the company. They need to not only consider client's satisfactions about software product, but also control the cost of development. Since minimizing the total cost of including new features into a software package and maximizing the total satisfaction of customers are contradictory objectives, the problem has a multi-objective nature. The sorting and selection problem of candidate software requirements is expressed as a series of feature subset selection problems applicable to software engineering based on search [4].

The developer department should portion a parameter as the development cost and another one as the rate of satisfactions in every release. Considering the presence of clients with different views and desires, the allocation of a crisp value as the satisfaction for each requirement is not logical [5–7]. It is unfeasible in actual application to use crisp number, because technical problems when the implementation requires the exact value of the allocation of the cost is not standard planned development requirements. In this situation, the developer team allocation of values of use fuzzy type of for satisfaction and cost for each requirement is more logical and possible.

In order to overcome above problem, we propose a new algorithm to solve software engineering requirement. We define this problem as fuzzy multi-objective problem for optimization—fuzzy multi-objective requirements based on particle swarm optimization (FMOPSO). Then particle swarm optimization (PSO) algorithm is used to adjust the convergence parameters of multiple object to search the optimal solution quickly. Finally, the results of the algorithm are evaluated by using three criteria of multi-objective problem through simulation data. Experimental results show that the algorithm is efficient and reliable, and can help developers make reasonable decisions.

2 Related Work

2.1 Software Engineering and Next Release Problem (NRP)

As an independent discipline, software engineering has developed for more than 40 years [8, 9]. In order to solve the problem of software crisis, the concept of "software engineering" was first proposed in 1968. In the era of rapid development of intelligence and information, software engineering technology has become an indispensable core component, and the market demand for software will only grow. Therefore, the accurate grasp of software development requirements is of vital importance to the future development of the software field [10].

With the rapid development of science and coding technology, the development trend of the software industry has become overwhelming, how to choose the best combination of requirements to meet as many users as possible has become the first difficulty for software developer to solve, which is called the next release problem (NRP). In large software projects, determining the requirements schedule for NRP is an important issue in requirements engineering [11].

2.2 Multi-objective Optimization

The optimization problem is one of the main problems in engineering practice and scientific research. Evolutionary multi-objective optimization is one of the research hotspots in the field of evolutionary computing.

In 1967, Rosenberg [12] proposed the use of evolutionary search to deal with multi-objective optimization problem. In 1975, Holland [13] put forward the genetic algorithm, which has important guiding significance for the research of subsequent evolutionary multi-objective optimization algorithms. It has attracted extensive attention from many scholars and a large number of research results [14, 15]. In recent years, ant colony algorithm, artificial immune system, distribution estimation algorithm and co-evolutionary algorithm have been used to solve multi-objective optimization problems. Ant colony algorithm [16] is a new bionic optimization algorithm inspired by the real ant foraging behavior in nature in the early 1990s by the Italian scholar Dorigo [17]. Its distributed computing mechanism and good integration with other algorithms make it rapidly developed and applied in a short time. However, there are still some shortcomings in the ant colony algorithm to be improved. The selection of key parameters and the setting of initial values in the algorithm are empirical and lack of scientific theoretical demonstration. Artificial immune system [18] is inspired by immunology, simulation of immunological functions, principles and models to solve complex problems. It has been successfully used in anomaly detection, computer security, data mining, optimization, etc., and some with artificial immune system algorithms for solving multi-objective optimization problems appear one after another. The non-dominated domain immune algorithm is one of the representative algorithms, this method has great advantage in solving high-dimensional multi-objective optimization problems. Especially, multi-objective optimization have many chance to solve practical problem in recent year [19, 20].

Due to the above algorithms have some shortcomings, this paper from the perspective of particle swarm optimization algorithm (PSO) [21] to study the multi-objective optimization algorithm. PSO is a swarm intelligent optimization algorithm proposed by Kennedy and Eberhart in 1995 [22]. It regards each individual in the population as a particle without volume and mass in the search space. These particles fly at a certain speed in the search space, and their speed is dynamically adjusted according to their own flight experience and the flight experience of the entire population. The process is simple and easy to implement, the parameters of the algorithm are simple and do not need to be adjusted by duplicating. PSO has already been rapidly applied in some original applications of genetic algorithm since it was proposed.

3 Fuzzy Multi-objective Requirement

It is important to choose the optimal subset of the proposed requirements for developer teams, which make requirements maximize client satisfactions and minimize company's cost and time. A good subset of requirements can help soft engineers make the right planning. In large projects with a great number of requirements, it is very difficult to select the optimal subset [23–28]. Also, optimization problem has multiple different requirement object. In this subset we will discuss some concept of multi-objective optimization include non-dominated, Pareto front and crowding distance.

Single-object problems only need to optimize one object. For multi-object problem, we assigned different weight to each object to main objective problem (F) as Eq. 1.

$$S = \min F(x) = \left[\alpha \times f_1(x) + \beta \times f_2(x) + \cdots \omega \times f_n(x)\right] \quad (1)$$

Where $f_1(x), f_2(x), \ldots, f_n(x)$ designate different n object fitness functions to make the minimum of $F(x)$, α, β, $\ldots$, ω denote the variable constraint of objects.

When the objective function is in a state of conflict, there is no optimal solution that can make all objective functions reach the maximum or minimum at the same time. But there is one set of solutions, none of which has an advantage over the others. This set of solutions is called a non-dominated solution (NDS).

For a nontrivial multi-objective optimization problem, there does not exist a single solution that simultaneously optimizes each objective. In that case, the objective functions are said to be conflicting, and there exists a (possibly infinite) number of Pareto optimal solutions. A solution is called nondominated, Pareto optimal, Pareto efficient or noninferior, if none of the objective functions can be improved in value without degrading some of the other objective values.

For example, there is a real problem which has 2 objectives; and $x = [x1, x2]$ and $y = [y1, y2]$ are two solution for this problem. Solution x dominates solution y only if for all x is better or equal to y, and x is exactly better than y in at least one objective.

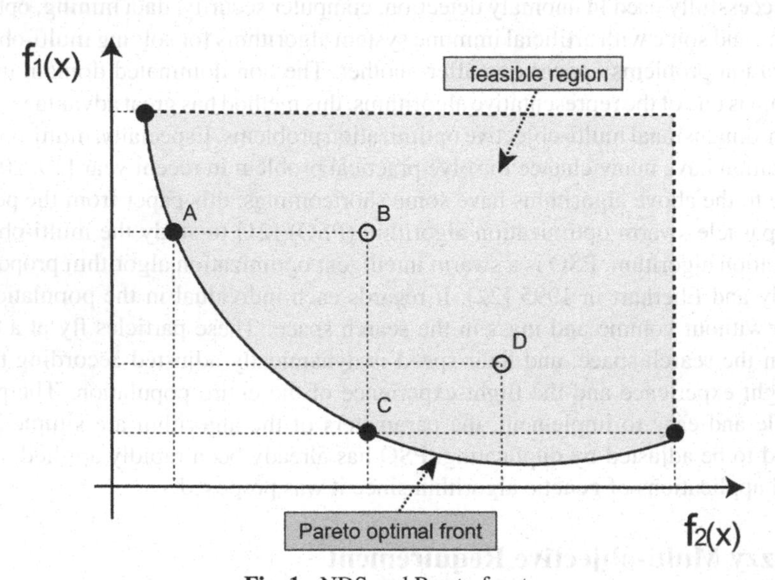

Fig. 1. NDS and Pareto front

The Fig. 1 shows the Pareto optimal select process, where solid points is the Pareto optimal solution based on $f1(x)$ and $f2(x)$ constraint. All Pareto-optimal solution constitute a Pareto optimal set. So, the value of objective function corresponding to the Pareto optimal solution is the pareto optimal front. For two objects problem, the Pareto optimal front is a line. But for multiple objects, the Pareto optimal front is usually a hypersurface. The non-dominated solutions are archived external and crowding distance is used to maintain diversity. Then the evaluation function can be defined to further evaluate the Pareto set and obtain the best satisfactory solution in Pareto solution set.

4 PSO Optimization for NRP

4.1 PSO

Particle swarm optimization (PSO), an evolutionary computing technique derived from studying the behavior of flocks of birds, it was first proposed by Dr Barnhart and Dr Kennedy in 1995. The relationships between particle swarm optimization and both artificial life and genetic algorithms are described. This is a population intelligence optimization algorithm. Based on the concept of group and evolution, PSO algorithm uses the synergy and competition among particles to finally get the optimal solution of search space. Algorithm process are as follow (Fig. 2, Table 1):

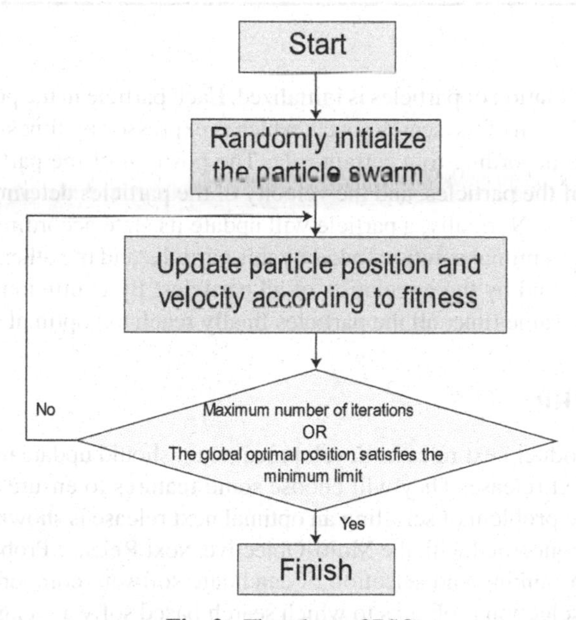

Fig. 2. Flowchart of PSO

Table 1. Algorithm of PSO

```
Algorithm : PSO
1   for each particle i
2   |   Initialize velocity V_i and position S_i for particle i;
3   |   Evaluate particle i and set pBest = X_i;
4   end
5   gBest = min{pBest};
6   while not stop
7   |   for i = 1 to N
8   |   |   Update the V_i and S_i;
9   |   |   if fit(X_i) < fit(pBest)
10  |   |       pBest = X_i
11  |   |   if fit(pBest) < fit(gBest)
12  |   |       gBest = pBest
13  |   end
14  end
15  print gBest
```

Firstly, the population of particles is initialized. Each particle in the population represents a feasible solution of the search space, which is expressed by fitness function. Then the particles move according to a certain rule. The position of the particles represents the performance of the particles, and the velocity of the particles determines the motion trend of the particles. Normally, a particle will update its state according to two values, one is the historical optimal solution found by this particle, and the other is the historical optimal solution found by the population of all particles. By continuously updating all the particles at the same time, all the particles finally reach the optimal solution.

4.2 Deal with NRP

In the software product next release, development team should update user requirement feature in their next release. They will choose some features to ensure the demands of their client. So, the problem of selecting an optimal next release is shown to be NP-hard. In this paper, we concerned with the Multi-Objective Next Release Problem (MONRP). It formulates both ranking and selection of candidate software components as a series of feature subset selection problems to which search based software engineering can be applied.

Using the non-dominated solution archive method and storing the suitable solutions found during each iteration in the archive. And the FMOPSO pseudo-code is shown to solve the problem of NRP in the algorithm 2 (Table 2).

Table 2. Algorithm of FMOPSO

```
Algorithm : FMOPSO
Input: Particles
Output: NDS-archive
1   Initialize parameters;
2   Particles = generate initial molecules;
3   Initialize velocity V_i and position P_i for each particle;
4   Evaluate pBest, gBest of particles;
5   gBest = min{pBest};
6   NDS_archive = empty;
6   while not stop
7     for i = 1 to N
8     |   Update the V_i and P_i of particles i;
9     |   Evaluate particle i
10    |   if fitness(X_i) < fitness(pBest)
11    |       pBest = X_i
12    end for
13  | Correct pBest of particles
14  | New_gBest = getBestBNSGA2(particles)
15  | if fitness(New_gBest) < fitness(gBest)
16  |     gBest = New_gBest
17  | front = getParetoFront(pBest)
18  | NDS_archive = updateNDS_archive(Front)
14  end while
```

5 Experiments

5.1 Dataset

In this paper, we use two different simulation datasets. Each data set has three dimensions, *r* represented requirement; *a* represented fuzzy satisfaction value obtained by survey, investigation and so on; *c* represented fuzzy cost value of requirements by the experts. For two datasets, all value of *a* and *c* are test after fuzzy treatment. In both datasets, the fuzzy number of how satisfying and cost each item requires. This dataset is much more complex than previous datasets. In addition, each requirement is scored on a scale of 1 to 10. For each software company, the client has a different level of importance. Customer importance levels can be considered clear or fuzzy numbers (Tables 3 and 4).

Above two datasets, there are four kinds of interrelationships between requirements that govern the next release of the problem. For example:

(1) $r4 \Longrightarrow r10$ is implication. If the requirement $r4$ does not develop, the requirement 10 still not develops.
(2) $r12 \oplus r15$ is combination. The requirements $r12$ and $r15$ must happen together or not happen.
(3) $r11 \otimes r30$ is exclusion. The requirements $r11$ and $r30$ can't develop at the same time.

(4) The last one is modification. However, the generally overlooked interaction have applied in our simulation datasets. And researcher divide this interaction into two distinct interaction as follows:

A. $r3(\mathbf{60\%}) \circledast r10$ The requirement $r3$ and $r10$ are developed, only 60% of satisfaction $r10$ is applied.

B. $r11(\mathbf{40\%}) \odot r6$ The requirement $r11$ develops and $r6$ does not, only 40% of satisfaction $r6$ is applied.

5.2 Analysis of Results

The following table shows the parameters of the experiment (Tables 5 and 6, Fig. 3).

We show the four situations about random experiments in dataset1. In the iteration, the results have different NDS search process shown in Fig. 4. On the overall trend, the image is increasing, and the image will converge in a certain range. But there is always a jitter in the iteration. And the number of NDS found in each iteration is also different in random experiments. The sub-Fig(a) is reasonable result in the experimental results is that the curve does not fall into the local optimum, and the number of NDS is also relatively average. The sub-Fig(c) the number of NDS is too much than average. The sub-Fig(c) is similar to sub-Fig(a) and suitable to our expect. The sub-Fig(d) may fell into the local optimum.

Table 3. Dataset1

	r1	*r2*	*r3*	*r4*	*r5*	*r6*	*r7*	*r8*	*r9*	*r10*	*r11*	*r12*
a_{1i}	1	2	2	1	2	2	1	3	2	1	2	1
a_{2i}	4	3	4	2	3	3	3	4	4	2	3	2
a_{3i}	5	4	5	4	5	4	5	5	5	3	4	4
c_{1i}	2	1	2	1	3	2	1	2	1	1	2	1
c_{2i}	3	3	3	2	4	3	2	3	3	3	3	3
c_{3i}	5	4	4	3	5	5	3	4	4	4	5	5
	r13	***r14***	***r15***	***r16***	***r17***	***r18***	***r19***	***r20***	***r21***	***r22***	***r23***	***r24***
a_{1i}	1	2	1	1	1	1	2	3	1	1	2	2
a_{2i}	2	3	2	2	3	3	3	4	4	4	4	3
a_{3i}	3	5	3	5	5	4	5	5	5	5	5	4
c_{1i}	2	1	2	1	3	1	3	1	1	2	2	2
c_{2i}	3	2	3	2	4	2	4	3	2	3	4	3
c_{3i}	4	3	5	4	5	4	5	5	4	4	5	4

$r4 \Rightarrow r10; r11 \Rightarrow r17; r14 \Rightarrow r17; r12 \oplus r15; r22 \oplus r19; r2 \oplus r9;\ r5 \oplus r24$

$r3(\mathbf{60\%}) \circledast r10; r4(\mathbf{70\%}) \circledast r7; r1(\mathbf{50\%}) \circledast r15; r11(\mathbf{40\%}) \odot r6; r21(\mathbf{25\%}) \odot r22$

Table 4. Dataset2

	r1	*r2*	*r3*	*r4*	*r5*	*r6*	*r7*	*r8*	*r9*	*r10*	*r11*	*r12*
a_{1i}	4	2	1	5	7	1	2	3	2	3	3	6
a_{2i}	7	6	4	7	8	3	5	6	7	4	5	8
a_{3i}	8	8	6	9	9	8	7	9	9	5	8	9
c_{1i}	5	2	1	1	5	3	2	3	6	6	3	4
c_{2i}	7	3	2	4	6	4	4	6	7	7	6	5
c_{3i}	9	4	3	6	8	6	6	7	9	8	9	6
	r13	*r14*	*r15*	*r16*	*r17*	*r18*	*r19*	*r20*	*r21*	*r22*	*r23*	*r24*
a_{1i}	3	4	2	3	1	3	2	4	6	7	1	1
a_{2i}	7	6	3	5	2	6	7	5	7	8	3	3
a_{3i}	8	8	4	8	3	9	9	6	9	9	8	5
c_{1i}	4	3	6	3	1	3	6	4	3	5	3	3
c_{2i}	5	5	7	6	2	6	7	5	4	6	4	5
c_{3i}	7	6	8	9	3	7	9	6	6	8	6	7
	r25	*r26*	*r27*	*r28*	*r29*	*r30*	*r31*	*r32*	*r33*	*r34*	*r35*	*r36*
a_{1i}	2	1	7	2	4	2	3	7	1	2	3	2
a_{2i}	6	4	8	5	5	3	5	8	3	4	6	7
a_{3i}	8	6	9	6	9	4	8	9	8	7	9	9
c_{1i}	2	1	6	4	6	6	3	5	3	2	3	6
c_{2i}	3	2	7	7	8	7	6	6	4	3	6	7
c_{3i}	4	3	8	9	9	8	9	8	6	4	7	9
	r37	*r38*	*r39*	*r40*	*r41*	*r42*	*r43*	*r44*	*r45*	*r46*	*r47*	*r48*
a_{1i}	2	1	5	3	2	2	1	4	2	3	7	6
a_{2i}	4	4	7	6	7	6	4	5	3	5	8	7
a_{3i}	6	6	9	9	9	8	6	8	4	8	9	9
c_{1i}	3	1	1	3	6	2	1	3	6	3	7	1
c_{2i}	5	2	4	6	7	3	2	4	7	6	8	2
c_{3i}	7	3	6	7	9	4	3	6	8	9	9	4
	r49	*r50*	*r51*	*r52*	*r53*	*r54*	*r55*	*r56*	*r57*	*r58*	*r59*	*r60*
a_{1i}	4	3	5	3	7	1	1	3	2	1	3	2
a_{2i}	5	4	6	5	8	3	2	5	6	4	6	7
a_{3i}	9	6	8	8	9	8	4	7	8	6	9	9
c_{1i}	3	2	6	3	5	3	1	1	2	1	3	6
c_{2i}	5	5	7	6	6	4	2	3	3	2	6	7
c_{3i}	6	7	8	9	8	6	4	5	4	3	7	9
	r61	*r62*	*r63*	*r64*	*r65*	*r66*	*r67*	*r68*	*r69*	*r70*	*r71*	*r72*
a_{1i}	1	7	1	1	5	2	3	7	3	2	2	1
a_{2i}	4	8	3	4	7	3	5	8	6	7	6	4
a_{3i}	8	9	8	6	9	4	8	9	9	9	8	6
c_{1i}	3	5	3	1	1	6	3	4	3	6	2	1
c_{2i}	5	6	4	2	4	7	6	5	6	7	3	2
c_{3i}	7	8	6	3	6	8	9	6	7	9	3	3

$r34 \Rightarrow r20; r21 \Rightarrow r47; r34 \Rightarrow r37;; r40 \Rightarrow r57$

$r12 \oplus r25; r32 \oplus r29; r43 \oplus r39;\ r52 \oplus r69$

$r11 \otimes r30; r23 \otimes r24; r55 \otimes r51;\ r56 \otimes r72$

$r33(\mathbf{50\%}) \circledast r38; r61(\mathbf{60\%}) \circledast r37; r20(\mathbf{55\%}) \circledast r51; r72(\mathbf{65\%}) \odot r64;$

$r17(\mathbf{30\%}) \odot r38; r30(\mathbf{40\%}) \odot r27\ ; r45(\mathbf{15\%}) \odot r66; r42(\mathbf{20\%}) \odot r64$

Table 5. Results of dataset1

Parameters	Values
$R_{min0\%}(cost, satifaction)$	(0,0)
$R_{max100\%}(cost, satifaction)$	(230,775)
$W_1 = clientsweight$	(1,2.8,4)
Initial particle	200
Iterations	500

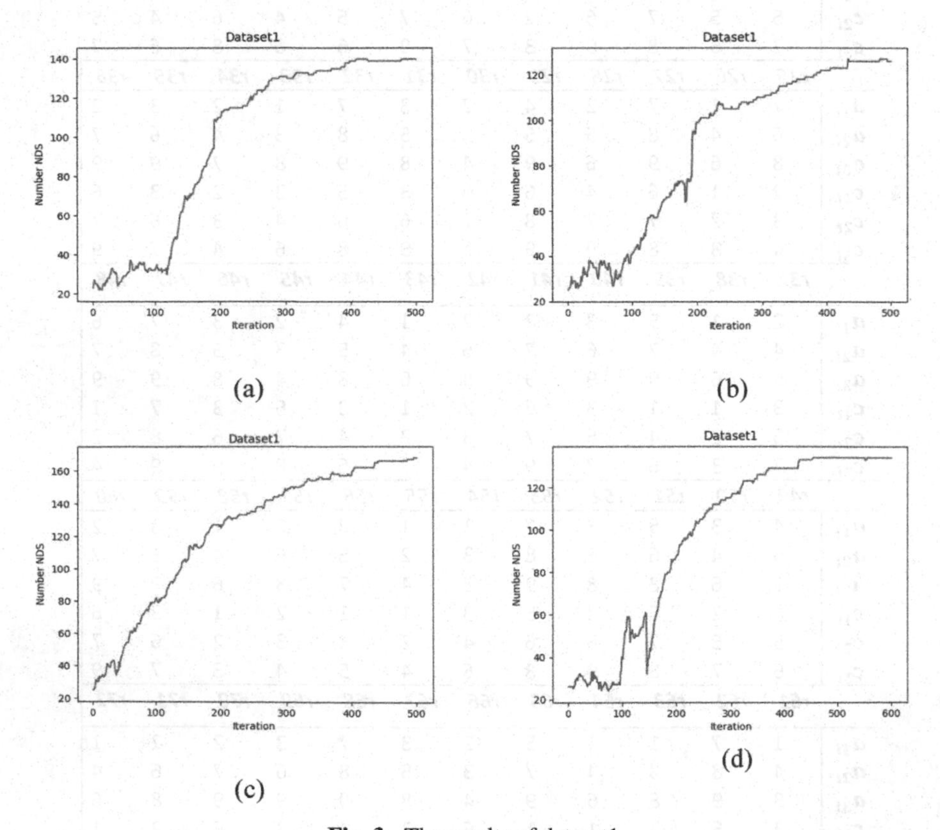

Fig. 3. The results of dataset1

We show the four situations about random experiments in dataset2. In the iteration, the results have different NDS search process shown in Fig. 4. On the overall trend, the image is increasing, and the image will converge in a certain range. Also, there is always a jitter in the iteration. And the number of NDS found in each iteration is also different in

Table 6. Results of dataset2

Parameters	Values
$R_{\min 0\%}(cost, satifaction)$	(0,0)
$R_{\max 100\%}(cost, satifaction)$	(1130,4293)
$W_1 = clientsweight$	(1,2.6,5)
Initial particle	800
Iterations	500

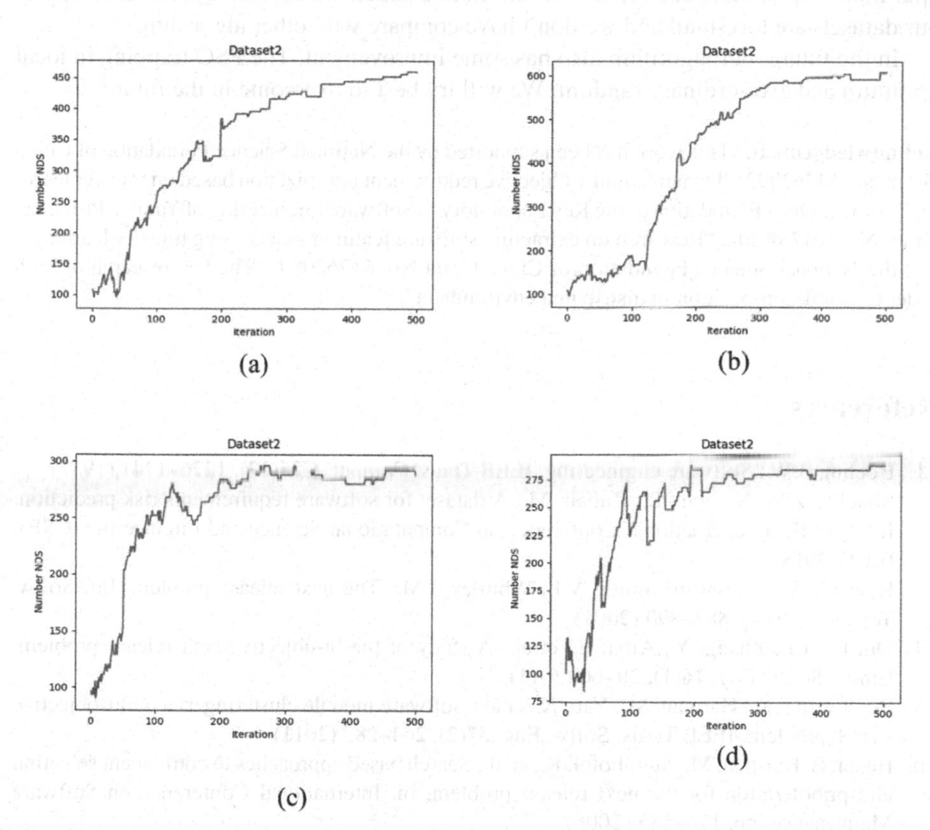

Fig. 4. The results of dataset2

random experiments. Because of big data of dataset, the sub-Fig(a) is reasonable result in the experimental results is that the curve has fell into the local optimum sometimes, and the number of NDS is also relatively average. The sub-Fig(b) and sub-Fig(c) show of sharp NDS fluctuation. The sub-Fig(d) the number of NDS is too much than average.

It can be seen from this that particles can expand the search space and have a faster convergence rate. However, due to the lack of local search, it is easier for complex problems to fall into local optimization than standard PSO. Because the number of NDS is different every test and we have different index to other experiment, this experiment doesn't have comparisons.

6 Conclusion

In this paper, we proposed a method for next release problem—fuzzy multi-objective requirements based on particle swarm optimization. The algorithm is efficient and reliable, and can help developers make reasonable decisions than experts. It can help developer make the suitable and reliable result. In this paper, we also have some advantages, our datasets are too small and we don't have compare with other algorithm.

In the future, our algorithm also has some improvement. The PSO trapping in local optimum and extraordinary random. We will try best to overcome in the future.

Acknowledgement. This work has been supported by the National Science Foundation of China Grant No. 61762092, "Dynamic multi-objective requirement optimization based on transfer learning", and the Open Foundation of the Key Laboratory in Software Engineering of Yunnan Province, Grant No. 2017SE204, "Research on extracting software feature models using transfer learning", and the National Science Foundation of China Grant No. 61762089, "The key research of high order tensor decomposition in distributed environment".

References

1. Boehm, B.W.: Software engineering. IEEE Trans. Comput. **C25**(12), 1226–1241 (1977)
2. Shaukat, Z.S., Naseem, R., Zubair, M.: A dataset for software requirements risk prediction. In: 2018 IEEE International Conference on Computational Science and Engineering (CSE). IEEE (2018)
3. Bagnall, A.J., Rayward-Smith, V.J., Whittley, I.M.: The next release problem. Inf. Softw. Technol. **43**(14), 883–890 (2001)
4. Durillo, J.J., Zhang, Y., Alba, E., et al.: A study of the bi-objective next release problem. Empir. Softw. Eng. **16**(1), 29–60 (2011)
5. Praditwong, K., Harman, M., Yao, X., et al.: Software module clustering as a multi-objective search problem. IEEE Trans. Softw. Eng. **37**(2), 264–282 (2011)
6. Baker, P., Harman, M., Steinhofel, K., et al.: Search based approaches to component selection and prioritization for the next release problem. In: International Conference on Software Maintenance, pp. 176–185 (2006)
7. Sun, Z., et al.: Designing and optimization of fuzzy sliding mode controller for nonlinear systems. Comput. Mater. Continua **61**(1), 119–128 (2019)
8. Wenkai, C.: Status and development trend in software engineering. Inf. Rec. Mater. **6**, 6–8 (2018)
9. Fan, X., Zhou, T.: Status and future of software engineering industry development strategy. Comput. Program. Skills Maint. **406** (04), 57–59 (2019)
10. Yanping, L.: Investigating the technical requirements for software development. Mod. Vocat. Educ. **36**, 210–211 (2017)

11. Chen, J.: Research versioning software searches for the next technology and implementation. Nanjing University of Posts and Telecommunications (2018)
12. Rosenberg, R.S.: Simulation of genetic populations with biochemical properties. Ph.D. Thesis. University of Michigan, Michigan (1967)
13. Holland, J.H.: Adaptation in Natural and Artificial Systems. The University of Michigan Press, Michigan (1975)
14. Goldberg, D.E.: Genetic Algorithm for Search, Optimization, and Machine Learning. Addison-Wesley Longman Pub lishing Co., Inc., Boston (1989)
15. Fonaeca, C.M., Fleming, P.J.: Genetic algorithm for multiobjective optimization: formulation, discussion and generation. In: Forrest, S., (ed.) Proceedings of the 5th International Conference on Genetic Algorithms, pp. 416–423. Morgan Kauffman Publishers, San Mateo (1993)
16. Zuo, L., et al.: A multi-objective optimization scheduling method based on the ant colony algorithm in cloud computing. IEEE Access **3**, 2687–2699 (2015)
17. Dorigo, M., Stützle, T.: Ant colony optimization: overview and recent advances. In: Gendreau, M., Potvin, J.-Y. (eds.) Handbook of Metaheuristics. ISORMS, vol. 272, pp. 311–351. Springer, Cham (2019). https://doi.org/10.1007/978-3-319-91086-4_10
18. Dasgupta, D. (ed.): Artificial Immune Systems and Their Applications. Springer, Heidelberg (2012)
19. Liu, Z., Xiang, B., Yuqing Song, H.L., Liu, Q.: An improved unsupervised image segmentation method based on multi-objective particle, swarm optimization clustering algorithm. Comput. Mater. Continua **58**(2), 451–461 (2019)
20. Liu, W., Tang, Y., Yang, F., Dou, Y., Wang, J.: A multi-objective decision-making approach for the optimal location of electric vehicle charging facilities. Comput. Mater. Continua **60**(2), 813–834 (2019)
21. Kennedy, J.: Swarm intelligence. In: Swarm intelligence. Morgan Kaufmann Publishers Inc. (2001)
22. Kennedy, J., Eberhart, R.: Particle swarm optimization. In: Proceedings of ICNN 1995 - International Conference on Neural Networks. IEEE (1995)
23. Alrezaamiri, H., Ebrahimnejad, A., Motameni, H.: Software requirement optimization using a fuzzy artificial chemical reaction optimization algorithm. Soft Comput. **23**(20), 9979–9994 (2018). https://doi.org/10.1007/s00500-018-3553-7
24. Ebrahimnejad, A., Tavana, M., Alrezaamiri, H.: A novel artificial bee colony algorithm for shortest path problems with fuzzy arc weights. Measurement **93**, 48–56 (2016)
25. Tajdin, A., Mahdavi, I., Mahdavi-Amiri, N., Sadeghpour-Gildeh, B.: Computing a fuzzy shortest path in a network with mixed fuzzy lengths using a-cut. Comput. Math Appl. **60**(2), 989–1002 (2010)
26. Hassanzadeh, R., Mahdavi, I., Mahdavi-Amiri, N., Tajdin, A.: A genetic algorithm for solving fuzzy shortest path problems with mixed fuzzy arc lengths. Math. Comp. Model. **57**(1–2), 84–99 (2013)
27. Mahdavi, I., Tajdin, A., Hassanzadeh, R., et al.: Genetic algorithm for solving fuzzy shortest path problem in a network with mixed fuzzy arc lengths. In: AIP Conference Proceedings, vol. 1337, p. 265 (2011)
28. Alrezaamiri, H., Ebrahimnejad, A., Motameni, H.: Software requirement optimization using a fuzzy artificial chemical reaction optimization algorithm. Soft Comput. - Fusion Found. Methodol. Appl. **23**, 9979–9994 (2019)

A Clustering Algorithm for Wireless Sensor Networks Using Geographic Distribution Information and Genetic Algorithms

Yu Song[1,2](✉), Zhigui Liu[1], and He Xiao[1,3]

[1] School of Information Engineering, South West University of Science and Technology, Mianyang 621010, Sichuan, China
3958613@qq.com

[2] Department of Network Information Management Center, Sichuan University of Science and Engineering, Zigong 643000, China

[3] School of Computer Science, Sichuan University of Science and Engineering, Zigong 643000, China

Abstract. WSN consists of a large number of micro sensor nodes with limited resources. The limited battery resources of these nodes have become an important bottleneck to the development of WSN. In order to improve the energy efficiency and prolong the network life cycle, we propose a clustering method GDGA based on improved genetic algorithm. In this method, we divide the sensor area into two parts: the first part is that the distance from the node to BS is less than the transmission threshold of the node. For the nodes in this area, we do not cluster but directly transmit the data to BS. The part beyond the threshold of BS is divided into the second region. The nodes in this region will be clustered using the improved genetic algorithm according to the characteristics of node distribution. Simulation results show that compared with other four protocols, GDGA has the highest energy efficiency, lower average energy consumption of cluster head and longer life cycle of the whole network.

Keywords: Genetic algorithm · Clustering · CH selection · Clustering beyond threshold

1 Introduction

WSN is composed of many miniature and cheap sensor nodes. These sensor nodes are usually deployed in the monitoring area and form a self-organizing network through wireless communication. Sensor nodes can sense, collect and process data and transmit the data to the BS [1]. In recent years, WSN has been widely used in military, environmental monitoring and disaster management [2]. The existing WSN mainly relies on a limited battery supply and its life cycle depends entirely on its battery power. Because WSNs are often deployed in remote or harsh environments, these batteries are almost impossible to replace. Therefore, the energy optimization of WSN has been a hot topic of research.

X. Sun et al. (Eds.): ICAIS 2020, LNCS 12240, pp. 156–165, 2020.
https://doi.org/10.1007/978-3-030-57881-7_14

The most popular solution among WSN's many energy-saving solutions is: clustering technology. This technique divides sensor nodes into multiple clusters. Each cluster in the network has a unique cluster head node. Ordinary nodes will send the perceived data to the cluster head node [3]. Then, the cluster head node sends the data to the BS after collecting and aggregating data. Clustering technology has many benefits, including scalability, energy efficiency, and reduced routing latency. In the past research, many clustering methods have been proposed. The common goal of these clustering technologies is to save energy. The operations in a clustering protocol are usually divided into three phases: CH selection, cluster formation, and data transmission. The main part of each method is the CH selection algorithm that defines the energy efficiency of the network [4].

1.1 Clustering

LEACH protocol-full name is "Low Energy Adaptive Clustering Hierarchy". It was proposed by Heinzelman et al. In 2000. This protocol is WSN's first dynamic clustering protocol and is considered the basis of other advanced clustering protocols in WSN. Generally, it is a layered, random, and distributed single-hop protocol. LEACH is executed continuously in rounds. Each round can be divided into two phases: the establishment phase of the cluster and the stable phase of transmitting data. The cluster establishment process can be divided into 4 stages: selecting the cluster head node, broadcasting the cluster head node, establishing the cluster head node, and generating a scheduling mechanism [5].

TEEN (Threshold sensitive Energy Efficient Sensor Network protocol) is different from all nodes in LEACH that always have data to transmit. TEEN is specifically designed for applications that should send data to the BS when certain events occur. TEEN is the first hierarchical WSN routing protocol for responsive networks. TEEN works basically the same way as LEACH, except that after each re clustering, the cluster head node needs to broadcast the following three parameters to the members in the cluster:

1) Characteristic value: The physical parameter of the data that the user cares about.
2) Hard Threshold (HT): The absolute threshold of the characteristic value of the monitoring data [6]. When the feature value monitored by the node exceeds this threshold, the transmitter is started to report this value to the cluster head node.
3) Soft Threshold (ST): Monitors a small range of characteristic values to change the threshold to trigger the node to start the transmitter to report data to the cluster head.

When the characteristic value of the node's sensing data exceeds its hard threshold for the first time, the node starts the transmitter to send the sensing data, and this characteristic value is also stored in the node's memory.

The node will start the next data transfer if and only if the following two conditions are met:

1) The feature value currently sensed is greater than the hard threshold.
2) The difference between the current feature value and the previous feature value is greater than or equal to the soft threshold.

However, there are some problems in TEEN. First, users will not get feedback from the area of interest until the threshold is reached. As a result, some nodes may die without the user being aware of their deaths because it has not received feedback.

1.2 Genetic Algorithm

Genetic algorithm is a kind of randomized search method that evolved from the evolutionary laws of the biological world (the genetic mechanism of survival of the fittest). Compared with traditional search algorithms, genetic algorithms have the following characteristics:

1) Genetic does not have too many mathematical requirements for the optimization problem to be solved. Genetic algorithms can handle any form of objective function and constraints, whether linear or non-linear, discrete or continuous, or even mixed search space.
2) The ergodicity of evolutionary operators makes genetic algorithms very efficient for global searches in the sense of probability, while traditional optimization methods transfer to better points by comparing neighboring points, thereby achieving a local search process that converges;
3) Using probabilistic optimization method, it can automatically obtain and guide the optimized search space, and adaptively adjust the search direction.
4) Genetic algorithms can provide great flexibility to mix domain-independent heuristics for various special problems, thereby ensuring the effectiveness of the algorithm [7].

2 System Model

2.1 Energy Model

The energy consumption of a classic wireless sensor network is mainly composed of two parts: information reception energy consumption and information transmission energy consumption.

$$E_{(Si)}(L, d) = \begin{cases} L \times E_{elec} + L \times E_{fs} \times d^2, (d < d_0) \\ L \times E_{elec} + L \times E_{mp} \times d^4, (d \geq d_0) \end{cases} \quad (1)$$

$$E_{(Ri)} = L \times E_{elec} \quad (2)$$

$$d_0 = \sqrt{E_{fs}/E_{mp}} \quad (3)$$

Where L is the total amount of data to be transmit or receive. E_{elec} is the energy consumed to send one bit of data; d is the distance between two sensor nodes. The signal energy consumption model is divided into two categories according to distance: free space model and multi-path attenuation model [8]. When the transmission distance is less than the threshold d_0, the free space energy consumption model is adopted in the communication mode, while the multi-path attenuation model is adopted in the other way. d_0 is a constant, and the value depends on the network environment.

2.2 System Model

We assume that there are N sensor nodes randomly distributed in the area of M × M. BS has infinite energy and is located in the center of the area.

1) Each node is equipped with a GPS device and knows its own coordinates.
2) After node deployment is completed, all nodes become static nodes. Each node can belong to only one cluster.
3) Each node has local information, including its own unique ID, CH node ID, information collection and transmission rounds, remaining energy level, and distance to its neighbors.
4) The nodes have the same structure, the same initial energy, the same computing and communication capabilities, and all have the ability to collect, calculate, store, and fuse data.

The whole operation of WSN is divided into several rounds according to time. In each round, the ordinary sensor node sends the sensed data to its cluster head [9].

3 Design and Analysis of Algorithms

3.1 Design of Clustering Algorithm

After the node arrangement is completed, the WSN forms C concentric circles with BS as the center. As shown in Fig. 1, the radius R of the innermost first circle of the concentric circles is equal to the threshold d_0. We call this circular area 0. In region 0, we believe that according to formula 1, nodes within the BS threshold (the threshold of BS in this article is equal to the threshold of the node) directly transmit data to the BS. The nodes in this area do not need to be clustered. Clustering instead wastes extra energy [10].

The radius of the 2nd to Cth concentric circles in the inner layer is 1/2 d_0. This is because during the data transmission phase, data will be transmitted from the nodes in the Cth concentric circle in the outermost layer to the nodes of the inner concentric circle in order. During data transmission, the transmission distance does not exceed the threshold d_0 [11].

Next, we determine the extent of each cluster within the area of each concentric circle. As shown in Fig. 1, in the first concentric circle (herein referred to as Zone 1, the concentric circles in the following and so on), the nodes in the area of points A and B form a new cluster. Point A is on the inner perimeter of the first concentric circle, and point B is on the outer perimeter of the first concentric circle. And the Euclidean distance between two points is d_0. Similarly, another cluster can be divided by points C and D, and the Euclidean distance between the two points is also d_0. In the same way, new clusters can be divided according to points E and F within the area of Zone 2. If the area is not enough after the division is completed, a small cluster is formed separately. Such as Cluster D in Zone 1 [12].

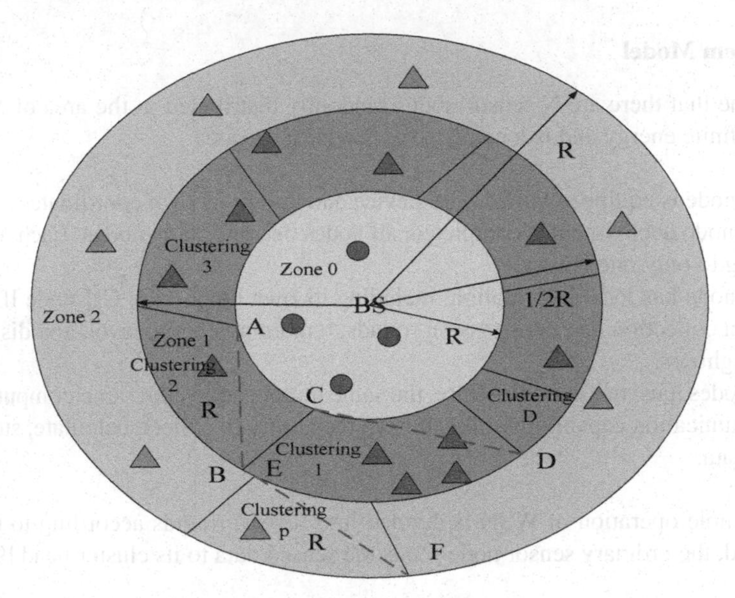

Fig. 1. Division of clusters

3.2 Cluster Head Selection Based on Genetic Algorithm

WSN clustering is a multi-objective optimization problem. This paper proposes a genetic algorithm based on the geographical distribution of nodes to solve the clustering optimization model [13].

Step 1: population initialization

Geographical Distribution Genetic Algorithm (GDGA) is a process to obtain the optimal chromosome through the process of chromosome selection, crossing, and mutation based on the initial population [14].

Geographical Distribution Genetic Algorithm (GDGA) is a process to obtain the optimal chromosome through the process of chromosome selection, crossing, and mutation based on the initial population. The algorithm uses two-dimensional gene coding. The number of chromosomes is first defined as Q, and each chromosome has P bits (the entire WSN is pre-divided into P clusters) in binary code. As shown in Table 1:

Table 1. Chromosome coding

Node number	1	2	3	4	5	6	....	N
CM or CH	1	0	0	0	0	0	...	1

One dimension of a chromosome is represented by node numbers 1 to N. Two-dimensional indicates whether this node is selected as the cluster head. 0 indicates that the node is a common node (CM), as represented by a cluster head (CH). For example, the first digit of the gene in Table 1 indicates that node 1 is selected as the cluster head [15].

3.3 Fitness Function

The calculation formula for the fitness function of the CH node selection needs to define some important parameters as the basis for the CH selection.

The fitness function of the CH node can be expressed as

$$F_t = \mu_1 \times E_{residual} + \mu_2 \times \frac{1}{D_{toCH_i}} + \mu_3 \times \frac{1}{D_{toRelay}} \tag{4}$$

$$\mu_1 + \mu_2 + \mu_3 = 1 \tag{5}$$

where μ_1, μ_2, and μ_3 are the control parameters of the three parts of the fitness function. In this algorithm, the equivalent values of the three control parameters represent the same effect of the three influencing factors. $E_{residual}$ represents the remaining energy of the node. The more the remaining energy, the more likely it is to be selected as the cluster head. D_{toCH_i} represents the distance from other nodes in the cluster to that node. The smaller the distance, the more likely it is to be selected as the cluster head. $D_{toRelay}$ represents the distance from this node to the previous relay node. The smaller the distance, the more likely it is to be selected as the cluster head [16].

3.4 Operation of Genetic Algorithm

Crossover and mutation operations are performed on the current genes to generate new genes. The operations are shown in Fig. 2 and Fig. 3.

Node number	1	2	3	4	5	6	......	N
CM or CH	1	0	0	0	0	0	......	1

Gene crossover

Node number	1	2	3	4	5	6	......	N
CM or CH	1	0	0	0	0	0	......	1

Fig. 2. Gene mutation

4 Simulation Experiment and Analysis

In order to verify the effectiveness of the GDGC algorithm, we used Matlab 2016 to compare its three algorithms, LEACH, ERP, and HEED. Assume that 100 wireless

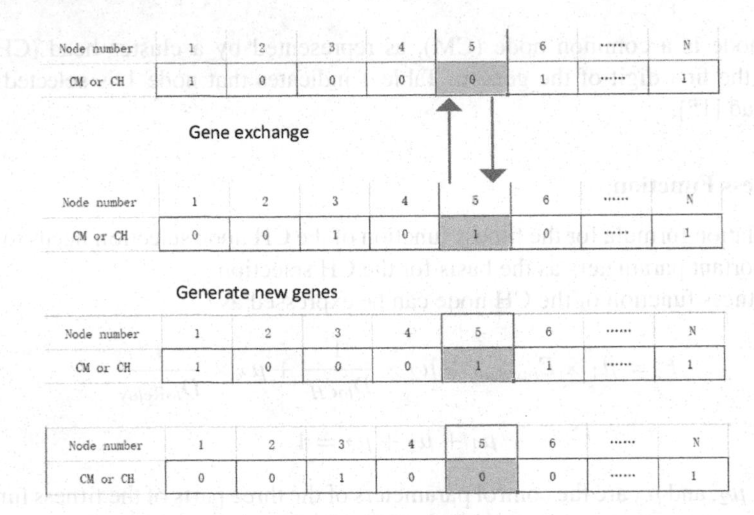

Fig. 3. Gene mutation

sensor nodes are randomly arranged in a 50 × 50 area. The position of BS is at the center coordinate of (100, 100). We compare the overall network energy consumption, node life, remaining energy (the number of remaining nodes) and other indicators [17].

The simulation parameters are shown in Table 2.

Table 2. Simulation input parameters

Parameters	Value
Sensor field region (m2)	(100 * 100)
BS location	(50, 50)
Number of sensors	100
Initial energy of the node (J)	200
Data packet length (bits)	2048
Number of iterations	100
E_{elec} (nJ/bit)	50
E_{amp} (pJ/bit)	0.0012
E_{fs} (pJ/bit)	10
d_{th} (m)	30

The simulation results are shown in Fig. 4. After the four protocols run the same 100, 300, and 500 rounds, the remaining energy of the WSN running the GDGA algorithm is higher than the other three algorithms.

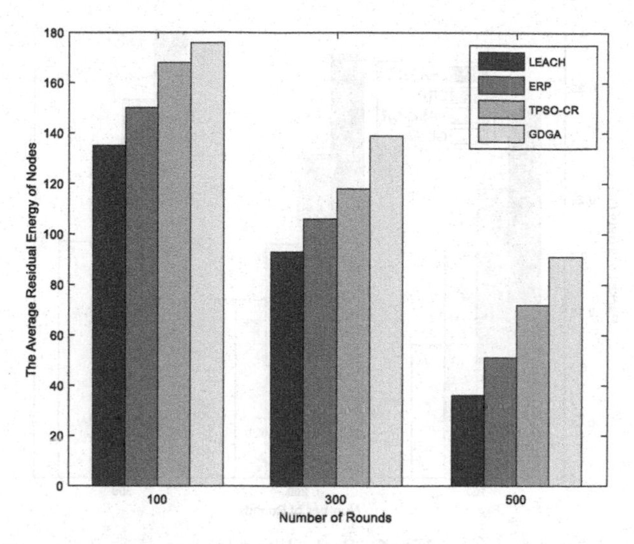

Fig. 4. The average remaining energy of the nodes when running the same round

As shown in Fig. 5, the number of nodes surviving is compared after running the four algorithms for several rounds. We can see that the GDGA algorithm has more nodes than the other three algorithms. This proves that our algorithm can extend the maximum life of WSN.

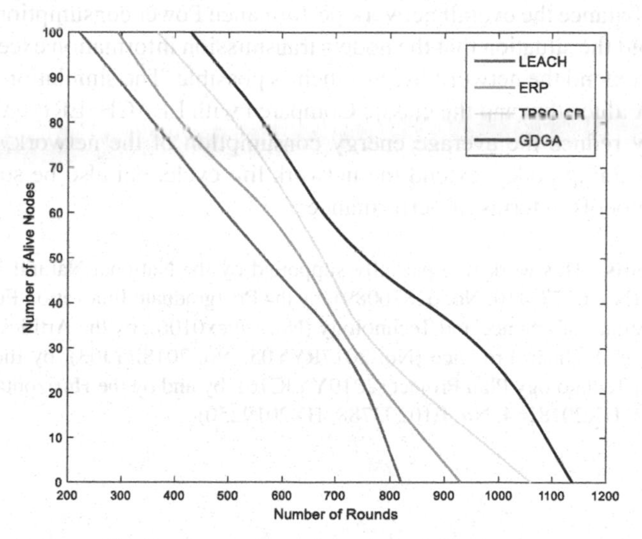

Fig. 5. Comparison of number of alive nodes

We cluster the four algorithms and then execute the greedy algorithm as the routing algorithm. This way we can compare the data transmission capabilities of the four algorithms (Fig. 6).

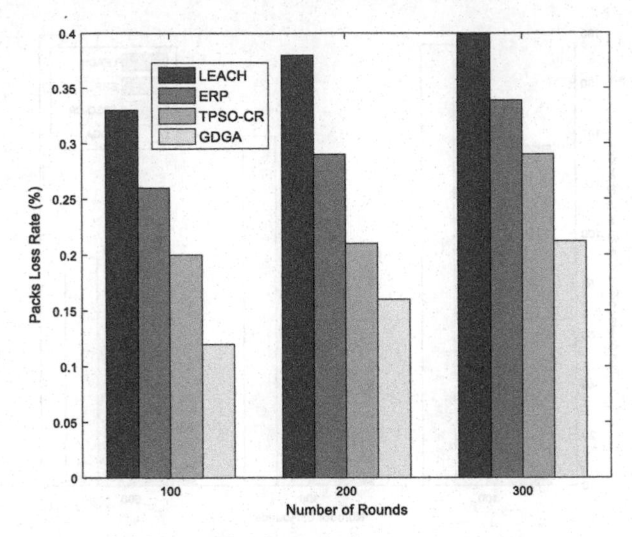

Fig. 6. Packs loss rate

5 Conclusion

In this paper, in the traditional clustering algorithm, a genetic algorithm based on the geographical distribution of sensor nodes is introduced to obtain the GDGA algorithm, which can cluster faster. At the same time, a fitness function based on multiple factors is constructed to balance the overall network performance Power consumption, to a certain extent, can avoid the situation that the node's transmission information exceeds the node threshold, and extend the network life as much as possible. The simulation results show that the GDGA algorithm and the classic Compared with LEACH, ERP, and TPSO-CR, it can not only reduce the average energy consumption of the network, increase the number of remaining nodes, extend the network life cycle, but also be superior to the other three protocols in terms of performance.

Acknowledgments. This work was partially supported by the National Natural Science Foundation of China (No. 61771410, No. 61876089), by the Postgraduate Innovation Fund Project by Southwest University of Science and Technology (No. 19ycx0106), by the Artificial Intelligence Key Laboratory of Sichuan Province (No. 2017RYY05, No. 2018RYJ03), by the Zigong City Key Science and Technology Plan Project (2019YYJC16), by and by the Horizontal Project (No. HX2017134, No. HX2018264, No. E10203788, HX2019250).

References

1. Akyildiz, I.F., Su, W., Sankarasubramaniam, Y., Cayirci, E.: Wireless sensor networks: a survey. Comput. Netw. **38**(17), 393–422 (2002)
2. Cardai, M., Du, D.Z.: Improving wireless sensor network lifetime through power aware organization. Wireless Netw. **3**, 333–340 (2005)
3. Chen, M.-T., Tseng, S.-S.: A genetic algorithm for multicast routing under delay constraint in WDM network with different light splitting. J. Inf. Sci. Eng. **21**(8), 85–108 (2005)

4. Bhondekar, A.P., Vig, R., Singla, M.L., Ghanshyam, C., Kapur, P.: Genetic algorithm based node placement methodology for wireless sensor networks. Proc. Int. Multiconf. Eng. Comput. Sci. **1**, 18–22 (2009)
5. Wang, P., He, Y., Huang, L.: Near optimal scheduling of data aggregation in wireless sensor networks. Ad Hoc Netw. **4**, 1287–1296 (2013)
6. Kulkarni, R.V., Forster, A., Venayagamoorthy, G.K.: Computational intelligence in wireless sensor networks: a survey. IEEE Commun. Surv. Tutor. **13**(1), 68–96 (2011)
7. Gazen, C., Ersoy, C.: Genetic algorithms for designing multihop lightwave network topologies. Artif. Intell. Eng. **13**, 211–221 (1999)
8. Jiang, H., Zhang, T., Zhao, X., et al.: Large data based anomaly detection mechanism for power information network traffic. Telecommun. Sci. **33**(3), 134–141 (2017)
9. Han, W., Tian, Z., Huang, Z., Zhong, L., Jia, Y.: System architecture and key technologies of network security situation awareness system YHSAS. Comput. Mater. Continua **59**(1), 167–180 (2019)
10. Li, R., Zhang, L., Li, H., et al.: Summary of network anomaly traffic detection based on entropy. Appl. Comput. Syst. **26**(6), 36–39 (2017)
11. Gu, Y., He, T.: Dynamic switching-based data forwarding for low-duty-cycle wireless sensor networks. IEEE Trans. Mob. Comput. **10**(12), 1741–1754 (2011)
12. Xu, G., Wang, Z., Zang, D., et al.: Data center network anomaly detection algorithm based on link state database. Comput. Res. Dev. **55**(4), 815–830 (2018)
13. Rout, R.R., Ghosh, S.K.: Adaptive data aggregation and energy efficiency using network coding in a clustered wireless sensor network: an analytical approach. Comput. Commun. **40**, 65–75 (2014)
14. Hong, M., Bei, Y.X.: Network anomaly data detection model based on intrusion feature selection. Modern Electron. Technol. **40**(12), 69–71 (2017)
15. Ying, W.: Wireless network traffic anomaly data detection simulation. Comput. Simu. **34**(9), 408–411 (2017)
16. Kaiwartya, O., Kumar, S., Abdullah, A.H.: Research on time synchronization method under arbitrary network delay in wireless sensor networks. Comput. Mater. Continua **61**(3), 1323–1344 (2019)
17. Zhang, H., Yi, Y., Wang, J., Cao, N., Duan, Q.: Analytical model of deployment methods for application of sensors in non-hostile environment. Wireless Person. Commun **97**, 389–399 (2017)

Dual Residual Global Context Attention Network for Super-Resolution

Jingjun Zhou[1,2], Jingbing Li[1,2](✉), Hui Li[1,2], Jing Liu[1], Qianning Dai[1], Saqib Ali Nawaz[1], and Jian Shen[1,2,3]

[1] College of Information and Communication Engineering, Hainan University, Haikou, Hainan, People's Republic of China
{juingzhou,lihui,shenjian}@hainanu.edu.cn, jingbingli2008@hotmail.com, jingliuhnu2016@hotmail.com, dqn0526@163.com, saqibsial20@gmail.com
[2] State Key Laboratory of Marine Resource Utilization in the South China Sea, Hainan University, Haikou, Hainan, People's Republic of China
[3] Dongguan ROE Technology Co., Ltd., Dongguan, Guangdong, People's Republic of China

Abstract. Recently, deep convolutional neural networks have enabled SISR to achieve amazing performance and visual experience. However, we note that most current methods fail to model effective long-distance dependence, either considering only channel-wise attention modeling or considering a self-attention mechanism to model long-range dependencies resulting in a number of parameters and memory costs have increased dramatically, hindering CNNs representation capabilities and deployment on edge devices. To solve this problem, we proposed a novel network named Dual Residual Global Context Attention Network (DRGCAN), which is lightweight and can effectively model global context information. Specifically, we proposed a dual residual in residual structure in which introduce dual residual learning in the residual block of each residual group of the traditional residual in residual (RIR) to fully capture the rich feature information. Furthermore, we further proposed a global context attention mechanism to insert it at the end of each residual group to effectively model long-distance dependence to improve the representation ability of the model. Extensive experiments shown that the proposed model is superior to state-of-the-art SR methods in both visual quality and memory footprint.

Keywords: Image super-resolution · Global context attention · Dual residual in residual

This work is supported by Hainan Provincial Natural Science Foundation of China [No. 2019RC018], and by the National Natural Science Foundation of China [61762033], and by the Natural Science Foundation of Hainan [617048, 2018CXTD333] and by Dongguan Introduction Program of Leading Innovative and Entrepreneurial Talents.

X. Sun et al. (Eds.): ICAIS 2020, LNCS 12240, pp. 166–178, 2020.
https://doi.org/10.1007/978-3-030-57881-7_15

1 Introduction

Single Image Super Resolution Technology (SISR) is a low-level computer vision technology problem that accurately reconstructs the corresponding HR output from a given LR input [7]. SISR is widely used in various fields, including remote sensing [19], medical imaging [8], and surveillance [25]. However, SISR is a natural ill-posed inverse problem because there is a one-to-many relationship between the input LR image and the output HR image. To solve this problem, a large number of researchers proposed many excellent methods, ranging from prediction-based [14] to learning-based [2]. Although many pioneers have done a lot of work, there are still several shortcomings in these work: 1). Most of these methods are based on image prior, and if there is a deviation, the reconstruction quality will be poor; 2). These methods have a slow optimization process and need to be improved.

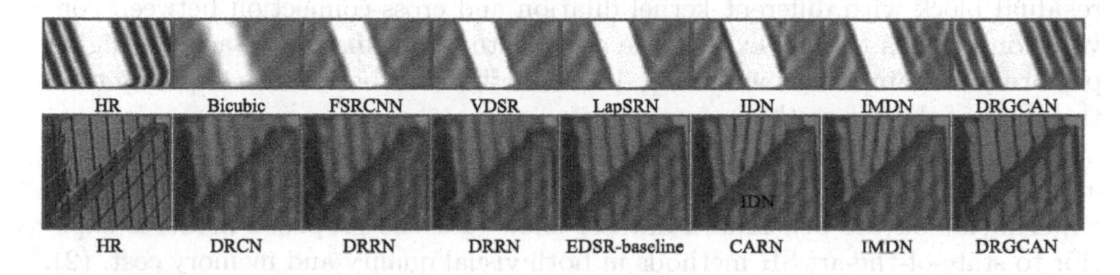

Fig. 1. Zoom visual results for 4× SR on "Dollgun" from Manga109 dataset and "img_033" from Urban100 dataset.

Recently, CNN has been widely used in many high level vision tasks with its powerful representation capability, and its performance has been significantly improved. Dong et al. [5] first proposed the CNN-based SR method, and then continued to appear CNN-based method to achieve a new level of SISR performance [6,15,16,18,20,27,34]. Most methods are learning a mapping function between the interpolated or LR input and the corresponding HR output, but most methods are explore the inherent statistical information of the image through the continuous stacking of convolution layers to achieve the state of the art performance. Although the performance of image SR has made unprecedented progress, these methods still have the following problems to be solved urgently: 1). Most CNN-based methods are large and deep, and they cannot be deployed on edge devices; 2). Most of the existing methods make it difficult to transmit information between different points due to simple stacking convolution layer, which is not enough to effectively model long-distance dependence, resulting in the failure of the model to learn more discriminative low level features, which hinders the representation ability of the model.

To solve these problems, we proposed a novel lightweight dual residual global context attention network (DRGCAN) to effectively model global context information. Specifically, we propose a global context attention mechanism for more efficient modeling of long-distance dependency and more powerful feature representation. Specifically, in the original paper of non-local block [9], different query positions are calculated according to different feature maps, but we have found that the feature maps of different query positions in the non-local block are almost the same, that is, all query positions are independent of the feature map [3]. Based on this finding, we present a simplified non-local block where all query positions are independent of the feature map. At the same time, it has been found that the simplified non-local block is similar in structure to the SE block [10] and have their own advantages. By combining the advantages of the simplified non local block and SE block [10], we further proposed a global context attention mechanism (GCA) for more efficient modeling of long-distance dependency. The proposed GCA is significantly lower than the non-local block parameter, and better performance is achieved compared to the SE block and the parameter amount is only increased a little. In addition, we also modified the single-branch residual block in the original RIR structure [34] to a double-branch residual block with different kernel dilation and cross-connection between convolutional layers to fully exploit the deep feature information to achieve higher performance gain. As shown in Fig. 1, our method has better visual quality than the state-of-the-art method.

In general, the main contributions of this paper are as follows: (1). We propose a lightweight global context attention network for accurate image SR. extensive experiments on the benchmark dataset show that the proposed model is superior to state-of-the-art SR methods in both visual quality and memory cost. (2). We propose a dual residual in residual structure to fully exploit deep feature information to achieve higher performance gains. (3). We propose a Global Context Attention Mechanism (GCA) to learn more discriminative features by more efficient modeling of long-distance dependencies, thereby improving the model's representation ability.

2 Related Work

In the past decade, a large variety of excellent SISR methods have been proposed, including patch-based, learning-based and CNN-based methods et al. Due to limited space, here we only discuss the CNN-based SR method and attention mechanism related to this study.

CNN for SR. The earliest introduction of CNN to Image SR was Dong et al. and proposed SRCNN [5], which consisted of only three layers of convolutional layers, which achieved significant performance improvements at the time. However, since SRCNN is the LR input interpolate to the desired size, thus leading to a large amount of computer memory spending. To solve this problem, the author further proposed FSRCNN [6], which performs convolution learning in the LR space and applies deconvolution at the end of the network to upscale it into the HR space. Then, in order to alleviate the difficulty of training the depth model, Kim et al. proposed a VDSR [15] with a residual structure and a DRCN [16] with a recursive structure, and the performance obtained was superior to the previous methods at that time. Since CNN-based SISR methods are unable to build deeper models, inspired by SRResNet [18], Lim et al. propose a deep and wide EDSR [20] model. EDSR has very large model parameters and significantly improves the performance of SR. Zhang et al. introduced SE block in SENet [10] to the image SR task and proposed RCAN [34] with channel-wise attention and up to 400 layers, which improves performance by modeling inter-channel relationships and is superior to all previous methods. In addition, in order to model global spatial dependence, Liu et al. introduces Non-Local block in NLNet [31] into recurrent network, and proposes NLRN [21], which models global spatial dependence according to self attention to improve model performance. Similarly, Zhang et al. [35] also introduced Non-Local block [31] into the trunk branches of their models to better guide feature extraction. Very recently, Liu et al. [22] Combines non-local operation with back project mechanism to learn the cross-correlation between features of different layers, so as to improve the quality of image SR.

For lightweight models, Ahn et al. [1] used a cascading mechanism on the residual network to make the model more efficient. Then Hui et al. [12] proposed to build cascaded information multi-distillation blocks and step-by-step to extract hierarchical features.

Attention Mechanism. The purpose of introducing attention mechanism into the model is to make it more focused on those more informative features, which are widely used in various tasks of computer vision. Hu et al. [10] proposed the squeeze-and-excitation (SE) block to model the interdependencies between channels and further enhance the model representation ability and validate it on image classification tasks. At the same time, Zhang et al. [34] also introduced the SE block into the image SR task. Specifically, the residual block in the RCAN introduces the SE block, which allows it to adaptively adjust the global context channel dependency. However, simply rescaling the feature fusion is not enough to effectively model the global context. In addition, He et al. [31]proposes NL block, which uses self attention mechanism to model the correlation of pixel level pairwise. Liu et al. [21] and Zhang et al. [35] also introduced NL block into the computational vision task of low level to model long-distance dependency. However, NL block is actually a feature map learning query independent for each query point, which wastes a lot of computing costs. Our DRGCAN absorbs the advantages of NL block [31] strong global context modeling ability and low computation of SE block [10]. Therefore, it can effectively global context modeling and DRGCAN achieved better performance.

3 Dual Residual Global Context Attention Network

3.1 Network Framework

As shown in Fig. 2, we proposed network model DRGCAN consists of four parts: shallow feature extraction, dual residual in residual deep feature extraction,

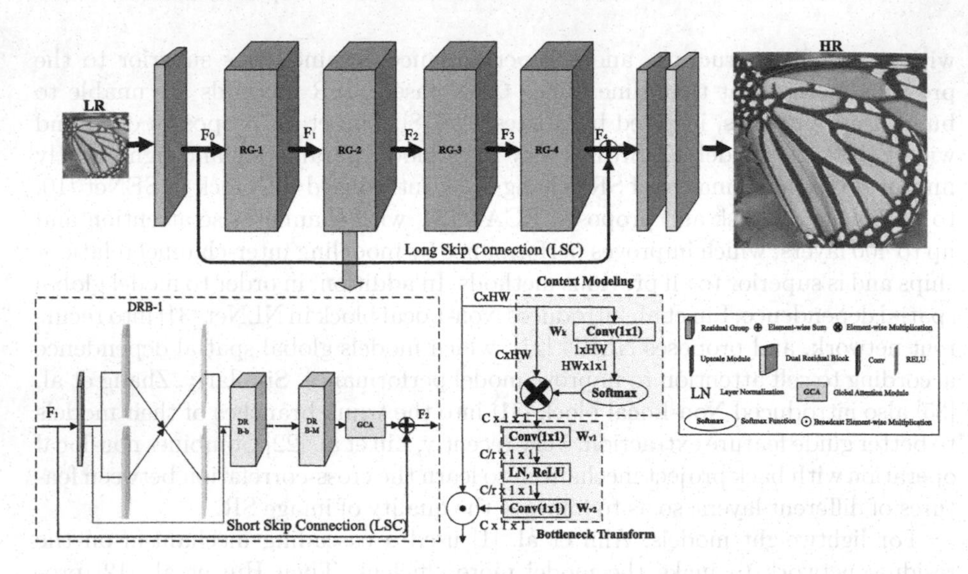

Fig. 2. The network structure of our dual residual global context attention network (DRGCAN)

upscale and reconstruction modules. First, for the sake of clarity, we assume that I_{LR} and I_{SR} are the inputs and outputs of DRGCAN, respectively. As with most studies [18,20], we only use a Conv layer to extract shallow features from LR input.

$$F_0 = H_{SF}(I_{LR}), \tag{1}$$

where $H_{SF}(\cdot)$ is a convolution operation, and the obtained F_0 is used for the deep feature extraction of the subsequent dual residual in residual module. Therefore, we can further obtain deep features,

$$F_{DF} = H_{DRIR}(F_0), \tag{2}$$

$H_{DRIR}(\cdot)$ represents our proposed dual residual in residual structure (DRIR), which contains 4 residual groups (RG). The proposed DRIR has a very large receptive field, which can fully extract the rich deep feature information F_{DF} and then upscaled it through the upscale module,

$$F_{UP} = H_{UP}(F_{DF}), \tag{3}$$

where $H_{UP}(\cdot)$ and F_{UP} are the corresponding upscale module and upscaled feature respectively. For upscale operation, we choose to use the sub-pixel convolution layer [27], and then through a convolution layer to complete the final output I_{SR} reconstruction

$$F_{SR} = H_{RECON}(F_{UP}) = H_{DRGCAN}(I_{LR}), \tag{4}$$

where $H_{RECON}(\cdot)$ and $H_{DRGCAN}(\cdot)$ represent the reconstruction module and the entire DRGCAN network structure, respectively.

Next, for the loss function of the DRGCAN model, we choose L1 Loss. Given a pair of N training sets, it is written as$\{I_{LR}^{i}, I_{HR}^{i}\}_{i=1}^{N}$ consisting of N LR and corresponding HR images. Our DRGCAN training optimization goal is to minimize the L1 loss function.

$$L(\theta) = \frac{1}{N}\sum_{i=1}^{N}\left\|H_{DRGCAN}\left(I_{LR}^{i}\right) - I_{HR}^{i}\right\|_{1}, \tag{5}$$

where θ is the parameter of the DRGCAN model, which will be optimized by the stochastic gradient descent method. For the DRGCAN parameter setting, we will give in Sect. 4.1.

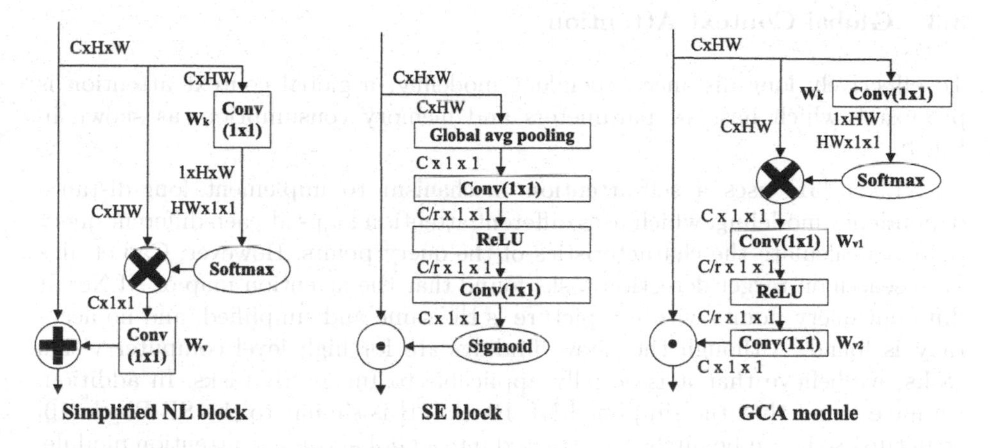

Fig. 3. Architecture of three different blocks. Where C × H × W denotes a feature map with channel number C, height H and width W.

3.2 Dual Residual in Residual (DRIR)

Now we give more details about the proposed DRIR structure, which consists of 4 residual groups (RGs). Due to limited space, we only introduce the differences from RIR [34], please refer to [34] for more details.

The main differences between us and RIR [34] are removing the CA module and introducing dual residual learning in the residual block, and embedding a GCA module at the end of each RG. Specifically, we first use a 1×1 convolutional layer in each dual residual block to halve the dimension, then use a two-branch convolutional layer with kernel dilation and two branches to cross-contract again, and finally concatenate the two branches and Output,

$$\begin{aligned}
F_{2,b1} &= H_{2,b,d2}(\tau(H_{2,b,d1}(F_{2,b-1}))) \\
F_{2,b2} &= H_{2,b,d1}(\tau(H_{2,b,d2}(F_{2,b-1}))) \\
F_{2,b} &= H_{2,b,cat}(F_{2,b1}, F_{2,b2}) + F_{2,b-1}
\end{aligned} \tag{6}$$

where $H_{2,b,d\_}(\cdot)$, $H_{2,b,cat}(\cdot)$ respectively represent convolutional layer functions with different kernel dilation and corresponding concatenate functions, $F_{2,b-1}$ and $F2,b$ respectively represent the bth of the 2nd RG the input and output of the dual residual block, τ represents the ReLU [26] activation function. This type of connection can fully exploit deep feature information to achieve higher performance gains.

Furthermore, in order to effectively model the long-distance dependence to improve the representation ability of the model, we further propose a global attention mechanism to insert it at the end of each residual group. More details about GCA, we will give in the next subsection.

3.3 Global Context Attention

To effectively long-distance-dependent modeling, a global context attention is proposed, which has low parameters and memory consumption, as shown in Fig. 3.

NLNet [31] uses a self-attention mechanism to implement long-distance dependency modeling, which uses different attention maps at each different query point to calculate the characteristics of the query points. However, Cao et al.'s [3] research on target detection tasks found that the attention map of NLNet at different query points within a picture is the same and simplified, and no accuracy is found. Although the above findings are for high level computer vision tasks, we believe that it is equally applicable to Image SR tasks. In addition, we have found that the simplified NL Block [31] is similar to the SE Block [10] structure, so it can be further abstracted into a Global context attention module.

The global context attention module consists of three parts, which first aggregate the features of all positions to construct a global context modeling,

$$Y_j = \sum_{j=1}^{N_p} \frac{e^{W_k \mathbf{x}_j}}{\sum_{m=1}^{N_p} e^{W_k \mathbf{x}_m}} \mathbf{x}_j, \tag{7}$$

where Y_j is the output of global attention pooling for context modeling, which is equivalent to the global average pooling in the SE block, N_p indicates the number of positions of the feature map H×W, and then the bottleneck transform to capture channel-wise dependencies, which is equivalent to the squeeze and excitation operation in the SE block,

$$T_i = W_{v2} ReLU\left(LN\left(W_{v1}(Y_j)\right)\right), \tag{8}$$

where T_i is the global context feature of the i-th feature map captured, we add layer normalization (LN) to simplify the training difficulty to speed up the optimization, and finally need to fuse it with the features of all locations.

$$Z_i = T_i * X_i, \tag{9}$$

where Z_i is the input X_i and the global context feature T_i for the broadcast element-wise multiplication to obtain the final output and complete the global context modeling.

3.4 Implementation Details

Now we need to specify the implementation details of the proposed model DRGCAN. First, we set the number 4 of RGs in DRIR and have M = 8 DRB blocks in each RG. The kernel size is 1×1 in GCA, and the kernel size of Conv layers of other modules without special instructions is 3×3. In addition to 1×1 Conv layers, other convolutional layers use zero-padding to ensure the same input and output size. Except for the Conv filters number C = 32 in the DRB, all other Conv filters number C = 64. In addition, the bottleneck radio of the bottleneck layer in the GCA is set to r = 16. For the upscale module $H_{UP}(\cdot)$, we follow [20,34] and use the sub-pixel layer in ESPCNN [27] for upsampling to achieve the coarse to fine feature, and finally Conv layer has 3 filters to reconstruct the output color image. At the same time, the proposed model is also applicable to gray images.

4 Experiments

4.1 Settings

Following [1,20,34], we selected the first 800 HD composite images of the DIV2K dataset [29] as the training dataset, and selected five benchmark datasets: Set5 [2], Set14 [32], BSD100 [23], Urban100 [11] and Manga100 [24], as test data sets for model evaluation and each data set has its own characteristics. We first use the Bicubic interpolation (BI) function in MATLAB to perform $\times 2$, $\times 3$, and $\times 4$ degradations on the data set. Then we obtain the experimental results and convert it to the YCbCr space, and only evaluate the Y channel using PSNR and SSIM. In addition, in order to obtain more training data sets, we randomly rotate the training data by 90°, 180°, 270° and horizontal flip during training, and randomly select 16 LR color paths of 48×48 as the input of each batch. We choose ADAM as the optimizer for our model and set its hyperparameters to $\beta_1 = 0.9$, $\beta_2 = 0.99$, and $\epsilon = 10^{-8}$. Our training learning rate is initialized to 10^{-4}, a total of 1000 epochs is trained, and the learning rate per 200 epochs will be reduced by half. We implemented the proposed DRGCAN model on Pytorch 1.0 and trained it on the Nvidia Tesla V100 GPU.

4.2 Model Analysis

In this section, we will study the effects of DRIR and GCA on SR performance in the DRGCAN model. First, we take the original RIR structure in [34] as the baseline, and then we test whether there are DRIR and GCA effects on the performance respectively. From Table 1, we can see that the performance of L_{base} is relatively low. When we add channel attention (CA), global context attention (GCA) or change their RIR to DRIR respectively, the performance of the model (L_{DRIR}, L_{CA}, L_{GCA}) is better than that of L_{base}. When DRIR and CA are added at the same time (denoted $L_{DRIR+CA}$), the performance of the model is increased from 37.42 dB to 37.54 dB, which also shows that DRIR and CA play a very important role in the model. Further, when the CA in $L_{DRIR+CA}$ is replaced by GCA (denoted $L_{DRIR+GCA}$), the performance of the model is further improved, and the parameters of CA and GCA are almost the same. Therefore, both DRIR and GCA are used in our network, and more results are given in Table 2.

Table 1. Effects of vary modules. We give the best PSNR value on Set5 (2×) in 50 epochs.

	L_{base}	L_{DRIR}	L_{CA}	L_{GCA}	$L_{DRIR+CA}$	$L_{DRIR+GCA}$
DRIR		✓			✓	✓
channel attention			✓		✓	
GCA				✓		✓
PSNR/Params	37.41/914k	37.42/1,638k	37.44/916k	37.45/916k	37.54/1640k	37.55/1640k

4.3 Comparison with State-of-the-art SR Methods

Quantitative Results. Our DRGCAN is compared to 13 state-of-the-art methods: SRCNN [5], FSRCNN [6], VDSR [15], DRCN [16],DRRN [28], LapSRN [17], SelNet [4], IDN [13], EDSR-baseline [20], SRMDNF [33],SRDenseNet [30], CARN [1] and IMDN [12]. Table 2 gives various methods for quantitative comparison of ×2, ×3 and ×4. We can note that although our DRGCAN parameters are slightly larger than other SR methods, their performance is significantly better than other state-of-the-art methods, especially in the case of ×3, ×4 scaling factor. This is mainly because the GCA in our DRGCAN can effectively model long-distance spatial information to ensure that DRGCAN has strong feature expression capabilities, while other methods only consider channel-wise attention modeling or treat channels equally.

Visual Quality. In Fig. 4, we give the zoom results of various methods, and we can see that most of them are seriously blurred. On the contrary, our DRGCAN can effectively alleviate artifacts and recover more real details. Taking "img_071" as an example, most of the early methods [15–17,28] fail to recover the line segments accurately, and have very serious blur. This is mainly because the networks of these methods are simply stacked and no correlation of features is explored. Some recent comparison methods [1,12,13,20] can better restore the

Table 2. Average PSNR/SSIM on five benchmark datasets. The best and the second best results are **highlighted** and underlined respectively.

Method	Scale	Params	Set5	Set14	BSD100	Urban100	Manga109
			PSNR/SSIM	PSNR/SSIM	PSNR/SSIM	PSNR/SSIM	PSNR/SSIM
Bicubic	×2	–	33.66/0.9299	30.24/0.8688	29.56/0.8431	26.88/0.8403	30.80/0.9339
SRCNN	×2	8K	36.66/0.9542	32.45/0.9067	31.36/0.8879	29.50/0.8946	35.60/0.9663
FSRCNN	×2	13K	37.05/0.9558	32.66/0.9088	31.53/0.8920	29.88/0.9020	36.67/0.9710
VDSR	×2	666K	37.53/0.9587	33.03/0.9124	31.90/0.8960	30.76/0.9140	37.22/0.9750
DRCN	×2	1,774K	37.63/0.9588	33.04/0.9118	31.85/0.8942	30.75/0.9133	37.55/0.9732
LapSRN	×2	251K	37.52/0.9591	33.08/0.9130	31.08/0.8950	30.41/0.9101	37.27/0.9740
SelNet	×2	974K	37.89/0.9598	33.61/0.9160	32.08/0.8984	–/–	–/–
IDN	×2	553K	37.83/0.9600	33.30/0.9148	32.08/0.8985	31.27/0.9196	38.01/0.9749
EDSR-baseline	×2	1,370K	37.99/0.9604	33.57/0.9175	32.16/0.8994	31.98/0.9272	38.54/0.9769
SRMDNF	×2	1,511K	37.79/0.9601	33.32/0.9159	32.05/0.8985	31.33/0.9204	38.07/0.9761
CARN	×2	1,592K	37.76/0.9590	33.52/0.9166	32.09/0.8978	31.92/0.9256	38.36/0.9765
IMDN	×2	694K	38.00/0.9605	33.63/0.9177	32.19/0.8996	32.17/0.9283	**38.88/0.9774**
DRGCAN(Ours)	×2	1,640K	**38.01/9606**	**33.64/0.9178**	**32.20/0.8998**	**32.22/0.9292**	38.73/0.9772
Bicubic	×3	–	30.39/0.8682	27.55/0.7742	27.21/0.7385	24.66/0.7349	26.95/0.8556
SRCNN	×3	8K	32.75/0.9090	29.30/0.8215	28.41/0.7863	26.24/0.7989	30.48/0.9117
FSRCNN	×3	13K	33.18/0.9140	29.37/0.8240	28.53/0.7910	26.43/0.8080	31.10/0.9210
VDSR	×3	666K	33.66/0.9213	29.77/0.8314	28.82/0.7976	27.14/0.8279	32.01/0.9340
DRCN	×3	1,774K	33.82/0.9226	29.76/0.8311	28.80/0.7963	27.15/0.8276	32.24/0.9343
LapSRN	×3	502K	33.81/0.9220	2979/0.8325	28.82/0.7980	27.07/0.8275	32.21/0.9350
SelNet	×3	1,159K	34.27/0.9257	30.30/0.8399	28.97/0.8025	–/–	–/–
IDN	×3	553K	34.11/0.9253	29.99/0.8354	28.95/0.8013	27.42/0.8359	32.71/0.9381
EDSR-baseline	×3	1,555K	34.37/0.9270	30.28/0.8417	29.09/0.8052	28.15/0.8527	33.45/0.9439
SRMDNF	×3	1,528K	34.12/0.9254	30.04/0.8382	28.97/0.8025	27.57/0.8398	33.00/0.9403
CARN	×3	1,592K	34.29/0.9255	30.29/0.8407	29.06/0.8034	28.06/0.8493	33.50/0.9440
IMDN	×3	703K	34.36/0.9270	30.32/0.8417	29.09/0.8046	28.17/0.8519	33.61/0.9445
DRGCAN (Ours)	×3	1,825K	**34.49/0.9278**	**30.41/0.8424**	**29.14/0.8063**	**28.36/0.8563**	**33.72/0.9455**
Bicubic	×4	–	28.42/0.8104	26.00/0.7027	25.96/0.6675	23.14/0.6577	24.89/0.7866
SRCNN	×4	8K	30.48/0.8628	27.50/0.7513	26.90/0.7101	24.52/0.7221	27.58/0.8555
FSRCNN	×4	13K	30.72/0.8660	27.61/0.7550	26.98/0.7150	24.62/0.7280	27.90/0.8610
VDSR	×4	666K	31.35/0.8838	28.01/0.7674	27.29/0.7251	25.18/0.7524	28.83/0.8870
DRCN	×4	1,774K	31.53/0.8854	28.02/0.7670	27.23/0.7233	25.14/0.7510	28.93/0.8854
LapSRN	×4	502K	31.54/0.8852	28.09/0.7700	27.32/0.7275	25.21/0.7562	29.09/0.8900
SRDenseNet	×4	2015K	32.02/0.8934	28.50/0.7782	27.53/0.7337	26.05/0.7819	–/–
SelNet	×4	1,417K	32.00/0.8931	28.49/0.7783	27.44/0.7325	–/–	–/–
IDN	×4	553K	31.82/0.8903	28.25/0.7730	27.41/0.7297	25.41/0.7632	29.41/0.8942
EDSR-baseline	×4	1,518K	32.09/0.8938	28.58/0.7813	27.57/0.7357	26.04/0.7849	30.35/0.9067
SRMDNF	×4	1,552K	31.96/0.8925	28.35/0.7787	27.49/0.7337	25.68/0.7731	30.09/0.9024
CARN	×4	1,592K	32.13/0.8937	28.60/0.7806	27.57/0.7349	26.07/0.7837	30.47/0.9084
IMDN	×4	715K	32.21/0.8948	28.58/0.7811	27.56/0.7353	26.04/0.7838	30.45/0.9075
DRGCAN (Ours)	×4	1,788K	**32.27/0.8958**	**28.61/0.7814**	**27.60/0.7372**	**26.19/0.7897**	**30.57/0.9089**

details. The main reason is that these network models are carefully designed and even partially modeled for local feature correlation, but only our DRGCAN can recover more results close to ground-truth. These observations also verify that our DRGCAN can effectively model long-distance dependency.

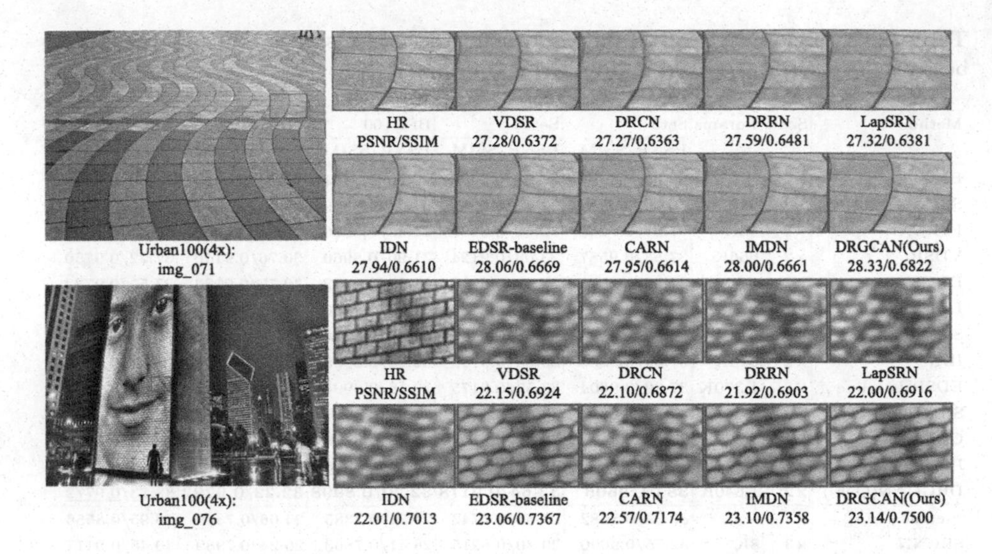

Fig. 4. Visual comparison of our DRGCAN with other SR methods on the Urban100 dataset.

5 Conclusions

In this paper, we proposed a lightweight dual residual global context attention network (DRGCAN) for accurately reconstructing image SR. Specifically, we proposed a dual residual in residual (DRIR) structure, that is, we introduce dual residual learning into the residual block in each residual group of the original RIR to fully exploit the deep feature information and achieve higher performance gains. In addition, in order to model long-distance dependencies more efficiently, we propose a global context attention mechanism (GCA) to learn more discriminative features to improve the model's representation ability. Extensive experiments on benchmark datasets with BI degradation models shown the effectiveness of our model DRGCAN in both visual quality and memory cost.

References

1. Ahn, N., Kang, B., Sohn, K.A.: Fast, accurate, and lightweight super-resolution with cascading residual network. In: Proceedings of the European Conference on Computer Vision (ECCV), pp. 252–268 (2018)
2. Bevilacqua, M., Roumy, A., Guillemot, C., Alberi-Morel, M.L.: Low-complexity single-image super-resolution based on nonnegative neighbor embedding (2012)
3. Cao, Y., Xu, J., Lin, S., Wei, F., Hu, H.: GCNet: non-local networks meet squeeze-excitation networks and beyond (April 2019). https://arxiv.org/abs/1904.11492v1
4. Choi, J.S., Kim, M.: A deep convolutional neural network with selection units for super-resolution. In: 2017 IEEE Conference on Computer Vision and Pattern Recognition Workshops (CVPRW), pp. 1150–1156 (July 2017). ISSN: 2160-7516

5. Dong, C., Loy, C.C., He, K., Tang, X.: Image super-resolution using deep convolutional networks. IEEE Trans. Pattern Anal. Mach. Intell. **38**(2), 295–307 (2015)
6. Dong, C., Loy, C.C., Tang, X.: Accelerating the super-resolution convolutional neural network. In: Leibe, B., Matas, J., Sebe, N., Welling, M. (eds.) ECCV 2016. LNCS, vol. 9906, pp. 391–407. Springer, Cham (2016). https://doi.org/10.1007/978-3-319-46475-6_25
7. Freeman, W.T., Pasztor, E.C., Carmichael, O.T.: Learning low-level vision. Int. J. Comput. Vis. **40**(1), 25–47 (2000)
8. Greenspan, H.: Super-resolution in medical imaging. Comput. J. **52**(1), 43–63 (2008)
9. He, K., Zhang, X., Ren, S., Sun, J.: Deep residual learning for image recognition. In: Proceedings of the IEEE Conference on Computer Vision and Pattern Recognition, pp. 770–778 (2016)
10. Hu, J., Shen, L., Sun, G.: Squeeze-and-excitation networks. In: Proceedings of the IEEE Conference on Computer Vision and Pattern Recognition, pp. 7132–7141 (2018)
11. Huang, J.B., Singh, A., Ahuja, N.: Single image super-resolution from transformed self-exemplars. In: Proceedings of the IEEE Conference on Computer Vision and Pattern Recognition, pp. 5197–5206 (2015)
12. Hui, Z., Gao, X., Yang, Y., Wang, X.: Lightweight Image Super-Resolution with Information Multi-distillation Network. arXiv preprint arXiv:1909.11856 (2019)
13. Hui, Z., Wang, X., Gao, X.: Fast and Accurate Single Image Super-Resolution via Information Distillation Network. arXiv:1803.09454 [cs] (March 2018)
14. Keys, R.: Cubic convolution interpolation for digital image processing. IEEE Trans. Acoust. Speech Sig. Process. **29**(6), 1153–1160 (1981)
15. Kim, J., Kwon Lee, J., Mu Lee, K.: Accurate image super-resolution using very deep convolutional networks. In: Proceedings of the IEEE Conference on Computer Vision and Pattern Recognition, pp. 1646–1654 (2016)
16. Kim, J., Kwon Lee, J., Mu Lee, K.: Deeply-recursive convolutional network for image super-resolution. In: Proceedings of the IEEE Conference on Computer Vision and Pattern Recognition, pp. 1637–1645 (2016)
17. Lai, W.S., Huang, J.B., Ahuja, N., Yang, M.H.: Deep Laplacian pyramid networks for fast and accurate super-resolution. In: Proceedings of the IEEE Conference on Computer Vision and Pattern Recognition, pp. 624–632 (2017)
18. Ledig, C., et al.: Photo-realistic single image super-resolution using a generative adversarial network. In: Proceedings of the IEEE Conference on Computer Vision and Pattern Recognition, pp. 4681–4690 (2017)
19. Li, S., Yin, H., Fang, L.: Remote sensing image fusion via sparse representations over learned dictionaries. IEEE Trans. Geosci. Remote Sens. **51**(9), 4779–4789 (2013). https://doi.org/10.1109/TGRS.2012.2230332
20. Lim, B., Son, S., Kim, H., Nah, S., Mu Lee, K.: Enhanced deep residual networks for single image super-resolution. In: Proceedings of the IEEE Conference on Computer Vision and Pattern Recognition Workshops, pp. 136–144 (2017)
21. Liu, D., Wen, B., Fan, Y., Loy, C.C., Huang, T.S.: Non-local recurrent network for image restoration. In: Advances in Neural Information Processing Systems, pp. 1673–1682 (2018)
22. Liu, Z.S., Wang, L.W., Li, C.T., Siu, W.C., Chan, Y.L.: Image Super-Resolution via Attention based Back Projection Networks. arXiv:1910.04476 (2019)
23. Martin, D., Fowlkes, C., Tal, D., Malik, J.: A database of human segmented natural images and its application to evaluating segmentation algorithms and measuring ecological statistics. In: ICCV, Vancouver (2001)

24. Matsui, Y.: Sketch-based manga retrieval using manga109 dataset. Multimedia Tools Appl. **76**(20), 21811–21838 (2016). https://doi.org/10.1007/s11042-016-4020-z
25. Mudunuri, S.P., Biswas, S.: Low resolution face recognition across variations in pose and illumination. IEEE Trans. Pattern Anal. Mach. Intell. **38**(5), 1034–1040 (2015)
26. Nair, V., Hinton, G.E.: Rectified linear units improve restricted Boltzmann machines. In: Proceedings of the 27th International Conference on Machine Learning, ICML-10, pp. 807–814 (2010)
27. Shi, W., et al.: Real-time single image and video super-resolution using an efficient sub-pixel convolutional neural network. In: Proceedings of the IEEE Conference on Computer Vision and Pattern Recognition, pp. 1874–1883 (2016)
28. Tai, Y., Yang, J., Liu, X.: Image super-resolution via deep recursive residual network. In: Proceedings of the IEEE Conference on Computer Vision and Pattern Recognition, pp. 3147–3155 (2017)
29. Timofte, R., Agustsson, E., Van Gool, L., Yang, M.H., Zhang, L.: Ntire 2017 challenge on single image super-resolution: methods and results. In: Proceedings of the IEEE Conference on Computer Vision and Pattern Recognition Workshops, pp. 114–125 (2017)
30. Tong, T., Li, G., Liu, X., Gao, Q.: Image super-resolution using dense skip connections. In: 2017 IEEE International Conference on Computer Vision (ICCV), Venice, pp. 4809–4817. IEEE (October 2017)
31. Wang, X., Girshick, R., Gupta, A., He, K.: Non-local neural networks. In: Proceedings of the IEEE Conference on Computer Vision and Pattern Recognition, pp. 7794–7803 (2018)
32. Zeyde, R., Elad, M., Protter, M.: On single image scale-up using sparse-representations. In: Boissonnat, J.-D., et al. (eds.) Curves and Surfaces 2010. LNCS, vol. 6920, pp. 711–730. Springer, Heidelberg (2012). https://doi.org/10.1007/978-3-642-27413-8_47
33. Zhang, K., Zuo, W., Zhang, L.: Learning a single convolutional super-resolution network for multiple degradations. In: Proceedings of the IEEE Conference on Computer Vision and Pattern Recognition, pp. 3262–3271 (2018)
34. Zhang, Y., Li, K., Li, K., Wang, L., Zhong, B., Fu, Y.: Image super-resolution using very deep residual channel attention networks. In: Proceedings of the European Conference on Computer Vision (ECCV), pp. 286–301 (2018)
35. Zhang, Y., Li, K., Li, K., Zhong, B., Fu, Y.: Residual non-local attention networks for image restoration. arXiv:1903.10082 (2019)

A Robust Zero-Watermarking Algorithm for Medical Images Using Curvelet-Dct and RSA Pseudo-random Sequences

Fengming Qin[1,2], Jingbing Li[1,2(✉)], Hui Li[1,2], Jing Liu[1,2], Saqib Ali Nawaz[1,2], and Yanlin Liu[1,2]

[1] College of Information and Communication Engineering, Hainan University, Haikou 570228, China
fengmingqin@hotmail.com, Jingbingli2008@hotmail.com, lihui@hainanu.edu.cn, jingliuhnu2016@hotmail.com, saqibsial20@gmail.com, yanlinliu567@163.com

[2] State Key Laboratory of Marine Resource Utilization in the South China Sea, Hainan University, Haikou 570228, China

Abstract. With the popularization of digital medicine, the protection of patients' privacy has become extremely important in the medical field. Embedding watermarks into medical images is one of the effective ways to protect patients' privacy. However, the existing watermarking algorithms have poor robustness under geometric attacks. To solve this problem, aiming at the particularity of medical image, a robust zero-watermarking algorithm using Curvelet-Dct and RSA pseudo-random sequence is proposed in this paper. First, extract the visual feature vectors of medical images where the energy is most concentrated via perform DCT transformation on the Coarse layer of Curvelet transformation. Then, the pseudo-random sequence generated by the RSA algorithm is used to encrypt the watermark which improved the security of the patients' privacy. Combined with the concept of zero-watermark, the experimental results show that the proposed algorithm has good robustness in both conventional attacks and geometric attacks.

Keywords: Curvelet transform · RSA algorithm · Pseudo-random sequence · Zero-watermark · Robustness

1 Introduction

With the rapid development of information society and the improvement of people's self-protection consciousness, people are paying more and more attention to the security of information and privacy. The digitization of information media facilitates the storage and transmission of information. Meanwhile, it may be stolen by malicious people or

This work is supported by Hainan Provincial Natural Science Foundation of China [No. 2019RC018], and by the National Natural Science Foundation of China [61762033], and by the Natural Science Foundation of Hainan [617048, 2018CXTD333].

X. Sun et al. (Eds.): ICAIS 2020, LNCS 12240, pp. 179–190, 2020.
https://doi.org/10.1007/978-3-030-57881-7_16

groups. It is generally considered that the implementation of communication security can be achieved by encryption. That is, the multimedia data files are first encrypted and then communicated, making it impossible for an illegal attacker to obtain confidential information from the ciphertext. Encrypted files hinder the propagation of multimedia information while it is easy to cause the curiosity and attention of attackers. Once the ciphertext is cracked, it is completely transparent. However, medical images should not only protect the privacy information of patients in the images, but also facilitate the storage and sharing of images to meet the communication between experts. Simple data encryption cannot solve this problem completely. Combined with the special requirements of medical image, this paper chooses to use information hiding technology when designing the algorithm scheme, and hides the patient's personal information in the form of zero-watermark in the ordinary medical image for transmission. It not only does not change the original image, but also can effectively protect the patient's personal privacy information.

Information hiding is fundamentally divided into digital watermarking [1, 2] and steganography [3]. Initially, digital watermarking is a technique for embedding watermark into multimedia data, such as images, sounds, videos, etc. for copyright protection [4]. In recent years, digital watermarking technology has developed rapidly and has been widely used in many aspects like fingerprint identification, digital signature, distance education and insurance companies, secure electronic voting system, protection of electronic Business and e-government, cultural relics protection [5] and so on. According to the data type of the adding watermarks, it can be divided into four types: text watermark, image watermark, audio watermark and video watermark [6]. Among them, image watermarking technology is mainly divided into spatial domain technology and transform domain technology [7]. The spatial domain method is to superimpose watermark directly into the spatial domain of image. Because of its visual invisibility, the watermark is usually embedded in the least significant pixel position in the image. Another type of method is to transform the image first, especially orthogonal transform. Then embed the watermark into the transform domain [8–11], such as DCT domain, wavelet transform domain, fourier-mellin domain, etc. Because medical images are an important basis for doctors to diagnose patients, it is best not to make any changes. Therefore, this paper chooses transform domain algorithm to process medical images. Recently, multi-scale geometric analysis methods in the transform domain have gradually attracted people's attention. Channapragada et al. proposed two algorithms based on Ridgelet Transform and Rubik's Cube technology to embed and extract watermarks into cover images [12]. Thanki et al. proposed a new watermarking technique consisting of finite ridgelet transform (Finite Ridgelet Transform) and discrete wavelet transform and singular value decomposition [13]. H. Sadreazami et al. proposed a novel watermark decoder for the contourlet domain [14]. Kim et al. proposed a robust watermarking technique in the Curvelet Domain to maintain the clarity of high-quality images. This technology maximized stealth while maintaining adequate robustness and data capacity [15]. L. Chen et al. proposed a conservative blind watermarking scheme based on Contourlet for rendering 3D images of depth images [16].

However, the existing watermark algorithms still have the problem of poor robustness under geometric attacks, and there are few studies focus on curvelet in the field of medical image. Thus, this paper proposed a robust zero-watermarking algorithm based on the combination of Curvelet-DCT and cryptography RSA algorithm for medical images. Curvelet transform is able to represent smooth and edge parts of image with sparsity which can provide more image information than wavelet transform. RSA encryption can be decrypted without directly transmitting the key, avoiding the risk of the watermark being cracked by passing the key directly. They are used together in the proposed algorithm to ensure the patient's privacy information, and to resist geometric attacks well.

2 Basic Theory

2.1 Curvelet Transform

As multi-scale ridgelet analysis has great redundancy [17], Candes and Donohoe proposed the first generation curvelet transform based on ridgelet transform. First, sub-band decomposition is performed on the image. Then different size sub-blocks are used for sub-band images of different scales. Finally, each block is subjected to ridgelet analysis [18]. The digital implementation of the first generation of curvelet is more complex, requiring a series of steps such as subbed decomposition, smooth partitioning, normalization and ridgelet analysis [19]. Therefore, Candes et al. proposed a second generation fast curvelet transform that is simpler and easier to understand. The construction idea of the first generation of curvelet is to look at the curve in each block by a sufficiently small block, and then use the local ridgelet to analyze its characteristics. While the second generation of curvelet has nothing to do with ridgelet theory. and its process does not need to use ridgelet. The similarities between them are only the abstract mathematical meanings of tight support and framework. There are two fast discrete curvelet transform implementation methods based on the second generation curvelet transform theory named two-dimensional FFT algorithm and Wrap algorithm for non-uniform spatial sampling [20].

The Curvelet coefficient can be obtained by the following equation, i.e. the inner product of the signal and the wavelet function:

$$C(j, l, k) = \langle f, \phi_{j,l,k} \rangle = \int_{R^2} f(x)\phi_{j,l,k}(x)dx \tag{1}$$

Here j denotes a scale, l denotes a direction, and k denotes a displacement.

Performing a curvelet decomposition on a medical image yields a low frequency sub band coefficient and a high frequency sub band coefficient in each direction of each scale. The low frequency sub band coefficient can well characterize the medical image, and the high frequency sub band coefficient reflects important information such as details and texture in all directions of the image. Taking the 512 × 512 spine image as an example, the decomposed Curvelet low frequency sub-band image is shown in Fig. 1(c).

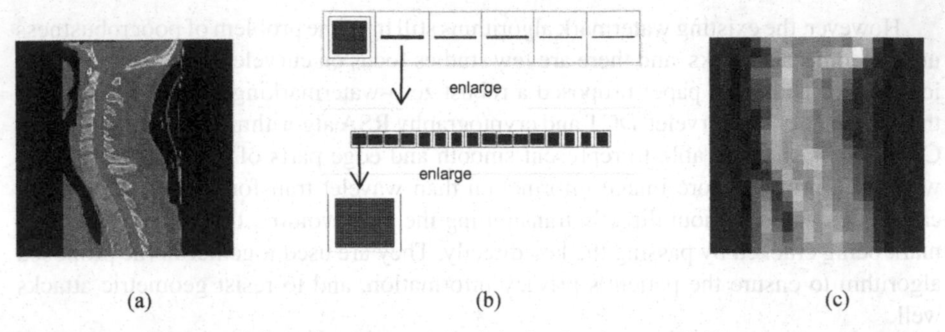

Fig. 1. Curvelet map of the spine image: (a) original spine image (b) Curvelet decomposition structure (c) low frequency image

2.2 Discrete Cosine Transform

Discrete cosine transform (DCT) works by dividing the image into different frequencies, including the low frequency, high frequency and medium frequency coefficients [21]. It is an orthogonal transform based on real numbers. DCT domain has a small amount of computation and a strong energy concentration characteristic. Most of the natural signals, including sound and image, are concentrated in the low frequency part of the discrete cosine transform, and it is easy to extract the visual feature vector. The two-dimensional discrete cosine transform (2D-DCT) formula (2) is:

$$F(u,v) = c(u)c(v)\sum\nolimits_{x=0}^{M-1}\sum\nolimits_{y=0}^{N-1} f(x,y)\cos\frac{\pi(2x+1)u}{2M}\cos\frac{\pi(2y+1)v}{2N} \tag{2}$$

Among them, $u = 0, 1, \ldots, M-1, v = 0, 1, \ldots, N-1$

$$c(u) = \begin{cases} \sqrt{\frac{1}{M}} & u = 0 \\ \sqrt{\frac{2}{M}} & u = 1, 2, \ldots, N-1 \end{cases} \qquad c(v) = \begin{cases} \sqrt{\frac{1}{M}} & v = 0 \\ \sqrt{\frac{2}{M}} & v = 1, 2, \ldots, N-1 \end{cases}$$

x, y is spatial sampling frequency domain; u, v are frequency domain samples, which are usually represented by a matrix of pixels in digital image processing, i.e. M = N.

2.3 RSA Algorithm

If a sequence can be predetermined, repeatedly produced and copied, and has the randomness of a random sequence, i.e. statistical characteristic, this sequence is called a pseudo-random sequence. Pseudo-random sequence is widely used in cryptography, communication ciphers, communication systems and radar signal design. RSA algorithm is one of the three major algorithms of cryptography (RSA, MD5, DES), which is an asymmetric encryption algorithm. Until now, RSA is still the most widely used asymmetric encryption algorithm. The asymmetric means the encryption key is inconsistent with the decryption key. The algorithm essentially uses a principle that large numbers are difficult to be encrypted by factorization. It includes three steps: generate a key, encrypt a password and decrypt the transform. The first step is key generation, which is used as a key to encrypt and decrypt data. The second step is encryption, in which the actual process is in the process of converting from m to cipher c.

The third step is decryption, in which the encrypted text c is restored to m.

3 Proposed Algorithm

3.1 Feature Extraction

We selected a 512 pixels × 512 pixels standard human brain slice map as our experimental object. The following content refers to this picture unless otherwise specified. As known that the low frequency sub band coefficients can well characterize the medical image, and the high frequency sub band coefficients reflect the details and texture of the image in all directions and other important information. Therefore, we extract the low-frequency sub-band coarse layer when extracting the visual features of the image. In this paper, we first perform curvelet transform, then perform DCT on its low-frequency sub-band coarse layer, which stands for the characteristics of energy concentration. After performing the Curvelet-DCT transform on the cover image and applying different attacks, we observed that the coefficient values changed after different attacks, but the symbols did not change.

In order to extract the visual feature vector of the image, we need to perform the perceptual hash binarization on the coefficients of the transform domain. If the coefficient is larger than the average value, we binarize it to "1". If it is smaller than the average value, we binarize it to "0". We set the average value as perceptual hash threshold, 32 bits as the visual feature vector bit length. In order to verify whether the image visual feature vector extracted by this method can distinguish different pictures, we randomly select six different test medical images and the "Lenna" image as shown in Fig. 2 to process and calculate the correlation between them, and get the Table 1 as follows.

We can observe that the correlation coefficients between different medical images are less than 0.5, which means that the visual feature vector we extracted can distinguish different medical images and the algorithm is feasible. In addition, it should be pointed out that from Fig. 2(h), we can find the algorithm can not only target medical images, but also suitable for normal images.

3.2 Watermark Encryption and Embedding

Watermark Encryption. As shown in Fig. 3, we first use RSA algorithm to generate a pseudo-random sequence *x(i)* of length $L = 2500$, $1 \leq i \leq 2500$. The *p*, *q* are random prime numbers and *e* is a random integer. Then we use the return value num function of the sort function and the original watermark *W(i ,j)* (32 pixels × 32 pixels) to perceive the binary matrix S(i, j) obtained by the hash binarization process. Finally, the combination of the two generate a scrambled encryption watermark *EW (i, j)*. The encryption of the RSA algorithm based on the large number decomposition mechanism and the scrambling of the num function space, the simultaneous use of the two, further ensure the security of the watermark.

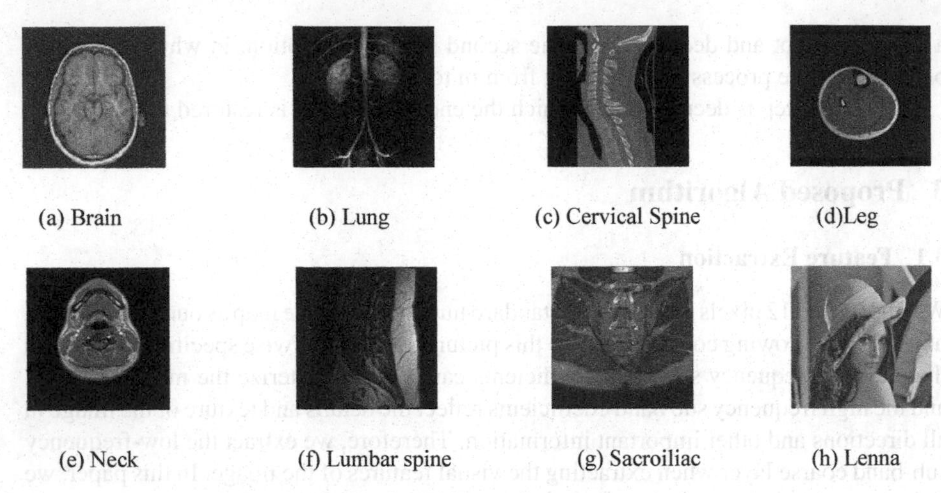

Fig. 2. Different test images

Table 1. Correlation of visual feature vectors of different images

Image	(a)	(b)	(c)	(d)	(e)	(f)	(g)	(h)
(a)	1.00	0.15	0.00	0.22	0.20	−0.13	0.28	0.07
(b)	0.15	1.00	0.00	0.00	0.22	−0.10	−0.04	0.11
(c)	0.00	0.00	1.00	0.06	−0.10	−0.10	−0.18	0.05
(d)	0.22	0.00	0.06	1.00	0.09	0.03	−0.04	−0.09
(e)	0.20	0.22	−0.10	0.09	1.00	−0.13	0.16	0.09
(f)	−0.13	−0.10	−0.10	0.03	−0.13	1.00	−0.09	0.13
(g)	0.28	−0.04	−0.18	−0.04	0.16	−0.09	1.00	0.04
(h)	0.07	0.11	0.05	−0.09	0.09	0.13	0.04	1.00

Watermark Embedding. The embedding method we use is shown in Fig. 4 XOR operation is performed on the visual feature vector *V(i)* of the original medical image and our encrypted watermark *EW (i, j)*. As we did not change the pixel values of the original image, this operate is a zero-watermark embedding.

3.3 Watermark Extraction and Recovery

Give the medical image 5° clockwise rotation, as shown in Fig. 5. The extraction and recovery process of the watermark is the inverse process of watermark embedding and encryption.

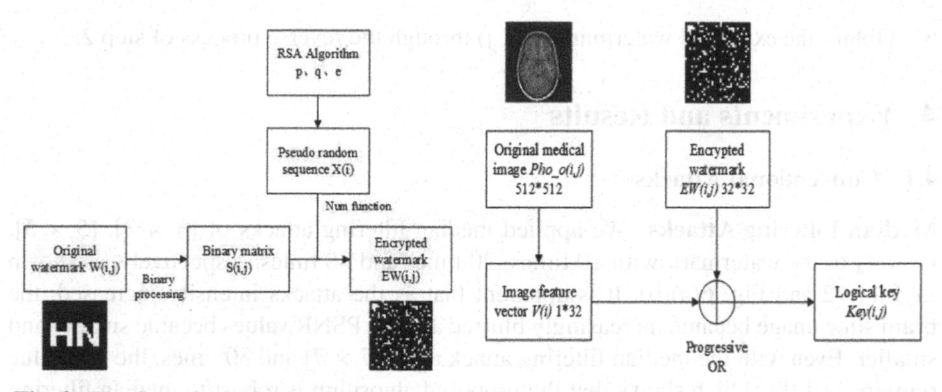

Fig. 3. The flow chart of watermark encryption

Fig. 4. Watermark embedding flow chart

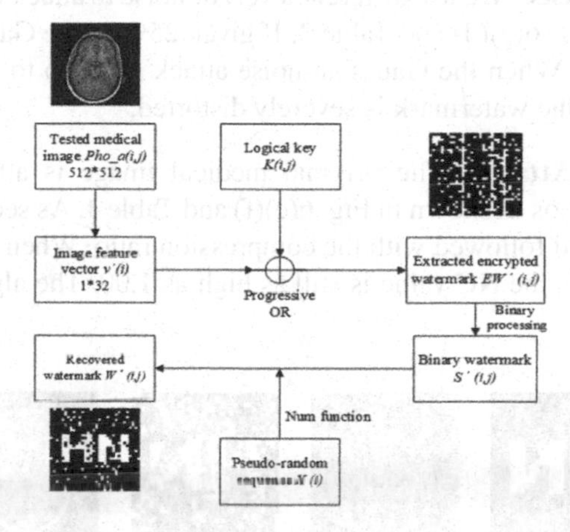

Fig. 5. Watermark extraction and recovery

3.4 Algorithm Summary

The entire algorithm is summarized as follows:

1. First, the original medical image pho_o (i, j) undergoes Curvelet-Dct transformation to generate a 32-bit image feature vector V (i);
2. The original watermark W (i, j) is subjected to binary processing and the pseudo-random sequence X (i) generated by the RSA algorithm is scrambled to obtain an encrypted watermark EW (i, j);
3. The V (i) generated in step 1 is XORed with the encrypted watermark EW (i, j) generated in step 2 to obtain the logical key key (i, j);
4. The attack test medical image through step one to generate a test image feature vector V'(i);
5. V′(i) and logical key key (i, j) XOR row by row to get the extracted encrypted watermark EW′ (i, j);

6. Obtain the extracted watermark w′(i, j) through the reverse process of step 2.

4 Experiments and Results

4.1 Conventional Attacks

Median Filtering Attacks. We applied median filtering attacks of [3 × 3], [5 × 5], [7 × 7] to the watermark with 10 times, 20 times and 30 times, respectively, as shown in Table 2 and Fig. 6(a)(b). It is apparent that as the attacks intensity increased, the brain slice image became increasingly blurred and the PSNR values became smaller and smaller. Even with the median filtering attack up to [7 × 7] and 30 times, the NC value remains at 1.00, which shows that the proposed algorithm is robust to median filtering attacks.

Gaussian Noise Attacks. We use different levels of noise to attack the original medical image, as shown in Fig. 6(c)(d) and Table 3. If given 25% of the Gaussian noise attack, the NC value is 0.80. When the Gaussian noise attack level up to 75%, the NC value drops below 0.5, and the watermark is severely distorted.

JEPG Compression Attacks. The original medical image is attacked by different JPEG compression ratios as shown in Fig. 6(e)(f) and Table 4. As seen from Table 5, the PSNR values increased followed with the compression ratio. When the JEPG compression rate reached 80%, the NC value is still as high as 1.00. The algorithm is robust to JEPG attacks.

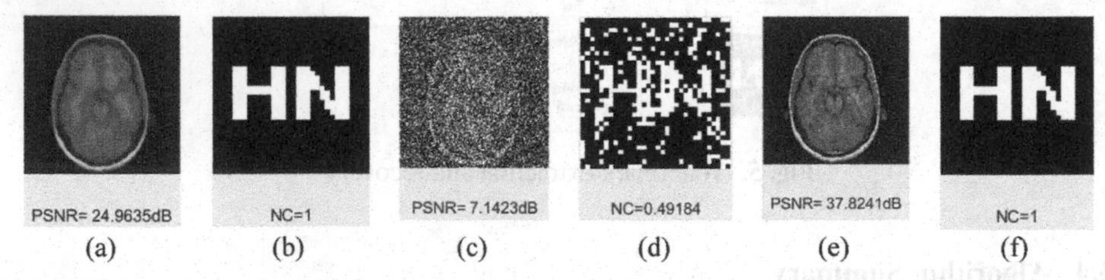

Fig. 6. Conventional attacks (a) Median filtering [7 × 7] 30 times (b) Extracted watermark under median filtering; (c)75% of Gaussian noise attacks (d) Extracted watermark under noise attacks; (e) JPEG compression 80% (f) Extracted watermark under JPEG compression.

Table 2. Data of median filtering attacks

Parameter	[3 × 3]			[5 × 5]			[7 × 7]		
Times	10	20	30	10	20	30	10	20	30
PSNR (dB)	34.00	33.82	33.78	28.56	27.46	26.92	26.40	25.45	24.96
NC	1.00	1.00	1.00	1.00	1.00	1.00	1.00	1.00	1.00

Table 3. Data of Gaussian noise attacks

Gaussian noise	1%	10%	25%	50%	75%
PSNR (dB)	20.43	11.89	9.20	7.77	7.15
NC	1.00	0.80	0.80	0.63	0.49

Table 4. The data of JEPG compression attacks

Compression quality	5%	10%	20%	40%	60%	80%
PSNR (dB)	28.44	31.29	33.81	35.46	36.42	37.82
NC	1.00	1.00	1.00	1.00	1.00	1.00

4.2 Geometric Attack

Rotation Attack. The original medical image is attacked by different angles of rotation, as shown in Fig. 7(a)(b) and Table 5. below. Although the NC values will drop below 0.5 after 30° clockwise rotation, the extracted watermark still clearly. It is also robust to rotational attacks.

Scaling Attacks. We selected scaling attacks with different coefficients to apply to the original medical image. The results are shown in Fig. 7(c)(d) and Table 6. We can find different coefficients. When the zoom factor is greater than 0.2, their NC values are all 1.00. Our extended attack algorithm is very powerful.

Translation Attacks. We move down different angles step by step, as shown in Fig. 7(e)(f) and Table 7. When down to 12% vertically, the extract watermark is still clearly visible, and the NC value is still as high as 0.79.

Y-direction Crop Attacks. In order to test the effect of the cropping attacks of the original medical image in Y direction, we gradually increased the crop angle from 3% to 56%, as shown in Fig. 7(g)(h) and Table 8. As shown, when cropping 50%, the NC value is still as high as 0.55.

4.3 Comparisons

A convenient way to fairly compare the performance of two watermarking algorithms is to compare the imperceptibility and robustness of the two methods on the premise that a common host image and a watermark of the same size are used [21]. So, for the following comparison, we used a host image with a size of 512 × 512 and a watermark image with a size of 32 × 32. Because the robustness and imperceptible performance of watermarked images in the face of geometric attacks are urgent problems to be solved, we only compare some common geometric attacks during comparison. First, by comparing the NC values in the Table 9, we can find that as the attack intensity increases, the NC

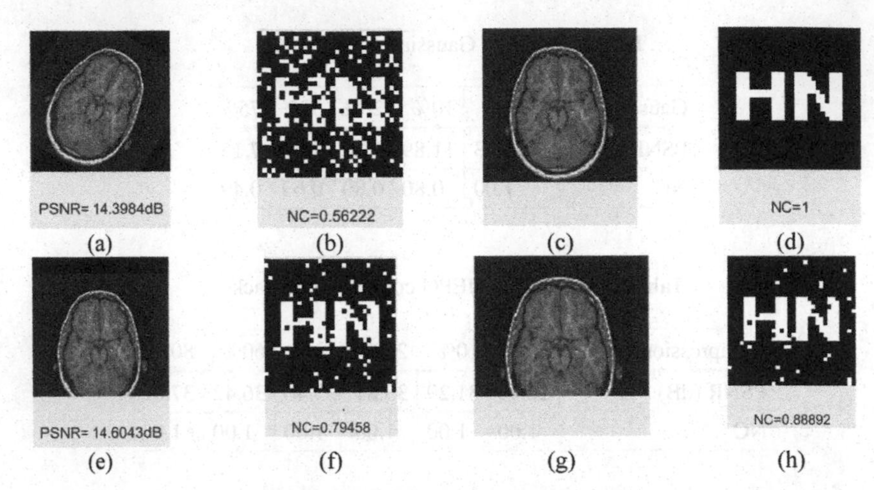

Fig. 7. Geometric Attacks (a) Rotate 30° clockwise (b) Extracted watermark under rotate attacks; (c) Scale 2 times (d) Extracted watermark under scale attacks; (e) Move down 12% vertically (f) Extracted watermark under translation attacks; (g) 20% in the Y direction (h) Extracted watermark under under translation attacks;

Table 5. The data of rotation attacks

Rotation (clockwise)	2°	4°	6°	10°	20°	30°	35°
PSNR (dB)	22.36	19.02	17.24	15.60	14.60	14.40	14.26
NC	0.89	0.79	0.71	0.63	0.63	0.56	0.45

Table 6. The data of scaling attacks

Zoom (×)	0.1	0.2	0.5	1.5	2.0
NC	0.79	0.89	1.00	1.00	1.00

Table 7. Translation attacks data

Move vertically (%)	2	6	8	10	12	16	20
PSNR	16.76	15.03	14.93	14.94	14.60	14.13	13.78
NC	1.00	1.00	1.00	0.89	0.79	0.57	0.45

value gradually decreases (except for telescopic attacks). Secondly, we can find that in the face of a shear attack in the Y direction and a vertical downward translation attack, our proposed algorithm is due to the Contourlet-Dct algorithm. The Dct algorithm is relatively small.

Table 8. The data of cropping attacks

Cropping (Y direction)	3%	9%	15%	20%	50%	56%
NC	1.00	1.00	1.00	0.89	0.55	0.48

Table 9. Comparison of different multi-scale algorithms

	Attacks type	Scaling attacks			Cropping attacks (Y direction)			Translation attacks (down)			Rotation attacks (clockwise)		
		x0.4	x0.8	x2.0	2%	8%	16%	1%	3%	4%	5°	10°	15°
Curvelet-Dct	NC	1.00	1.00	1.00	0.89	0.89	0.71	1	0.89	0.79	0.89	0.70	0.55
Contourlet-Dct [7]	NC	0.76	0.76	1.00	1.00	0.67	0.58	0.77	0.64	0.52	0.95	0.82	0.60

5 Conclusions

This paper presents a new, robust, zero-watermarking medical image algorithm. It used a pseudo-random sequence, Curvelet-DCT and RSA encryption algorithms to generate a key for embedding and extracting the watermark. The experimental results show that the proposed algorithm not only improved the security of watermark information for medical images, but also has good robustness under conventional attacks and geometric attacks. It does not affect the quality of original medical images, also suitable for normal images and can be widely used in various fields such as digital medical, copyright protection, cultural relic protection, digital signature, insurance companies, etc.

References

1. Liu, R., Tan, T.: A review of digital image watermarking. J. Commun. **21**(8), 39–48 (2000)
2. Wang, J., Lian, S., Shi, Y.: Hybrid multiplicative multi-watermarking in DWT domain. Multidim. Syst. Signal Process. **28**(2), 617–636 (2017)
3. Xiong, L., Xu, Z., Shi, Y.: An integer wavelet transforms-based scheme for reversible data hiding in encrypted images. Multidim. Syst. Signal Process. (2017). https://doi.org/10.1007/s11045-017-0497-5
4. Jiang, X.: Digital watermarking and its application in image copyright protection. In: 2010 International Conference on Intelligent Computation Technology and Automation (2010)
5. Zhou, M., Geng, G., Wu, Z.: Digital Preservation Technology for Cultural Heritage. Springer, Heidelberg (2012). https://doi.org/10.1007/978-3-642-28099-3
6. Shih, F.Y.: Digital Watermarking and Steganography: Fundamentals and Techniques. CRC Press, Boca Raton (2017)
7. Wu, X., et al.: Contourlet-DCT based multiple robust watermarking for medical images. Multimed. Tools Appl. **78**, 8463–8480 (2018). https://doi.org/10.1007/s11042-018-6877-5
8. Jayashree, N., Bhuvaneswaran, R.S.: A robust image watermarking scheme using Z-transform, discrete wavelet transform and bidiagonal singular value decomposition. Comput. Mater. Continua **58**(1), 263–285 (2019)

9. Liu, J., Li, J., Chen, Y., Zou, X., Cheng, J., Liu, Y., Bhatti, U.A.: A robust zero-watermarking based on SIFT-DCT for medical images in the encrypted domain. Comput. Mater. Continua **61**(1), 363–378 (2019)
10. Liu, J., Li, J., Cheng, J., Ma, J., Sadiq, N., Han, B., Geng, Q., Ai, Y.: A novel robust watermarking algorithm for encrypted medical image based on DTCWT-DCT and chaotic map. Comput. Mater. Continua **61**(2), 889–910 (2019)
11. Zhou, Y., Jin, W.: A novel image zero-watermarking scheme based on DWT-SVD. In: IEEE International Conference on Multimedia Technology, pp. 2873–2876 (2009)
12. Channapragada, R.S.R., Prasad, M.V.N.K.: Digital watermarking based on magic square and ridgelet transform techniques. In: Mohapatra, D.P., Patnaik, S. (eds.) Intelligent Computing, Networking, and Informatics. AISC, vol. 243, pp. 143–161. Springer, New Delhi (2014). https://doi.org/10.1007/978-81-322-1665-0_14
13. Thanki, R., Borisagar, K., Borra, S.: Speech watermarking technique using the finite ridgelet transform, discrete wavelet transform, and singular value decomposition. Advance Compression and Watermarking Technique for Speech Signals. SECE, pp. 27–45. Springer, Cham (2018). https://doi.org/10.1007/978-3-319-69069-8_3
14. Sadreazami, H., Ahmad, M.O., Swamy, M.N.S.: Multiplicative Water630 mark decoder in contourlet domain using the normal inverse Gaussian distribution. IEEE Trans. Multimed. **18**(2), 196–207 (2016). https://doi.org/10.1109/TMM.2015.2508147
15. Kim, W.H., Nam, S.H., Kang, J.H., et al.: Multimed. Tools Appl. (2019). https://doi.org/10.1007/s11042-018-6879-3
16. Chen, L., Zhao, J.: Robust contourlet-based blind watermarking for depth-image-based rendering 3D images. Signal Process. Image Commun. **54**, 56–65 (2017)
17. Candes, E.J., Guo, F.: New multiscale transforms, minimum total variation synthesis: Application to edge-preserving image reconstruction. Sig. Process. **82**, 1519–1543 (2002)
18. Candes, E.J., Donoho, D.L.: Curvelets, multiresolution representation, and scaling laws. In: Proceedings of SPIE 4119, Wavelet Applications in Signal and Image Processing VIII, 4 December 2000
19. Starck, J.L., Candes, E.J., Donoho, D.L.: The curvelet transform for image denoising. IEEE Trans. Image Process. **11**, 670–684 (2002)
20. Candès, E., Demanet, L., Donoho, D., Ying, L.: Fast discrete curvelet transforms. Multiscale Model. Simul. **5**(3), 861–899 (2006)
21. Mohammad, M., Gholamhossein, E.: A new DCT-based robust image watermarking method using teaching-learning-based optimization. J. Inf. Secur. Appl. **47**, 28–38 (2019)

Encryption Algorithm for TCP Session Hijacking

Minghan Chen, Fangyan Dai, Bingjie Yan, and Jieren Cheng(✉)

Longjuan Wang Hainan University, Haikou 570228, Hainan Province, China
mh.chen@hainanu.edu.cn, fangyeee@163.com, beiyuouo@foxmail.com,
cjr22@163.com, 40552382@qq.com

Abstract. Distributed network of the computer and the design defects of the TCP protocol are given to the network attack to be multiplicative. Based on the simple and open assumptions of the TCP protocol in academic and collaborative communication environments, the protocol lacks secure authentication. In this paper, by adding RSA-based cryptography technology, RSA-based signature technology, DH key exchange algorithm, and HAMC-SHA1 integrity verification technology to the TCP protocol, and propose a security strategy which can effectively defend against TCP session hijacking.

Keywords: TCP protocol · Session hijacking · Security countermeasure · Three-time handshake

1 Introduction

The transmission of network information is inseparable from the most essential network protocol of the computer. The problem of information transmission affects the size of the cluster and the size of the throughput. The connection-oriented TCP protocol (Transmission Control Protocol) is often used at the transport layer. As the most popular protocol in the world today, the TCP protocol can establish a connection through the three-way handshake method, provide a reliable end-to-end byte transmission stream for network information transmission and prevent packet loss. The purpose of the three-way handshake is to synchronize the serial numbers and acknowledgments of both parties and exchange TCP window size information. However, since protocols such as TCP were formed in the early 1980s, more practical aspects of the protocol were considered during design. Therefore, the TCP protocol data stream is transmitted in plain text, which lacks encryption and authentication of data. There are insufficient considerations in network security, such as: unable to provide reliable identity verification, unable to effectively prevent information leakage, and not providing reliable complete verification of information. And the protocol has no means to occupy and allocate resources, making the TCP protocol fragile and limited. Attacks such as IP spoofing and TCP session hijacking using the TCP protocol are increasing. Therefore, this article studies the problem of TCP plaintext transmission and lack of data encryption. Using the RSA asymmetric encryption algorithm and DH key exchange algorithm, the existing TCP session process

X. Sun et al. (Eds.): ICAIS 2020, LNCS 12240, pp. 191–202, 2020.
https://doi.org/10.1007/978-3-030-57881-7_17

adds identity verification and encryption to ensure a higher algorithm safety. At the same time, this paper improves the single authentication of the original server to the client to achieve double-ended authentication to confirm the authenticity of the data source. In the third handshake process, this article chooses to use HAMC-SHA1 with a hash operation with a key, and performs HMAC-SHA1 integrity verification for each subsequent transmission. Compared with MD5 without a key, In this paper, identity verification is performed while encrypting, which makes the session process more secure and can solve the problems of an attacker disguising as an attacker and communicating with the server.

Compared to other attacks, TCP session hijacking has many benefits, such as sniffing passwords in IP datagrams, especially when using advanced authentication and authentication techniques. However, since all of these advanced authentication techniques occur at the time of the connection, they will not provide protection after this. Therefore, an attacker can enter the system simply by hijacking a legitimate connection. This allows an attacker to appear as a legitimate user in the operating system security mechanism [1].

From 90 mid-year introduction of cookies since the use of cookies loopholes session hijacking attacks has been a security problem. In the article, Vissiago [2] describes the use of Web application design flaws ED for session management attacks and discusses the "current prevalence" of these defects. This survey emphasizes that cookies are vulnerable to session hijacking and are not a source of ability to maintain web authentication. Because of this, many researchers began to study network security protocols. Wannes Meert and some other people in 2001 publish a paper "SessionShield:Lightweight Protection Against the Session Hijacking" [3] which presented SessionShield - a lightweight client protection mechanisms. Prevent session hijacking by using session identifier values not used by legitimate client scripts. Dacosta and some other researcher [4] proposed a more robust session authentication scheme for one-time cookies (OTC) as a solution for session hijacking. Each user request is signed by using a session secret that is securely stored in the browser. Each token is only allowed to be used once. One cookie can significantly improve the security of the web application with minimal impact on performance.

However, cookies have a serious design flaw that limits their security. In particular, cookies cannot provide session integrity for an attacker capable of hosting content on the relevant domain. Therefore, Bortz [5] proposed a new concept of related domains that are vulnerable to session hijacking attacks. They proposed a solution to suppress attacks and lightweight expansion of cookies which are secure against related domains and network attackers. By this way, they result in session integrity.

The simplest and most effective defense against IP spoofing, TCP spoofing and TCP session hijacking are those that provide Internet access. If all of these organizations have sufficient responsibility to prevent IP datagrams coming from outside the network from reaching the network, the above attacks cannot be performed. Unfortunately, many networks offer unregulated Internet access. Therefore, other means of preventing fraud and hijacking attacks must be used. The easiest and most effective way for an organization is to block all IP datagrams from the network, and a properly configured firewall can be used to enforce such policies. Therefore, Elie [6] exploited the features of a web browser to develop a secure login system-Session Juggler, from an untrusted terminal and provided a secure logout mechanism. In 2012, Asif and some researchers [7] found that the efficient identity management system OpenID is easily hijacked by the session. In order to solve the theft of user identity information caused by session hijacking, ELie and other researchers proposed a two-factor authentication method. It authenticates the user by checking the credentials and PIN code stored on the ID server. Even if the session between the user and the identity provider is hijacked by an intruder, the intruder may not be able to access the PIN due to the two-factor authentication system. Burgers and other researchers [8] made a new way to prevent theft of the session, via the secure communication channel negotiation bound to the application user authentication, a server is introduced independently of the client and server software running a reverse proxy, the mechanism Established on the communication channel of security negotiation. This is achieved by establishing a server-side reverse proxy. It runs independently of the client and server software.

It is also important to assess threats. For example, an attacker is less likely to encounter hijacking an anonymous FTP session. In this regard, Stango and other researchers [9] used threat analysis to show how potential opponents can use system weaknesses to achieve their goals. It identifies threats and defines risk mitigation strategies for specific architectures, features and configurations. Desmet [10] analyzed the threats that might be associated with using web services in web applications. In session hijacking, the intruder mimics the identity of the victim and uses the same resource access rights as the victim. The consequences can be catastrophic because it can lead to the loss of critical information. Therefore, session hijacking has always been the focus of researchers, and they have proposed strategies to prevent and mitigate session hijacking.

2 Background and Relevant Work

2.1 Attack on TCP

SYN Attack. A SYN attack is a denial of service attack against the three-way handshake of the TCP protocol. In the server receives from the client SYN after a request packet will be responded by ACK, and reserve resources for the client until the client from listening to the ACK packet start communicating parties. If the client sends a large number of SYN packets in a short interval, the limited resources of the server will be exhausted, the server will not continue to work, and the network hacker successfully implements the denial of service attack. (complementary map).

Session Hijacking. The TCP protocol stipulates that authentication between two hosts takes place during the connection establishment phase (Handshake stage), after which no authentication is required. The lack of security for TCP session authentication gives network hackers a chance. After the client establishes a connection with the server, the attacker can send a fake RST message to the client, causing the client to terminate the session connection with the server. At this point, the network hacker forges the data packet to communicate with the server based on the obtained packet identifier (the client's IP address, the client's port, the server's IP address, the server's port, the serial number, and the acknowledgment number).

3 Defense Mechanism for TCP Session Hijacking

3.1 Improved TCP Handshake Scheme

Problem Solving Process

The first handshake. Client by client and server issuing common certification authority to a third party server obtaining the digital certificate server public key S+ ; client to select a large prime number P and P is a primitive root of A, and the selected private XC(XC < P); client according DH secret key exchange algorithm and P, a computing client public key YC; client with RSA encryption algorithm to S+ is a secret key of large prime numbers P||A||YC|| bits Description the combined data is encrypted ciphertext C. client will SYN set a randomly selected sequence number seq is J, the SYN packet header splicing C sent to the Server.
Second handshake. After server receiving the data packet, the data portion client and server mutual recognition of third party certification authority's private key S- to C for RSA decryption algorithm, to obtain a large prime number P, the original root A and YC; server selected private key XS(XS < P); the DH exchange algorithm and key P, a calculation server public key YS; the DH key exchange algorithm and YC, P calculate the session key K; server to client, and server third-party certification authority's private key common authentication S- as a key in accordance with RSA algorithm, YS signing obtain S; server will SYN set to 1, the ACK is set 1, randomly selected sequence number seq is I, the acknowledgment number ack For J, the ACK message header is spliced as data S to send the data packet to the client.
The third handshake. After the client receives the data packet, it first checks whether the ACK is 1, and confirms whether the ack is J. If not, it discards it. If it is, it uses S+ as the key and S is the message input. According to the RSA algorithm, it is obtained. The server's public key YS; client uses the YS and DH key exchange algorithm P, to calculate the session key K; the client uses the session key K as the key, according to the HAMC-SHA1 algorithm to generate the message hash value HAMC; client will ACK Set to 1, ack is I, with the HAMC as the data, send the packet to the server.

After receiving the data packet, the server first checks whether the ACK is 1, and whether ack is I. If it is not discarded, if yes, the session key K is used as the key, and the HAMC' is calculated according to the HAMC-SHA1 algorithm. If HAMC' = HAMC, Then the connection is successfully established. 4. In the data transmission phase, every 100 data packet senders use the session key K as a key, and according to the HAMC-SHA1 algorithm, calculate the HAMC; the sender splicing data packets are sent together with the HAMC;

After receiving the data packet, the receiver uses the session key K as the key and calculates the HAMC' according to the HAMC-SHA1 algorithm. Only when HAMC = HAMC', the transmission can continue (Fig. 1).

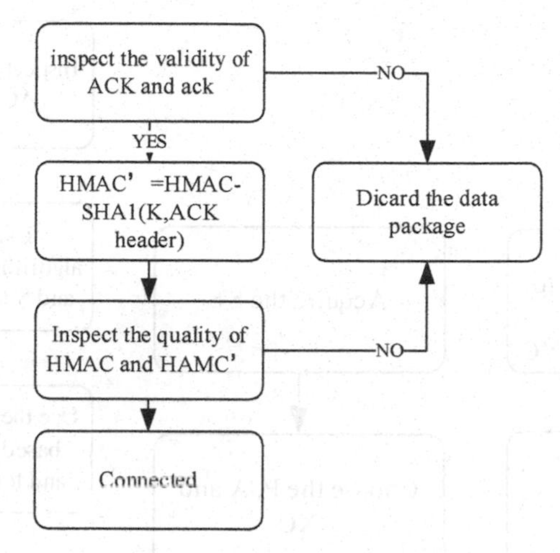

Fig. 1. Third shake of server

4 Experiment Analysis

4.1 TCP Session Hijacking Experiment

TCP Session Hijacking Experiment Process. The server creates a socket object, binds the IP address and port number (192.168.0.105, 49999), and then listen on the 49999 port number. The client creates socket object, binds IP address and port number (192.168.0.104, 59999), then send a connection request to the server (192.168.0.105 49999). The server receives the client's connection request and creates a thread to handle the connection verification of the client's TCP protocol. The server and client perform a TCP protocol three way handshake to verify identity. The attacker use Ettercap to implement ARP attack to make the client mistakenly think it is a gateway and send packets to the attacker. Then the attacker capture the packets that are transferred between the server and the client, as shown in the picture below. The attacker uses the information in the TCP packet. The required information is as follows: using TCP protocol; Time To Live: 128; Source IP address: 192.168.0.104; Destination IP address: 192.168.0.105; Source Port: 59999; Destination Port: 49999; Next Sequence number: 2775375568; Acknowledgment number: 2356415172.

The attacker uses the Spoof Ip4Tcp packet option in the Netwag tool and enter the command "sudo passwd root" that attacker wants to send to the server to generate the Netwox command (Figs. 2, 3 and 4).

```
40 --ip4-ttl 128 --ip4-protocol 6 --ip4-src 192.168.0.104 --ip4-dst 192.168.0.105 --
tcp-src 59999 --tcp-dst 49999 --tcp-seqnum 2775375568 --tcp -acknum
2356415172 --tcp-ack --tcp-psh --tcp-window 128 --tcp-data "'sudo passwd root'"
```

The attacker uses the Netwox tool to generate and send fake TCP packets to the server.

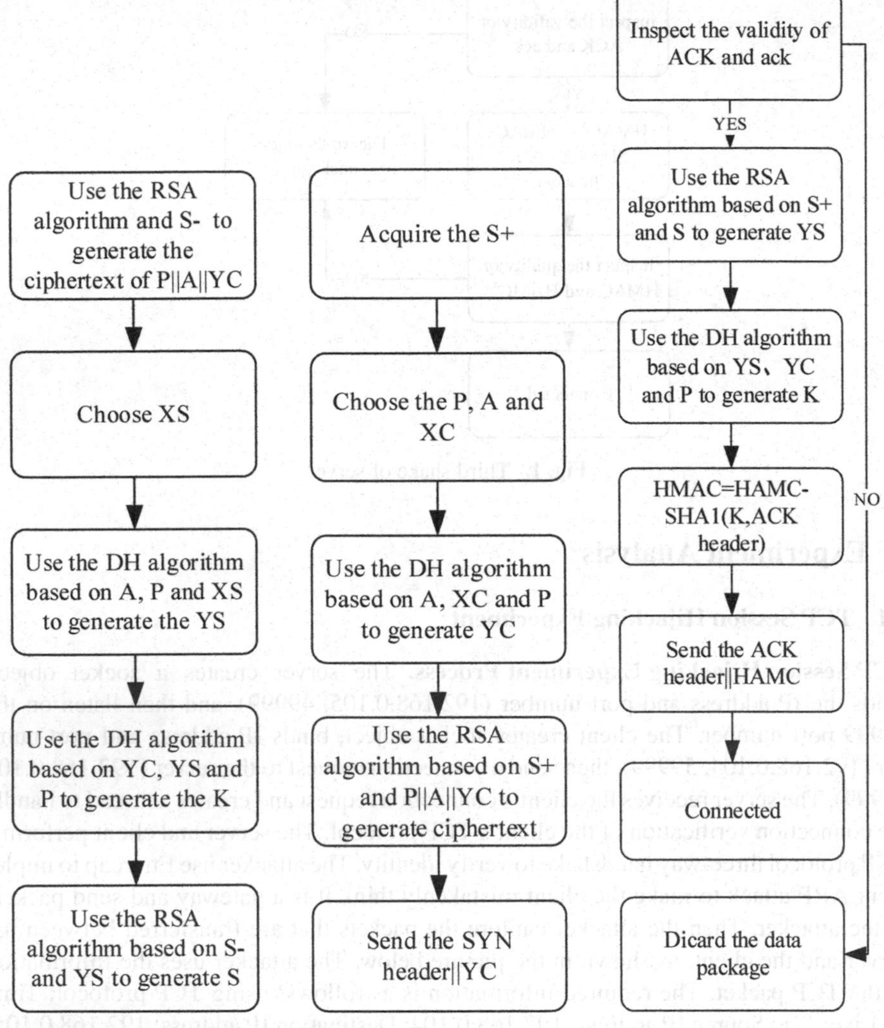

Fig. 2. First shake of client **Fig. 3.** Second shake of server **Fig. 4.** Third shake of client

The server receives the data packet sent by the attacker. After verification, it is mistaken that it is the data packet sent by the client, correctly receives the data packet. and execute the commands. The output on the server side is as follows.

```
Listening…
Accept!
Connection addr: (192.168.0.104, 59999)
Recv: sudo passwd root
```

You can see that the server received the packet forged by the attacker and thought that the packet came from the client (192.168.0.104: 59999). Then the attacker forges the server to send the packet containing RST command to the client to let the client go offline. After that, the attacker gets all the permissions of the client on the server.

Encryption Experiment Process. Note: For the convenience of experiment, the number of prime p used in the DH key exchange protocol used in the experiment is small, but it can be recommended according to RPC 3626 in the actual environment should use a larger prime number, so that the attacker can not hack the server's or client private's key in the DH key exchange protocol in a limited time.

Specific Experimental Process. In the first handshake session, the client obtains the server's public key S+ from the digital certificate issued by the third-party authority authenticated by the client and the server to the server:

```
S+:(728947908706520468702710835614097609778934317790345395720490862
5740114208903632340542215628999136015415388380418420294863244335815
571945495531508812116349,65537)
```

and selects the prime number p = 97, the original root A = 5, the private key XC = 36 in the DH key exchange protocol, calculate the public key YC = A^{XA}(mod p) = 50 in the DH key exchange protocol. The information is stitched in the order of the order negotiated with the server 'pAXA$', and the data is obtained by encrypting with the RSA encryption algorithm:

```
b'v\xbf\x8f\x85+\x9f\xb4\xcd|\x17\x81\xe9y\xcd\x96\xa6\xd8\xbe\x80\x89\x04l\
x0e\xb6\\\xc2y\x93iO\xb0sBL\x84D\x13\xf33E]\xf8\xe4\x97\t\xb1\xb0\xb4\xb5\
x0e\xfd9\xa0]\xf6\x19\xb7\xedh\xf6\x84G\x8e'
```

At the same time, the client sets SYN to 1, randomly selects the serial number seq as J, and packs it with other information into a TCP packet and sends it to the server.

The server receives the TCP packet sent by the client, and uses the RSA private key corresponding to the public key when the digital certificate issued by the client and the server co-certified by the server and the server authenticates the public key (Fig. 5):

```
S-:(7289479087065204687027108356140976097789343177903453957204908625
7401142089036323405422156289991360154153883804184202948632443358155
71945495531508812116349,65537)
```

Then the server decrypts the information sent by the client, and the information can only be decrypted by the server. After that the server can obtain the prime number p, the original root A and the client public key YC in the DH key exchange protocol. The server selects the private key XS = 58 in the DH key exchange protocol, calculates the public key YS = A^{XB} (mod p) = 44 in the DH key exchange protocol, and then calculates the shared key K = YC^{XS} mod p = 50^{58} mod 97 = 75 in the DH key exchange protocol.

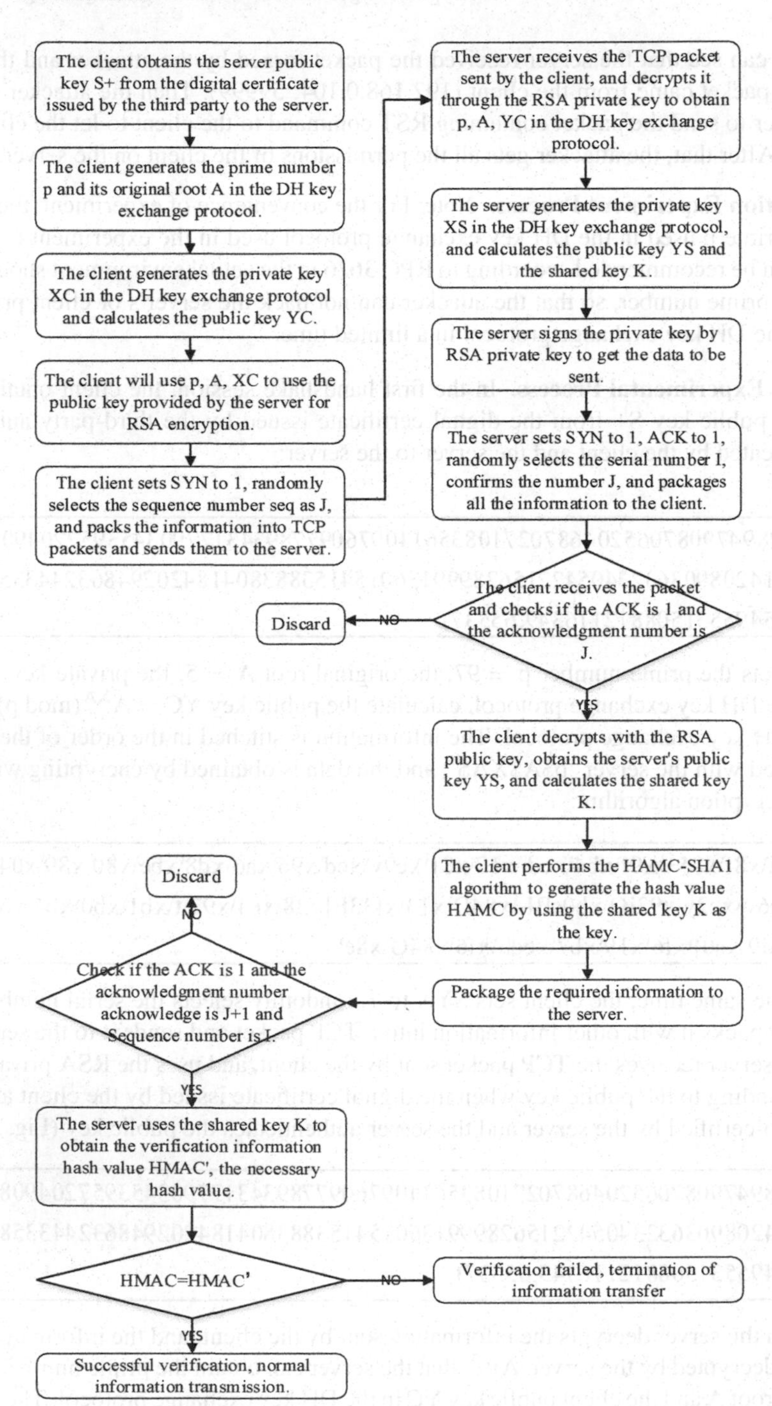

Fig. 5. Improved three handshakes of TCP

The server encrypts the public key YS in its own DH key exchange protocol by its own RSA private key to obtain the data:

b'e\xe3J\xbdF\x93^\x86\xa9\xba\xe3d\xc6\x94R\xd3_\xaf\xb1\xefL_%\x8a\x0e\
xdb\x19\xc7\x8a>\x8c\xea\x8d\xf1)P\xe0q\xa5\t\xa5\xa9+{\x8fd0<\x8b\x19b\xbe\
x86\xb0\x86yf\xcd\xa0@\xf7\xe8C3'

The server sets SYN to 1, ACK to 1, randomly selects the sequence number seq as I, confirms the acknowledgment number as J, and packs it and other required information into a TCP packet and sends them to the client.

After the client receives the TCP packet sent to him by the server, it first checks that the ACK is 1, the confirmation acknowledgment number is J. Then, the data packet is decrypted by using the RSA public key S+ on the server digital certificate as the key, and the data the public key YS in the server DH key exchange protocol contained in the data packet is obtained.

The client then calculates the shared key K using the DH key exchange protocol YS and according to $K = XC^{YS} = 50^{58} \bmod 97 = 75$. The client generates a mutual authentication message = 'message' hash value:

$$HAMC = ada668f4688e906e157d8613dc4408ce00de1cf0$$

According to the HAMC-SHA1 algorithm using the shared key K in the DH key exchange protocol as the key.

The client sets ACK to 1 and acknowledgment number to I, which packs the HAMC and other TCP packet headers and sends them to the server.

After receiving the data packet, the server first checks that the ACK is 1, and the acknowledgment is I. Then, using the shared key $K = 75$ as the secret key.

$$HAMC' = ada668f4688e906e157d8613dc4408ce00de1cf0$$

It is calculated according to the HAMC-SHA1 algorithm to verify whether

$$HAMC' = HAMC.$$

After the two parties successfully establish a connection, it is required to perform verification once every 100 data packets. The $K = 75$ in the DH key exchange protocol is used as the key. According to the HAMC-SHA1 algorithm, the HAMC is calculated, and the data to be sent is sent together with the HAMC data, and the party that accepts the verification uses the DH key exchange protocol in its own. $K = 75$, calculate HAMC' for verification. Message transmission can continue if and only if HAMC = HAMC'.

Experimental Results. After experiment, this paper used the modified encrypted TCP protocol verification, because it is impossible to brute force all possible private keys in the theoretical time complexity, and both parties will perform message verification every time period, which can prevent the attacker from pairing the client. The session hijacking of the server cannot obtain the operation permission of the client to the server, and the injection operation of the attacker to the server is extremely effectively avoided.

5 Discuss

5.1 Limitations

The defense mechanism based on TCP session hijacking proposed in this paper can effectively solve the session hijacking that occurs during the TCP three-way handshake phase. However, in the data transmission process, the HAMC-SHA1 signature and integrity verification for each data packet will spend a lot of system resources.

5.2 Solution

In order to ensure integrity verification and reduce overhead, we currently propose a window value of 100. When the number of packets sent by the client reaches 100, the 100th data packet is signed and the counter is cleared to 0. After receiving the data packet with the HAMC signature, the server performs integrity verification according to the negotiated HAMC-SHA1 algorithm and the session key K. Therefore, it is necessary to obtain a reasonable window value through a large number of experimental and scientific mathematical proofs, and achieve the best balance between resource consumption and security.

5.3 Advantages

HAMC-SHA1 Algorithm. Currently used integrity algorithms are MD5, MD4, SHA1 and others. SHA1 is one of the most widely used algorithms. HMAC is a key-related hash operation message authentication code. The HMAC algorithm using hashing algorithm and using a message and a key as input, generates a message digest as output. We use the HAMC-SHA1 algorithm to use the session key as the key to perform identity verification and complete the authentication.

Window Value Based Integrity Verification. To completely prevent the session be hijacked, it is the most reliable method for verifying the integrity of each data packet based on the HAMC-SHA1 and session secret key. Therefore we have proposed a more scientific window value. When the number of transmitted packets reaches the window value, both parties perform integrity verification and the counter is reset to 0. It can find a relative balance between security and resource consumption.

6 Conclusion

The design flaw of the TCP protocol is an objective cause of hacker attacks, and there are many technologies that can protect the network from attacks at present. This paper proposes a security mechanism based on TCP three-way handshake to defend against a well-known attack method - TCP session hijacking. First we show how to implement session hijacking, and then we add cryptography and authentication techniques to the TCP protocol for session hijacking. Through the three-way handshake phase of TCP protocol combined with cryptography to enhance identity verification, we propose a security countermeasure for session hijacking based on TCP protocol. This security countermeasure not only can effectively identify malicious packets constructed by network attackers, but also protect the protocol security. At the same time, it can also sign the authentication label of the data packet (client IP address, client port, server IP address, server port, serial number and confirmation number). Once these identifiers are changed, they are considered to be fake packets constructed by the attacker. If the packets can not pass the integrity verification on the server side, the server will discard the packets. The server only processes the data in the packet after the integrity verification is passed, and the integrity verification is based on the HAMC-SHA1 algorithm, where the secret key is the session key of the first handshake and the second handshake. In this way, in the case of relatively small resource consumption, integrity verification and authentication are implemented, and TCP session hijacking attacks are effectively blocked.

Future work includes a large number of cyberattack experiments and mathematical proofs, taking reasonable window values and find a balance between security and resource consumption. When the window value is smaller, the success rate of successfully intercepting TCP session hijacking is higher, but the resource consumption of the server and client will become very large; when the window value is larger, it is possible for a network attacker to take control of the communication details of the packet before successful hijacking until the next integrity check, and the security is reduced. Therefore, finding a scientific and reasonable window value through mathematical proof and a large number of experimental tests is the key to the practical application of this strategy.

References

1. Harrisa, B., Huntb, R.: TCP/IP security threats and attack methods. Comput. Sci. **22**, 885–897 (1999)
2. Visaggio, C.: Session management vulnerabilities in today's web. IEEE Secur. Priv. **8**(5), 48–56 (2010)
3. Nikiforakis, N., Meert, W., Younan, Y., Johns, M., Joosen, W.: SessionShield: lightweight protection against session hijacking. In: Erlingsson, Ú., Wieringa, R., Zannone, N. (eds.) ESSoS 2011. LNCS, vol. 6542, pp. 87–100. Springer, Heidelberg (2011). https://doi.org/10.1007/978-3-642-19125-1_7
4. Dacosta, I., Chakradeo, S., Ahamad, M., Traynor, P.: One-time cookies: Preventing session hijacking attacks with stateless authentication tokens. ACM Trans. Internet Technol. **12**(1), 1 (2012)
5. Bortz, A., Barth, A., Czeskis, A.: Origin cookies-: session integrity for web applications. In: Proceedings of the Web 2.0 Security and Privacy Workshop (W2SP) (2012)
6. Bursztein, E., Soman, C., Boneh, D., Mitchell, J.C.: Sessionjuggler: secure web login from an untrusted terminal using session hijacking. In: Proceedings of the 21st International Conference on World Wide Web, pp. 321–330. ACM (2012)
7. Asif, M., Tripathi, N.: Evaluation of OpenID-based DoubleFactor authentication for preventing session hijacking in web applications. J. Comput. **7**(11), 2623–2628 (2012)
8. Burgers, W., Verdult, R., van Eekelen, M.: Prevent session hijacking by binding the session to the cryptographic network credentials. In: Riis Nielson, H., Gollmann, D. (eds.) NordSec 2013. LNCS, vol. 8208, pp. 33–50. Springer, Heidelberg (2013). https://doi.org/10.1007/978-3-642-41488-6_3

9. Stango, A., Prasad, N.R., Kyriazanos, D.M.: A threat analysis methodology for security evaluation and enhancement planning. In: Third International Conference on Emerging Security Information, Systems and Technologies, SECURWARE 2009, 18–23 June, pp. 262–267 (2009)
10. Desmet, L., Jacobs, B., Piessens, F., Joosen, W.: Threat modelling for web services based web applications. In: Chadwick, D., Preneel, B. (eds.) CMS 2004. ITIFIP, vol. 175, pp. 131–144. Springer, Boston (2005). https://doi.org/10.1007/0-387-24486-7_10
11. Sahu, D.R., Tomar, D.S.: Strategy to handle end user session in web environment. In: Proceedings of National Conference on Computing Concepts in Current Trends, NC4T 2011, 11th and 12th August 2011 (2011)
12. Liu, B., Wang, L., Liu, M.: Lifelong federated reinforcement learning: a learning architecture for navigation in cloud robotic systems. IEEE Robot. Autom. Lett. **4**(4), 4555–4562 (2019)
13. Liu, B., Wang, L., Liu, M., Xu, C.: Federated imitation learning: a privacy considered imitation learning framework for cloud robotic systems with heterogeneous sensor data arxiv:1909.00895 (2019)
14. Kaddi, M., Benahmed, K., Omari, M.: An energy-efficient protocol using an objective function & random search with jumps for WSN. Comput. Mater. Continua **58**(3), 603–624 (2019)
15. Centonze, P.: Security and privacy frameworks for access control big data systems. Comput. Mater. Continua **59**(2), 361–374 (2019)
16. Tian, W., et al.: Defense strategies against network attacks in cyber-physical systems with analysis cost constraint based on honeypot game model. Comput. Mater. Continua **60**(1), 193–211 (2019)

A Covert Information Transmission Scheme Based on High Frequency Acoustic Wave Channel

Huiling Li, Yayi Zou, Wenjia Yi, Ziyi Ye, and Yi Ma(✉)

Software College, Northeastern University, Shenyang 110169, China
may@swc.neu.edu.cn

Abstract. With the development of Internet and communication technology, the security threats of mobile communication network emerge one after another. Using sound wave and covert channel, it's possible to find a covert security communication method based on high frequency sound waves at short distance. Initially, we use technologies such as channel coding and decoding, message authentication, signal synchronization, signal modulation and demodulation. Then we designed a multi-thread based sound wave acquisition and playback scheme. We also constructed a one-way P2P high frequency acoustic wave covert channel with good basic performance. On this basis, we introduced the error control mechanism and other related technologies and designed a two-way alternating information sharing scheme. Finally, the information security application of high frequency acoustic wave covert channel is realized on mobile devices.

Keywords: Covert security communication · High frequency acoustic wave · Information security

1 Introduction

After Pwn2Own Hacker Challenge founder Dragos Ruiu discovered the badBIOS virus [1], the existence of covert channels attracted people's attention. A covert channel is a communication channel that is not designed to transmit information [2]. Using this channel for communication can not only guarantee the concealment of communication content, but also hide the traces of communication [3]. If combined with encryption communication technology, another layer of protection can be added. Therefore, covert channel communication can become an effective means to prevent intruders from sniffing and monitoring based on traditional computer networks [4].

In computer system, acoustic wave channel has low cost and supports many kinds of equipment. Not affected by wireless communication and electromagnetic interference, acoustic wave communication is a wireless data transmission based on acoustic wave signals [5]. With the extensive application of near-field communication technology in mobile electronic equipment, the application of air channel communication technology that uses sound waves for information transmission becomes diversified in various fields and attracts more and more attention from scholars.

X. Sun et al. (Eds.): ICAIS 2020, LNCS 12240, pp. 203–214, 2020.
https://doi.org/10.1007/978-3-030-57881-7_18

Due to the development of mobile Internet and the endless emergence of network security threats, people urgently call for the emergence of a near-field covert communication technology. The research on acoustic wave covert communication has great potential in this field.

This paper presents a covert communication system architecture based on high frequency acoustic wave. Using the human ear's insensitivity to high frequency components, and acoustic wave can communicate with other equipment by its audio input and output equipment (microphone and speaker) without network, we proposed high frequency acoustic wave communication technology to realize safe and concealed transmission in short distance. The main work is listed below:

(1) Based on the principle of acoustic wave communication, we study and construct P2P high frequency acoustic wave covert communication channel.
(2) For the limitations of peer-to-peer communication, we designed a secure two-way alternating information sharing scheme.
(3) We have realized two-way alternating high frequency acoustic covert communication. Meanwhile, combined with the technology of information security, the system application scenarios of mobile devices are designed and implemented. Through testing and verification, the feasibility of the system is guaranteed.

2 Overview

In this paper, a covert communication system architecture based on high frequency acoustic waves is designed. As shown in Fig. 1, it is divided into five layers: Application Layer, Interface Layer, Logical Layer, Functional Layer and Storage Layer.

(1) Application Layer: It's the implementation of software at the application level. Users communicate through the system. The carrier of communication is the high frequency acoustic wave generated by software.
(2) Interface Layer: This layer implements the code interface of the software. It contains Configuration interface, Encryption and decryption interface, Signal processing interface, High frequency acoustic wave playback and receiving interface, Password authentication and key agreement interface.
(3) Logic Layer: The layer implements code logic for methods in the interface. It includes System configuration, Encryption algorithm, Modulation and demodulation, Sound wave playback and receiving configuration, Acoustic FFT transform, Password authentication and key agreement, Two-way alternating communication.
(4) Functional Layer: We realized specific system functions according to the code in the Logical Layer. It can realize covert information transmission based on high frequency acoustic channel and key agreement protocol based on password authentication.
(5) Storage Layer: This layer implements local storage of information. The high frequency acoustic wave communication system software can store the public information of the users in the local database.

Figure 2 shows the overall process of high frequency acoustic wave covert channel communication system.

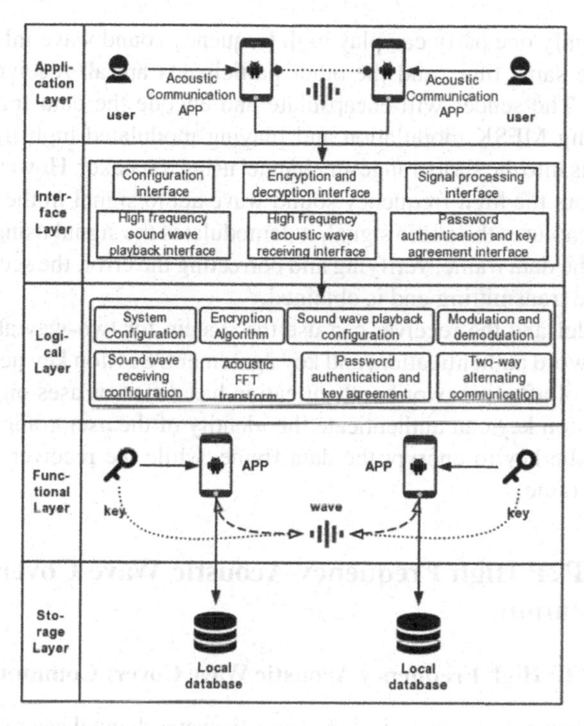

Fig. 1. The architecture design of high frequency acoustic wave covert communication system

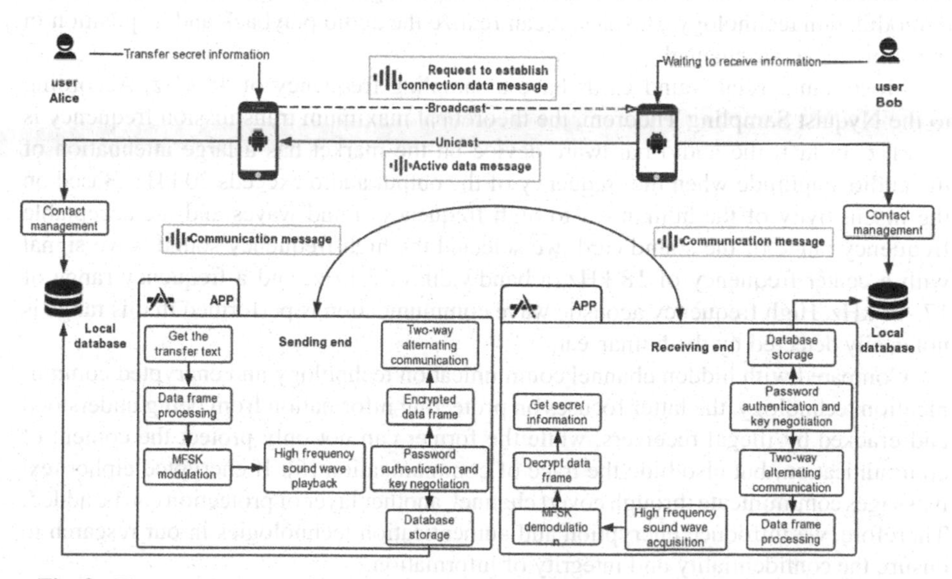

Fig. 2. The overall process of high frequency acoustic wave covert communication system

Within the system, the sender end and receiver end will perform different operations at the same time. Since the communication mode of the system is two-way alternating communication, only one party can play high frequency sound wave information to the air channel at the same time, and the other participants are all receiving ends in the monitoring state. The sender will encapsulate and encode the data frame, generating audio signals using MFSK modulation and playing modulated high frequency sound wave audio signals into the short-range air channel using a speaker. However, the receiver constantly monitors the high frequency sound wave audio signal in the short-range air channel. After it captures the audio signal, it demodulates the signal using MFSK. Then, after decrypting the data frame, verifying and correcting the error, the secret information transmitted by the transmitting end is obtained.

Both the sender and the receiver can use the system for two-way alternating communication, password authentication, and key agreement. Session key negotiated by key agreement protocol after password authentication has different uses on the sender and receiver. The session key can authenticate the identity of the user communicating. The sender still uses the key to encrypt the data frame, while the receiver uses the key to decrypt the data frame.

3 One-Way P2P High Frequency Acoustic Wave Covert Channel Communication

3.1 One-Way P2P High Frequency Acoustic Wave Covert Communication Model

The information transmission based on the acoustic wave channel can realize the mutual conversion of the data information and the audio signal through the modulation and demodulation technology. Besides, it can realize the audio playback and acquisition in the short-range air channel.

Most commercial sound cards have a sampling frequency of 48 kHz. According to the Nyquist Sampling Theorem, the theoretical maximum transmission frequency is 24 kHz. In fact, the audio hardware device on the market has a large attenuation of the audio amplitude when the frequency of the output audio exceeds 20 kHz. Based on the insensitivity of the human ear to high frequency sound waves and the acceptable frequency range of the sound card, we selected the high frequency sound wave signal with a center frequency of 18 kHz, a bandwidth of 3 kHz, and a frequency range of 17–19 kHz. High frequency acoustic wave communication is performed in this range is not easily detected by the human ear.

Compared with hidden channel communication technology and encrypted communication technology, the latter focuses on protecting information from being understood and cracked by illegal receivers, while the former can not only protect the content of communication, but also hide the trace of communication [6]. If encrypted cipher text messages communicate through covert channel, another layer of protection can be added. Therefore, we introduce encryption and authentication technologies in our research to ensure the confidentiality and integrity of information.

The advantage of the covert information transmission based on the acoustic channel is that sender and receiver only need to rely on the speaker and microphone and cooperate with the corresponding software to complete the covert information transmission with high frequency acoustic signal as the carrier [7]. According to the general model of acoustic communication, we designed a communication system model of one-way P2P high frequency acoustic wave covert channel as shown in Fig. 3.

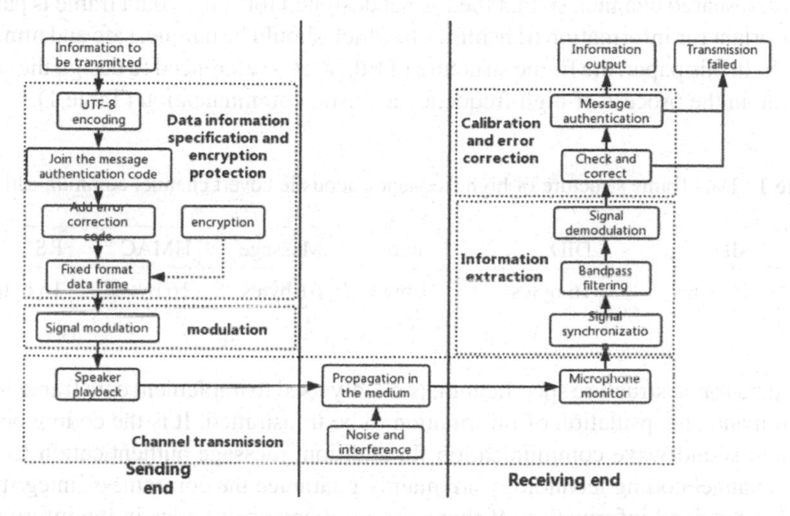

Fig. 3. One-way P2P high frequency acoustic wave covert communication system model

Step 1. Preparation for transmission of information. Get the information to be transmitted from the interface, which is stored in bytes.
Step 2. Information encoding and data frame format normalization. The sender encodes the text to be transmitted in UTF-8, adds the part used to identify the data frame, and generates the data frame in fixed format for information transmission. In addition, the session key is used to encrypt the data frame to ensure the confidentiality of the transmitted information.
Step 3. Modulation generating audio signal. Using appropriate modulation technology to modulate the data frame and band-pass filtering, audio signal is generated.
Step 4. Secure transmission of covert channel. Sender use the speaker to play the audio, which is transmitted in the short-range air channel and is subject to noise and interference. The receiver uses a microphone to monitor and capture the audio signal.
Step 5. Information extraction. Adopting appropriate signal synchronization technology to reduce the effect of multipath effect and Doppler effect on signal transmission. Then, using demodulation technology to obtain the bypass information.
Step 6. Check and correction of information. Check and correct the demodulated information. If there are too many error codes, the transmission fails. Moreover, if the correction is successful, then authenticating the message. However, if authentication failed, transmission fails. The received information can be output to the receiver's application interface if it is certified.

3.2 Specification and Encryption Protection of Data Information

The air channel on which high frequency acoustic covert channel communication depends is a shared channel, so that the format designed for source data frame is particularly important for information transmission, which should be unique, safe and minimize the length. In this paper, the frame structure of Ethernet is referenced to design the source data frame in the process of high frequency acoustic communication (Table 1).

Table 1. Data frame structure of high frequency acoustic covert channel communication

Type	SID	DID	Start	Message	HMAC	RS
1 byte	16 bytes	16 bytes	1 byte	61 bytes	16 bytes	16 bytes

The data frame structure specification is mainly used to implement data frame format generation and encapsulation of information to be transmitted. It is the coding basis of subsequent sound wave communication. In addition, message authentication technology and channel coding technology are mainly guarantee the correctness, integrity and security of received information. If there are too many error codes in the information, the transmission fails. And if error correction and authentication are passed, then the correct and unchanged information is obtained.

Obviously, the transmission of information in plaintext in the channel is extremely insecure and vulnerable to malicious attacks leading to the disclosure of information. Therefore, an appropriate and secure encryption algorithm, SM4 algorithm is selected in this paper. Cipher text is obtained by encrypting the plaintext information to be transmitted. The data information exists in the channel in the form of cipher text. And the cipher text data is converted into signals. Encryption algorithm ensures the confidentiality and security of the data transmission in the channel, also ensures the security of the data information.

3.3 Signal Processing, Modulation and Demodulation

The purpose of modulation is to convert the baseband signal generated by the signal source that is not suitable for transmission into the signal not only suitable for transmission in the channel but also meet communication requirements [8]. Furthermore, modulation also improves the signal's resistance to channel noise. In frequency domain, modulation is the shift of signal spectrum [9]. Considering the characteristics of air channel, complexity, variability and limited bandwidth, concealment of communication process, limitation of audio output equipment, and reliable transmission of information with a certain data rate, MFSK was selected as the modulation scheme of the system.

The processing of acoustic signal mainly includes noise reduction, anti-interference and synchronization, which is the preparation for the subsequent demodulation and decoding. The noise reduction and anti-interference algorithms mainly aim at the distortion caused by channel fading and the loss of main information in the process of signal transmission to reduce the bit error rate caused by channel transmission. While the purpose of synchronization is to detect whether the transmitted signal is valid and find the initial position of the information transmission [10]. Due to the time-varying of acoustic channel, frequency deviation would occur in the transmission process of acoustic signal, such as Doppler frequency shift, or frequency deviation between sender and receiver caused by multipath effect, which will both affect the performance of the system. Therefore, by referring to the signal synchronization principle of OFDM, the idea of cyclic prefix is introduced to realize the signal synchronization of this system.

3.4 Acoustic Wave Collection and Playback Based on Multithreading

In the process of communication in high frequency acoustic covert channel, the input text and audio are modulated and converted before they can be played and received. However, there are several problems with using the Android API alone, mainly the limited cache problem. The problem of limited cache could be described as follows, the sending end continuously generates audio signals. For the receiving end, after parsing to the beginning syllable frequency in the audio information, scans back and forth to determine the best position to start parsing, that is, to align the syllables. In this case, the rate of sound production is much higher than the rate of sound consumption (Fig. 4).

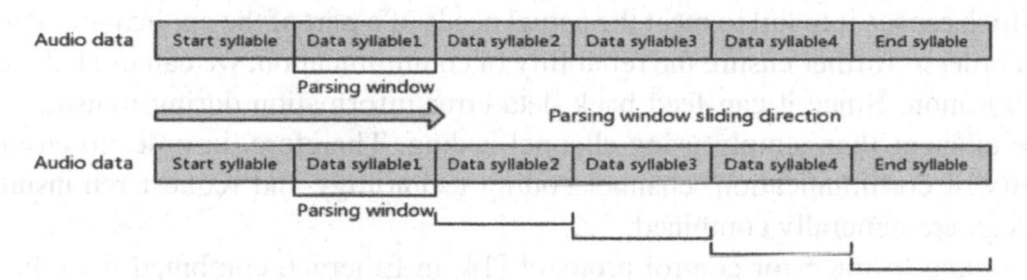

Fig. 4. The audio parsing process after aligning syllables

To solve the limited cache problem, it is necessary to balance the rate between the audio recording thread and parsing threads to prevent data loss and improve information transmission efficiency. Therefore, based on the solution of the classical "Producer-Consumer" limited cache problem in the operating system, this paper introduces the concurrent cooperation model into the system and solves the problem of playback blocking by means of multi-threading. LinkedBlockingQueue is a common method that is conducive to coordinating the balance between producer and consumer threads. In this paper, a thread-safe and first-in-first-out LinkedBlockingQueue is adopted to realize the producer and consumer model.

4 Two-Way Alternating Information Sharing Based on Identity Authentication

4.1 Key Agreement Protocol Based on Password Authentication

Key agreement protocol based on password authentication means that participants can conduct identity authentication through password and negotiate a high-entropy session key so as to realize the safe transmission of data [11]. The password, as the long-term key of the communication participants, is used to establish the session key without being leaked or stolen and used by the communication party to authenticate the other party. Only after the authentication is passed can the session key be obtained through key negotiation. If the authentication is not passed and the security threat is detected, the security early warning scheme will be started, and the communication will be interrupted. Depending on the difficulty of CDL issue and CDH issue [12], the protocol can ensure that the session key negotiated by password authentication has strong security, can also resist various attack techniques such as Denning-Sacco attack, impersonation attack, online and offline dictionary attack, and man-in-the-middle attack. In addition, any party in the terminal device doesn't store a complete session key locally. So even if attacked, the malicious attacker cannot obtain its session key store from the local cache or database of the attacked terminal device, which solves the key management problem of private key cryptography technology.

4.2 Error Control Method for Information Sharing

When one-way P2P communication has an error during transmission and the receiving end cannot correct it by channel coding technology, this means that the communication fails, which causes it to fail to meet the actual needs of a part of the application scenarios [13]. In order to further ensure the reliability of communication, we can use half-duplex communication. Since it can feed back data error information during transmission, it is more efficient than simply using channel coding. Therefore, in order to ensure the reliability of communication, channel coding technology and request retransmission technology are generally combined.

According to the error control protocol [14] in Ethernet, combined with the characteristics of acoustic communication itself, the error control protocol in acoustic communication is given. The working principle is shown in Fig. 5.

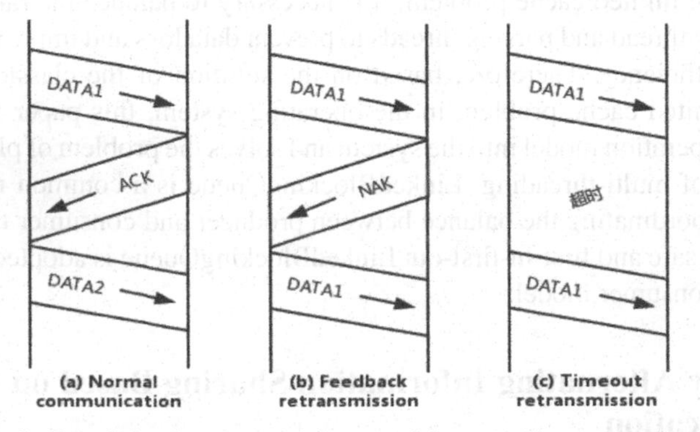

Fig. 5. Principle of error control protocol for acoustic communication

Among them, Fig. 5(c) describes the time-out retransmission in the communication process, which can be divided into three cases:

1. The data packet of the sending end is lost. The receiving end does not receive any information, and it waits for the sending end to time-out retransmission.
2. The data packet sent by the sending end is correctly received, but the feedback information of the receiving end is lost. The sending end performs time-out retransmission after exceeding the waiting time interval;
3. When an error occurs in the process of receiving the feedback packet, the sending end performs time-out retransmission.

4.3 Decentralized Two-Way Alternating Information Sharing Scheme

The two-way information sharing of the high frequency acoustic covert channel is based on the information sharing of the decentralized idea. Decentralization makes the role of the server no longer exist, and there is also no longer a communication protocol based on TCP/IP protocol between terminal devices. Nevertheless, the system still needs the idea based on TCP/IP protocol [15] to design a communication protocol suitable for high frequency acoustic covert channels. To this end, this paper establishes a two-way information sharing model based on identity authentication.

Two-way alternating communication is also known as half-duplex communication. Both sides of communication can send data, but they cannot send and receive at the same time. This type of communication is mutual, one end sends and the other receives.

In the two-way communication of high frequency acoustic covert channels, it combines various technologies to form a complete communication process and specification. The two-way communication of high frequency acoustic waves is based on P2P communication, introduces a two-way alternating communication reliability guarantee mechanism to ensure the effectiveness of data transmission and improve communication efficiency. By listening to user events and determining the type of data frame, there is an interface interaction between the user and the system. Communication between users is accomplished by mutual conversion of data frames and audio, so the ability to perform data frame type determination and encapsulated communication protocol control class is the underlying basis of P2P communication. In order to ensure the security, authenticity and integrity of the communication, a key agreement protocol based on password authentication is introduced. The protocol can perform identity authentication according to the password and negotiate the session key for the encryption of the communication message in the process of communication at both ends. The encryption algorithm adopts the SM4 algorithm.

This paper designs the communication scheme into four parts, namely "Establishing connection - Identity authentication and key agreement - Sharing information - Disconnection". Figure 6 is the information transmission timing of the connection requesting end A and the connected end B.

Therefore, the whole process of information sharing between the connection requesting end A and the connected end B is as follows:

(1) When A wants to send and receive messages with people nearby, it is necessary to search for a subject that can be connected nearby first. Then A will broadcast the request connection message 'Request Connect [$Name_A$]' with the identity flag.

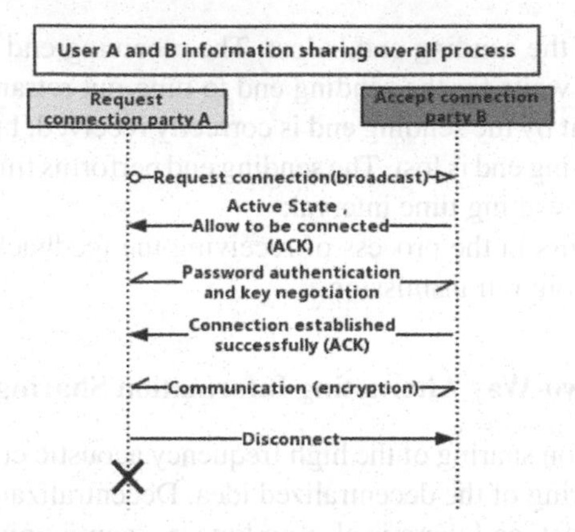

Fig. 6. Two-way alternating information sharing timing

(2) After receiving the broadcasted data packet, the active B obtains the user name $Name_A$ from the data packet and displays the information of A on the software interface. Then B sends a message 'Active State [$Name_B$]' to A, A obtains the information of the nearby subject B, and also displays it on the interface.
(3) If A initiates a communication connection with B first, then the data packet whose destination ID is $Name_B$ is sent, and then enters the waiting process.
(4) After B agrees and sends the corresponding ACK packet, A and B will continuously change the identity of the sending end and receiving end and perform password authentication and key agreement.
(5) After A and B pass the authentication and complete the key agreement, the connection is formally established and enters the communication process.
(6) The packets in the communication process of A and B are encrypted by using the negotiated session key and then sent.
(7) When either end of the communication clicks the disconnect button, the communication connection is broken. At this time, the 'Connect Shut Down' message is sent. It can exit directly without receiving a response. After the disconnected end receives the message, it also exits directly.

5 Experiments

According to the above high frequency acoustic covert communication model, a covert secure communication system based on high frequency acoustic waves is built. Sonic communication requires no additional hardware support and has a natural advantage on mobile devices, so this paper applies the system to text transmission between mobile devices. The microphones and speakers of the two Android mobile devices are used as the receiving end and the sending end.

The experimental scenario is an open environment. The relative position of the receiving end and the sending end when sending the message is kept unchanged during the experiment. The experimental process is based on the two-way alternating information sharing model. First, the communicating parties search for nearby contacts and display them on the interface. One party requests to establish a connection with the other one. When the invitee agrees to establish a connection and successfully passes the identity authentication and key negotiation, both parties enter the communication interface. Finally, the two sides will conduct two-way communication of high frequency acoustic waves. Through many experiments, the basic functions of the system are complete, and the overall performance is stable.

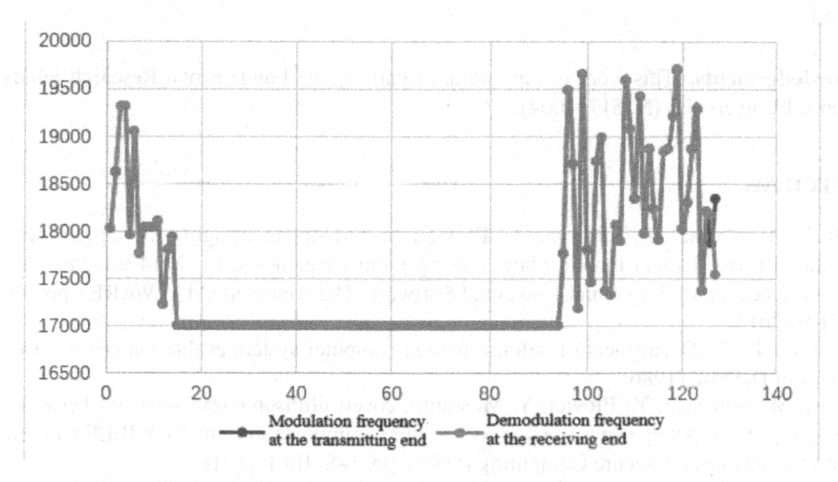

Fig. 7. Comparison of frequency between sending end and receiving end

In order to test the modulation and demodulation performance, the frequency modulated by the sending end is compared with the frequency demodulated by the receiving end, which is displayed in the form of a line graph. As can be seen from Fig. 7, the system can achieve fidelity and anti-interference of audio information.

Meanwhile, the system conducts performance tests such as communication accuracy, transmission distance and transmission time to verify the reliability and stability of the system during operation. The main implementation methods of the test are the control variable method and the comparative verification method. Through the test, it can be concluded that the system supports Chinese, English, Chinese characters, expressions and other characters, and the transmission accuracy is 99%. The linear distance between the sending end and the receiving end affects the decoding rate of the data message and the decoding rate can reach 98% or more within 3 meters. The complexity of transmitting data does not affect the transmission time, and it remains at around 3 s.

6 Conclusion

In recent years, with the deepening of research on covert channels, the rapid development of mobile Internet, more and more cyber security threats and privacy protection issues, the information security of mobile devices has become increasingly prominent. This paper focuses on the research of secure and reliable communication between devices for high frequency acoustic covert channels, designs a covert communication system architecture based on high-frequency acoustic waves. The system combines acoustic wave communication technology, key agreement, message authentication, identity authentication, data encryption and other technologies. Based on the successful construction of P2P high frequency acoustic covert channel, the two-way alternating information sharing scheme is designed and implemented. Finally, high frequency acoustic covert channel communication with effectiveness, confidentiality, integrity and authenticity is realized.

Acknowledgements. This work is supported, in part, by the Fundamental Research Funds for the Central Universities (N181704004).

References

1. Guri, M., Kedma, G., Kachlon, A., Elovici, Y: AirHopper: bridging the air-gap between isolated networks and mobile phones using radio frequencies. In: 2014 9th International Conference on Malicious and Unwanted Software: The Americas (MALWARE), pp. 58–67. IEEE (2014)
2. Latham, D. C.: Department of defense trusted computer system evaluation criteria. Department of Defense (1986)
3. Guri, M., Solewicz, Y., Elovici, Y: Mosquito: covert ultrasonic transmissions between two air-gapped computers using speaker-to-speaker communication. In: 2018 IEEE Conference on Dependable and Secure Computing (DSC), pp. 1–8. IEEE (2018)
4. Zhiguo, S.: Computer Network Security Tutorial. Tsinghua University Press, Beijing (2004)
5. Hanspach, M., Keller, J: On the implications, the identification and the mitigation of covert physical channels. In: 9th Future Security Conference, pp. 563–570 (2014)
6. Wayner, P.: Disappearing Cryptography: Information Hiding: Steganography and Watermarking. Morgan Kaufmann, Burlington (2009)
7. Xuejie, D., Bin, L., Di, W., Meng, Z., Degang, S.: Key technologies of covert information transmission based on acoustic channel. Inf. Secur. Res. **2**(2), 131–136 (2016)
8. Gao, N.: Shimamoto, S: Amplitude and phase modulation for ultrasonic wireless communication. Int. J. Wirel. Mob. Netw. **6**(2), 1 (2014)
9. Tu, Y., Lin, Y., Wang, J., Kim, J.-U.: Semi-supervised learning with generative adversarial networks on digital signal modulation classification. Comput. Mater. Continua **55**(2), 243–254 (2018)
10. Pushpa, K., Kishore, C. N., Yoganandam, Y: A new technique for frame synchronization of OFDM systems. In: 2009 Annual IEEE India Conference, pp. 1–5. IEEE (2009)
11. Zhou, Z., Peng, C.: Improvement and design of password-based key agreement protocol. Inf. Netw. Secur. 14 (2014)
12. Zhou, F., Liu, X., Yan, H., Chang, G: Cryptanalysis and improvement of EC2C-PAKA protocol in cross-realm. In: 2008 IFIP International Conference on Network and Parallel Computing, pp. 82–87. IEEE (2008)
13. Wang, Y., Zhang, X., Zhang, Y.: Joint spectrum partition and performance analysis of full-duplex D2D communications in multi-tier wireless networks. Comput. Mater. Continua **61**(1), 171–184 (2019)
14. Zorzi, M., Rao, R.R., Milstein, L.B.: ARQ error control for fading mobile radio channels. IEEE Trans. Veh. Technol. **46**(2), 445–455 (1997)
15. Graniszewski, W., Krupski, J., Szczypiorski, K.: The covert channel over HTTP protocol. In: Photonics Applications in Astronomy, Communications, Industry, and High-Energy Physics Experiments 2016. International Society for Optics and Photonics (2016)

Privacy Preserving Mining System of Association Rules in OpenStack-Based Cloud

Zhijun Zhang[1,3](✉), Zeng Shou[2,3], Zhiyan Ning[1,3], Dan Wang[1,3], Yingjian Gao[1,3], Kai Lu[1,3], and Qi Zhang[2,3]

[1] China NARI Group Corporation (State Grid Electronic Power Research Institute), Nanjing 210061, China
zzj519@163.com

[2] State Grid Liaoning Electronic Power Supply Co. Ltd., Shenyang 110004, China

[3] Software College, Northeastern University, Shenyang 110169, China

Abstract. As an efficient data analysis tool, data mining can discover the potential association and regularity of massive data, and it has been widely used and played an important role in business decision, medical research and so on. However, the data mining technology is also a double-edged sword, in bringing convenience at the same time, will also cause the user's privacy leak problem. In order to solve the problem, the symmetric searchable encryption technology is introduced into the association rule mining system to protect the privacy, and privacy preserving mining system of s (PP-MSAR) in OpenStack-based Cloud environment is designed. In order to solve the problem that the existing data mining algorithms can't deal with large-scale data, this paper uses the computational power of Hadoop platform and add the global pruning technique to the existing algorithm based on MapReduce association rules, so that the counting of frequent item sets get reduced. At the same time, this paper add frequent matrix storage method into the distributed association rules algorithm and realize he algorithm of mining association rules for frequent matrix storage based on MapReduce. In addition, the introduction of symmetric searchable encryption technology to support the cloud server-side ciphertext retrieval, on the one hand to ensure that users stored in the database information will not be leaked to the outside for others, on the other hand also to ensure that the user data for the system staff confidential. Finally, we test the system, and the results show that the system can carry out association rules mining under the premise of protecting user privacy, and provide the correlation degree between data, which has certain practical significance and application value.

Keywords: Cloud computing · Association rules · Privacy preserving

1 Introduction

With the rapid development and increasing popularity of cloud computing [1, 2], mobile computing and the Internet of Things, the era of big data has arrived. Because big data has the characteristics of large volume, diversity, low value density, and high speed, traditional data processing technology is difficult to effectively process big data in effective

X. Sun et al. (Eds.): ICAIS 2020, LNCS 12240, pp. 215–224, 2020.
https://doi.org/10.1007/978-3-030-57881-7_19

time [3, 4]. In order to solve the processing problems of big data, big data technology is gradually emerging. Data mining is an important part of big data technology. Through the mining of data, it is possible to predict the trend of events and make adjustments, which is of great significance to the economic development and progress of the future society. With the continuous development of data mining technology, the development of data mining technology at domestic and foreign is becoming more and more mature, and is widely used in the fields of banking, insurance, astronomy, molecular biology, etc., combining cloud computing platform and data mining to solve the traditional Data mining calculates the problem of insufficient storage capacity and can reduce the cost of mining.

Association rules are one of the most active research directions in data mining [5–8]. It was proposed by Agrawa et al. [9, 10], the main idea is to mine the frequent k-item set from the candidate frequent k-item set. Among them, it is necessary to scan the transaction set multiple times. When the data size is large, the cost of the practice and resources consumed by the association rule mining of the data will far exceed the value of the obtained information, which is the main reason for restricting the development of data mining technology. After that, there have been a lot of researches on the mining of association rules by many researchers. DDM (Distribution Data Mining) focus on the application of the best available data resources in the distributed computing environment. With the wide application of the Internet and the maturity of distributed technologies, and Google's breakthrough development of distributed file systems, it has opened up new research ideas for distributed association rules mining. Google has proposed the MapReduce computing model. The MapReduce model can be used in multi-core machine learning. At the same time, it has the advantage of solving communication and synchronization problems. In addition, MapReduce will automatically handle faults, while also hiding some of the programmers. The complexity of fault tolerance, the association rule algorithm can use the advantages of this model to improve mining efficiency. Many scholars at home and abroad have proposed a variety of data mining methods and platforms to improve the processing power of big data. However, at present, the research on association rule mining that supports privacy protection has not been perfected, and the user's private data is easily leaked in existing data mining technology, which brings security risks. Therefore, how to efficiently obtain the required information and knowledge from the data while protecting user privacy becomes a challenge.

Combined with the existing research, this paper adds the global pruning technology based on the existing MPAriori (MapReduce Apriori Algorithm) algorithm, which makes the count of counting frequent itemsets decrease again. At the same time, the storage method of frequent matrix is added to the distributed association rule algorithm, and the MAPDAP (Map Frequent Matrix Distributed Apriori Algorithm) [9] algorithm for the frequent matrix storage of MapReduce calculation model is proposed. The algorithm only needs to traverse the data set once, as long as the data is stored in the matrix, the rest of the work is to operate on the matrix. Based on the algorithm and the characteristics of OpenStack-based Cloud environment, this paper proposes an association mining system that supports privacy protection. Under the premise of protecting user privacy, the association rules mining data.

2 Building Blocks

2.1 Symmetric Searchable Encryption

Table 1 is the relevant symbols that will be used in the algorithm.

Table 1. Description of the symbol

Symbols	Meaning
$C = (F_1, F_2, \cdots, F_l)$	A collection of l encrypted documents
$W = (w_1, w_2, \cdots, w_n)$	Collection of keywords in document set C
$S_1, S_2, \cdots, S_l$	Pseudo random value sequence
$x \oplus y$	XOR x and y
$f : K \times X \longrightarrow \lambda$	Pseudorandom function
$\{T_w\}$	The Trapdoor of keyword w
$<x, y>$	connection of x and y
ID_{file}	Document marking
(C_i, ID_{file})	File-based index
$T_i = <S_i, F_{K_i}(S_i)>$	A bit string generated from a sequence of pseudo-random values

Symmetric searchable encryption based on MapReduce can be described as a six-tuple *SSE* = (*KeyGen*, *Enc*, *BuildIndex*, *Trapdoor*, *Search*, *Dec*) which is defined on word $\Delta = \{w_1, w_2, \ldots, w_p\}$.

- $K = KeyGen(k)$: The user executes this algorithm to build the system, the input is the security parameter k, and the output is a randomly generated key. These keys are used in subsequent encryption algorithms and trapdoor generation algorithms;
- $C_i = X_i \oplus T_i$: The user encrypts the plaintext uploaded to the cloud database. First, each keyword is pre-encrypted by AES encryption algorithm, and each word w in the plaintext is pre-encrypted. Then, the ciphertext is generated by using the pseudo-random sequence and the pre-encrypted result XOR, and finally the ciphertext is uploaded to the cloud server;
- *BuildInde*(*key*, *value*): The cloud server side uses MapReduce technology to parallel integration these ciphertexts C_i. First use the Map function, list each C_i in the cloud's document collection and its corresponding ID_{file}. For each C_i,sort into format (C_i, ID_{file}) Then use the Reduce function, Organize files that have the same ID_{file} (the same keyword), finally established an index table based on (C_i, ID_{file});
- *Trapdoor*: The user executes this algorithm to generate the keyword trapdoor to be queried, input is keyword w, Firstly, compute $X = E_{K''}(w)$. Then divide the results into two parts, the first $n - m$ bits and the last m bits. Then compute $K_i = f_{k'}(L_i)$ and $C_i = X_i \oplus T_i$, send C_i to the cloud.

- *Search*: The cloud server executes this algorithm to retrieve the document containing the request keyword w. The cloud database retrieves the index table (C_i, ID_{file}) according to C_i, and return the corresponding file to the user.
- *Dec*: This step is performed by the user to decrypt the resulting document. The user generates a S_i using a pseudo-random generator, Then XOR the first $n - m$ bits of S_i and C_i, Can be recovered L_i. Finally the value of L_i can let user compute k_i, and complete the entire decryption process.

The process of Symmetric Searchable Encryption is given as follows:

Step 1. Encryption process

Step 1.1. The user uses the AES encryption algorithm to pre-encrypt the keywords in the plaintext, i.e. $X_i = E_{k''}(W_i)$.
Step 1.2. Split X_i into left and right parts L_i and R_i. Where L_i corresponds to the first $n - m$ bits of X_i and R_i corresponds to the last $n - m$ bits of X_i.
Step1.3. User generate $K_i = f_{k'}(L_i)$, Pseudo-random function $f : K_F * \{0, 1\}^* \rightarrow K_F$.
Step1.4. The user generates a pseudo-random value sequence $S_1, S_2 \ldots S_i$ using the stream cipher, each of them is 10 $n - m$ bits. Define T_i is the connection of S_i and $F_{K_i}(S_i)$. $F_{K_i}(S_i)$ is also a pseudorandom function, input is a $n - m$-bit word, output is a m-bit word, K_i is a pseudo random number generated in the left part L_i.
Step1.5. Encrypted output ciphertext, $C_i = X_i \oplus T_i$. $X_i = E_{k''}(W_i)$, T_i is the connection of S_i and $F_{K_i}(S_i)$, $K_i = f_{k'}(L_i)$.
Because as long as the data owner can be generated as a random stream $T_i(i = 1, 2, \cdots, n)$, others can not decrypt it, ensuring the security of the data.

Step 2. Search process
During retrieval, The cloud server XORs X in $<X, k>$ with C_i in the database. If the XOR result corresponds to a certain $<S_i, F_{K_i}(S_i)>$, the verification relationship is satisfied, and the keyword information is returned, and the retrieval is completed.
Step 3. Decryption process
When the user needs to decrypt the keyword, since the user knows that the random number generates the seed, and can generate it by using the stream cipher generator, then XOR S_i and the first $n - m$ bits of C_i, and L_i can be recovered, and k_i can be computed by L_i, the entire decryption operation can be completed, the specific process is as follows:

Step 3.1. $C_i = <C_{i,l}, C_{i,r}>$: Divide cyphertext C_i Divide ciphertext A into left and right parts $C_{i,l}$ and $C_{i,r}$, $C_{i,l}$ is left,and $C_{i,r}$ is right.
Step 3.2. XOR the left part of the ciphertext $C_{i,l}$ and stream sequence S_i.
Step 3.3. $k_i = f_{k'}(X_{i,l})$: Take $X_{i,l}$ as a parameter and k' as a random function of the key.
Setp 3.4. Connect S_i and $F_{K_i}(S_i)$ to get T_i.
Step 3.5. XOR ciphertext C_i and T_i.
Decrypt X_i by the AES decryption algorithm.

2.2 Mining Algorithm of Association Rules

The association rule mining algorithm includes a matrixing process and frequent item set generation, as follows.

The pseudo code of the frequent matrix MapReduce association rule algorithm is as follows:

```
(1) Matrixing process:
public void addTransaction
{   if (Collection is empty) return;
        Define a BigInteger class to store matrix elements;
        Store the collection in a table in Hbase;
            For (Traversing the entire table)
            { Define an int variable to find if the item exists in all_item;
              If (Attribute already exists)
              { Support count ++ }
                  Else{ Record the location of the new attribute and add it to
      all_item. The initial value of the support of attribute is 1.}
              Set the project location to 1;
            }
            Store the transformed a transaction in the list of transactions;;
      }
(2) Frequent item set generation
Private Map<BigInteger,Integer>get CandidateCollection(
            Map<BigInteger,Integer>itemmap{
            The definition collectionCollection is used to store the candidate
      K+1 item set;
            Set two item sets Item1 and item2 for self-joining;
            For (Traversing item1) {
              For (Traversing item2) {
                XOR an element in item1 with an element in item2;
                 The result is stored in tmp_candidateCollection;
                If (The number of 1 in tmp_candidateCollection is not equal
      to the number of 1 in item1 plus 1)
                    Indicates that it is not a k+1 item set, filtered;
              }
            }
    Return k+1 order candidate set candidateCollection}
```

3 System Design

3.1 Overview

The PP-MSAR system is divided into two parts: the client and the cloud. The cloud is based on the HBase database, combined with MapReduce technology to achieve multi-dimensional association rules mining of data. The client implements a privacy protection function, introduces symmetric searchable encryption, encrypts data, protects user data privacy, and provides user interaction functions. The specific structure of the cloud and client is shown in Fig. 1.

The system uses a three-tier architecture, including the data layer, business layer, and application layer.

(1) Data layer: Provides the business layer with access to data and access to the cloud and local servers through data access.
(2) Business layer: Centralized to provide logic implementation of various functions of the association rule mining system supporting privacy protection in the cloud computing environment, and encapsulate various functional modules, including data interaction module, registration and login module, keyword encryption and generation trapdoor module, file encryption upload module, trapdoor retrieval module, MapReduce preprocessing module and association rule mining module.
(3) Application layer: The user completes the information entry by registering, and logs in for identity verification. The user can request to encrypt the keyword, and encrypt the file and upload it to the cloud and request the association rule mining. After receiving the files and keywords, the cloud performs MapReduce preprocessing and stores them in the HBase database. At the same time, the cloud retrieves the encrypted data through trapdoor retrieval and mines the association rules, and returns the mining results to the user interface.

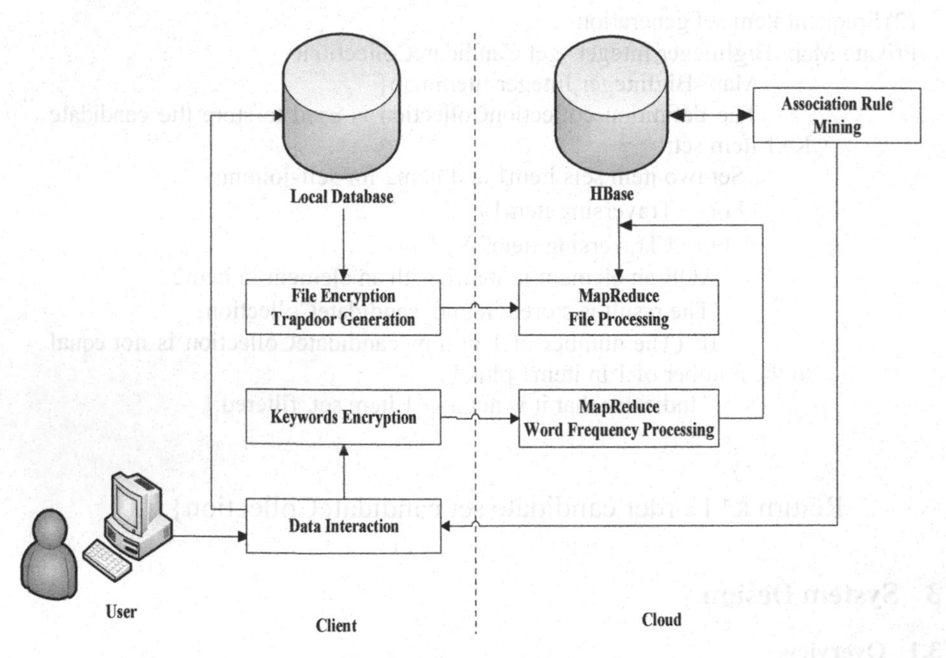

Fig. 1. System architecture

3.2 Keyword Encryption and Trapdoor Generation

This function includes generation of a random number $k^{'}$, generation of a retrieval voucher *Trapdoor* and etc. Encrypt the keyword to get the ciphertext form of the keyword. Simultaneous generation of trapdoors. When the user wants to retrieve the corresponding keyword, the user does not trust the cloud server, and cannot decrypt the data on the cloud server side for searching, so the trapdoor is used to solve the problem. Suppose the user wants to retrieve the keyword w, firstly performs AES encryption into the result X_i, then divides X_i into the left and right parts L and R, and then calculate $K_i = f_k(L_i)$, where the function f is a random function, and the X_i and $K_i = f_k(L_i)$ connections are generated to generate trapdoors and sent to HBase. On the database side, the trapdoor will be used for the subsequent trapdoor search module.

3.3 Uploading of Encrypted Files

This function encrypts the document that the user wants to encrypt and uploads it to the cloud HBase database. Assume that the document information to be encrypted by the user contains a keyword set: $w_1, w_2, \cdots, w_n$. This function module uses a pseudo-random generator to generate a pseudo-random bit string of a special structure, and then XORs it with the plaintext. The result of this XOR operation allows the cloud server to search for keywords without Disclosure of any information about plaintext and keywords to external systems.

3.4 Trapdoor Retrieving

The main function of the trapdoor retrieval module is to retrieve the encrypted keywords in the encrypted data in the database, laying the foundation for the subsequent association analysis. The process is that the cloud server XORs the E_i in the trapdoor with the C_i in the index table, and if the XOR result corresponds to the form of a certain $<S_i, F_{K_i}(S_i)>$, the verification is satisfied. In the cloud, because the encrypted S_i is different, the generated ciphertext is also different, so it is necessary to obtain different ciphertext forms of the same disease by searching.

3.5 Mining of Association Rules

The association rule mining module is mainly used for association analysis of data. The MapReduce association rule algorithm (MFMDAP) of the frequent matrix is used. The algorithm only traverses the data set once, and only needs to store the thing set in the transaction matrix. It is the operation of the matrix. When the transaction matrix and the weight matrix are established, a frequent episode is generated at the same time, and finally the pruning method is used to extract the data items that do not satisfy the min_sup. The specific steps are as follows:

Step 1. Input data fragmentation: Use MapReduce to control data fragmentation and segment data.

Step 2. The local data slice is transformed into a matrix: the data is sent to different computing nodes, and the sequence is converted into a matrix. The first pass scans the data, and generates a frequent episode while generating a local frequent matrix.
Step 3. Generation of local frequent itemsets: Frequent binomia itemset is generated through frequent itemset, calculate supportiveness techniques at the same time, generate multiple sets, trim at the same time, remove unnecessary items, and generate final local frequent matrix itemsets at each DataNode node.
Step 4. The local frequency frequent matrix generates a local frequent item set: the calculated matrix of each node is converted into a local frequent item set according to the conversion rule.
Step 5. Correlation analysis: Find out the non-empty true subset for each frequent item set, calculate the confidence by using the grouping method for each rule, and obtain all the strong association rules.

4 Performance Analysis

The association rule mining system that supports privacy protection in the cloud computing environment uses a symmetric searchable scheme for encryption. The encrypted data source is a local database, and the data is encrypted and uploaded while the database is being read. Due to the large amount of data, the amount of data can greatly affect performance.

First, the encryption upload test, mainly including ordinary AES encryption and symmetric searchable encryption, after testing, each encryption consumption time is shown in Fig. 2, as can be seen from the figure, each encryption takes less than a few seconds Below, therefore, for the user, the encryption speed is within an acceptable range.

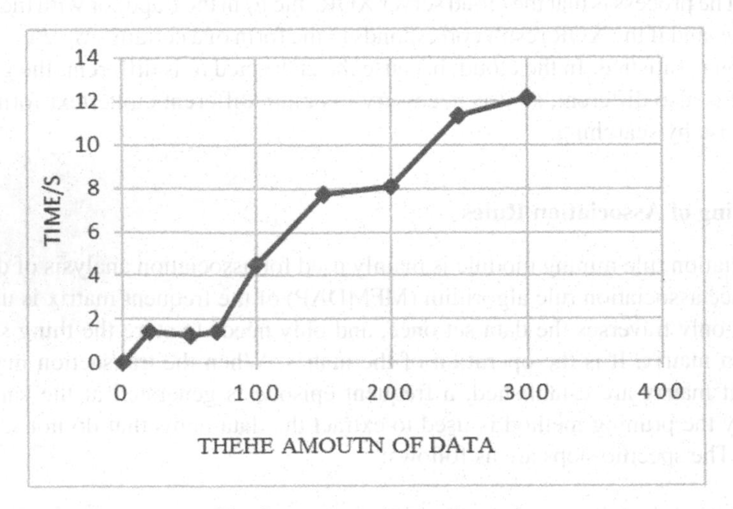

Fig. 2. Encrypted upload speed

Test preprocessing and correlation analysis need to add data, the main influencing factors are: local database read speed, cloud database write speed, MapReduce processing speed, specific data shown in Fig. 3, Fig. 4.

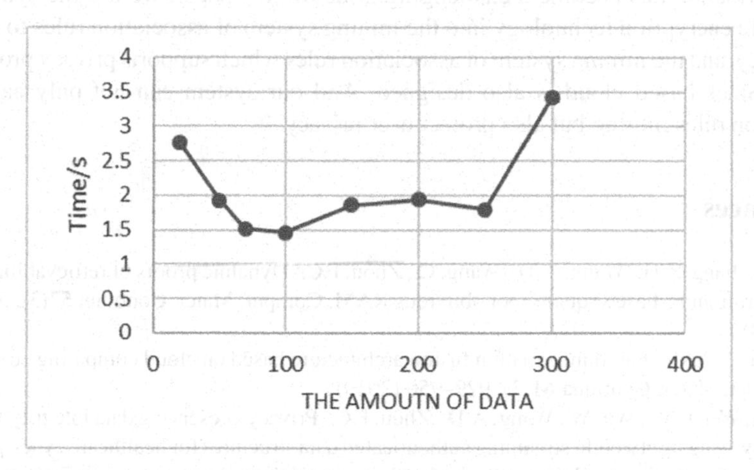

Fig. 3. Pretreatment speed

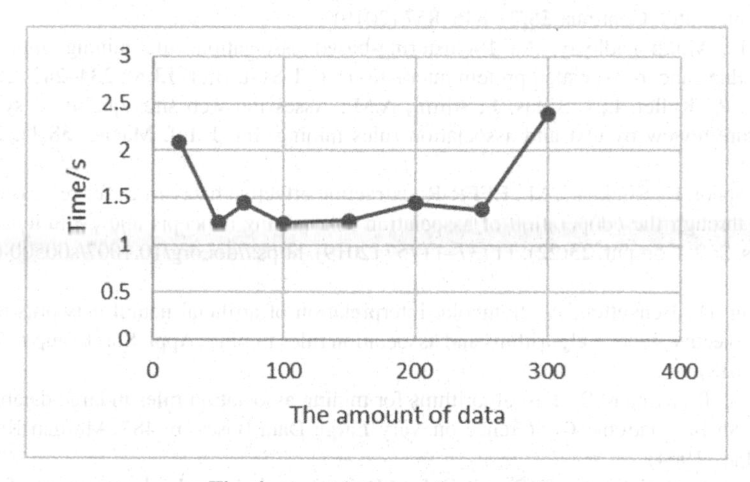

Fig. 4. Association analysis speed

From the above analysis, the correlation analysis and preprocessing are quite fast. However, due to the limitation of the hard disk read/write speed and the influence of the Hadoop mechanism, the speed of data is reduced when writing HBase, but the speed of association analysis and correlation analysis that users care about is not affected much.

5 Conclusions

Data mining technology can discover the potential association and regularity of massive data, and it has been widely used and played an important role in business decision, medical research and so on. However, this technology is also a double-edged sword, in bringing convenience at the same time, will also cause the user's privacy leak problem. So how to protect user privacy under the premise of efficient analysis and mining of useful data information has become a challenging issue. In this paper, we use the symmetric searchable encryption technology into the mining system of association rules to protect the privacy, and the mining system of association rules which supports privacy protection in OpenStack-based cloud is also designed. And our system can not only carry out association rules mining but also protect user privacy.

References

1. Xu, J., Jiang, Z.H., Wang, A.D., Wang, C., Zhou, F.C.: Dynamic proofs of retrievability based on partitioning-based square root oblivious RAM. Comput. Mater. Continua **57**(3), 589–602 (2018)
2. Hsu, I.C.: XML-based information fusion architecture based on cloud computing ecosystem. Comput. Mater. Continua **61**(3), 929–950 (2019)
3. Xu, J., Wei, L.W., Wu, W., Wang, A.D., Zhou, F.C.: Privacy-preserving data integrity verification by using lightweight streaming authenticated data structures for healthcare cyber-physical system. Future Gener. Comput. Syst. (2018). https://doi.org/10.1016/j.future.2018.04.018
4. Wang, C., Liu, M.J.: Dynamic trust model based on service recommendation in big data. Comput. Mater. Continua **58**(3), 845–857 (2019)
5. Dey, L., Mukhopadhyay, A.: Biclustering-based association rule mining approach for predicting cancer-associated protein interactions. IET Syst. Biol. **13**(5), 234–242 (2019)
6. Rekik, R., Kallel, I., Casillas, J., Alimi, A.M.: Assessing web sites quality: a systematic literature review by text and association rules mining. Int. J. Inf. Manag. **38**(1), 201–216 (2018)
7. Aghaeipoor, F., Eftekhari, M.: EEFR-R: extracting effective fuzzy rules for regression problems, through the cooperation of association rule mining concepts and evolutionary algorithms. Soft. Comput. **23**(22), 11737–11757 (2019). https://doi.org/10.1007/s00500-018-03726-1
8. Yedjour, D., Benyettou, A.: Symbolic interpretation of artificial neural networks based on multiobjective genetic algorithms and association rules mining. Appl. Soft Comput. **72**, 177–188 (2018)
9. Agrawal, R., Srikant, R.: Fast algorithms for mining association rules in large databases. In: The 20th International Conference on Very Large Data Bases, p. 487. Morgan Kaufmann Publ Inc. (1994)
10. Agrawal, R., Imielinski, T., Swanmi, A.: Mining association rules between set of items in large databases. ACM SIGMOD Rec. **22**(2), 207–216 (1993)

Unsupervised Data Transmission Scheduling in Cloud Computing Environment

Gui Liu(✉), Wei Zhou, Yongyong Dai, Haijiang Xu, and Li Wang

Jiangnan Institute of Computing Technology, Wuxi, Jiangsu, China
lgzhy@163.com

Abstract. How to make a reliable and efficient replica is an important aspect of data replication management, which involves data transmission, including bandwidth application, mobility and bandwidth release. Although data transmission is an important aspect of data intensive applications, there are still some problems in data transmission processing technology.

In view of the problems existing in the transmission process of data replication, this paper studies some popular data transmission services, points out the defects in the massive information processing business process in the cloud computing environment, and then proposes an unsupervised data transmission scheduling model, which adopts a transmission strategy called DRFT to realize the automatic division of data transmission process, and Each stage is managed by "job" scheduling. Finally, the difference of reliability and performance between DRFT and GridFTP is analyzed through comparative experiments.

Keywords: Data transmission · DRFT · Unsupervised data transmission

1 Relevant Research

In order to transmit data reliably, effectively and quickly, researchers have adopted many technologies:

GridFTP (Grid File Transfer Protocol)
Although GridFTP [1] has many features compared with FTP, the GridFTP client is required to remain open until the transmission is completed. Although GridFTP supports reliable data transmission and data retransmission, in the case of losing the connection with the server, it must be manually involved in restarting the transmission.

RFT: Reliable File Transfer Service
RFT [2] is a service of grid based on WSRF(Web Services Resource Framework) [3], It allows the byte stream to be transmitted in a reliable way, which ensures the reliability of the transmission task.

X. Sun et al. (Eds.): ICAIS 2020, LNCS 12240, pp. 225–235, 2020.
https://doi.org/10.1007/978-3-030-57881-7_20

However, both GridFTP and RFT services have not completely solved the scheduling problem of data transmission, and the processing of data transmission is still in the stage of human intervention.

2 Problems in Copying Massive Data in Cloud Computing Environment

In the cloud computing environment, a large amount of data needs to be copied [4, 5]. The usual business process is as follows: first, establish several data centers to store the original data received from various channels in these data centers. Then, according to the demand, the data needed for processing are copied to the cluster computer, supercomputer, supercomputer and other computing nodes, and then these data are processed with massive information. Here, the computing node and the data center may be physically located in the same place, or they may be located in different places and connected through the network (Fig. 1).

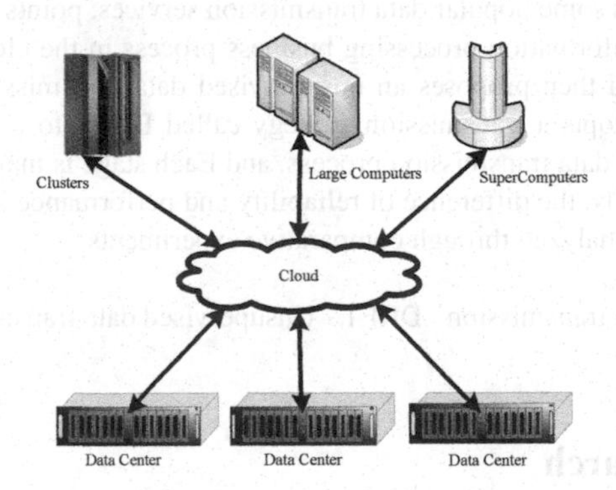

Fig. 1. Massive information processing business process in cloud computing environment

The business process of this data transmission mainly has the following shortcomings:

No intelligent automatic processing: the data can only be copied to multiple computing nodes (microcomputers, cluster computers, supercomputers, etc.) in a manually specified way, and the data can not be copied automatically and intelligently according to the needs of users;

There is no scheduling: there is no automatic scheduling process between copying data and processing these data, which is to copy a large number of data to each computing node under human intervention;

No effective fault-tolerant measures: if there is no effective fault-tolerant measures in case of network disconnection, insufficient disk space, machine restart and other situations, only the data can be manually retransmitted after the error is found. Obviously, this method of data replication is not suitable for the current development direction.

3 Unsupervised Data Transmission Scheduling Model

According to the process characteristics of massive information processing business in cloud computing environment, this paper proposes an unsupervised data transmission scheduling method with automatic scheduling function. This method abstracts every step in the business process, and expresses their dependency in the way of directed graph, and then schedules them through the automatic scheduler, thus realizing the data transmission method without human intervention and automatic scheduling.

Data transfer scheduling is not the same as calculation job scheduling. For example, if a large file fails to be transferred, we are not willing to simply restart the job to retransmit the whole file, but rather only transfer the remaining unfinished parts. Similarly, if we fail to use one protocol at a time, we can try other protocols. In the process of data transmission, the network parameters can also be changed dynamically or the number of concurrent computing can be calculated, but the computing job scheduling does not deal with these situations.

3.1 Data Transmission Steps

In the process of data transmission, which is composed of N transmission steps, there may be some dependency between the transmission steps.

Define 1 data transmission scheduling step set $t = \{t_1, t_2, \ldots, t_n\}$, consisting of n meta-steps. The input data of transmission scheduling is the data to be copied, and the output data is the data to be sent to the specified node according to the user's requirements.

Specifically, a data transmission includes the following steps:

Determine replication content: determine the content to be replicated, as well as the source and target nodes;

Bandwidth preallocation: before data replication, first calculate the optimal transmission channel according to the current network traffic, and allocate the required bandwidth;

Data transmission: reliable data transmission in blocks between source and target nodes;

Release data bandwidth: after data transmission, the occupied bandwidth needs to be released in time.

3.2 Representation of Data Transmission Steps

The steps are as follows:

```
{
transfer_step="Determine replication content";
src_url ="10.0.0.20/test/1.dat";
size="200MB";
}
{
transfer_step="Bandwidth preallocation";
dest_host="10.0.0.21";
size="200MB";
duration="60 minutes";
allocation_id=1;
}
{
transfer_step="Data transmission";
src_url ="10.0.0.20/test/1.dat";
dest_url ="10.0.0.21/home/1.dat";
size="200MB";
threads="5";
}
{
transfer_step="Release data bandwidth";
dest_host = "10.0.0.21";
allocation_id = 1;
}
```

Define 2 dependency set $E = \{e_1, e_2, \ldots, e_m\}$, consisting of M dependencies. E_{UV} indicates that there is a dependency between the transmission step t_u and the transmission step t_v.

It is defined that TSG = t, is a directed graph. Where T is the set of nodes $T = \{t_1, t_2, t_n\}$, E is the set of edges $e = \{e_{ij} \mid 0 < i, j \leq n; i \neq j\}$. The node t_u in the graph represents the identification of each step of a data transmission, the directed edge e_{uv} represents the dependency between the scheduling step t_u and t_v, and the c_{uv} represents the size of the data to be transmitted between t_u and t_v.

Among them, the content that needs to be replicated can be determined by the algorithm described in "Multi-stage Replica Consistency Algorithm based on Activity" [6], so that the replication of the replica can avoid human intervention to the greatest extent and has high intelligence.

In the whole process, no storage devices or computing resources will be occupied repeatedly, and computing tasks will be started at the right time (for example, the required input data has been prepared), so as to maximize the utilization of resources.

For example, there are two data transmission tasks, job_a and job_b, which need to occupy the data transmission channel between node A and node B. in order to make the description simple, we assume that the data transmission channel can only transmit one data at the same time (this assumption simplifies the discussion of bandwidth limitation. In the actual data transmission, due to the limitation of system load and network bandwidth, the two points can often advance at the same time Row data transmission is limited. In addition, in order to make full use of bandwidth resources, each data transmission task is often carried out in a multi-threaded way. In fact, the number of data transmission tasks that can be carried out at the same time between two points is very limited, and too many tasks will lead to bandwidth contention, which will be described in the later content.

In the case of limited channels, there are a variety of policy options:

1. Transfer the data of job_a first, and then carry out job_b task after job_a is finished;
2. To estimate job $_a$ and job_b, the first task is to take a short time, and then another task;
3. Block job_a and job_b, and transfer each block in turn.

The above three strategies can be respectively expressed as the following situations by adopting the data transmission process description method described above.

1. Sequential execution

{
Dag_id = v001
Dag_type = "Allocated bandwidth"
Data_size = 1G
Host = A
Duration = 2 minutes
ReserveID = 2
}
	{
	Dag_id = v002
	Dag_type = "Data transmission"
	Src_url = D1://from/data/path
	Dest_url = A://to/data/path
	}

{
Dag_id = v003
Dag_type = "Release data bandwidth"
Host = A
ReserveID' = 2
}
	{
	Dag_id = v004
	Dag_type = "Allocated bandwidth"
	Data_size = 500M
	Host = B
	Duration = 2 minutes
	ReserveID = 3
	}

{
Dag_id = v005
Dag_type = "Data transmission"
Src_url = A://from/data/path
Dest_url = B://to/data/path
}
	{
	Dag_id = v006
	Dag_type = "Release data bandwidth"
	Host = A
	DataID' = 3
	}

2. Optimized execution

{
Dag_id = v001
Dag_type = “Time consuming”
Data_sizeA = 1G
Data_sizeB = 500M
DurationA = 2 minutes
DurationB = 1 minutes
}
	{
	Dag_id = v002
	Dag_type = “Allocated bandwidth”
	Data_size = 1G
	Host = A
	Duration = 2 minutes
	ReserveID = 2
	}

{
Dag_id = v003
Dag_type = “Data transmission”
Src_url = D1://from/data/path
Dest_url = A://to/data/path
}
	{
	Dag_id = v004
	Dag_type = “Release data bandwidth”
	Host = A
	ReserveID’ = 2
	}

{
Dag_id = v005
Dag_type = “Allocated bandwidth”
Data_size = 500M
Host = A
Duration = 2 minutes
ReserveID = 3
}
	{
	Dag_id = v006
	Dag_type = “Data transmission”
	Src_url = A://from/data/path
	Dest_url = B://to/data/path
	}

{
Dag_id = v007
Dag_type = “Release data bandwidth”
Host = A
DataID’ = 3
}	

3. Block execution

{
Dag_id = v001
Dag_type = "Blocking"
Data_size = 1G
Duration = 2 minutes
Threads=5
}
	{
	Dag_id = v002
	Dag_type = "Blocking"
	Data_size = 500M
	Duration = 1 minutes
	Threads=5
	}

{
Dag_id = v003
Dag_type = "Allocated bandwidth"
Data_size = 200M
Host = A
Duration = 24 seconds
ReserveID = 1
}
	{
	Dag_id = v004
	Dag_type = "Data transmission"
	Src_url = A://from/data/path
	Dest_url = B://to/data/path
	}

{
Dag_id = v005
Dag_type = "Release data bandwidth"
Host = A
DataID'= 1
}
	{
	Dag_id = v006
	Dag_type = "Allocated bandwidth"
	Data_size = 100M
	Host = A
	Duration = 12 seconds
	ReserveID = 2
	}

{
Dag_id = v007
Dag_type = "Data transmission"
Src_url = A://from/data/path
Dest_url = B://to/data/path
}
	{
	Dag_id = v008
	Dag_type = "Release data bandwidth"
	Host = A
	DataID' = 2
	}

{
Dag_id = v003
Dag_type = "Allocated bandwidth"
Data_size = 200M
Host = A
Duration = 24 seconds
ReserveID = 3
}
	……

3.3 Distribute Reliable File Transfer: DRFT

When there are errors in the process of data transmission, such as the target disk is full or a node suddenly fails, the transmission management system must be able to adjust the strategy in time, so that the data transmission can continue. Because the network environment of a large number of data providers (DPS) is usually very different, transmission failures often occur in the process of data replication. In this paper, a distributed reliable file transfer (DRFT) is proposed. By using this strategy, the transmission management system can automatically handle the errors that may occur in the transmission process, so as to realize the reliable data transmission in the unsupervised environment.

Based on the reliable data transfer (RFT), the distributed reliable file transfer strategy can capture the abnormal events such as network disconnection, machine restart, server crash and network timeout, and form the corresponding retry mechanism. Different from RFT, in DRFT, customer's file transfer request is not treated as a task, but a threshold value is obtained by calculating the bandwidth status and system load of the data provider, according to which the files to be transferred are divided into blocks, and then each block is treated as a task. Through block processing, on the one hand, it can improve the concurrency of transmission, on the other hand, it can effectively restrain the impact of fault.

When selecting the nodes for data transmission, DRFT is also different from RFT. DRFT adopts an adjacent algorithm, which allows multiple adjacent nodes to transmit data with DP at the same time. Each node completes different blocks, and the completed blocks are shared in the nodes. This strategy is based on the assumption that the reliability of DP is far less than that of the central node. In fact, this assumption can also be established.

4 Performance Analysis and Experimental Results

In order to verify the validity of the data transmission model proposed in this paper, two groups of experiments are designed, the first group uses GridFTP, the second group uses DRFT strategy proposed in this paper. Both groups adopt the same node configuration and network connection, and the specific network structure diagram is shown in Fig. 2:

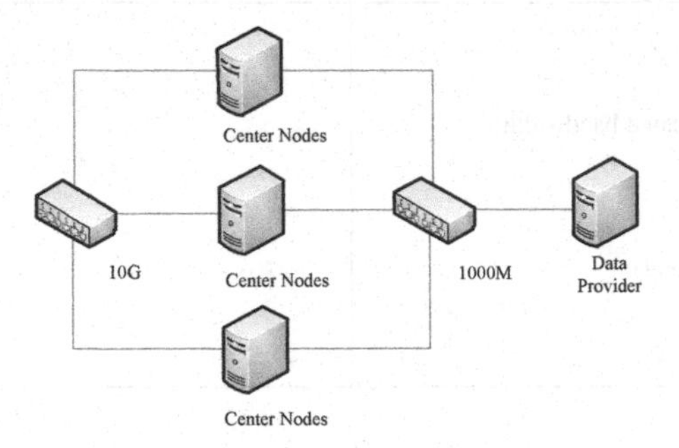

Fig. 2. Network structure

One machine on the right is the data provider DP, and three machines on the left are the central node cluster. They are interconnected by 10 Gigabit network. Each central node and DP are connected by a hub with Gigabit bandwidth. The data to be transmitted is three files on DP, with file sizes of 5g, 10g and 20G.

In the experiment, firstly, the stability and self recovery ability of data transmission is verified by the artificial network disconnection and DP restart, at the same time, the impact of each fault on the total transmission time is recorded for comparison; secondly, the performance between the two transmission systems is compared by recording the time of completing three file transmission.

4.1 Reliability Verification

In this experiment, two kinds of faults are designed, one is network disconnection, the other is DP restart. Network disconnection can be divided into two situations: the first is to disconnect DP, and the second is to disconnect a central node from DP randomly.

In the first case, the nature of network disconnection is the same as that of DP restart. Since all the connections between the central node and DP are disconnected, data transmission is immediately interrupted. When reconnected, the two groups using GridFTP and DRFT can immediately resume data transmission.

In the second case of network disconnection, because there is still a connection between the central node and the DP, the DRFT strategy group still continues to carry out data transmission, and because the change of bandwidth is detected, the other two central nodes increase the number of concurrent, ensuring the full use of bandwidth, and the total data transmission time is not affected. A group of GridFTP uses the Striped multi server parallel transmission mode. In this mode, interrupting a central node will not affect the data transmission of other nodes. However, due to the lack of effective monitoring of bandwidth, the other two central nodes will not improve the concurrency, so the total data transmission time is significantly longer. Table 1 shows the comparison of the results of the two groups of experiments:

Table 1. Comparison of two groups of experiments

Transmission strategy	File size (G)	Transmission time (s)	Transmission speed (MB/s)
GridFTP (Three center nodes)	5	68.7	72.8
GridFTP (Two center nodes)	5	105.7	47.3
GridFTP (Single center node)	5	178.6	28.0
DRFT (Three center nodes)	5	70.2	71.2
DRFT (Two center nodes)	5	72.0	69.4

4.2 Performance Verification

In the performance verification, two groups of comparative experiments are still used, one group uses GridFTP, the other group uses DRFT. The difference is that in the group using DRFT, the first suitable strategy, the maximum or minimum suitable strategy and the best suitable strategy of genetic algorithm are used for the test. Each theory test uses three files of 5g, 10g and 20g for transmission, respectively statistical transmission Enter the time of completion. The test results of each round are listed in Table 2, Table 3, Table 4 and Table 5 respectively:

Table 2. GridFTP test results

Transmission strategy	File size (G)	Transmission time (s)	Transmission speed (MB/s)
GridFTP	5	68.7	72.8
GridFTP	10	139.7	71.6
GridFTP	20	284.5	70.3

Table 3. Test results of the earliest fit strategy (FF)

Transmission strategy	File size (G)	Transmission time (s)	Transmission speed (MB/s)
DRFT with FF	5	104.6	47.8
DRFT with FF	10	234.7	42.6
DRFT with FF	20	490.2	40.8

Table 4. Test results of LFSF

Transmission strategy	File size (G)	Transmission time (s)	Transmission speed (MB/s)
DRFT with LFSF	5	107.5	46.5
DRFT with LFSF	10	224.2	44.6
DRFT with LFSF	20	459.8	43.5

Table 5. Test results of the best fit strategy (gabf) of genetic algorithm

Transmission strategy	File size (G)	Transmission time (s)	Transmission speed (MB/s)
DRFT with GABF	5	70.1	71.3
DRFT with GABF	10	139.9	71.5
DRFT with GABF	20	277.4	72.1

When using FF strategy, only the number of concurrent physical nodes is considered, so when the number of physical nodes is small, the degree of concurrency is not enough, resulting in the waste of transmission bandwidth, thus reducing the speed of data transmission. When using LFSF strategy, although the size of blocks is considered, in most cases, the size of blocks is fixed, so In fact, this strategy and FF strategy are similar. From the experimental results, it can be seen that these two strategies obtain similar data transmission rates. However, the BF strategy using GA algorithm can dynamically allocate the concurrency of each node according to the bandwidth resources, because of considering the bandwidth constraints, so it can make full use of the effective transmission bandwidth. From the experimental results, this strategy can have The transmission rate is relatively stable, which is the same as GridFTP with parallel transmission, and has basically reached the limit of physical bandwidth.

5 Concluding Remarks

In terms of performance index, under the limited bandwidth condition, GridFTP and DRFT based on GA algorithm can make full use of the bandwidth and reach the limit of almost physical bandwidth, while DRFT based on FF and LFSF strategy is slightly deficient in performance due to the lack of dynamic deployment capability for concurrency. However, considering the simplicity of implementation of these two strategies, it can be achieved through Dynamic adjustment of the number of central nodes involved in data transmission to indirectly adjust the degree of concurrency, so these two strategies still have application value.

References

1. GT 4.0 GridFTP. http://www.globus.org/toolkit/docs/4.0/data/gridftp/index.pdf
2. GT 4.0 Reliable File Transfer (RFT) Service. http://www.globus.org/toolkit/docs/4.0/data/rft/index.pdf
3. Czajkowski, K., et al.: From open grid services infrastructure to WS-resource framework. In: Refactoring & Evolution, pp. 88–101 (2004)
4. Yun, D., Wu, C.Q.: An integrated transport solution to big data movement in high-performance networks. In: International Conference on Network Protocols, pp. 460–462. IEEE (2015)
5. Yun, D., Wu, C.Q., Rao, N.S.V., et al.: Profiling transport performance for big data transfer over dedicated channels. In: International Conference on Computing, Networking and Communications, pp. 858–862. IEEE (2015)
6. Liu, G., Xu, H.: Multi-stage replica consistency algorithm based on activity. In: The Second International Cognitive Cities Conference, pp. 221–231, IC3 (2019)

DUMPLING: Cross-Domain Data Security Sharing Based on Blockchain

Mingda Liu(✉) and Yijuan Shi

Jiangnan Institute of Computing Technology, Wuxi 214083, China
happyliumd@163.com

Abstract. Blockchain is an effective solution for data sharing. The usual practice is to write metadata and the hash of raw data to the blockchain. This design does not take the confidentiality of metadata into account. The data provider wants to freely decide whether to share data through the data applicant's search conditions, without exposing what data they have. This paper presents a Cross-Domain data security sharing model called DUMPLING, which is based on consortium blockchain. It achieves trusted data transaction evidence and privacy protection search. Analysis shows that the system has good security and efficiency.

Keywords: Blockchain · Data sharing · Metadata confidentiality

1 Introduction

In the era of big data, data has become an important asset [1]. Big data mining can greatly increase the value of data. However, the problem of information silos is objective. Everyone needs data, and everyone lacks enough data. Therefore, it is crucial to establish a secure and trusted data sharing mechanism.

Recently, researchers have proposed methods based on blockchain [2–6] technology to achieve data sharing, including the data transaction, data validation, data determination. For example, DESC [7] implements automatic decision-making for data sharing based on smart contracts. It only implements access judgment on the chain and does not implement chain decision. The design of data confidentiality protection is also weak. There are also researches which combine secure mutiparty computing(SMC) [8], and other cryptographic [9, 10] techniques with blockchain, such as Dong's research [11]. But The performance of these cryptographic techniques is not high enough to be practical on a large scale. The other direction is the combination of blockchain and TEE to implement a privacy-protected data security computing platform, such as Ekiden [12]. But it needs special hardware, meanwhile TEE also has many security risks. The basic architecture of the data exchange market based on blockchain is shown in Fig. 1. Each organization has the same blockchain ledger, which in turn maintains the same data directory. This architecture enables the same data view across the entire network, and each participant can retrieve the data they want and get the data by paying or applying.

In the actual application scenario, data owners sometimes don't want data applicants to be able to retrieve their own data. In other words, they want to choose their own

X. Sun et al. (Eds.): ICAIS 2020, LNCS 12240, pp. 236–245, 2020.
https://doi.org/10.1007/978-3-030-57881-7_21

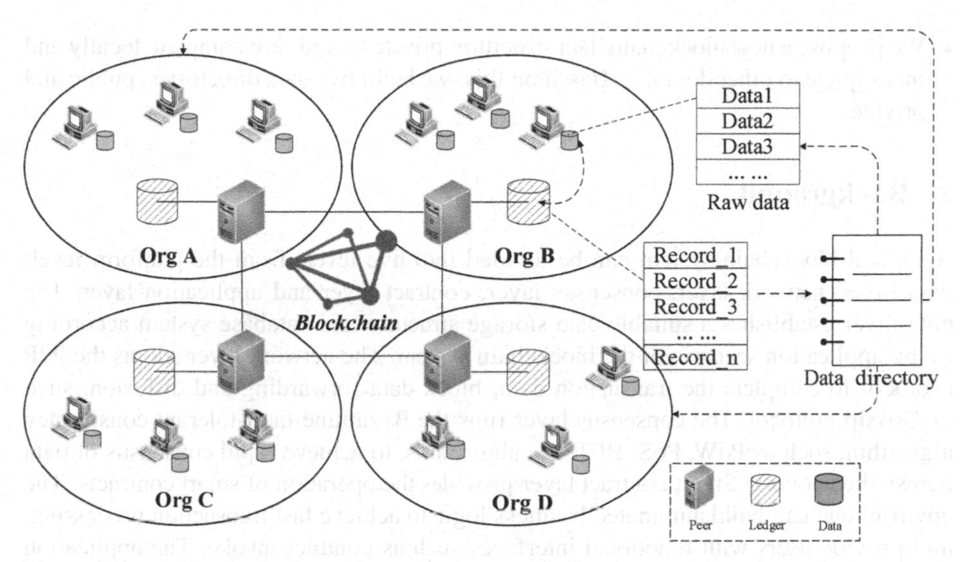

Fig. 1. Basic architecture of data market based on blockchain

buyers, because they don't expose some kind of data. Sometimes they even don't want others to know that they already have some data. The metadata itself may be valuable information which cannot be published on the whole network. For example, I got a correlation analysis report on the social relationship of 1000 important people on Twitter. But I just want to sell the data to the people who need, and I don't want more people to know that I have it. The reason may be that I don't want others to know I focus on data analysis of social networks, or that I don't want hackers to stare at my computer. In summary, data owners need a security mechanism that enables authorized retrieval to protect the confidentiality of metadata. This is also a balance between security and sharing. However, current practical blockchain-based data sharing researches do not take this into account.

In fact, searchable encryption can achieve what we need. All data stored on the blockchain is encrypted, but the user can retrieve the data he needs based on the conditions. However, searchable encryption cannot support large-volume data retrieval and is not compatible with distributed deployment environments. In addition, homomorphic encryption can implement a complete ciphertext retrieval mechanism, but the performance of current homomorphic cryptographic algorithms cannot meet the requirements of large-scale data retrieval and sharing.

Therefore, this paper proposes a blockchain-based cross-domain data security sharing method, called *DUMPLING*. The following are our contributions:

- The data transaction market based on blockchain does not consider the confidentiality of metadata. We propose a practical data security sharing mechanism that can protect the confidentiality of metadata.
- We implemented a full process record of data transactions based on the consortium chain architecture. The blockchain application not only guarantees the privacy needs of the data owner, but also prevents his repudiation.

- We propose a new blockchain data structure: private records are complete locally and incomplete in other domains. Based on this, we build two data directories, public and private.

2 Background

A typical blockchain system can be divided into five levels from the platform level: data layer, network layer, consensus layer, contract layer and application layer. The data layer establishes a suitable data storage structure and database system according to the application scenario of the blockchain system. The network layer adopts the P2P protocol to complete the transaction data, block data forwarding and diffusion, such as Gossip protocol. The consensus layer runs the Byzantine fault-tolerant consistency algorithm, such as PoW, PoS, BFT-like algorithms, to achieve rapid consensus of data across the network. Smart contract layer provides the operation of smart contracts. The environment can build automated business logic to achieve fast transaction processing, and provide users with functional interfaces such as contract invoke. The application layer implements basic business systems based on smart contracts, such as financial services and copyright management.

The process of blockchain generation can be summarized as: creating transactions, broadcasting transactions, verifying transactions, and adding blocks. The user first creates a transaction, which can be a transfer, data record or other information according to the actual logic of the business. Then the blockchain node broadcasts the transaction to the P2P network. Each consensus node verifies the legitimacy of the transaction and checks the user status to be verified conclusion. The last new block will be added to the blockchain and cannot be tampered or deleted.

At present, the blockchain ledger uses a full account synchronization mechanism. It means participants are able to obtain a full history of the blockchain, which is critical to the fact that the blockchain cannot be tampered. However, full historical ledger will reveal the privacy of users, especially when applied to the field of data sharing, in which the blockchain used as a data retrieval ledger. Therefore, this paper has modified the data organization form of the blockchain to meet different trust domains with different data directory views.

3 Design of DUMPLING

3.1 Architecture of DUMPLING

DUMPLING achieves two goals, one is to build a data sharing platform based on blockchain, and the other is to establish a secure and trusted data flow auditing system. In order to make the scheme practically feasible, we did not choose advanced cryptography such as homomorphic and searchable encryption, but based on blockchain nodes for fine-grained data access control.

As can be seen in Fig. 2, every organization holds a peer, which representing the organization in the blockchain network. Peer is responsible for maintaining the blockchain ledger and running the smart contract. Each participant of *DUMPLING* has an identity and is identified by a certificate. Users in each organization connect to the blockchain via Peer. Peer is responsible for five main functions: 1. Data registration. 2 access control. 3. Data retrieval, it returns relevant data according to search conditions. 4. Hold the ledger. 5. Execute smart contracts.

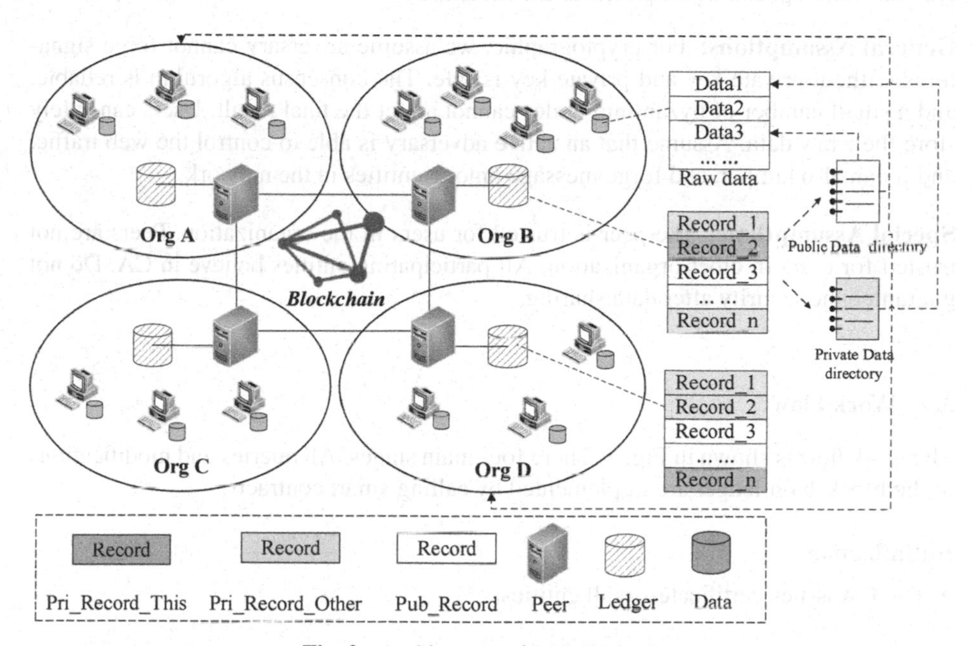

Fig. 2. Architecture of DUMPLING

Each organization can be data owner and data transferee. As data owner, it produces data, provides data, and earns revenue. As data transferee, it retrieves data, requests data, and obtains data. In consortium blockchain, the credentials to get the data can be either paid or user rights. Further, the user can obtain the data by payment, or can request the data by providing legal identity information.

The data is divided into two types: raw data and metadata. Metadata is a structured description of the raw data, containing the keyword information of the raw data. The raw data is stored and protected by the data producer, while the protection methods could adopt traditional encryption, access control and other methods. Since the metadata itself may also be sensitive, *DUMPLING* provides confidentiality protection for sensitive metadata,so the metadata storage of the system is designed in a special way, which will be described in detail. When data is registered, for sensitive metadata, two kinds of metadata are generated for the same raw data: complete metadata called ***Pri_Record_This***, partly metadata of sensitive information called ***Pri_Record_Other***. If the metadata is fully publicly available, we get ***Pub_Record***.

Each blockchain node maintains two types of databases, a blockchain database and a state database,the latter is a real-time analysis of the former state,which is used for fast retrieval of blockchain ledger.

3.2 Threat Model and Assumptions

This section presents a threat model and makes basic security assumptions. General assumptions is the basic security assumption of blockchain system. Special assumptions are additional special assumptions in DUMPLING.

General Assumptions: For cryptographic, we assume adversary cannot forge signatures without private key and private key is safe. The consensus algorithm is reliable, and a small number of byzantium nodes cannot affect the final result. Users can safely store their raw data. Assume that an active adversary is able to control the web traffic, and it can also tamper, and forge message among entities in the network.

Special Assumptions: The peer is trusted for users in the organization. Peers are not trusted for users in other organization. All participating entities believe in CA. Do not guarantee the security after data sharing.

3.3 Work Flow

The work flow is shown in Fig. 3. There four main stages. All queries and modifications to the blockchain ledger are implemented by calling smart contracts.

Initialization

0. The CA issues certificates to all entities.

Data registration

1A. Data owner sets the data access control policy and register data to the blockchain system.
1B. The regional center peer check the policy, classify the newly registered data, and determine whether the complete metadata can be shared to the entire network.
1C. According to the strategy, the smart contract is called to broadcast the data information.
1D. Run a consensus agreement.
1E. Consensus reached. Each peer updates the blockchain and update the local database.

Data query

2A. The user sets the search condition according to the data view and initiates a data query.
2B. Retrieve the domain data (***Pri_Record_This***) and the public data (***Pub_Record***) of the whole network to obtain the search results.
2C. Metadata that is owned by other domains (***Pri_Record_Other***) requires cross-domain data query.

2C-1 The consensus peer invokes the smart contract and provides the identity information of the data applicant as a parameter.
2C-2 Run consensus agreement to update the blockchain.
2C-3 The peer of other areas checks the user rights, and parses the search conditions to filter out the metadata that the user has permission to access.
2C-4 Encrypt the metadata list and send it to the querying user.

2D. The user parses the obtained data list and filters out the data that he wants to obtain.

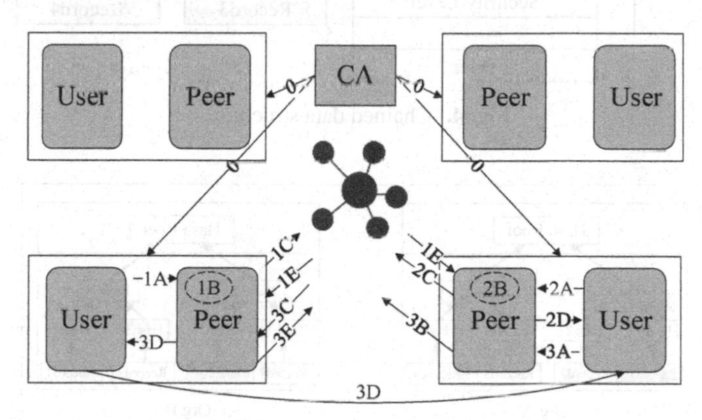

Fig. 3. Workflow of DUMPLING

Data sharing
3A. User invokes a smart contract to apply for data.
3B. The smart contract checks the permissions of the user and generates an access token if the permissions meet the requirements, such as identity and fee. If it does not meet the requirements, the request is rejected and the behavior is recorded on the chain.
3C. The smart contract sends the token to the data owner.
3D. The data owner verifies the token and sends the data encrypted to the applicant.
3E. The data transmission process records the deposit on the blockchain.

3.4 Data Structure

Blockchain is a chained data structure, so as DUMPLING, seen in Fig. 4.

Every block has a head and a body. The block headers include Pre_BlockHash, Leader and Root Hash. PrevBlockHash is the hash of the previous block to ensure that the order is consistent. Hash Root is actually the root of the Merkle tree. The above records are stored in the block, and the number of records in one block depends on the size of our design. In summary, once the record is written to the blockchain, it will no longer be tampered with by others.

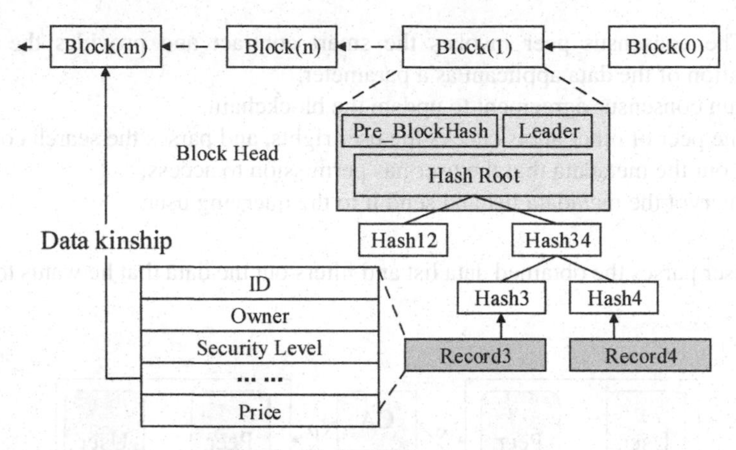

Fig. 4. Chained data structure

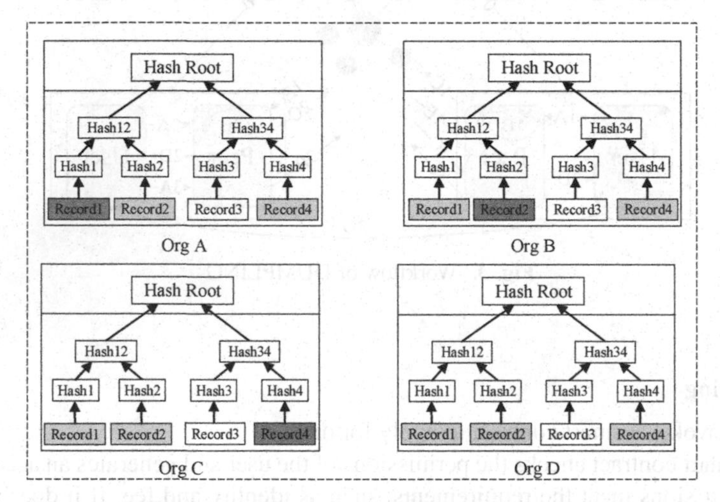

Fig. 5. Data structure in different organization

In addition, in the application scenario of data sharing, the data production and storage based on the chain structure can establish the data blood relationship in the data production and processing process.

Data structure is a special design of DUMPLING. An example is shown in Fig. 5, described as follows. It shows the data structure of one block in different Orgs. All peers hold a complete hash tree, but the leaf node is different. Record1, belongs to A. Record2, belongs to B. Record3, belongs to all. Record4, belongs to C. For example, record1 is ***Pri_Record_This*** data type, while it is ***Pri_Record_Other*** in other organization. The metadata format is as follows:

Pri_Record_This:{*ID, Owner, Security Level, Access Strategy, Storage Location, Key Words, Data Description, Other Description, Price*}

Pri_Record_Other:*{ID, Owner, Security Level, Access Strategy, Other Description, Price}*
Pub_Record:*{ID, Owner, Security Level, Access Strategy, Storage Location, Key Words, Data Description, Other Description, Price}*

This data structure design implements the privacy protection of metadata, and sensitive metadata is visible only within the organization. Data description entries are dynamically extensible and can support diverse data queries. Obviously, the structure of the metadata is only an example, and the specific format design needs to be performed according to the actual application scenario.

4 Analysis

4.1 Security Analysis

Security analysis under the Dolev-Yao threat model allows an attacker to eavesdrop, acquire, and tamper with protocol messages, and can pretend to be a legitimate subject participating protocol. The underlying cryptographic algorithm is secure, and random and private keys cannot be compromised. Assume that more than 2/3 of the compute nodes are honest and trustworthy, and the attacker cannot control more than 1/3 of the nodes. *DUMPLING* inherits the security features of the blockchain, has decentralization, traceability, unforgeable, and cannot be tampered.

- Decentralization

The main threat comes from the Sybil Attack, the adversary try his best to control peers and then throw an erroneous conclusion. His purpose is to modify the block to illegally obtain data. But And the adversary cannot control more 1/3 nodes in blockchain, so the final conclusion cannot be controlled by him.

- Traceable

Malicious nodes are often accompanied by malicious behavior. Trusted audits can expose malicious behaviors and discover malicious nodes. The blockchain is a chained storage structure marked by hash. Once the data is written into the blockchain, it can be along The blockchain traces back to the history. Traceability has two levels of meaning, and the evidence written into the blockchain can eventually be retrieved, and the retrieved data is credible and non-repudiation.

- Tamper-resistant

If an attacker wants to tamper with the blockchain database, he must control most of the peers in the network, which is inconsistent with the security assumptions in this paper. Suppose the adversary attempts to modify one of the blockchain systems involved in the *DUMPLING*. If the block is modified, then the hash value of the block will change, and all blocks behind the block must be changed. The blockchain is a distributed database system, only on most peers. The corresponding modification will make the attack take effect.

• Prevent data repudiation

Private metadata is the core design of this project. Owners of private data may use a low-value data to spoof data applicants to get more revenue. In the design of this article, although the metadata is private, the evidence of its integrity and the description of the foundation will be shared throughout the blockchain network. Therefore, the data applicant can determine whether he has obtained the desired data through the blockchain, thereby preventing data repudiation.

4.2 Efficiency Analysis

The efficiency analysis of the blockchain system mainly includes two aspects, throughput and latency. Throughput determines the scale of services, and the number of users and data sizes that can be supported on this system. Latency reflects the quality of service and determines the time delay introduced by blockchain system.

The data organization structure introduced in this paper obviously brings a little performance overhead, mainly reflected in the data registration and data query parts.

As for throughput, because the core idea of this solution is to limit the spread of data according to access rights. So this scheme does not increase the number of blockchain messages, no additional impact on throughput. The throughput performance metrics are primarily determined by the performance of the blockchain base platform.

As for latency, this scheme increases the latency of the blockchain system during the data query phase. Because in the cross-domain query, a process of consensus is actually added, causing the latency from the additional consensus. In addition, the processor needs to parse the data type and access rights, then send the corresponding location of the data. Therefore, the solution of this paper increases the computational overhead of some processors, but this has little effect on modern processors.

Besides, the performance impact of additional processing is different under different blockchain platforms. For example, under Hyperledger Fabric, whose throughput is around 1000 to 3000TPS and latency around 800 ms [13], our solution will increase the latency from 800 ms to 1600 ms. It is obviously unbearable. But for Hotstuff, whose throughput is around 100KTPS and latency around 5–20 ms [14],our solution will increase the latency from 20 ms to 40 ms. This is completely unfeelable.

5 Conclusion

In order to solve the data sharing problem under the metadata privacy protection requirements, we propose DUMPLING, a practical data transaction platform based on blockchain. This paper proposes the architecture design, workflow and data structure design of DUMPLING. The analysis proves that DUMPLING is safe, efficient and can be used in actual scenarios.

References

1. Naimi, A.I., Westreich, D.J.: Big data: a revolution that will transform how we live, work, and think. Math. Comput. Educ. **47**(17), 181–183 (2014)

2. Eyal, I., Gencer, A.E., Sirer, E.G., Renesse, R.V.: Bitcoin-NG: a scalable blockchain protocol. In: Proceedings of the 13th Usenix Conference on Networked Systems Design and Implementation, pp. 45–59. USENIX Association Berkeley, CA, USA (2015)
3. Deng, Z.L., Ren, Y.J., Liu, Y.P., et al.: Blockchain-based trusted electronic records preservation in cloud storage. Comput. Mater. Continua **58**(1), 135–151 (2019)
4. Song, R., Song, Y., Liu, Z., Tang, M., Zhou, K.: GaiaWorld: a novel blockchain system based on competitive PoS consensus mechanism. Comput. Mater. Continua **60**(3), 973–987 (2019)
5. Jiang, X., Liu, M., Yang, C., Liu, Y., Wang, R.: A blockchain-based authentication protocol for WLAN mesh security access. Comput. Mater. Continua **58**(1), 45–59 (2019)
6. Pass, R., Seeman, L., Shelat, A.: Analysis of the blockchain protocol in asynchronous networks. In: Coron, J.-S., Nielsen, J.B. (eds.) EUROCRYPT 2017. LNCS, vol. 10211, pp. 643–673. Springer, Cham (2017). https://doi.org/10.1007/978-3-319-56614-6_22
7. Liang, J., Han, W., Guo, Z., et al.: DESC: enabling secure data exchange based on smart contracts. Sci. China (Inf. Sci.) **61**(4), 049102 (2018)
8. Zyskind, G., Nathan, O.A.: Decentralizing privacy: using blockchain to protect personal data. In: 2015 IEEE Security and Privacy Workshops (SPW). IEEE Computer Society (2015)
9. Kosba, A., Miller, A., Shi, E., et al.: Hawk: the blockchain model of cryptography and privacy-preserving smart contracts. In: 2016 IEEE Symposium on Security and Privacy (SP). IEEE (2016)
10. Hu, S., Cai, C., Wang, Q., et al.: Searching an encrypted cloud meets blockchain: a decentralized, reliable and fair realization. In: IEEE INFOCOM 2018 - IEEE Conference on Computer Communications. IEEE (2018)
11. Dong, X.Q., Guo, B., Shen, Y., et al.: An efficient and secure decentralizing data sharing model. Chin. J. Comput. **425**(5), 55–70 (2018)
12. Cheng, R., Zhang, F., Kos, J., et al.: Ekiden: A Platform for Confidentiality-Preserving, Trustworthy, and Performant Smart Contract Execution (2018)
13. Androulaki, E., Barger, A., Bortnikov, V., et al.: Hyperledger Fabric: A Distributed Operating System for Permissioned Blockchains (2018)
14. Yin, M., Malkhi, D., Reiter, M.K., et al.: HotStuff: BFT consensus with linearity and responsiveness. In: ACM Symposium on Principles of Distributed Computing (PODC 2019) (2019)

A Semi-quantum Group Signature Scheme Based on Bell States

Jinqiao Dai, Shibin Zhang(✉), Yan Chang, Xueyang Li, and Tao Zheng

Chengdu University of Information Technology, Chengdu 610225, Sichuan, China
cuitzsb@cuit.edu.cn

Abstract. In this paper, we propose a group signature scheme based on Bell states, which restricts the group member Alice to transmit the signature message to a fully quantum ability group manager TP through semi-quantum communication, then TP sends the signature message to an outsider verifier Bob to verify the signature. The signature message is sent by TP which can ensure Alice's anonymous in group signature scheme. Besides, the limited Alice is a "classic" user in semi-quantum communication and can only measure, prepare, reorder and send quantum states only in the classical basis $\{|0\rangle, |1\rangle\}$, while the quantum TP can perform arbitrary quantum operations. Besides, the "classical" user Alice does not require quantum memory. Finally, the security analysis shows that the scheme can not only resist the repudiation of the signer and denial of the verifier, but also detect the forgery of the attacker, and the efficiency of the scheme is slightly higher than most of the previous signature schemes.

Keywords: Quantum cryptography · Semi-quantum · Group signature · Bell states

1 Introduction

Quantum signature is the quantum counterpart of classical signature, the first classic signature was introduced by Diffie and Hellman in 1976 [1], since then it has been playing a very critical role in classical cryptography and many other applications, s, for example, electronic government affairs, electronic medical and electronic payment. However, classical signature schemes are insecure in quantum environment [2, 3]. Therefore, quantum digital signature is proposed, it is extended by classical digital signatures, unlike classical signature schemes, the security of quantum signature schemes is guaranteed by quantum non-cloning theorem [4] and Heisenberg's uncertainty principle [5], and is independent of the attacker's computing power. It can be theoretically unconditional security in front of a powerful quantum computer.

Quantum group signature is a special signature method. Quantum group signature consists of many group members and one group manager. In this group, any group member can represent the whole group to sign the message and the process of signing message is completely anonymity. The verifier cannot know the specific signer but can check the validity of the signature. The group manager, considered as the arbitrator, can know

X. Sun et al. (Eds.): ICAIS 2020, LNCS 12240, pp. 246–257, 2020.
https://doi.org/10.1007/978-3-030-57881-7_22

the specific identity of the signer. When dispute happens, the group manager can open the signature to reveal the identity of the signer. In 2008, Yang and Wen [6] suggested a multi-proxy quantum group signature, which achieved threshold shared verification. Then in 2010 [7], they proposed a new quantum group scheme based on quantum teleportation and improved by Su et al. [8] in 2014 for its insecurity to the inside attacker during the eavesdropping process after the transmission. After that, many researchers have been used and improved quantum group signatures in many practical applications. Such as inner bank quantum proxy blind signature [9], quantum communication-based online shopping mechanism [10]. And in 2017, Zhang [11] et al. proposed a third-party e-payment protocol based on quantum group blind signature by using four-qubit entangled state.

However, all of the above signature schemes have the disadvantage of quantum resource. The quantum state they used mostly are GHZ state or multi-qubit entangled state which cannot be implemented with the existing technology and they require all users in the signature have fully quantum ability, the actual application cost of these schemes will be expensive. To solve the difficulty of quantum resource and the cost of signature scheme, we consider using Bell states and semi-quantum communication technology as new directions of implementing in quantum group signature. "Semi-quantum" is a novel concept introduced by Boyer et al. [12], and Bell states is widely used in many quantum protocols and easier to implement in practice. In semi-quantum communication, one of the communication parties has complete quantum capability and can perform arbitrary quantum operations; the other party does not have a quantum memory and can only perform limited quantum operations, which is called 'classic' party. Due to its requirement for quantum resources in quantum communication, it has attracted the interest of many researchers and applied it to many quantum protocols. Such as SQKD protocols [13–16], semi-quantum direction protocols [17, 18], semi-quantum secret sharing [19–21], semi-quantum private comparison and semi quantum dialogue [22] etc. In addition, many other studies have provided many novel ideas for the development of quantum encryption [23, 24].

Based on the above analysis, we propose a semi-quantum group signature scheme based on Bell states, where the group member is classical party and can only limited quantum operations, and allowing the classical party to send the signature message to a fully quantum group manager. The proposed scheme can resist eavesdropping attacks and can the repudiation of the signer and verifier. In addition, by comparing our scheme with the previous scheme, our scheme has higher qubits efficiency.

The rest of this paper is outlined as follows. In Sect. 2, we briefly introduce the theory of quantum mechanics and semi-quantum communication to transmit signature message which will be used in the next section. And then in Sect. 3, we proposed the semi-quantum signature scheme based on Bell states. Then, we analysis the security and efficiency of the proposed protocol in Sect. 4 and 5. And the conclusion is drawn in Sect. 6.

2 Preliminary Theory

Here is some basic knowledge of quantum mechanics to facilitate the readers to understand the principles of protocol implementation.

The basic single photon measurement basis: Z basis $\{|0\rangle, |1\rangle\}$.

The description of Bell states (EPR pairs) is as follows:

$$|\phi^{+}\rangle = \frac{1}{\sqrt{2}}(|00\rangle + |11\rangle)$$

$$|\phi^{-}\rangle = \frac{1}{\sqrt{2}}(|00\rangle - |11\rangle)$$

$$|\psi^{+}\rangle = \frac{1}{\sqrt{2}}(|01\rangle + |10\rangle)$$

$$|\psi^{-}\rangle = \frac{1}{\sqrt{2}}(|01\rangle - |10\rangle) \tag{1}$$

In the formula (1), if we use Z basis to measure the first particle of Bell states $|\phi^{+}\rangle$, there is a probability of 1/2 that the measurement result is $|0\rangle$ or $|1\rangle$, and the second particle will collapse into the same state as the first one. If we use Z basis to measure the first particle of Bell states $|\psi^{+}\rangle$, the probability of measuring $|0\rangle$ or $|1\rangle$ is also 1/2 but the measurement results of the second particle is opposite.

In semi-quantum communication, one of the communication parties has complete quantum capability and can perform arbitrary quantum operations; the other party does not have a quantum memory and can only perform limited quantum operations, which is called 'classic' party. In the process of semi-quantum communication, the 'classic' party can only perform the following operations:

- Reflecting the received particles to the sender, defined as REFLECT operations;
- Measuring the received particles with Z basis, defined as MEASURE operations;
- Preparing single particles identical or opposite to the measurement results of MEASURE operations.
- Reorder all the received particles and return them to the sender.

3 The Proposed Scheme

In this scheme, we propose a group signature scheme using Bell states and semi-quantum communication technology. The communication between the group manager and group members are semi-quantum communication. Figure 1 depicts the generalization steps of the proposed scheme, which will be detailed as follows.

3.1 Initializing Phase

(I1) The group manager TP shares his secret key Ka with group member Alice (signer of the signature) and then shares his secret key Kb with verifier of the signature Bob. These keys are distributed by QKD protocols [25–27].

(I2) Alice wants to generate a signature, she firstly prepares n bits classic signature message M = (m(1), m(2), …, m(n)) and then sends a requirement to the group manager TP. TP prepares 2n pairs of Bell states particles where n is the length of the signature message and the prepared particles are randomly in states $\{|\phi^{+}\rangle, |\phi^{-}\rangle, |\psi^{+}\rangle$ or $|\psi^{-}\rangle\}$.

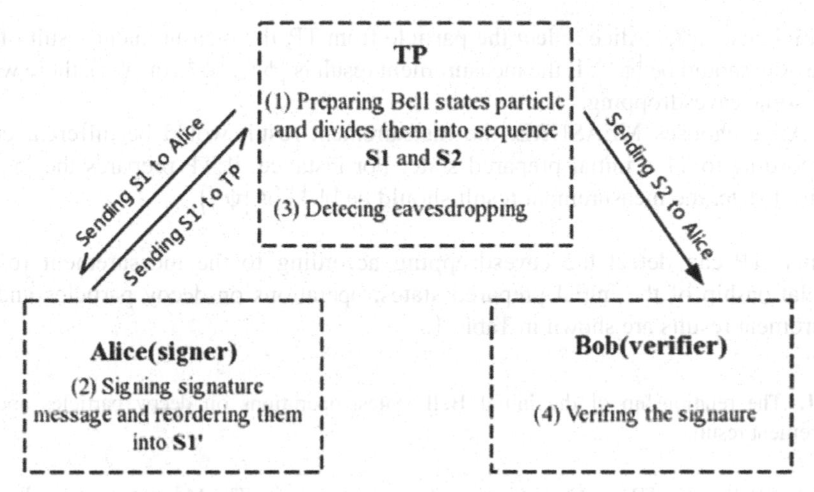

Fig. 1. The generalization steps of the proposed scheme.

(I3) TP divides these Bell states particles into two sequence, all the first particles are ordered into sequence S1 and all the second particles are ordered into sequence S2, both S1 and S2 contain 2n qubits particles. Finally, TP sends S1 to Alice and keeps S2.

3.2 Signing Phase

(S1) In order to resist the Trojan horse attack [28–30], a photon number splitter (PNS) and a wavelength filter are added before Alice's devices. After receiving the particles from TP, Alice randomly selects a half of particles from the sequence S1 as decoy particles which are used to check eavesdropping. For each decoy particle, Alice randomly selects either to measure it with Z basis (MEASURE) or to reflect it (REFLECT). The remaining n qubits particles are used to transmit signature message, for each remaining particle, Alice firstly measure it with Z basis, and prepares different single particle state according to her signature message M; if m(i) = 0, Alice prepares the same particle state as the measurement result, if m(i) = 1, she prepares t prepares the particle state opposite to the measurement result. And Bob randomly reorders the 2n particles and generates a rearranged particle sequence $S1'$ [12, 17]. Alice does not require quantum memory to implement this scheme. She sends $S1'$ to TP. The order of the rearranged sequence is completely secret to others but Alice herself.

(S2) After TP receives the sequence $S1'$ from Alice, he informs Alice to help him detecting eavesdropping. Alice tells TP the rearranged order of sequence $S1'$ and the operations on decoy particles through classic encrypted channel using the pre-shared secret key Ka. Based on the information Alice offers, TP reorders sequence $S1'$and performs a Bell measurement on the decoy particle with his own one, then he can detect the eavesdropping. In detail:

a) If Alice chooses REFLECT, the measurement result should be the same as the Bell state TP prepared in the initialing phase. For instance, if TP prepares the $|\phi^+\rangle$ as initial state, after Alice reflect the particle from TP, the measurement result of this particle should be $|\phi^-\rangle$. If the measurement result is $|\phi^-\rangle$, $|\psi^+\rangle$ or $|\psi^-\rangle$, there would be some eavesdropping.

b) If Alice chooses MEASURE, the measurement result would be different either according to TP's initial prepared state. For instance, if TP prepares the $|\phi^+\rangle$ as initial state, the measurement result should be $|\psi^+\rangle$ or $|\psi^-\rangle$.

Thus, TP can detect the eavesdropping according to the measurement results. The relationship of the initial prepared states, operations on decoy particles and the measurement results are shown in Table 1.

Table 1. The relationship of the initial Bell states, operations on decoy particles and the measurement results.

The initial Bell states TP prepares	Operations on decoy particles	The Measurement result of TP's own particles
$\|\phi^+\rangle$	REFLECT MEASURE	$\|\phi^+\rangle$ $\|\phi^+\rangle/\|\phi^-\rangle$
$\|\phi^-\rangle$	REFLECT MEASURE	$\|\phi^-\rangle$ $\|\phi^+\rangle/\|\phi^-\rangle$
$\|\psi^+\rangle$	REFLECT MEASURE	$\|\psi^+\rangle$ $\|\psi^+\rangle/\|\psi^-\rangle$
$\|\psi^-\rangle$	REFLECT MEASURE	$\|\psi^-\rangle$ $\|\psi^+\rangle/\|\psi^-\rangle$

Then TP calculates the corresponding error rate after the eavesdropping detection. If the total error rate is lower than the threshold, TP will continue the communication and proceed to the next step. Otherwise, the communication will be terminated.

(S3) TP measures all the remaining particles with Z basis to obtain Alice's signature message M. In addition, TP encodes the Bell states information he prepared in the initialing phase into Mp. Finally, he encrypts M and Mp with the secret key Kb, which can be descried as $E_{Kb}\{M, Mp\}$, and sends it along with the particle sequence S2 to Bob. Since the encrypted message is sent from TP, Bob cannot recognize the identity of the singer.

3.3 Verification Phase

(V1) After receiving $E_{Kb}\{M, Mp\}$ and S2 from TP, Bob firstly decrypts $E_{Kb}\{M, Mp\}$ with the secret key Kb to get the message M and Mp. Then Bob uses Z basis to measure all the particles in the sequence S2 and Bob can deduce Alice's signature message according to the measurement results and the Bell states information TP prepared in the initialing phase Mp. The details of deducing Alice's signature are shown in Table 2. The deduced signature message is denoted as M′ = (m(1)′, m(2)′, …, m(n)′). For instance, if the initial state TP prepares is $|\phi^{+}\rangle$ and the measurement result of S2 is $|0\rangle$, the signature message of m(i)′ is classic bit 0.

(V2) Bob compares M with M′. If M = M′, he will accept the signature, otherwise, he will reject it.

Table 2. The details of deducing Alice's signature message.

The initial Bell states TP prepares (Mp)	The measurement results of S2	The deduced signature message m′(i)
$\|\phi^{+}\rangle/\|\phi^{-}\rangle$	$\|0\rangle$ $\|1\rangle$	0 1
$\|\psi^{+}\rangle/\|\psi^{-}\rangle$	$\|0\rangle$ $\|1\rangle$	1 0

4 Security Analysis

4.1 Impossibility of Disavowal by the Signatory

Due to group member Alice is a limited semi-quantum ("classical") user and has no ability of quantum memory. Thus, she has to choose REFLECT or MEASRUE for each particle she receives. Besides, the information about the order of the rearranged sequence and operations on decoy particles are encrypted by the secret key Ka, the secret key Ka is distribute via QKD protocols, it is difficult for other members to get the secret key Ka. In step (S2) and (S3) of signing phase, TP detects eavesdropping and measures all the remaining particles to obtain Alice's signature M. If Alice If wants to Alice disavow her signature, which means she did not perform step (S1) of signing phase, and the scheme will not continue after (S1). However, after Bob accept the signature, the entire scheme has been completed, which means someone had used Ka to send information to TP. Since Ka is distribute via QKD protocols, only TP and Alice knows the secret key Ka. TP can confirm Alice has signed the signature. If Alice's disavowal engenders a dispute with Bob, TP can ask Alice to announce her operations on the particle sequence, i.e. which particles are reflected and which particles are measured, and the rearranged sequence. Thus, TP and Bob can deduce the signature M according to the message Alice announced, if the signature M is the same as the signature M′ Bob accepted in step (V2) of verification phase. The signature must have been signed by Alice.

4.2 Impossibility of Denial by the Receiver

If Bob cannot deny his receiving of the signature, the scheme is termed to be undeniable and this feature plays a very important role in many practical applications. In this scheme, if Bob tempts to deny his receiving of the signature, he cannot get any benefit. Because

Bob needs the assistant of TP to gain Alice's full signature during the verifying phase, in addition, the message M and Mp sent by TP are encrypted with the secret key Kb. Therefore, the message can only be decrypted by Bob and he cannot deny his receiving of the signature once he accepts the signature. On the other hand, if Bob wants to deny the integrity of the signature, which means Bob recognizes some received signature but disaffirms correctness of the signature, that is, Bob declares $M' \neq M$ during the verifying phase, even if $M' = M$. However, this kind of denial has no benefit for Bob either. Once Bob claims $M' \neq M$, in the view of TP, he will consider that the signature has been intercepted by the eavesdropper. Thus, TP will invalidate the signature and terminate this communication. The signature Bob obtained will be useless.

4.3 Impossibility of Forgery

Assume that there is a malicious attracter Eve attempts to obtain and forge Alice's signature message. We analyze some common attack methods as follow.

- Intercept-measure-resend attack

Assume that Eve attempts to forge Alice's signature by using intercept-measure-resend attack, he needs to intercept the particle sequence $S1'$ Alice sent back to TP and measure the remaining n qubits particles to gain the signature message. However, the sequence $S1'$ was randomly rearranged and the order of the rearranged $S1'$ is completely secret to others but Alice herself. In order to gain Alice's signature message, Eve has to measure all the particles in $S1'$ with Z basis. If Eve measures decoy particles with Z basis, for the MEASURE particles, this kind of operation will not disturb the state of these particles, as for the reflected ones, the state of these particles is randomly in Bell states. When Eve measures Bell states in the Z basis, according to the theory of quantum mechanics, the particle will collapse in to $|0\rangle$ or $|1\rangle$ with equal probability. After these particles return to TP, TP will use the original generation basis to measure these reflected particles to calculate the corresponding error rate during the eavesdropping detection. If the initial states of decoy particles are $|\phi^+\rangle$, after Eve's attack, the measurement results would be $|\phi^+\rangle$ or $|\phi^-\rangle$ instead of $||\phi^+\rangle$, which could be detected by TP and the error rate introduced by Eve is 50%. For each decoy particle, the possible of Alice choosing REFLECT or MEASURE is 50%. Thus, the possibility of Eve successfully pass the eavesdropping detection defined as Pe.

$$P_e = \left(\frac{1}{2} \times (\frac{1}{2} \times \frac{1}{2} + \frac{1}{2} \times \frac{1}{2})\right) = \frac{1}{4} \tag{2}$$

Eve's possibility of forgery can be quantitatively assessed. Assume that for the reflected particles containing k quits, there are n quits are decoy particle. The possibility of Eve successfully pass the eavesdropping is Pe. The possibility that Eve successfully forged a signature is defined as Ps, which is depended on binomial distribution.

$$P_s = \binom{n}{k}(P_e)^k(1-P_e)^{n-k}$$
$$= \binom{n}{k}\left(\frac{1}{4}\right)^k\left(\frac{3}{4}\right)^{n-k} \quad (3)$$

The Binomial coefficient can be defined as.

$$\binom{n}{k} = \frac{n!}{k!(n-k)!} \quad (4)$$

It can be seen from (2) and Fig. 2. There is a maximum probability value for each different values of m. Moreover, the peak value of Ps decreases as n increases. It can be deduced that Ps will be quite small when n is large enough. Therefore, it is impossible for Eve to use intercept-measure-resend attack to forge Alice's signature message.

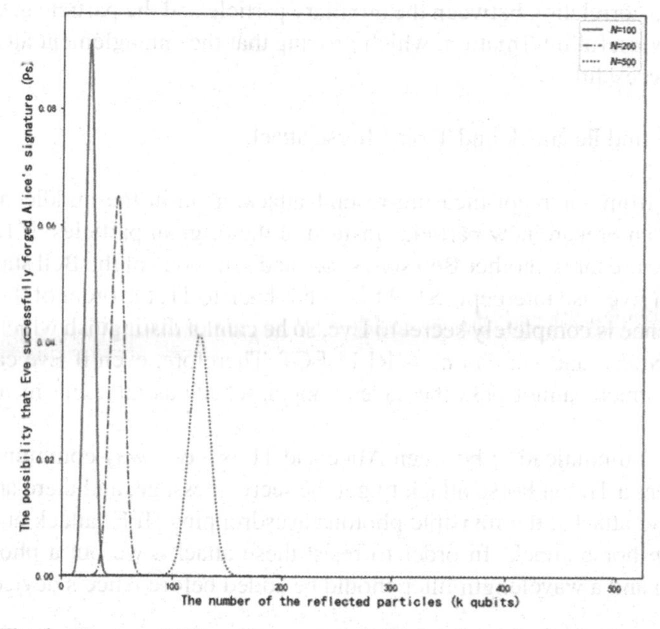

Fig. 2. The possibility of Eve successfully forging Alice's signature.

On the other hand, if Eve attempts to intercept sequence S2 and measures all the particles with Z basis during the step (S3) of the signing phase, he cannot obtain Alice's signature message either. Because the measurement results of S2 are not Alice's signature and Eve has to gain Mp to deduce the final signature message. However, Mp is encrypted by TP using the secret key Kb which can only be decrypted by TP and Bob. Thus, Eve cannot obtain Alice's signature by intercepting S2.

- Entanglement attack

Another strategy for Eve to forgery Alice's signature is to prepare an intermediate state e. When Eve intercept S1 or S1′, he entangle each particle of the quantum sequence with the pre-prepared intermediate state. Without loss of generality, Eve's unitary operation Ue can be described as

$$U_e|0\rangle|E\rangle = a_e|0\rangle|e_{00}\rangle + b_e|1\rangle|e_{01}\rangle \tag{5}$$

$$U_e|1\rangle|E\rangle = c_e|0\rangle|e_{10}\rangle + d_e|1\rangle|e_{11}\rangle \tag{6}$$

Where $|e_{00}\rangle$, $|e_{01}\rangle$, $|e_{10}\rangle$, $|e_{11}\rangle$ are pure states and $|a_e|^2 + |b_e|^2 = 1$, $|c_e|^2 + |d_e|^2 = 1$. For particles using transmit signature message, the particles are all in state $|0\rangle$or $|1\rangle$, and it can be seen from the above equations, when the state of the particle is $|0\rangle$, the state of the target bit particle does not change, which means Eve's entanglement attack will not cause any disturb. However, in this case, the total particles must be related to Eve's direct state of the auxiliary quantum states. However, the stochastic state indicates that there is no correlation between the auxiliary particle and the particle g, thus Charlie cannot get any useful information, which proving that the entanglement attack strategy will not be successful.

- Man-in-the-middle attack and Trojan horse attack

Different from intercept-measure-resend attack, man-in-the-middle assumes the attacker Eve can prepare new particles instead of the original particles. If Eve captures S1 from TP, he prepares another Bell states $|\phi_e\rangle$ and sends one of the Bell states particles to Alice. Then Eve also intercepts S1′ Alice sends back to TP, the order of the rearranged particle sequence is completely secret to Eve, so he cannot distinguish which decoy particle is MEASURE and which one is REFLECT. Therefore, even if Eve catches these particles, his attack cannot pass the eavesdropping check as intercept-measure-resend attack.

Since the communication between Alice and TP is a two-way communication, Eve may implement a Trojan horse attack to get the secret message, and there are two kinds of Trojan horse attacks: the invisible photon eavesdropping (IPE) attack and the delay-photon Trojan horse attack. In order to resist these attacks, we put a photon number splitter (PNS) and a wavelength filter should be added before Alice's devices.

5 Efficiency Analysis

The efficiency of the proposed semi-quantum group signature scheme can be calculated with the definition [31] $\eta = \frac{b_s}{q_t + b_t}$. Where bs is the total number of transmitted classical bits, qt is the total number of qubits utilized in the protocol, and bt expresses the number of classical bits utilized to decode the message. (Classical communications used for checking of eavesdropping is not counted.)

In our proposed protocol, TP prepares 2n pairs of Bell states particles and sends on of the particles to Alice, Alice chooses n qubits as decoy particles and n qubits to transmits n bit classic signature message then send 2n qubits particles back to TP. After TP detects eavesdropping, he and measures particles, he sends n qubits particles to Bob along with 2n bit classic message to help Bob deduce the signature message and verify with Alice's original signature message. Thus, the efficiency of our proposed scheme is $\eta = n/(3n + n) = 25\%$. The efficiency comparison of the proposed scheme with the previous protocols are demonstrated in Table 3.

Table 3. Efficiency and other features comparing with previous schemes.

Protocol	Ref. [9]	Ref. [11]	Ref. [32]	Our scheme
$\eta(\%)$	18	16.7	20	25
Participant attribute	Fully quantum	Fully quantum	Fully quantum	Semi-quantum
Quantum states used in the scheme	three-particle entangled states	Four-particle cluster states	Bell states	Bell states

It is explicit that the efficiency of our scheme is slightly higher than most of the previous schemes. Furthermore, the quantum states our protocol prepares are Bell states particles, the communication between group manager and members are through semi-quantum communication.

6 Conclusion

In this paper, we propose a semi-quantum group signature scheme based on Bell states, which enable the classical group member Alice to send the signature message to a quantum TP directly, and TP sends the signature to a verifier Bob to ensure Alice's anonymous. Compared with previous quantum signature schemes, the semi-quantum scheme requires less quantum resource and have higher efficiency. In addition, the security analyses shows that is also secure against the intercept-measure-resend attack, man-in-the-middle attack, entanglement attack and Trojan horse attacks.

Acknowledgement. This work is supported by the National Natural Science Foundation of China (No. 61572086, No. 61402058), the Key Research and Development Project of Sichuan Province (No. 20ZDYF2324, No. 2019ZYD027, No. 2018TJPT0012), the Innovation Team of Quantum Security Communication of Sichuan Province (No. 17TD0009), the Academic and Technical Leaders Training Funding Support Projects of Sichuan Province (No. 2016120080102643), the Application Foundation Project of Sichuan Province(No. 2017JY0168), the Science and Technology Support Project of Sichuan Province (No. 2018GZ0204, No. 2016FZ0112).

References

1. Diffie, W.: New directions in cryptography. IEEE Trans. Inf. Theory **22**, 644–654 (1976)
2. Shor, P.W.: Polynominal-time algorithms for prime factorization and discrete logarithms on a quantum computer. SIAM J. Sci. Statist. Comput. **26**, 1484–1509 (1997)

3. Grover, L.K.: Quantum mechanics helps in searching for a needle in a haystack. Phys. Rev. Lett. **79**(2), 325 (1997)
4. Barnum, H., Crepeau, C., Gottesman, D., et al.: Authentication of quantum messages, pp. 449–58. IEEE Computer Society, Washington, DC (2002)
5. Busch, P., Heinonen, T., Lahti, P.: Heisenberg's uncertainty principle. Phys. Rep. **452**(6), 155–176 (2007)
6. Yang, Y., Wen, Q.: Sci. China Ser. G Phys. Mech. Astron. **51**, 1505 (2008)
7. Wen, X., et al.: A group signature scheme based on quantum teleportation. Phys. Scr. **81**(5), 055001 (2010)
8. Su, Q., Li, W.-M.: Improved group signature scheme based on quantum teleportation. Int. J. Theor. Phys. **53**(4), 1208–1216 (2013). https://doi.org/10.1007/s10773-013-1917-4
9. Wen, X.J., Chen, Y.Z., Fang, J.B.: An inter-bank E-payment protocol based on quantum proxy blind signature. Quantum Inf. Process. **12**(1), 549–558 (2013)
10. Chou, Y.-H., Lin, F.-J., Zeng, G.-J.: An efficient novel online shopping mechanism based on quantum communication. Electron. Commer. Res. **14**(3), 349–367 (2014). https://doi.org/10.1007/s10660-014-9143-6
11. Zhang, J.Z., Yang, Y.Y., Xie, S.C.: A third-party e-payment protocol based on quantum group blind signature. Int. J. Theor. Phys. **56**(9), 2981–2989 (2017)
12. Boyer, M., Kenigsberg, D., Mor, T.: Quantum key distribution with classical Bob. Phys. Rev. Lett. **99**(14), 140501 (2007)
13. Zou, X.F., Qiu, D.W., Li, L.Z., Wu, L.H., Li, L.J.: Semi-quantum key distribution using less than four quantum states. Phys. Rev. A **79**(5), 052312 (2009)
14. Sun, Z.W., Du, R.G., Long, D.Y.: Quantum key distribution with limited classical Bob. Int. J. Quant. Inform. **11**(1), 1350005 (2013)
15. Zou, X., Qiu, D., Zhang, S., Mateus, P.: Semiquantum key distribution without invoking the classical party's measurement capability. Quantum Inf. Process. **14**(8), 2981–2996 (2015). https://doi.org/10.1007/s11128-015-1015-z
16. Zhang, W., Qiu, D.W.: A single-state semi-quantum key distribution protocol and its security proof. arXiv:quant-ph/161203087 (2017)
17. Boyer, M., Gelles, R., Kenigsberg, D., Mor, T.: Semiquantum key distribution. Phys. Rev. A **79**(3), 032341 (2009)
18. Zou, X.F., Qiu, D.W.: Three-step semi-quantum secure direct communication protocol. Sci. China Phys. Mech. Astron. **57**(9), 1696–1702 (2014)
19. Li, Q., Chan, W.H., Long, D.Y.: Semi-quantum secret sharing using entangled states. Phys. Rev. A **82**(2), 022303 (2010)
20. Li, L.Z., Qiu, D.W., Mateus, P.: Quantum secret sharing with classical Bobs. J. Phys. A: Math. Theor. **46**(4), 045304 (2013)
21. Yan, L.L., Chang, Y., Zhang, S.B.: Measure-resend semi-quantum private comparison scheme using GHZ class states. Comput. Mater. Continua **61**(2), 877–887 (2019)
22. Shukla, C., Thapliyal, K., Pathak, A.: Semi-quantum communication: protocols for key agreement, controlled secure direct communication and dialogue. Quantum Inf. Process. **16**, 295 (2017)
23. Chang, Y., Zhang, S.B., Yan, L.L.: A quantum authorization management protocol based on EPR-pairs. Comput. Mater. Continua **59**(3), 1005–1014 (2019)
24. Zhang, S.B., Chang, Y., Yan, L.L.: Quantum communication networks and trust management: a survey. Comput. Mater. Continua **61**(3), 1145–1174 (2019)
25. Shor, P.W., Preskill, J.: Simple proof of security of the BB84 quantum key distribution protocol. Phys. Rev. Lett. **85**(2), 441–444 (2000)
26. Mayers, D.: Unconditional security in quantum cryptography. J. Assoc.: Comput. Math. **48**(1), 351–406 (2001)

27. Inamon, H., Lutkenhaus, N., Mayers, D.: Unconditional security of practical quantum key distribution. Eur. Phys. J. D **41**(3), 599–627 (2007)
28. Cai, Q.Y.: Eavesdropping on the two-way quantum communication protocols with invisible photons. Phys. Lett. A **351**(1–2), 23–25 (2006)
29. Li, X.H., Deng, F.G., Zhou, H.Y.: Improving the security of secure direct communication based on the secret transmitting order of particles. Phys. Rev. A **74**(5), 054302 (2006)
30. Deng, F.G., Li, X.H., Zhou, H.Y., Zhang, Z.J.: Improving the security of multiparty quantum secret sharing against Trojan horse attack. Phys. Rev. A **72**(4), 044302 (2005)
31. Cabello, A.: Quantum key distribution in the Holevo limit. Phys. Rev. Lett. **85**, 5635 (2000)
32. Zhang, K., Song, T., Zuo, H., et al.: A secure quantum group signature scheme based on Bell states. Phys. Scr. **87**(4), 045012 (2013)

An Improved Quantum Identity Authentication Protocol for Multi-party Secure Communication

Peilin He[1], Yuanyuan Huang[1(✉)], Jialing Dai[2], and Shibin Zhang[1,2]

[1] Department of Network Engineering, Chengdu University of Information Technology, Chengdu 610225, China
hy@cuit.edu.cn

[2] Department of Information Management and Information System, Beijing University of Chemical Technology, Beijing 100020, China

Abstract. Quantum secure identity authentication is a gripping part in the field of quantum communication which has an obvious superiority to makes the identity authentication system more secure. There are many existing multi-party quantum authentication protocols whereas not impeccable, some potential insecurity for some multi-party communication in those protocols can lead to be eavesdropped. Therefore, this paper aims to proposed an improved multi-party quantum identity authentication for secure communication scheme, by coalescing two advancing methods from Hong et al. [23] and Xiong et al. [24] and introduces a simple protocol that incorporates quantum identity authentication technique into multi-party communication medley with Greenberger-Horne-Zeilinger states as well as photons in Bell entangled states can be safely measured by legitimate parties. This protocol literally not only smooth over the insecure problems but also easy to realize. With this improved protocol, the authentication method can defend many attacks in the real world. In addition, effective protection against various attacks like eavesdropping or entangled-and-measure attack can also be achieved via this improved protocol.

Keywords: GHZ states · Multi-Party communication · Quantum identity authentication (QIA)

1 Introduction

There is no doubt in stating that Quantum Cryptography (QC) could revolutionize todays classical cryptography with solving problems such as Photon Number Splitting attack, Man-in-the-Middle attack, etc. The field of quantum cryptography is growing fast and is one of the most active field of modern science, mainly including: QKD (quantum key distribution), QSS (quantum secret sharing), QIA (quantum identity authentication) and so on. The quantum cryptography idea started since 1969. Inspired by this idea, Bennett in IBM and Brassard in the University of Montreal in Canada proposed the concept of QKD and the first QKD protocol (BB84 protocol) [Bennett and Brassard (1984)] in 1984, which marked the beginning of the study of quantum cryptography [1, 2]. Particularly,

X. Sun et al. (Eds.): ICAIS 2020, LNCS 12240, pp. 258–266, 2020.
https://doi.org/10.1007/978-3-030-57881-7_23

there ensures the safety of quantum communication on quantum channel for quantum key distribution (QKD) protocols [3].

Since the classical authentication method is not unconditional secure, it cannot effectively prevent the attacker to pretend to be a legitimate correspondent. If the attacker controls the quantum channel and classical channel, he can succeed in middle-man attack. Therefore, identity authentication between both sides in one communication is imperative [4]. There are some fundamental protocols in quantum key distribution [5–8] which strongly support the growth of identity authentication protocols.

In recent years, researchers have proposed a variety of quantum identity authentication (QIA) schemes [9–20]. In 2011, Yi et al. proposed a spin Squeezing of Superposition of Multi-Qubit GHZ State and W State protocol [21]. In 2014, Yuan et al. put forward a QIA scheme based on non-entanglement "Ping-Pong" technology [12]. In 2016, Ma et al. proposed continuous-variable quantum identity authentication based on quantum teleportation [13]. In 2019, Zawadzki proposed a quantum identity authentication protocol without entanglement [22]. From the above analysis, we know that quantum identity authentication plays an important role in quantum communication. It serves as a bridge which can connect quantum mechanics to quantum cryptography. These methods can create a more secure environment that different parties in such environment can communicate freely and openly. Different from the above QIA protocols, Hong et al. proposed a new idea in regard to simulate quantum identity authentication with single photon [23]. This protocol can create an environment and improve the assurance of multi-party communication. Take Xiong et al.'s [24] attacking scheme as an example, assuming that if there exist an eavesdropping, secret key sequence is restored with limited trust, Eve must eavesdrop on multiple seemingly identical authentication instances to make her guesses less or more likely. Instead, Bob's probe gives confidence in the value of two shared secrets. The proposed version of the protocol uses codes that are immune to general conditions.

Nonetheless, most quantum identity authentication protocols are still in the form of the point to point authentication scheme which have low efficiency and wasting of classical and quantum resources. We are also pleased to see that some scholars also paid close attention to these problems and put forward multi-party quantum identity authentication protocols.

This paper proposed an improved protocol of quantum identity authentication (QIA) for multi-party secure communication. Based on the contribution of the papers which author are Hong et al. and Xiong et al., we obtained a vast bulk of ideas such as by using GHZ states to create multi-party communication environment, or using binary bit to represent quantum bits and so on to substantiate our protocol. GHZ states can represent three party's communication, with this protocol, we can incorporate it into Bell states, creating a relatively stable environment which can provide three or multi party communication more reliable. With some certain technology which can find in quantum laboratory, our protocol can be testified and realized in some fields such as multi-party communication integrated network, communication between large enterprises even in the domain of blockchain security. This paper is organized as follows. In Sect. 2, we indicate those schemes are not comprehensively represent when multiple users communicating. In Sect. 3, we proposed an improvement protocol. In Sect. 4, Based on some possible attacks, the security analysis of the improved protocol is discussed. Lastly, the summary is given in Sect. 5.

2 Analyze These Two Protocols

2.1 Analyze Hong et al.'s Scheme

In this paper, quantum identity authentication method can surely guarantee the security in intercommunication. We assume that as there is an eavesdropper who name is Eve, and Eve wants to eavesdrop the message from quantum channel, point-to-point identity authentication will verify the sequence that sent by Alice to Bob. Eve does not know the authentication key sequence because the key sequence is not published. Only Alice and Bob know that sequence that they told each other. Eve can only guess the sequence and send quantum states randomly to Bob. The probability of correction sequence is 1/4 because there are four Bell states [35-36].

As Hong et al. said that: 'Eve will pass every particle of the security mode with the probability $P_{pass} = \frac{3}{4}$, and with probability P_{pass}, Eve can pass the security mode and the authentication mode. When the number of the security mode particles reaches S, the probability that Eve succeeds in passing the security mode is $P_{Tpass} = (\frac{3}{4})^S$. As long as S is large enough, the success probability of Eve becomes minuscule. Therefore, Eve's impersonation attack is impossible without being detected'. Therefore, this growth is exponential so that the probability of a message being tapped can drop to almost zero. However, this protocol does not satisfy multi-party communication, as if there existing the quantity of users want to send messages to other parties, this protocol should be improved.

2.2 Analyze Xiong et al.'s Scheme

In Xiong's paper, [25] they proposed a protocol which illustrates the QIA protocol on the basis of entanglement. The two different papers reveal one question that is per to per communication with single-photon and entanglement. Here I will brief on introducing Xiong's protocol for further illustrating my improved protocol.

1. Firstly, Alice has her binary identity string $ID_A = \{C_{12}^1, C_{12}^2, \ldots, C_{12}^{2n}\}$, Bob has his binary identity string $ID_B = \{C_{34}^1, C_{34}^2, \ldots, C_{34}^{2n}\}$, besides, Alice and Bob share ID_A and ID_B.
2. Then go to the next step. Alice prepares particle sequence ID_A and Bob does the same thing that sequence called ID_B. Here is the rule of preparing particle sequence blew.

$$\begin{aligned} |\phi^+\rangle &\to 00 \\ |\phi^-\rangle &\to 01 \\ |\varphi^+\rangle &\to 10 \\ |\varphi^-\rangle &\to 11 \end{aligned} \quad (1)$$

3) The two communicators exchange their particles in Bell states and each gets a new sequence S'_{ID_A} and S'_{ID_B}. Then they measure each other's sequences by using Bell basis. Finally, entanglement exchange is fully finished.
4) After that, the results are represented by binary digit. The presentation rules can consult those equations above. Here will not redundantly repeat.
5) The last step comes. The binary of S'_{ID_A} is C'_{ID_A}, the binary of S'_{ID_B} is C'_{ID_B}, then verify it, if $ID_A \oplus ID_B = C'_{ID_A} \oplus C'_{ID_B}$, those two communicator are legitimate users. If not, it needs to recertification.

Also we know that entangled exchange between Alice and Bob finally got the binary sequence C'_{ID_A} and C'_{ID_B}, if the result is correspond to the equation $ID_A \oplus ID_B = C'_{ID_A} \oplus C'_{ID_B}$, then after Alice and Bob announced their authentication key, we call it completed. None of eavesdropper will not know the authentication key because the key is randomly generated, there's only a 1/4 chance of being able to guess it correctly without authentication, once Eve generates a probe to impact quantum tunnel, he will be detected, then end up communication. However, it still not suitable for multi-users communication. Therefore, we proposed an scheme in order to cater for multi-party on the basis of Hong et al.'s protocol mixing with Xiong et al.'s theory, it will be a newly design.

3 Improved Protocol Scheme

Up to now scientists were dealing with entanglement of only two subsystems, based on GHZ states [21, 26–29], it combines quantum identity authentication protocol with GHZ states and can be realized in multi-party communication.

a) Assuming that there have three communicators Alice, Bob, Charlie, respectively. They want to send messages to others. For example, Alice sends message to Bob meanwhile Bob wants to send message to Charlie, Charlie wants to send message to Alice. Alice, Bob, Charlie prepare three Bell states photons with horizontal $|H\rangle$ or vertical $|V\rangle$ polarizations in the form:

$$|\phi\rangle_{GHZ} = \frac{1}{\sqrt{2}}(|H\rangle_1|H\rangle_2|H\rangle_3 + |V\rangle_1|V\rangle_2|V\rangle_3) \tag{2}$$

Then taking GHZ into account of three photons in linear polarization base (*H/V, states* $|H\rangle$ *and* $|V\rangle$*):*

$$|\phi\rangle_{GHZ} = \frac{1}{\sqrt{2}}(|H\rangle_1|H\rangle_2|H\rangle_3 + |V\rangle_1|V\rangle_2|V\rangle_3) \tag{3}$$

Then writing vectors as:

$$|H\rangle = \begin{pmatrix}1\\0\end{pmatrix} |V\rangle = \begin{pmatrix}0\\1\end{pmatrix} \tag{4}$$

The measurements in base H'/V' that be considered results are rotated relate to H/V approximately 45°. New bases states are shown up:

$$\left|H'\right\rangle = \frac{1}{\sqrt{2}}(|H\rangle + |V\rangle) \tag{5}$$

$$\left|V'\right\rangle = \frac{1}{\sqrt{2}}(|H\rangle - |V\rangle) \tag{6}$$

After adding and subtracting sides Eqs. (4) and (5), new equations can be induced as:

$$|H\rangle = \frac{1}{\sqrt{2}}\left(\left|H'\right\rangle + \left|V'\right\rangle\right) \tag{7}$$

$$|V\rangle = \frac{1}{\sqrt{2}}\left(\left|H'\right\rangle - \left|V'\right\rangle\right) \tag{8}$$

Now based on those equations, simplifying common elements:

$$|\phi\rangle_{GHZ} = \frac{1}{4}\left((H_1' + V_1')(H_2' + V_2')(H_3' + V_3') + (H_1' - V_1')(H_2' - V_2')(H_3' - V_3')\right)$$

$$= \frac{1}{2}\left(H_1'H_2'H_3' + H_1'V_2'V_3' + V_1'V_2'H_3' + V_1'H_2'V_3'\right) \tag{9}$$

b) Now we take the similar circumstance of measurement into account, if we measure single photon in base H'/V', we find two situations as blew:

1. If the outcome is in $|H\rangle$ state, the left two photons will present opposite circular polarizations.
2. If the outcome is in $|V\rangle$ state, the left two photons will present the same circular polarizations.

c) Now let's generalize it to the general case:

Firstly, we change states $|H\rangle$ and $|V\rangle$ to $|0\rangle$ and $|1\rangle$ respectively, generalized GHZ state is as follows:

$$|\phi\rangle_{GHZ} = \frac{1}{\sqrt{N}}(|0\rangle_1|0\rangle_2\cdots|0\rangle_N + |1\rangle_1|1\rangle_2\cdots|1\rangle_N) \tag{10}$$

So, one particle can come across different symbol:

$$|n\rangle_1|n\rangle_2|n\rangle_3\cdots|n\rangle_N = |n\cdots n\rangle_N = |n\rangle^{\otimes N} \tag{11}$$

N belongs to 0 and 1, besides, that state is extremely entangled. A system that reduces three qubits to two qubits (tracking the operation of a third qubit) results in a completely non-entangled state. Thus, the nature of entanglement is extremely sensitive to the loss of one of these particles.

d) According to the above proof, we incorporate that results onto quantum identity authentication protocol with strongly entanglement. Assuming that there exist three parties in order to represent multi-party communication. Alice, Bob and Charlie prepare particle sequence separately as S_a, S_b, S_c. Of course, Alice has binary string ID_a, Bob has ID_b, Charlie has ID_c.

e) Alice sends entangled particles to Bob and Charlie, meanwhile Bob sends entangled particles to Charlie, now the three of them correspond with each other. After exchanged particles, Alice got sequence S_{ID_b} and S_{ID_c}, in a similar way, Bob got S_{ID_a} and S_{ID_c}, Charlie got S_{ID_a} and S_{ID_b}. Now Alice measures S_{ID_a} by using Bell base, S_{ID_b} and S_{ID_c} will at entangled state at once. On the contrary, Bob and Charlie do the same measurement by measuring S_{ID_b} and S_{ID_c}, now the entanglement swapping is completed.

f) Using binary digit to represent the results, here is the rule:

$$|0\rangle = 00$$
$$|1\rangle = 01$$
$$|+\rangle = \frac{1}{\sqrt{2}}(|0\rangle + |1\rangle) = 10$$
$$|-\rangle = \frac{1}{\sqrt{2}}(|0\rangle - |1\rangle) = 11 \tag{12}$$

S'_{ID_a} will represent as C'_{ID_a} in binary digit.
S'_{ID_b} will represent as C'_{ID_b} in binary digit.
S'_{ID_c} will represent as C'_{ID_c} in binary digit.

g) Now they doing identity authentication, just notifying Bob and Charlie announce C'_{ID_b} and C'_{ID_c} via the classic channel, then verifying them whether

$$ID_A \otimes ID_B \otimes ID_C = C'_{ID_A} \otimes C'_{ID_B} \otimes C'_{ID_C} \tag{13}$$

h) After that, if the results are corresponding to that equation, authentication succeed, they are legitimate users. On the contrary, they should resume from the beginning which is similar to the original protocol.

4 Security Analysis

4.1 The Man-in-the-Middle Attack

1) Sender (Alice) prepares photons in two orthogonal bases (rectilinear and pivoted by 45° diagonal) in random order. In both she can encrypt "1" and "0". Receiver (Bob and Charlie) detects photons also in random bases. Measurement in wrong basis does not provide any information, so Alice exchanges information about used bases (in classical channel) and sift bits with Bob and Charlie, accepting only 100% correct ones.
2) Eve performs man-in-the-middle attack because she is unable to copy bits due to no-cloning theorem [40]. Unknown bases used to encrypt photons enforce Eve to guess the proper ones. This implies 50% chance of success like in a coin tossing. In

Table 1. Error detection

Alice Random Bits	**0**	**1**	**1**	**0**	**1**	**0**	**0**	**1**
Alice Sending Basis	+	×	+	×	×	+	×	+
Alice Polarization	→	↖	↑	↗	↖	→	↗	↑
Eve basis measurements	+	+	×	×	+	×	×	+
Polarization Eve measures and sends	→	→	↗	↗	↑	↗	↗	↑
Bob basis measurement	+	×	×	+	×	×	+	+
Polarization Bob measures	→	↗	↗	→	↖	↗	→	↑
Charlie basis measurement	+	×	×	+	×	×	+	+
Polarization Charlie measures	→	↗	↗	→	↖	↗	→	↑
Shared Secret key	0	0	--	--	--	--	--	1
Error Generated	✓	×	--	--	--	--	--	√

the case of wrong guess, Bob and Charlie have 50% chance to detect bad bit value besides good basis choice. Because of the fact of eavesdropping errors arise (25% of sifted key as in the example in Table 1).

3) Then, Alice sends the first qubit of S_{ID_a} in Bell entangled states to Bob and Charlie, because of Eve does not know the sequence of ID_b and ID_c, thus she prepares two sequences S_{ID_e} randomly and resend to Alice. After swapping, Alice gets particle sequence S_{ID_a}, Eve gets sequence S'_{ID_b}. Then they both do entanglement swapping, Alice presents the results as binary sequence C'_{ID_a}, Eve presents the results as binary sequence C'_{ID_e}.

4) If Alice wants to do identity authentication with Bob and Charlie, she needs to verify the identification of them. Because of Eve intercepted their qubits so that entanglement swapping cannot be completed. Then Alice verifies identification according to (13), according to $ID_aID_bID_c \neq C'_{ID_a}C'_{ID_b}C'_{ID_c}$, Eve cannot pass the authentic process. Therefore, eavesdropping will be detected, Alice, Bob and Charlie will terminate communication immediately.

5 Conclusion

To sum up, presented above propositions for a family of protocols for extended quantum identity authentication are based on multi-party qubits entanglement properties. We initially proposed a feasible QIA protocol with multi-party communication in GHZ state and used the formulas to derive the correct result. Then we simulated some possible attacks to prove the security of the protocol that can hold back such as Man-in-the-middle (MITM) attack, Photon Number Splitting (PNS) and even maybe Trojan Horse (light injection attack). What is more, the implementation of the proposed QIA protocol for multi-party only needs to utilize EPR pairs and Quantum Random Number Generator (QRNG). Therefore, our protocol is feasible with the present techniques.

Acknowledgement. This work is supported by the National Key Research and Development Project of China (No. 2017YFB0802302), the Key Research and Development Project of Sichuan Province (No. 20ZDYF2324, No. 2019ZYD027, No. 2018TJPT0012), the Science and Technology Support Project of Sichuan Province (No. 2018GZ0204, No. 2016FZ0112, No. 2018RZ0072), the Science and Technology Project of Chengdu (No. 2017RK0000103ZF), the Innovation Team of Quantum Security Communication of Sichuan Province (No. 17TD0009), the Academic and Technical Leaders Training Funding Support Projects of Sichuan Province (No. 2016120080102643), the Application Foundation Project of Sichuan Province (No. 2017JY0168), the Foundation of Chengdu University of Information Technology (No. J201707) and the College Students' innovation project (No. S201910621082).

References

1. Einstein, A., Podolsky, B., Rosen, N.: Can quantum-mechanical description of physical reality be considered complete? Phys. Rev. **47**, 777 (1935)
2. Bennett, C.H., et al.: Teleporting an unknown quantum state via dual classical and EPR channels. Phys. Rev. Lett. **70**, 1895–1899 (1993)
3. Bennett, C.H., Brassard, G.: Quantum cryptography: public key distribution and coin tossing. In: Proceedings of the IEEE International Conference on Computers, Systems and Signal Processing (1984)
4. Bennett, C.H., Bessette, F., Brassard, G., Salvail, L., Smolin, J.: Experimental quantum cryptography. J. Cryptol. **5**(1), 3–28 (1992). https://doi.org/10.1007/BF00191318
5. Jacak, M., Martynkien, T., Jacak, W., et al.,: Quantum cryptography: quantum mechanics as foundation for theoretically unconditional security in communication. Institute of Physics, Wrocław University of Technology, Wyb. Wyspiańskiego 27, 50-370 Wrocław, Poland
6. Goldenberg, L., Vaidman, L.: Quantum cryptography based on orthogonal states. Phys. Rev. Lett. **75**, 1239–1243 (1995)

7. Sun, Y., Wen, Q.-Y., Gao, F., Zhu, F.-C.: Robust variations of the Bennett-Brassard 1984 protocol against collective noise. Phys. Rev. A **80**, 032321 (2009)
8. Song, T.-T., Wen, Q.-Y., Guo, F.-Z., Tan, X.-Q.: Finite-key analysis for measurement-device independent quantum key distribution. Phys. Rev. A **86**, 022332 (2012)
9. Curty, M., Santos, D.J.: Quantum authentication of classical messages. Phys. Rev. A **64**, 062309 (2001)
10. Liu, B., Gao, Z.F., Xiao, D., Huang, W., Liu, X., Xu, B.: Quantum identity authentication in the orthogonal-state-encoding QKD system. Quantum Inf. Process. **18**(137), (2019)
11. Shi, B.-S., Li, J., Liu, J.-M., Fan, X.-F., Guo, G.-C.: Quantum key distribution and quantum authentication based on entangled state. Phys. Lett. A **281**, 83–87 (2001)
12. Yuan, H., Liu, Y.-M., Pan, G.-Z., Zhang, G., et al.: Quantum identity authentication based on ping–pong technique without entanglements. Quantum Inf. Process. **13**, 2535–2549 (2014)
13. Ma, H., Huang, P., Bao, W., Zeng, G.: Continuous-variable quantum identity authentication based on quantum teleportation. Quantum Inf. Process. **15**(6), 2605–2620 (2016). https://doi.org/10.1007/s11128-016-1283-2
14. Wang, J., Zhang, Q., Tang, C.-J.: Multiparty simultaneous quantum identity authentication based on entanglement swapping. Chin. Phys. Lett. **23**, 2360–2363 (2006)
15. Ljunggren, D., Bourennane, M., Karlsson, A.: Authority-based user authentication in quantum key distribution. Phys. Rev. A **62**, 022305 (2000)
16. Gong, J., Zhang, W., Deng, Y. Q., Liu, B.,: The BB84 protocol with identity authentication. Chin. Core J. **04**(03) (2011)
17. Inamori, H., Lütkenhaus, N., Mayers, D.: Unconditional security of practical quantum key distribution. Eur. Phys. J. D **41**, 599 (2007)
18. Yan, D., Wang, X.G., Song, L.J., Zong, Z.G.: Cent. Eur. J. Phys. **5**, 367 (2007)
19. Barnum, H., Knill, E., Ortiz, G., Somma, R., Viola, L.: Phys. Rev. Lett. **92**, 107902 (2004)
20. Lewenstein, M., Kraus, B., Cirac, J.I., Horodecki, P.: Phys. Rev. A **62**, 052310 (2000)
21. Yi, X.J., Wang, J.M.: Spin squeezing of superposition of multi-qubit GHZ state and W state. Int. J. Theor. Phys. **50**, 2520–2525 (2011)
22. Zawadzki, P.: Quantum identity authentication without entanglement. Quantum Inf. Process. **18**(1), 1–12 (2018). https://doi.org/10.1007/s11128-018-2124-2
23. Hong, C., Heo, J., Jang, J.G., Kwon, D.: Quantum identity authentication with single photon. Quantum Inf. Process. **16**(10), 1–20 (2017). https://doi.org/10.1007/s11128-017-1681-0
24. Ekert, A., Rarity, J., Tapster, P., Palma, G.M.: Practical quantum cryptography based on two-photon interferometry. Phys. Rev. Lett. **69**, 1293 (1992)
25. Xiong, J.X., Chang, Y., Zhang, S.B.: Quantum identity authentication protocol based on Bell states and entanglement swapping. Appl. Res. Comput. **36**(4) (2019)
26. Bouwmeester, D., Pan, J.-W., Daniell, M., Weinfurter, H., Zeilinger, A.: Observation of three-photon Greenberger-Horne-Zeilinger entanglement. PRL **82**, 1345 (1999)
27. Werner, R.F.: Quantum states with Einstein-Podolsky-Rosen correlations admitting a hidden-variable model. Phys. Rev. A **40**, 4277 (1989)
28. Bouwmeester, D., Ekert, A., Zeilinger, A.: The Physics of Quantum Information. Springer, Heidelberg (2000). https://doi.org/10.1007/978-3-662-04209-0
29. Zurek, W., Wootters, W.: A single quantum cannot be cloned. Nature **299**, 802–803 (1982)
30. Yan, L., Chang, Y., Zhang, S., Wang, Q., Sheng, Z., Sun, Y.: Measure-resend semi-quantum private comparison scheme using GHZ class states. Comput. Mater. Continua **61**(2), 877–887 (2019)
31. Zhao, X., et al.: A high gain, noise cancelling 3.1–10.6 GHz CMOS LNA for UWB application. Comput. Mater. Continua **60**(1), 133–145 (2019)

A Mixed Mutual Authentication Scheme Supporting Fault-Detection in Industrial Internet of Things

Gongxin Shen(✉)

School of Information Engineering, Nanjing Polytechnic Institute, Nanjing 210048, China
sgx0554@163.com

Abstract. In the Industry Internet of Thing (IIoT), industrial robots and employees play very important roles in manufacturing process. However, a secure and efficient authentication scheme for industrial robot and employee is under research and challengeable. For instance, the industrial robot and employee are vulnerable to impersonation and desynchronization attack in the authentication phase. Meanwhile, it is hard for control center to find the disobedient industrial robot. Hence, in this paper, we set a threshold of defect rate to find and identity the source of error. In addition, by introducing the PUF to authentication between robot and server, our scheme can effectively resist the desynchronization attack. Finally, through our security analysis and experimental simulation, our proposed scheme can ensure secure and efficient in authentication phase.

Keywords: IIoT · Fault detection · Authentication · Threshold-based

1 Introduction

With the advancement of manufacturing sector informatization, network and IoT related technologies are increasingly wide utilization in all fields of industrial production system, which tremendously lead to gains in productivity and economic efficiency. However, when taking the characteristics of manufacturing industry into consideration (e.g., closure, high-value data, decentralization), it's unavoidable to tempt malicious attackers to eavesdrop or destroy valuable manufacturing data by taking advantage of virus, vulnerability of network protocol, so on and so forth [1–3]. The unending attacks may lead to catastrophic results, like leakage of core technology of industry or the illegal manipulation of the network terminal, which will cause enormous loss of life and property [4–6]. In addition, the most fatal threats not only come from the external hacker or malicious competitor, but also stem from potential internal attacker including disgruntled former employees who may have permissions to access control center. By various different attack ways, the normal production activities will be distributed unconsciously.

X. Sun et al. (Eds.): ICAIS 2020, LNCS 12240, pp. 267–279, 2020.
https://doi.org/10.1007/978-3-030-57881-7_24

1.1 Potential Threats and Attacks

The potential threats and attacks are generally fall into two categories, the one is coming from external and the other is stemming from internal. Meanwhile, these threats can be viewed in three perspectives, including authentication, communication and eviction

(1) Impersonation Attacks
The adversary can execute impersonation attacks to pass himself off a legitimate robot or user [7] but also can impersonate the valid SCDAC [8] to other entities.
(2) Desynchronization attack
The adversary's aim is to desynchronize the state of two parties in the phase of authentication. Attacker will block updating secret information and counterfeit the wrong state to victimized party [9, 10].
(3) Eavesdropping attack
Because of the insecurity of wireless channel, the adversary can execute eavesdropping attack to acquire the identify information in the authentication phase.
(4) Attack Against Anonymity
The industrial robots are monitored by SCDAC in real time for requirements of manufacturing industry, which should be anonymous to adversary and can be identified by SCDAC. As to the adversary, they will try to reveal the identity information in authentication phase and data transmission phase.
(5) Inside Attack
As to inside attack, which has been seen as the maximal threats to industrial production. The internal employees who are discontented to the company, maybe have authority to access the SCDAC. Without doubt, it is very dangerous to close the production line.

1.2 Related Works and Problem Definition

Recently, researches on authentication in IoT are seem to focus on three aspects: 1) public cryptography system depended on algorithm of RSA, ECC [11, 12]. 2) lightweight authentication schemes, especially hash based [13, 14] or PUF-based [15, 16] schemes. 3) ultralightweight authentication protocols are basically simply logic operations like AND, OR, XOR. Hence, public crypto based and lightweight authentication schemes will be the key point of the review.

Since 2004, Lauter [17] claimed that elliptic curve cryptography (ECC) can be used into resource-constrained equipment, the research on applying ECC into IoT authentication has a great process. In 2015, vanga proposed a scheme [18] supporting the multi-server and biometric authentication that can resist known session temporary information attack and device loss attack, compared with he-wang's scheme [19]. Later, Amin proposed [20] in 2018, which address the aforesaid security weakness, but without protecting impersonation attack and privileged-inside attack.

As to lightweight authentication protocol in IoT, 2008, Godor proposed a SLAP (Simplest Lightweight Authentication Protocol) [21], which can resist the well-known attack with a theoretical analysis. Nevertheless, Mete [22] pointed out the impersonation attack to his protocol and fixed the problem without increasing any extra computation and memory cost. However, almost every hash-based scheme has an inherent deficiency that is vulnerable to physical attack. Hence, PUF based schemes with the tamper-evident feature which can remedy this deficiency has gotten more attention. In 2008, bringer [23] firstly proposed a mutually tree-based authentication protocol using PUF with improve security. However, actually the protocol was suffering from the DoS and impersonation attack [24]. Lately, Akgun [25] presented a scalable PUF based authentication protocol, which notwithstanding high lever privacy and scalability in RFID system gope [27] assert that scheme is impractical when executing exhaustive search to identifying a tag.

1.3 Contributions

A mixed mutual authentication scheme is proposed by us firstly, which including ECC-based and PUF-based authentication. Subsequently, this scheme is security enough to withstand several attacks. Finally, a simulation and comparison with other related scheme is presented to show the superiority of our scheme. Following are our summarized core contributions:

1. Proposing an Improved authentication scheme, which satisfy several security requirements, including eavesdropping-resistance, withstanding desynchronization attack, Dos attack, limited anonymity and resisting inside attack.
2. Setting a threshold for defect rate of production to detect the corrupted robot. Valid industrial robots will adhere to the standard instructions, if the defect rate of one batch of products are abnormal, SCDAC will challenge the all industrial robots and identity the corrupted one.
3. Adding a table of evocation to prevent inside attack from malicious employees who have authority to access the system.

1.4 Outline

The rest of our paper is structured as follows: Related preliminaries and definition of system model are presented in Sect. 2. In Sect. 3, our protocol is detailed introduced in five phases. Security proof and the simulation of our scheme are shown in Sects. 4 and 5, respectively. Note that the notations and functions are presented in Table 1.

Table 1. Notations used in this paper.

Notations	Description
G1,G2	Multiplicative subgroup of *G*
||	Concatenation operation
XOR	Exclusive-OR operation
Ps,ms	Public key and private key of SCDAC
P_u,*PW*	Verifier value and password of employee
$h(*)$	One-way hash function $\{0, 1\}* \rightarrow \{0, 1\}^K$
PID_i,TID_i	Pseudo identifier and temporary *ID*
SK_1,SK_2 *A. ECC*	Verifier value calculated by employee

2 Preliminaries

2.1 Elliptic Curve Discrete Logarithm

In finite filed F_p, there is a non-singular elliptic curve $y^2 = x^3 + ax + b$, where for any constant satisfy $4a^3 + 27b^3$ not equal 0, together with all pairs of affine coordinates $(x,y) \in Fp$ and a special point O called infinity point. The set of points which on the elliptic curve forms an abelian group with O as the additive identity [26, 27].

2.2 Bilinear Pairing

Let's define that G_1, G_2 and G_t are multiplicative cyclic group of order *p*. If $e: G_1 \times G_2 \rightarrow G_t$ satisfies following properties, the map is known as bilinear map.

1) Bilinearity: for all $W,Q,Z \in G_1$, we have $e(W,Q+Z) = e(W,Q)e(W,Z)$.
   Hence, for any $x,y \in Z_p^*$, we have $e(xW,yQ) = e(W,Q)^{xy} = e(yW,xQ)$.
2) Computability: there is an efficient algorithm computing $e(W,Q)$ for $W,Q \in G1$.
3) Non-degeneracy: for all $P \in G_1, e(P,P) \neq 1_{G2}$, where 1_{G2} is the identity element of *G2*.

Finally, in our paper it's a symmetrical bilinear pairing for $G1 = G2$.

2.3 Physical Unclonable Function

Physical Unclonable Function (PUF): The industrial robot chips consist microprocessor with PUF, which mark the only identity of the robot. PUF is depended on the inner physical characteristics and will output the only value regardless of temperature, voltage, time so on and so forth. Any attempt to detect or observe PUF operations will change the underlying circuit characteristics. PUF as challenge response micro-structure, which defined as $R = PUF_T(C)$, is virtually impossible to predict the response.

3 Adversary Model

3.1 System Model

An authentication model in IIoT is presented in Fig. 1, there are three entities in the system model, including industrial robot (IR), supervisory control and data acquisition center (SCDAC) and employee. Under this model, IR and Employee will authenticate mutually with SCDAC. After authenticating successfully, IR will transmit the production data to SCDAC regularly, while employee will upload the work data and send the instructions to IR. Note that IR's password should be regularly changed for security. Finally, for avoiding malicious employee's attack, SCDAC will maintain an eviction list to prohibit the access of malicious employee. In such architecture, in the phase of authentication between IR and SCDAC, computation cost and memory cost should be considered for IR is resource-constrained device. Meanwhile, when it comes to authentication between employee and SCDAC, security should take more attention than computation cost and memory overhead.

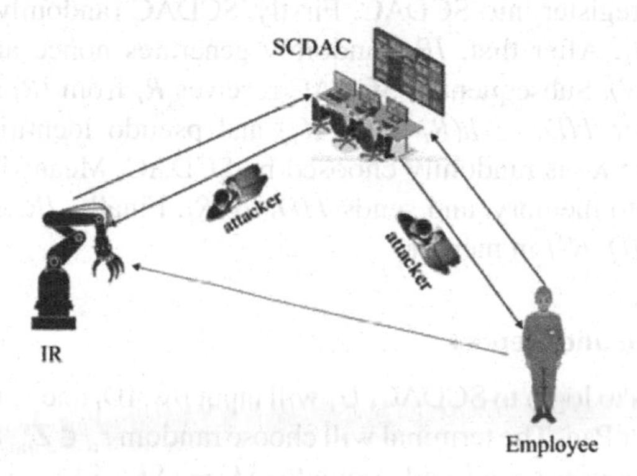

Fig. 1. Modern Industrial Internet of Things

3.2 Adversary Model

Under the security model, we assume that SCDAC is fully trusted, IR is trusted but easy to be attacked by side channel attack. Employee is semi-trusted and will try to steal confidential data or destroy the normal production activity to send error instructions. Adversary not only can eavesdrop the secretive information between the open channel in the phase of authentication, but also can retrieve the information stored in the IR memory. Another attack we should deserve is that adversary can block the information from SCDAC and vice verse.

4 Proposed Scheme

In this section, the detailed scheme will be explicated in four phases, including setup and register, authentication, communication, password changed phase and user eviction phase. Generally speaking, our authentication protocol is ECC based and PUF-based for employee and IR, respectively. Additionally, our protocol has error detection and correction mechanism that can find corrupted IR and evict employees whose permission is expired

4.1 Setup and Register

(1) SCDAC chooses a non-singular elliptic curve E_p on a finite field F_p, where G with order n is basic point such that $nG = 0, n \in Z_p^*$. Then SCDAC selects $m_s \in Z_p^*$ as master secret key, $G, Z \in G_1$ and computes $P_s = m_s \times G$. After that, SCDAC chooses an anti-collision hash function $h(*)$ to Z_p^* and releases parameters $\{E_p, F_p, G, Z, n, P_s, h(*)\}$.
(2) Employee will register an account to access SCDAC and send instructions to IR. Employee U_i registers the ID_i from SCDAC and the employee's biological information like fingerprint information f_i and computers $P_u = h(f_i || pw || ID_i) \times G$. Then SCDAC stores P_u and corresponding ID_i into database. And marked 1, which represents it's an active account. On the contrary, this is an inactivated account.
(3) Every IR will register into SCDAC. Firstly, SCDAC randomly generate the challenge C_i to IR_i. After that, IR_i randomly generates nonce and calculates $R_i = PUF(C_i \; XOR \; N)$. Subsequently, SCDAC receives R_i from IR_i and calculates temporary identifier $TID_i = h(R_i \; XOR \; K_i)$ and pseudo identities $PID_i = h(TID_i \; XOR \; R_i)$, where K_i is randomly choosed by SCDAC. Meanwhile, SCDAC stores $\{TID_i, PID_i\}$ into memory, and sends TID_i to IR_i. Finally, IR_i calculates PID_i and stores $\{TID_i, PID_i, N'\}$ in memory.

4.2 Authentication and Report

(1) When U_i wants to login to SCDAC, U_i will input pw, ID_i and f_i into terminal device like computer or Pad. The terminal will choose random $r_a \in Z_p^*$, and calculate. Next, U_i randomly chooses r $\in Z_p^*$ and computes $SK = (SK_1, SK_2) = (h(f_i || pw || ID_i) \times Z + r \times U, r_a \times G)$. Finally, the terminal device sends SK to SCDAC.
(2) SCDAC verifies the legitimacy of user identity by checking the following equation holds or not. $e(SK_1, G) = e(Z, P_u)e(U, SK_2)$
(3) If U_i passes the verification, the employee can access into SCDAC and check the production process. After that, IR_i tries to authenticate with server before starting production. IR_i sends authentication request, TID_i and count to SCDAC. SCDAC checks if TID_i is in the database. If yes, SCDAC generates random number N_s, and calculates $N_s^* = R_i \; XOR \; N_s$, and sends C_i, N_s^* along with $VER_s = h(PID_i || R_i | N_s^* || count + 1)$ to IR_i. Hereafter, IR_i verify the correctness of VER_s to check the identity of SCDAC.
(4) If SCDAC is certified as legitimate, IR_i calculates, $R_{i+1} = PUF(C_{i+1} \; XOR \; (N' + 1))$ and $VERr = h(R_{i+1} || C_{i+1} || N_s)$, and sends$\{C_{i+1}, R_{i+1}, VERr\}$ to SCDAC. Subsequently, SCDAC will check $VERr$ to verify the identity of IR_i because only valid IR_i can get the correct N_s. After that, SCDAC calculates, $PID_{i+1} = h(TID_{i+1} \; XOR \; R_{i+1})$ and update $\{TID_{i+1}, PID_{i+1}\}$. Meanwhile, SCDAC calculates $Vs = h(TID_{i+1} || PID_{i+1} || count + 3 || N_s)$ and sends $\{TID_{i+1}, PID_{i+1}, V_s\}$ to IR_i. Finally, if V_s is verified correctly, IR_i will updates $\{TID_{i+1}, PID_{i+1}, N' + 1\}$ in memory.

(5) If all industry robots pass the authentication, the realtime production data will generated by industry robots to transmit to SCDAC.

4.3 Password Update

(1) When the employee U_i wants to change the old password pw to new password *pw'*. U_i will update the $h(f_i||pw||ID_i)$ to new $h(f_i||pw'||ID_i)$ after login the SCDAC. U_i utilizes the old password pw to login the SCDAC and send the password updating request.
(2) U_i computes $h(f_i||pw'||ID_i)$, $P_unew = h(f_i||pw'||ID_i) \times G$ and SK_{new} by *pw'*.

4.4 Threshold-Based Fault Detection (TFD)

Within a certain time period t_i, the defect rate d of products must no higher than SCDAC's instruction d. If in that time, the defect rate is no higher than d, the production log would not be called to check. Conversely, if the defect rate is abnormally higher than instruction d, there is a doubt that the production parameter of some industry robots have been modified by adversary or internal employee. Hence, SCDAC calls the production log during that time period and detect the fault.

(1) SCDAC will define the instruction d and broadcast it to all employee. Firstly, SCDAC chooses random value $p_1 \in Z_p^*$ and computes $W = p_1 \times G, f = H(d||W||t)$ and $s_1 = p_1 - f \times m_s$. Hereafter, SCDAC broadcasts the tuple $\{d, s1, f, t\}$ to all employees.
(2) Employee checks the validity of instruction d by using the elements in tuple $\{d, s_1, f, t\}$. Employee first computes $W' = f \times p_s + s_1 \times G$ and checks the validity of equation $f = H(d||W'||t)$. If the equation holds, employee will compare the defect rate *d'* with the instruction *d* during one certain period t_i. If *d'* is obviously above d, the employee will send warning report $u = d'||t_i$ to SCDAC, where d' is the defect rate during a period t_i, and otherwise ignoring it.
(3) Upon receiving the warning information from employee, SCDAC will call the production log in period t_i and fix the source of fault to make sure whether the industry robot has been corrupted by adversary.

4.5 User Eviction

For prevent the attack from the expired or malicious employees, SCDAC will maintain an eviction table to prohibit them to access SCDAC just as shown in Table 2. If the employee send the error instruction to industry robots, SCDAC will delete the account in verifier table and move the employee's account to eviction table after the accident was deliberately caused by employee.

Table 2. Eviction table

Identity	Password-verifier	Status
ID_1	$P_{u1} = h(f_1 \| pw \| ID_1) \times G$	Blocked
ID_2	$P_{u3} = h(f_3 \| pw \| ID_3) \times G$	Blocked

5 Security Proof

The security analysis of our proposed scheme is presented in this section. It is worth noting that, the potential threats and attacks are shown in Sect. 1.1, and our security analysis consists of five parts which is based on potential threats and attacks, including resistance of impersonation attack, desynchronization attack, eavesdropping attack, and inside attack, and supporting anonymity.

5.1 Resistance of Impersonation Attack

In our proposed scheme, adversary will try to simulate a valid employee or industry robot to login the SCDAC. Firstly, the adversary attempt to forge new $\{h(f_i||pw||ID_i)guess \times Z + r_a \times U, r_a \times G\}$ to impersonate employee to authenticate with SCDAC, however, it is infeasible to guess the correct under a hash collision-resistant assumption and intractability of elliptic curve discrete logarithm problem(ECDLP). More concretely, it is hard for adversary to calculate $h(f_i||pw||ID_i)$ for $P_u = h(f_i||pw||ID_i) \times G$, so that adversary cannot impersonate employee to pass the authentication. Secondly, adversary may impersonate industry robot to authenticate with SCDAC by conducting the BAN Logic [A logic of authentication] to analyze the security of authentication between industry robot and SCDAC. A direct proof is given to present that only valid industry robot can pass identity authentication.

5.2 Resistance of Desynchronization Attack

As far as we know, the resource-constrained devices are vulnerable to desynchronization attack. Consequently, we only consider the security of anti-desynchronization attack to industry robot.

In IIoT system, synchronization is essential for two parties of communication. In our proposed scheme, adversary may tamper to block the message $\{TID_{i+1}, PID_{i+1}, V_s\}$ to break the synchronization between SCDAC to IR. However, SCDAC has gotten C_{i+1} and R_{i+1} so that SCDAC can calculate pseudo identity $PID_{i+1} = h(TID_{i+1}\ XOR\ R_{i+1})$. Hereafter, SCDAC and IR share the same updated pseudo identity PID_{i+1} so that they can authenticate successfully in the next authentication.

5.3 Resistance of Eavesdropping Attack

If adversary executes eavesdropping attack to acquire the confidential information in authentication phase and data reporting phase. In our proposed scheme, adversary can acquire SK, Pu between employee and SCDAC. However, it is no way for adversary to pass the authentication because of the infeasible of ECDLP, so that adversary cannot figure out the correct password pw. As to the authentication between SCDAC and IR, adversary may steal TID_i and PID_i but without R_i, hence, adversary cannot impersonate the valid IR to authenticate with SCDAC. Additionally, in the phase of data reporting, our proposed protocol is based on which can achieve the strong anonymity and prevent the leakage of reporting data. As such, our scheme can withstand eavesdropping attack.

5.4 Resistance of Reply Attack

If adversary can intercept and resend the authentication data in the next session, the temporary identifier TID, pseudo identity PID, counter count and random number N' will be different in every session. Meanwhile, the challenge C_i and response R_i will update as $C_{i+1}= h(R_i\ XOR\ count + 2)$, $R_{i+1}= PUF(C_{i+1}\ XOR\ (N' + 1))$. Evidently, adversary is incapable to execute the reply attack in the next session by using last authentication message.

5.5 Resistance of Inside Attack

In our security model, we assume that the inside malicious employee may have permission to access the SCDAC and send wrong commend. Firstly, SCDAC will call the production log from industry robot if the abnormal defect rate is detected, and indicates that one employee U_i has changed the production parameter. SCDAC will ensure the identity of employee ID_i and remove it to eviction table to prevent U_i accessing the server again. In addition, by setting life cycle of employee, SCDAC can only be accessibility in that time which can efficiently monitor the operations by employee and prevent the insider attack previously.

From aforesaid security analysis, it is apparently shown that our proposed scheme can withstand all security threats and attacks mentioned in Sect. 1.2. To present the superiority of our scheme in security features, our proposed authentication scheme is compared with some related and state-of-the-art works [28–31] and the comparison results are shown in Table 3, where Anti-imper represents resistance of impersonation attack, resistance of desynchronization attack denotes as Antidesy, Anti-eva represents resistance of eavesdropping attack, Anti-rep means resistance of reply attack, Anti-ins indicates resistance of inside attack and L-anony represents the feature of limited anonymity. Note that, although A's scheme can achieve the anonymity, it cannot identify the corrupted tag as same as C's scheme. In addition, on other schemes can resist the inside attack except our scheme and TAI. Hence, our proposed scheme is superiority to other related scheme in terms of security features.

Table 3. Security comparision.

	[28]	[29]	[30]	[31]	Our
Anti-imper	√	√	√	√	√
Anti-desy	√	√	√	×	√
Anti-eva	×	√	√	√	√
Anti-rep	×	√	√	√	√
Anti-ins	√	×	×	×	√
L-anony	√	×	×	√	√

6 Performance Analysis and Comparison

6.1 Communication Cost Analysis

Firstly, let the length of temporary and pseudo identity, the size of output length of secure hash and PUF, the length of random nonce and challenge, and the random value K_i are 32 bits, 128 bits, 120 bits and 128 bits, respectively. In addition, the comparison is consist of three phased including IR setup phase, SCDAC setup phase and authentication between IR and SCDAC phase. Specifically, in the setup phase, SCDAC sends the challenge Ci, and the temporary identity TIDi to IR, while IR sends Ri to SCDAC. Hence, the communication cost is 152 bits to SCDAC and 128 bits to IR in setup phase. As to authentication phase, the total four messages sizes, $\{C_{i+1},R_{i+1},VERr\}$, $\{TID_{i+1},PID_{i+1},V_s\}$ are 32 + 32 + 32 + 120 + 120 + 128 + 120 + 120 ++128 + 32 + 32 + 128 = 1024 bits.

6.2 Computation Cost Analysis

When it comes to computation cost comparison, firstly cryptographic operations in authentication phase will be counted in our and compared works, then each of operation is simulated on the desktop installed Ubuntu16.04 with Intel Core i3-3120 CPU and 4 GB memory except XOR operation and Random number generation operation for their extremely short running time. The simulation experiment is executed by C language and pairing-based cryptography(PBC) library. Hence, let us start to define the symbol notations as follows.

T_{Hash} *represents the time of one way hash function*
T_{PUF} *denotes the time of one PUF operation*
T_{XOR} *is remarked as the time of XOR operation*
T_{RAN} *denotes the time of random number generation*
$T_{Sys-e/d}$ *denotes the time of symmetric encryption and decryption*

We can clearly observe that the computation overhead of our proposed scheme is less than other related works on authentication phase, which is 4THash + 3TXOR + TPUF for SCDAC and 3THash + 3TXOR + TPUF for IR. Meanwhile, the running time of one way secure hash is 0.00048 s which is approximately equally to one PUF operation. In conclusion, the total running time of IR and SCDAC is about 0.00432 s which is the least compared with related work in Table 4.

Table 4. Compasion on computation cost.

	User	Server
[7]	4THash+3TPUF+TRAN	4THash+2TRAN
[18]	3TSys−e/d+12THash	3TSys−e/d+12THash
[25]	4THash+2TPUF+TRAN	4THash+2TRAN
Ours	3THash+3TXOR+TPUF	4THash+3TXOR+TPUF

7 Conclusion

In our work, we first survey the three main kinds of authentication and compare their strengths and weaknesses. According the different roles in the industrial internet of thing, a mixed authentication scheme is proposed to address the security issues. More specifically, an ECC-based authentication scheme and a PUF-based authentication are designed for employee and IR, respectively according the different security threats. Moreover, we design a threshold-based fault detection scheme to detect the disobedient the robot by broadcasting the defect rate to employee. As to corrupted or malicious employee, we add the eviction table to prevent them to access the SCDAC. In addition, by the security analysis in term of impersonation attack, desynchronization attack, eavesdropping attack, and inside attack, we ensure that our proposed scheme is superiority to other related schemes. Finally, our scheme is proved efficient in communication and computation overhead. In our future work, we will try to design more efficient and secure authentication scheme between IoT devices using blockchain.

References

1. Grammatikis, P.I.R., Sarigiannidis, P.G., Moscholios, I.D.: Securing the Internet of Things: challenges, threats and solutions. Internet of Things **5**, 41–70 (2019)
2. Deogirikar, J., Vidhate, A.: Security attacks in IoT: a survey. In: 2017 International Conference on I-SMAC (IoT in Social, Mobile, Analytics and Cloud) (I-SMAC), pp. 32–37. IEEE (2017)
3. Abdul-Ghani, H.A., Konstantas, D.: A comprehensive study of security and privacy guidelines, threats, and countermeasures: an IoT perspective. J. Sens. Actuator Netw. **8**(2), 22 (2019)
4. Chen, D., Chang, G., Sun, D., Li, J., Jia, J., Wang, X.: TRM-IoT: a trust management model based on fuzzy reputation for Internet of Things. Comput. Sci. Inf. Syst. **8**(4), 1207–1228 (2011)
5. Jabbar, W.A., Shang, H.K., Hamid, S.N., Almohammedi, A.A., Ramli, R.M., Ali, M.A.: IoT-BBMS: Internet of Things-based baby monitoring system for smart cradle. IEEE Access **7**, 93791–93805 (2019)

6. Paţa, S.D., Milici, D.L., Poienar, M., Cenuşă, M.: Management system for the control of the forklifts activity in a factory. In: 2019 8 th International Conference on Modern Power Systems (MPS), pp. 1–4, May 2019
7. Gope, P., Lee, J., Quek, T.Q.: Lightweight and practical anonymous authentication protocol for RFID systems using physically unclonable functions. IEEE Trans. Inf. Forensics Secur. **13**(11), 2831–2843 (2018)
8. Song, B.: Server impersonation attacks on RFID protocols. In: 2008 The Second International Conference on Mobile Ubiquitous Computing, Systems, Services and Technologies, pp. 50–55. IEEE (2008)
9. Lee, J., Yu, S., Park, K., Park, Y.: Secure three-factor authentication protocol for multi-gateway IoT environments. Sensors **19**(10), 2358 (2019)
10. Shuai, M., Liu, B., Yu, N., Xiong, L.: Lightweight and secure three factor authentication scheme for remote patient monitoring using on-body wireless networks. Secur. Commun. Netw. **2019**, 1–14 (2019)
11. Lohachab, A., et al.: ECC based inter-device authentication and authorization scheme using MQTT for IoT networks. J. Inf. Secur. Appl. **46**, 1–12 (2019)
12. Gopinath, V., Bhuvaneswaran, R.S.: Design of ECC based secured cloud storage mechanism for transaction rich applications. Comput. Mater. Continua **57**(2), 341–352 (2018)
13. Shi, C.: A novel ensemble learning algorithm based on D-S evidence theory for IoT security. Comput. Mater. Continua **57**(3), 635–652 (2018)
14. Kou, L., Shi, Y., Zhang, L., Liu, D., Yang, Q.: A lightweight three-factor user authentication protocol for the information perception of IoT. Comput. Mater. Continua **58**(2), 545–565 (2019)
15. Chatterjee, U., Chakraborty, R.S., Mukhopadhyay, D.: A PUF-based secure communication protocol for IoT. ACM Trans. Embed. Comput. Syst. (TECS) **16**(3), 67 (2017)
16. Chatterjee, U., et al.: Building PUF based authentication and key exchange protocol for IoT without explicit CRPS in verifier database. IEEE Trans. Dependable Secure Comput. **16**(3), 424–437 (2018)
17. Lauter, K.: The advantages of elliptic curve cryptography for wireless security. IEEE Wirel. Commun. **11**(1), 62–67 (2004)
18. Odelu, V., Das, A.K., Goswami, A.: A secure biometrics-based multi-server authentication protocol using smart cards. IEEE Trans. Inf. Forensics Secur. **10**(9), 1953–1966 (2015)
19. He, D., Wang, D.: Robust biometrics-based authentication scheme for multiserver environment. IEEE Syst. J. **9**(3), 816–823 (2014)
20. Amin, R., Kumar, N., Biswas, G., Iqbal, R., Chang, V.: A light weight authentication protocol for IoT-enabled devices in distributed cloud computing environment. Future Gener. Comput. Syst. **78**, 1005–1019 (2018)
21. Godor, G., Antal, M., Imre, S.: Mutual authentication protocol for low computational capacity RFID systems. In: IEEE GLOBECOM 2008, 2008 IEEE Global Telecommunications Conference, pp. 1–5. IEEE (2008)
22. Akgun, M., Caglayan, M.U.: Server impersonation attacks and revisions to slap, RFID lightweight mutual authentication protocol. In: 2010 Fifth International Conference on Systems and Networks Communications, pp. 148–153. IEEE (2010)
23. Bringer, J., Chabanne, H., Icart, T.: Improved privacy of the tree-based hash protocols using physically unclonable function. In: Ostrovsky, R., De Prisco, R., Visconti, I. (eds.) SCN 2008. LNCS, vol. 5229, pp. 77–91. Springer, Heidelberg (2008). https://doi.org/10.1007/978-3-540-85855-3_6
24. Kardaş, S., Çelik, S., Yıldız, M., Levi, A.: PUF-enhanced offline RFID security and privacy. J. Netw. Comput. Appl. **35**(6), 2059–2067 (2012)
25. Akgün, M., Çağlayan, M.U.: Providing destructive privacy and scalability in RFID systems using PUFs. Ad Hoc Netw. **32**, 32–42 (2015)

26. Hankerson, D., Menezes, A., Vanstone, S.: Guide to Elliptic Curve Cryptography. Springer, Heidelberbg (2004). https://doi.org/10.1007/b97644. 332 p., ISBN 0-387-95273-X
27. Stallings, W.: Cryptography and Network Security: Principles and Practice. Pearson, Upper Saddle River (2017)
28. Sui, Z., Niedermeier, M., et al.: Tai: a threshold-based anonymous identification scheme for demand-response in smart grids. IEEE Trans. Smart Grid **9**(4), 3496–3506 (2016)
29. Fan, K., Jiang, W., Li, H., Yang, Y.: Lightweight RFID protocol for medical privacy protection in IoT. IEEE Trans. Ind. Inf. **14**(4), 1656–1665 (2018)
30. Aghili, S.F., Ashouri-Talouki, M., Mala, H.: Dos, impersonation and de-synchronization attacks against an ultra-lightweight RFID mutual authentication protocol for IoT. J. Supercomput. **74**(1), 509–525 (2018). https://doi.org/10.1007/s11227-017-2139-y
31. Li, X., Niu, J., Kumari, S., Wu, F., Choo, K.-K.R.: A robust biometrics based three-factor authentication scheme for global mobility networks in smart city. Future Gener. Comput. Syst. **83**, 607–618 (2018)

Multi-party Semi-quantum Secret Sharing Scheme Based on Bell States

Xue-Yang Li, Yan Chang(✉), and Shi-Bin Zhang

School of Cyberspace Security,
Chengdu University of Information Technology, Chengdu 610225, Sichuan, China
cyttkl@cuit.edu.cn

Abstract. Based on the semi-quantum theory, we propose an easy-to-implement multi-party semi-quantum secret sharing scheme using Bell states, which reduces the construction cost of quantum secret sharing network and completes the multi-party secret sharing. The quantum Alice uses Bell states as the initial resource and has the abilities of performing Bell basis measurement and storing qubits, the participant Bob_i has limited quantum capabilities. The secret sharing network based on semi-quantum theory completes the secret sharing between the secret distributor and multiple participants. Security analysis shows that the scheme can resist internal attacks and external entanglement attacks and is safe and feasible for current technologies.

Keywords: Quantum cryptography · Quantum secret sharing · Semi-quantum theory · Bell states

1 Introduction

Quantum secret sharing is an important branch of quantum cryptography. It is a combination of classical secret sharing and quantum theory. It enables secret information (classical information or quantum-encoded information) to be distributed, transmitted and restored through quantum operations. The security of quantum secret sharing is based on the fundamental principles of quantum mechanics, which makes quantum secret sharing safer than traditional secret sharing.

The earliest quantum secret sharing scheme was proposed by Hillery et al. in 1999 [1], which uses the Greenberger-Horne-Zeilinger (GHZ) entangled particles to complete the secret sharing. Since then, more and more quantum schemes based on Bell entangled states or multi-particle entangled states have been proposed [2–4], including quantum secret sharing schemes [5–8]. However, the difficulty in the preparation of Bell states or multi-particle entangled states suggests that quantum secret sharing schemes based on entangled states are not worthwhile in some cases. After all, practicability is an important pursuit of quantum information theory. These technical barriers make such quantum secrets. The practicality of the sharing scheme is greatly reduced. In this regard, Guo et al. proposed a quantum secret sharing scheme without entanglement in 2003 [9], which uses the single particle to complete the secret sharing of classical information.

X. Sun et al. (Eds.): ICAIS 2020, LNCS 12240, pp. 280–288, 2020.
https://doi.org/10.1007/978-3-030-57881-7_25

Since then, Yan et al. proposed a quantum secret sharing scheme between multi-party and multi-party without entanglement in 2005 [10], but then the literature [11] pointed out that this scheme has security risks in particle transmission, resulting in the disclosure of secret information. And the corresponding improvement measures are given. Although such quantum secret sharing schemes do not use the entanglement characteristics of entangled particles to complete secret sharing, it is difficult to ensure the security of particle transmission. Most of the existing quantum secret sharing protocols require communication parties to have complete quantum capabilities. The limitation of cost and quantum resources seriously hinders the commercialization and popularization of quantum secret sharing.

Semi-quantum cryptography is a research branch of quantum communication, which refers to communication between communicators with complete quantum capabilities and quantum capabilities. It does not require both parties to have complete quantum capabilities, but it also enhances the security of the communication process through quantum mechanical properties, while reducing the dependence on quantum device resources. In 2007, Boyer et al. [12] proposed the definition and application of semi-quantum protocols. Based on the semi-quantum idea, they proposed the first semi-quantum cryptographic protocol based on BB84. Since then, researchers have begun to study quantum cryptography based on semi-quantum ideas, applying the concept of semi-quantum cryptography to quantum cryptography tasks such as quantum key distribution, quantum direct communication, quantum privacy comparison, and quantum secret sharing [13–16]. In 2010, Long et al. extended the semi-quantum idea to quantum secret sharing, and proposed two semi-quantum secret sharing schemes based on the GHZ-like state [17]. Subsequently, Wang et al. [18] proposed a semi-quantum secret sharing scheme based on two-particle entangled states in 2012. In 2013, Li et al. proposed a semi-quantum secret sharing scheme for two-particle product states [19], which uses the state as the initial state to complete the three-party secret sharing, which makes the quantum secret sharing scheme more practical and reduces quantum. The consumption of resources. Since then, Xie et al. proposed a semi-quantum secret sharing protocol based on GHZ-like state in 2015 [20]. Ye et al. proposed a single-photon-based ring semi-quantum secret sharing protocol in 2018 [21]. It can be seen that semi-quantum communication is a practical communication scheme, which greatly reduces the dependence on quantum resources while ensuring communication security. Inspired by semi-quantum cryptography, we propose a single-photon-based multi-party semi-quantum secret sharing scheme, which uses only single particles to complete the secret sharing between multiple parties, and reduces the dependence on quantum devices, which is easy to implement in practice.

Therefore, this paper proposes a semi-quantum secret sharing scheme based on Bell states and can be extended to multiple parties.

The rest of this manuscript is outlined as follows. Section 2 introduces the preliminaries. Section 3 introduces the proposed scheme. Section 4 analyzes the security of the proposed scheme. Finally, Sect. 5 concludes the work.

2 Preliminaries

2.1 Quantum Secret Sharing

Quantum secret sharing is a combination of classical secret sharing and quantum cryptography. It is based on the characteristics of quantum mechanics to improve the security of secret sharing. In the secret sharing, the secret distributor divides the classical information coding into quantum states. After the participants receive the quantum state through quantum communication, they recover part of the secret information through quantum operations. Each participant can only recover through honest cooperation. Original secret information.

2.2 Semi-quantum Cryptography Communication

Semi-quantum cryptography means that one of the two parties has complete quantum power (quantum side) and the other has limited quantum power (classic side), which stipulates that the classical side can only perform the following operations:

1. Measuring particles with a Z-based.
2. Not measuring the particles, directly reflecting the particles to the quantum side.
3. Preparing a particle based on the Z group and sending it to the quantum side.
4. Rearrange and reorder the received particle sequence.

Semi-quantum cryptography communication does not strictly require communication parties to have complete quantum capabilities, reduce the dependence on quantum resources, but has the characteristics of quantum cryptography, which improves security.

3 The Proposed Protocol

It is assumed that the secret distributor Alice is prepared to complete the secret information sharing of length M with the n recipients $\mathrm{Bob_i}$. Alice has quantum capabilities, and $\mathrm{Bob_i}$ has only classic capabilities. In this section, to accomplish the task of sharing Alice keys with Bobi, we construct a multi-party semi-quantum secret sharing scheme using Bell states.

We define two operations that the classic side $\mathrm{Bob_i}$ has:

1. Measure the received particles by Z-basis ($\{|0\rangle, |1\rangle\}$), and a new particle of the same quantum state is prepared and sent to Alice. (Referred to as MEASURE).
2. Simply return the particles back to Alice without disturbance. (Referred to as REFLECT)

For the sake of simplicity, in this paper, we record the measurement basis $\{|0\rangle, |1\rangle\}$ as Z basis, the measurements of particles $|0\rangle$ and $|1\rangle$ represent classical bits 0 and 1 respectively, the four Bell states can be denoted as

$$|\phi^+\rangle = \frac{1}{\sqrt{2}}(|0\rangle|0\rangle + |1\rangle|1\rangle) \tag{1}$$

$$|\phi^-\rangle = \frac{1}{\sqrt{2}}(|0\rangle|0\rangle - |1\rangle|1\rangle) \quad (2)$$

$$|\psi^+\rangle = \frac{1}{\sqrt{2}}(|0\rangle|1\rangle + |1\rangle|0\rangle) \quad (3)$$

$$|\psi^-\rangle = \frac{1}{\sqrt{2}}(|0\rangle|1\rangle - |1\rangle|0\rangle) \quad (4)$$

The specific steps of the plan are as follows.

Step 1: Secret distributor Alice prepares $M + T$ pairs of Bell states where $M + T$ is the length of secret message, the entire sequence is recorded as $|S\rangle$. Each of the Bell states $|S_i\rangle$ is randomly placed in one of four Bell states. Alice divides these states into two ordered sequences $|S1\rangle$ which is all the first qubits and $|S2\rangle$ which is all the second qubits. Finally, Alice keeps $|S1\rangle$, and sends $|S2\rangle$ to Bob_1 by the block transmission technology [22, 23]. $|S1\rangle$ ($|S2\rangle$) contains $M + T$ qubits.

Step 2: In order to resist the Trojan horse attack [56–58], a photon number splitter (PNS) is needed for Bob. After receiving the particles from Alice, Bob_1 randomly selects M particles for MEASURE and performs REFLECT on the remaining T particles. Bob_1 represents the measurement results of M particles as classic information, which is recorded as K_{B1}.

Step 3: After Alice confirming the receiption of Bob_1's $M + T$ particles, Bob_1 announces the location of the particles he chooses to MEASURE and the particles he selects to REFLECT.

Step 4: Alice starts eavesdropping detection. Since Alice knows the positions of the particles that Bob_1 chooses to MEASURE and REFLECT, she could distinguish which particles are reflected back and which are measured. For the T reflected particles, that is, the particles for which Bob_1 performs REFLECT, Alice combines the corresponding positional particles $|S1_i\rangle$ and $|S2_i\rangle$ in her hands, and uses Bell measurement basis to measure them. Alice compares the measurement results with the Bell state particles $|S_i\rangle$ she prepared in Step 1 and calculates the error rate of the particles being REFLECT, if the error rate is below the threshold, Alice proceeds to the next step. Otherwise, Alice resumes communication with Bob_1. For example, suppose Alice prepares the initial state $|S_i\rangle = |\psi^+\rangle$, the Bell measurement result of $|S1_i\rangle$ and $|S2_i\rangle$ should be $|\psi^+\rangle$ if there is no eavesdropping.

Step 5: Alice uses the Z basis to measure the remaining M particles $|S2_i\rangle$, and expresses the measurement results as classic information K_{B1}'. Then Alice uses the Z basis to measure the corresponding particles $|S1_i\rangle$, if the message is not eavesdropped, the relationship among the initial Bell state, Z basis measurement results of $|S1_i\rangle$ and $|S2_i\rangle$ and K_{B1}' is as shown in the following Table 1. So far, Alice has completed her secret sharing with Bob_1, $K'_{B1} = K_{B1}$.

Step 6: For other participants, repeat Steps 1–5, Alice can completes the secret sharing with Bob_2, Bob_3, Bob_4 … Bob_n.

Record the classic secret information obtained by Bob_2, Bob_3, Bob_4 … Bob_n as K_{B2}, K_{B3}, K_{B4} … K_{Bn}, and the classic secret information obtained by Alice's hand as K'_{B2}, K'_{B3}, K'_{B4} … K'_{Bn}. If the message has not been eavesdropped, then $K'_{Bi} = K_{Bi}$.

Alice performs an XOR operation on K'_{B1}, K'_{B2}, K'_{B3} … K'_{Bn} to obtain secret information K_A at the length of M, $K_A = K'_{B1}$ XOR K'_{B2} XOR K'_{B3} XOR … XOR K'_{Bn}. At

Table 1. The relationship among the initial Bell state, Z basis measurement results and K'_{B1}

Initial Bell state	Z basis measurement results of $\|S1_i\rangle$ and $\|S2_i\rangle$	K'_{B1}
$\|\phi^+\rangle$ or $\|\phi^-\rangle$	$\|0\rangle_1\|0\rangle_2$	0
	$\|1\rangle_1\|1\rangle_2$	1
$\|\psi^+\rangle$ or $\|\psi^-\rangle$	$\|0\rangle_1\|1\rangle_2$	1
	$\|1\rangle_1\|0\rangle_2$	0

this point, Alice established her secret sharing relationship with Bob_1, Bob_2, Bob_3 … Bob_n. Only when Bob_1, Bob_2, Bob_3 … Bob_n work together honestly, can they jointly recover Alice's secret information K_A.

4 Security Analysis

This protocol can effectively resist internal participant attacks and external attacks, and ensure the security of quantum secret sharing.

4.1 Internal Participant Attacks

- Intercept–resend attack by Bob^*

Any internal participant Bob^* can not gain benefits by intercept/resend attacks.

First case, suppose Bob^* intercepts the particle sequence sent by Alice to Bob_i, and then Bob_i prepares a new sequence of particles at the length of $M + T$ and sends it to Bob_i according to its own interests. Since Bob^* does not know where Bob_i chooses MEASURE and REFLECT, Alice can detect the anomaly by eavesdropping after receiving Bob_i's particles. Bob_i's particle sequence is not the same as the particle sequence prepared by Alice, Alice can detect anomalies by checking Bob_i's execution of REFLECT particles.

Second case, suppose Bob^* intercepts the particle sequence that Bob_i sends to Alice, then Bob^* prepares a new sequence of particles at the length of $M + T$ and sends it to Alice. Similarly, Bob^* does not know where Bob_i choose MEASURE and REFLECT. After that, Alice can detect anomalies in eavesdropping detection after measuring the particles sent by Bob_i. In addition, in this case, Alice and Bob_i can not establish a secret sharing relationship because the K_{Bi}' measured by Alice is not equal to K_{Bi}.

- Intercept-measure-resend attack by Bob^*

Any internal participant Bob^* can not steal the classic information K_{Bi} of other participants' measurements by intercept-measure-resend attacks.

Suppose Bob^* intercepts and measures the particle sequence that Bob_i sends to Alice and resends the measured particle sequence to Alice in an attempt to determine K_{Bi} after Bob_i announces the location of the particles that MEASURE and REFLECT. But before this, Bob^* didn't know Bob_i's choice for MEASURE and REFLECT, Alice could detect the exception by checking Bob_i' REFLECT particles.

In addition, any internal participant Bob* cannot infer Alice's complete secret information K_A by guessing the classic information of other participant measurements.

The secret information K_A of length M is obtained by bitwise XOR of K_{B1}, K_{B2}, K_{B3} … K_{Bn}. For each XOR value, assuming that Bob* has a 50% probability of successfully guessing the XOR value of the classical information of other participants' measurements, the probability P_i of Bob* correctly inferring the entire message secret information K_A can be quantitatively evaluated based on the statistical data.

$$P_i = \binom{M}{k}\left(\frac{1}{2}\right)^k\left(\frac{1}{2}\right)^{M-k} \tag{5}$$

Where k represents the total number of XOR values correctly guessed by Bob*, and M represents the length of the entire secret information K_A. The probability P_i conforms to the binomial distribution and the binomial coefficient.

$$\binom{M}{k} = \frac{M!}{k!(M-k)!} \tag{6}$$

By calculating $M = 256, M = 512, M = 1024, M = 2048$, the probability P_i of Bob* correctly guessing XOR values under the total number of k, knowing that for different M, P_i has its maximum value within the interval $(0, k)$. ($P_{max}(M = 256) \approx 0.0575$, $P_{max}(M = 512) \approx 0.0407$, $P_{max}(M = 1024) \approx 0.0288$, $P_{max}(M = 2048) \approx 1.4804*10^{-102}$), and decreases as M increases. Therefore, any internal participant Bob* can not guess the complete secret information K_A.

4.2 External Attacks

External eavesdropper Eve or any internal participant cannot gain benefits through entanglement attacks.

Suppose Bob* intercepts the particles $|S2\rangle$ that Alice sent to Bob_i during transmission and uses unitary operation E to entangle the new particles e with $|S2\rangle$ to form a bigger Hilbert space.

$$E \otimes |0e\rangle = a|0e_{00}\rangle + b|1e_{01}\rangle \tag{7}$$

$$E \otimes |1e\rangle = b'|0e_{10}\rangle + a'|1e_{11}\rangle \tag{8}$$

$$\begin{aligned} E \otimes |{+}e\rangle &= \frac{1}{\sqrt{2}}(a|0e_{00}\rangle + b|1e_{01}\rangle + b'|0e_{10}\rangle + a'|1e_{11}\rangle) \\ &= \frac{1}{2}[|{+}\rangle(a|e_{00}\rangle + b|e_{01}\rangle + b'|e_{10}\rangle + a'|e_{11}\rangle) \\ &+ |{-}\rangle(a|e_{00}\rangle - b|e_{01}\rangle + b'|e_{10}\rangle - a'|e_{11}\rangle)] \end{aligned} \tag{9}$$

$$E \otimes |{-e}\rangle = \frac{1}{\sqrt{2}}(a|0e_{00}\rangle + b|1e_{01}\rangle - b'|0e_{10}\rangle - a'|1e_{11}\rangle)$$
$$= \frac{1}{2}[|+\rangle(a|e_{00}\rangle + b|e_{01}\rangle - b'|e_{10}\rangle - a'|e_{11}\rangle)$$
$$+ |-\rangle(a|e_{00}\rangle - b|e_{01}\rangle - b'|e_{10}\rangle + a'|e_{11}\rangle)] \tag{10}$$

The unitary operation matrix E is expressed as

$$E = \begin{pmatrix} a\, b' \\ b\, a' \end{pmatrix} \tag{11}$$

where $e_{i,j}$ decided by operator E satisfy the normalization condition

$$\sum_{i,j\in\{0,1\}} \langle e_{ij} | e_{ij} \rangle = 1 \tag{12}$$

Since $EE^* = 1$, a, b, a', b' satisfy the following relationship

$$|a|^2 + |b|^2 = 1,\ |a'|^2 + |b'|^2 = 1,\ ab^* = (a')^* b' \tag{13}$$

We can get the result

$$|a|^2 = |a'|^2,\ |b|^2 = |b'|^2 \tag{14}$$

Suppose Alice prepares Bell states $|\phi^+\rangle$ and sends the second particle to $\mathrm{Bob_i}$, after Eve performing the attack operator E, the state of the composed system becomes

$$|\varphi\rangle_{Eve} = \frac{1}{\sqrt{2}}[|0\rangle_{S1}(a|0e_{00}\rangle + b|1e_{01}\rangle)_{S2,E} + |1\rangle_{S1}(b'|0e_{10}\rangle + a'|1e_{11}\rangle)_{S2,E}]$$
$$= \frac{1}{\sqrt{2}}[a|00e_{00}\rangle + b|01e_{01}\rangle + b'|10e_{10}\rangle + a'|11e_{11}\rangle]_{S1,S2,E}$$
$$= \frac{1}{\sqrt{2}}[|0\rangle_{S2}(a|0e_{00}\rangle + b|1e_{01}\rangle)_{S1,E} + |1\rangle_{S2}(b'|0e_{10}\rangle + a'|1e_{11}\rangle)_{S1,E}] \tag{15}$$

If Eve chooses to MEASURE, and the particle e is in an entangled state, Eve's entanglement attack will inevitably introduce an error rate P_{error}

$$P_{error} = |b|^2 = 1 - |a|^2 = |b'|^2 = 1 - |a'|^2 \tag{16}$$

If Eve tries to achieve the eavesdropping without being detected, the transmitted qubits and Eve's auxiliary particles are in a tensor-product state. However, in direct state there is no correlation between the auxiliary particle e and the whole system, Eve could not get any useful information, thus proving that the entanglement attack is futile.

5 Conclusion

In this paper, we analyze the previous quantum-secret sharing scheme based on entangled states and semi-quantum secret sharing scheme, and propose a multi-party semi-quantum secret sharing scheme based on Bell state, which uses the Bell state to complete the secret sharing between the quantum side and multiple semi-quantum sides. The secret sharing between the square and multiple semi-quantum squares can be applied to a more practical quantum communication network, such as Alice as a quantum party, which is served by a network information service provider, Bob_i and other classic parties represent ordinary customers in the network. To achieve safe and reliable multi-party secret sharing.

Unlike previous quantum secret sharing schemes, the advantages of our approach are summarized below. First, our protocol does not require the classical side to have complete quantum capabilities, reducing the need for quantum device resources. Second, we have completed the secret sharing between the secret distributor and the parties, not just the secret sharing between the three parties. In addition, the final security analysis shows that our solution is resistant to internal attacks and external entanglement attacks and is safe and feasible under current technology.

Acknowledgments. This work is supported by the National Natural Science Foundation of China (No. 61572086, No.61402058), the Key Research and Development Project of Sichuan Province (No. 20ZDYF2324, No. 2019ZYD027, No. 2018TJPT0012), the Innovation Team of Quantum Security Communication of Sichuan Province (No. 17TD0009), the Academic and Technical Leaders Training Funding Support Projects of Sichuan Province (No. 2016120080102643), the Application Foundation Project of Sichuan Province (No. 2017JY0168), the Science and Technology Support Project of Sichuan Province (No. 2018GZ0204, No. 2016FZ0112).

References

1. Hillery, M., Buzek, V., Berthiaume, A.: Quantum secret sharing. Phys. Rev. A **59**(3), 1829–1834 (1999)
2. Chang, Y., Zhang, S.B., Yan, L.L., et al.: A quantum authorization management protocol based on epr-pairs. Comput. Mater. Continua **59**(3), 1005–1014 (2019)
3. Yan, L.L., Chang, Y., Zhang, S.B., et al.: Measure-resend semi-quantum private comparison scheme using GHZ class states. Comput. Mater. Continua **61**(2), 877–887 (2019)
4. Zhang, S.B., Chang, Y., Yan, L.L., et al.: Quantum communication networks and trust management: a survey. Comput. Mater. Continua **61**(3), 1145–1174 (2019)
5. Bai, C.M., Li, Z.H., Xu, T.T., et al.: A generalized information theoretical model for quantum secret sharing. Int. J. Theor. Phys. **55**(11), 4972–4986 (2016)
6. Dou, Z., Xu, G., Chen, X.B., et al.: Rational non-hierarchical quantum state sharing protocol. Comput. Mater. Continua **58**(2), 335–347 (2019)
7. Yang, Y.-G., Gao, S., Li, D., Zhou, Y.-H., Shi, W.-M.: Three-party quantum secret sharing against collective noise. Quantum Inf. Process. **18**(7), 1–11 (2019). https://doi.org/10.1007/s11128-019-2319-1
8. Karlsson, A., Koashi, M., Imoto, N.: Quantum entanglement for secret sharing and secret splitting. Phys. Rev. A **59**(1), 162–168 (1999)
9. Guo, G.P., Guo, G.C.: Quantum secret sharing without entanglement. Phys. Lett. A **310**(4), 247–251 (2003)
10. Yan, F.L., Gao, T.: Quantum secret sharing between multiparty and multiparty without entanglement. Phys. Rev. A **72**(1), 012304 (2005)
11. Han, L.F., Liu, Y.M., Shi, S.H., et al.: Improving the security of a quantum secret sharing protocol between multiparty and multiparty without entanglement. Phys. Lett. A **361**(1), 24–28 (2007)

12. Boyer, M., Kenigsberg, D., Mor, T.: Quantum key distribution with classical bob. Phys. Rev. Lett. **99**(14), 140501 (2007)
13. Boyer, M., Gelles, R., Kenigsberg, D., et al.: Semiquantum key distribution. Phys. Rev. A **79**(3), 032341 (2009)
14. Shukla, C., Thapliyal, K., Pathak, A.: Semi-quantum communication: protocols for key agreement, controlled secure direct communication and dialogue. Quantum Inf. Process. **16**(12), 1–19 (2017). https://doi.org/10.1007/s11128-017-1736-2
15. Yan, L.L., Sun, Y.H., Chang, Y., Zhang, S.B., Wan, G.G., Sheng, Z.W.: Semi-quantum protocol for deterministic secure quantum communication using Bell states. Quantum Inf. Process. **17**(11), 1–12 (2018). https://doi.org/10.1007/s11128-018-2086-4
16. Sun, Y.H., Yan, L.L., Chang, Y., et al.: Two semi-quantum secure direct communication protocols based on Bell states. Mod. Phys. Lett. A **34**(1), 1950004 (2019)
17. Long, D.Y., Chan, W.H., Li, Q.: Semi-quantum secret sharing using entangled states. Phys. Rev. A **82**(82), 2422–2427 (2010)
18. Wang, J., Zhang, S., Zhang, Q., et al.: Semi-quantum secret sharing using two-particle entangled state. Int. J. Quantum Inf. **10**(05), 1250050 (2012)
19. Li, L., Qiu, D., Mateus, P.: Quantum secret sharing with classical Bobs. J. Phys. A: Math. Theor. **46**(4), 045304 (2013)
20. Xie, C., Li, L., Qiu, D.: A novel semi-quantum secret sharing scheme of specific bits. Int. J. Theor. Phys. **54**(10), 3819–3824 (2015). https://doi.org/10.1007/s10773-015-2622-2
21. Ye, C.Q., Ye, T.Y.: Circular semi-quantum secret sharing using single particles. Commun. Theor. Phys. (12) (2018)
22. Long, G.L., Liu, X.S.: Theoretically efficient high-capacity quantum-key-distribution scheme. Phys. Rev. A **65**(3), 032302 (2002)
23. Deng, F.G., Long, G.L., Liu, X.S.: Two-step quantum direct communication protocol using the Einstein-Podolsky-Rosen pair block. Phys. Rev. A **68**(4), 042317 (2003)

A Quantum Proxy Arbitrated Signature Scheme Based on Two Three-Qubit GHZ States

Tao Zheng, Shi-Bin Zhang(✉), Yan Chang, and Lili Yan

School of Cyber Security, Chengdu University of Information Technology, Chengdu 610225, Sichuan, China
cuitzsb@cuit.edu.cn, sunsunk@foxmail.com

Abstract. In this paper, we proposed a quantum proxy arbitrated signature scheme based on two three-qubit GHZ states. In the scheme, we chose a trust third party (TP) as proxy signer, which can finish the signature process without the original signer's authority. The scheme uses the physical characteristics of quantum mechanics to implement delegation, signature and verification. It could guarantee not only the unconditionally security but also the anonymity of the message owner. We prove that our scheme can resist the eavesdropping attack, and the security analysis shows the scheme satisfies the security features of proxy arbitrated signature, singers cannot disavowal his signature while the signature cannot be forged by others, and the message owner can also be traced.

Keywords: Proxy arbitrated quantum signature · Quantum teleportation · Two three-qubit GHZ states

1 Introduction

Quantum technology is a research hotspot in the 21st century. Since Bennett and Brassard put forward the first QKD protocol in 1984 [1], according to different application scenarios, scholars have proposed a large number of quantum information schemes [2–10].

Quantum signature can be used in electronical voting or paying system. In 2001, Gottesman and Chuang proposed the pioneering quantum digital signature protocol based on weak quantum one-way functions [11]. With the development of quantum signature technology, two branches of QS scheme have been established: true quantum signature (TQS) and arbitrated quantum signature (AQS). In 2002, Zeng et al. [12] proposed the first AQS protocol with Green-Horne-Zeilinger (GHZ) states. Lee et al. [13] proposed two quantum signature schemes with message recovery relying on the availability of an arbitrator. After that, many AQS scheme had been proposed [14–20] in the past decade. In 2019, Feng et al. [21] proposed an AQS scheme with quantum walk-based teleportation, which did not need to prepare entangled particles in advance. Recently, Li et al. [22] showed that two different quantum channels of two three-qubit GHZ states and the six-qubit entangled state can be used for quantum teleportation of

X. Sun et al. (Eds.): ICAIS 2020, LNCS 12240, pp. 289–297, 2020.
https://doi.org/10.1007/978-3-030-57881-7_26

an arbitrary two-qubit state deterministically. Only Bell state measurements, Projective measurements and Unitary operation are needed to recover the arbitrary two-qubit state.

With the development of quantum communication network, the importance of quantum signature scheme shall be increase. Inspired by Feng et al. [21] and Li et al. [22], we proposed a quantum proxy arbitrated signature scheme with quantum teleportation. Based on Li et al.'s research, this quantum teleportation scheme was proved to be secure under internal and external attacks, and is robust against decoherence. The summary process of their scheme is: Alice performs BSMs on her qubit pairs (A, 1) and (B, 6), and Charlie performs a two-qubit projective measurement on the qubit pairs (3, 5) with X-basis. Based on both Alice and Charlie's measurement outcomes, Bob performs an appropriate unitary on the qubit pairs (2, 4) and he can reconstruct the signature quantum state $|\psi\rangle_{AB}$.

The rest of this paper is organized as follows. In Sect. 2, we introduce quantum teleportation scheme based on two three-qubit GHZ states. A quantum proxy arbitrated signature scheme with quantum teleportation is proposed in Sect. 3 and the security analysis is discussed in Sect. 4. Finally, a conclusion is drawn in Sect. 5.

2 Quantum Teleportation Based on Two Three-Qubit GHZ States

In the reference of [22], Li et al. show the implementation of the scheme with two cases. **Case1** uses a six-qubit as the quantum channel, and Alice performs the BSMs on her qubit pairs, and Charlie performs the BSMs operation on his qubits as well. On the other hand, **Case2** uses two three-qubit GHZ states as quantum channel and Alice performs BSMs operation but Charlie performs a two-qubit projective measurement in the X-basis on his qubits. We selected **Case2** as the theoretical basis of our proposed scheme. The process of quantum teleportation protocol based on two three-qubit GHZ states is as follow.

Alice, Bob and Charlie share two three-qubit GHZ states. Alice possesses qubits 1 and 6, Bob possesses qubits 2 and 4 and Charlie possesses qubits 3 and 5 respectively. The six-qubit cluster state is described as follows:

$$|\psi\rangle_{123456} = \frac{1}{2}(|000000\rangle + |000111\rangle + |111000\rangle + |111111\rangle)_{123456}$$

Alice holds the arbitrary two-qubits $|\psi\rangle_{AB} = (\alpha|00\rangle + \beta|10\rangle + \gamma|01\rangle + \eta|11\rangle)_{AB}$, which is the qubits to be transmitted. Now the whole system can be written as:

$$\begin{aligned}
|\Theta\rangle_{AB123456} = |\psi\rangle_{AB} \otimes |\psi\rangle_{123456} &= (\alpha|00\rangle + \beta|10\rangle + \gamma|01\rangle + \eta|11\rangle)_{AB} \\
&\otimes \frac{1}{2}(|000000\rangle + |000111\rangle + |111000\rangle + |111111\rangle)_{123456} \\
&= \frac{1}{2}[\alpha|00000000\rangle + \alpha|00000111\rangle + \alpha|00111000\rangle + \alpha|00111111\rangle \\
&+ \beta|10000000\rangle + \beta|10000111\rangle + \beta|10111000\rangle + \beta|10111111\rangle \\
&+ \gamma|01000000\rangle + \gamma|01000111\rangle + \gamma|01111000\rangle + \gamma|01111111\rangle \\
&+ \eta|11000000\rangle + \eta|11000111\rangle + \eta|11111000\rangle + \eta|11111111\rangle]_{AB123456}
\end{aligned}$$

Alice performs BSMs on her qubit pairs (A, 1) and (B, 6), then she can get 16 kinds of measurement results with equal probability. After Alice's measurement operations, the qubit pairs (2, 3, 4, 5) will also have 16 kinds of collapse results to the corresponding states $|\psi\rangle_{2345}$. Alice informs the measurement results to Bob and Charlie through a classical communication channel. Then Charlie performs a two-qubit projective measurement on the qubit pairs (3, 5), and Charlie sends the measurement outcomes to Bob. After that, Bob performs an appropriate unitary on the qubit pairs (2, 4) and he can reconstruct the state $|\psi\rangle_{AB}$, which means that the qubit $|\psi\rangle_{AB}$ holds by Alice has been transmitted to Bob with quantum teleportation scheme.

3 The Proposed Scheme

In this section, we propose a quantum proxy arbitrated quantum signature scheme based on two three-qubit GHZ states. Four participants are in this protocol: Alice is the original owner of message M, trust third party (TP) is the message's proxy signer, Bob is the verifier and Charlie is the arbitrator. The whole system of our proposed scheme is shown as Fig. 1.

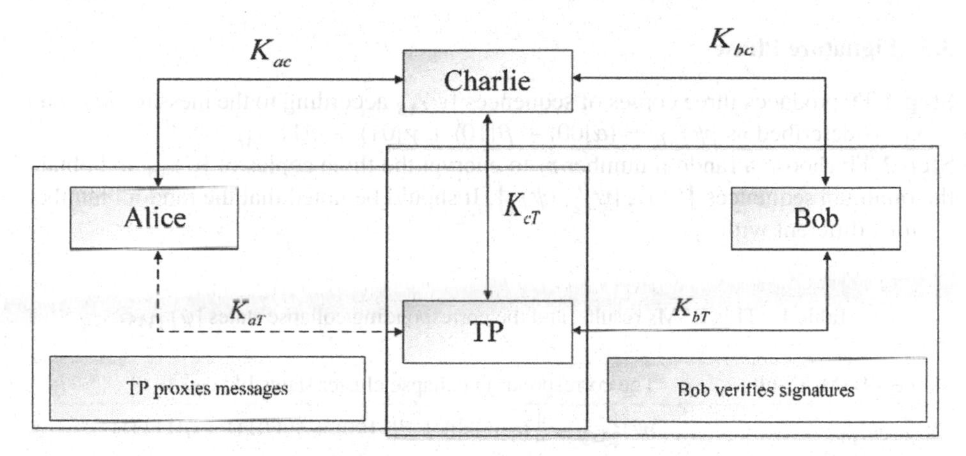

Fig. 1. The process of our scheme.

3.1 Initial Phase

Step 1 Charlie prepares N ordered two three-qubit GHZ states, each of which is in the state: $|\psi\rangle_{123456} = \frac{1}{2}(|000000\rangle + |000111\rangle + |111000\rangle + |111111\rangle)_{123456}$. Charlie takes one and six particles from each state to compose S_{16} sequence. The remaining particles compose S_{35} and S_{24} according to the same rules.

Step 2 With the quantum key distribution system [23–25], TP shares K_{aT} with Alice, and K_{bT} with Bob. Charlie shares K_{ac} shared with Alice, and K_{bc} with Bob. The shared keys can be written as: $K_{ij} = \left\{K_{ij}^{1}, K_{ij}^{2}, \ldots, K_{ij}^{2n}\right\}$ and the length of these keys is related to the message M.

Step 3 Charlie holds the qubit sequence S_{35}, and sends the qubit pairs and quantum key (S_{16}, K_{ac}) to Alice, and sends (S_{24}, K_{bc}) to Bob.
Step 4 Alice or Bob applies to Charlie to set up the arbitrated quantum signature system.

3.2 Proxy Phase

Step 1 Alice and TP finish the identity authentication process with one-way hash function as follow.

1. Both Alice and TP calculate hash function, then Alice obtains $M_A = hash(K_{aT})$ and TP obtains $M_T = hash(K_{aT})$.
2. Alice and TP publish their results, then they can judge the identity of each other.

Step 2 Alice informs Charlie that TP is the legal proxy agent. Charlie shares K_{cT} with TP.
Step 3 Alice uses a random number r_1 to encrypt original message $M: M_T = E_{r_1}(M)$, and sends M_T and S_{16} to TP.

3.3 Signature Phase

Step 1 TP produces three copies of sequences $|\psi\rangle_{AB}$ according to the message M_T, and $|\psi\rangle_{AB}$ is described as $|\psi\rangle_{AB} = (\alpha|00\rangle + \beta|10\rangle + \gamma|01\rangle + \eta|11\rangle)_{AB}$.
Step 2 TP choose a random number r_2 to encrypt the three copies of $|\psi\rangle_{AB}$ and obtain the quantum sequences $\{|\psi\rangle'_1, |\psi\rangle'_2, |\psi\rangle'_3\}$. It should be noted that the random number r_2 must different with r_1.

Table 1. TP's BSMs results, and the corresponding collapse states $|\psi\rangle_{2345}$

Alice's BSMs result	The corresponding collapse cluster states $|\psi\rangle_{2345}$
$|\phi\rangle^+_{A1}|\phi\rangle^+_{B6}$	$|\psi\rangle^1_{2345} = \frac{1}{4}(\alpha|0000\rangle + \beta|1100\rangle + \gamma|0011\rangle + \eta|1111\rangle)$
$|\phi\rangle^+_{A1}|\phi\rangle^+_{B6}$	$|\psi\rangle^3_{2345} = \frac{1}{4}(\alpha|0011\rangle + \beta|1111\rangle + \gamma|0000\rangle + \eta|1100\rangle)$
$|\phi\rangle^+_{A1}|\psi\rangle^-_{B6}$	$|\psi\rangle^4_{2345} = \frac{1}{4}(\alpha|0011\rangle + \beta|1111\rangle - \gamma|0000\rangle - \eta|1100\rangle)$
$|\phi\rangle^-_{A1}|\phi\rangle^+_{B6}$	$|\psi\rangle^5_{2345} = \frac{1}{4}(\alpha|0000\rangle - \beta|1100\rangle + \gamma|0011\rangle - \eta|1111\rangle)$
$|\psi\rangle^-_{A1}|\phi\rangle^+_{B6}$	$|\psi\rangle^7_{2345} = \frac{1}{4}(\alpha|0011\rangle - \beta|1111\rangle + \gamma|0000\rangle - \eta|1100\rangle)$
$|\psi\rangle^-_{A1}|\phi\rangle^+_{B6}$	$|\psi\rangle^8_{2345} = \frac{1}{4}(\alpha|0011\rangle - \beta|1111\rangle - \gamma|0000\rangle - \eta|1100\rangle)$
$|\psi\rangle^+_{A1}|\phi\rangle^+_{B6}$	$|\psi\rangle^9_{2345} = \frac{1}{4}(\alpha|1100\rangle + \beta|0000\rangle + \gamma|1111\rangle + \eta|0011\rangle)$
$|\psi\rangle^-_{A1}|\phi\rangle^+_{B6}$	$|\psi\rangle^{13}_{2345} = \frac{1}{4}(\alpha|1100\rangle - \beta|0000\rangle + \gamma|1111\rangle - \eta|0011\rangle)$

Step 3 TP encrypts $|\psi\rangle_1'$ with key K_{cT} to get $|S\rangle_a = E_{K_{cT}}\left(|\psi\rangle_1'\right)$. Then he uses Li et al.'s protocol [22] to transmit the qubit pairs $|\psi\rangle_2'$, and the detailed steps are as follow. TP holds the qubit sequence S_{16} and the qubit pairs $|\psi\rangle_2'$, Then TP performs BSMs on his qubit pairs (A, 1) and (B, 6) respectively. TP will get 16 kinds of possible measurement results R_a with equal probability 1/4. TP picks a random number r_3 to encrypt R_a, then obtains $M_a = E_{r_3}(R_a)$. TP generates the quantum state sequence $|S\rangle = \left\{|S\rangle_a, |\psi\rangle_1', M_a, |\psi\rangle_3'\right\}$, and TP encrypts $|S\rangle$ with K_{bT} to obtain $|S\rangle'$, then sends it to Bob. The corresponding collapse states $|\psi\rangle_{2345}$ is illustrated as Table 1.

3.4 Verification Phase

Step 1 Bob decrypts $|S\rangle'$ with K_{bT}, and he uses K_{bc} to encrypts $|S\rangle_a$ and $|\psi\rangle_1'$ to get $|\psi\rangle_B = E_{K_{bc}}\left(|S\rangle_a, |\psi\rangle_1'\right)$, then sends $|\psi\rangle_B$ to Charlie.
Step 2 Charlie decrypts $|\psi\rangle_B$ and obtains $|S\rangle_a$ and $|\psi\rangle_1'$. Charlie uses K_{cT} to encrypt $|\psi\rangle_1'$ and gets $|S\rangle_t$: $|S\rangle_t = E_{K_{cT}}\left(|\psi\rangle_1'\right)$. Charlie gets λ by comparing the values of $|S\rangle_t$ with $|S\rangle_a$: He sets $\lambda = 1$ When $|S\rangle_t = |S\rangle_a$, and $\lambda = 0$ when $|S\rangle_t \neq |S\rangle_a$.
Step 3 Charlie performs a two-qubit projective measurement on S_{35} with $\left\{|+\rangle = \frac{1}{\sqrt{2}}(|0\rangle + |1\rangle), |-\rangle = \frac{1}{\sqrt{2}}(|0\rangle - |1\rangle)\right\}$, and gets the measure result R_c.
Step 4 Charlie decrypts $|S\rangle_t$ with K_{cT} to recovers $|\psi\rangle_1'$, and encrypts $|\psi\rangle_1'$, $|S\rangle_a$ R_c, λ with key K_{bc} to obtain $|\psi\rangle_{BC} = E_{K_{bc}}\left(|\psi\rangle_1', |S\rangle_a R_c, \lambda\right)$. Charlie transmits $|\psi\rangle_{BC}$ to Bob.
Step 5 After receiving $|\psi\rangle_{BC}$, Bob decrypts it and obtains $|\psi\rangle_1'$, $|S\rangle_a$, R_c λ. If $\lambda = 0$, it means the signature is illegal and Bob rejects it immediately. If $\lambda = 1$, Bob informs TP to publish r_3 via a public channel, but he can only confirm the quantum state $|S\rangle_a$ prepared by TP is accurate, and needs further verification.
Step 6 Bob decrypts M_a with r_3 to get TP's BSMs result R_a. With R_a and R_c, Bob can recover the teleported state $|\psi\rangle_{2(out)}'$ by applying local unitary operation U on his qubit pairs S_{24}. If $|\psi\rangle_{2(out)}' \neq |\psi\rangle_3'$, Bob rejects the signature. If $|\psi\rangle_{2(out)}' = |\psi\rangle_3'$, Bob notifies TP to publish the random number r_2. The relationship between Bob's unitary operation and Charlie's measurement outcomes R_c and the corresponding particle state obtained by Bob is shown in Table 2.
Step 7 Alice publishes r_1 and Bob contains $|\psi\rangle_{AB}$ by decrypting $|\psi\rangle_{2(out)}'$ or $|\psi\rangle_3'$, and confirms ($|S\rangle_a$, r_2,K_{aT},K_{bT}) as TP's proxy signature for message M_T. It can be noted that the proxy signature has been established.

4 Security Analysis

4.1 Impossibility of Disavowals

Bob cannot deny his receiving of $|S\rangle'$ because $|S\rangle'$ is encrypted by TP with K_{bT}. Moreover, Bob encrypts $|S\rangle_a$ and $|\psi\rangle_1'$ with K_{bc} to obtain $|\psi\rangle_B$, and Charlie can judge Bob's disavowals with theirs shared key K_{bc}. Similarly, Proxy signer TP also cannot deny his proxy signature with $K_{cT}(|S\rangle = E_{K_{cT}}\left(|\psi\rangle_{AB}\right))$ and K_{bT}.

Table 2. The relationship of Charlie's result and Bob's unitary operation.

Charlie's result	(2, 4) collapse state	Bob's unitary operation
$\vert +\rangle_3 \vert +\rangle_5 = \frac{1}{2}(\vert 00\rangle + \vert 01\rangle + \vert 10\rangle + \vert 11\rangle)_{35}$	$\vert\psi\rangle_{24}^{1} = \frac{1}{16}(\alpha\vert 00\rangle + \beta\vert 10\rangle + \gamma\vert 01\rangle + \eta\vert 11\rangle)$	$I \otimes I$
$\vert +\rangle_3 \vert -\rangle_5 = \frac{1}{2}(\vert 00\rangle - \vert 01\rangle + \vert 10\rangle - \vert 11\rangle)_{35}$	$\vert\psi\rangle_{24}^{2} = \frac{1}{16}(\alpha\vert 00\rangle + \beta\vert 10\rangle - \gamma\vert 01\rangle - \eta\vert 11\rangle)$	$I \otimes \sigma_Z$
$\vert -\rangle_3 \vert +\rangle_5 = \frac{1}{2}(\vert 00\rangle + \vert 01\rangle - \vert 10\rangle - \vert 11\rangle)_{35}$	$\vert\psi\rangle_{24}^{3} = \frac{1}{16}(\alpha\vert 00\rangle - \beta\vert 10\rangle + \gamma\vert 01\rangle - \eta\vert 11\rangle)$	$\sigma_Z \otimes I$
$\vert -\rangle_3 \vert -\rangle_5 = \frac{1}{2}(\vert 00\rangle - \vert 01\rangle - \vert 10\rangle + \vert 11\rangle)_{35}$	$\vert\psi\rangle_{24}^{4} = \frac{1}{16}(\alpha\vert 00\rangle - \beta\vert 10\rangle - \gamma\vert 01\rangle + \eta\vert 11\rangle)$	$\sigma_Z \otimes \sigma_Z$
$\vert +\rangle_3 \vert +\rangle_5 = \frac{1}{2}(\vert 00\rangle + \vert 01\rangle + \vert 10\rangle + \vert 11\rangle)_{35}$	$\vert\psi\rangle_{24}^{5} = \frac{1}{16}(\alpha\vert 00\rangle + \beta\vert 10\rangle - \gamma\vert 01\rangle - \eta\vert 11\rangle)$	$I \otimes \sigma_Z$
$\vert +\rangle_3 \vert -\rangle_5 = \frac{1}{2}(\vert 00\rangle - \vert 01\rangle + \vert 10\rangle - \vert 11\rangle)_{35}$	$\vert\psi\rangle_{24}^{6} = \frac{1}{16}(\alpha\vert 00\rangle + \beta\vert 10\rangle + \gamma\vert 01\rangle + \eta\vert 11\rangle)$	$I \otimes I$
$\vert -\rangle_3 \vert +\rangle_5 = \frac{1}{2}(\vert 00\rangle + \vert 01\rangle - \vert 10\rangle - \vert 11\rangle)_{35}$	$\vert\psi\rangle_{24}^{7} = \frac{1}{16}(\alpha\vert 00\rangle - \beta\vert 10\rangle - \gamma\vert 01\rangle + \eta\vert 11\rangle)$	$\sigma_Z \otimes \sigma_Z$
$\vert -\rangle_3 \vert -\rangle_5 = \frac{1}{2}(\vert 00\rangle - \vert 01\rangle - \vert 10\rangle + \vert 11\rangle)_{35}$	$\vert\psi\rangle_{24}^{8} = \frac{1}{16}(\alpha\vert 00\rangle - \beta\vert 10\rangle + \gamma\vert 01\rangle - \eta\vert 11\rangle)$	$\sigma_Z \otimes I$
$\vert +\rangle_3 \vert +\rangle_5 = \frac{1}{2}(\vert 00\rangle + \vert 01\rangle + \vert 10\rangle + \vert 11\rangle)_{35}$	$\vert\psi\rangle_{24}^{9} = \frac{1}{16}(\alpha\vert 00\rangle + \beta\vert 10\rangle + \gamma\vert 01\rangle + \eta\vert 11\rangle)$	$I \otimes \sigma_x$
$\vert +\rangle_3 \vert -\rangle_5 = \frac{1}{2}(\vert 00\rangle - \vert 01\rangle + \vert 10\rangle - \vert 11\rangle)_{35}$	$\vert\psi\rangle_{24}^{10} = \frac{1}{16}(-\alpha\vert 00\rangle - \beta\vert 10\rangle + \gamma\vert 01\rangle + \eta\vert 11\rangle)$	$I \otimes -i\sigma_y$

4.2 Impossibility of Forgeries

According to the analysis of the protocol process, TP must complete the identity authentication process with Alice, and he shall publish the random number r_2 and r_3. It's impossible for others to forge these right random numbers. In signature phase, only TP and Bob, Charlie have the secret key K_{bT}, K_{cT}. Based on the above analysis, only the proxy signer TP can produce the legal proxy signature.

4.3 Traceability

In a proxy signature scheme, the original signer should be able to correctly judge the specific identity information of the proxy signer from a proxy signature. In our scheme, original signer Alice must be able to trace the proxy signer TP's identity information.

In verification phase, Bob confirms ($|S\rangle_a$, r_2, K_{aT}, K_{bT}) as TP's proxy signature for message M_T. Alice can trace and confirm TP with r_2 and K_{aT}.

4.4 Distinguishability

In a proxy signature protocol, the trust-third-party's proxies the original signer Alice's signature must be distinguishable, which means that the proxy signature and the original signer's signature can be distinguished by the arbitrator. In this scheme, others must can distinguish the signature is signed by TP or by Alice. In the verification phase, after Alice's announcement of r_1, Bob, Charlie and others can distinguish the signature is signed by the original signer Alice or the proxy signer TP.

4.5 Analysis of the Pre-shared Keys

Due to the unconditional security of quantum key distribution [26–30], only Alice and Charlie know the secret key K_{ac}, Alice and TP share the key K_{aT}, TP shares key K_{bT} and K_{cT} with Bob and Charlie. According to the analysis of the protocol flow, K_{cT} is used to encrypt messages in the signature phase, and in the verification phase, Bob or external attackers are unable to obtain any information related to K_{cT}. Assuming that the information of K_{cT} is tempered by external attacker and the value of λ is changed, Bob will refuse to accept this signature. The analysis of the other keys is the same as that of K_{cT}. when a failure occurs in the eavesdropping check or when the secret keys are used for a long period of time does, the new secret keys have to be shared again.

5 Conclusion

In this paper, we proposed a quantum proxy arbitrated signature scheme based on two three-qubit GHZ states. Our scheme does not require the complex measurement operation, only Bell states Measurement and two-qubit projective operation can realize the whole system. It is easy to implement in practical situations based on current technologies. The security of the protocol is guaranteed by the pre-shared keys and random numbers. The secret keys which are used in the scheme, and one-time pad algorithm, all of which have been proven unconditionally secure. According to the security analysis, the scheme can meet the security requirements of the impossibility of disavowals, impossibility of forgeries, and it also meets the traceability and distinguish.

Acknowledgments. This work is supported by the National Natural Science Foundation of China (No. 61572086, No.61402058), the Key Research and Development Project of Sichuan Province (No. 20ZDYF2324, No. 2019ZYD027, No. 2018TJPT0012), the Innovation Team of Quantum Security Communication of Sichuan Province (No. 17TD0009), the Academic and Technical Leaders Training Funding Support Projects of Sichuan Province (No. 2016120080102643), the Application Foundation Project of Sichuan Province (No. 2017JY0168), the Science and Technology Support Project of Sichuan Province (No. 2018GZ0204, No.2016FZ0112).

References

1. Bennett, C.H., Brassard, G.: Public key distribution and coin tossing. In: Proceedings of the IEEE International Conference on Computers, Systems and Signal Processing, Bangalore, pp. 175–179. IEEE, New York (1984)
2. Liu, W.-J., Gao, P.-P., Yu, W.-B., Qu, Z.-G., Yang, C.-N.: Quantum relief algorithm. Quantum Inf. Process. **17**(10), 1–15 (2018). https://doi.org/10.1007/s11128-018-2048-x
3. Qu, Z.-G., Wu, S.-Y., Wang, M.-M., et al.: Effect of quantum noise on deterministic remote state preparation of an arbitrary two-particle state via various quantum entangled channels. Quantum Inf. Process. **16**(306), 1–25 (2017)
4. Chang, Y., Zhang, S.-B., Yan, L.-L., et al.: A quantum authorization management protocol based on EPR-pairs. Comput. Mater. Continua. **59**(3), 1005–1014 (2019)
5. Yan, L.-L., Chang, Y., Zhang, S.-B., et al.: Measure-resend semi-quantum private comparison scheme using GHZ class states. Comput. Mater. Continua **61**(2), 877–887 (2019)
6. Zhang, S.-B., Chang, Y., Yan, L.-L., et al.: Quantum communication networks and trust management: a survey. Comput. Mater. Continua **61**(3), 1145–1174 (2019)
7. Tao, Z., Chang, Y., Zhang, S., Dai, J., Li, X.: Two semi-quantum direct communication protocols with mutual authentication based on bell states. Int. J. Theoret. Phys. **58**(9), 2986–2993 (2019). https://doi.org/10.1007/s10773-019-04178-5
8. Yang, Y.-G., Sun, S.-J., Xu, P., Tian, J.: Flexible protocol for quantum private query based on B92 protocol. Quantum Inf. Process. **13**(3), 805–813 (2013). https://doi.org/10.1007/s11128-013-0692-8
9. Yang, Y.-G., Zhang, M.-O., Yang, R.: Private database queries using one quantum state. Quantum Inf. Process. **14**(3), 1017–1024 (2014). https://doi.org/10.1007/s11128-014-0902-z
10. Zheng, T., Zhang, S.-B., Gao, X., Chang, Y.: Practical quantum private query based on Bell state. Modern Phys. Lett. A **34**, 1950196 (2019)
11. Gottesman, D., Chuang, I.: Quantum digital signatures. arXiv preprint arXiv:quant-ph/0105032 (2001)
12. Zeng, G., Keitel, C.H.: Arbitrated quantum-signature scheme. Phys. Rev. A **65**(4), 042312 (2002)
13. Lee, H., Hong, C., Kim, H., Lim, J., Yang, H.J.: Arbitrated quantum signature scheme with message recovery. Phys. Lett. A **321**(5–6), 295–300 (2004)
14. Curty, M., Lütkenhaus, N.: Comment on arbitrated quantum-signature scheme. Phys. Rev. A **77**(4), 046301 (2008)
15. Li, Q., Chan, W.H., Long, D.Y.: Arbitrated quantum signature scheme using Bell states. Phys. Rev. A **79**(5), 054307 (2009)
16. Zou, X., Qiu, D.: Security analysis and improvements of arbitrated quantum signature schemes. Phys. Rev. A **82**(4), 042325 (2010)
17. Yang, Y.-G., Zhou, Z., et al.: Arbitrated quantum signature with an untrusted arbitrator. Eur. Phys. J. D **61**(3), 773–778 (2011)
18. Yang, Y.-G., Lei, H., Liu, Z.-C., Zhou, Y.-H., Shi, W.-M.: Arbitrated quantum signature scheme based on cluster states. Quantum Inf. Process. **15**(6), 2487–2497 (2016). https://doi.org/10.1007/s11128-016-1293-0
19. Zhang, L., Sun, H.-W., Zhang, K.-J., Jia, H.-Y.: An improved arbitrated quantum signature protocol based on the key-controlled chained CNOT encryption. Quantum Inf. Process. **16**(3), 1–15 (2017). https://doi.org/10.1007/s11128-017-1531-0
20. Yang, Y.-G., Liu, Z.-C., Li, J., Chen, X.-B., Zuo, H.-J., Zhou, Y.-H., Shi, W.-M.: Theoretically extensible quantum digital signature with starlike cluster states. Quantum Inf. Process. **16**(1), 1–15 (2016). https://doi.org/10.1007/s11128-016-1458-x

21. Feng, Y., Shi, R., Shi, J., Zhou, J., Guo, Y.: Arbitrated quantum signature scheme with quantum walk-based teleportation. Quantum Inf. Process. **18**(5), 1–21 (2019). https://doi.org/10.1007/s11128-019-2270-1
22. Li, D., Wang, R., Baagyere, E.: Quantum teleportation of an arbitrary two-qubit state by using two three-qubit GHZ states and the six-qubit entangled state. Quant. Inf. Process. **18**(5), 1–15 (2019). https://doi.org/10.1007/s11128-019-2252-3
23. Su, X.: Applying Gaussian quantum discord to quantum key distribution. Chin. Sci. Bull. **59**(11), 1083–1090 (2014). https://doi.org/10.1007/s11434-014-0193-x
24. Wang, L., et al.: Correction to: New scheme for measurement-device-independent quantum key distribution. Quantum Inf. Process. **18**(1), 12 (2019)
25. Yang, Y.-G., Liu, Z.-C., Chen, X.-B., Cao, W.-F., Zhou, Y.-H., Shi, W.-M.: Novel classical post-processing for quantum key distribution-based quantum private query. Quantum Inf. Process. **15**(9), 3833–3840 (2016). https://doi.org/10.1007/s11128-016-1367-z
26. Cabello, A.: Quantum key distribution in the Holevo limit. Phys. Rev. Lett. **85**, 5635 (2000)
27. Banerjee, A., Shukla, C., Thapliyal, K., Pathak, A., Panigrahi, P.K.: Asymmetric quantum dialogue in noisy environment. Quantum Inf. Process. **16**(2), 1–23 (2017). https://doi.org/10.1007/s11128-016-1508-4
28. Yang, Y.G., Sun, S.J., Zhao, Q.Q.: Trojan-horse attacks on quantum key distribution with classical bob. Quantum Inf. Process. **14**, 681 (2015)
29. Jouguet, P., Kunz-Jacques, S., Leverrier, A.: Long-distance continuous-variable quantum key distribution with a Gaussian modulation. Phys. Rev. A **84**(6), 062317 (2011)
30. Cai, H., Long, C.M., DeRose, C.T., Boynton, N., Urayama, J., Camacho, R., Pomerene, A., Starbuck, A.L., Trotter, D.C., Davids, P.S., Lentine, A.L.: Silicon photonic transceiver circuit for high-speed polarization-based discrete variable quantum key distribution. Opt. Express **25**(11), 12282–12294 (2017)

Quantum Key Distribution Protocol Based on GHZ Like State and Bell State

Ji-Zhong Wu(✉) and Lili Yan(✉)

Chengdu University of Information Technology, Chengdu 610255, Sichuan, China
564446870@qq.com, yanlili@cuit.edu.cn

Abstract. In this paper, we propose a multi-party quantum key distribution protocol using Bell state particles and GHZ like particles. The protocol adds Charlie between Alice and Bob. Alice and Bob can establish a secret key with Charlie at the same time. Through security analysis, Alice and Bob can't eavesdrop on each other's key. The three parties can not cheat each other. The protocol can also resist eavesdroppers. The advantage of this protocol is that two different keys can be established among three parties at the same time.

Keywords: Multi-party quantum key distribution protocol · Bell state · GHZ like particles

1 Introduction

Since the first quantum cryptography protocol was proposed by Bennett and brassard in 1984. Quantum cryptography attracts researchers' interest because of its unconditional security. At present, many quantum cryptography protocols like Quantum digital signature [1–3] Quantum secret sharing (QSS) [4–9] Quantum secure direct communication(QSDC) [5, 10–15] have been published. At the same time, quantum key distribution (QKD) [16–24] is also a very popular research direction.

The concept of quantum key distribution is that two parties can establish a key safely through the characteristics of quantum mechanics. The IBM company first realized the QKD protocol in 1989. It makes BB84 [17] protocol move from theory to practice. After that, a variety of QKD protocols with outstanding points in security and practicability were published. But the deviation of the experimental device will affect the security of the protocol. In view of the deviation of the device itself, Lo proposes the QKD protocol independent of the device. It not only avoids the eavesdropping caused by the imperfect equipment, but also improves the distance of secure communication. Various international scholars have put forward a variety of experimental schemes on this protocol [25–28].

Phoenix [29] proposed a more easily implemented QKD protocol named PBC00 in 2000. This protocol adds a third non orthogonal polarization state. Compared with B92 protocol [30], it is more secure and it can defend the eavesdropper more comprehensively. Four years later, Renes and others [31] proposed a protocol called R04. It is an improvement of PBC00 protocol. In the protocol, spherical encoding is used to enhance

X. Sun et al. (Eds.): ICAIS 2020, LNCS 12240, pp. 298–306, 2020.
https://doi.org/10.1007/978-3-030-57881-7_27

the anti eavesdropping ability and the utilization of key. In 2008, Boileau and others [32] proved that R04 protocol and PBC00 protocol really have unconditional security. In 2016, Schiavon [33] Proved that R04 protocol is feasible in experiments. Although its efficiency is lower than that of BB84 protocol, R04 also has implementation value because its implementation difficulty is lower than that of BB84 protocol.

Based on the QKD protocol, we proposes a protocol to establish two security keys among the three parties. In this agreement, third party Charlie is added between Alice and Bob. Allow Alice and Bob to establish a secret key with Charlie at the same time. At the same time, ensure that the key is not eavesdropped by any third party.

2 Multiparty Quantum Key Distribution Protocol

Different from the previous QKD protocol, there are three communication parties in my protocol. In order to realize the secret key distribution of three parties, a group of particles in the Bell state and a group of GHZ-like particles need to be prepared. At the same time, in order to ensure the security of this protocol, the receiver and the sender also need to conduct integrity detection, only to ensure that there is no possibility of deception between the two parties. We can say that the protocol is safe.

First of all, there are three communication parties: Alice, Bob and Charlie. They want to communicate with each other, but they can't know the communication key between the other parties. This is the purpose of this protocol. Therefore, in the three parties, Bob prepared a group of $N + K$ pair particles in the Bell state, while Alice prepared a group of particles in the GHZ-like [34–37] state at the same time. When an interactive key is needed, Bob sends a part of the particles in the Bell state to Alice. Alice uses the particle to perform a CNOT operation on a part of the particles in the GHz like state. Then Alice sends the particle for the CNOT operation to Charlie. Bob also sends the remaining particles to Charlie. Charlie compares the information of both sides, and then obtains two different secret keys. The specific implementation process is as follows:

Step1
Alice and Bob both want to establish a secure channel with Charlie. But their keys are not the same, and they can't eavesdrop on other people's key.

First of all, we need to define the measurement basis between Alice Bob and Charlie. Here we choose Z-basis and X-basis:

$$Z = \{|0\rangle, |1\rangle\}$$

$$X = \{|+\rangle, |-\rangle\}$$

After the measurement basis has been selected, Bob prepares a group of $N + K$ pair particles in the Bell state:

$$|\psi^+\rangle = \frac{1}{\sqrt{2}}(|01\rangle + |10\rangle)$$

Among them, the N pair particles in Bell state are used to generate the key, while the other K-pair particles in Bell state are used for checking eavesdropping. After the preparation, Bob will form the first particle of each group in P_1 and the second particle in P_2:

$$P_1 = \left\{x_1^1, x_1^2, \ldots\ldots, x_1^{n+k}\right\}$$

$$P_2 = \left\{x_2^1, x_2^2, \ldots\ldots, x_2^{n+k}\right\}$$

Step2

Bob randomly selects Z or X measurement basis for the particles in P_1 to measure. When the measurement is completed, Bob sends the second particle in each group to Alice and randomly adds K decoy particles in $|0\rangle$, $|1\rangle$, $|+\rangle$, $|-\rangle$ in P_2:

$$P_{2K} = \left\{x_2^1, x_2^2, \ldots\ldots, x_2^{n+2K}\right\}$$

Meanwhile, Alice prepares N-group particles in *GHZ*-like state:

$$|\psi\rangle = \frac{1}{2}\{|010\rangle + |100\rangle + |001\rangle + |111\rangle\}$$

Alice needs to analyze the channel security after receiving P_2 from Bob.

Step3

Bob did not publish the measurement basis after randomly selecting Z-basis or X-basis for measurement. When Alice receives P_2, she doesn't know the state of all particles because she doesn't know the measurement basis selected during Bob's preliminary measurement. It will cause the initial state of particles to be destroyed when Alice chooses the wrong measurement basis measuring particles. This causes some of the particles' states to be unrecoverable. This is an important part of the key security between Bob and Charlie. Alice needs to carry out monitoring detection first before she makes the measurement. Bob will tells Alice the location and measurement basis of the K measured particles. When Alice knows the test particles' positions, she measures the tested particles, and then publishes the test particle status:

$$P_k' = \{x_1, \ldots\ldots, x_k\}$$

Bob analyzes P_k'. For Bob, if there is no listener, the state of the particles in $|0\rangle$, $|1\rangle$, $|+\rangle$ and $|-\rangle$ in the test particle should remain the same. If there is a problem in the comparison result, it indicates that Alice is dishonest or there is a eavesdropper in the channel between Alice and Bob. At this time, communication fails. If there is no eavesdropper. Then Alice, who does not know the original measurement basis, measures N particles with Z-basis. After the measurement, P_2' is obtained. At this time, there are still K undetected particles.

Step4

After Bob passed the monitoring and detection, Alice performed a controlled non gate operation (CNOT) [38–40] on $|\psi\rangle_{345}$, taking the particle in P_2 as the control bit and the second particle in $|\psi\rangle_{345}$ as the target qubit. After the operation, Alice retained the third and fifth particle:

$$P_3 = \left\{x_3^1, x_3^2, \ldots\ldots, x_3^n\right\}$$

$$P_5 = \left\{x_5^1, x_5^2, \ldots\ldots, x_5^n\right\}$$

Send the fourth particle and K undetected particles to Charlie at the same time:

$$P_4' = \left\{x_4^1, x_4^2, \ldots\ldots, x_4^{n+k}\right\}$$

The definition of CNOT is to operate two qubits. The second qubit can only be not operated when the first qubit is $|1\rangle$, otherwise it will remain unchanged.

As shown in the figure below:

3-1 CNOT operation process 1

$$C|00>=\begin{pmatrix}1&0&0&0\\0&1&0&0\\0&0&0&1\\0&0&1&0\end{pmatrix}\left(\begin{pmatrix}1\\0\end{pmatrix}\otimes\begin{pmatrix}1\\0\end{pmatrix}\right)=|00>$$

3-2 CNOT operation process 2

$$C|00>=\begin{pmatrix}1&0&0&0\\0&1&0&0\\0&0&0&1\\0&0&1&0\end{pmatrix}\left(\begin{pmatrix}0\\1\end{pmatrix}\otimes\begin{pmatrix}1\\0\end{pmatrix}\right)=|11>$$

3-3 Final result of CNOT operation

GHZ-1 / P_2	001	010	100	111
0	001	010	100	111
1	011	000	110	101

Step5

Finally Bob sends P_1 to Charlie. After Bob sends P_1 to Charlie, he publishes the selected measurement basis. At this time, Charlie and Bob share the particles measured with X as the measurement basis. Bob and Charlie Think of $|+\rangle$ as 1 and $|-\rangle$ as 0. This group serves as the secret key between Bob and Charlie. Charlie and Alice share the particles measured with Z as the basis. Charlie decodes P_4 through P_1 and obtains P_4. Alice can also know which particles use z-based measurements. Alice and Charlie use the particles measured by Z-based particles for CNOT operation as the Key generation tool, and know the relationship between P_3P_5. Output 1 when P_3P_5 are the same, otherwise output 0. Charlie and Alice use the relationship between P_3P_5 as the secret key.

3-4 restore P_4'

P_4' \ P_1	0	1
0	1	0
1	0	1

According to the properties of *GHZ*-like states. The satus of GHZ-like states may be $|010\rangle$ or $|111\rangle$ When the initial state of P_4' is $|1\rangle$. Then P_3P_5 is $|00\rangle$ or $|11\rangle$. At this time, Alice and Charlie record that this bit key is 1, otherwise it is 0.

3 Protocol Analysis

3.1 Outside Eavesdropper

There are three Communication sides in this protocol. For this protocol, as long as it can ensure that there is no eavesdropper between the three interactions, and the three parties can not cheat each other during the communication. It can ensure that the key is safe.

First, the security analysis of three communication processes:

(1) Bob sends P_2 to Alice

When Bob sends P_2 to Alice, he will add K randomly detected particles in the state of $|0\rangle$, $|1\rangle$, $|+\rangle$, $|-\rangle$. At the same time, Alice did not know the measurement basis selected for Bob's preliminary measurement. After Alice determines to receive the particles, Bob will announce the position and state of the detected particles inserted by himself. Alice uses Z-based measurements for particles in the $|0\rangle$, $|1\rangle$ States, and X-based measurements for particles in the $|+\rangle$, $|-\rangle$ states. If there is no listener between Alice and Bob, the state of these K detection particles should be unchanged. That is, Alice's measurement results should be consistent with Bob's published results:

$$P_{2k} = \{x_1, \ldots\ldots, x_k\}$$

When Alice and Bob determine that the measurement results are consistent at the same time, the protocol can proceed to the next step. When the K value is larger, the probability that the listener or dishonest person will be found is higher. Because Bob randomly selects Z or X measurement basis for measurement, only *(1/2)K* probability of dishonest person can not be found.

(2) Bob sends P_1 to Charlie

After Bob sends *P1* to Charlie, Alice publishes k particles in $P2'$ that are not involved in the CNOT operation to serve as the particles for security monitoring. Bob then publishes his own measurement base. If at this time Charlie finds that the particles published in Alice's hands do not match the state of the particles he has obtained, then there is a listener between Alice and Charlie or Alice is a deceiver.

(3) Alice sends P'_4 to Charlie

Because the safety monitoring of 1 and 2 is established. So when we only talk about external monitors, P'_2 can guarantee its security even if it is not detected.

3.2 Inside Eavesdropper

The second is the safety inspection of three communication parties. That is to say, there are betrayers inside, only to ensure that the three parties can not eavesdrop each other can guarantee the security of the agreement. The following are necessary conditions to ensure internal Honesty:

(1) Bob's credibility:

If Bob listens for communication between Alice and Charlie. Because Bob knows the secret key used to encrypt P_4 between Alice and Charlie. So to ensure that Bob can't cheat, Bob needs to be unable to crack even if he gets this group of information. Because there is a group of particles in the state of $|0\rangle$, $|1\rangle$, $|+\rangle$, $|-\rangle$ in the particles sent by Alice to Charlie. If Bob intercepts the information of Alice and Charlie in the middle, the state of the particles in $|+\rangle$, $|-\rangle$ will be destroyed when Bob detects them. Bob will be exposed when Alice tells Charlie the state and position of the k-test particles.

(2) Alice's credibility:

The possibility of Alice cheating is that Alice listens for the key between Bob and Charlie. At this time, Alice makes z-based measurement on P_2, and then performs random CNOT gate operation on P'_4 to be sent to Charlie. Because Charlie didn't know the original state of P'_4. So even after obtaining Bob's information, it's impossible to identify whether Alice is cheating. So Charlie needs to ask Alice about the original state of K particles at random after getting P'_4. When Bob announced the measurement basis, Charlie screened the particles. Where P_1 is in the state of $|0\rangle$ and $|1\rangle$ should be able to directly analyze the original state of P'_4corresponding position. However, the particles originally in the $|+\rangle$, $|-\rangle$ state should be able to restore the original state of particles close to half of the corresponding position of P'_4. For the detection particle K between Alice and Charlie, if $|0\rangle$ in P_1 after restoration and the detection particle corresponding to the particle in $|1\rangle$ state cannot be restored, it means that Alice is dishonest and will end this communication.

3.3 Efficiency Analysis

Quantum efficiency formula:

$$\eta = \frac{r_c}{r_q}$$

r_c is the number of classic bits in the comparison. r_q is the number of qubits consumed. For the correspondents of Bob and Charlie. Bob prepared $N + K$ group Bell state particles and K detection particles ($2N + 3K$). The key is $N/2$ bits in total.

The quantum communication efficiency between Bob and Charlie is shown in the formula:

$$\eta_{BC} = \frac{N}{2(2N + 3K)} \quad (1)$$

During Alice's communication with Charlie. Alice prepared N-group GHZ-like states. There are *3N* particles for the secret key exchange between Charlie and Alice in total. Because half of the particles in P_2 that perform the CNOT operation are different from the measurement basis selected by Bob. So only half of P_2 can be used effectively. The secret key is determined according to the relationship between P_3P_5. And there are k particles in P_2 that are consumed as detection particles, that is, the number of particles as key is $(N - K)/2$.

The quantum communication efficiency between Alice and Charlie is shown in the formula:

$$\eta_{AC} = \frac{(N - K)}{6N} \quad (2)$$

4 Conclusion

In this paper, we propose a quantum key distribution protocol, which can construct two key channels between three parties through a group of Bell state particles and a group of GHz like states.

Through the above analysis, we can ensure the correctness and security of this protocol. Compared with the traditional QKD protocol, th [2–4] is protocol can establish multiple keys at one time, but the disadvantage is that it is not efficient and entangled state is not easy to achieve.

References

1. Lee, H., Hong, C., Kim, H., Lim, J., Yang, H.J.: Arbitrated quantum signature scheme with message recovery. Phys. Lett. A **321**(5–6), 295–300 (2004)
2. Li, Q., Chan, W.H., Long, D.: Arbitrated quantum signature scheme using Bell states. Phys. Rev. A **79**(5), 54307 (2009)
3. Yang, Y.G., Wen, Q.Y.: Circular threshold quantum secret sharing. Chin. Phys. B **17**(2), 419 (2008)
4. Hillery, M., Nek, V.B., Berthiaume, A.: Quantum secret sharing. Phys. Rev. A **59**(3), 1829 (1998)
5. Xiao, L., Long, G.L., Deng, F., Pan, J.: Efficient multi-party quantum secret sharing schemes. Physics **69**(5), 521–524 (2004)
6. Tittel, W., Zbinden, H., Gisin, N.: Experimental demonstration of quantum secret sharing. Phys. Rev. A **63**(4), 42301 (2001)
7. Gottesman D.: On the theory of quantum secret sharing. Phys. Rev. A, **61**(4) (1999)
8. Deng, F.G., Long, G.L., Zhou, H.Y.: An efficient quantum secret sharing scheme with Einstein-Podolsky-Rosen pairs. Phys. Lett. A **340**(1–4), 43–50 (2005)

9. Guo, G.P., Guo, G.C.: Quantum secret sharing without entanglement. Phys. Lett. A **310**(4), 247–251 (2002)
10. Boström, K., Felbinger, T.: Deterministic secure direct communication using entanglement. Phys. Rev. Lett. **89**(18), 187902 (2002)
11. Deng, F.G., Long, G.L., Liu, X.: Two-step quantum direct communication protocol using the Einstein-Podolsky-Rosen pair block. Phys. Rev. A **68**(4), 42317 (2003)
12. Wang, J., Zhang, Q., Tang, C.: Quantum secure direct communication based on or der rearrangement of single photons. Phys. Lett. A **358**(4), 256–258 (2006)
13. Li, X.H.: Quantum secure direct communication. Acta Phys. Sin-Ch Ed, **64**(16) (2015)
14. Jie, S., Ai-Dong, Z., Shou, Z.: Quantum secure direct communication protocol with blind polarization bases and particles' transmitting order. Chin. Phys. **16**(3), 621–623 (2007)
15. Gao, F., Qin, S., Wen, Q., Zhu, F.: Cryptanalysis of multiparty controlled quantum secure direct communication using Greenberger–Horne–Zeilinger state. Opt. Commun. **283**(1), 192–195 (2010)
16. Bennett, C.H., Brassard, G.: Quantum cryptography: public key distribution and coin tossing. Theor. Comput. Sci. **560**, 7–11 (2014)
17. Bennett, C.H., Brassard, G., Mermin, N.D.: Quantum cryptography without Bell's theorem. Phys. Rev. Lett. **68**(5), 557 (1992)
18. Ekert, K.A.: Quantum cryptography based on Bell's theorem. Phys. Rev. Lett. **67**(6), 661–663 (1991)
19. Lütkenhaus, N.: Security against individual attacks for realistic quantum key distribution. Phys. Rev. A **61**(5), 52304 (2000)
20. Boyer, M., Ran, G., Dan, K., Mor, T.: Semiquantum key distribution. Phys. Rev. A **79**(3), 032341 (2009)
21. Shor, P.W., Preskill, J.: Simple proof of security of the BB84 quantum key distribution protocol. Phys. Rev. Lett. **85**(2), 441–444 (2000)
22. Scarani, V., Bechmann-Pasquinucci, H., Cerf, N., Du Ek, M., Lütkenhaus, N., Peev, M.: The security of practical quantum key distribution. Rev. Mod. Phys. **81**(3), 1301–1350 (2009)
23. Hoi-Kwong, L.O.: Decoy state quantum key distribution. Phys. Rev. Lett. **03**(1), 143 (2008)
24. Mi, J.L., Wang, F.Q.: Practical non-orthogonal decoy state quantum key distribution with heralded single photon source. Chin. Phys. B **17**(4), 1178–1183 (2008)
25. Hughes, R., Morgan, G., Glenpeterson, C.: Quantum key distribution over a 48 km optical fibre network. Opt. Acta Int. J. Opt. **47**(2–3), 533–547 (2000)
26. Ma, X., Razavi, M.: Alternative schemes for measurement-device-independent quantum key distribution. Phys. Rev. A **86**(6), 3818–3821 (2012)
27. Jouguet, P., Kunz-Jacques, S., Leverrier, A.: Long distance continuous-variable quantum key distribution with a gaussian modulation. Phys. Rev. A **84**(6), 062317 (2011)
28. Zhu, F., Wang, Q.: Quantum key distribution protocol based on heralded single photon source. Acta Opt. Sinica **34**(6), 0627002 (2014)
29. Phoenix, S.J.D., Barnett, S.M., Chefles, A.: Three-state quantum cryptography. J. Mod. Opt. **47**(2–3), 507–516 (2000)
30. Zhang, Q., Tang, C.J., Zhang, S.Q.: Modification of B92 protocol and the proof of its unconditional security. Acta Phys. Sinica –Chin. Edn.- **51**(7), 1446–1447 (2002)
31. Renes, M.J.: Spherical-code key-distribution protocols for qubits. Phys. Rev. A **70**(5), 52314 (2004)
32. Boileau, J.C., Tamaki, K., Batuwantudawe, J., Laflamme, R., Renes, J.M.: Unconditional security of three state quantum key distribution protocols. Phys. Rev. Lett. **94**(4), 40503 (2004)
33. Schiavon, M., Vallone, G., Villoresi, P.: Experimental realization of equiangular three-state quantum key distribution (2016)

34. Dai, H.Y., Chen, P., Li, C.: Probabilistic teleportation of an arbitrary two-particle state by a partially entangled three-particle GHZ state and W state. Opt. Commun. **231**(1–6), 281–287 (2004)
35. Dong, L., Xiu, X., Gao, Y., Ren, Y., Liu, H.: Controlled three-party communication using GHZ-like state and imperfect Bell-state measurement. Opt. Commun. **284**(3), 905–908 (2011)
36. Tsai, C.W., Hwang, T.: Teleportation of a Pure EPR state via GHZ-like state. Int. J. Theor. Phys. **49**(8), 1969–1975 (2010)
37. Hassanpour, S., Houshmand, M.: Efficient controlled quantum secure direct com munication based on GHZ-like states. Quantum Inf. Process. **14**(2), 739–753 (2015)
38. O'Brien, J.L., Pryde, G.J., White, A.G., Ralph, T.C., Branning, D.: Demonstration of an all-optical quantum controlled-NOT gate. Nature **426**(6964), 264–267 (2003)
39. Huang, Y.F., Ren, X., Zhang, Y., Duan, L., Guo, G.: Experimental teleportation of a quantum controlled-NOT gate. Phys. Rev. Lett. **93**(24), 240501 (2004)
40. Wang, H.F., Zhu, A.D., Zhang, S., Yeon, K.H., Wen, J.J.: Deterministic CNOT gate and entanglement swapping for photonic qubits using a quantum-dot spin in a double-sided optical microcavity. Phys. Lett. A **377**(40), 2870–2876 (2013)
41. Xiao, D., Liang, J., Ma, Q., Xiang, Y., Zhang, Y.: High capacity data hiding in encrypted image based on compressive sensing for nonequivalent resources. Comput. Mater. Continua **58**(1), 1–13 (2019)
42. Luo, M., Wang, K., Cai, Z., Liu, A., Li, Y., Cheang, C.F.: Using imbalanced triangle synthetic data for machine learning anomaly detection. Comput. Materi. & Continua **58**(1), 15–26 (2019)
43. Xia, Z., Lu, L., Qiu, T., Shim, H.J., Chen, X., Jeon, B.: A privacy-preserving image retrieval based on ac-coefficients and color histograms in cloud environment. Comput. Mater. Continua **58**(1), 27–43 (2019)

Design and Implementation of Heterogeneous Identity Alliance Risk Assessment System

Jianchao Gan, Zhiwei Sheng(✉), Shibin Zhang, and Yang Zhao

School of Cybersecurity, Chengdu University of Information Technology, Chengdu 610225, China
997725846@qq.com

Abstract. Heterogeneous identity alliance consists of multiple identity management platforms cross architectures and applications, provides unified identity management and services. This paper focuses on the heterogeneous alliance risk assessment framework, The network behavior and identity attributes of users in the alliance system are chosen as the main evaluated factors. Combined with the information security risk assessment, the risk assessment system of the alliance system is designed and implemented by Web. Finally, The development of the system shows the results of the risk assessment in a visual way, which assistants the regulators to visually find out the dangerous factors of the alliance system and take corresponding measures.

Keywords: Heterogeneous identity alliance · Risk assessment · Visualization

1 Related Work

1.1 Information Security Risk Assessment Methods

In modern world, information systems are becoming more and more complex, meanwhile, there are increasingly security risks. How to accurately conduct risk assessment is a hot topic. In the risk assessment process, how to choose the appropriate assessment method is critical to the correctness of the assessment results. Various risk assessment methods have emerged nowadays [1–3]. To sum up, it can be divided into three categories: qualitative assessment methods, quantitative assessment methods, and comprehensive assessment methods combining qualitative and quantitative [4].

Qualitative assessment method:

The qualitative assessment method is mainly a method that evaluators utilize knowledge and experience. It needs high requirements for evaluators, A risk assessment expert will be better. The method is more subjective, and the evaluator pays more attention to the consequences of the risk, while ignoring the probability of the event happens. In the process of qualitative evaluation, there are generally many factors that cannot be quantified, so the evaluated results also have uncertainty. This method is widely used in incomplete data collection. Common qualitative analysis methods include logic analysis, Delphi method, historical comparison method and factor analysis method [5].

X. Sun et al. (Eds.): ICAIS 2020, LNCS 12240, pp. 307–317, 2020.
https://doi.org/10.1007/978-3-030-57881-7_28

The quantitative assessment method:

The quantitative assessment method is an evaluation method based on the mathematical indicator system. First, mathematical modeling is performed on the collected statistics, and the values obtained by the modeling are evaluated. Common quantitative evaluation methods include cluster analysis, time series model, regression model, factor analysis method, decision tree method, etc. [6]. Its advantage is to use the intuitive data to present the evaluation results, making the evaluation results more scientific and rigorous. However, the quantitative evaluation method may lead the evaluator simplify the complicated data, so that the data becomes ambiguous and the risk factor confirmation has certain difficulties to dig out.

Comprehensive assessment method combining qualitative and quantitative:

Qualitative assessment methods are subjective and short of data support, while the quantitative assessments are lack of subjectivity. Not all factors can be quantified in complex systems, forcing quantification can lead to inaccurate evaluation results. Therefore, the combination of the two can improve the accuracy of risk assessment. Analytic Hierarchy Process [7] (AHP) is a multi-objective decision analysis method combining quantitative and qualitative [8], which was proposed by T.L in 1970s. The idea is to quantify the experience judgment of decision makers, which provides decision makers with a quantitative basis for decision making [9].

1.2 Steps on Information Security Risk Assessment

Information security risk assessment is a systematic analysis towards threats and vulnerabilities of networks and information systems from the perspective of risk management, using scientific methods and means to assess the hazards and losses that may occur in the event of a security incident [10]. In order to prevent and reduce information security risks, corresponding defense measures are formulated to ensure the security of information system to a large extent [11]. The information security risk assessment process is shown in Fig. 1.

The specific process mainly includes the following steps:

1) Risk assessment preparation activities:
   The preparatory activities mainly include the determination of risk assessment objectives, risk assessment scopes, establishment of risk assessment management team and assessment implemented team, systematic research, determination assessment basis, and obtaining the support of top management for risk assessment [12].
2) Asset, threat, vulnerability identification:
   Asset identification refers to the measurement and evaluation of the confidentiality, integrity and availability of assets. In practical, assets are usually divided into data, software, hardware, documents, services, personnel and others. In the assignment of assets, we can not only consider the economic value of assets, but also attention the security of assets. According to the three indicators, the corresponding marks—very low, low, medium, high and very high, and the final comprehensive evaluated score is 1–5 (points) [13].
   Asset identification refers to the measurement and evaluation of the confidentiality, integrity and availability of assets. In practical evaluation, assets are usually divided

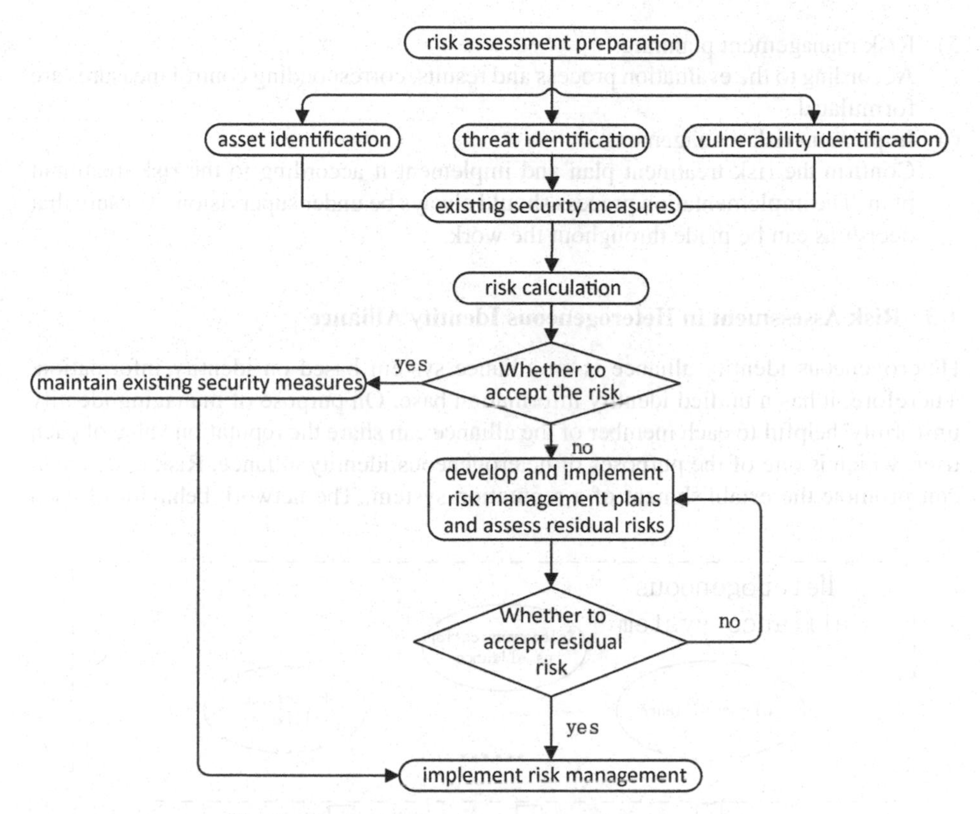

Fig. 1. Steps on information security risk assessment

into data, software, hardware, documents, services, personnel and others. In the assignment of assets, we can not only consider the economic value of assets, but also the security of assets. According to the three indicators, the corresponding marks—very low, low, medium, high and very high, and the final comprehensive evaluated score is 1–5 (points).

Vulnerability identification is the most vital part of risk assessment. Generally, vulnerability identification is based on assets, using tools detection, manual audit, penetration test and other methods to identify.

3) Confirmation of existing safety measures:
   At the same time of vulnerability identification, the existing security measures should be tested, and the use of security measures will reduce the vulnerability of information system.
4) Risk analysis and calculation:
   Risk is defined as the possibility of security incidents caused by threat using vulnerability. After the asset identification, threat identification, vulnerability identification and existing security measures are confirmed, appropriate methods are used for risk analysis and calculation. The commonly used methods are matrix method and multiplication, and finally the risk level is obtained.

5) Risk management planning:
   According to the evaluation process and results, corresponding control measures are formulated.
6) Implement risk management:
   Confirm the risk treatment plan and implement it according to the risk treatment plan. The implementation process should always be under supervision to ensure that decisions can be made throughout the work.

1.3 Risk Assessment in Heterogeneous Identity Alliance

Heterogeneous identity alliance is an alliance system based on identity information. Therefore, it has a unified identity information base. On purpose of managing identity uniformly. helpful to each member of the alliance can share the reputation value of each user, which is one of the purposes of heterogeneous identity alliance. Risk assessment can promote the establishment of a reputation system, The network behavior of each

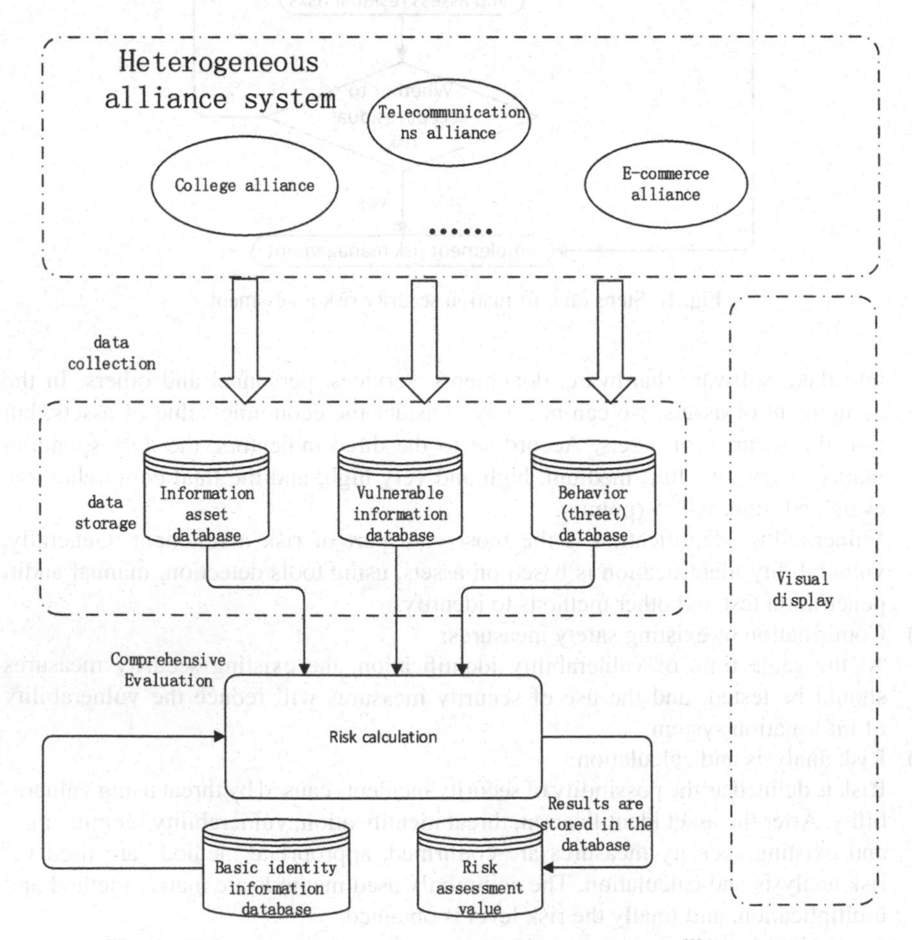

Fig. 2. Risk assessment structural graph of heterogeneous alliance system

user in the alliance is an important factor in risk assessment. Therefore, we replace the traditional threat identification with the user network behavior identification and scoring. Taking the network behavior and identity attribute of users in the alliance system as the main evaluation factors, combined with information security risk assessment, the risk assessment structure chart of heterogeneous identity alliance is designed. Its structure is shown in Fig. 2.

The first layer is the evaluated object, including the whole heterogeneous alliance system, the evaluation data are from the heterogeneous alliance system. Data collection objects include alliance system information assets, brittleness of information assets, and network behavior data of federated users (threat data). On the basis of this data, we carry out asset valuation, vulnerability valuation, behavior (threat) scoring and combined with the comprehensive assessment of basic identity information base to calculate the behavior risk value of each user and the risk value of the alliance to be stored in the database. From the data collection and storage to the statistical analysis of risk value, in this paper, we accomplish the visualization display and operation, and improve the efficiency of evaluation.

2 Systematic Requirements and Design

2.1 Requirement Design

Risk assessment includes three elements: asset, vulnerability and threat [14], so it is necessary to assess the risk of heterogeneous identity alliance system. The system needs to manage these parts visually, and then carry out relevant risk assessment and visual display [15]. It mainly includes the following requirements:

1) Visual management of assets.
2) Visual management of vulnerability.
3) Visual management of threats (behaviors).
4) Comprehensive risk calculation.
   In accordance with user behavior, the threat and the vulnerability of the system, it evaluates the risk value of a single user to the system. According to the risk value of each alliance member, a total risk value is calculated.
5) Risk distribution statistic.
   Risk distribution statistic includes risk statistics of each alliance member, regional risk distribution of alliance users and regional statistics of alliance users. Among them, the risk statistics of alliance members can be visualized by dashboard, which is convenient for managers to understand the risks of each alliance system intuitively; the risk distribution of alliance user areas can be visualized by map thermal map, managers can directly see the risk distribution of users all over the country; the risk level statistics of behavior can be visualized by line chart.

2.2 System Framework Design

The overall structure of risk assessment visualization system is shown in Fig. 3.

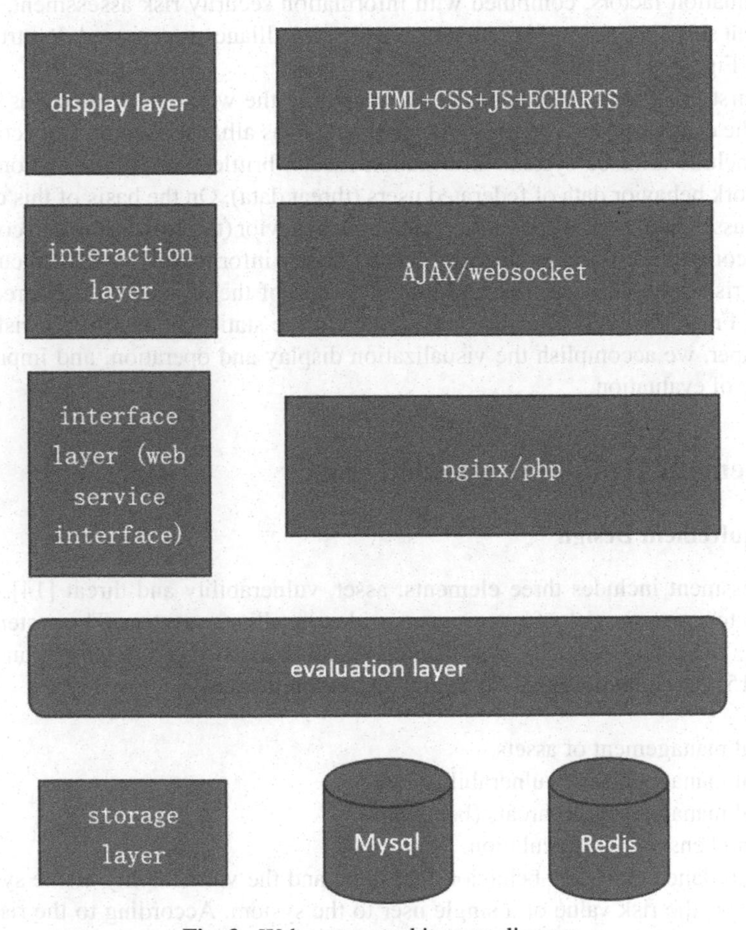

Fig. 3. Web system architecture diagram

1. Display layer

The display layer control the visualization. It is proposed to use the Web to display. The web mode has low requirements for hardware equipment. Only the client computer needs to be installed with a browser, and no client software needs to be installed. With the help of the mature visualization tool ECHARTS, web pages can be developed quickly and efficiently [16–18].

2. Interaction layer

The interaction layer is a module which is responsible for the interaction between users and the system. Ajax asynchronous loading mode is proposed to reduce the pressure of server request. When user needs data, trigger the interaction layer to request the data by clicking the corresponding display layer or other ways. The output data is, then, handed over to the display layer for showing. It avoids the repeated refresh of the page and enhances embodiment quality. If function needs to get real-time data, we can use WebSocket—long connection to get. The server has new data that can be pushed to the browser without repeated Ajax queries.

3. Interface layer

The interface layer is responsible for all requests and data return. It is proposed to adopt the restful api style to encapsulate all functional interface modules into a unified format and return json data format.

4. Evaluation layer

Assessment layer is the core part of the system, and the engine module of risk assessment, one of the core sources of risk statistical analysis data. It is responsible for system risk assessment and user behavior assessment. It is a layer of real-time and independent operation. It calculates the corresponding risk value in real-time or regularly according to the change of user behavior and updates and stores the corresponding data.

5. Storage layer

The storage layer manages storing all the original data and risk assessment data, including user information, alliance member information, alliance information, system asset information, user behavior information, etc., as well as the data calculated by risk assessment, including personal risk value, alliance risk value, etc. It is the basis of visualization. We plan to use MySQL database to store data, which can be combined with Redis to store high-frequency access data.

3 Systematic Research and Practice

3.1 Permission Module

In the process of risk assessment, the work of different stages needs different assessors, and different member units of the alliance have independent data management. In order to ensure the information security of the system, we use access control technology to separate different modules.

In practice, we use RBAC (Role Based Access Control) model to achieve access control. Combining functions into roles and assigning roles to users means that roles are a collection of functions. Its advantage is to decouple users and functions, reduce operation error rate, and reduce the tedious degree of function permission assignment.

3.2 Asset Visualization Module

The asset visualization module includes the input, evaluation (assignment), deletion and other functions of assets. The inputs of assets are based on each alliance member. Each asset is assigned according to confidentiality requirements, integrity requirements and availability requirements. Assignment is defined as high security requirement = 5, high security requirement = 4, medium security requirement = 3, low security requirement = 2, low security requirement = 1.

3.3 Vulnerability Assessment Module

Vulnerability assessment module includes vulnerability classification management and vulnerability assignment. Vulnerability classification management is to create different vulnerability classifications for each alliance unit. Some common information system threat classifications are built in the system for using. After the classification is created, the work of threat identification is recorded in the system, and assigned according to the asset and vulnerability.

3.4 Behavior (Threat) Assessment Module

User behavior is the operation trace left by users in the information system, including normal behavior and some abnormal behaviors. We need to evaluate (score) these behaviors for the convenience of subsequent risk calculation. The evaluation is conducted by expert scoring and rule matching. For some behaviors that conform to the rules, such as XSS and SQL injection behavior system, direct scoring is required. For some behaviors that cannot be directly determined by the system, artificial scoring is required, and corresponding system vulnerabilities (threats) need to be found. This behavior risk value is used as the system threat value to evaluate the system risk. At the same time, the behavior risk value of each user can be calculated synthetically.

3.5 Risk Calculation Module

According to the above system assets, vulnerability of system assets, user behavior and user basic information base, the risk can be calculated as follows:

1. Calculation of risk value of individual behavior

The risk value of a single behavior is the difference between the threat value and the vulnerability value of the behavior, so that the risk value of the behavior is r_b, behavior threat value is r_t, behavior vulnerability value is r_v, the following formula can be obtained:

$$r_b = r_t - r_v$$

2. User static attribute trusted value calculation

User static attribute trusted value calculation method:

Step 1: check whether the user fills in the required fields (name, gender, age, ID card, etc.). If it is not filled in, the static attribute directly sets the trusted value to 0. If it is filled in, the initialization attribute is 5. Continue with the following steps.
Step 2: for each additional attribute, the trust value will increase by 0.5.
Step 3: add the initial trust value and the score of attribute increase to get the users static attribute trust value, with a full score of 10.

For example (all identity properties are not listed in the Table 1):

Table 1. Trusted value calculation example table

Name	Gender	Age	ID card	Household registration	Address	Score
Bom	Male	20		XXXX	XXXX	0
Abel	Male	22	XXXX	XXXX	XXXX	6
Baron	Female	23	XXXX		XXXX	5.5

(1) calculation of individual comprehensive risk value

The calculation of individual comprehensive risk value is composed of user static attribute credible weighted value and behavior average risk weighted value. The behavior average risk value is obtained by dividing the sum of all individual risk values by the total number of risk behaviors, so that the individual comprehensive risk value isr_p, user static attribute trusted value is t_p, the sum of individual risk behaviors is r_c, the total number of personal risk behaviors is〖r〗_n, the following formula can be obtained:

$$r_p = 0.4 * t_p + 0.6 * \left(\frac{r_c}{r_n}\right)$$

(2) calculation of individuals risk value for an alliance

The risk value of an individual to an alliance is composed of the s static attribute trust weighted value and the average risk weighted value of the alliance behavior. The average risk value of the alliance behavior is obtained by dividing the sum of all the risk behavior values of the individual in the alliance by the total number of the risk behavior within the alliance energy, so that the risk value within the individual alliance is〖r〗_ap, user static attribute trusted value is t_p, the sum of risk behaviors in individual alliance is〖r〗_ac, the total number of risk behaviors within individual alliance is〖r〗_an, the following formula can be obtained:

$$r_{ap} = 0.4 * t_p + 0.6 * \left(\frac{r_{ac}}{r_{an}}\right)$$

3.6 Risk Statistics Analytic Module

According to the calculated risk data, risk statistics can be carried out. In this module, Charts technology is used to visually display the data of various risk results and factors, As shown in Fig. 4, including the following six parts:

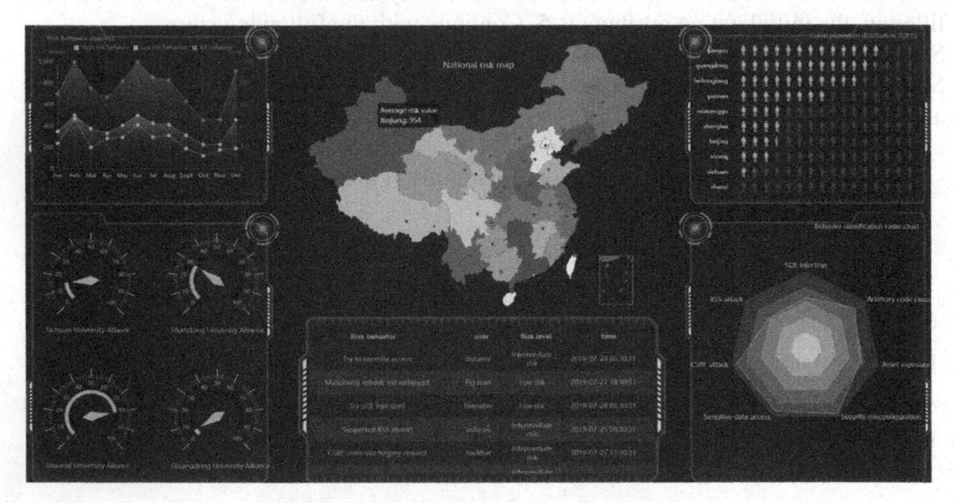

Fig. 4. Visualization of risk statistics

1) behavior risk is classified into high-risk behavior and low-risk behavior according to the time and classification statistics, and the occurrence times of various behaviors are calculated according to the time gradient.
2) The risk value of each alliance member system is displayed in the form of a dashboard
3) The average risk value is calculated according to the group of alliance member provinces and the risk distribution is displayed in the way of map.
4) Scroll through user behavior data in a list.
5) Top 10 provinces with the largest number of alliance users.
6) Display behavior radar map according to user behavior classification.

Its purpose is to enable regulators to more intuitively detect user risk value and related risk factors, and then, make corresponding decisions.

4 Conclusion

In the era of complex network environment, the unified identity authentication management becomes more popular, the framework of heterogeneous identity alliance can be applied for all kinds of systems. The risk assessment system of heterogeneous identity alliance designed can effectively realize the risk assessment of user behavior and alliance systems, and realize the visual display, which is conductive to promote the reputation of heterogeneous alliance establishment of the department.

Acknowledgments. The authors would like to thank the reviewers for their detailed reviews and constructive comments, which have helped improve the quality of this paper. This work is supported by the National Key Research and Development Project of China (No. 2017YFB0802302), the Key Research and Development Project of Sichuan Province (No. 20ZDYF2324, No. 2019ZYD027, No. 2018TJPT0012), the Science and Technology Support Project of Sichuan Province (No. 2018GZ0204, No. 2016FZ0112), the Science and Technology Project of Chengdu (No. 2017-RK00-00103-ZF).

References

1. Liu, J., Xu, C.: Comparative analysis of information security risk assessment methods. J. Central Univ. Natl. (Nat. Sci.) **21**(02), 91–96 (2012)
2. Rodríguez, A., Ortega, F., Concepción, R.: A method for the evaluation of risk in IT projects. Expert Syst. Appl. **45**, 273–285 (2016)
3. Shameli-Sendi, A., Aghababaei-Barzegar, R., Cheriet, M.: Taxonomy of information security risk assessment (ISRA). Comput. Secur. **57**, 14–30 (2016)
4. Feng, D., Zhang, Y., Zhang, Y.: Overview of information security risk assessment. Trans. Commun. **25**(7), 10–18 (2004)
5. Zhang, L., Peng, J., Du, Y., Wang, Q.: A review of comprehensive evaluation methods for information security risk assessment. J. Tsinghua Univ. (Sci. Technol.) **52**(10), 1364–1369 (2012)
6. Zhao, Y., Zhang, S., Yang, M., He, P., Wang, Q.: Research on architecture of risk assessment system based on block chain. Comput. Mater. Continua **61**(2), 677–686 (2019)
7. Satty, T.L.: The Analytic Hierarchy Process. McGraw-Hill, New York (1980)
8. Zhang, K.: Research on information security risk assessment based on AHP and BP. Hebei University of Engineering (2016)
9. Fan, G., Zhong, D., Yan, F., Yue, P.: A hybrid fuzzy evaluation method for curtain grouting efficiency assessment based on an AHP method extended by D numbers. Expert Syst. Appl. **44**, 289–303 (2016)
10. Chi-Qian, J., Dong-Qin, F.: Security assessment for industrial control systems based on fuzzy analytic hierarchy process. J. Zhejiang Univ. (Eng. Sci.) **50**(4), 759–765 (2016)
11. U. S. Government Accountability Office. Information Security Risk Assessment: Practices of Leading Organizations, pp. 1086–1089 (1999)
12. Lin, Y., Chen, D., Peng, Y., Wang, H.: Information security risk assessment of power plant industrial control system based on DS evidence theory. J. East China Univ. Sci. Technol. **40**, 505
13. Wang, Y.: Design and implementation of data center information security risk assessment system. Hunan University (2015)
14. Huang, G.: Design and implementation of remote information security risk assessment system. Shandong University (2018)
15. Liu, F.: Information visualization technology and application research. Zhejiang University (2013)
16. Chang, Y.: Visualization and analysis of web data in the background of big data. Inf. Syst. Eng. (01), 148+150 (2018)
17. Shen, T., Nagai, Y., Gao, C.: Improve computer visualization of architecture based on the bayesian network. Comput. Mater. Continua **58**(2), 307–318 (2019)
18. Han, S.W., Seo, J.Y., Kim, D.-Y., Kim, S.H., Lee, H.M.: Development of cloud based air pollution information system using visualization. Comput. Mater. Continua **59**(3), 697–711 (2019)

A Blockchain Based Distributed Storage System for Knowledge Graph Security

Yichuan Wang(✉), Xinyue Yin(✉), He Zhu(✉), and Xinhong Hei(✉)

Xi'an University of Technology, Xi'an, China
chuan@xaut.edu.cn, 767495461@qq.com, 745394370@qq.com, 59277100@qq.com

Abstract. With the rapid rise of artificial intelligence, knowledge graph has also been widely concerned, and the security of knowledge graph has also followed. In the process of knowledge graph construction, process files are easy to be tampered and forged by malicious, so that the reasoning results are biased. The purpose of this experiment is to explore the availability and effectiveness of knowledge graph based on blockchain and distributed system storage and traceability. This paper proposes a scheme of process file storage and traceability of knowledge graph based on blockchain and distributed file storage system. Firstly, the process files of building knowledge graph are preprocessed by distributed file system to improve the storage efficiency and save resources; then the processed files are stored in the blockchain network to ensure the security and integrity of the files.

Keywords: Blockchain · Distributed file system · Knowledge graph · Decentralized

1 Introduction

With the development and application of artificial intelligence technology, knowledge graph has become an important branch of artificial intelligence, and knowledge engineering has been successfully applied in big data environment [1]. Knowledge graph is a knowledge base that represents concepts and entities in the objective world and their relationships in the form of graph. Building knowledge graph is an iterative process, which includes three stages, such as information extraction, knowledge fusion and knowledge processing [2]. Because the key application of knowledge graph lies in the query and reasoning of knowledge, the modeling stage of knowledge graph becomes particularly important. If the following problems occur in the modeling process, it will lead to wrong reasoning results. For example, before the knowledge update, the process file of building the graph is illegally tampered with or lost, which makes the security of the file not guaranteed.

In this paper, a solution is proposed to store the process file of building the knowledge graph into the blockchain to ensure the security and traceability of the process file. In the case of error in reasoning or security problems in files, process files of a certain period can be obtained from the chain on demand to avoid repetitive work such as rebuilding the graph, which effectively solves the above problems.

X. Sun et al. (Eds.): ICAIS 2020, LNCS 12240, pp. 318–327, 2020.
https://doi.org/10.1007/978-3-030-57881-7_29

2 Background

2.1 Knowledge Graph

Knowledge graph, a structured semantic knowledge base, is used to rapidly describe the concepts and their relationships in the physical world. By reducing the data granularity from the document level to the data level and aggregating a large number of knowledge, it can achieve rapid response and reasoning of knowledge. The basic unit of knowledge graph is the triple of "Entity-Relationship-Entity", which is also the core of knowledge graph. Building knowledge graph is an iterative updating process. According to the logic of knowledge acquisition, each iteration includes three stages [3]: (a) Information extraction: Extract entities, attributes and relationships among entities from various types of data sources, and then form ontology knowledge representation; (b) Knowledge fusion: After acquiring new knowledge, it needs to be integrated to eliminate contradictions and ambiguities, for example, some entities may have multiple expressions, a specific title may correspond to multiple different entities, etc.; (c) Knowledge processing: For new knowledge after integration, only after quality assessment can qualified parts be added to the knowledge base to ensure the quality of the knowledge base. Because of this iterative process, once is tampered by malicious people in the process of updating, it will directly affect the output of the knowledge graph. Therefore, we should protect the knowledge graph from tampering and forgery, which is the original intention of this article.

2.2 Blockchain

With the rapidly development of information and communication technology, the Internet has penetrated into almost all aspects of people's life, such as production, trade, communication, learning, entertainment and so on, which has made revolutionary changes in many fields [4]. Blockchain is the underlying technology of Bitcoin. It was proposed by Satoshi Nakamoto in Bitcoin: A peer-to-peer electronic cash system as a decentralized tamper-proof and trust-free peer-to-peer trading system [5]. Although Bitcoin, a digital currency, has received different reviews, as the core technology of Bitcoin, blockchain technology has received increasing attention. Its characteristics can be explained as follows [6]: (1) Decentralization, which means not focusing on any party involved in the transaction. Decentralization can increase efficiency and reduce cost. (2) Trust. Assuming that any node participating in the transaction is untrustworthy, and force all parties to abide by the integrity by recording the non repudiation transaction information. Therefore, blockchain has solved the problem of integrity, the pain point of the current Internet. (3) Distributed structure. Blockchain can be regarded as a kind of distributed database. Its core lies in how to automatically form a consensus mechanism in a distributed, untrusted node environment [7].

Figure 1 shows a blockchain network, it is consisted of two components which include blockchain nodes and blockchain database. The blockchain database is a distributed database and storage the data as a block. Each blockchain node can only read the information. The blocks are connected together as a blockchain due to the hash value. Each block has an unique hash value and the value will be recorded in its previous block.

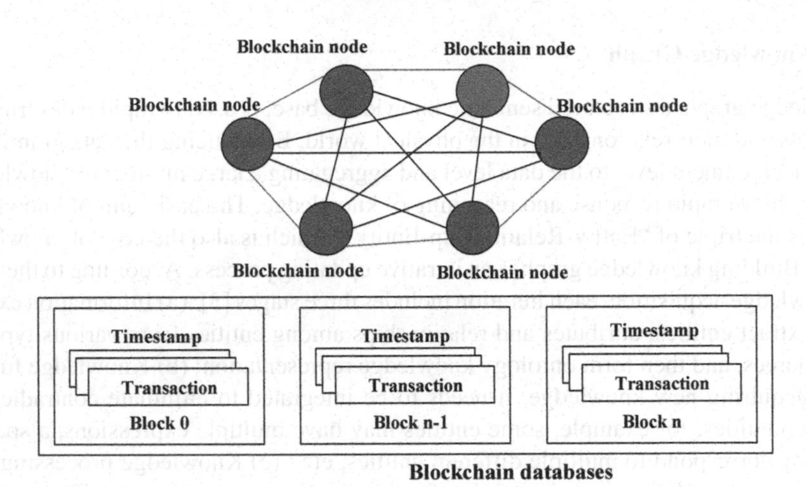

Fig. 1. Blockchain network architecture

Digital Signature. Suppose A sends an encrypted message to B using B's public key [8]. A declares that it is a in the message, because anyone can send encrypted message to B using B's public key, so B can't verify whether the message comes from A. To verify the sender of the message, you can use RSA to sign the message. For example, A sends A message with its own private key signature to B. First, the message needs to be hashed. Assuming the hash value is set to h, the signature message can be expressed as

$$c = h^d \ mod \ n \tag{1}$$

And attach it as a signature to the sent message. When B receives a signed message, use the formula:

$$c^e \ mod \ n = h^{e \times d} \ mod \ n \tag{2}$$

Perform signature verification. Compare the validated value with the hash value in the message. If the two are the same, B can confirm that the sender has the private key of A, and the message has not been changed during sending.

Elliptic Curve Encryption Algorithm (**ECC**)**.** ECC is a public key encryption algorithm based on the algebraic operation of elliptic curves over infinite fields [9]. Compared with RSA algorithm, ECC has higher security, faster processing speed, less space, and more difficult to crack secret key. ECC is used in transaction transmission among peers is Fabric1.0.0. This encryption principle as follows:

G is a point on the elliptic curve, *k* is the private key, *K* is the public key, then:

$$K = kG \tag{3}$$

Select a random number *r* and generate ciphertext from message *M*, and the public key encryption algorithm is shown in (4):

$$C = \{rG, M + rK\} \tag{4}$$

The algorithm of private key decryption is shown in formula (5):

$$M + rK - k(rG) = M + r(kG) - k(rG) = M \tag{5}$$

2.3 Distributed File System

Distributed file system is a file system fixed in a certain location, extended to any number of locations, multiple file systems, many nodes constitute a file system network, each node can be distributed in different locations, through the network for communication and data transmission between nodes, its technical architecture is mainly based on model client/server [10, 11]. At present, the storage system using blockchain technology mainly refers to the point-to-point distributed file system, which adopts the relevant blockchain technology in the incentive layer and can provide users with a storage system similar to cloud. The blockchain storage systems that have been formally used or are being developed mainly include Storj, Sia, IPFS and Maidsafe [12, 13].

3 Experiment Architecture and Implement

3.1 System Architecture

In this system, we have four components which is blockchain network, IPFS core, knowledge graph and database (Fig. 2).

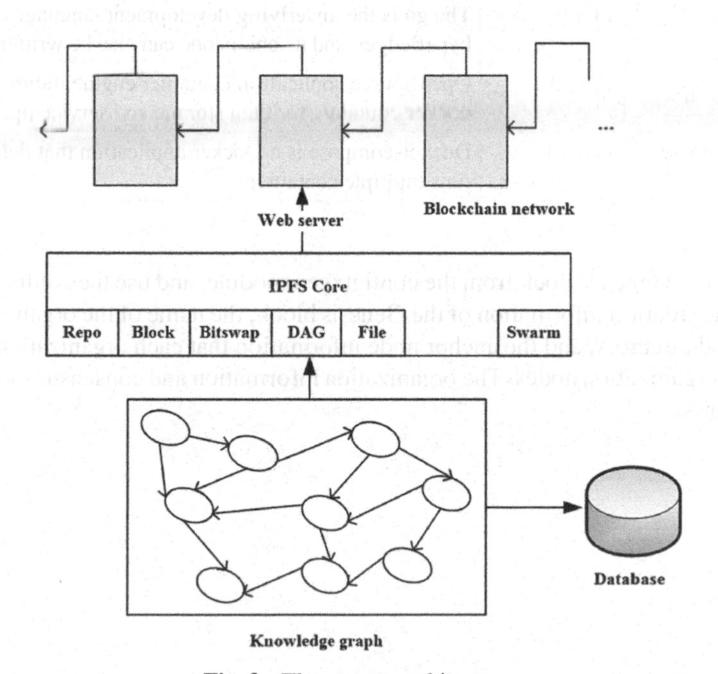

Fig. 2. The system architecture

In system architecture, the process files of knowledge graph are exported from the graph database and stored in IPFS and local database. After IPFS returns the file index hash value, the web service obtains the file index hash value and coexists in the blockchain.

3.2 Deploying the Underlying Blockchain

In this experiment, The graph database of this experiment adopts neo4j, the distributed file system is IPFS (Distributed file system) [14], and the blockchain adopts the Consortium blockchains-Hyperledger Fabric as an example. The environment of Hyperledger Fabric is two organizations, including four Peers (One is Endorser peer used to verify that the transaction is legal, endorsing and signing the legal transaction; the others are Committer peer, which checks the Orderer peer's sorted transactions, performs legitimate transactions, and writes to the store) one Orderer peer and one CA peer. The version of Fabric is 1.0.0, and experiment uses a single machine and multinode deployment (Table 1).

Table 1. Implementation environment

Environment	Version	Notes
Operating system	Ubuntu18.04	
Hyperledger Fabric	Fabric 1.0.0	Blockchain applications
Go	v1.13	The go is the underlying development language of the hyperledger, and its chaincode can also be written in the go
Docker	v19.03.1	Open source application container engine, fabric uses docker container for data storage and service operations
Docker-compose	v1.10.1	Docker-compose is a Docker application that defines and runs multiple containers

Generate a Genesis block from the configtxgen module, and use the configtx.yaml to describe the creation information of the Genesis block, the name of the organization, the MSP ID and directory, and the anchor node information that each organization interacts with other organization nodes. The organization information and consensus mechanisms are as follows:

```
Organizations:
-&OrdererOrg
    Name: OrdererOrg
    ID: OrdererMSP
    MSPDir: crypto-config/ordererOrganizations/example.com/msp
 &Org1
    Name: Org1MSP
    ID: Org1MSP
    MSPDir: crypto-config/peerOrganizations/org1.example.com/msp
    AnchorPeers:
        - Host: peer0.org1.example.com
                peer1.org1.example.com
          Port: 7051
 &Org2
    Name: Org2MSP
    ID: Org2MSP
    MSPDir: crypto-config/peerOrganizations/org2.example.com/msp
    AnchorPeers:
        - Host: peer0.org2.example.com
                peer1.org2.example.com
          Port: 7051
```

3.3 Design and Implementation of Chaincode

The smart contract in Hyperledger Fabric is called chaincode. The chaincode in this project is written in go language. When certain conditions are triggered, the contract code is automatically executed and is not subject to human control.

Because chaincode in this experiment was originally intended to upload and read index hashes of files, there are no initialization parameters in the Init interface. To meet the requirements of the experiment, writeHash, queryHash, deleteHash, getHash, and cleanHash are defined in the Invoke interface. In writeHash design, when the same time to upload multiple files index hash value to blockchain, in order to satisfy the couchDB is key - value form and the uniqueness of the key, the function sets the system time stamp to set key, set the value to graph structure, key in the graph for the first visit to store values, the value for a certain storage file index hash value (Table 2).

Table 2. Main function of these two API chaincode

API	Fuction	Notes
Init		Parameter initialization
Invoke	writeHash	Write hash to fabric ledger
	queryHash	Query hash form fabric ledger
	deleteHash	Delete hash of fabric ledger
	gethash	Acquire hash from the file and show on blockchain
	cleanhash	Clean hash of the file

3.4 Overview of Experimental Process

The experimental process of this experiment is roughly described as follows: after the knowledge graph process file is exported from the graph database, it is stored in the IPFS interplanetary file system, and the index hash value of the file generated by IPFS is stored in the Hyperledger Fabric. The flow chart is shown in Fig. 3.

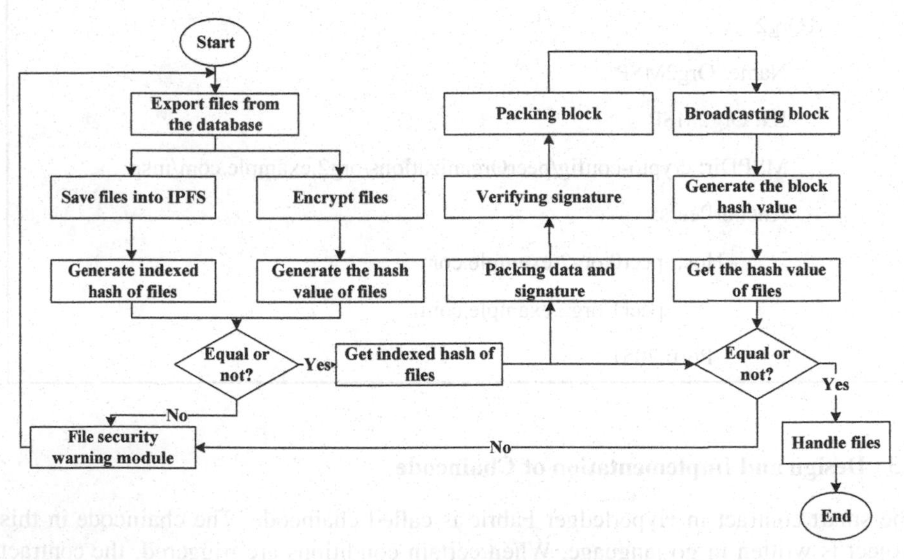

Fig. 3. Experimental operation flow chart

4 Experiment Analysis and Comparison

In this experiment, we implement the environment simulation experiment of single machine and multi node, and set up two organizations, one orderer node and four peer nodes. After several groups of experimental tests, the experimental data after statistics is taken as the average value. The result is shown in Fig. 4. The final experimental results show that the overall trend of the time when the file index hash value is stored in the blockchain is on the rise, and the time for storing data less than 8 KB is less than 1 ms. When the 16 KB data is stored, there is a jump increase, but the overall data storage is less than 10 ms. The time of data storage in the blockchain is fast, which meets the demand of fast upload and read hash value in the blockchain (Fig. 5).

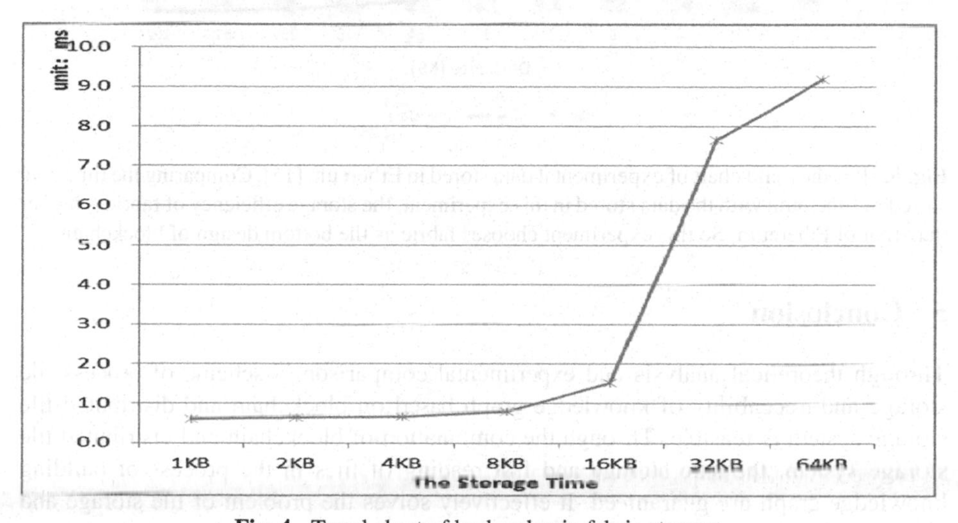

Fig. 4. Trend chart of hash value in fabric storage

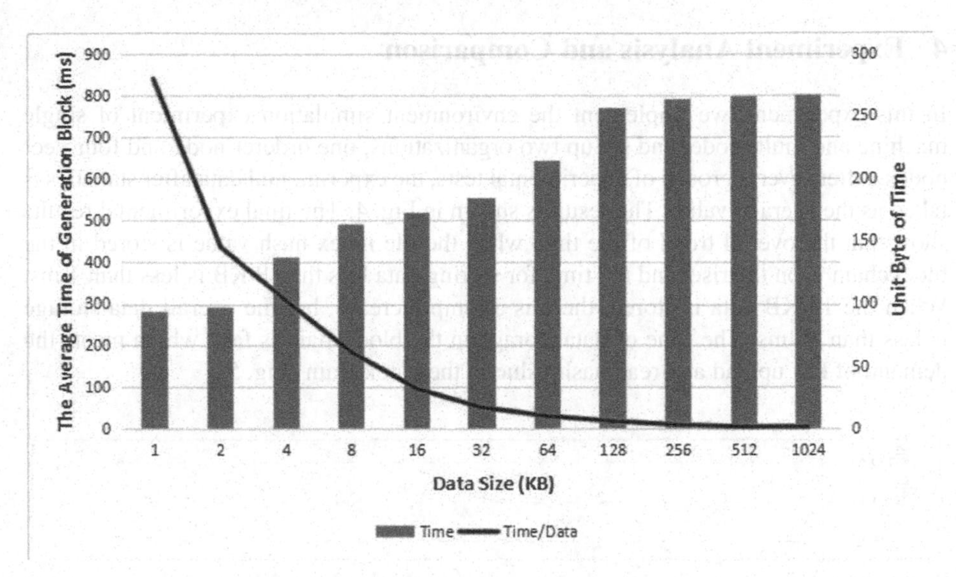

Fig. 5. It is the trend chart of experimental data stored in Ethereum [15]. Comparing the time data stored in Ethereum with the data stored in this experiment, the storage efficiency of fabric is higher than that of Ethereum. So this experiment chooses fabric as the bottom design of blockchain.

5 Conclusion

Through theoretical analysis and experimental comparison, a scheme of process file storage and traceability of knowledge graph based on blockchain and distributed file storage system is feasible. Through the combination of blockchain and distributed file storage system, the safe storage and fast reading of files in the process of building knowledge graph are guaranteed. It effectively solves the problem of file storage and anti tampering in the process of building knowledge graph.

Acknowledgement. This research work is supposed by the National Key R&D Program of China (2018YFB1201500), National Natural Science Founds of China (61602376, 61773313, 61602374, 61702411), National Natural Science Founds of Shaanxi (2017JQ6020, 2016JQ6041), Key Research and Development Program of Shaanxi Province (2017ZDXM-GY-098, 2019TD-014), Science Technology Project of Shaanxi Education Department (16JK1573, 16JK1552).

Reference

1. Sisson, P., Ryan, J.J.C.H.: A knowledge concept map: structured concept analysis from systematic literature review. J. Entrep. Manag. Innov. **13**(3), 29–69 (2017)
2. Lin, F., Hsueh, C.: Knowledge map creation and maintenance for virtual communities of practice (2003)
3. Niu, B., Huang, Y.: An improved method for web text affective cognition computing based on knowledge graph. Comput. Mater. Continua **59**(1), 1–14 (2019)
4. Zyskind, G., Zekrifa, D.M.S., Alex, P., et al.: Decentralizing privacy: using blockchain to protect personal data. In: 2015 IEEE Security and Privacy Workshops (SPW). IEEE (2015)

5. Hazazi, M., Almousa, A., Kurdi, H., Al-Megren, S., Alsalamah, S.: A credit-based approach for overcoming free-riding behaviour in peer-to-peer networks. Comput. Mater. Continua **59**(1), 15–29 (2019)
6. Croman, K., et al.: On scaling decentralized blockchains. In: Clark, J., Meiklejohn, S., Ryan, P.Y.A., Wallach, D., Brenner, M., Rohloff, K. (eds.) FC 2016. LNCS, vol. 9604, pp. 106–125. Springer, Heidelberg (2016). https://doi.org/10.1007/978-3-662-53357-4_8
7. Zheng, Z., Xie, S., Dai, H., et al.: An overview of blockchain technology: architecture, consensus, and future trends. In: 6th IEEE International Congress on Big Data. IEEE (2017)
8. Underwood, S.: Blockchain beyond bitcoin. Commun. ACM **59**(11), 15–17 (2016)
9. Androulaki, E., Barger, A., Bortnikov, V., et al.: Hyperledger fabric: a distributed operating system for permissioned blockchains (2018)
10. Benet, J.: IPFS - Content Addressed, Versioned, P2P File System (DRAFT 3). eprint arXiv (2014)
11. Kelly, M., Alam, S., Nelson, M.L., Weigle, M.C.: InterPlanetary wayback: peer-to-peer permanence of web archives. In: Fuhr, N., Kovács, L., Risse, T., Nejdl, W. (eds.) TPDL 2016. LNCS, vol. 9819, pp. 411–416. Springer, Cham (2016). https://doi.org/10.1007/978-3-319-43997-6_35
12. Patsakis, C., Casino, F.: Hydras and IPFS: a decentralised playground for malware. Int. J. Inf. Secur. **18**(6), 787–799 (2019). https://doi.org/10.1007/s10207-019-00443-0
13. Deng, Z., Ren, Y., Liu, Y., Yin, X., Shen, Z., Kim, H.-J.: Blockchain-based trusted electronic records preservation in cloud storage. Comput. Mater. Continua **58**(1), 135–151 (2019)
14. Zheng, Q., Li, Y., Chen, P., et al.: An innovative IPFS-based storage model for blockchain. In: 2018 IEEE/WIC/ACM International Conference on Web Intelligence (WI). ACM (2018)
15. Zhu, H., Wang, Y., Hei, X., et al.: A blockchain-based decentralized cloud resource scheduling architecture. In: 2018 International Conference on Networking and Network Applications (NaNA). IEEE Computer Society (2018)

A Fragile Watermarking Algorithm Based on Audio Content and Its Moving Average

Xizi Peng, Jinquan Zhang(✉), and Shibin Zhang

School of Cybersecurity, Chengdu University of Information Technology, Chengdu 610225, China

zhjq@cuit.edu.cn

Abstract. A fragile audio watermarking scheme for content authentication was proposed in the paper. First, two proper positive integers are chosen. Then, the audio signal is segmented. Two moving averages are computed according to the chosen integers in each segment. The watermark information obtained from audio segment is embedded at crosses of the two moving averages. The watermark can be blindly extracted without original signal when authentication is needed. Subjective and object tests show that the fragile watermarking scheme is inaudible and sensitive\e against MP3 compression, additive noise, re-sampling, re-quantization, low-pass filtering, etc. Especially, it can detect and localize the tampered regions effectively.

Keywords: Audio watermarking · Fragile watermarking · Moving average

1 Introduction

Watermarking is a widely used solution to the problems of authentication and copyright protection of digital media especially for images, videos, and audio [1]. Although image watermarking in various fields has always been a research hotspot [2, 3], in recent years, copyright protection for audio works has been widely concerned by researchers [4–6]. A way to protect the copyright of audio works is to embed robust watermark into it. On the other hand, embedding (semi-) fragile watermarking into an audio works to authenticate the integrity is also received research's attention.

Recently, copyright protection for audio works has been widely concerned by researchers [4–6]. A way to protect the copyright of audio works is to embed robust watermark into it. On the other hand, embedding (semi-) fragile watermarking into an audio works to authenticate the integrity is also received research's attention.

In literature [7], Zhao etc. located the tampered position based on the change of binary watermark image extracting from the watermarked audio. In literature [8], authors partitioned the audio signal into frame, then extracted the energy of the critical bands from audio frame as the feature, and embedded the feature into frame by modifying coefficients in DCT domain. In literature [9], Wang etc. computed the centroid of each audio frame and quantized it as watermark. They embedded watermark in the frame in hybrid domain. It is a semi-fragile watermark fragile. In literature [10], Lei et al. proposed

X. Sun et al. (Eds.): ICAIS 2020, LNCS 12240, pp. 328–340, 2020.
https://doi.org/10.1007/978-3-030-57881-7_30

a semi-fragile audio watermarking scheme. They used binary image as fragile watermark, and embedded the watermark signal into the average value of the wavelet coefficients. A speech content authentication algorithm based on Bessel–Fourier moments was proposed in [11]. The watermarking scheme is imperceptible and has excellent ability of tamper detection. A fragile watermarking scheme for speech content authentication was also proposed in literature [12].

If the watermark is generated from the audio signal, when the watermark is embedded into the audio signal, the signal usually was modified, so regeneration of the watermark will be influenced in extracting phase. In literature [8], the author noticed this problem.

The impairment to audio signal by embedding fragile watermark should be negligible. Literature [8, 9, 13] embedded the watermark in transform domain, so, it is difficult to control the distortion of waveform, especially in hybrid domain. Literature [10, 14] paid attention to the distortion of waveform due to embedding watermark, and indicated that audible noise may be brought out by improper embedding rule.

In order to keep the stability of audio signal, our algorithm preprocesses the audio clip. At the same time, the watermark is embedded in time domain, and the distortion of waveform can be predicted, so the algorithm keeps the waveform of the audio signal well. In the algorithm, the embedding strength is very small, just like adding a weak echo signal with a delay of 0, so, it has less impact on auditory sense.

2 Embedding and Extraction of Fragile Watermark

In this section, we describe the definition of moving average, the preprocessing to the audio signal, the generation of fragile watermark, the rule of embedding and extracting the watermark, and locating the tampered position.

2.1 Definition of Moving Average in Audio Signal

Assume an audio signal X has L samples. They are denoted as x1, x2, ..., xL. Two positive integer a and b are chosen. In this paper, we always set $a < b$. We can obtain the moving average M_A as Eq. (1).

$$M_{A_k} = \frac{1}{a}(x_k + x_{k+1} + \cdots + x_{k+a-1}) = \frac{1}{a}\sum_{i=k}^{k+a-1} x_i, k = 1, 2, \ldots, L - a + 1 \quad (1)$$

M_A is used to represent $(M_{A_1}, M_{A_2}, \ldots, M_{A_{L-a+1}})$. Similarly, M_B is used to represent $(M_{B_1}, M_{B_2}, \ldots, M_{B_{L-b+1}})$. From Eq. (1), we know that M_A is the low frequency components of the audio signal X.

M_A and M_B are discrete. For a point M_{B_i}, a corresponding point $M_{A_{i+b-a}}$ in M_A, and a corresponding sample x_{i+b-1} in X is at the same time in time axis. If

$$(M_{B_i} - M_{A_{i+b-a}})(M_{B_{i+1}} - M_{A_{i+b-a+1}}) \leq 0, i > b \quad (2)$$

We called that there is a cross between M_A and M_B at x_{i+b-1}.

2.2 Preprocessing

Amplitude of samples will be modified when message is embedded into an audio signal. So, this may change the characteristics of some segment. In order to obtain relatively stable features, it is necessary to pretreat the original signal.

(1) Partition the audio signal X into frame, $X = (X_1, X_2, \ldots)$. There are l samples in each frame. Then computing the sum of absolute value of amplitude of samples in each frame as Eq. (3), we get $T = (T_1, T_2, \ldots)$.

$$T_i = \sum_{k=(i-1)\times l+1}^{i\times l} |x_k| \tag{3}$$

(2) Choose a positive number $q > 0$. For T_i, compute $r = T_i \bmod q$, $0 \leq r < q$. If $r \leq q \times 0.1$ or $r \geq q \times 0.9$, preprocessing is not necessary. If not,

(i) If $q \times 0.1 < r \leq q \times 0.5$, for x_i in X_i, if $x_i \geq 0$, x_i is set as $x_i - r/l$, otherwise x_i is set as $x_i + r/l$.

(ii) If $q \times 0.5 < r < q \times 0.9$, for x_i in X_i, if $x_i \geq 0$, x_i is set as $x_i + r/l$, otherwise x_i is set as $x_i - r/l$.

For an audio frame X_i, after the preprocessing, set the sum of amplitude is S_i. As can be seen from the processing procedure, S_i is almost integral multiple of the q.

It is necessary to preprocess the audio signal, as described in Sect. 5.1.

2.3 Fragile Watermark Generation

After the preprocessing, the S_i is a relatively stable characteristic of X_i. Calculate $\lfloor (S_i + q/2)/q \rfloor$, $\lfloor \cdot \rfloor$ means floor function. Then converting the result into binary, we get the watermark w.

For each sample, the amplitude is less than 1. That is, $|x_i| \leq 1$. So, $S_i < l$. For $q > 0$, $S_i/q < l/q$. Then

$$\log_2^{T_i^{'}/q} < \log_2^{l/q} \tag{4}$$

That is, the length of w is less than $\log_2^{l/q}$ bits. Usually, $|x_i| \leq 0.5$ and $S_i \leq 0.5l$. In this case, the length of w is less than $\log_2^{l/q} - 1$ bits. The difference of their length is only one bit. So we won't reiterate it here.

Then a public stream cipher algorithm is used to encrypt w. Here, we choose ZUC algorithm. w can be concatenated itself any number of times. A key K with suitable length is chosen for each frame. Then we compute

$$W_C = (w\|\ldots\|w) \oplus K \tag{5}$$

where $\oplus$ means XOR operation. W_C will be embedded into X_i as described in Sect. 2.4.

2.4 Embedding Watermark

For an audio signal, two positive integers a and b are chosen. Assume the embedding strength is s, $s > 0$. Calculate M_A and M_B according to Eq. (1) in each frame.

The embedding phase is as follows.

(1) cnt = 0, i = 1.
(2) Search a cross of MA and MB until the end of the frame X_i.
do while $\left(M_{B_i} - M_{A_{i+b-a}}\right)\left(M_{B_{i+1}} - M_{A_{i+b-a+1}}\right) > 0$ {
i = i + 1,
cnt = cnt + 1,
}
(3) If cnt <= b, go to step (2). If not, go to step (4) to embed one bit message.
(4) Assume $u = M_{B_{l+1}}$. For the kth bit of $W_c(k)$, the embedding rule is as follows:

$$u' = \begin{cases} \lfloor (u + s/2)/s \rfloor \times s, W_C(k) = 1 \\ \lfloor u/s \rfloor \times s + s/2, W_C(k) = 0 \end{cases} \tag{6}$$

Set c = u′ − u. Notice the ith point of the sequence MB corresponds to the (i + b − 1)th sample of the audio in time axis. The amplitude of each sample from former cross to the sample xi + b adds c.
(5) Set cnt = 0. Goto step (2) to embed the next bit.

In the embedding rule, since the amplitude of the original audio is directly modified, and the parameter cnt is larger than b, the position of the crosses of MA and MB will not change.

2.5 Detecting Watermark

Assume all parameters are obtained, including: l, a, b, q, s, K. For a certain frame X_i, calculate the moving average M_A and M_B, Then extract message in frame X_i.

(1) cnt = 0, i = 1.
(2) Search a cross of MA and MB until the end of current frame.
do while $\left(M_{B_i} - M_{A_{i+b-a}}\right)\left(M_{B_{i+1}} - M_{A_{i+b-a+1}}\right) > 0$ {
i = i + 1,
cnt = cnt + 1,
}
(3) If cnt <= b, go to step (2). If not, go to step (4) to extract one bit message.
(4) Assume $u' = M_{B_{i+1}}$. The detecting rule is as follows:

$$w^* = \begin{cases} 0, s/4 < u' \text{mod} < 3s/4 \\ 1, others \end{cases} \tag{7}$$

(5) Set cnt = 0. Goto step (2) to search the next cross.
Assume all message extracting in this frame is W_C'. Compute

$$w* = W_C' \oplus K \tag{8}$$

2.6 Locating the Tampered Position

From the frame length l and parameter q, we get $S' = (S_1', S_2', \ldots)$ according to Eq. (3). Then we will compare w', the fragile watermark generated from watermarked frame, with $w*$, the extracted watermark from the watermarked frame.

Define the authentication sequence as follows:

$$e(i) = w'(i \bmod t') \oplus w * (i), i \in [1, t] \tag{9}$$

t′ is the length of w′. t is the length of w*. For a certain i, if $e(i) \neq 0$, it means samples in this frame which watermark bit is embedded into in the watermarked audio is changed to a certain extent.

3 Analysis of the Algorithm

In this section, we will describe why the sum of amplitude in a frame changes small when the watermark is embedded, derive the correlation among parameters, and analyze the imperceptibility of the proposed algorithm.

3.1 Effect of Embedding Algorithm to S_i

For a certain frame X_i, set the difference is c between the marked and the original audio when one bit watermark is embedded. Then, According to the embedding algorithm, since the number of sample is large for the audio clip, we may presume the difference obeys uniform distribution. That is, $c \sim U\left(-\frac{s}{2}, \frac{s}{2}\right)$.

For a certain audio clip, regardless of which distribution its amplitude obeys, it can be presumed that the mean of amplitude of its samples is zero [15]. Assume the number of positive and negative amplitude is about 50% respectively in the audio signal. For two samples x_1 and x_2, if the sign of their amplitude is opposite, and if the sign of $x_1 - c$ and x_1 is the same, the sign of $x_2 - c$ and x_2 is the same, the following equation can be satisfied.

$$|x_1 - c| + |x_2 - c| = |x_1| + |x_2| \tag{10}$$

From the embedding rule, the maximum of c is s/2. As can be seen in Sect. 4.2, s is very small, so the condition that sign of $x_1 - c$ and x_1 is the same and sign of $x_2 - c$ and x_2 is the same is easily satisfied.

So, for a certain audio frame, the sum of absolute value of amplitude has little change due to embedding into watermark. It can be achieved that the change is less than $0.5q$ by adjusting q.

3.2 Correlation Among Parameters

From the embedding algorithm, we know that the embedding capacity is mainly affected by the choice of the parameter a and b, especially b. Here we assume the embedding capacity is in the range $(l/nb, l/b)$. That is, the ceiling means there is one bit in each group of b samples, and the floor means there is one bit in each group of nb samples.

We assume the w is embedded once at least, that is

$$\log_2^{l/q} < l/nb \tag{11}$$

For the frame X_i, from the embedding phase of w, if amplitude of all samples decreased or increased by $s/2$, the change of S_i should be larger than $q/2$. If not, it means that q is too large, and the algorithm isn't fragile enough. That is,

$$q/2 < ls/2 \tag{12}$$

From Eq. (11), we get

$$l/nb > \log_2^{l/q} > \log_2^{1/s} \tag{13}$$

That is,

$$l > -nb\log_2^s \tag{14}$$

According to Eq. (14), it is helpful for the choice of parameters.

3.3 Analysis of Necessity of Preprocessing

If preprocessing isn't performed, assume the sum of i th frame X_i is T_i. If $r = T_i \bmod q$, $0 \leq r < q$, r obeys uniform distribution. That is, $r \sim U(0, q)$. Assume the sum of ith frame X_i is $T_i^{'}$ after watermark had been embedded into. When the audio signal is long, No matter how much the parameter q is, there must be $r \approx 0.5q$. Then, $\left\lfloor (T_i^{'} + 0.5q)/q \right\rfloor$ may not be equal to $\lfloor (T_i + 0.5q)/q \rfloor$. So feature of the corresponding frame changes after watermark is embedded into.

So, it is necessary to preprocessing the audio signal as described in Sect. 2.2. The verification is presented in Sect. 5.1.

3.4 Analysis of Inaudibility

In audio watermarking, the SNR is a difference indicator between the watermarked and the original audio. The definition of SNR is shown as follows:

$$SNR = 10 \times \lg\left(\frac{\sum_{i=1}^{L} x_i^2}{\sum_{i=1}^{L} (x_i - x_i^{'})^2}\right) \tag{15}$$

where x_i and $x_i^{'}$ are the original and watermarked audio signal respectively. L is the number of samples in audio signal.

In the algorithm, for each sample, the maximal change of amplitude is the sum of two numbers. One is the change, $(q/2)/l$, in preprocessing. The other is half of the embedding strength, $s/2$, in embedding phase. That is,

$$\left|x_i - x_i^{'}\right| < (q/2)/l + s/2 = (q + ls)/2l \tag{16}$$

According to Eq. (12),

$$\left|x_i - x_i'\right| < s \tag{17}$$

So,

$$SNR \geq 10 \times (\lg \sum_{i=1}^{L} x_i^2 - \lg Ls^2) \tag{18}$$

In fact, Eq. (16) is the worst case. So, SNR of the experimental result is larger than the derived one from Eq. (18).

For SNR doesn't take the characteristics of human auditory into account, perceptual evaluation of audio quality (PEAQ), an assessment tool recommended by ITU BS1387, is also used in objective evaluation tests. It utilizes software to simulate perceptual properties of the human ear and then, integrate multiple model output variables into a single metric. At the same time, Mean Opinion Score (MOS) is used in the subjective listening test.

4 Choice of Parameters

In this section, according to correlation among parameters based on part 3 and experimental result, the choice of parameters in the algorithm is described. In experiments, different audio clips including pop, light, march, piano, jazz and rock are tested. All music clips are in WAV format, mono, 16 bits/sample, 8 s, and 44.1 kHz sampling frequency. The experimental results are similar for all audio files. We report the results in the paper with a few music clips.

4.1 Choice of Parameter a and b

For a fragile watermark algorithm, the parameter a often ranges from 20 to 50, and b is 1.5–3 times as much as a. The smaller the b is, the larger the embedding capacity is. From Eq. (14), the smaller the b is, the smaller the l is, and the more accurately locating the tampered position is.

4.2 Choice of Parameter s

In experiments, we test our algorithm with different audio clips. The experimental results are similar for all audio clips. That is, when the embedding strength $s \leq 0.00006$, no matter how much a and b is, the BER isn't equal to 0. When $s = 0.00007$, the message can be detected correctly.

The embedding strength s is very small, which will be beneficial to keep the stability of the feature for each frame.

4.3 Choice of Parameter l and q

From the embedding algorithm, we know that the embedding capacity is mainly affected by the choice of the parameter a and b. Experimental results show that the embedding capacity is always in $(l/2b, l/b)$. L is the number of samples of audio signal. That is, $n = 2$ is acceptable in Eq. (14).

Take march music for example. Set $a = 20$, $b = 30, 40, 50 \ldots, 90$, we can see that the assumption and experimental result are shown in Fig. 1.

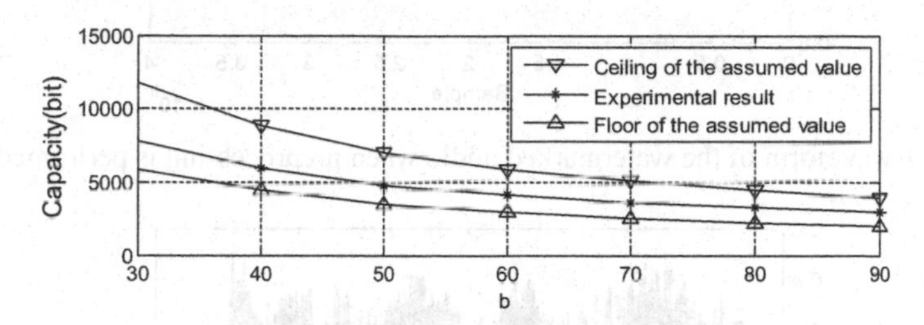

Fig. 1. Test of embedding capacity

Many other music clips are chosen to make this experiment, and a and b are also changed. The capacity is always in $(l/2b, l/b)$.

For s, n and b are known as described above, l can be computed according to Eq. (14). From Eq. (12), $0 < q < ls$. We can let q gradually increase from 0 until all the watermark information in each frame can be correctly extracted.

5 Experiments and Analysis

In this section, the necessity of preprocessing to audio signal is verified, locating the tampered position is test, and the experimental results about imperceptibility and fragility of the watermarked audio are presented.

For all music clips, $n = 2$, $s = 0.00007$. For light music, from Eq. (14), $l > 1662$. We set $l = 2048$. Other parameters for light music, $(a, b, q) = (30, 60, 0.02)$. Parameters for pop music, $(a, b, l, q) = (50, 80, 2560, 0.06)$. Parameters for march music, $(a, b, l, q) = (50, 80, 2048, 0.04)$.

5.1 Necessity of Preprocessing to Audio Signal

Take light music for example, whether preprocessing is performed or not, waveform of the watermarked audio is similar, as shown in Fig. 2(a) and Fig. 2(b). When preprocessing isn't performed, the extracted message is inconsistent with the watermark generated from the watermarked audio, as shown in Fig. 2(c), and the BER = 0.008. When preprocessing is performed, the extracted message is consistent with the watermark generated from the watermarked audio, and the BER = 0. When preprocessing isn't performed, and other parameters are unchanged, q is equal to 0.03, 0.05, 0.08 and 0.1 respectively, BER are not 0. Fox example, when $q = 0.1$, BER = 0.001, as shown in Fig. 2(d).

BER is not equal to 0, which means features of some frames are changed when message was embedded.

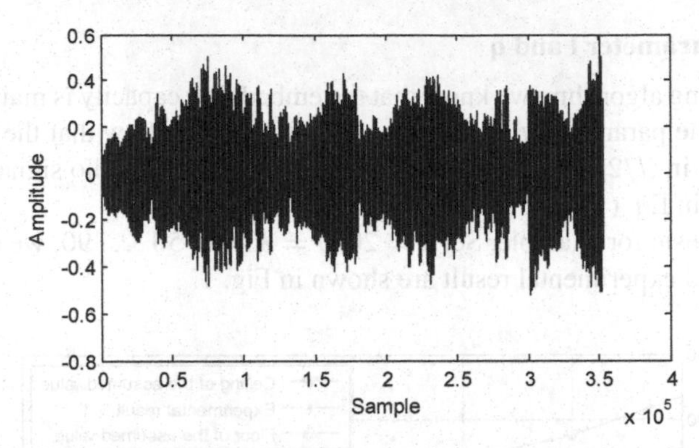

(a) waveform of the watermarked audio when preprocessing is performed

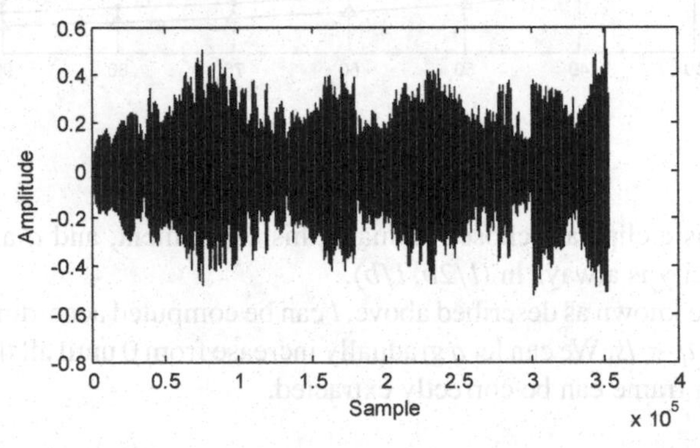

(b) waveform of the watermarked audio when preprocessing isn't performed

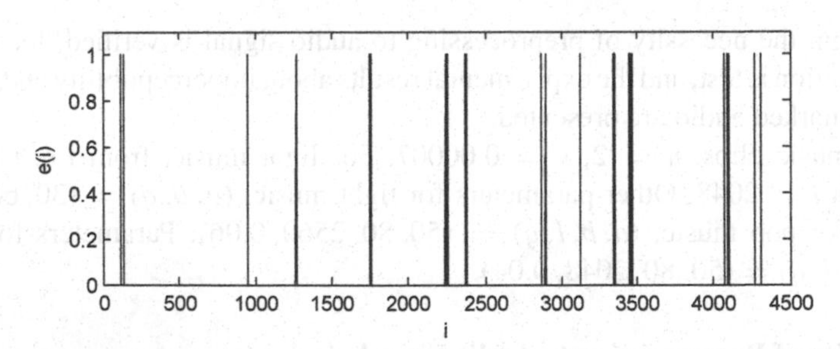

(c) False alarm when preprocessing isn't performed and $q = 0.02$

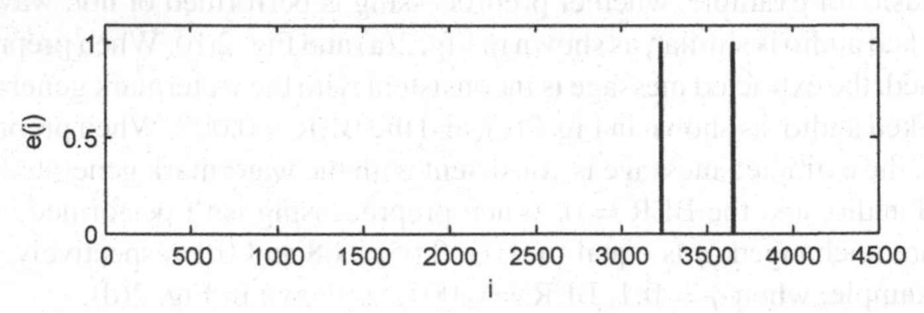

(d) False alarm when preprocessing isn't performed and $q = 0.1$

Fig. 2. Necessity of preprocessing to audio signal

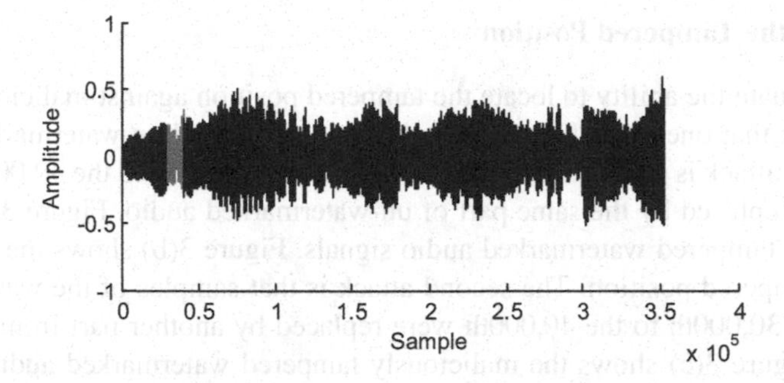

(a) Waveform of the watermarked audio signal after the first attack

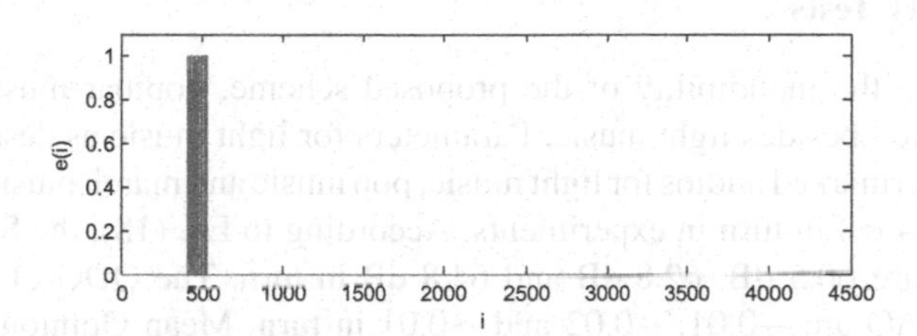

(b) Locating the tampered position after the first attack

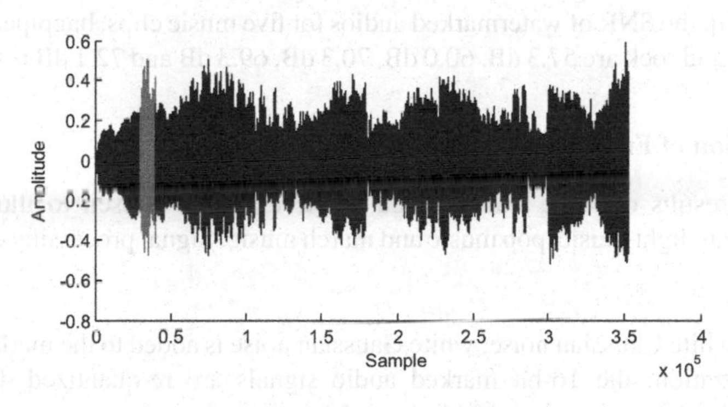

(c) Waveform of the watermarked audio signal after the second attack

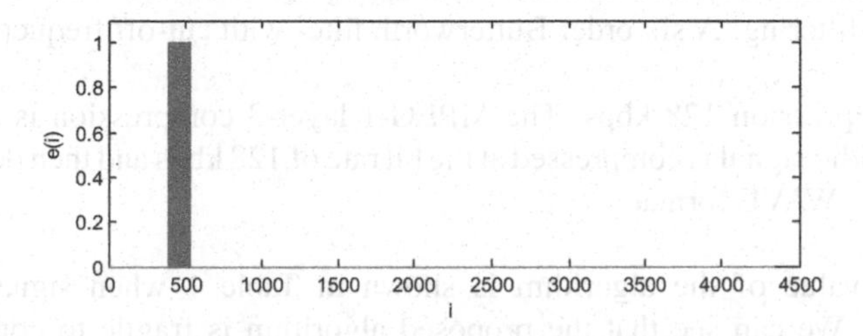

(d) Locating the tampered position after the second attack

Fig. 3. Verification of locating the tampered position

5.2 Locating the Tampered Position

In order to evaluate the ability to locate the tampered position against malicious operations, the attack that one part of audio is replaced is performed on watermarked audio signal. The first attack is that samples of the watermarked audio from the 30,000th to the 40,000th were replaced by the same part of un-watermarked audio. Figure 3(a) shows the maliciously tampered watermarked audio signals. Figure 3(b) shows the results of locating the tampered position. The second attack is that samples of the watermarked audio from the 30,000th to the 40,000th were replaced by another part from 70,000th to 80,000th. Figure 3(c) shows the maliciously tampered watermarked audio signals. Figure 3(d) shows the results of locating the tampered position.

5.3 Inaudibility Tests

When evaluating the inaudibility of the proposed scheme, popular music and march music are reported besides light music. Parameters for light music as described above. The SNR of watermarked audios for light music, pop music and march music are 77.3 dB, 73.3 dB and 75.4 dB in turn in experiments. According to Eq. (18), the SNR of watermarked audios are 66.5 dB, 62.8 dB and 64.8 dB in turn. The ODG of watermarked audios with PEAQ are −0.01, −0.02 and −0.01 in turn. Mean Opinion Score in the subjective listening test are all equal to 0.

From the SNR of the algorithm, the proposed algorithm is better than literature [10]. In literature [10], the SNR of watermarked audios for five music clips, bagpipe, classical, country, dance and rock are 57.3 dB, 60.0 dB, 70.3 dB, 69.3 dB and 72.1 dB respectively.

5.4 Verification of Fragility

Experimental results of three watermarked music clips are chosen to illustrate the fragility. They are light music, pop music and march music. Signal processing operations include:

(1) Additive white Gaussian noise: white Gaussian noise is added to the marked signal.
(2) Re-quantization: the 16-bit marked audio signals are re-quantized down to 8 bits/sample and then back to 16 bits/sample.
(3) Resampling: The marked signal, originally sampled at 44.1 kHz, is re-sampled at 22.05 kHz, and then restored back by sampling again at 44.1 kHz.
(4) Low-pass filtering: A six order Butterworth filter with cut-off frequency 21.4 kHz is used.
(5) MP3 Compression 128 kbps: The MPEG-1 layer-3 compression is applied. The marked audio signal is compressed at the bit rate of 128 kbps and then decompressed back to the WAVE format.

The BER value of the algorithm is shown in Table 1 when signal operations are performed. We can see that the proposed algorithm is fragile to common signal operations.

Table 1. Verification of fragility

Signal operation	Light music	Pop music	March music
Re-quantization	0.493	0.502	0.509
Resampling	0.465	0.495	0.508
Additive noise	0.501	0.508	0.489
Low-pass filtering	0.505	0.481	0.509
MP3 compression	0.502	0.500	0.509

6 Conclusions

A fragile watermarking algorithm for content authentication of audio signal is proposed in this paper. We introduce the definition of moving average in audio signal. Then, the generation, embeddedness and extraction of watermark are presented. By the extracted message, the tampered position can be located if the marked audio is maliciously modified. We analyze why the feature of the frame is stable, the necessity of preprocessing and inaudibility of the algorithm, introduce the choice of parameters, and verify them with experiments.

One of the shortcomings of the proposed algorithm is that the algorithm needs a good synchronization code scheme to locate the position of the frame. This is also an open problem in the industry.

Acknowledgments. The authors would like to thank the reviewers for their detailed reviews and constructive comments, which have helped improve the quality of this paper. This work was supported in part by the National Key Research and Development Project of China (No. 2017YFB0802302), the Science and Technology Support Project of Sichuan Province (No. 2016FZ0112, No. 2017GZ0314, No. 2018GZ0204), the Academic and Technical Leaders Training Funding Support Projects of Sichuan Province (No. 2016120080102643), the Application Foundation Project of Sichuan Province (No. 2017JY0168), the Science and Technology Project of Chengdu (No. 2017-RK00-00103-ZF, No. 2016-HM01-00217-SF).

References

1. Jayashree, N., Bhuvaneswaran, R.S.: A robust image watermarking scheme using z-transform, discrete wavelet transform and bidiagonal singular value decomposition. Comput. Mater. Continua **58**(1), 263–285 (2019)
2. Liu, J.: A robust zero-watermarking based on SIFT-DCT for medical images in the encrypted domain. Comput. Mater. Continua **61**(1), 363–378 (2019)
3. Gong, D., Chen, Y., Haoyu, L., Li, Z., Han, Y.: Self-embedding image watermarking based on combined decision using pre-offset and post-offset blocks. Comput. Mater. Continua **57**(2), 243–260 (2018)
4. Hua, G., Huang, J., Shi, Y.Q., Goh, J., Thing, V.L.L.: Twenty years of digital audio watermarking-a comprehensive review. Signal Process. **128**, 222–242 (2016)

5. Hua, G., Goh, J., Thing, V.L.L.: Time-spread echo-based audio watermarking with optimized imperceptibility and robustness. IEEE/ACM Trans. Audio Speech Lang. Process. **23**, 227–239 (2015)
6. Fallahpour, M., Megias, D.: Audio watermarking based on fibonacci numbers. IEEE/ACM Trans. Audio Speech Lang. Process. **23**(8), 1273–1282 (2015)
7. Zhao, H., Shen, D., Zhu, Y.: A Semi-fragile audio watermarking against shearing. Acta Automatica Sinica **34**(6), 647–651 (2008)
8. Gulbis, M., Muller, E., Steinebach, M.: Content-based authentication watermarking with improved audio content feature extraction. In: Proceedings of International Conference on Intelligent Information Hiding and Multimedia Signal Processing, Harbin, China, pp. 620–623. IEEE (2008)
9. Wang, H., Fan, M.: Centroid-based semi-fragile audio watermarking in hybrid domain. Sci. China Inf. Sci. **53**(3), 619–633 (2010)
10. Lie, W., Chang, L.: Robust and high-quality time-domain audio watermarking based on low-frequency amplitude modification. IEEE Trans. Multimedia **8**(1), 46–59 (2006)
11. Liu, Z., Wang, H.: A novel speech content authentication algorithm based on bessel-fourier moments. Digit Signal Process. **24**, 197–208 (2014)
12. Qian, Q., Wang, H.-X., Hu, Y., Zhou, L.-N., Li, J.-F.: A dual fragile watermarking scheme for speech authentication. Multimedia Tools Appl. **75**(21), 13431–13450 (2015). https://doi.org/10.1007/s11042-015-2801-4
13. Mingquan, F., Hongxia, W.: Content-based fragile audio watermarking in hybrid domain. J. China Railway Soc. **32**(1), 118–122 (2010)
14. Megías, D., Serra-Ruiz, J., Fallahpour, M.: Efficient self-synchronised blind audio watermarking system based on time domain and FFT amplitude modification. Signal Process. **90**(12), 3078–3092 (2010)
15. Mendes, R.S., Ribeiro, H.V., Freire, F.C.M., Tateishi, A.A., Lenzi, E.K.: Universal patterns in sound amplitudes of songs and music genres. http://pre.aps.org/abstract/PRE/v83/i1/e017101. Accessed 13 July 2012

An AHP/DEA Methodology for the Public Safety Evaluation

Li Mao[1], Naqin Zhou[2](✉), Tong Zhang[1], Wei Du[1], Han Peng[1], and Lina Zhu[1]

[1] Guangdong Police College, No. 118, Wenshengzhuang Road, Baiyun District, Guangzhou, China

[2] Cyberspace Institute of Advanced Technology, Guangzhou University, Guangzhou, China
439657699@qq.com

Abstract. Public safety may be endangered by events that could lead to huge losses including human life, injuries or property damage, such as crime or disaster. In addition, there are some factors that affect public safety, such as the ratio of temporary residents, per capita GDP, etc. This paper aims at integrating Non-discretionary variable in data envelopment analysis (DEA-NDSC) and analytic hierarchy process (AHP) to evaluate the public safety of cities in China. The proposed AHP/DEA methodology uses the AHP to determine the weights of criteria to assess each item under each criterion, and the DEA-NDSC method to evaluate the efficiency of safety services. A cities' case of public safety perception is employed to illustrate the proposed method. The results show that the method can effectively and reasonably evaluate the public safety of the city. This research also offers a simple enough and applicable approach to a number of multiple criteria decision making problems.

Keywords: Analytic hierarchy process (AHP) · Data envelopment analysis (DEA) · Public safety · Integration

1 Introduction

One of the functions of governments is the public security, which ensures the protection of persons in their territory against threats to their well-being. To meet the increasing challenges in the public security, Chinese government try to address possible threats in advance using information technology in the field of big data [1–3]. The government optimizes the efficiency of public security services, and always evaluates the public security to prevent public crises.

Public safety evaluation is essentially a multiple criteria decision making problem, which involves multiple evaluation criteria such as crime and facts, security inputs, and environmental factors. Therefore, analytic hierarchy process (AHP) approach can be used for public safety evaluation. [4] developed an improved AHP method for assessing social vulnerability of Beijing. [5] investigated the relative importance of crime safety assessment indicators in outdoor public places and conducted AHP analysis. Furthermore, the efficiency of public safety service is an important consideration in public safety

X. Sun et al. (Eds.): ICAIS 2020, LNCS 12240, pp. 341–352, 2020.
https://doi.org/10.1007/978-3-030-57881-7_31

evaluation. Chinese government is struggling with tight budgets while trying to maintain stability. Given a set of inputs used and outputs achieved in the public security sector, efficiency of public security can be estimated using data envelopment analysis (DEA) technique. Paper [6] and [7] applied the three-stage DEA model to measure the performance of police forces. For a more accurate assessment of public safety, we combine the three-stage DEA with AHP and propose an integrated DEA-NDSC/AHP methodology in this paper.

The paper proceeds as follows. In Sect. 2, we give the brief descriptions of the AHP and the three-stage DEA to provide a ground for the later developed method. In Sect. 4, elucidation of the integrated DEA-NDSC/AHP approach in this analysis is given. The next section presents an application of the proposed DEA-NDSC/AHP method to public safety assessment, where the proposed DEA-NDSC/AHP method results were compared with the method of three-stage used in the DEA-NDSC for handling city alternatives. Discussion and conclusions are offered in Sect. 5.

2 Overview of the Methodologies

2.1 AHP Process

In this study,the AHP method is applied for Security situation perception. The AHP method is commonly used among multi-criteria decision making techniques developed by Thomas Saaty [8], which provides an approach to calibrating the numeric scale for the measurement of quantitative and qualitative performances, based on mathematics and psychology. AHP requires following main steps:

Development of the Hierarchical Structure. Stating the problem goal and structuring the hierarchy from the top (decision goal) through the intermediate levels (criteria and subcriteria) to the lowest level (alternatives).

Definition of Criteria Matrix and Determination of the Relative Weights. Constructing a series of pair-wise comparison matrices at each level of the hierarchy. Comparisons are made between the indicators by using relative scale measurement which is shown in Table 1, in the light of Saaty's semantic scale [9].

Let $C = \{C_j | j = 1, 2, \cdots, n\}$ be the set of criteria, and n is the number of criteria. A $n \times n$ pairwise comparison matrix can be made between n criteria, each element a_{ij} represents the intensity of importance of the i th criterion relative to the j th criterion.

$$A = \begin{bmatrix} a_{11} & a_{12} & \cdots & a_{1n} \\ a_{21} & a_{22} & \cdots & a_{2n} \\ \vdots & \vdots & \ddots & \vdots \\ a_{n1} & a_{n2} & \cdots & a_{nn} \end{bmatrix} \tag{1}$$

Where $a_{ii} = 1$, $a_{ji} = 1/a_{ij}$, $a_{ij} > 0$.

Denote the relative weights of each criterion by $w = (w_1, w_2, \cdots, w_n)^T$, and ω corresponding to the largest eigenvalue ($\lambda_{\max}$) can be computed by the equation:

$$Aw = \lambda_{\max} w \tag{2}$$

Besides, the quality of the judgements and the consistency ratio of matrix must be taken into account. The judgements of the pairwise comparison matrix are perfectly consistent so long as

$$a_{ij} \cdot a_{jk} = a_{ik},\ or\ (w_i/w_j)(w_j/w_k) = (w_i/w_k),\ \text{with}\ i, j, k = 1, 2, \cdots, n \tag{3}$$

The consistency index (CI) is defined by Saaty as following:

$$CI = \frac{\lambda_{\max} - n}{n - 1} \tag{4}$$

For different size of matrix, random matrices are generated and their mean value, called the random index (RI), is tabulated Table 1 below.

Table 1. Random Consistency Index (RI) for different matrix size.

1	2	3	4	5	6	7	8	9
0.00	0.00	0.58	0.90	1.12	1.24	1.32	1.41	1.45

The consistency ratio (CR) proposed by Saaty is as shown in a comparison.

$$CR = \frac{CI}{RI} \tag{5}$$

The measurement of the consistency ratio (CR) can allow to conclude whether the evaluations are sufficiently consistent, which is calculated as a comparison between Consistency Index (CI) and the random index (RI). The value of Consistency Ratio

Table 2. Scale for pairwise comparisons.

Importance intensity	Definition
1	Equal importance
3	Moderate importance of one over another
5	Strong importance of one over another
7	Very strong importance of one over another
9	Extreme importance of one over another
2, 4, 6, 8	Intermediate values
Reciprocals	Reciprocals for inverse comparison

(CR) less than 0.1 is acceptable. If it is greater than 0.1, then the comparison matrix should be revised and the consistency ratio recomputed to assure consistency.

In Table 2, where it is assumed that the i th criterion is more or less or equally important than the j th criterion according to Saaty's semantic scale [9].

Finally, Synthesizing the Total Priority for the Alternatives. After the local weights of decision alternatives with respect to decision criteria are generated by using pairwise comparison, the total priority (overall weight) can be aggregated by using the following simple additive weighting (SAW) [10] method:

$$W_{A_i} = \sum_{j=1}^{m} W_{ij} W_j, i = 1, \ldots, n \tag{6}$$

In Eq. (6), where $W_j (j = 1, \ldots, m)$ are the decision criterion weights determined by the AHP methodology, $W_{ij} (j = 1, \ldots, m)$ are the weights of decision alternatives with respect to criterion j, and $W_{A_i} (i = 1, \ldots, n)$ are the overall weights (total priority) of the n alternatives.

2.2 Non-discretionary Variable in Data Envelopment Analysis (DEA-NDSC)

Data envelopment analysis (DEA) is a linear programming method to measure productive efficiency value of a set of decision making units (DMUs) that change several inputs into several yields. As pointed out by paper [11] in their reviews in the history of DEA, the oldest paper provided the background for DEA that they have found from 1957 [12]. The original DEA model was proposed by Charnes et al. [13] to estimate extremal relations from observational data. The most basic type of DEA is CCR known by the initials of its developers - CCR (Charnes, Cooper, and Rhodes). The CCR model allows each participating unit being measured the efficiency to objectively determine optimal weights by reference to the observational data for the multiple outputs and inputs in the model. Besides CCR model, there are several representing models of DEA, such as BCC, SBM, RAM, ST, FG, [14] that have been applied to diverse fields.

A non-discretionary variable model allows non-discretionary inputs that impact production. [15] The production technology transforming multiple inputs $x = (x_1, \cdots, x_M) \in \mathbb{R}_+^M$ into multiple outputs $y = (y_1, \cdots, y_S) \in \mathbb{R}_+^S$ for $j = 1, \cdots, J$ firms are represented by the input set $L(y) = \{x : (y, x) \text{ is feasible}\}$.

For each output vector y, $L(y|z)$ has isoquant Isoq $L(y|z) = \{x : x \in L(y|z), \lambda x \notin L(y|z), \lambda \in [0, 1]\}$, and efficient subset Eff $L(y|z) = \left\{x : x \in L(y|z), x' \notin L(y|z), x' \leq x\right\}$.

The Ruggiero input-oriented (variable returns to scale) efficiency measure for production possibility (y^0, x^0) can be defined as a three-stage procedure.

$$
\begin{aligned}
F(y^0, x^0) &= \min \lambda \\
&\text{s.t.} \sum_{j=1}^{J} \theta_j y_i^j \geq y_i^0 \, \forall i = 1, \cdots, S, \\
&\sum_{j=1}^{J} \theta_j x_k^j \leq \lambda x_k^0 \, \forall k = 1, \cdots, M, \\
&\sum_{j=1}^{J} \theta_j = 1, \\
&\theta_j \geq 0 \, \forall j = 1, \cdots, J
\end{aligned} \tag{7}
$$

And the second-stage regression is specified as

$$
FS = \alpha + \beta_1 z_1 + \cdots + \beta_R z_R + \varepsilon \tag{8}
$$

After constructing z, the third-stage linear program is solved as follows:

$$
\begin{aligned}
R3S(y^0, x^0) &= \min \lambda \\
&\text{s.t.} \sum_{j=1}^{J} \theta_j y_i^j \geq y_i^0 \, \forall i = 1, \cdots, S, \\
&\sum_{j=1}^{J} \theta_j x_k^j \leq \lambda x_k^0 \, \forall k = 1, \cdots, M, \\
&\theta_j = 0 \text{ if } z^j > z^0, \\
&\sum_{j=1}^{J} \theta_j = 1 \\
&\theta_j \geq 0 \, \forall j = 1, \cdots, J
\end{aligned} \tag{9}
$$

3 Integrated DEA-NDSC/AHP Approach

In this study, we propose an integrated DEA-NDSC/AHP method to help the government to evaluate the relative scores of public security in Chinese cities. Figure 1 shows the schematic of the proposed method.

Step 1: Identify criteria for evaluating public safety in China city

In public safety evaluation, analyzing the selection criteria is an essential part of the process. Together with a comprehensive search from related issues [6, 16, 17], we select i + j + k criteria (i input criteria, j non-discretionary input criteria, k output criteria) for the safety evaluation of Chinese cities in this study.

Step 2: Data collection

An evaluation of public safety can use both quantitative and qualitative data. Data collection should include characteristics of crime rates and event rates: violent crimes, property crimes, other crimes, traffic accidents and fire accidents. The two discretionary inputs used in the analysis were the Police force ratio, and public safety expenditure. In addition, four variables were considered as environmental factors affecting the public safety.

Step 3: Obtain AHP score by computing weights of available cities on safety criteria.

Step 4: Obtain DEA-NDSC score of public security inputs efficiency of each given city.

Step 5: Compute combined scores of each city

In this step, we calculate combined score by adding up each DEA-NDSC score and AHP score to for evaluating the public security of city. The public security with high combined score means that the city is safe and effective.

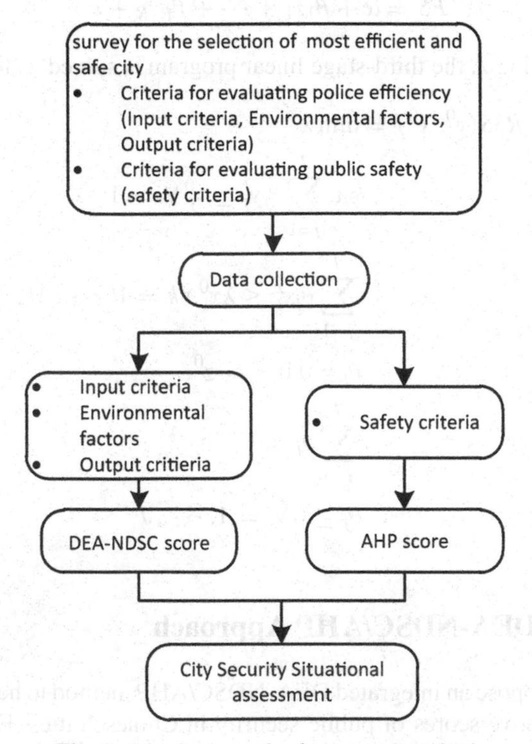

Fig. 1. The framework of proposed method

4 Empirical Application

China's hierarchical classification is well known as tier-city, which contains 338 cities ranked on 6 tiers: tier 1, new tier 1, tier 2, tier 3, tier 4, and tier 5. We select 10 cities from tier 1, new tier 1 and tier 2 to evaluate the public security in this research. We collect data on public security from the city yearbook Statistics of China. Data regarding environmental factors (Per capita GDP, Temporary population ratio, low guarantor ratio, Residents' satisfaction with social security ratio) that may influence the public safety were taken from the city Budget and investigation report.

4.1 AHP Evaluation of Public Security

The hierarchy for public security prioritization is shown in Fig. 2. The decision hierarchy has three levels. The goal of the decision process is the first level of the hierarchy. The second level is consisted of three important criteria for public security evaluation. The criteria include the following:

C1: Crime and facts: This criterion comprises main indicators of public safety, including crime rates, road safety indicator, and fire safety indicator.

C2: Security inputs: It is generally believed that the government's public safety expenditures and the number of police officers have a significant impact on public safety.

C3: environmental factors: This criterion is not a significant factor direct affecting public safety.

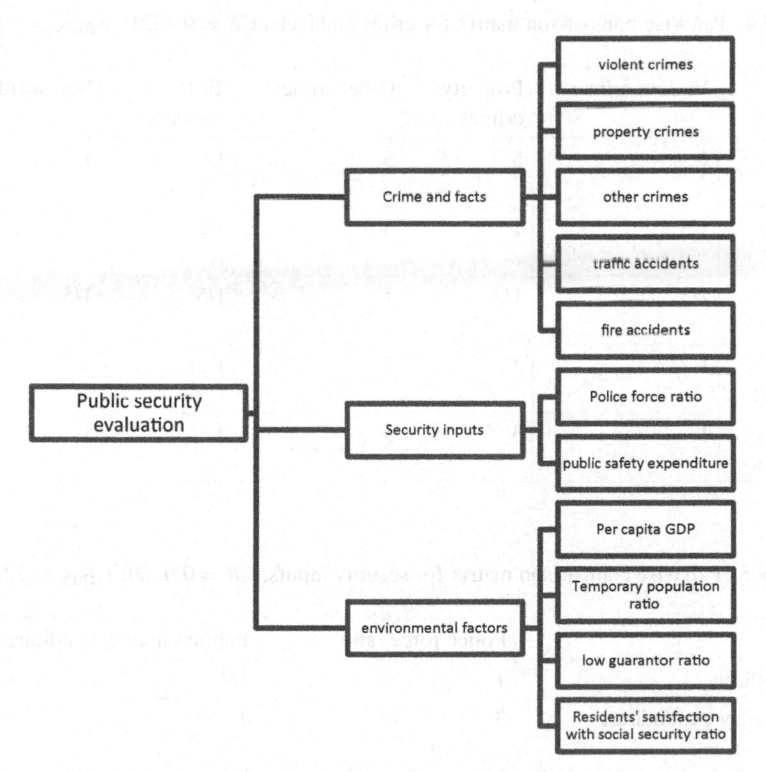

Fig. 2. The hierarchical structure of public security evaluation

Weight Calculation. Security expert rating was used for AHP analyzing to get synthetic weights. After determining the first level criteria, experts make pairwise comparisons of the criteria in Table 3. The decision matrixes of "crime and facts", "security inputs" and "environmental factors" were shown in Table 4, Table 5, and Table 6. The Consistency ratio of the pairwise comparison matrix was calculated using formula (5) as listed was less than 0.1. Thus, the weights of each level's factors were consistent and could be calculated according to the formula (3). The final outcome of weights was obtained and shown in Table 7.

Table 3. Pairwise comparison matrix for the first level criteria, $CR = 0.0176$; $\lambda_{max} = 3.0183$

	Crime and facts	Security inputs	Environmental factors
Crime and facts	1	2	3
Security inputs	1/2	1	1
Environmental factors	1/3	1	1

Table 4. Pairwise comparison matrix for crime and facts, $CR = 0.0221$; $\lambda_{max} = 5.0991$

	Violent crimes	Property crimes	Other crimes	Traffic accidents	Fire accidents
Violent crimes	1	5	5	1	1
Property crimes	1/5	1	3	1/3	1/3
Other crimes	1/5	1/3	1	1/5	1/5
Traffic accidents	1	3	5	1	1
Fire accidents	1	3	5	1	1

Table 5. Pairwise comparison matrix for security inputs, $CR = 0.0000$; $\lambda_{max} = 2.0000$

	Police force ratio	Public safety expenditure
Police officers	1	1/3
Public safety expenditure	3	1

Table 6. Pairwise comparison matrix for environmental factors, $CR = 0.0888$; $\lambda_{max} = 4.2370$

	Per capita GDP	Temporary population ratio	Low guarantor ratio	Residents' satisfaction with social security ratio
Per capita GDP	1	1/3	3	3
Temporary population ratio	3	1	7	5
low guarantor ratio	1/3	1/7	1	3
Residents' satisfaction with social security ratio	1/3	1/5	1/3	1

Table 7. Public security index system and weight distribution

Level 2 category	Local weight	Level 3 variable	Local weight	Weight with respect to category
Crime and facts	0.5499	Violent crimes	0.3100	0.1705
		Property crimes	0.0957	0.0526
		Other crimes	0.0495	0.0272
		Traffic accidents	0.2724	0.1498
		Fire accidents	0.2724	0.1498
Security inputs	0.2402	Police officers	0.7500	0.0601
		Public safety expenditure	0.2500	0.1802
Environmental factors	0.2098	Per capita GDP	0.2337	0.0490
		Temporary population ratio	0.5785	0.1214
		low guarantor ratio	0.1161	0.0244
		Residents' satisfaction with social security ratio	0.0718	0.0151

4.2 DEA Evaluation of Public Security Inputs Efficiency

The data of input, environmental and output variables of this paper were applied to 2017 data and were taken from the City yearbook. The descriptive statistics are provided in Table 8.

Table 8. Descriptive statistics

Variable	Maximum	Minimum	Mean	Std. Dev.
Security inputs				
Police officers	74275	3937	32854	23065.579
Public safety expenditure	356.12	11.24	121.8785	109.4878
Outputs (Crime and facts)				
Violent crimes	0.550307	0.037014	0.18452	0.157177
Property crimes	0.68773	0.082941	0.326799	0.175531
Other crimes	1.632689	0.009102	0.718716	0.425213
Traffic accidents	0.246012	0.017193	0.130204	0.082757
Fire accidents	0.437168	0.154185	0.216836	0.077241
Environmental factors				
Per capita GDP	187000	51015	116445.5	42908.7
Temporary population ratio	1.10374	0.322898	0.720982	0.1967
low guarantor ratio	0.035275	0.000497	0.011085	0.010089
Residents' satisfaction with social security ratio	0.96	0.84	0.894	0.042942

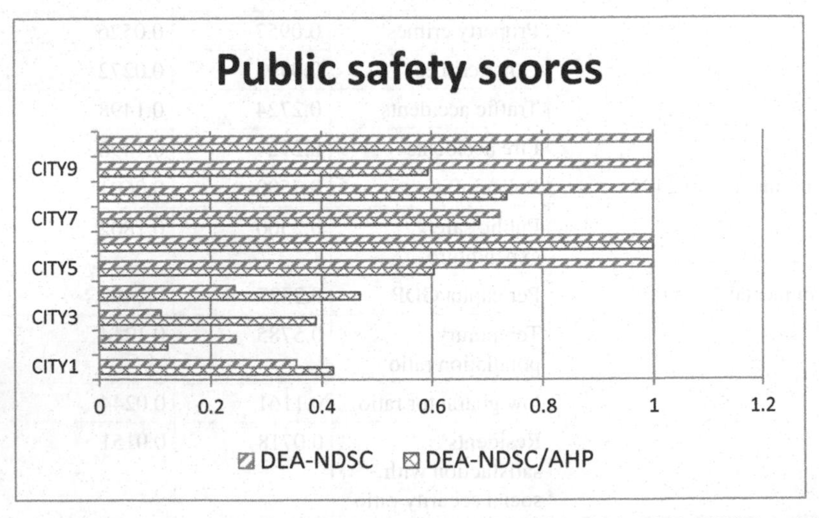

Fig. 3. Comparison of scores of each city security evaluation with DEA-NDSC and DEA-NDSC/AHP

We apply DEA-NDSC and DEA-NDSC/AHP model in sample cities to generate public safety scores. The scores for all cities were obtained and the public safety evaluation comparison between two methods is shown in Fig. 3. The public safety score is always the-bigger-the-better. We observe that the inefficient cities (city1, city2, city3, city4, city7) represent that the local governments have invested a lot of manpower and resources while they can not provided adequate security services. In relation to the ranking of public safety, city6 with the highest ranking is consistent between the two models. However, the lowest ranked city2 using DEA-NDSC/AHP model (penultimate ranking using DEA-NDSC model) because of its Security service inefficiency and high crime rate. Meanwhile, Scores with single DEA-NDSC method lack discrimination, because 5 cities get the same score.

5 Conclusions

The main aim of this paper was to perceive security posture by integration of AHP and DEA-NDSC analysis in the context of city public safety. The results are useful in improving the government understanding the trends of the public safety in Chinese cities. The public security situation analysis needs to consider a number of security situation indicators. Usually, because the analysis results of individual indicators are incompatible, direct analysis is more difficult, and the integration analysis model can effectively solve this problem. In addition, the proposed method not only can sense public safety, but also compare the efficiency of urban security services. In Sect. 4, half of the local governments are inefficient relative to their peers. This means that there are opportunities for cost cutting via contraction of Police officers and Public safety expenditure. The research also offers a simple enough and applicable approach to a number of multiple criteria decision making problems. The control part of public security (such as early warning, tracking and management of public security) needs to study in the future research.

Acknowledgements. This work is supported by Guangdong Science and Technology Plan Project (2018KJYZ009), the young teachers training of Guangdong police officer college (2018QNGG06), and Key research platforms and projects of universities in Guangdong province (2018KTSCX157).

References

1. Wu, C., Lee, V., McMurtrey, M.E.: Knowledge composition and its influence on new product development performance in the big data environment. Comput. Mater. Continua **60**(1), 365–378 (2019)
2. Xiao, B., Wang, Z., Liu, Q., Liu, X.: SMK-means: an improved mini batch k-means algorithm based on mapreduce with big data. Comput. Mater. Continua **56**(3), 365–379 (2018)
3. Wang, B., et al.: Research on hybrid model of garlic short-term price forecasting based on big data. Comput. Mater. Continua **57**(2), 283–296 (2018)
4. Zhang, N., Huang, H.: Social vulnerability for public safety: a case study of Beijing, China. Chin. Sci. Bull. **58**(19), 2387–2394 (2013)
5. Byun, G., Ha, M.-K.: An analysis of crime safety evaluation indicators in urban outdoor public space by using AHP. J. Archit. Inst. Korea Plann. Design **35**(5), 11–20 (2019)
6. Wu, T.-H., Chen, M.-S., Yeh, J.-Y.: Measuring the performance of police forces in Taiwan using data envelopment analysis. Eval. Program Plann. **33**(3), 246–254 (2010)

7. Sinuany-Stern, Z., Alper, D.: Factors affecting police station efficiency: DEA in police logistics. Int. J. Logist. Syst. Manag. **34**(1), 75–101 (2019)
8. Wind, Y., Saaty, T.L.: Marketing applications of the analytic hierarchy process. Manag. Sci. **26**(7), 641–658 (1980)
9. Saaty, T.L.: The Analytic Hierarchy Process. McGraw Hill, New York (1980)
10. Hwang, C.-L., Yoon, K.: Methods for multiple attribute decision making. In: Hwang, C.-L., Yoon, K. (eds.) multiple attribute decision making. Lecture Notes in Economics and Mathematical Systems, vol. 186, pp. 58–191. Springer, Heidelberg (1981). https://doi.org/10.1007/978-3-642-48318-9_3
11. Mardani, A., et al.: A comprehensive review of data envelopment analysis (DEA) approach in energy efficiency. Renew. Sustain. Energy Rev. **70**, 1298–1322 (2017)
12. Farrel, J.M.: The measurement of productive efficiency (1957)
13. Charnes, A., Cooper, W.W., Rhodes, E.: Measuring the efficiency of decision making units. Eur. J. Oper. Res. **2**(6), 429–444 (1978)
14. Cooper, W.W., Seiford, L.M., Zhu, J.: Data envelopment analysis. In: Handbook on Data Envelopment Analysis, pp. 1–39. Springer, Heidelberg (2004)
15. Ruggiero, J.: Non-discretionary inputs in data envelopment analysis. Eur. J. Oper. Res. **111**(3), 461–469 (1998)
16. Gorman, M.F., Ruggiero, J.: Evaluating US state police performance using data envelopment analysis. Int. J. Prod. Econ. **113**(2), 1031–1037 (2008)
17. Sun, S.: Measuring the relative efficiency of police precincts using data envelopment analysis. Socio-Econ. Plann. Sci. **36**(1), 51–71 (2002)

Detection and Information Extraction of Similar Basic Blocks Used for Directed Greybox Fuzzing

Chunlai Du[1], Shenghui Liu[1], Yanhui Guo[2(✉)], Lei Si[2], and Tong Jin[1]

[1] School of Information Science and Technology, North China University of Technology, Beijing 100144, China

[2] Department of Computer Science, University of Illinois Springfield, Springfield, IL, USA

yguo56@uis.edu

Abstract. Directed gray-box fuzzing generates input samples with the objective of reaching a given set of target program locations efficiently so that improves the fuzzy efficiency and reduces the time cost. This Scheme can find well the vulnerabilities hided in update patch so that relies heavily on feature extraction of target blocks. Whether there are other basic blocks with similar features in the target program to speed up the efficiency of vulnerability fuzzing becomes the starting point of this paper. Our main work focuses on the static analysis of the target program to find feature similar blocks. We proposed a similarity feature discovery model of blocks by designing basic feature description vector of block. Standard feature extraction of malicious basic block from lava dataset by which we can quickly fuzz these basic blocks with similar characteristics and possibly potential threats in the target program. Through experiments, we find other basic blocks similar to malicious basic blocks and add them into dataset so that speed up the effectiveness of vulnerability fuzzing in directed gray-box fuzzing mode.

Keywords: Directed gray-box fuzzing · Static analysis · Deep learning

1 Introduction

In recent years, with the rapid development of the network, the demand for software in various industries is growing rapidly, which inevitably leads to a sharp increase in the number and types of software. At the same time, software development no longer has the characteristics of lightweight code and single function, but pursues the completeness and practicability of function, which inevitably leads to the increase of software complexity. For network security researchers, the increase of software complexity brings about the increase of software vulnerability. In the field of Cyberspace Security, the confrontation between attacker and defender is becoming more and more serious [1–4]. It is an important method for attackers to launch attacks by using the vulnerabilities of the target software or operation system. Just in the first half of 2019, the number of security vulnerabilities has been kept above 1000 every month, and the number of high-risk vulnerabilities is growing synchronously. the number and type of vulnerability

X. Sun et al. (Eds.): ICAIS 2020, LNCS 12240, pp. 353–364, 2020.
https://doi.org/10.1007/978-3-030-57881-7_32

CVEs (Common Vulnerabilities and Exposures) released each year from 1999 to 2019 is shown in Fig. 1 and Fig. 2 [5, 6].

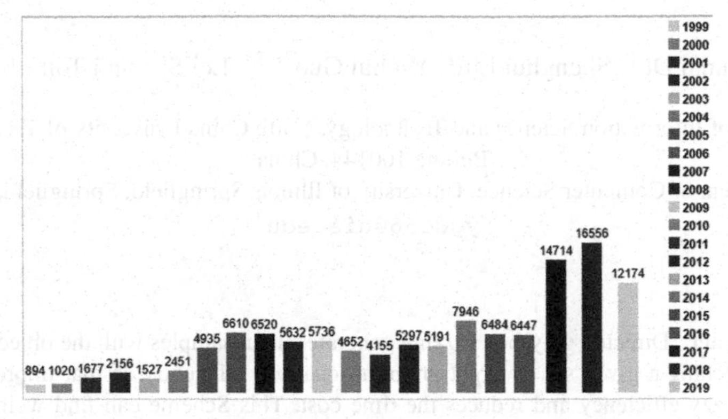

Fig. 1. Vulnerabilities by year

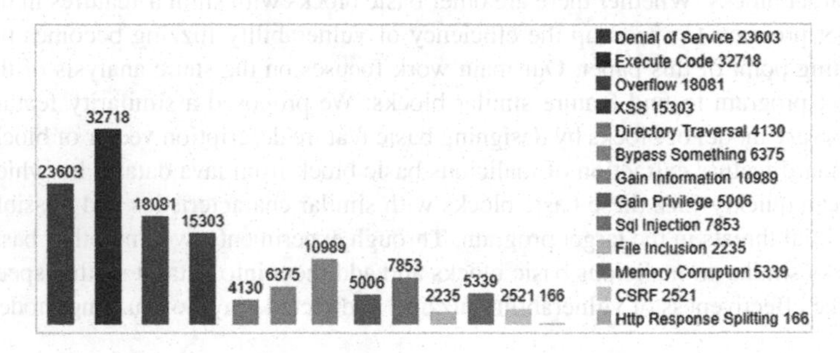

Fig. 2. Vulnerabilities by type

Although security researchers are constantly exploiting vulnerabilities to reduce the security threat, the technology of attackers is also constantly updated. software security is always an inevitable battlefield of confrontation. Therefore, it is of great significance to analyze software vulnerabilities, innovate detection technology and improve detection accuracy.

The key contributions of our work are shown as follows:

- designed a feature description vector of basic block.
- proposed a detection and extraction algorithm of suspiciously malicious basic blocks whose features are similar to the known malicious basic blocks.
- the set of suspiciously malicious basic blocks can be used for Directed Greybox Fuzzing to speed up the effectiveness as one of the inputs.

The rest of the paper is organized as follows. Section 2 show related work. In Section 3, we describe the proposed similarity detection model of static basic block. Section 4 shows the experimental results. Section 5 presents the conclusions and future work and acknowledgement given in acknowledgement section.

2 Related Work

Traditional software vulnerability detection can be divided into content-based mode and behavior-based mode according to its data source. The content-based mode is called static analysis. Researchers analyze program function through open-source code or disassembly code of binary execution files, and find out program vulnerability through manual analysis. The behavior-based mode is the so-called narrow sense dynamic analysis (i.e. excluding dynamic pin insertion, stain analysis, symbolic execution and so on). It is necessary to execute the program through actual mode or virtual mode, and observe the execution of the program to find the program vulnerability. Through manual analysis, researchers need to have a keen insight into code with deep experience in code analysis. But with the increase of software complexity, the vulnerability fuzzing not only is lack of enough researchers, but also requires less time consumption. As a result, automated vulnerability fuzzing methods are gradually emerging and developing rapidly.

Program execution process can be represented by CFG (control flow graph) and CG (call graph), which can contribute to reveal the basic blocks and their relations in the program [7]. The basic block of program is a sequence of statements executed sequentially with only one entry and one exit. From the view of assembly-level code, the program vulnerability is essentially triggered by the crash of some basic blocks. Therefore, it is an import way to exploit vulnerabilities by fuzzing and analyzing the dangerous basic blocks of the program. Iván [8] point out that most programmers spend more time modifying code than writing new software functions. Accordingly, different versions of the same software maybe contain a lot of the same code or similar code. Once one version of the software exposed some vulnerability, the other versions maybe have the same vulnerability accordingly.

In static analysis mode, program control flow is an important analysis method of program execution process. Zhou [9] proposed a vulnerability detection algorithm based on the definition of syntax rules for each type of vulnerability which includes unused variable, use_of_uninitialized_variable and UAF (use_after_free). Feng [10] proposed a method called Bintaint which performs static taint analysis and generates a TCFG (taint control flow graph). Combined with two-way search technology and path guidance algorithm, all known vulnerabilities can be detected without false negatives, and the problem of path explosion and computing cost can be effectively alleviated. Jihun Kim [11] proposed the execution path flow of program by analyzing the execution process, and adds the path flow to the binary program analysis process by using BFS (breath first search) algorithm. Program's assembly instructions and registers are also the focus of security researchers. Analyzing the assembly instructions of programs and the operation of registers is an import work, such as modifying or replacing some values maybe lead the program to crash. Through this operation, potential program vulnerabilities can be found effectively. Ghiasi [12] proposed a model to analyze the operation of binary program and find out different influences on the values in program registers. By detecting this effect, program vulnerabilities with similar behaviors can be found. Tian [13] proposed a model to identify the entry instructions of function module through static binary analysis, and implemented dynamic detection through memory security in combination with virtual machine. It mainly detects buffer overflow vulnerabilities. Chao [14] proposed a model to analyze the risk function of binary program stack overflow, model the function behavior of stack overflow, and design and implement a method to detect stack overflow vulnerability. Jeon [15] proposed a static binary analysis tool crashfiler, which classifies the crashes in the fuzzing process according to the risk level. Because of the memory location analysis and intraprocedural analysis, its accuracy and analysis scope have been greatly improved. Liang [16] proposed a collaborative analysis method which based on the unified defect pattern set to detect vulnerability. Enas [17] firstly used static analysis tools to classify error information, and then uses support vector machine (SVM), k-nearest neighbor (KNN), random forest and repeated increment algorithm to compare performance.

AFL (American Fuzzy Lop) [18] is a brute-force fuzzer which uses a modified form of edge coverage to effortlessly pick up subtle, local-scale changes to program control flow. Chen [19] proposed a creation and selection model of sample seed with deep learning theory. Rawat [20] used the smart static analysis of learning the key constant bytes in program to reduce the generation of invalid fuzzing samples. Du [21] proposed a grey-box fuzzing model which uses simulated annealing algorithm to gain optimal sample seeds. In order to find those seeds having capacity of executing low frequency of path, Marcel [22] proposed a model Aflfast that Markov chain can be used to specify the probability of execution path i to execution path j.

Directed fuzzing is used to exploit these vulnerabilities towards problematic changes or patches, towards critical system calls or dangerous locations, or towards functions in the stack trace of a reported vulnerability. Marcel [23] proposed a model AFLgo which design a simulated annealing-based power schedule to select seeds closer to the target locations. Directed fuzzer spends more time in searching specific target location than other. Chen [24] proposed a model Hawkeye. Hawkeye precisely collects the information such as the call graph, function and basic block level distances to the targets, evaluates exercised seeds based on both static information and the execution traces to generate the dynamic metrics, which are useful to achieve better directedness and gravitate towards the target.

Directed grey-box fuzzing must firstly obtain the locations of targeted code area, then focus on how to generate and select the better sample seeds which help to execute these paths to the target area as soon as possible. Therefore, directed grey-box fuzzing heavily depends on the number of locations of targeted code area. the more locations, the more gain. Therefore, we proposed a model **SBFK** (Similar Basic Block Found by KNN) to find and output the suspiciously malicious basic blocks of the program.

3 Proposed Model

Our research focuses on the static analysis of the program to provide the directed grey-box fuzzing with more location of suspiciously malicious basic blocks. In order to finish the aim, we search and determine some famous datasets containing a lot of vulnerability. We achieve the crashes by some tools, for example AFL, Vuzzer, etc. According to these achieved crashes, we further achieve the basic blocks which trigger crash due to uncompliant inputs. We extract the features of the malicious basic blocks by designing vector expression and build the dataset. Based this dataset, we check the target program by **KNN** (K-nearest neighborhood) to find such basic blocks that maybe have the similar vulnerability features. Based on those found basic blocks, we output the location used for directed grey-box fuzzing tools, such as AFLgo.

In this section, we first introduce the model **SBFK**. Because KNN Algorithm is used in our model, we then make a brief introduction to KNN theory. The rest introduce the modules of **SBFK** in order.

3.1 Model-SBFK

The model **SBFK** (Similar Basic Block Found by KNN) mainly includes data set building module, crash module of basic block, extraction module of basic block feature, training module of KNN and detection module. Each module can work independently and provide data input and output for other modules (see in Fig. 3).

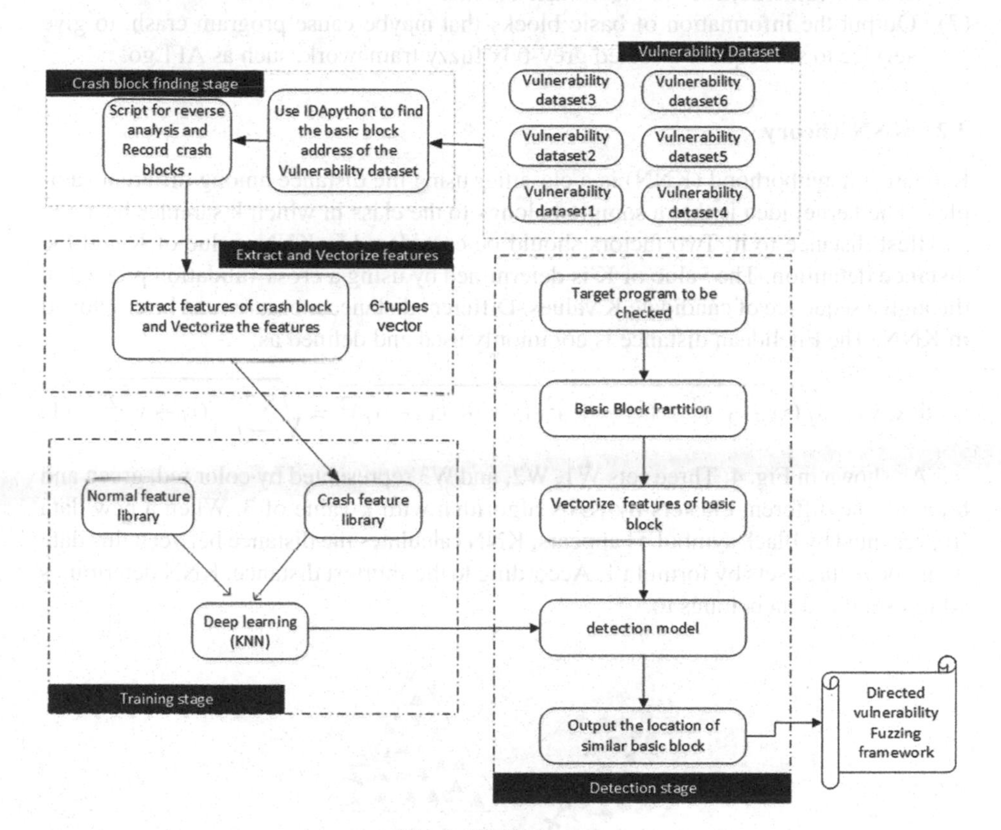

Fig. 3. SBFK model

This research mainly serves for the directed grey-box fuzzing tools, and provides the location of the target basic block and similar basic blocks in binary program. The main work includes:

(1) Determine the dataset source whose vulnerability blocks can be triggered.
(2) Fuzzing the programs from vulnerability dataset by AFL tools.
(3) Record the position of crash.
(4) Analyze and extract the features of these crash blocks by six-dimension vector expression.
(5) In order to apply KNN to train detection module more accurate, we also introduce some features of these basic block not triggering vulnerabilities.
(6) Detect and find the similar binary basic block of targeted program which maybe contain vulnerabilities in these basic blocks.
(7) Output the information of basic blocks that maybe cause program crash, to give service to subsequent directed grey-box fuzzy framework, such as AFLgo.

3.2 KNN Theory

K-nearest neighborhood (KNN) is a classifier using the distance among different samples. The kernel idea is that a sample belongs to the class in which k samples have the smallest distance to it. Two factors should be considered in KNN: value of K and the distance definition. The value of K is determined by using a cross validation procedure through a sequence of candidate K values. Different distance measures can be employed in KNN. The Euclidean distance is commonly used and defined as:

$$d(x, y) = \sqrt{(x_1 - y_1)^2 + (x_2 - y_2)^2 + \cdots + (x_n - y_n)^2} = \sqrt{\sum\nolimits_{i=1}^{n} (x_i - y_i)^2} \quad (1)$$

As shown in Fig. 4, Three sets W1, W2, and W3 represented by color red, green and blue are the different clusters by KNN algorithm with k-value of 3. When a new data (represented by black symbol +) appears, KNN calculates the distance between this data with above three sets by formula 1. According to the shortest distance, KNN determines which set this data belongs to.

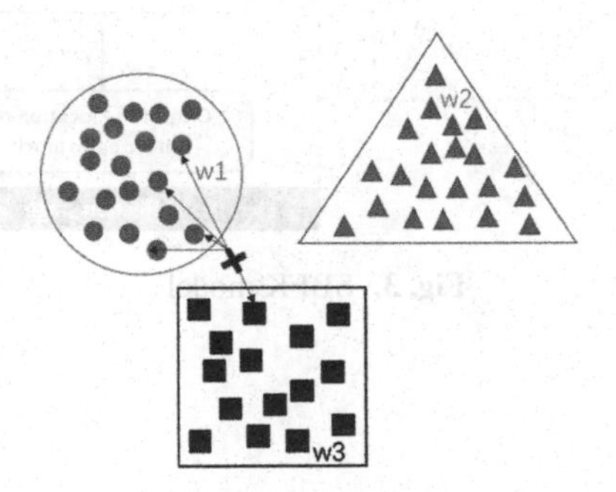

Fig. 4. Example

3.3 Dataset Construction

The system selects the LAVA data set [25, 26] as the training set. LAVA is a set of software used to evaluate the vulnerability detection tool. The vulnerability information is injected into the binary program through manual construction. LAVA generates and validates the validity and availability of the dataset. Currently, although there are many public corpora that explain very well on the web, they will soon become obsolete due to their release. LAVA, based on pre-training, considers new exploitable vulnerabilities in its process, thus overcoming the shortcomings of the public corpus. The vulnerabilities in LAVA are synthetic, but in a sense, they are embedded in the depths of the program, triggered by real input, and are still realistic.

For each vulnerability that injects a binary file, LAVA provides the corresponding input file that triggers the vulnerability and the corresponding back-traces file. The LAVA data set contains two parts, LAVA-1 and LAVA-M. LAVA-M includes four programs, namely base64, md5sum, uniq and who. This paper mainly performs static analysis on LAVA-M.

3.4 Dataset Preprocess

Manual analysis of the four binary programs in the LAVA-M data set, using the input files provided by LAVA to dynamically debug the binary program, find the crash point corresponding to each vulnerability, extract the corresponding malicious basic blocks and define it as the crash blocks. Due to the injection of more vulnerabilities, in order to ensure the timeliness of data extraction, the automated basic block extraction module of triggering binary program to crash was developed, and the extracted results were saved to the JSON file to prepare for the subsequent work.

Figure 5 show a snapshot of basic block when the program crash. Figure 6 show the process of locating the crash address and achieving the data of the crash block.

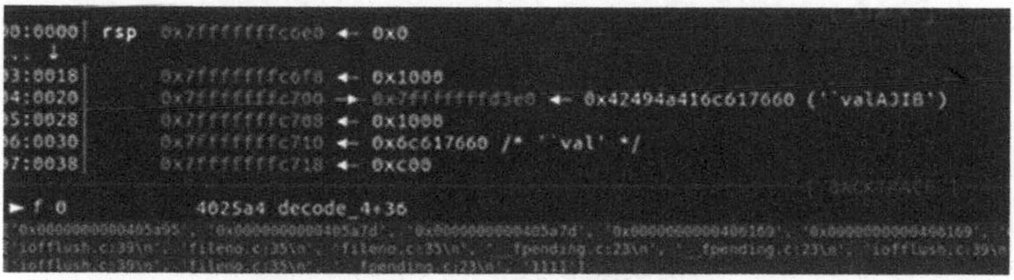

Fig. 5. Snapshot of crashed program

3.5 Extracting Feature Data

Binary program and disassembled text processing methods are a key issue in the study of binary program reverse analysis. We adopt vector expression of six-dimension tuple by analyzing the basic block structure of binary programs. The composition of the six-dimension tuple vector is show as follow.

<register, memory, Immediate, Call, Jxx, Total>

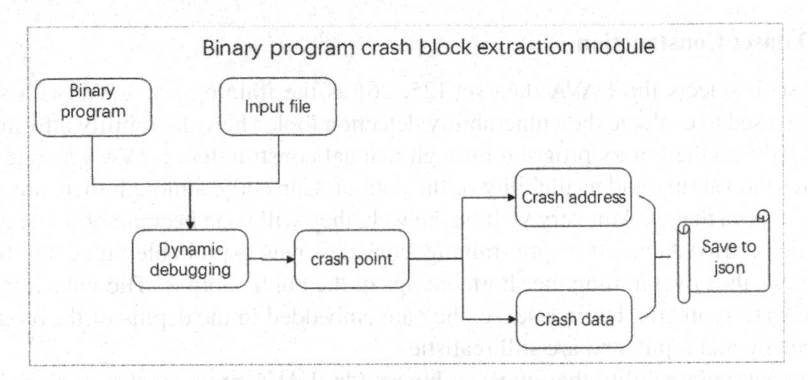

Fig. 6. Crash block extraction module

The meaning of each factor is shown as follow:

- **Register** - the number of the target register operation instruction in crash block.
- **Memory** - the number of the target memory operation instruction in crash block.
- **Immediate** - the number of the source operation code being immediate number in crash block.
- **Call** - the number of call instructions in crash block.
- **Jxx** - the number of Jump instruction in crash block, such as JMP, JNZ, JZ, JL, JA, JNA, JL, and so on.
- **Total** - the total number of instructions in crash block.

The six-dimension tuple takes full account of the structure of basic block and its relation while extracting features of the crashed block. Figure 7 shows an example of feature extraction of basic block and constructs a six-dimension tuple **<4, 0, 1, 0, 1, 5>** as result to output. Based on the extracted six-dimension tuple, the crash feature dataset is constructed.

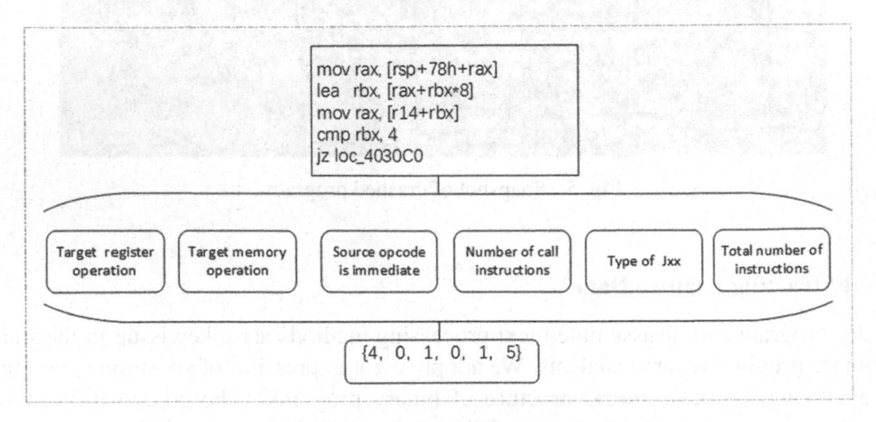

Fig. 7. Extraction of feature vector

In order to detect these crash blocks and similar blocks better, we also construct the six-dimension tuple of normal blocks with the same method.

3.6 Classifier

Based on the dataset of six-dimension tuple, a feature vector with six attributes is used to describe all crash blocks and normal blocks. Statistical analysis labeled the crash blocks as A and normal ones as B, and then their feature vectors are normalized to reduce the effect of data distribution. This dataset is split with 90% for the training data and 10% for the testing data.

KNN is used for classifier to extract feature in training phase and detect feature in testing phase. The process is shown in Fig. 8.

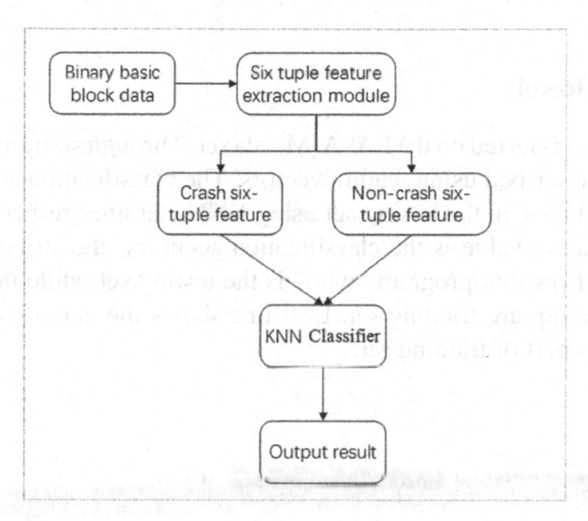

Fig. 8. KNN classifier

3.7 Output of Similar Block

Based on the KNN classifier, Our Model SBFK can detect the crash blocks and similar blocks. Firstly, we make the target program into basic blocks with static analysis and extract the six-dimension tuple of each basic block. Then, we use the trained KNN classifier to detect whether the target program include crash blocks or suspicious blocks. Once detected crash blocks or similarly suspicious blocks,we add them into output data set. The output set can be an input of directed gray-box fuzzing tool, such as AFLgo, to improve the speed of vulnerability exploitation.

4 Experiment

4.1 Experiment Environment

The experiment environment is shown in Table 1.

Table 1. Experimental environment.

Configure	Versions
Operating system	Ubuntu 16.04.1 64-bit
Kernel version	4.8.0-58-generic
Processor	Intel Xeon(R)
Memory	36 GB
Python	py3-v5.0.0-linux64
IDA	6.8
IDAPython	1.7.2
Dataset	LAVA-M

4.2 Experiment Result

The proposed model is tested on the LAVA-M data set. Through static analysis, all blocks are extracted and described using feature vectors. The classification accuracy is 96% in the training set and 88% in the testing set using KNN classifier respectively.

In Fig. 9, Abscissa value is the classification accuracy. the string "base64 + md5 + uniq who" descripts that program "who" is the testing set while the others, such as base64, md5 and uniq, are training set. Last line shows the classification accuracy of testing set which is part of training set.

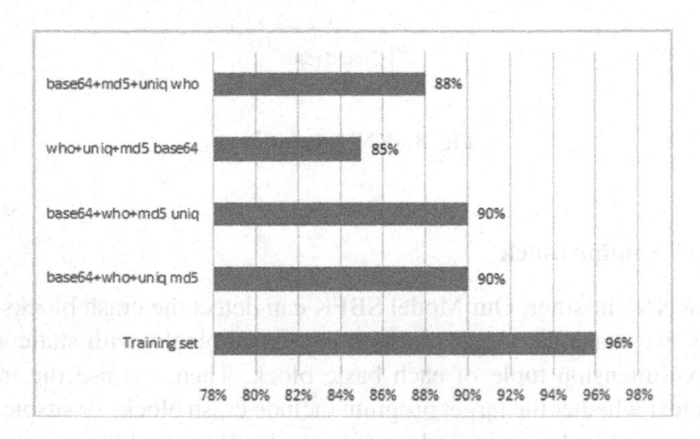

Fig. 9. Experiment results

5 Conclusion

This work proposed a method to identify the potentially crashed blocks used for directed vulnerability fuzzing on the targeted programs. It mainly includes Fuzzing of basic blocks, location of crash address and feature extraction of crash blocks. The features are extracted from the crash blocks by six-dimension tuple. The KNN algorithm is trained to identify the crash blocks or suspiciously crash blocks in the targeted program. The proposed method not only presents a way to efficiently identify potentially threatening basic blocks in targeted program, but also provides inputs for the directed fuzzing tools in our next work.

Acknowledgement. This work was supported by Natural Science Foundation of China (61702013), Joint of Beijing Natural Science Foundation and Education Commission (KZ201810009011), Science and Technology Innovation Project of North China University of Technology (19XN108).

References

1. Arif, M.S., Raza, A., Shatanawi, W., et al.: A stochastic numerical analysis for computer virus model with vertical transmission over the internet. Comput. Mater. Continua **61**(3), 1025–1043 (2019)
2. Tian, Z., Shi, W., Wang, Y., et al.: Real-time lateral movement detection based on evidence reasoning network for edge computing environment. IEEE Trans. Ind. Inf. **15**(7), 4285–4294 (2019)
3. Yu, X., Tian, Z., Qiu, J., et al.: An intrusion detection algorithm based on feature graph. Comput. Mater. Continua **61**(1), 255–274 (2019)
4. Zhao, W., Li, P., Zhu, C., et al.: Defense against poisoning attack via evaluating training samples using multiple spectral clustering aggregation method. Comput. Mater. Continua **59**(3), 817–832 (2019)
5. Vulnerabilities by Date. https://www.cvedetails.com/browse-by-date.php. Accessed 22 Dec 2019
6. Vulnerabilities by Type. https://www.cvedetails.com/vulnerabilities-by-types.php. Accessed 22 Dec 2019
7. Toman, M.: LLVM IR service for Fedora, Masaryk University, USA (2013). https://is.muni.cz/th/n9bfn/dp.pdf. Accessed 22 Dec 2019
8. García-Ferreira, I., Laorden, C., Santos, I.: Static analysis: a brief survey. Logic J. IGPL **24**(6), 871–882 (2016)
9. Zhou, M., et al.: A method for software vulnerability detection based on improved control flow graph. Wuhan Univ. J. Nat. Sci. **24**(2), 149–160 (2019). https://doi.org/10.1007/s11859-019-1380-z
10. Feng, Z., Wang, Z., Dong, W., et al.: Bintaint: a static taint analysis method for binary vulnerability mining. In: International Conference on Cloud Computing, Big Data and Blockchain (ICCBB), Fuzhou, China, pp. 1–8. IEEE (2018)
11. Kim, J., Youn, J.M.: Malware behavior analysis using binary code tracking. In: International Conference on Computer Applications and Information Processing Technology (CAIPT), Kuta Bali, Indonesia, pp. 1–4. IEEE (2017)
12. Ghiasi, M., Sami, A., Salehi, Z.: Dynamic VSA: a framework for malware detection based on register contents. Eng. Appl. Artif. Intell. **44**, 111–122 (2015)
13. Tian, D., Xiong, X., Changzhen, H., et al.: Defeating buffer overflow attacks via virtualization. Comput. Electr. Eng. **40**(6), 1940–1950 (2014)
14. Feng, C., Zhang, X.: A static taint detection method for stack overflow vulnerabilities in binaries. In: International Conference on Information Science and Control Engineering (ICISCE), ChangSha, China, pp. 110–114. IEEE (2017)

15. Jeon, H.-G., Mok, S.-K., Cho, E.-S.: Automated crash filtering using interprocedural static analysis for binary codes. In: Annual Computer Software and Applications Conference (COMPSAC), Turin, Italy, pp. 614–623. IEEE (2017)
16. Liang, X., Cui, B., Lv, Y., et al.: Research on the collaborative analysis technology for source code and binary executable based upon the unified defect mode set. In: International Conference on Innovative Mobile and Internet Services in Ubiquitous Computing, Blumenau, Brazil, pp. 260–264. IEEE (2015)
17. Alikhashashneh, E.A., Raje, R.R., Hill, J.H.: Using machine learning techniques to classify and predict static code analysis tool warnings. In: International Conference on Computer Systems and Applications (AICCSA), Aqaba, Jordan, pp. 1–8. IEEE (2018)
18. google/AFL. https://github.com/google/AFL. Accessed 23 Dec 2019
19. Chen, L., Yang, C., Liu, F., et al.: Automatic mining of security-sensitive functions from source code. Comput. Mater. Continua **56**(2), 199–210 (2018)
20. Rawat, S., Jain, V., Kumar, A., et al.: Vuzzer: application-aware evolutionary fuzzing. In: The Network and Distributed System Security Symposium (NDSS), San Diego, California, pp. 1–14. Internet Society (2017)
21. Du, C., Tan, X., Guo, Y.: A gray-box vulnerability discovery model based on path coverage. In: Sun, X., Pan, Z., Bertino, E. (eds.) ICAIS 2019. LNCS, vol. 11635, pp. 3–12. Springer, Cham (2019). https://doi.org/10.1007/978-3-030-24268-8_1
22. Böhme, M., Pham, V.-T., Roychoudhury, A.: Coverage-based greybox fuzzing as markov chain. In: ACM SIGSAC Conference on Computer and Communications Security, Vienna, Austri, pp. 1032–1043. ACM (2016)
23. Böhme, M., Pham, V.-T., Nguyen, M.-D., et al.: Directed greybox fuzzing. In: ACM SIGSAC Conference on Computer and Communications Security (CCS), Dallas, USA, pp. 2329–2344. ACM (2017)
24. Chen, H., Xue, Y., Li, Y., et al.: Hawkeye: towards a desired directed grey-box fuzzer. In: The ACM SIGSAC Conference on Computer and Communications Security (CCS), Toronto, ON, Canada, pp. 2329–2344. ACM (2018)
25. Dolan-Gavitt, B., Hulin, P., Kirda, E., et al.: Lava: large-scale automated vulnerability addition. In: Symposium on Security and Privacy, San Jose, USA, pp. 110–121. IEEE (2016)
26. Panda-re/lava. https://github.com/panda-re/lava. Accessed 22 Dec 2019

Summary of Research on Information Security Protection of Smart Grid

Li Xu and Yanbin Sun(✉)

Cyberspace Institute of Advanced Technology,
Guangzhou University, Guangzhou 510006, China
{2111906094,sunyanbin}@gzhu.edu.cn

Abstract. With the continuous development of smart grid, information security has become an important guarantee to ensure the efficient and stable operation of smart grid. In recent years, researches on the security level of the smart grid has not made some mature architectures, or the existing work has only solved a very small number of security problems. This paper will explain the concept of smart grid and its threats in information security, on this basis we will put forward some corresponding security protection and make detailed description of the status of security research. We hope that this paper can contribute to the safe and stable operation of the smart grid.

Keywords: Smart grid · Information security · Security threats · Security protection

1 Introduction

With the gradual development of science and technology, smart grid has combined the traditional grid with sensory measurement technologies, intelligent cloud computing and wireless communication technologies. These technologies help to solve some problems about the smart grid's power generation, transmission and distribution [1]. Advanced sensory measurement technology, data acquisition and processing systems work together to enable people can obtain the real-time operating status of the smart grid. The comprehensive application of modern communication technology and management technology establishes an intelligent and two-way service model. The smart grid embodies a high degree of intelligence in its operation and truly benefits thousands of families.

However, with the advancement of the smart grid, its information security issues have become increasingly prominent. And the continuous occurrence of the smart grid security incidents has sounded the alarm for people. For example, the US-Mexico blackout that occurred in 2011 was caused by some internal employees who implanted viruses and malicious codes into the substation system. They directly led to the smashing of transmission lines between the United States and Mexico. The Ukrainian smart grid blackout occurred in 2015 that caused a series of serious consequences. The reason was the internal computers of the smart grid were broken which caused by a malicious software called "Black-Energy". It led to a wide range of blackout in western Ukraine for

X. Sun et al. (Eds.): ICAIS 2020, LNCS 12240, pp. 365–379, 2020.
https://doi.org/10.1007/978-3-030-57881-7_33

half an hour. In 2016, the Israel Electricity Authority suffered a grievous network attack [2]. Because the workers could not shut down the virus-infected computers in the grid, which made the smart grid inoperable. It can be seen that the problems of smart grid security become very imminent.

To achieve the protection of the smart grid's security, we must understand its concept, architecture, which type of attack, and some security requirements firstly. In the second chapter, we will introduce the structure and basic technologies of the smart grid. In the third chapter, we will analyze some attacks and threats of smart grid in detail. In the fourth chapter, we will propose the corresponding smart grid's security protection strategies and analyze those measures' advantages and disadvantages. We will also summarize some security risks that faced by the smart grid. In the fifth chapter we will finally summarize the work which carried out in the paper and then look into the smart grid's future.

2 Smart Grid

2.1 Concept

The concept of the smart grid was firstly proposed by the American Electric Power Research Institute [3]. It is a combination of the advanced equipment technology, the advanced network technology and the physical grid. Its characteristics are self-healing, resisting to the external attacks, and highly intelligence. The smart grid applies a series of advanced technologies to generate, transport and distribute electricity. It brings high-quality electricity to our users and promotes the economic development of the society. It also meets the needs of people about the green, healthy, intelligent and highly integrated smart grid. As the smart grid is shown in Fig. 1.

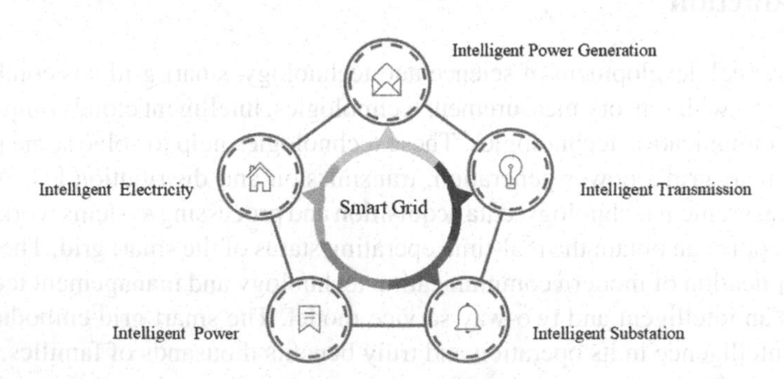

Fig. 1. The smart grid

2.2 Smart Grid's Architecture

Generally speaking, the architecture of the smart grid consists of four layers: the basic hardware layer, the perceptual measurement layer, the information communication layer, and the scheduling operation and maintenance layer. The smart grid architecture is shown in Fig. 2.

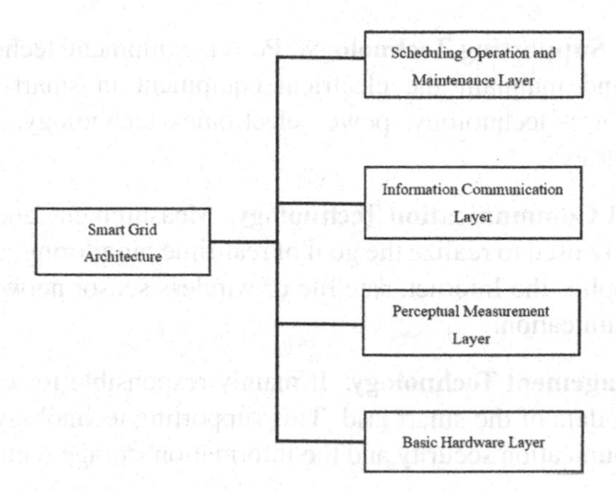

Fig. 2. Smart grid architecture

Basic Hardware Layer. The smart grid's basic hardware layer consists of the physical devices and the communication devices. First of all, the physical equipment mainly includes some advanced power generation facilities, power transmissions, distribution equipment, and some measuring equipment. And the communication equipment comprises some information processing systems, some data storage devices, and the computer processing systems which are responsible for the dispatching of smart grid.

Perceptual Measurement Layer. The primary function of the sensing layer is intelligent sensing, monitoring the operational status of some power equipment, and controlling some decision. Especially, the use of these advanced communication technologies can achieve data docking between the sensory measurement layer and the communication network layer.

Information Communication Layer. The information communication layer is mainly composed of a power backbone communication network and a personal user network. The power backbone communication network adopts communication protocol groups such as IEC60870 and IEC61970. The personal user network usually adopts the Open HAN protocol and the Zigbee protocol which based on IEEE802.15.4 [4].

Scheduling Operation and Maintenance Layer. China divides the large smart grid into small parts according to the region, and sets one or more control centers in each regional smart grid. It also adopts data sharing, distributed methods and artificial intelligence technologies to realize the control and scheduling of the overall business of the smart grid.

2.3 Supporting Technology of Smart Grid

People divide the supporting technology of smart grid into four aspects: basic power equipment, measurement and communication technology, information management technology and decision- controlling technology [5].

Power equipment Supporting Technology. Power equipment technology is mainly used to operate and maintain the electrical equipment in smart grid. It includes high-efficiency energy technology, power electronics technology, transmission and distribution technology.

Measurement and Communication Technology. Measurement and communication technology is mainly used to realize the goal of real-time monitoring, alarming on smart grid, and it also applies the Internet, satellite or wireless sensor network to achieve the smart grid's communication.

Information Management Technology. It mainly responsible for collecting, analyzing and processing data of the smart grid. This supporting technology ensures the data security, the communication security and the information storage security of smart grid.

Decision-Controlling Technology. Decision-controlling technology mainly includes the high-performance computing technology, the real-time distribution system and the database-designing technology. It plays a decision-making role in smart grid's operation. The smart grid's structure hierarchy is shown in Fig. 3.

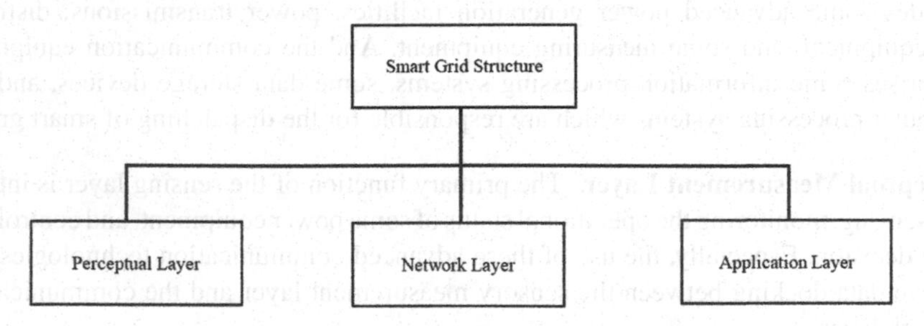

Fig. 3. Smart grid structure hierarchy

3 Security Threats and the Status of Smart Grid

3.1 Smart Grid Security Threat Events

To some extent, the smart grid is a large-scale and complex network, which means that its design and logic are prone to loopholes [6]. As a typical industrial control system, the smart grid is also vulnerable to suffer malicious attacks on its physical devices and communication protocols. The following Table 1 shows some smart grid security incidents all over the world.

3.2 Security Threats to the Smart Grid

The security problems of smart grid can be divided into two categories. The first category is artificially unintentional or maliciously attacking the smart grid. The second type is caused by some natural disasters causing damage to the power facilities and resulting in grid operation failure.

Table 1. Smart grid security incidents

Blackout accident case	2006-11-4 Western Europe blackout	2009-11-10 Brazil blackout	2011-9-8 US Mexico blackout	2012-7-30 India blackout	2015 Ukrainian blackout
Cause of power failure	The power supply company did not strictly enforce the N-1 standard and the coordination between the power companies was improper	Faulty current caused by lightning or heavy rain, power equipment parameter setting error	Failure of employee operation caused the substation monitoring system to malfunction	Protection failed to operate normally, abnormal operation was performed	Power plant computer infected with virus and stopped working
Network attack scenario reappears	The power system is injected into the false data by the attackers, hindering or delaying the communication, and maliciously tampering with the power data	Bypassing the authentication mechanism to tamper with power device parameters through an opening remote configuration interface	Implanting malicious code or viruses in the substation system by internal employees	Using a man-in-the-middle attack interception scheduling instruction to send a false trip command to the relay protection device	The attacker put a virus called "Black Energy" into the computer and caused it to malfunction

Man-Made Malicious Attack Type. Most of the attacks on the smart grid are caused by humans on purpose. Human malicious attacks can be divided into three categories: device attacks, network attacks, and data attacks.

Device Attacks. Device attacks against the smart grid can be divided into four types: hardware attacks, software attacks, impersonation and access restrictions, data tampering and energy theft [7].

(1) *Hardware attack*

Attackers can use some opening source software tools or hardware tools to gain the internal data of the power devices or inject some virus into the devices through an external interface [8].

Some grid system processors may have defects in board making, drawing processes, or software loopholes in production. Developers even can implant chips with some malicious program when they develop processors. For example, some logic bombs might be planted in the program. Once the program is triggered by a specific logic that will cause a malicious attack which resulting in system defects, data leakage and other problems [9].

(2) *Software attack*

Attackers often plant virus or malicious program into the grid system software, so that the power devices cannot operate normally. The attackers can also invade the control system of some power equipment, perform malicious operations on the equipment. They usually affect the normal operation of the power grid to obtain some illegal benefits.

(3) *Impersonation and access restrictions*

Attackers can destroy the smart grid transmission module by making the servers or users lose the authentication ability when they communicate with each other. When the attackers hold or destroy the transmission module, which will become some initiator or victim of the fake message attack [10]. Attackers also impersonate the users or servers and uses wireless noise information to interfere with the smart grid's communication process.

(4) *Data tampering and energy theft*

Attackers can hold or invade a power device and modify the historical data or real-time data in the smart meter. For example, they can invade a smart meter and modify the amount of data about the electricity that used by users. Thereby, it causes greatly reducing the cost of electricity. The data modification of the power equipment by the attackers can make them easily steal the energy and cause huge economic losses of the power company.

Network Attacks. At present, smart grid has suffered from enormous security threats in cyberspace. Generally speaking, network attacks in the smart grid can be divided into protocol, operating system vulnerabilities, spoofing attacks, DoS (denial of service) attacks and so on.

(1) *Protocol and operating system vulnerabilities*

The communication network of the smart grid, the information interaction and data communication between the devices are mostly based on the TCP/IP protocol. Attackers may detect or intercept the serial numbers and connect the target host to transmit some false data, which make TCP serial number spoofing.

Most TCP/IP protocols apply plain text to transmit data passwords or information. And the plain texts are very easy to be detected and intercepted. If attackers capture the transmitted plain texts, they can tamper with the source IP address and transmit the fake data, thereby threaten the corresponding session and service computers.

An attacker can exploit vulnerability in the operating system of smart grid to implant malware, trojans, or viruses, who will control or destroy the power system and steal some important data from the system.

(2) *DoS (Denial of Service) attack*

The attackers will broadcast a large amount of spam, which will cause network congestion and occupy a large amount of network resources. This approach will interfere with the normal operation of the smart grid. In a general way, the denial of service attacks make users unable to perform normal information interaction, which greatly reduces the scheduling capability of the smart grid and hinders the smart grid from providing normal power services to users.

(3) *Proofing attack*

Attackers will take advantage of the characteristic about somc devices fail to perform identity authentication or version update in timc, and then they send a large amount of false information to the network. They will destroy the redundancy and congestion of the network, which is also called "witch attack".

(4) *Middleman attack*

The attackers access the power devices through technologies such as DNS spoofing, SMB session hijacking and ARP spoofing. They intercept the normal network communication between data and traffic. Attackers will also tamper with the transmission data or transmits a false encryption key to decrypt. This method will destroy the normal information communication of the power grid.

(5) *Replay attack*

Attackers can make use of the network to intercept and resend data. They use the internal malicious node to resend the information to the power system in order to achieve the purpose of deceiving the system without losing the certificate.

(6) *Illegal crack*

Attackers utilize some relevant technologies to invade some power equipment information or control systems in the smart grid. They will also brute force the encryption key of the data inside. The attackers can disrupt the operation of smart grid by destroying the communication method of the smart grid.

Data Attack. The data security of smart grid is a necessary condition for its normal operation. It is related to the operation and maintenance, marketing, business services and user security of the power system. The types of data attacks on the smart grid can be divided into false data injection attacks and user privacy leaks.

(1) *False data injection attack*

False data injection attacks can be divided into two categories. One category is random attack: attackers will find a highly sparse attack vector that satisfies the condition.

However, this form of attack could not attack the specific state variable or measurement point. And the other type is a specific attack, which is possible to modify certain power flow measurements or state variables to meet their attack intentions.

(2) *Users' privacy leakage*

Smart meters can bring very large privacy and security threats to our users. The users' information leakage may occur when they collect users' information. Attackers can use a certain technical means to invade smart meters and obtain sensitive information of a large number of users.

3.3 Brief Summary

Through the previous section, we have divided and introduced the security issues that the smart grid may suffer. The types of attacks that smart grid suffers can be classified into three types: device attacks, network attacks, and data attacks. Attackers use different offensive technologies to attack some hardware devices, communication networks, and data information of smart grid, which will destroy the normal operation of the smart grid.

4 Smart Grid Security Protection Method

The smart grid is in an extremely severe situation. According to the characteristics of the smart grid and the security threats, people propose the security area division of the smart grid and the corresponding security protection methods. The smart grid safe zoning plan is shown in the Fig. 4.

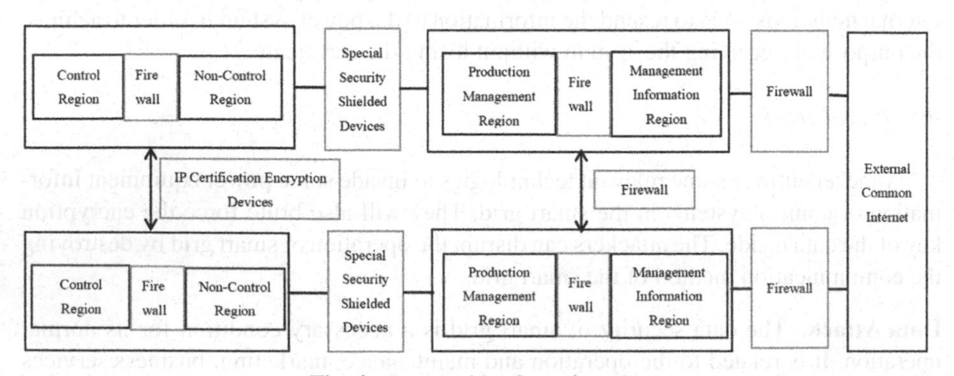

Fig. 4. Smart grid safe zoning plan

4.1 Physical Security

The physical security protection of the smart grid mainly refers to the protection of the power equipment to prevent the physical equipment from being externally invaded. This approach will avoid causing the equipment operation to malfunction or the power system to stop serving.

Power Equipment Safety Assessment. We have been inspired by some security work of the Internet of Vehicles, such as the evaluation of the Internet of Vehicles reputation management based on evolutionary game theory proposed by Z. Tian et al. [10]. And a new reputation framework for identifying DoS attacks of the Internet of Vehicles proposed by S. Su et al. [11]. However, the security protection system of the smart grid is relatively fragile. The power equipment is extremely vulnerable to be attacked and damaged, which in turn causes paralysis of the smart grid. Therefore, many researchers have conducted deep research on the protection of power equipment in the smart grid. Zhengwei Jiang et al. [12] constructed a safety assessment model for power equipment terminals. This model can evaluate the safety status of hardware, software and applications in power equipment. It can also identify high-risk power equipment, and then provide targeted protection.

Nan Quan et al. [13] proposes a strategy for smart grid protection from the network layer of power equipment. By studying the access control of power equipment and performing corresponding identity authentication. The strategy can enter the network and protect the power devices. Therefore, the security protection of power equipment is achieved on purpose.

Side channel Analysis Technology Application. The side channel technology refers to the study and analysis of some information, which related to its own operating state leaked during the operation of the power equipment. The leaked information includes various power information and physical information. Therefore, side channel technology is widely used in the monitoring and analysis of power grid equipment.

We refer to the detection of edge computing environment based on evidence-based reasoning network proposed by Y. Sun and N. Guizani et al. [14]. A data-driven model for future Internet routing decision modeling proposed by M. Guizani and X. Yu et al. [15]. Many scholars have proposed a number of safety monitoring methods for power equipment by using the acquisition and analysis of side channel information about power equipment. For example, OVuagnoux M et al. [16] proposes a collection and analysis of the bypass information of the power equipment, so as to locate the equipment accordingly. Once the equipment attack of the power grid occurred, the source equipment would be quickly found; Xiongwei Li et al. [17] studies In-depth study of the bypass information about the equipment, and proposed a hardware Trojan detection system which based on power equipment bypass information. This method can accurately detect the hardware Trojan in the FPGA chip, effectively prevent the hardware Trojan in the power equipment. Because these malicious softwares can damage to the operation of smart grid.

Intrusion Detection Technology. We refer to a secure digital evidence system using blockchain proposed by M. Qiu, Y. Sun et al. [18]. We were also inspired by Q. Tan, Y. Gao et al. [19] for an Eclipse attack that fully understands Tor's hidden services.At present, many scholars have conducted in-depth research on grid intrusion detection. Xin Fang et al. [20] proposes an intrusion detection method that based on protocol parsing technology. It can detect some attacks against power devices which appear in power systems. Zijian Cao et al. [21], after researching industrial intrusion detection system (IDS) and information network protection wall, he proposed an intrusion detection strategy combining IDS and information network protection.

Yahui Yang et al. [22] extends a GHSOM neural network model to construct an intrusion detection system that can learn new types of attack behavior. Qiang Du et al. [23] proposed an improved clustering analysis algorithm, which can effectively solve some local problems and further optimize the performance of the intrusion detection system.

4.2 Network Security Protection Method

The dispatching data network and management information network of the smart grid are mainly responsible for power electricity metering, power dispatch management and power information management. Therefore, the smart grid has high requirements on the performance of the network, and it is necessary to establish a comprehensive, secure and reliable power network security protection strategies.

Security Awareness and Assessment. In order to predict the development trend of the smart grid's operating state. Ge Jun et al. proposed to construct a network situation assessment model for the smart grid [24]. The smart grid's security perception situation model is shown in the Fig. 5.

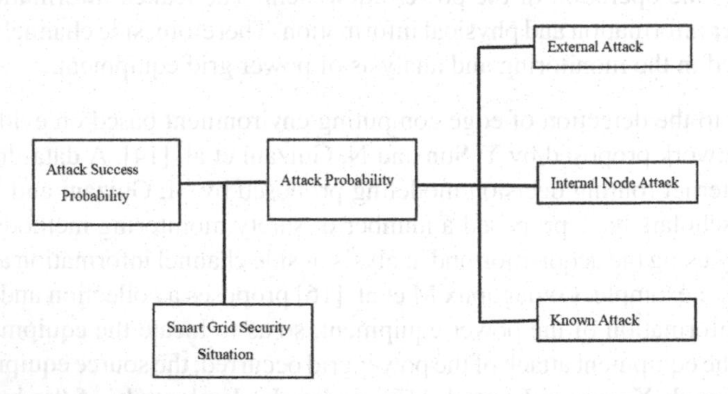

Fig. 5. Smart grid's security perception situation model

The smart grid's security perception situation model combines extreme learning machine, least squares support vector machine and auto-regressive prediction model to accurately detect and analyze the security situation of the smart grid, so as to prevent the security problem of the smart grid.

Dynamic State Estimation About DOS Attack. For solving the problems of data losses caused by denial of service (DOS) attacks. Li Xue et al. proposed an improved unscented Kalman filter (UKF) method for DOS attacks [25].

They firstly reconstruct the dynamic model of the grid system to compensate for the packet loss caused by the denial of service attack. Then, the state equation of the data compensation information is established, which is based on the unscented Kalman filter method and the two-parameter exponential smoothing method. At last, the UKF-based state estimation model is derived. And the algorithm steps are shown in the Fig. 6.

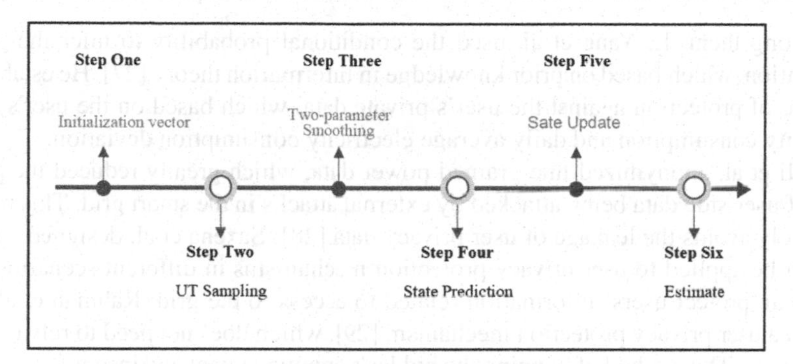

Fig. 6. The UKF algorithm steps

Through a large number of experiments, this improved UKF-based dynamic estimation method can effectively solve the problems of grid's data losses caused by DOS attack, and the computational efficiency of this model is very perfect.

Vulnerability Mining Model Based on Data Mining Technology. There are also many vulnerabilities in the smart grid, which threaten the safe operation of the grid. In order to improve the accuracy and efficiency of vulnerability detection, Niu Wennan et al. proposed a grid vulnerability mining model based on data mining technology [26].

The principle of data mining based on grid vulnerability mining model: Firstly, the model defines the parameters related to the vulnerability data, the vulnerability authority and the vulnerability related transaction set. The corresponding attribute division of the historical data and the running data in the grid. And then, Vulnerability parameters are linked with attributes to establish an association rule. Finally, the vulnerability identification matrix is combined with the associated rule equation to obtain a vulnerability mining model for the grid.

Experiments show that the data mining based on security vulnerability mining model can identify the vulnerabilities in the smart grid with high efficiency and high precision. Compared with the traditional method of analyzing the vulnerability behavior, the false positive rate of the model and the rate of missed investigations has been significantly reduced. And it has strong practicability.

5 Data Security Protection Method

In order to prevent data information in smart grid from being stolen or forged by attackers, we need to protect the data of the power grid. Only by ensuring the data security of the smart grid, which can operate more healthily, efficiently and stably.

Data Protection from a Privacy Perspective. The smart grid collects and processes a large amount of data and information during operation. And all the data includes important power data from the station side and from the user side. User-side data includes user-sensitive private information such as many users' names, residential addresses or phone numbers. Therefore, in order to prevent the theft of user privacy data in smart grid, many scholars at home and abroad have conducted in-depth research.

Among them, L. Yang et al. used the conditional probability to infer the private information, which based on prior knowledge in information theory [27]. He established a degree of protection against the user's private data, which based on the user's actual electricity consumption and daily average electricity consumption deviation.

Balli et al. anonymized fine-grained power data, which greatly reduced the probability of user-side data being attacked by external attacks in the smart grid. This method effectively avoids the leakage of user privacy data [28]; Saxena et al. designed a model that can be applied to user privacy protection mechanisms in different scenarios. This model can protect users' information related to access to the grid; Rahman et al. constructed a user privacy protection mechanism [29], which does not need to rely on other third parties. The method of winning the bidder's announcement and incentive statement ensures the confidentiality of the privacy data of the grid users.

Data Aggregation. As a common means of protecting important data of smart grids, LU et al. proposed a data aggregation scheme EPPA to protect the data in smart grid. They used the super-increment sequence to upgrade the power data, and then applied the homorphic paillier cryptography to encrypt the data after the dimension [30].

Yu Yong et al. [31] proposed an efficient data aggregation method which using bilinear mapping in verification. They use the DGK cryptogenic system to encrypt the sensitive data of power users to protect the power data. FAN et al. [32] proposed a user consumption data aggregation scheme, which can effectively resist the internal attacks frequently occurring in the smart grid and achieve the role of strengthening data protection.

In addition, some scholars have also proposed to build a data aggregation model based on the multilevel network and multidimensional data. This model can extend the grid data aggregation technology to multidimensional angle and multilevel gateway model, which greatly improving the accuracy of data aggregation.

Data integrity Protection. Most of the data attacks on the smart grid are disrupting the availability, integrity or confidentiality of the data. Therefore, in order to avoid the destruction of the internal data about smart grid during the operation, many scholars have proposed corresponding defense methods.

First, Clark and Wilson proposed the Clark-Wilson integrity strategy in 1987, which provided a reference for the data integrity protection mechanism of smart grids. Zhouyi Chou et al. [33] proposed a hybrid mandatory integrity model based on Clark-Wilson and Biba strategies. The model applies the Clark-Wilson model based on the Biba integrity strategy to protect the grid data.

In the database, the integrity of data is characterized by transactions. And the recovery of power data is equal to the recovery of transactions. For transaction recovery, researchers at home and abroad have proposed a large number of log models and recovery mechanisms. Gary et al. [34] summarized and elaborated on the methods of transaction recovery in the database. When the concept of trusted recovery was developed into intrusion tolerance, Liu et al. proposed to build a database architecture based on intrusion tolerance. However, the architecture is easily bypassed by attackers and does not have a strong defense effect. Therefore, Ragothaman et al. [35] proposed a scheme for improving database logs, which utilizes the dependency of power data and performs fragmentation processing before they are written to the log, thereby improving the data recovery speed of the power system.

M. Yu et al. [36] proposed a data integrity recovery method based on multilevel layer architecture, which can effectively solve the uneven size of log fragments caused by the convergence of data fragments, and further improve the Protection of grid data integrity.

6 Conclusion

As the most important infrastructure in the 21st century, the smart grid has brought great convenience to people's lives and promoted a new round of energy revolution. However, smart grid also suffers from huge security threats. Therefore, it is very necessary to carry out security protection for the smart grid. From the perspective of the attackers, we deeply analyze the possible security problems of the smart grid and propose corresponding solutions.

When we propose solutions to grid security problems, we should formulate a systematic and comprehensive security protection system, in order to strive to build a safe, flexible and reliable smart grid that can bring a better life to human beings.

Acknowledgments. This work is funded by the National Key Research and Development Plan (No. 2018YFB0803504), the National Natural Science Foundation of China (No. 61702220, 61702223, 61871140, U1636215, 61602210), the Guangdong Basic and Applied Basic Research Foundation (No. 2020A1515010450), the Science and Technology Planning Project of Guangdong (2017A040405029, 2018KTSCX016, 2019A050510024), the Science and Technology Planning Project of Guangzhou (201902010041), the Fundamental Research Funds for the Central Universities (21617408, 21619404), Guangdong Province Universities and Colleges Pearl River Scholar Funded Scheme (2019), and the Opening Project of Shanghai Trusted Industrial Control Platform (TICPSH202003014-ZC).

References

1. Liang, G., et al.: The 2015 ukraine blackout: implications for false data injection attacks. IEEE Trans. Power Syst. **32**(4), 3317–3318 (2016)
2. Chakrabortty, A., Bose, A.: Smart grid simulations and their supporting implementation methods. Proc. IEEE **105**(11), 2220–2243 (2017)
3. Uzunoğlu, B., Ülker, M.A.: Maximum likelihood ensemble filter state estimation for power systems. IEEE Trans. Instrum. Meas. **67**(9), 2097–2106 (2018)
4. Ghosal, M., Rao, V.: Fusion of multirate measurements for nonlinear dynamic state estimation of the power systems. IEEE Trans. Smart Grid **10**(1), 216–226 (2017)

5. Befekadu, G.K., Gupta, V., Antsaklis, P.J.: Risk-Sensitive control under Markov modulated denial-of-service (DoS) attack strategies. IEEE Trans. Autom. Control **60**(12), 3299–3304 (2015)
6. Lu, R., Liang, X.: EPPA: an efficient and privacy-preserving aggregation scheme for secure smart grid communications. IEEE Trans. Parallel Distrib. Syst. **23**(9), 1621–1631 (2012)
7. Marksteiner, S., Vallant, H., Nahrgang, K.: Cyber security requirements engineering for low-voltage distribution smart grid architectures using threat modeling. J. Inf. Secur. Appl. **49**, 102389 (2019)
8. Shan, X., Zhuang, J.: A game-theoretic approach to modeling attacks and defenses of smart grids at three levels. Reliab. Eng. Syst. Saf. **195**, 106683 (2019)
9. Davis, M.: Is It Dumb to Be Smart in Grid Security? Transmission & Distribution World (2019)
10. Tian, Z., Gao, X., Su, S., Qiu, J., Du, X., Guizani, M.: Evaluating reputation management schemes of internet of vehicles based on evolutionary game theory. IEEE Trans. Veh. Technol. **68**(6), 5971–5980 (2019)
11. Tian, Z., Su, S., Yu, X., et al.: Vcash: a novel reputation framework for identifying denial of traffic service in internet of connected vehicles. IEEE Internet Things J. **7**(5), 3901–3909 (2019)
12. Jiang, Z., Wang, D., Wang, H.: Safety evaluation model of power intelligent terminal. Comput. Eng. Des. **35**(1), 6–10 (2014)
13. Quan, N., Lei, Y., Huang, B.: Research on power terminal communication access architecture in smart grid. Power Syst. Commun. **33**(1), 74–77 (2012)
14. Tian, Z., et al.: Real time lateral movement detection based on evidence reasoning network for edge computing environment. IEEE Trans. Ind. Inf. **15**(7), 4285–4294 (2019)
15. Tian, Z., Su, S., Shi, W., Du, X., Guizani, M., Yu, X.: A data-driven model for future internet route decision modeling. Future Gener. Comput. Syst. **95**, 212–220 (2019)
16. OVuagnoux, M., Pasini, S.: Compromising electromagnetic emanations of wired and wireless key boards. In: USENIX Security Symposium (2009)
17. Li, X., Xu, X., Zhang, Y.: Hardware trojan detection method based on electromagnetic bypass analysis. Comput. Eng. Appl. **49**(12), 97–100 (2013)
18. Tian, Z., Li, M., Qiu, M., Sun, Y., Su, S.: Block-DEF: a secure digital evidence system using blockchain. Inf. Sci. **491**, 151–165 (2019)
19. Tan, Q., Gao, Y., Shi, J., Wang, X., Fang, B., Tian, Z.: Toward a comprehensive insight to the eclipse attacks of tor hidden services. IEEE Internet Things J. **6**(2), 1584–1593 (2019)
20. Fang, X., Wan, Y., Wen, X.: Research on DDoS attack in network intrusion detection system based on protocol analysis technology. Inf. Netw. Secur. **4**, 112–123 (2012)
21. Cao, Z., Cao, Y., Rong, X.: Design of network intrusion detection and firewall linkage platform. Inf. Netw. Secur. **1**, 86–95 (2012)
22. Yang, Y., Huang, H., Shen, Q.: Research on intrusion detection based on incremental GHSOM neural network model. Chin. J. Comput. **37**(5), 1216–1224 (2014)
23. Du, Q., Sun, M.: Research on intrusion detection system based on improved cluster analysis algorithm. Comput. Eng. Appl. **47**(11), 106–108 (2011)
24. Ge, J., Meng, G., Chou, D.: Construction of network situation assessment model for smart grid. Autom. Instrum. (4) (2019)
25. Li, X., Li, W., Du, D.: Research on dynamic state estimation of smart grid based on UKF under Denial of Service attack. J. Autom. **45**, 122–133 (2019)
26. Niu, W., Bao, P., Tang, H.: A data mining model for smart grid security vulnerability. Power Technol. **331**(4), 126–129 (2018)
27. Yang, L., Chen, X., Zhang, J., Poor, H.V.: Optimal privacy-preserving energy management for smart meters. In: Proceeding of IEEE INFOCOM, p. 513 (2014)

28. Balli, M., Uludag, S., Tavli, B.: Distributed multi-unit privacy assured bidding (PAB) for smart grid demand response program. IEEE Trans. Smart Grid **9**(5), 4119–4127 (2017)
29. Saxena, N., Chou, B.J., et al.: Authentication and authorization scheme for various user roles and devices in smart grid. IEEE Trans. Inf. Forensics Secur. **11**(5), 907–921 (2016)
30. Fu, Y., et al.: Droop control for DC multi-microgrids based on local adaptive fuzzy approach and global power allocation correction. IEEE Trans. Smart Grid **10**(5), 5468–5478 (2018)
31. Yu, Y., Ye, Y., Huang, L.: Data aggregation protocol for privacy protection for smart grid. Small Microcomput. Syst. **10**(5), 666–675 (2016)
32. Fan, D.Z., Mi, W.: Data aggregation scheme for safety measurement in smart grid. Electr. Appl. **13**(4), 50–57 (2019)
33. Chou, Z., He, Y., Liang, H.: Hybrid mandatory integrity based on Biba and Clark-Wilson strategiesmodel. J. Softw. **21**(1), 98–106 (2010)
34. Gray, J., Reuter, N.: A Transaction Processing, Concepts and Techniques. Morgan Kaufmann (1993)
35. Chen, C., Feng, D., Xu, Z.: Research on database transaction recovery and intrusion response model. Comput. Res. Dev. **47**(10), 1797–1804 (2010)
36. Zheng, J., Qin, X., Sun, J.: Research on malicious transaction repair algorithm in survivable DBMS based on SPN model. Chin. J. Comput. **29**(8), 1480–1486 (2006)

Framework Design of Environment Monitoring System Based on Machine Learning

Lingxiao Meng[1], Shudong Li[1(✉)], Xiaobo Wu[2(✉)], and Weihong Han[1]

[1] Cyberspace Institute of Advance Technology, Guangzhou University, Guangzhou 510006, China
lishudong@gzhu.edu.cn

[2] School of Computer Science and Cyber Engineering, Guangzhou University, Guangzhou 510006, China
happywxb@gzhu.edu.cn

Abstract. With the gradual deepening of industrialization process, a series of problems brought about by industrial pollution. Urban development has exerted tremendous pressure on the ecological environment and has resulted in a serious impact on the lives and residents. Therefore, real-time monitoring of air quality and accurate forecasting are very important for the full utilization of the environmental data index. With an original platform, this paper deeply considers the actual needs and the performance defects of users, and build the platform of system: (1) The data map function has been added. Real-time data is identified on the map, and different colors are displayed in different intervals for PM2. 5; (2) Alarm function is added. When data exceeds a certain threshold value, logos will be displayed on the map; (3) Trend analysis functions are added, and BP artificial neural network is used to predict the data of environmental data in the future. The data scrolls playback on the map over time; (4) Reconstructing the change curve of data makes the system more intuitive, simple and easy to use. Based on the theory of BP artificial neural network, the influence of temperature, humidity, PM2.5, atmospheric pressure and illumination intensity on one of five factors above are carefully analyzed respectively. And then a reliable and suitable environmental index for predicting the five factors above is obtained. Finally, the forecast results are shown on a heat map for better visualization.

Keywords: Environment monitoring · Prediction · BP neural network

1 Introduction

Ecological environment is the material basis for human beings, and it is one of the basic components for all living things to reproduce. The quality of the environment is directly related to people's physical condition and the level of a industry. Especially, in the era of serious industrial pollution, it is even more

X. Sun et al. (Eds.): ICAIS 2020, LNCS 12240, pp. 380–392, 2020.
https://doi.org/10.1007/978-3-030-57881-7_34

important. With the deepening of industrialization process, a series of problems caused by industrial pollution. And urban development put great pressure on the ecological environment. Since the reform and opening up, the initial extensive development mode of our country has made us pay a heavy price, and China has become a country with relatively serious environmental pollution. With PM2.5 exceeding the standard, the beijing-tianjin-hebei region has been plagued by smog for a long time. According to the 2016 China environment bulletin, 84 of the 338 cities of China above the prefectural level met the air quality standards, accounting for 24.9% of the total. Air quality in 254 cities exceeded the standard, accounting for 75.1 % of the total [1,26].

Under the situation, the convenient, practical and real-time environmental monitoring cloud platform plays an important role. At present, although some institution has applied such systems. Their functions are mainly to monitor the data sent by the sensor in real time and manage the equipment. Intelligent system should have intuitive visual function: users do not need leave the house, they just need to open the browser to facilitate the situation of a certain location; The existing data should also be used to estimate the future time point to achieve the purpose of disaster prevention. Relevant departments can prevent disasters through early warning information and control environmental conditions in advance. Therefore, real-time monitoring of air quality, temperature, humidity, light, air pressure and other environmental indexes, visual trend analysis and accurate prediction are useful for the full utilization of environmental data indexes and comprehensive control of the overall environment. We built a novel monitoring system for the Internet of things. Great progress has been made in functionality and user experience. This system can get the current weather conditions on the map. When an index exceeds a threshold, the system alerts the users. And it is a very powerful tool system for observers. We studied bp artificial neural network and had a detailed understanding of its principles and algorithms. Based on the data of the original environmental monitoring system of the Internet of things (located in China university of geosciences), temperature, humidity, air pressure, light and PM2. 5 were selected as training samples. And appropriate intermediate layer number and weight were selected to obtain a feasible prediction model.

2 Related Work

At the end of the last century, artificial neural network was gradually adopted by scholars. It is very suitable for such nonlinear prediction, because of being trained to approach the actual state scenarios [2,10,14,17,21]. The researchers mainly made the following attempts: Claudio Carnevale et al. predicted daily AQI by neural network based on principal component analysis (PCA) [9]. Kumar et al. compared two neural networks, SOM and MLP, and predicted the NO2 of Stockholm [8]. In the design of prediction model, Ruobo Xin applied LM algorithm, simulated annealing algorithm and early stop algorithm for BP network [24]. Jehng-jung Kao used traditional time series method and neural network to predict ambient air quality [7]. By using improved BP artificial neural network, Ghong established a basic model to predict PM10 in guangzhou [16]. Abd Rahman compared the Box Jenkins method of ARIMA, ANN and three fuzzy time series (FTS) by means of mean absolute percentage error, mean absolute error, mean square error and root mean SQ [20]. Ioannis Kyriakidis applied neural network model to test the good performance of Helsinki [22]. P. GOYAL proposed the development of an integrated model based on air diffusion model and neural network (NN), which optimized the performance of each method [3]. In terms of air quality or environmental data prediction, China started late but developed rapidly. Since the beginning of the new century, environmental forecasting has begun to rise in the country [5,11–13,23]. Domestic scholars have mainly carried out the following work: Zhang tian et al. used chi-square test to determine air quality impact factors and BP artificial neural network to predict them [27]. On the basis of training neural network algorithm, Xin introduced genetic algorithm to optimize the initial weight of neural network, and introduced bayesian normalization algorithm to directly improve the generalization ability of neural network [25].

3 BP Artificial Neural Network

BP artificial neural network is a relatively effective model for predicting air quality. This is due to its non-feedback, multi-layered, and non-connected structure. It consists of at least three layers: input, hidden layer and output. The number of hidden layer is at least one, which can be expanded as needed. There is no connection between the nodes of each layer, but there is correlation between the layer and the perceptron of the layer. The propagation direction can only be from input to hidden layer and then to output. The actual output is out of line with the calculated result. The algorithm needs to be corrected according to the deviation between the learning result and the real value. Correction is done by adjusting the weight of each connection. This is called "reverse propagation error reduction". Error back propagation algorithm is shown in Algorithm 1 [19].

4 Logical Design of Environmental Monitoring Platform

4.1 Demand Analysis of Platform

Environmental data prediction: prediction is the grasp of future environmental conditions. BP artificial neural network algorithm is used. First, the data in the several periods of time were inputed into the prediction model as training input, and the data of one time point after each group of time points were taken as output. The data obtained by this algorithm is visualized in the website. Environmental danger alarm: prediction is invalid information for some non-professional users who do not understand the meaning behind the data and do not know the range of dangerous situation. Therefore, the detection point in the map with a prominent marker to remind the user that a certain threshold is

1. Error back propagation algorithm

: Training set $D = \{(\mathbf{x}_k, \mathbf{y}_k)\}_{k=1}^m$; Learning rate η
: Connect multilayer feedforward neural networks
function $BP(D, \eta)$
Randomly initialize all connection weights and thresholds in the network in range of $(0, 1)$
repeat
for all$(\mathbf{x}_k, \mathbf{y}_k) \in D$ **do**
Calculate the output of current sample $\hat{\mathbf{y}}_k = f(\beta_j - \theta_j)$
Calculate the gradient of output neurons

$$\begin{aligned} g_j &= -\frac{\partial E_k}{\partial \hat{y}^k} \cdot \frac{\partial \hat{y}^k}{\partial \beta_j} \\ &= -(\hat{y}_j^k - y_j^k) f'(\beta_j - \theta_j) \\ &= \hat{y}_j^k (1 - \hat{y}_j^k)(\hat{y}_j^k - y_j^k) \end{aligned}$$

Calculate hidden neuron gradients

$$\begin{aligned} e_h &= -\frac{\partial E_k}{\partial b_h} \cdot \frac{\partial b_h}{\alpha_h} \\ &= -\sum_{j=1}^{\ell} \frac{\partial E_k}{\partial \beta_j} \cdot \frac{\partial \beta_j}{b_h} f'(\alpha_h - \gamma_h) \\ &= \sum_{j=1}^{\ell} w_{hj} g_j f'(\alpha_h - \gamma_h) \\ &= b_h (1 - b_h) \sum_{j=1}^{\ell} w_{hj} g_j \end{aligned}$$

Update weight

$$\Delta w_{hj} = \eta g_j b_h$$
$$\Delta v_{ih} = \eta e_h x_i$$

Update threshold

$$\Delta \theta_j = -\eta g_j$$
$$\Delta \gamma_h = -\eta e_h$$

end for
until Stop condition is reached
end function

exceeded at the time. Such information is useful and convenient. Environmental index trend analysis: the data analysis of a single point is very limited. Using dynamic simulation on a map can help users understand the curve. At the same time, we use the heat map, which is one of the most intuitive forms [15].

4.2 Frame Design of Platform

Function and Framework. The platform is innovative in following four aspects:

Four lines in one: four graphs are merged together. The types represented in the lines are plentiful. The novel drawing library is used.

Data map: we use the API interface to display data in real time where the sensor is, and use different labels and colors to represent different data types and degrees. Trend analysis: past trend and future predictions are integrated in this function. The images (of each time point) scrolls in the map as heat maps, and the predicted value are displayed at the top of the map.

Environment alarm: we set the threshold value for various indicators. And after exceeding the threshold, they will flicker and jump on the map (Fig. 1).

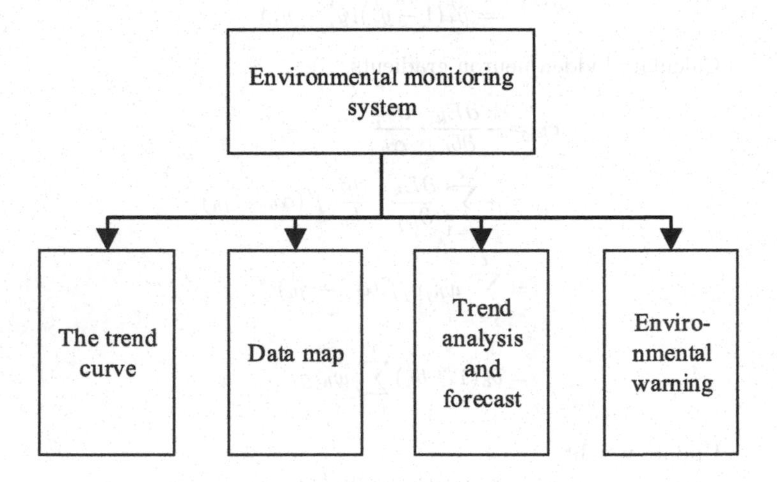

Fig. 1. Platform frame

Website Interface. The page of the column is divided into four parts: Real-time monitoring, Data map, Trend analysis, Environmental warning. The function of this part is placed in HTML. The main realization is about the site information and four buttons corresponding to the four links. In order not to display its subordinate content code, the iframe is used to store links in the control. And the pages in the link complete the function in the current box.

5 Detail Design of Environmental Monitoring Platform

5.1 Real-Time Monitoring

This function is implemented by a file named 'Change'. Ihe main task of it is building the graph, querying from the database in real time, and filling the data into the corresponding position of the graph. In the process of using the Echarts library, it is necessary to configure it with JavaScript code. Cs files access the database to read the data, and the front-end JavaScript calls the background function to display the data in the line.

5.2 Data Map

The realization of data map is the process of data visualization based on baidu map API. From the implementation of view, the highlight of work is: (1) calling the background database (2) displaying level of color changes, label changes and label display.

5.3 Trend Analysis

Trend analysis is the core function of the platform. This part applys the combination of echarts and Baidu map. The animation effect of the heat maps is realized by Echarts. The data is provided not only from the database, but also from the BP artificial neural network algorithm. In addition, at the top of the map, the predicted value is displayed.

5.4 Environmental Early Warning

There are many similarities between the realization of environmental warning function and data map. For example, baidu map API is used to display the current environmental index. However, the most prominent highlight of this function is setting the threshold of each index. Once the index exceeds the preset threshold, there will be a marked jumping dynamic figure, which has strong practicability.

5.5 Other Development Details

In the process of building website, because the .Net framework is used, it is necessary to use Microsoft's IIS as the website management tool. Clicking on the publishing site in visual studio can easily push the tested web application. When using Baidu map, users need to get the key to use the interface. Map resources can be called by writing them to the program.

6 Experiment and Evaluation

6.1 The Results

When a user opens the interface there will be the title and four function selection boxes at the top of the page. Legends represent the various colors and shapes on the map, which show all the sensors' data in time. ICONS is used to differentiate each sensor. The PM2.5 index is shown in different colors [18] (Fig. 2).

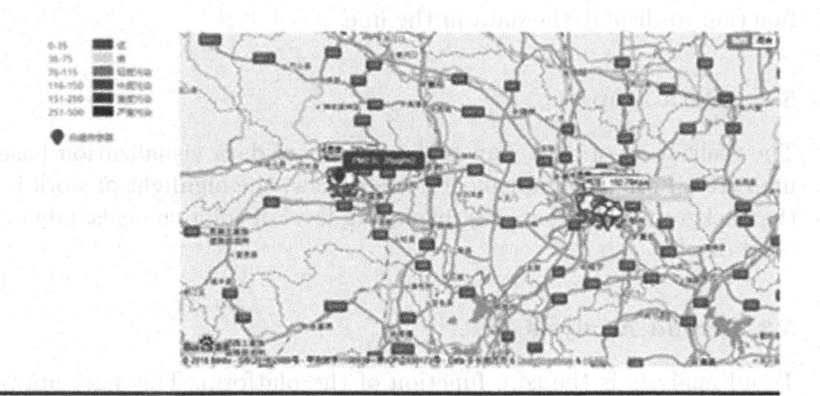

Fig. 2. Home page display effect

When the user click the real-time display button, a trend curve will appear. The top column is the same as the home page, and the bottom column shows the graph line in real time, including curves of temperature, humidity, air pressure, illumination and PM2.5 (Fig. 3).

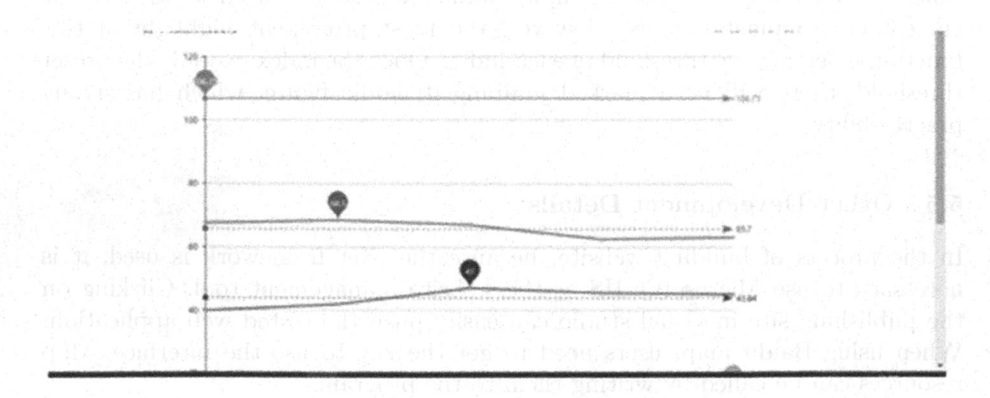

Fig. 3. Display of real-time monitoring function

When users click the trend analysis button, the heat diagram interface will appear. In addition to the illustration above, four buttons have been added to switch the content of the heat map on the right side. Below the heat map, there is a time point diagram. It is automatically played, showing the previous five hours and the current data, and finally making predictions for the next time point. The top of the image shows the predicted value. When users click environment alert, the last function image appears [6] (Fig. 4).

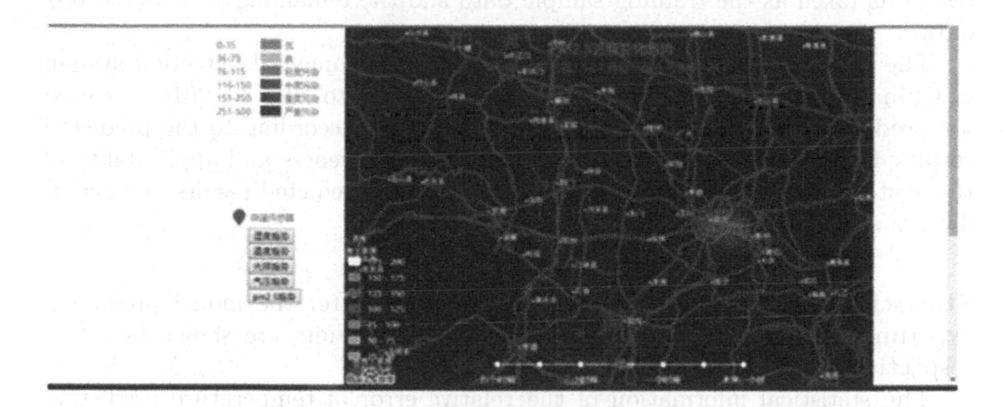

Fig. 4. Display of trend function

Different from the real-time monitoring, this function does not be changed in terms of ICONS, but adds the marking jumping. When the data of the dot exceeds the threshold set by the program, an alarm will be triggered, and the dot will appear jumping animation (Fig. 5).

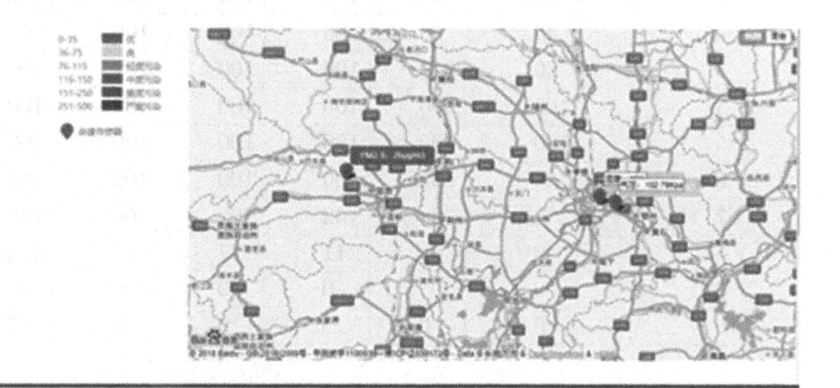

Fig. 5. Display of environmental warning

6.2 Evaluation

In this paper, the number of nodes in the input layer is 5, the number of nodes in the output layer is 1, and the number of nodes in the hidden layer is within [3, 5]. After a lot of tests, the number of nodes in the hidden layer is set as 13. The transfer function from the input layer to the hidden layer is Sigmoid function, the learning rate is 0.004, the number of iterations is 5000, and the convergence criterion is 0.0001. A total of 10,000 sets of sample data were collected. The first 90% was taken as the training sample data and the remaining 10% as the test data.

The environmental index collected by the environmental detection system of China university of geosciences from May 1, 2016 to July 30, 2016 was used for prediction. The predicted results were analyzed according to the predicted results[4]. This experiment is used to test the effectiveness and applicability of the system. The comparison between experimental predicted results and actual data is shown in the Fig. 6.

Statistical Analysis of Prediction Results. After the model prediction experiment of BP neural network, the predicted results are shown in tables respectively.

The statistical information of the relative error of temperature prediction results is shown in Table 1. The proportion of annual relative error less than 20% is 9.90%, and the average relative error of prediction results is 10.44%.

Table 1. Statistical information on the proportion of relative error in temperature prediction results (%)

Period of time	$\delta < 8.5\%$	$\delta < 9\%$	$\delta < 10\%$	$\delta < 11\%$	$\delta < 12\%$	$\delta < 17\%$
1	13	23	100	100	100	100
2	43	48	86	100	100	100
3	60	79	100	100	100	100
4	100	100	100	100	100	100
5	100	100	100	100	100	100
6	84	100	100	100	100	100
7	0	3	62	100	100	100
8	0	0	11	27	52	100
9	0	0	0	0	0	100
10	0	0	0	0	0	67

The statistical information of the relative error of humidity prediction results is shown in Table 2. The proportion of annual relative error less than 20% is 59.70%, and the average relative error of prediction results is 19.56%.

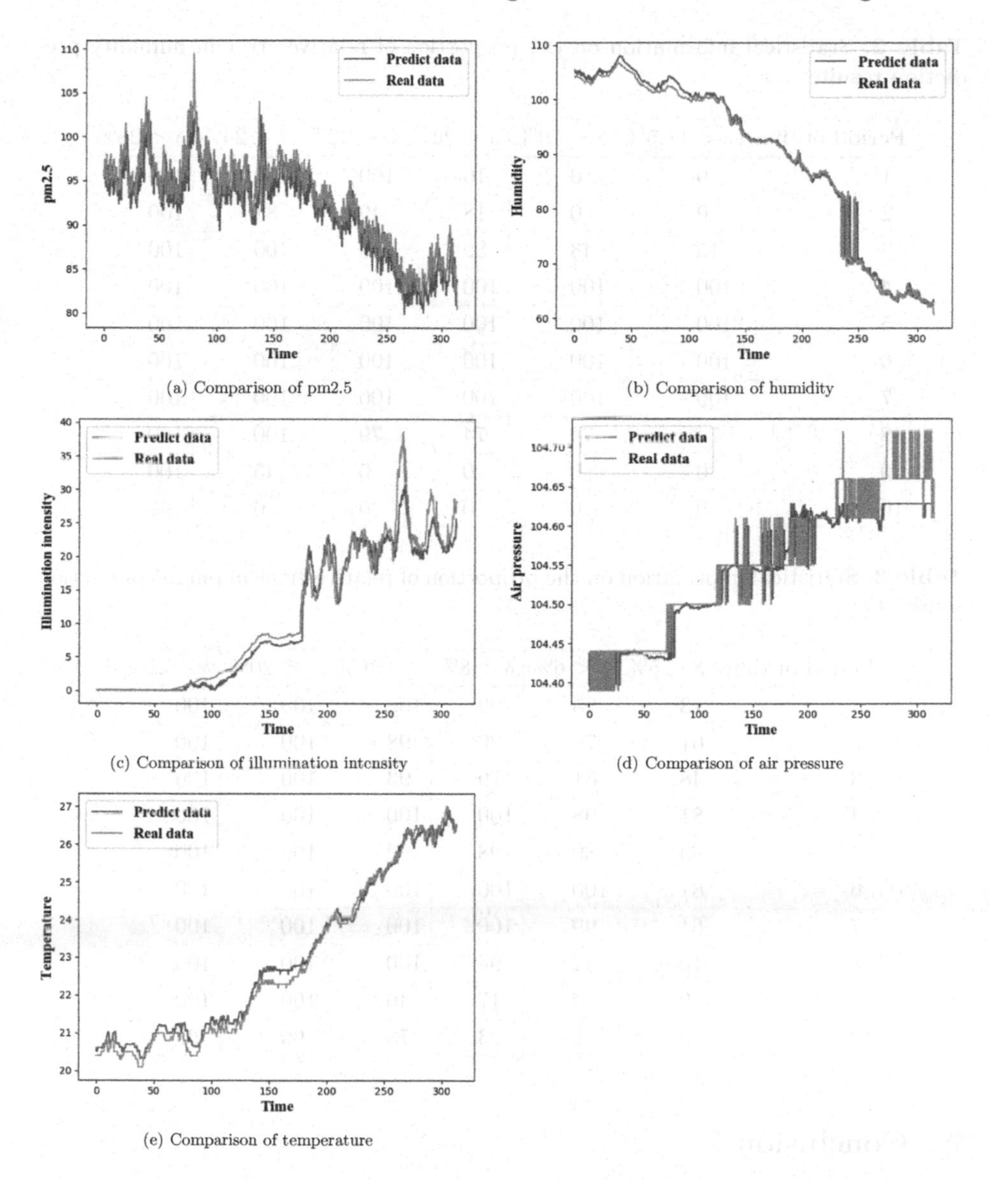

(a) Comparison of pm2.5

(b) Comparison of humidity

(c) Comparison of illumination intensity

(d) Comparison of air pressure

(e) Comparison of temperature

Fig. 6. Comparison

The statistical information of the relative error of humidity prediction results is shown in Table 3. The proportion of annual relative error less than 20% is 99.90%, and the average relative error of prediction results is 6.12%.

The result can meet the prediction requirements. Compared with the research results of others, the prediction results of this experiment are relatively accurate. The average relative error appears to be large because the real value is relatively small.

Table 2. Statistical information on the proportion of relative error in humidity prediction results (%)

Period of time	δ < 18.5%	δ < 19%	δ < 20%	δ < 22%	δ < 24%	δ < 28%
1	0	0	16	100	100	100
2	0	0	18	49	88	100
3	15	43	89	100	100	100
4	100	100	100	100	100	100
5	100	100	100	100	100	100
6	100	100	100	100	100	100
7	100	100	100	100	100	100
8	74	74	74	79	100	100
9	0	0	0	0	45	100
10	0	0	0	0	0	92

Table 3. Statistical information on the proportion of relative error in pm 2.5 prediction results (%)

Period of time	δ < 5%	δ < 6%	δ < 8%	δ < 10%	δ < 20%	δ < 25%
1	73	95	99	100	100	100
2	61	76	90	98	100	100
3	48	61	79	93	100	100
4	81	98	100	100	100	100
5	63	89	98	99	100	100
6	81	100	100	100	100	100
7	61	99	100	100	100	100
8	10	42	94	100	100	100
9	0	0	17	46	100	100
10	0	1	23	73	99	99

7 Conclusion

In this paper, we design a novel framework of environment detection system, and add four functions to the system.

(1) The data mapping function: Real-time data is identified on the map, and different logo on behalf of different environment indexes;
(2) Alarm function: when data exceeds a certain threshold value, alarm logos will be displayed on the map;
(3) Trend analysis functions: BP artificial neural network is used to predict the data of each hour in the future. The data scrolls playback on the map over time;

(4) The change curve of data: The curve appears on the page in a novel form. Based on the theoretical basis of BP artificial neural network algorithm, we analyze various environmental indicators in-depth and obtain a reliable and applicable environmental index for this system. The algorithm uses the current environment data as the inputvector of the training sample, and operates on the input and output with normalization and de-normalization respectively to obtain the parameters of BP artificial neural network.

We conducted a lot of experiments and comparisons. Above these works, we found that the BP neural network performed well under the input conditions. This way, the proposed algorithm considerably endow the capability of predicting new variants of environment.

Acknowledgment. This work was supported in part by the National Natural Science Foundation of China (NSFC) under Grant 61672020, Grant U1803263, Grant 61972106, Grant U1636215, and Grant 61972105, in part by the National Key Research and Development Program of China under Grant 2019QY1406, and in part by the Key Research and Development Program of Guangdong Province under Grant 2019B010136003.

References

1. Bulletin on the state of China's environment (excerpt). The environmental protection 45, 35–47 (2017)
2. Carnevale, C., Finzi, G., Pisoni, E., Volta, M.: Neuro-fuzzy and neural network systems for air quality control. Atmos. Environ. **43**(31), 4811–4821 (2009)
3. Goyal, P., Kumar, A.: Air quality forecasting throught integrated model using air dispersion model and neural network, pp. 219–224. Latest advances in systems science and computational intelligence, WSEAS, WSEAS LLC pp (2012)
4. Han, W., Tian, Z., Huang, Z., Li, S., Jia, Y.: Bidirectional self-adaptive resampling in imbalanced big data learning. Multimed. Tools Appl. (2018)
5. Hou, P., Lei, Q.: Air quality monitoring system based on wireless sensor network. Industrial instrumentation and automation, pp. 110–113
6. Hua, Y., Chen, B., Yuan, Y., Zhu, G., Ma, J.: An influence maximization algorithm based on the mixed importance of nodes. CMC Comput. Mater. Continua **59**(2), 517–531 (2019)
7. Kao, J.J., Huang, S.S.: Forecasts using neural network versus box-Jenkins methodology for ambient air quality monitoring data. J. Air Waste Manag. Assoc. **50**(2), 219–226 (2000)
8. Kolehmainen, M., Martikainen, H., Ruuskanen, J.: Neural networks and periodic components used in air quality forecasting. Atmos. Environ. **35**(5), 815–825 (2001)
9. Kumar, A., Goyal, P.: Forecasting of air quality index in delhi using neural network based on principal component analysis. Pure Appl. Geophys. **170**(4), 711–722 (2013)
10. Kumar, A., Singh, I.P., Sud, S.K.: Energy efficient air quality monitoring system
11. Li, L., Zou, C., Yu, J., Wang, B.: Research progress of air quality monitoring network design at home and abroad. China Environ. Monit. **28**, 57–64 (2012)
12. Li, L., Fu, Q.: Design of urban air quality monitoring system based on distributed wireless sensor network. Comput. Meas. Control **22**, 706–708 (2014)

13. Li, S., Wu, X., Zhao, D., Li, A., Tian, Z., Yang, X.: An efficient dynamic ID-based remote user authentication scheme using self-certified public keys for multi-server environments. PLoS ONE **13**(10), e0202657 (2018)
14. Li, S., Zhao, D.L., Wu, X., Tian, Z., Li, A., Wang, Z.: Functional immunization of networks based on message passing. Appl. Math. Comput. **366**, 124728 (2020)
15. Liu, C., Guo, H., Li, Z., Gao, X., Li, S.: Coevolution of multi-game resolves social dilemma in network population. Appl. Math. Comput. **341**, 402–407 (2019)
16. Liu, Y., Zhu, Q., Yao, D., Xu, W.: Forecasting urban air quality via a back-propagation neural network and a selection sample rule. Atmosphere **6**(7), 891–907 (2015)
17. Ma, Y., Yang, S., Huang, Z., Hou, Y., Cui, L., Yang, D.: Hierarchical air quality monitoring system design (2015)
18. Qu, Z., Cao, B., Wang, X., Li, F., Xu, P., Zhang, L.: Feedback LSTM network based on attention for image description generator. CMC Comput. Mater. Continua **59**(2), 575–589 (2019)
19. Qu, Z., Wu, S., Liu, W., Wang, X.: Analysis and improvement of steganography protocol based on bell states in noise environment. Comput. Mater. Continua **59**(2), 607–624 (2019)
20. Rahman, N.H.A., Lee, M.H., Suhartono, L.M.T.: Artificial neural networks and fuzzy time series forecasting: an application to air quality. Qual. Quant. **49**, 2633–2647 (2015)
21. Saad, S.M., Saad, A.R.M., Kamarudin, A.M.Y., Zakaria, A., Shakaff, A.Y.M.: Indoor air quality monitoring system using wireless sensor network (WSN) with web interface. In: 2013 International Conference on Electrical, Electronics and System Engineering (ICEESE) (2013)
22. Voukantsis, D., Karatzas, K., Kukkonen, J., Rsnen, T., Karppinen, A., Kolehmainen, M.: Intercomparison of air quality data using principal component analysis, and forecasting of pm10 and pm2.5 concentrations using artificial neural networks, in thessaloniki and helsinki. Science of the Total Environment 409, 1266–1276
23. Wu, X.: Design of air quality monitoring system based on wireless sensor network. Ph.D. thesis, Beijing university of posts and telecommunications (2014)
24. Xin, R.B., Jiang, Z.F., Li, N., Hou, L.J.: An air quality predictive model of Licang of Gingdao city based on BP neural network. In: Advanced Materials Research, vol. 756, pp. 3366–3371. Trans Tech Publ (2013)
25. Xin, R.: Air quality prediction based on genetic optimization and Bayesian normalization neural network. Ph.D. thesis, Shandong University (2013)
26. Yang, H., Li, S., Wu, X., Lu, H., Han, W.: A novel solutions for malicious code detection and family clustering based on machine learning. IEEE Access **7**, 148853–148860 (2019)
27. Zhang, T., Mao, Y., Mao, J., Wang, H., Ren, P.: An air quality prediction method based on BP neural network. Inf. Commun. 72–74 (2017)

Compression Detection of Audio Waveforms Based on Stacked Autoencoders

Da Luo[1(✉)], Wenqing Cheng[1], Huaqiang Yuan[1], Weiqi Luo[2], and Zhenghui Liu[3]

[1] Dongguan University of Technology, Dongguan, People's Republic of China
luoda@dgut.edu.cn
[2] Sun Yat-sen University, Guangzhou, People's Republic of China
[3] Xinyang Normal University, Xinyang, People's Republic of China

Abstract. With the easy acquisition of digital recordings, the field of audio forensics has become increasingly prominent. Detection of audio compression history is an important issue in the field of audio forensics. In this paper, a detection framework is proposed to detect whether a given audio waveform is an original waveform or a decompressed one. We extract the spectrum features from the frequency domain and then adopt a stacked autoencoder to effectively detect the frame-level audio fragments to distinguish between the original audio frames and the decompressed audio frames. Then, a majority voting algorithm is applied to make the final decision for an audio clip. Our analysis focuses on multi-time compressed audio, including single compression, double compression, triple compression and even four-time compression in three kinds of compression formats. The experimental results show that the proposed framework can effectively detect multi-time compressed audio. Furthermore, the proposed framework can also estimate the compression bitrate.

Keywords: Audio compression detection · Statistical features · Multi-time compression

1 Introduction

The compression history analysis of the audio waveform is one of the research fields in audio forensics, especially for identification of the originality of the legal evidence.

Original audio is a kind of raw data recorded and stored in the form of waveform, which is uncompressed audio data. They are usually saved as compressed audio by an audio encoder such as MP3 to save storage space. Generally speaking, if a tamper want to modify a compressed audio such as MP3, OGG or WMA, he must first decompress the audio into audio waveform and then perform tampering operations in the waveform. From the perspective of forensic evidence, there should be no compression trace on the original audio waveform. Conversely, if an audio waveform has a compressed trace, it should be suspected that the audio has been tampered with. That's the reason why the compression history analysis is important.

Another situation is double compression, which means a decompressed audio waveform is compressed a second time. For example, after the above tamper has modified

X. Sun et al. (Eds.): ICAIS 2020, LNCS 12240, pp. 393–404, 2020.
https://doi.org/10.1007/978-3-030-57881-7_35

the waveform, compressing again will form a doubly compressed audio. Multi-time compressed audio is created by compressing one or more times, which usually will be in the case of forgery case. A decompressed/decoded audio means that its originality is questioned and the double compressed or even multi-time compressed audio are the same.

In the past decade, researchers have begun to study compression detection problems in the field of multimedia. Initially, scholars focused on the compression detection problems in images such as double JPEG compression detection [1, 2] and the estimation of compression parameters [3]. In recent years, there has also been some audio compression detection research.

In the beginning, researchers mainly focused on MP3 compression detection. Yang [4] first proposed the method of recompression detection of the MP3 audio signal. The author found that the difference between a false MP3 and a normal MP3 exists in the small amplitude coefficient of the MDCT (modified discrete cosine transform) and analyzed the reason for this phenomenon through the mechanism of MP3 encoding. In [5], a method based on Benford's law is proposed that can be used to detect MP3 audio recompression. However, the detection rate is not sufficient for the case in which the second compression bitrate is lower than the first compression bitrate. Later, Qiao [6] proposed a method to detect MP3 recompressed audio by using multiple statistics of subband MDCT coefficients as features, which achieve good performance in the above problem. Liu [7] also utilized the number of subband MDCT coefficients that approach zero, which can also be used as a good indicator to detect MP3 recompressed audio. Ma [8] analyzed MP3 recompression quantitatively from the Huffman code table index. Bießmann [9] estimated the MP3PRO parameters, and Yan [10] also detected the compression history for MP3 audio files from the scale factor and the Huffman code table index.

Scholars have also studied compression detection in other common audio formats as well as the identification of the codec. Jenner [11] proposed a method to identify several speech codecs. Hiçsönmez [12] proposed a method for audio codec identification through payload sampling. They built statistical models through the randomness and chaotic nature of encoded audio. They also identify compression traces by means of audio quality measures [13]. Hennequin [14] proposed a method for codec independent lossy audio compression detection. Luo [15] proposed a method using MDCT coefficients and MFCCs to detect the compression history of audio waveforms. The AAC format is also considered by Huang [16], who detected AAC double compression based on the difference of the scale factor. An earlier approach to detect double AMR compression was proposed by Shen [17], which achieved a detection accuracy of 87%. Luo [18, 19] also proposed AMR recompression detection based on a deep learning technique. Real-scenario application is also considered in compression history analysis. Gärtner [20] proposed an efficient cross-codec framing grid analysis for audio tampering detection and released a corresponding database for evaluating audio tampering detection methods [21].

In this paper, we concentrate on the audio forensics problem of detecting whether an audio waveform is an original (uncompressed) waveform or a decompressed one. We have focused on multi-time compressed audio (especially triple compression, four-time compression) since few studies have reported beyond the scope of single and double compressed audio. We propose a feature set that extracts from spectrum of audio waveform to characterize the statistics of both the original and decompressed audio frames. Then, a trained stacked autoencoder network is used for frame-level detection, and the final decision of an audio clip will be made by majority voting.

The paper is organized as follows. Section 1 is the introduction. Section 2 describes the proposed framework for detecting the decompressed audio waveform. Section 3 describes the experimental results, and the conclusion is drawn in Sect. 4.

2 Proposed Method

We focus on detecting whether an audio waveform is an original (uncompressed) waveform or a decompressed one. Traditional machine learning methods include two stages of feature extraction and classification. However, the multi-time compression audio waveforms have almost the same statistical characteristics. Figure 1 (a)–(d) show the spectrograms of original audio and different compression format audio, while the Fig. 1 (e)–(h) show spectrograms of decoded waveform of multi-time compressed MP3. We can see that these spectrograms are of high similarity, make it hand to design discriminative hand-crafted features for this problem.

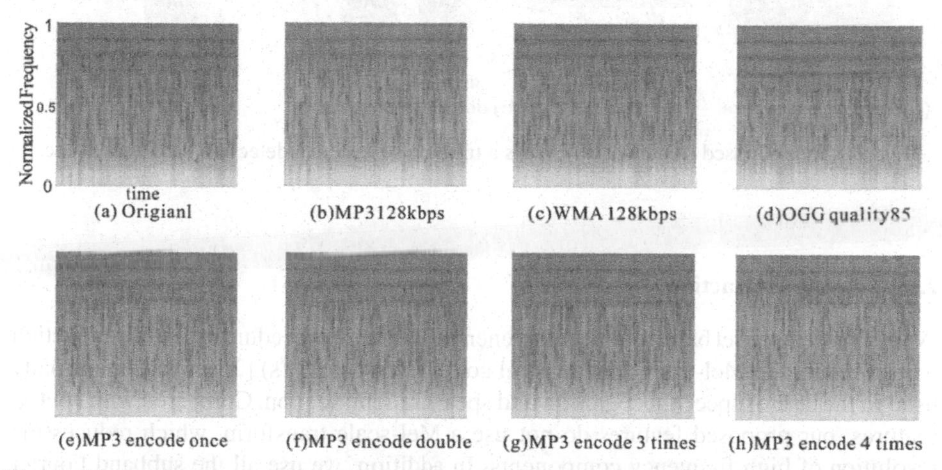

Fig. 1. The spectrograms of original waveform and decoded waveform of compressed audio. The 1st row (a)–(d) show spectrograms of original audio and different compression formats. The 2nd row (e)–(h) show spectrograms of decoded waveform of multi-time compressed MP3.

Recent techniques usually perform convolutional neural network (CNN) on audio spectrogram for detection problem. However, we find that the more obvious clues are the energy patterns that exist horizontally across the time dimension. Therefore, we proposed a detection framework based on spectrum scanning (of audio frame) to extract the subtle differences of multi-time decompressed audio. For classification, we adopt stack autoencoders (SAE), which is a kind of deep learning technique to help for solving this problem.

Figure 2 illustrates the proposed framework for discriminating decompressed audio waveforms from original ones. In the training stage, an SAE network is trained by frame-level samples of two kinds of waveforms (original and decompressed). The detection procedure involves frame scanning, spectrum feature extraction and a SAE network for frame-level classification. For an audio clip that may have a number of frames, a majority voting technique could be used to make the final decision. The category of an audio clip is defined as the category with the largest number of frames is classified by SAE.

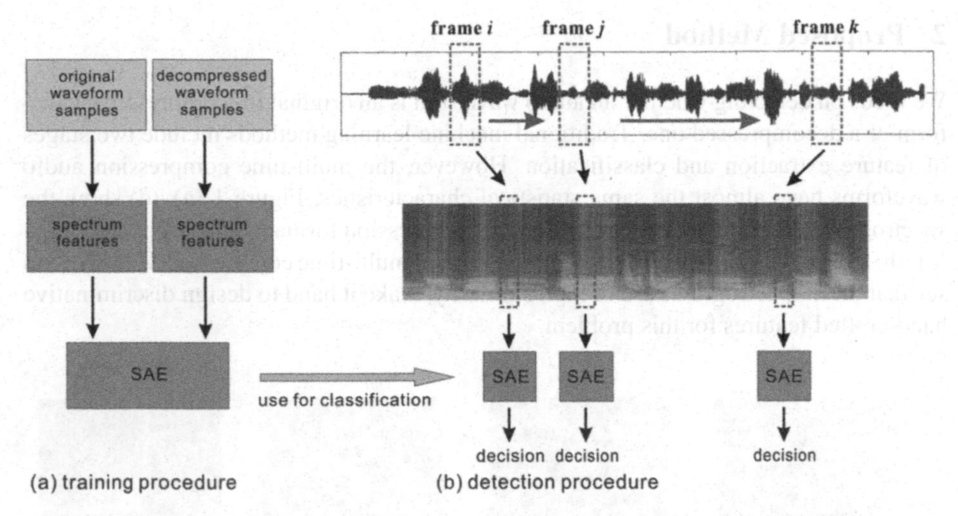

Fig. 2. The proposed framework includes a training stage and a detection procedure stage.

2.1 Feature Extraction

We use the feature set based on spectrum energy. The main procedure of feature extraction is inspired by the Mel-frequency cepstral coefficients (MFCCs) [23], which are usually used in the task of speech recognition and speaker identification. Compared with MFCC features, our proposed features do not use a Mel-scale transform, which reduces the resolution of high frequency components. In addition, we use all the subband Fourier coefficients because the high frequency components that the MFCCs discarded are crucial for detecting compressed audio. The steps to extract feature set from an audio frame are as follows:

- Adding a Hamming window to the audio frame
- Performing a fast Fourier transform and obtaining subband energy coefficients
- Performing triangular window filtering on the subband energy coefficients
- Obtaining the log energy of each subband energy coefficients

In all the experiments, we use a frame size of 1024, which is a normal length for audio processing and is appropriate for performing a fast Fourier transform. There are

513 (=1024/2 + 1) effective bins due to FFT spectrum symmetry. Triangular filtering uses a window size of 15 with half overlapping (advance 8 bins each time). Therefore, the proposed feature set has 64 (≈513/8) coefficients.

2.2 SAE Network Training

A stacked autoencoder (SAE) is a kind of deep neural network, which has been used in many kinds of applications [25, 26]. Its training procedure is shown in Fig. 3. An autoencoder (AE) is the main functional unit that can learn data representation in an unsupervised way. An AE is a fully connected 3-layer neural network. The first layer is used as input, while the third layer is used as the output layer. An AE artificially sets the training target of the output layer to be consistent with the input data (the meaning of "autoencoder"). Therefore, the training procedure of an AE is in an unsupervised manner. An AE is trained like a conventional neural network via back-propagation. After an AE finishes training, the hidden layer in the middle of AE can be regarded as another representation (i.e., feature) of the input data. The output of hidden layer will be transmitted to next AE as its input (for training the next AE).

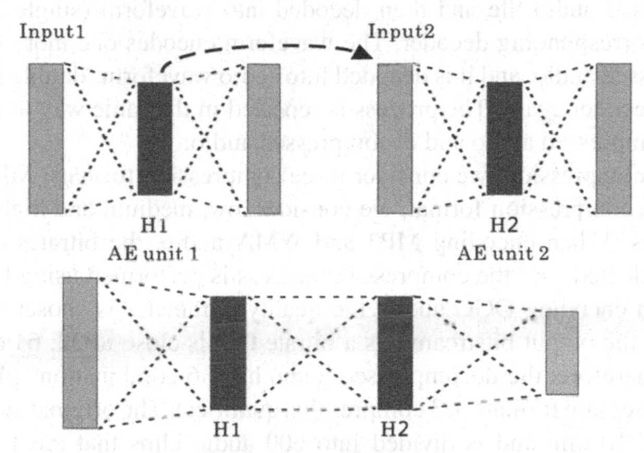

Fig. 3. Illustration of the SAE network construction. Each AE unit is trained, and all hidden layers are combined together to form the SAE. (Color figure online)

To construct an SAE network, the hidden layer of each AE (blue in Fig. 3) is connected together, and a classification layer of N nodes (one-hot type) is added to the last layer (yellow in Fig. 3). The SAE network is designed as a classifier for classifying the N categories. A fine-tuned stage is carried out in a supervised manner for the SAE network by back-propagation with the data of N categories and their corresponding labels.

The feature extraction mentioned above produces a feature set of 64 coefficients for each frame. Therefore, the first layer of the SAE network has 64 nodes. A common SAE network design that can achieve better results is to raise the dimension first and then reduce the dimension. In our experiments, we use 2 hidden layers in the SAE, with the first and second hidden layers containing 100 and 60 nodes, respectively. The node number of the classification layer (the last added layer) is determined by how many categories to be distinguished for a given task.

2.3 Majority Voting

The SAE network uses as a classifier and applies to frame-level audio data. In detection procedure, decision will be obtained by SAE for each frame. Generally, an audio clip contains several frames. After using the SAE to detect all frames, a category of audio clip will be determined by majority voting. The category of the largest number of frames is recorded as the category of audio clips.

3 Experiments

3.1 Dataset

Recently, little research focus on multi-time compression data. Therefore, we create a large dataset that containing 6.2 h recording and have a storage capacity of 4 GB. We first recorded a long audio waveform mono sound signal with a sampling rate of 44100 Hz and saved it as a WAV file. This audio file is regarded as the original audio. The audio waveform is compressed by the encoder (e.g., MP3 or WMA) to obtain a single compressed audio file and then decoded into waveform (single decompressed audio) by the corresponding decoder. The waveform encodes one more time to obtain double compressed audio, and it is decoded into audio waveform (double decompressed audio) by the decoder again. The process is repeated in the same way to produce triple or four-time compressed audio and decompressed audio.

During the compression, we consider three compression formats: MP3, WMA and OGG. For each compression format, we consider low, medium and high compression bitrates/qualities. When encoding MP3 and WMA audio, the bitrates of 32, 64 and 128 kbps are selected, and the compression process is performed using FormatFactory software. When encoding OGG audio, the quality parameter is chosen as 20, 40 and 85, with which the output bitstream has a bitrate that is close to 32, 64 and 128 kbps, respectively. Therefore, the decompressed audio has 36 combinations (4 compression times $\times$ 3 compression formats $\times$ 3 compression qualities). The original audio waveform has a length of 10 min and is divided into 600 audio clips that are 1 s long. Their corresponding 36 decompressed audio clips are obtained. The dataset has totally $600 \times (36 + 1) = 22{,}200$ audio clips, which could form more than 900,000 audio frames for experimental evaluation.

3.2 Discriminative Power of Feature Set

In this section, we analyze the extracted feature set by the t-SNE [22], which is a good dimension reduction technology and a tool for high-dimensional data visualization. Given a batch of data, t-SNE first performs unsupervised dimension reduction to 2 dimensions, which can be easily visualized. Figure 4 shows the t-SNE results of the feature set from selected frames (both original audio and decompressed audio). We can see that even when using only unsupervised learning, the two categories of data have centralized into two different centers, showing an initial discriminative power.

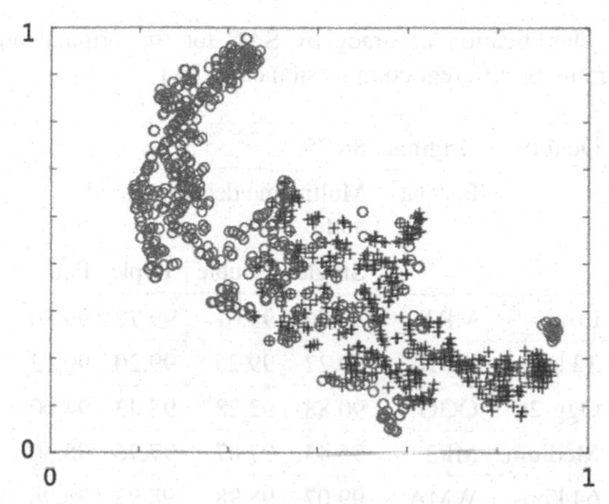

Fig. 4. The t-SNE visualization of selected frames to show the discriminative ability of the extracted feature set (blue: original audio frames red: decompressed audio frames). The t-SNE converts each sample feature to a two-dimension point (shown in normalized X- and Y-axis). (Color figure online)

3.3 Detection of the Decompressed Audio

In this section, our goal is to identify whether an audio waveform is an original uncompressed or a decompressed signal. To achieve this goal, we need to train an SAE network that can distinguish between two categories (original audio frame and decompressed audio frame). We randomly select 25% of original audio clips from the dataset, along with decompressed audio clips from the 36 combinations. Then these audio clips are divided into frames and features are extracted from the frames to train SAE network. The remaining 75% audio clips are used as testing data. Frame scanning is performed for each audio clip. The extracted features of the audio frames fed into the SAE network to produce a category decision. Majority voting is carried out through decisions of all of the frames to finally determine the category of a clip.

Note that in this experiment, we train only one model, and it is used for discriminating between original audio and decompressed audio of 3 formats and different bitrates. Table 1 shows the frame-level classification results. The testing data include 10,000 original audio frames and 10,000 of each kinds of 36 compression cases. Therefore, the evaluation is tested on 370,000 (= 10,000 $\times$ 37) audio frames. We can see that the decompressed audio frames of medium and low quality can achieve good classification accuracy. Most classification errors occur on OGG format due to the confusion between the original and the decompressed with a high bit rate.

Next, we evaluate the ability of our framework to classify audio clips. For an audio clip, the majority vote of each frame outputs the final category. The SAE network accumulates the classification effect of each frame so that it improves final detection rate. Our experiments also conducted on the original audio clips and the decompressed audio clips of different cases (see Table 2). It demonstrated that our method can effectively identify whether an audio waveform is original or decompressed.

Table 1. Frame-level classification accuracy by SAE for the original audio frame and the decompressed audio frame of different compression cases [%]

Quality	Original	88.75			
	Formats	Multi-time decompressed audio			
		Single	Double	Triple	Four
Low	MP3	99.56	99.61	99.72	99.80
32 kbps	WMA	99.27	99.26	99.20	99.12
Ogg-20	OGG	90.88	92.28	93.43	94.00
Medium	MP3	96.43	97.07	97.76	98.30
64 kbps	WMA	99.07	98.98	98.93	98.96
Ogg-40	OGG	77.76	77.03	77.10	78.10
High	MP3	72.91	77.40	81.47	84.52
128 kbps	WMA	98.41	98.39	98.34	98.31
Ogg-85	OGG	56.02	55.24	54.99	56.79

Table 2. Classification accuracy of the original and the decompressed audio clip of different cases [%].

Quality	Original	100			
	Formats	Multi-time decompressed audio			
		Single	Double	Triple	Four
Low	MP3	100	100	100	100
32 kbps	WMA	100	100	100	100
Ogg-20	OGG	100	100	100	100
Medium	MP3	100	100	100	100
64 kbps	WMA	100	100	100	100
Ogg-40	OGG	100	100	100	100
High	MP3	97.08	99.16	100	100
128 kbps	WMA	100	100	100	100
Ogg-85	OGG	70.83	71.25	69.16	73.75

3.4 Discussions

MFCCs are common features and effective on many audio detection tasks. Although the features we use is inspired by the idea of MFCCs [23], the experimental results are much better than those of the MFCCs (the reason is stated in Sect. 2.1). We compared the results for the task of detecting the original and the decompressed audio clips. Table 3 shows the results mainly on medium- and high-quality audio (MFCCs perform well on low-quality audio, therefore not shown in the table). Note that the MFCC features have a detection accuracy of approximately 60–70% when identifying high quality MP3 and OGG files. Therefore, our proposed features significantly improve the detection results.

Table 3. Classification accuracy improvement of our method compared with the known MFCC feature [%].

Quality	Original	+12.7			
	Formats	Multi-time decompressed audio			
		Single	Double	Triple	Four
Medium	MP3	0	0	0	0
64 kbps	WMA	0	0	0	0
Ogg-40	OGG	+32.2	+31.5	+27.5	+24.5
High	MP3	+22.5	+22.4	+21.9	+15.3
128 kbps	WMA	0	0	0	0
Ogg-85	OGG	+13.8	+10.2	+0.8	+1.6

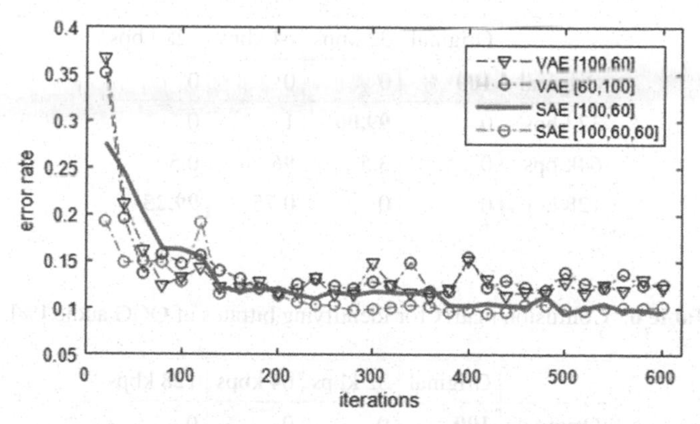

Fig. 5. Comparison with variational auto encoding (VAE) on frame-level detection.

We also discuss and compare another deep learning technique, variational auto encoding (VAE) [24]. VAE is a kind of deep learning technology that can be used to learn the mean and variance of features for signal reconstruction, and it has been widely concerned in recent years. We use it to build a classifier replace SAE for comparison. The experiments are conducted to evaluate the frame-level detection on the original audio or decompressed ones. Figure 5 shows the error rate of frame-level detection results by

SAE and VAE. In general, the performance of the two methods is not much different. However, the error rate of SAE is slightly (≈2%) lower than VAE.

3.5 Identification of the Compression Bitrate

In this section, we further evaluate the compression bitrate of the decompressed audio. The data consist of the original audio and the decompressed audio of low, medium and high bitrate/quality. We also considered single, double, triple and four-time compression, and during the multi-time compression for one individual audio, we use a fixed bitrate/quality. We trained 3 models for identifying MP3, WMA, and OGG audio, respectively. Table 4, 5 and 6 show the confusion matrixes of different formats.

Table 4. Confusion matrix for identifying bitrates of MP3 audio [%].

	Original	32 kbps	64 kbps	128 kbps
Original	**99.75**	0	0	0.25
32 kbps	0	**100**	0	0
64 kbps	0	0	**99.50**	0.50
128 kbps	0	0	0	**100**

Table 5. Confusion matrix for identifying bitrates of WMA audio [%].

	Original	32 kbps	64 kbps	128 kbps
Original	**100**	0	0	0
32 kbps	0	**99.00**	1	0
64k bps	0	3.5	**96**	0.5
128 kbps	0	0	0.75	**99.25**

Table 6. Confusion matrix for identifying bitrates of OGG audio [%].

	Original	32 kbps	64 kbps	128 kbps
Original	**100**	0	0	0
32 kbps	0	**100**	0	0
64 kbps	0.5	0	**99.5**	0
128 kbps	20.5	0	0	**79.5**

The results show that the proposed method can also achieve good performance in estimating the compression bitrate.

4 Conclusion

In this paper, we proposed a framework for compression history detection of audio waveforms, and we focused on detecting whether an audio waveform is original or decompressed one. Spectrum features are extracted from audio frames, then SAE network is trained to detect frame-level data and final decisions are made by majority voting for an audio clip. The experimental results show that the proposed method can effectively identify between original and decompressed audio from various combinations of compression formats (codecs) and compression levels (single to four-time compression). Moreover, the proposed method can also effectively recognize the compression bit rate even in multi-time compression situations. In the future, we will analyze more audio codecs, and we will also try to solve the problem of cross-format compression, which is a more complex case.

Acknowledgments. The work presented in this paper was supported in part by the NSFC (61602318, 61672551, 61631016, 61972090), the Guangzhou Science and Technology Plan Project under Grant 201707010167, the Science and Technology planning project of Guangdong Province (2014A010103039, 2015A010103022).

References

1. Barni, M., Chen, Z., Tondi, B.: Adversary-aware, data-driven detection of double JPEG compression: how to make counter-forensics harder. In: IEEE International Workshop on Information Forensics and Security (2016). https://doi.org/10.1109/wifs.2016.7823902
2. Pevný, T., Fridrich, J.: Detection of double-compression in JPEG images for applications in steganography. IEEE Trans. Inf. Forensics Secur. **3**(2), 247–258 (2008)
3. Galvan, F., Puglisi, G., Bruna, A., Battiato, S.: First quantization matrix estimation from double compressed JPEG images. IEEE Trans. Inf. Forensics Secur. **9**(8), 1299–1310 (2014)
4. Yang, R., Shi, Y., Huang, J.: Defeating fakequality MP3. In: Proceedings of ACM Workshop Multimedia Security, Princeton, NJ, USA, pp. 117C-124 (2009)
5. Yang, R., Shi, Y., Huang, J.: Detecting double compression of audio signal. In: Procedings of SPIE Electronic Imaging. International Society for Optics and Photonics, 75410 K-75410 K-10 (2010)
6. Qiao, M., Sung, A., Liu, Q.: Revealing real quality of double compressed MP3 audio. In: Proceedings of the International Conference on Multimedia, pp. 1011–1014 (2010)
7. Liu, Q., Sung, A., Qiao, M.: Detection of double MP3 compression. Cognit. Comput. **2**(4), 291–296 (2010)
8. Ma, P., Wang, R., Yan, D., Jin, C.: A Huffman table index based approach to detect double MP3 compression. In: Shi, Y.Q., Kim, H.J., Pérez-González, F. (eds.) IWDW 2013. LNCS, vol. 8389, pp. 258–271. Springer, Heidelberg (2014). https://doi.org/10.1007/978-3-662-43886-2_19
9. Bießmann, P., et al.: Estimating MP3PRO encoder parameters from decoded audio. In: Proceedings of GI-Jahrestagung, 2841C–2852 (2013)
10. Yan, D., Wang, R., Zhou, J., Jin, C., Yang, Z.: Compression history detection for MP3 audio. KSII Trans. Internet Inf. Syst. **12**(2), 662–675 (2018)
11. Jenner, F., Kwasinsk, A.: Highly accurate non-intrusive speech forensics for codec identifications from observed decoded signals. In: Proceedings of International Conference Acoustic, Speech and Signal Process, Kyoto, Japan, pp. 1737–1740 (2012)

12. Hiçsönmez, S., Sencar, H., Avcibas, I.: Audio codec identification through payload sampling. In: Proceedings of Workshop Information Forensics Security (2011). https://doi.org/10.1109/wifs.2011.6123128
13. Hiçsönmez, S., Uzun, E., Sencar, H.: Methods for identifying traces of compression in audio. In: Proceedings of 1st International Conference Communication, Signal Processing and Application, Sharjah, United Arab Emirates (2013). https://doi.org/10.1109/iccspa.2013.6487284
14. Hennequin, R., Royo-Letelier, J., Moussallam, M.: Codec independent lossy audio compression detection. In: IEEE International Conference on Acoustics, Speech and Signal Processing (ICASSP), pp. 726–730 (2017)
15. Luo, D., Luo, W., Yang, R., Huang, J.: Identifying compression history of wave audio and its applications. ACM Trans. Multimed. Comput. Commun. Appl. (TOMCCAP), **10**(3), 30 (2014)
16. Huang, Q., Wang, R., Yan, D., Zhang, J.: AAC double compression audio detection algorithm based on the difference of scale factor. Information **9**(7), 161 (2018)
17. Shen, Y., Jia, J., Cai, L.: Detecting double compressed AMR-format audio recordings. In Proceedings of 10th Phonetics Conference of China (2012)
18. Luo, D., Yang, R., Huang, J.: Detecting double compressed AMR audio using deep learning. In: IEEE International Conference on Acoustics, Speech and Signal Processing (ICASSP), pp. 2688–2692 (2014)
19. Luo, D., Yang, R., Li, B., Huang, J.: Detection of double compressed AMR audio using stacked autoencoder. IEEE Trans. Inf. Forensics Secur. **12**(2), 432–444 (2017)
20. Gärtner, D., Dittmar, C., Aichroth, P., Cuccovillo, L., Mann, S., Schuller, G.: Efficient cross-codec framing grid analysis for audio tampering detection. In: Proceedings of Audio Engineering Society Convention, p. 136 (2014)
21. Gärtner, D., Cuccovillo, L., Mann, S., Aichroth, P.: A multi-codec audio dataset for codec analysis and tampering detection. In: Proceedings of 54th Audio Engineering Society Conference on Audio Forensics (2014)
22. Maaten, L., Hinton, G.: Visualizing data using t-SNE. J. Mach. Learn. Res. **9**, 2579–2605 (2008)
23. Reynolds, D., Quatieri, T., Dunn, R.: Speaker verification using adapted gaussian mixture models. Digit. Signal Process. **10**(1–3), 19–41 (2000)
24. Kingma, D.: Max welling: auto-encoding variational Bayes. arXiv:1312.6114 [stat.ML] (2013)
25. Kan, X., et al.: Snow cover mapping for mountainous areas by fusion of MODIS L1B and geographic data based on stacked denoising auto-encoders. Comput. Mater. Continua **57**(1), 49–68 (2018)
26. Zhao, X., Jiaxin, W., Zhang, Y., Shi, Y., Wang, L.: Fault diagnosis of motor in frequency domain signal by stacked de-noising auto-encoder. Comput. Mater. Continua **57**(2), 223–242 (2018)

A Deep Reinforcement Learning Framework for Vehicle Detection and Pose Estimation in 3D Point Clouds

Weipeng Wang[1,2], Huan Luo[1,2(✉)], Quan Zheng[1,2], Cheng Wang[3], and Wenzhong Guo[1,2]

[1] Fujian Provincial Key Laboratory of Network Computing and Intelligent Information Processing, Fuzhou University, Fuzhou 350003, China
hluo@fzu.edu.cn
[2] Key Laboratory of Spatial Data Mining and Information Sharing, Ministry of Education, Fuzhou 350003, China
[3] Fujian Key Laboratory of Sensing and Computing for Smart City, Xiamen University, Xiamen 361005, FJ, China

Abstract. As for autonomous driving in urban environments, it is of significance to accurately capture the position and pose of vehicles. Those information can assist self-driving system in making right decisions to avoid potential risks. Currently, 3D point clouds captured by laser scanners are widely used in self-driving systems to sense the real environment. Therefore, in this paper, we propose a deep reinforcement learning framework for vehicle detection and pose estimation by using 3D point clouds. Specifically, to estimate the pose of vehicles, we propose to design a rotation action in Deep Q Network (DQN). In addition, by considering the whole detection procedure as Markov Decision Process (MDP), our intermediate detected results can further improve the detection performance of our proposed method. The evaluations are carried on outdoor point cloud scenes captured by VMX450 laser scanning system. The experimental results demonstrate the satisfied performance on vehicle detection and pose estimation.

Keywords: Vehicle detection · Deep Q Network · Point cloud

1 Introduction

With fast development of deep learning, self-driving related techniques have been flourishing in recent years. Automatically locating and recognizing on-road objects as an essential task to self-driving has been attracted wide attentions. In practice, not only object locations but also accurate object poses provide important information to self-driving system. This is because object pose information assist in predicting and analyzing on-road object behavior, which gives clues to guide the decisions of self-driving system.

Nowadays, 3D point cloud acquired by the laser scanning system has been introduced into self-driving system. Compared with traditional optical images,

X. Sun et al. (Eds.): ICAIS 2020, LNCS 12240, pp. 405–416, 2020.
https://doi.org/10.1007/978-3-030-57881-7_36

3D point cloud has exhibited many advantages. Specifically, 3D point cloud can provide 3D geometric information which reflects the real state of objects in outdoor scenario. Moreover, because of laser scanner belonging to active sensor, collection of 3D point is not influenced by illumination conditions. Last but not least, point cloud is invariant under transformation. Therefore, this paper focuses on detecting objects and estimating their poses in 3D point cloud scenes.

Most existing deep learning approaches achieved superior results in image area [1,2]. Works on object detection has been explored for years. Faster R-CNN [3] detects objects based on proposals. The proposals are generated by another neural network, RPN. YOLO [4] worked on images and could handle real-time jobs to some extent. Recently a well-designed fully convolutional network [5] has been proposed to detect digital image region forgery. However, one common essential issue is that they were not applicable to the pose estimation. For real-world vehicle detection problem, it makes sense to estimate a contour with pose information on account of the complexity of urban environments. This is because making a prediction about the movements of vehicles, based on their poses, is meaningful practically. In [6,7], a method based on Hough Forest (HF) was proposed to detect vehicles in 3D point cloud. In [8], Liu et al. also proposed a method using 3DCNN-DQN-RNN framework to do point cloud semantic parsing task. In [9], YOLOv2 and Euler-Region Proposal Network was used to estimate the pose of objects. PointRCNN [10] was an upgrade of 2D R-CNN series for point cloud. Some methods [11–14] operated on point cloud also took the pose of objects into consideration. There was no doubt that their works leveraged the field of vehicle detection. However, the pose estimation was not exploited in the methods mentioned above.

We believe that the pose of vehicle can indeed assist with the detection and further improve the detection results. By dividing detection into sequential decision process, we reduce the number of training data labeled manually by a wide margin. Deep Q Network (DQN) is active in making sequential decisions to gain fine accumulated rewards. It is frequently used in game playing field such as playing Atari [15]. Notably this model was under the spotlight when the breakthrough, AlphaGo, beat the human champion player in 2016. In [16], Mathe et al. designed a reinforcement learning approach to fix evidence regions during detection process. Another effort was to invoke an eye window. This allowed the agent to imitate the human beings and capture objects. In [17], Bueno et al. provided an approach to estimate an appropriate bounding box for objects. In [18], Li et al. utilized a pre-trained model to estimate the location of aircrafts with the apprenticeship learning. However, the pose of object which is important information for real application was omitted in those methods.

In this paper, we propose a new approach based on DQN in detecting vehicles and determining the vehicle poses in 3D point cloud. To estimate the pose of vehicles, we propose an approach to estimate the pose of vehicles by designing a rotation action. Meanwhile by considering the process as the Markov Decision Process, our intermediate results can further assist to improve our detection.

Experimental results on VMX-450 dataset support our point of view. Comparative studies also demonstrate the superiority of our proposed approach.

2 Method

2.1 Overview of the Algorithm

Our method can be generally divided into two stages: the generation and the detection. In the generation stage, the 3D point cloud dataset $P = \{p_i = (x, y, z), i = 1, 2, ...\}$ is processed and projected into 2D grid images in three channels based on height information. Through this procedure, our network model can estimate a rotated bounding box $(x, y, w, h, angle)$ that sufficiently represents the pose of a vehicle, where x and y are coordinates, w and h are respectively width and height, and $angle$ is the azimuth angle relative to the x-axis. In the detection stage, at first, we construct a set of proposals with simple priori knowledge. And our problem is considered as a task of reinforcement learning aiming at detection and pose estimation. During the training process, the agent is invoked to learn a detection policy through a sequential of trial and error. We design a feasible reward function for the agent and strike a balance between exploration and exploitation by using $\epsilon - greedy$ policy. During the test process, the trained model are utilized to estimate vehicle poses. Exploration is not required in the test step for a steady detection results.

In the next few subsections, approaches are described in detail. Section 2.2 talks about the construction of bird's eye views. Section 2.3 shows the mechanism of our detector and define the elements of reinforcement learning. Section 2.4 shows how the DQN works on the detection task.

2.2 Construction of Bird's Eye View

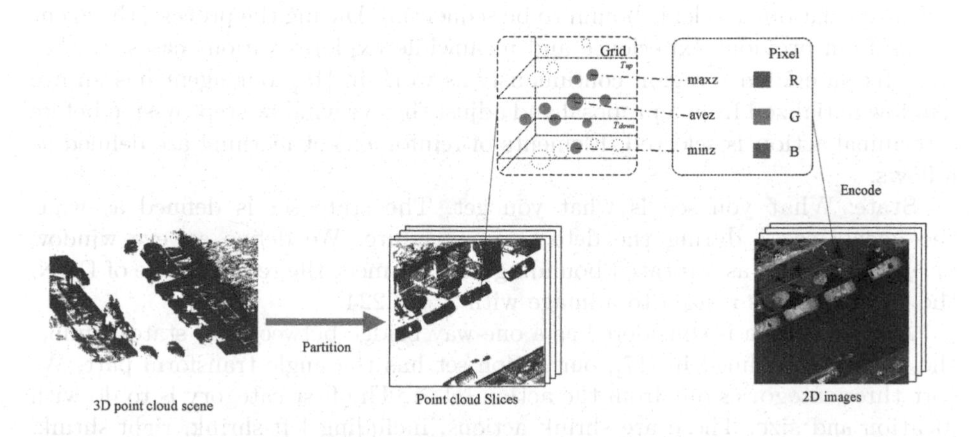

Fig. 1. The construction of bird's eye view.

To avoid the large computational consumption to handle the raw large-scale point cloud, we construct bird's eye views from point clouds as [19–22] do. The image set $I = \{i_j = (x_j, y_j, c_j), j = 0, 1, ...\}$ of 3D point cloud $P = \{p_j = (x_j, y_j, z_j, l_j)\}$ is constructed in this stage. To gain more useful slices, we utilize the PCA algorithm in the horizontal aspect ($xy - plane$) to transform x and y while z remains unchanged. Large-scale point cloud scenes are then partitioned into slices S. Each slice $S_i \in S$ corresponding to an image is further divided into $w \times h$ grids. For each grid $G \in S_i$, we encode height information $(max(p_z), ave(p_z), min(p_z))$ into pixels and finally normalized to 0–255 based on Eq. (1)

$$norm(p_z) = \frac{p_z}{maxz - minz} \tag{1}$$

where $p \in G$ and $maxz(minz)$ is the max (min) z value in S_i. However, point cloud scenes in civilian environments can be complicated and layered in vertical. To avoid overlays such as trees in $xy - plane$, we define an upper threshold and lower threshold T_{up} and T_{down}. If p_z exceeds these thresholds, it is set to the boundary. For those $G = \emptyset$, values are set to zero. The whole procedure is illustrated in Fig. 1.

We employ a simple method of priori knowledge i.e. dense points tends to exist objects, to generate our proposals. After encoding height information into the bird's eye views, ground and overlays are removed. We applied the pixel-wise Euclidean Cluster in the bird's eye view images and generate rectangular proposals.

2.3 Sequential Vehicle Detector

We define a detector for visual civilian environments based on Deep Q Network (DQN). The detection task is defined as a goal that the agent learns a fine way to accumulate the max reward instead of traditional one-step regression such as Faster R-CNN [3]. Hence the agent ought to make diverse decisions, which indicates that our model is bound to be sequential. During the process, the agent learns from previous experience and meanwhile explores various cases. Rather than fix an evidence region continuously as work in [16], our agent has an eye window initialized from a proposal and adjust the eye window step-by-step before a terminal action is selected. Elements of reinforcement learning are defined as follows.

State: What you see is what you get. The state set is defined as what the agent "sees" during the detection procedure. We define an eye window $(x, y, w, h, angle)$ as a rotated bounding box. To meet the required size of DQN, the eye window is resized to a image with 224×224.

Action: Action is considered as a one-way bridge between two states. Unlike the action set defined in [17], our action set has the angle transform part. We sort three categories out from the action set A. The first category is to do with location and size. These are shrink actions, including left-shrink, right-shrink, top-shrink and bottom-shrink. Given a shrink variance δ, once a certain action is selected, the eye window shrinks δ towards a specific direction e.g. bottom-shrink, which means that given a current eye window $s_n = (x, y, w, h)$, the next eye window is $s_{n+1} = (x, y+\delta, w, h-\delta)$. To make our model feasible and flexible, we define a shrink rate as ξ. The relationship between δ and size is shown in Eq. (2).

$$\delta = \begin{cases} \xi h, & action = a_{top}, a_{bottom}, \\ \xi w, & action = a_{left}, a_{right}. \end{cases} \tag{2}$$

The second type of action is the angle-plus action. Once the action is determined, the eye window rotates angle $\Delta\theta$. In practice, we set $\Delta\theta$ to be 10. The last type is the action which sends a terminal signal to the agent. During the searching progress, the agent simultaneously observes the following rules:

$$\begin{cases} w \geq T_w, \\ h \geq T_h, \\ \theta \leq T_{angle}, \end{cases} \tag{3}$$

where T_w, T_h and T_{angle}) indicate the threshold of width, height and rotation angle, respectively. When the eye window is about to exceed, the search progress are forced to be terminated passively, which indicates that there is no objects. Furthermore, for the eye window, we do not leave a fallback such as expansion or angle-minus in case of being trapped in a local optima. The period in detail is shown in Fig. 2.

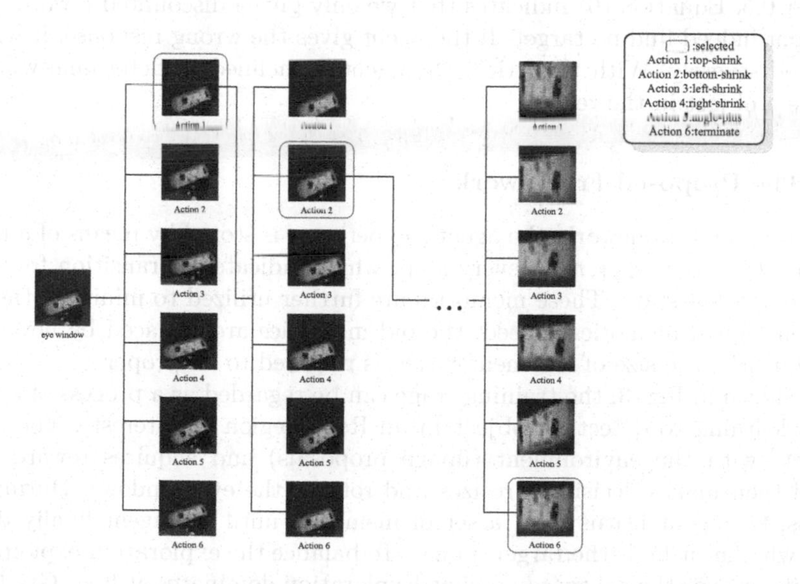

Fig. 2. Example of the designed detection agent making sequential decisions.

Reward: Reward is the feedback given by the environment. The Intersection-over-Union (IoU) is tightly linked to the detection result. So reward function consists of two parts. If the action is not the terminal, the IoU will be encouraged to be higher than the previous state. The relationship is formulated in Eq. (4)

$$reward = \omega sgn(IoU_{s_n} - IoU_{s_{n+1}}), \tag{4}$$

where ω is a constant. In practice, we set ω to be 1.6. When the action is terminal, the IoU determines the final reward. The relationship is formulated in Eq. (5)

$$reward = \begin{cases} +\varphi + \eta(IoU_{terminal} - \tau_{IoU}), & IoU_{terminal} \geq \tau_{IoU}, \\ -\varphi + \eta(IoU_{terminal} - \tau_{IoU}), & IoU_{terminal} \leq \tau_{IoU}, \end{cases} \tag{5}$$

where τ_{IoU} is a threshold of IoU. φ represents base reward. In practice, we set τ_{IoU} and φ to be 0.5 and 2.0, respectively. $\eta(IoU_{terminal} - \tau_{IoU})$ is an additional reward for obtaining better detection performance. In our experiment, we set $\eta = 20$. There remains one possibility that if the given proposal contains no any objects of interest, the searching progress is demanded to end forcedly. This part of reward is defined in (6)

$$reward = \begin{cases} +\beta(\varphi + \eta(1.0 - \tau_{IoU})), & Objects \cap Proposal = \emptyset, \\ -\dfrac{\varphi + \eta(1.0 - \tau_{IoU})}{\beta}, & otherwise, \end{cases} \tag{6}$$

where β is a balance factor to balance the primary and secondary jobs. Here, we set it at 0.5. Equation (6) indicates that we only give a discounted reward when the agent indeed find no target. If the agent gives the wrong response, it will be punished severely. With this trick, the agent is inclined to determine whether there is a object in the region.

2.4 The Proposed Framework

In the proposed framework, the agent's experience is stored by means of a memory set $M = (s_t, a, s_{t+1}, r)$ in every step, which indicates a transition from one state to another state. These memories are further utilized to minimize the loss function. When memories exceed, the old memories are replaced by new ones. Consequently, the size of the memory set is required to be proper.

As shown in Fig. 3, the training stage can be regarded as a process of human beings learning to detect an object in an RoI (Region of Interest). The agent interacts with the environment (image proposals) and acquires reward from it, and then makes decisions (resizes and rotates the eye window). During the process, the agent learns from a set of memories until the agent finally determines whether it finds the target object. To balance the exploration-exploitation dilemma, we use the $\epsilon - greedy$ policy. Exploration dominates at first. Gradually when phenomena are generally explored, exploitation takes over. The agent still remains possibilities of exploring. Hence ϵ descents from 1.0 to 0.1 in the training stage. Algorithm of the training stage is shown as follows: Following the work in [23], our network consists of two parts. One is the target network. Another

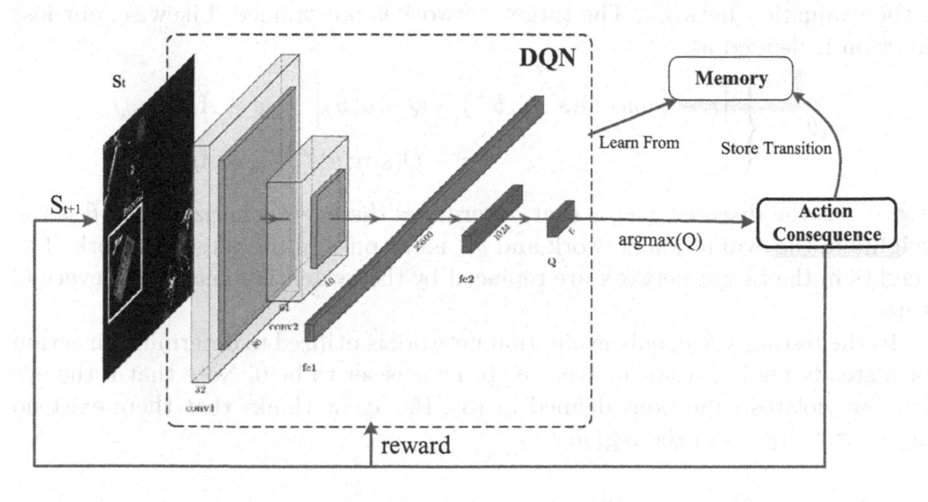

Fig. 3. The illustration of our proposed framework.

Algorithm 1. Training Stage

---

initialize memory set M where $|M| = 10000$, $\epsilon = 1.0$, DQN parameters θ and θ^-
for $e = 1$ to $TOTAL\_EPOCH$ **do**
  update $\epsilon = \epsilon - 0.001$ if $\epsilon > 0.1$ else $\epsilon = 0.1$
  update θ^- each C steps
  for p in *proposals* **do**
    initialize the eye window $(x, y, w, h, angle)$ and initial state s_t
    while True **do**
      if random $< \epsilon$ **then**
        $action$ = select randomly
      else
        $action = argmax(s_t; \theta)$
      end if
      step $action$, gain $reward$ and next observation s_{t+1}
      store transition $(s_t, action, reward, s_{t+1}) \rightarrow M$
      if $e \geq |M|$ **then**
        minimize loss $L(\theta)$ using SGD policy to learn from experience
      end if
      if $action = a_{terminal}$ **then**
        break
      end if
      $s_t = s_{t+1}$
    end while
  end for
end for

---

is the evaluation network. The target network is not trained. Likewise, our loss function is defined as:

$$L(\theta) = \begin{cases} \left[r + \gamma \max_{a'} Q(s', a'; \theta^-) - Q(s, a; \theta)\right]^2, & a \notin A_{terminal}, \\ \left[r - Q(s, a; \theta)\right]^2, & a \in A_{terminal}, \end{cases} \tag{7}$$

where γ is the discount factor that determines the agent's horizon [23]. θ is the weights in the evaluation network and θ^- is the ones in the target network. The weights in the target network are replaced by the evaluation network in every C steps.

In the testing stage, only evaluation network is utilized to determine an action for a steady testing result in every step, i.e. ϵ is set to be 0. Note that if the eye window violates conditions defined in (3), the agent thinks that there exist no any target objects in the region.

3 Experiment

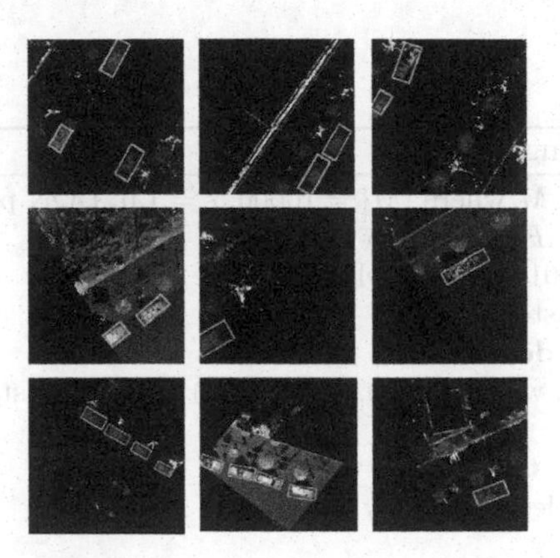

Fig. 4. Visual results demonstrate that our proposed approach can effectively estimate poses of vehicles.

Table 1. Comparison of Hough Forest and our proposed method.

	Precision	Recall
Hough forest	.910	.805
Our approach	.888	.912

Our experiment was conducted on a dataset collected by a RIEGL VMX-450 mobile LIDAR system on Xiamen Island, China, which contains annotated vehicles in point cloud. In order to handle the large number of points in our dataset, we construct bird's eye views from point clouds in advance. Here our point cloud scene is partitioned into slices with a scale of 20 m × 20 m. In addition, the slices are converted into 224 × 224 images with $T_{up} = 3.0$ and $T_{down} = 0.8$ to our priori knowledge. The shrink rate $\xi = 0.25$. Proposals are generated by means of the approach shown in Sect. 2.2. The total number of bird's eye views is 47 with 133 vehicles.

To evaluate the performance of our proposed framework on detecting vehicles in 3D point cloud, we split all the views into a training data set with 6 views containing 20 vehicles and a test data set with 41 views containing 113 vehicles. The VMX-450 dataset contains vehicles variously placed in orientation. Our method aims at determining a fine rotated bounding box for the vehicles. Figure 4 shows the visual result using our designed model.

To quantitatively analyze the accuracy and precision of the detection results on our test set, we choose Precision and Recall to estimate. The definition of Precision and Recall is as follows:

$$precision = \frac{TP}{TP + FP} \tag{8}$$

$$recall = \frac{TP}{TP + FN} \tag{9}$$

Moreover, an object is considered to be detected when the IoU of the terminate eye window with ground truth is greater than 0.5, which was defined by the Pascal VOC challenge [24]. As shown in Fig. 5, the Precision-Recall curves are generated under the 100th, 400th and 700th training epoch.

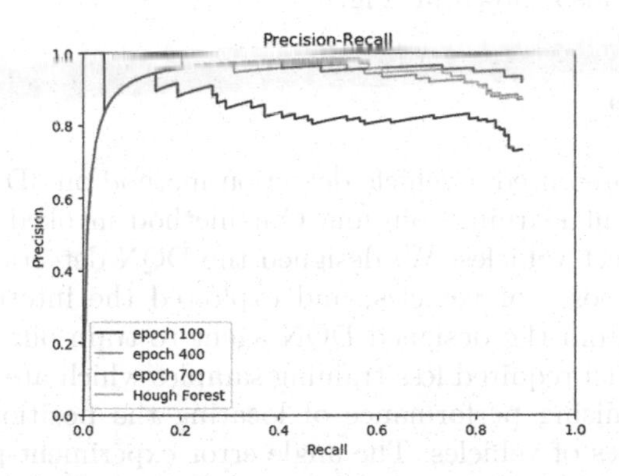

Fig. 5. The Precision and Recall curves under varying training epochs.

To validate that our proposed approach has the ability to estimate the pose of vehicles, we design an experiment to statistically analyze the performance of our framework. As shown in Fig. 6, we estimate the angle of the detected bounding box compared with the corresponding ground truth. The experimental result

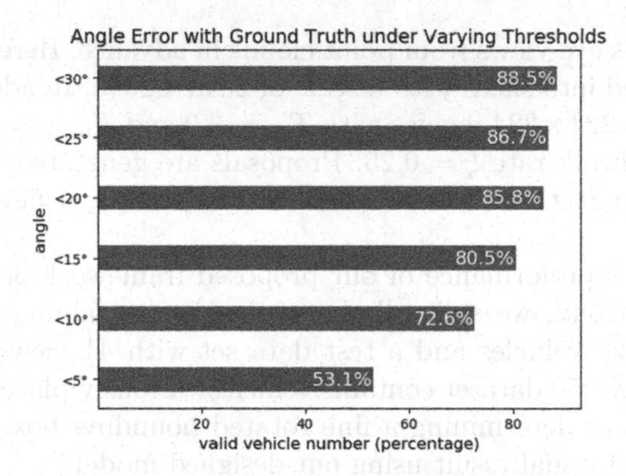

Fig. 6. The angle error of the detected bounding box with the corresponding ground truth under varying angle thresholds.

shows that the accuracy of our proposed method can achieve 88%. Generally, the angle errors between detection results and ground truths are 30° at most.

To further analyze the effectiveness of our proposed framework in vehicle detection when the number of training set is small, we compare with the Hough Forest algorithm [6]. As shown in Table 1, our proposed approach achieves a precision of 88.8% and Hough Forest achieves 91.0% while the recall of ours is better. The radius of vehicle in Hough Forest algorithm is set to 2.1 m to our priori knowledge. The number of clusters used to estimate in Hough Forest is sampled and equal to the number of proposals in our experiment. The P-R curve of Hough Forest is also shown in Fig. 5.

4 Conclusion

In this paper, we presented a vehicle detection method on 3D point cloud using a deep reinforcement learning solution. Our method invoked an agent with an eye window to detect vehicles. We designed the DQN detection agent to detect and estimate the poses of vehicles, and exploited the intermediate detection results generated from the designed DQN agent to train our model iteratively. Therefore, our model required less training samples which are labeled manually, and achieved promising performance of locating the position of vehicles and estimating the poses of vehicles. The angle error experiment presents that 88% vehicles are detected within a 30° error, which has demonstrated that our proposed method can be used to estimate vehicle poses. In spite of the truth that our training set is much smaller than the test set, our proposed method achieved the precision of vehicle detection at 88.8%. In conclusion, we have provided an effective solution to accurately detect vehicles and estimate their poses in 3D point clouds.

References

1. Guo, Y., Li, C., Liu, Q.: R2N: a novel deep learning architecture for rain removal from single image. Comput. Mater. Continua **58**(3), 829–843 (2019)
2. Pan, L., Qin, J., Chen, H., Xiang, X., Li, C., Chen, R.: Image augmentation-based food recognition with convolutional neural networks. Comput. Mater. Continua **59**(1), 297–313 (2019)
3. Ren, S., He, K., Girshick, R., Sun, J.: Faster R-CCNN: towards real-time object detection with region proposal networks. In: Advances in Neural Information Processing Systems, pp. 91–99 (2015)
4. Redmon, J., Divvala, S., Girshick, R., Farhadi, A.: You only look once: unified, real-time object detection. In: Proceedings of the IEEE Conference on Computer Vision and Pattern Recognition, pp. 779–788 (2016)
5. Zhang, J., Li, Y., Niu, S., Cao, Z., Wang, X.: Improved fully convolutional network for digital image region forgery detection (2019)
6. Wang, H., et al.: Object detection in terrestrial laser scanning point clouds based on hough forest. IEEE Geosci. Remote Sens. Lett. **11**(10), 1807–1811 (2014)
7. Qi, C.R., Litany, O., He, K., Guibas, L.J.: Deep hough voting for 3D object detection in point clouds. arXiv preprint arXiv:1904.09664 (2019)
8. Liu, F., et al.: 3DCNN-DQN-RNN: a deep reinforcement learning framework for semantic parsing of large-scale 3d point clouds. In: Proceedings of the IEEE International Conference on Computer Vision, pp. 5678–5687 (2017)
9. Simon, M., Milz, S., Amende, K., Gross, H.-M.: Complex-YOLO: an Euler-region-proposal for real-time 3D object detection on point clouds. In: Leal-Taixé, L., Roth, S. (eds.) ECCV 2018. LNCS, vol. 11129, pp. 197–209. Springer, Cham (2019). https://doi.org/10.1007/978-3-030-11009-3_11
10. Shi, S., Wang, X., Li, H.: PointRCNN: 3D object proposal generation and detection from point cloud. In: Proceedings of the IEEE Conference on Computer Vision and Pattern Recognition, pp. 770–779 (2019)
11. Qi, C.R., Liu, W., Wu, C., Su, H., Guibas, L.J.: Frustum pointnets for 3D object detection from RGB-D data. In: Proceedings of the IEEE Conference on Computer Vision and Pattern Recognition, pp. 918–927 (2018)
12. Yang, B., Luo, W., Urtasun, R.: Pixor: real-time 3D object detection from point clouds. In: Proceedings of the IEEE conference on Computer Vision and Pattern Recognition, pp. 7652–7660 (2018)
13. Lang, A.H., Vora, S., Caesar, H., Zhou, L., Yang, J., Beijbom, O.: Pointpillars: Fast encoders for object detection from point clouds. In: Proceedings of the IEEE Conference on Computer Vision and Pattern Recognition, pp. 12 697–12 705 (2019)
14. Behl, A., Paschalidou, D., Donné, S., Geiger, A.: Pointflownet: learning representations for rigid motion estimation from point clouds. In: Proceedings of the IEEE Conference on Computer Vision and Pattern Recognition, pp. 7962–7971 (2019)
15. Mnih, V., et al.: Playing atari with deep reinforcement learning. arXiv preprint arXiv:1312.5602 (2013)
16. Mathe, S., Pirinen, A., Sminchisescu, C.: Reinforcement learning for visual object detection. In: Proceedings of the IEEE Conference on Computer Vision and Pattern Recognition, pp. 2894–2902 (2016)
17. Bueno, M.B., Giró-i Nieto, X., Marqués, F., Torres, J.: Hierarchical object detection with deep reinforcement learning. In: Deep Learning for Image Processing Applications, vol. 31, no. 164, p. 3 (2017)

18. Li, Y., Fu, K., Sun, H., Sun, X.: An aircraft detection framework based on reinforcement learning and convolutional neural networks in remote sensing images. Remote Sens. **10**(2), 243 (2018)
19. Yang, B., Liang, M., Urtasun, R.: HDNET: exploiting HD maps for 3D object detection. In: Conference on Robot Learning, pp. 146–155 (2018)
20. Ku, J., Mozifian, M., Lee, J., Harakeh, A., Waslander, S.L.: Joint 3D proposal generation and object detection from view aggregation. In: 2018 IEEE/RSJ International Conference on Intelligent Robots and Systems (IROS), pp. 1–8. IEEE (2018)
21. Zeng, Y., et al.: RT3D: Real-time 3-D vehicle detection in lidar point cloud for autonomous driving. IEEE Robot. Autom. Lett. **3**(4), 3434–3440 (2018)
22. Yu, S.-L., Westfechtel, T., Hamada, R., Ohno, K., Tadokoro, S.: Vehicle detection and localization on bird's eye view elevation images using convolutional neural network. In: 2017 IEEE International Symposium on Safety, Security and Rescue Robotics (SSRR), pp. 102–109. IEEE (2017)
23. Mnih, V., et al.: Human-level control through deep reinforcement learning. Nature **518**(7540), 529 (2015)
24. Everingham, M., Van Gool, L., Williams, C.K., Winn, J., Zisserman, A.: The pascal visual object classes (VOC) challenge. Int. J. Comput. Vision **88**(2), 303–338 (2010)

A Network Security Situation Prediction Algorithm Based on BP Neural Network Optimized by SOA

Ran Zhang(✉), Min Liu, Qikun Zhang, and Zengyu Cai

School of Computer and Communication Engineering, Zhengzhou University of Light Industry, Zhengzhou 450002, China

ranranzh@sina.com

Abstract. The current cybersecurity situation is getting worse. In order to improve the accuracy of network security situation prediction, a network security situation prediction method based on BP neural network optimized by Seeker Optimization Algorithm (SOA) is proposed. The algorithm uses the four behavioral characteristics of SOA: the self-interest, altruism, pre-action and uncertainty reasoning to determine the search strategy, find the best fitness individual, obtain the optimal weight and threshold, and then assign value to the random initial weights and thresholds of BP neural network. After training the neural network, the predicted values are obtained. Finally, it is compared with the predicted values obtained by other two optimization algorithms. The experimental results show that this prediction algorithm has higher accuracy, smaller error and better stability.

Keywords: BP neural network · Seeker Optimization Algorithm · Network security · Situation prediction

1 Introduction

With the wide application of computer networks, multi-layer network security threats and security risks are increasing, and the network security situation is becoming more and more serious. Network attack is becoming more and more distributed, large-scale and complex. The traditional intrusion detection system, firewall and other network security defense means can not meet the current high-speed, intelligent and multi-source network security demands. We need more advanced and optimized technical means and methods to prevent the occurrence of network security events. In 1999, Bass first proposed the concept of network security situation awareness (NSSA) [1]. Security situation awareness was first used in the field of aviation and military affairs to quickly make decisions and deal with complex aviation and military affairs. Later, it was widely used in the field of network security by researchers. Researchers found that situation awareness can not only be used for network security situation assessment, but also for network security situation prediction, so that the original passive defense can be turned into active defense, which can largely solve the problem of network security defense, so it has become a hot research direction. The main work of this article is as follows:

X. Sun et al. (Eds.): ICAIS 2020, LNCS 12240, pp. 417–427, 2020.
https://doi.org/10.1007/978-3-030-57881-7_37

(1) A network security situation prediction method based on the SOA_BP neural network is proposed. The improved neural network model training generates a security situation value to predict the future network security situation.
(2) Apply the seeker optimization algorithm to BP neural network to optimize its connection weights and thresholds, and improve the accuracy of network security situation prediction based on neural networks.
(3) Experimental comparisons using different optimization algorithms show that our proposed network security situation prediction method based on the SOA_BP neural network has the smallest error, the highest accuracy, which can effectively predict the network security situation changes.

At present, the research of network security situation prediction has been very extensive. The research in this field is relatively early abroad. Fava et al. [2] proposed a variable-length Markov model (VLMM), which captures the sequential properties of the attack trajectory, so that it can predict the ongoing attack. Fachkha et al. [3] proposed a method based on time series fluctuation analysis and prediction to realize the prediction of distributed denial of service (DDoS) activities. Thonnard et al. [4] proposed a method based on prior knowledge and data mining to predict the behavior of attackers. Ramaki et al. [5] used the concept of Bayesian networks to construct attack scenarios and predict the next target of the attacker by creating attack prediction rules. In recent years, a lot of research has been done in the field of network security situation prediction in China, and a lot of network security situation prediction models and methods have been put forward. Many researchers try to introduce the method of artificial intelligence into the field of situation assessment to improve the accuracy of situation awareness and prediction. Lu studied an Elman prediction model based on genetic algorithm in his master's thesis [6]. Chen et al. proposed an improved BP neural network evaluation model using simulated annealing algorithm [7]. Zhang et al. proposed a network security situation prediction model based on improved convolution neural network by using the characteristic of weight sharing of convolution neural network and the convolution technology of depth separability and decomposability [8]. Qian and others proposed a prediction method based on improved particle swarm optimization algorithm to optimize BP neural network [9]. Zhao proposed a network intrusion detection algorithm which combines crowd search algorithm with support vector machine [10]. Cheng et al. proposed a method for predicting characteristic sequence of abnormal network flow suitable as a DDoS attack detector in a big data environment [11]. Zhang et al. proposed a network security situation awareness method based on random games in the cloud computing environment, which uses the effects of both players to quantify the network security situation value [12]. In addition, Han Weihong et al. studied a YHSAS system, which includes three steps of the network security situation; the acquisition, evaluation, and prediction of security elements. However, the system is difficult to accurately predict the occurrence and development trend of major cyber attack events, and further research is needed [13].Compared with traditional methods, these improved methods improve the effect of evaluation and prediction to a certain extent, but these methods are not ideal in accuracy and efficiency for the massive data collected, and can not adapt to the dynamic and changing network security demands.

To solve the problem that the accuracy of security situation prediction is not high at present, based on the existing methods and models of network security situation prediction, this paper combines SOA algorithm with BP neural network to study and simulate the network security prediction method.

2 Network Security Situation Prediction Based on BP Neural Network Optimized by SOA

BP (back propagation) neural network was proposed by a scientific team led by Rumelhart and McCelland in 1986 [14]. BP neural network is one of the most widely used neural network prediction models because of its simple structure, many adjustable parameters, many training algorithms, strong robustness and self-learning ability, and good operability. At present, most neural network models are based on BP neural network structure [15]. BP neural network mainly uses back propagation algorithm to adjust the weights and thresholds of the network repeatedly until the optimal weights and thresholds are obtained. After continuous learning and training, it makes the output data consistent with the true value as much as possible. Finally, when the sum of error squares is less than the specified error, the training is completed and the optimal network weights and thresholds are saved. However, due to the limitations of its many iterations and low operation speed, most of the research is to combine BP neural network with other optimization algorithms, such as using particle swarm optimization (PSO) to optimize BP neural network [16], using genetic algorithm (GA) to optimize BP neural network [17], and so on. In this paper, the seeker optimization algorithm (SOA) and BP neural network are combined to make up for the limitations of BP neural network and improve the efficiency and accuracy of the security situation prediction algorithm.

2.1 Seeker Optimization Algorithm

Seeker Optimization Algorithm (SOA) is a new heuristic random search algorithm [18]. It is to analyze the random search behavior of human, relies on the social experience of human beings, combines with the thought of evolution, takes searching the best position as the core, through the four search strategy behavior, the self-interest, altruism, pre-action and uncertainty, to model, determine the direction and step size of the crowd search, and then constantly update the position to obtain the best solution. The advantages of SOA algorithm are clear concept, easy to understand, fast convergence and high precision. The calculation steps mainly include:

(1) Determination of search step

When determining the step size, first arrange the individual optimal fitness value in descending order, and give index number to each individual as the input of fuzzy reasoning. In this paper, Gauss linear membership function is used to represent the output of the fuzzy variable of search step size, which can well map the optimal fitness value of the i-th individual to the minimum and maximum membership. The mapping formula is as follows:

$$u_i = U_{max} - (sizepop - Indexfitnessgbest(i)) * \frac{U_{max} - U_{min}}{sizepop - 1} \quad (1)$$

$$u_{ij} = u_i + (1 - u_i) * rand(j = 1, 2, 3, \ldots D) \quad (2)$$

where u_i is the membership degree of the i-th individual; $Indexfitnessgbest(i)$ is the index number of the optimal fitness value of the i-th individual; $sizepop = 30$ is the population size; $U_{max} = 0.95$ and $U_{min} = 0.0111$ respectively represent the maximum and the minimum function membership degree; u_{ij} is expressed as the membership degree corresponding to the objective function value i in the j-dimensional exploration space; according to formulas (1) and (2), the membership degree corresponding to the best fitness value individual is obtained, and then according to the formula (3) determine the step size:

$$a_{ij} = \delta_{ij}\sqrt{-\log(u_{ij})} \quad (3)$$

where a_{ij} is the search step length of the i-th individual in the j-dimensional search space, and δ_{ij} is the parameter of the Gaussian membership function, whose value is determined by the following formula:

$$\delta_{ij} = H(t) * |zbest - 5 * rands(1, 10)| \quad (4)$$

$$H(t) = \frac{maxgen - t}{maxgen} \quad (5)$$

where $zbest$ is the global best; $rands(1, 10)$ is the random real number between [1, 10]; $H(t)$ is the weight function value of the t-th iteration, which is constantly changing in the process of iteration. It is affected by the maximum number of iterations and the current number of iterations, where maxgen = 100.

(2) Determination of search direction

When determining the search direction, it is determined whether the search direction is self-interest, altruism or pre-action according to the individual best and the whole best compared with the current individual.

$$\vec{d}_{i,ego}(t) = \vec{g}_{i,best} - \vec{x}_i(t) \quad (6)$$

$$\vec{d}_{i,alt}(t) = \vec{z}_{i,best} - \vec{x}_i(t) \quad (7)$$

$$\vec{d}_{i,pro}(t) = \begin{cases} \vec{D}_i & F_{i,best} < F_{x_i} \\ -\vec{D}_i & F_{i,best} \geq F_{x_i} \end{cases} \quad (8)$$

where $\vec{x}_i(t)$ is the best position of the i-th individual at the t-th iteration, $\vec{g}_{i,best}$ is the current best position of the i-th individual, $\vec{z}_{i,best}$ is the global best position, $F_{i,best}$ is the fitness value of the position of $\vec{g}_{i,best}$, and F_{x_i} is the fitness value of the position of $\vec{x}_i(t)$.

The direction of self-interest is determined by subtracting the current best position of the individual from the optimal position of the individual at the t-th iteration. The determination of the altruistic direction is based on the global optimal position and the best position of the individual at the t-th iteration. The determination of the pre-action direction is based on the comparison of individual's current fitness value and individual's best fitness value.

In this experiment, the search direction is determined by the random weighted geometric mean of three directions. The calculation formula is as follows:

$$\vec{d_{ij}}(t) = sign\left(W\vec{d}_{ij,pro} + \varphi_1\vec{d}_{ij,ego} + \varphi_2\vec{d}_{ij,alt}\right) \tag{9}$$

$$\mathrm{W} = W_{max} - t * \frac{W_{max} - W_{min}}{maxgen} \tag{10}$$

where W is the inertia weight, φ_1 and φ_2 are constants, and their range of values is a uniformly distributed constant in [0, 1]. t is the current number of iterations, the value range is an integer between [2, maxgen], $W_{max} = 0.9$ is the maximum weight, and $W_{min} = 0.1$ is the minimum weight.

(3) Location update

After calculating the direction and step length of the individual exploration, the position of the individual should be updated. The location update formula is as follows:

$$\Delta x_{ij}(t+1) = \Delta x_{ij}(t) + a_{ij}(t) * \vec{d_{ij}}(t) \tag{11}$$

2.2 Prediction Algorithm Based on BP Neural Network Optimized by SOA

The main steps of network security situation prediction algorithm based on BP neural network optimized by SOA (abbreviated as SOA_BP neural Network) are as follows:

(1) *Initialization.* Initialize the population size, the maximum number of iterations, the spatial dimension, the minimum and maximum membership degree, and the minimum and maximum value of the weight, initialize the population individuals, and calculate the fitness of each individual.
(2) *Look for individuals with the best fitness.* Find the global best, individual best, individual best fitness value and global best fitness value. In this step the design of the fitness function is most important. In this experiment, the absolute value of the error between the predicted value and the real value obtained by training the BP neural network with the training data is used as the individual fitness value.
(3) *Iterative optimization.* First, the empirical gradient direction, the search step length and direction, and the parameter δ_{ij} of the Gaussian function are initialized. Then it enters into the iterative loop, and the termination condition of the loop is the double-layer loop of the maximum iteration number and the population size. The search direction is determined according to formulas (6)–(7); the direction of empirical

gradient is determined according to formula (9); the parameter δ_{ij} of the Gaussian function can be calculated according to formulas (4)–(5), and the search step length can be calculated according to formula (3); based on the calculated step and direction, the position, the individual optimal and optimum fitness values can be updated according to formula (10). In addition, the global optimal fitness values also should be updated. The iterative loop doesn't end until the termination condition is met, then the global optimal fitness value is output.

(4) *Get the best network weights and thresholds.* The optimal weights and thresholds are assigned to random initial values for network security prediction.

(5) *BP neural network training.* Initialize the network parameters, create a neural network, call the traingdm algorithm to train the BP network, use BP neural network for simulation and prediction, calculate the error, and analyze the results.

Flow Chart of network security situation prediction based on BP neural network optimized by SOA is shown in Fig. 1.

3 Simulation and Analysis of Network Security Situation Prediction Based on SOA_BP Neural Network

In this paper, the BP neural network algorithm optimized by SOA algorithm is applied to network security situation prediction to improve the accuracy of network security situation prediction. The experiment takes the data published in the network security information and the dynamic weekly report as the experimental data. It mainly takes the number of hosts infected with virus, the total number of websites tampered with, the total number of websites implanted in the back door, the number of phishing pages of domestic websites and the number of new information security holes as the evaluation indexes, which can reflect the situation of modern network security comprehensively and can be used as the evaluation indexes of weekly basic network security situation. In this experiment, the network security data information from the first period of 2015 to the seventh period of 2017 was selected. For the convenience of experimentation, we converted five security levels into digital levels, as shown in Table 1.

3.1 Data Preprocessing

According to the characteristics of neural network, too many training samples will increase the training time, too few will reduce the accuracy, so this paper chooses 101 training samples and 10 test samples. In order to increase the accuracy and convergence of the experiment, the data were preprocessed and normalized. There are generally two methods for data normalization, one is normalization to [0, 1], the other is normalization to [−1, 1]. The latter one is used in this experiment. The normalization formula is shown in formula (12), and the normalized result is shown in Fig. 2.

$$y = 2 \times \frac{x_i - x_{min}}{x_{max} - x_{min}} + (-1) \tag{12}$$

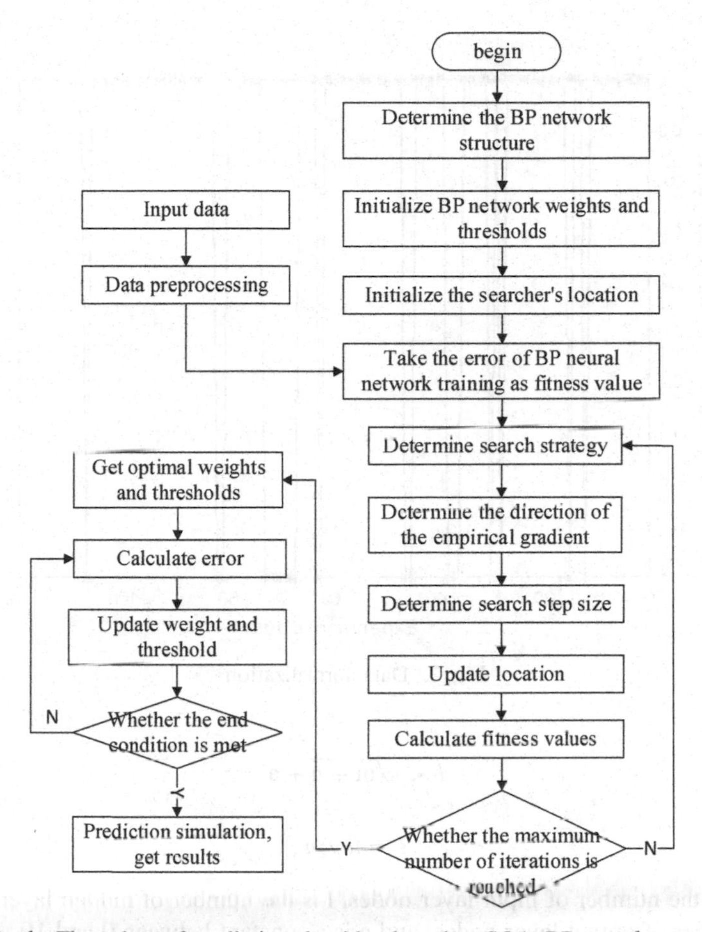

Fig. 1. Flow chart of prediction algorithm based on SOA_BP neural network

Table 1. Network security situation value conversion table

Excellent	Good	Normal	Bad	Dangerous
5	4	3	2	1

3.2 Analysis of Prediction Results

(1) Determine the Network Structure of the BP Neural Network. There are five evaluation indexes of network security situation known from the above, which should be converted into a security level finally. Therefore, this experiment has five input parameters and one output parameter, and then the number of hidden layer nodes is determined according to formula (13)–(15).

$$l < n - 1 \tag{13}$$

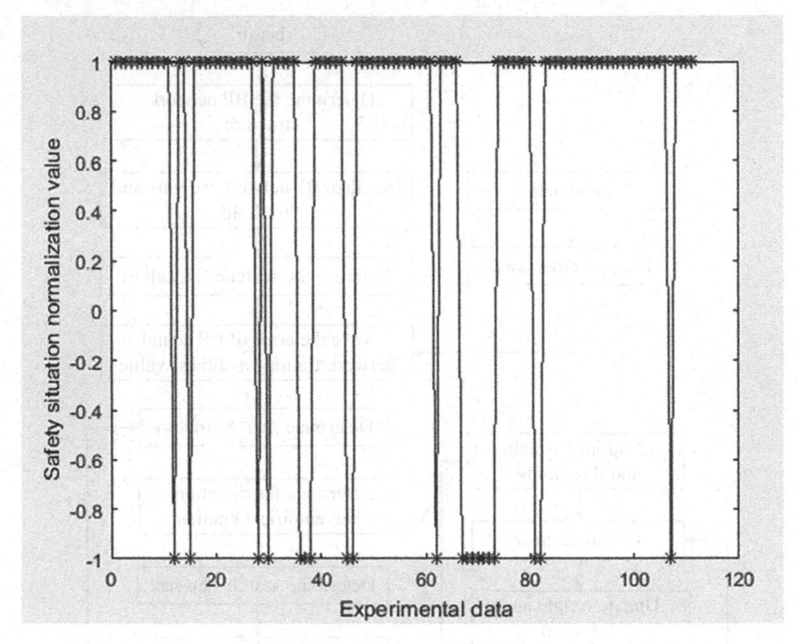

Fig. 2. Data normalization

$$l < \sqrt{m + n} + a \tag{14}$$

$$l = log_2 n \tag{15}$$

where n is the number of input layer nodes, l is the number of hidden layer nodes, m is the number of output layer nodes, and a is a constant between 0 and 10. According to the trial and error method, the number of hidden layer nodes in this experiment is determined to be 8, and the network structure of this experiment is finally determined to be 5-8-1.

(2) Verify the Accuracy of the SOA Algorithm. Mean square error (MSE) is used to measure the difference between the true value and the predicted value. It is a simple method to measure the average error, which can well reflect the change degree of data. The smaller MSE is, the more accurate the predicted data is, and vice versa. The calculation formula of MSE is as shown in formula (16).

$$\text{MSE} = \frac{1}{n} \sum_{t=1}^{n} \left(x_t - x_t^*\right)^2 \tag{16}$$

Where x_t and x_t^* is the true value and the predicted value respectively.

(3) Compared with the Prediction Results of other Optimization Algorithms. Because the currently used global optimization algorithms are GA and PSO, and in many articles, the SOA algorithm is compared with the PSO algorithm and the GA algorithm. So the comparison of experimental results of PSO optimization, GA optimization

and SOA optimization BP neural network is shown in Fig. 3. In the graph, the difference between the true value and the predicted value of each optimization algorithm can be clearly seen.

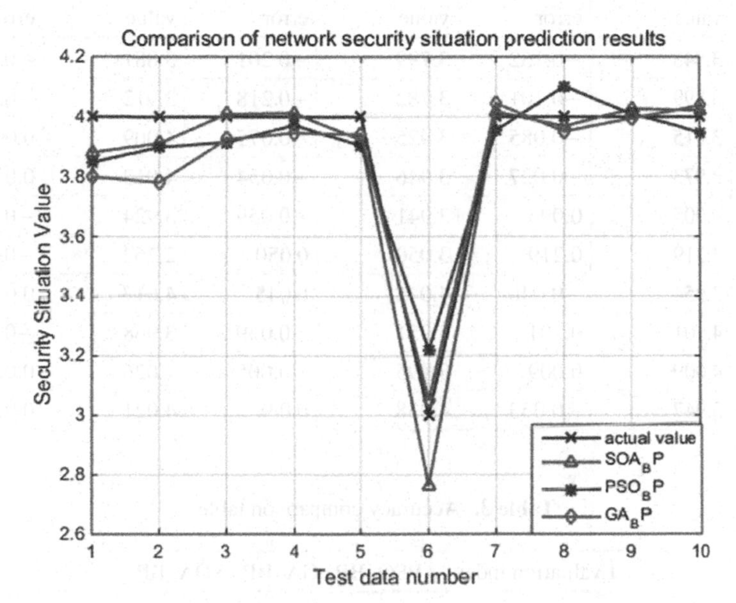

Fig. 3. Comparison of experimental results

It can be seen from Fig. 3 that the error between the predicted value and the true values of BP neural network optimized by SOA is smaller than that of PSO Optimized BP neural network and GA Optimized BP neural network, and the error between the predicted value obtained from SOA optimization and the true value is larger at the beginning, but the latter ones tend to be stable gradually, close to the true value, and the curve fluctuation is relatively smaller. This shows that the algorithm of SOA Optimized BP neural network is more stable and accurate than the other two algorithms.

Table 2 shows the ten test values obtained by three different optimization algorithms, the relative error between the predicted value and the true value. Table 2 shows from the microscopically that the SOA_BP algorithm has higher accuracy than that of the PSO_BP and the GA_BP.

Table 3 calculates the mean square error of the three algorithms. From Table 3, it can be seen macroscopically that the mean square error between the predicted value and the true value obtained by the SOA_BP optimization algorithm is the smallest. This also shows that the SOA optimization algorithm has higher accuracy and certain superiority.

Table 2. Prediction data analysis table

Actual value	PSO Predictive value	PSO Absolute error	GA Predictive value	GA Absolute error	SOA Predictive value	SOA Absolute error
4	3.848	−0.152	3.799	−0.201	3.881	−0.119
4	3.899	−0.101	3.782	−0.218	3.912	−0.088
4	3.915	−0.085	3.925	−0.075	4.009	0.009
4	3.973	−0.027	3.946	−0.054	4.012	0.012
4	3.905	0.095	3.941	−0.059	3.924	−0.076
3	3.219	0.219	3.050	0.050	2.762	−0.238
4	3.954	−0.046	4.045	0.045	4.017	0.017
4	4.101	0.101	3.951	−0.049	3.968	−0.032
4	4.009	0.009	3.995	−0.005	4.026	0.026
4	3.947	−0.053	4.038	0.038	4.021	0.021

Table 3. Accuracy comparison table

Evaluation index	PSO_BP	GA_BP	SOA_BP
Mean square error	0.0113	0.0108	0.00871

4 Conclusion

For the problem of low accuracy of current network security situation prediction, this paper proposes a network security situation prediction method based on SOA_BP Neural network, which uses SOA to optimize the weight and threshold of BP neural network, and compares it with PSO_BP optimization algorithm and GA_BP optimization algorithm. Experiments show that this algorithm can predict the future network security situation more accurately. However, SOA tends to fall into local optimum or "premature" phenomenon in the later stage of search. In the next work we will further improve it to solve such defects and further improve its global search ability and accuracy of prediction. In addition, the comparative experiments in this paper are not enough. It needs to be compared with more prediction methods, and the convergence of the algorithm needs to be studied and analyzed.

Acknowledgements. This paper is sponsored by key foundation of Science and Technology Development of Henan Province (No. 142102210081) and National Natural Science Foundation of China (No. 61772477).

References

1. Tim, B.: Multi-sensor data fusion for next generation distributed intrusion detection sytem. In: Proceedings of 1999 IRIS National Symposium on Senor and Data Fusion, pp. 1–6. The Johns Hopkins University (1999)
2. Fava, D.S., Byers, S.R., Yang, S.J.: Projecting cyberattacks through variable-length markov models. IEEE Trans. Inf. Forensics Secur. **3**(3), 359–369 (2008)
3. Fachkha, C., Bou-Harb, E., Debbabi, M.: Towards a forecasting model for distributed denial of service activities. In: Proceedings of the 12th IEEE International Symposium on Network Computing and Applications, pp. 110–117. IEEE, Cambridge (2013)
4. Thonnard, O., Dacier, M.: Actionable knowledge discovery for threats intelligence support using a multi-dimensional data mining methodology. In: Proceedings of the International Conference on Data Mining Workshops, 154–163. IEEE, Pisa (2008)
5. Ramaki, A.A., Khosravi-Farmad, M., Bafghi, A.G.: Real time alert correlation and prediction using Bayesian networks. In: Proceedings of the 12th International Iranian Society of Cryptology Conference on Information Security and Cryptology (ISCISC), pp. 98–103. IEEE, Rasht (2015)
6. Lu, D.: Research on network situation prediction model based on GA-Elman neural network. Zhejiang University (2017)
7. Chen, W., Zang, Z., Guo, J., Yu, Q., Tong, J.: Security assessment of network space situational awareness system based on improved bp neural network. Comput. Sci. **45**(S2), 335–337+341 (2018)
8. Zhang, R., Zhang, Y., Liu, W., Fan, Y.: A network security situation prediction method based on improved convolutional neural network. Comput. Eng. Appl. **55**(06), 86–93 (2019)
9. Qian, Z., Jiang, J., Yu, H.: BP neural network prediction method based on improved particle swarm optimization algorithm. Electron. Test (20), 39–40+25 (2015)
10. Zhao, W.: Research on network intrusion detection method based on SOA-SVM. Autom. Instr. (01), 39–42 (2015)
11. Han, W., Tian, Z., Huang, Z., Zhong, L., Jia, Y.: System architecture and key technologies of network security situation awareness system YHSAS. Comput. Mater. Continua **59**(1), 167–180 (2019)
12. Tang, X., Zheng, Q., Cheng, J., Sheng, V.S., Cao, R., Chen, M.: A DDoS attack situation assessment method via optimized cloud model based on influence function. Comput. Mater. Continua **60**(3), 1263–1281 (2019)
13. Cheng, J., Xu, R., Tang, X., Sheng, V.S., Cai, C.: An abnormal network flow feature sequence prediction approach for DDoS attacks detection in big data environment. Comput. Mater. Continua **55**(1), 095–119 (2018)
14. Rumelhart, D.E., Hinton, G.E., Williams, R.J.: Learning representations by back-propagating errors. Nature **323**(3), 533–536 (1986)
15. Zhu, D., Shi, H.: Principle and application of artificial neural networks (2006)
16. Li, C., Liu, X.: An improved PSO-BP neural network and its application to earthquake prediction. In: Proceedings of the Chinese Control and Decision Conference (CCDC), pp. 3434–3438. IEEE, Yinchuan (2016)
17. Ding, S., Su, C.: Application of optimizing BP neural networks algorithm based on genetic algorithm. In: Proceedings of the 29th Chinese Control Conference, pp. 2425–2428. IEEE, Beijing (2010)
18. Dai, C., Zhu, Y., Chen, W.: Seeker optimization algorithm. In: Wang, Y., Cheung, Y.-m., Liu, H. (eds.) CIS 2006. LNCS (LNAI), vol. 4456, pp. 167–176. Springer, Heidelberg (2007). https://doi.org/10.1007/978-3-540-74377-4_18

Research on User Preference Film Recommendation Based on Attention Mechanism

Lei Zhu, Yufeng Liu(✉), Wei Zhang, and Kehua Yang

College of Computer Science and Electronic Engineering, Hunan University, Changsha 410082, China
fx_yfliu@163.com

Abstract. Due to the influence of different factors such as environment, age, and interest, everyone has a different taste and appreciation of the movie. These factors can be used to make more personalized recommendation calculations for movie recommendations, but many traditional methods do not integrate these factors well. Therefore, how to effectively extract key information from multiple directions in a recommendation system is still a challenging problem.

In this paper, we use user, film, and movie scoring data as input, and use the CNN-BLSTM deep learning model that incorporates the multi-head attention mechanism as a training, and finally combine the output features to calculate user preferences for recommendation. The convolutional neural network and LSTM are used to extract the user and movie feature information from the matrix, and the multi-head attention mechanism can also extract the key information from the data. The comparative experimental analysis of different models shows that our proposed user preference model based on attention mechanism can obtain better performance than traditional extraction methods.

Keywords: Multi-head attention · Recommendation system · User preference · Deep learning

1 Introduction

Material life is constantly improving, and watching movies one by one is an entertainment way for people to relax daily. In movie recommendation, the traditional collaborative filtering recommendation algorithm mainly uses users' ratings for movies as the basis for recommendation. And other characteristics of users and movies, such as the user's personal attributes, movie attribute categories, and the relationship between user history and movies. Most of them are not fully utilized, and the disadvantage of this is the lack of fine positioning of user behavior preferences.

The study of deep learning shows great potential in effective learning. Deep learning can be used to automatically learn features, such as learning effective feature representations from text content, and then extracting valid information from the features.

X. Sun et al. (Eds.): ICAIS 2020, LNCS 12240, pp. 428–439, 2020.
https://doi.org/10.1007/978-3-030-57881-7_38

However, many existing models and methods still fail to extract key information from the data.

For deep learning models, such as RNN [1], this type of method does not get rid of the limitation of timing, that is, it cannot be parallelized, which leads to speed and efficiency problems on large data sets. Another example is CNN [2], which is convenient for parallelism and easy to capture some global structural information, but only has the advantage of feature detection but lacks feature understanding.

In order to overcome these difficulties, this paper proposes a user preference learning model that integrates the attention mechanism. Mainly have the following advantages: (1) Use deep learning models to automatically learn deep-seam semantic information without relying on any manually designed feature extraction rules. (2) We introduce a multi-head attention mechanism into the model, which can adaptively combine context information and extract text features from it, thereby improving the accuracy of key information acquisition. (3) In multiple dimensions, try to combine the movie's profile information to better extract features at a deep level.

The rest of this article is organized as follows. The related work is described in Sect. 2. Section 3 details the user preference learning model based on the attention mechanism. Section 4 demonstrates the model through experiments. The detailed conclusions and expectations are summarized in Sect. 5.

2 Related Work

In the current recommendation algorithm research, the collaborative filtering algorithm is one of the most enthusiastic and recommended algorithms by researchers, but the traditional collaborative filtering algorithm will inevitably face cold start and data sparsity [3]. In contrast, deep learning-based recommendation systems lack extensive attention and comment. Deep neural network models are now increasingly used in a variety of tasks, including text summaries, information retrieval, and relationship recognition [4, 5, 32].

In the study of deep learning, Socher et al. [1, 6] used recurrent neural networks to learn the logical relationships between sentences, and performed sentiment analysis and sentence semantic relationship recognition. Kim [2] uses the word vector trained by word2vec to map the sentence sequence into a two-dimensional feature matrix, and uses the convolutional neural network to extract the features of the sentence. Tang et al. [7] realized text-level sentiment classification based on the composition principle and relationship of text semantics, combined with convolutional neural networks and cyclic neural networks. Xie et al. [8] proposed the DKRL model, which uses entity description to represent entity vectors, and uses continuous word bag model and convolutional neural network to encode the semantics of entity descriptions, but this model does not consider the screening of entity information.

With the continuous development of deep learning, the use of attention mechanisms to optimize neural networks [9, 11] has become a hot topic. Bahdanau et al. [10] applied the attention mechanism to the NLP field for the first time, and simultaneously translated and aligned on the machine translation task, and finally achieved good performance. Later, Luong et al. [12] proposed a global attention mechanism and a local attention mechanism, providing a method for calculating the extended attention. Xu et al. [13] proposed the joint representation model of BLSTM, using the attention mechanism to select the relevant information in the entity description, and designing the gate mechanism to control the weight of the structural entity information. This method has a significant improvement in performance, but the hidden state of the BLSTM model needs to be generated in order, resulting in no parallel processing during training, thus reducing efficiency.

Although, deep learning has a good ability to capture features, but different models have different limitations. We propose a user preference learning model that integrates the attention mechanism. It not only considers the screening of the entity information, but also captures the feature information. It is enough to build a long-distance dependency between sentences, to solve the limitation of distance on features, and to have parallelization characteristics. Through experimental verification, our proposed method performs better than the previous method.

3 Model Building

This chapter proposes a user preference learning model that integrates the attention mechanism. The model mainly includes CNN-BLSTM feature extraction module, Multi-head attention module and recommendation table generation module. A schematic diagram of the model is shown in Fig. 1. Firstly, the user's features are extracted by multi-layer neural network. Then the feature matrix is extracted by CNN-BLSTM, and the scoring matrix is reduced in dimension, placed together in the embedding layer. Next, the user features and the movie features processed by the attention mechanism are further processed in the fully connected layer. Finally, the vector obtained by the joint calculation is used to calculate the similarity, thereby implementing the recommended task.

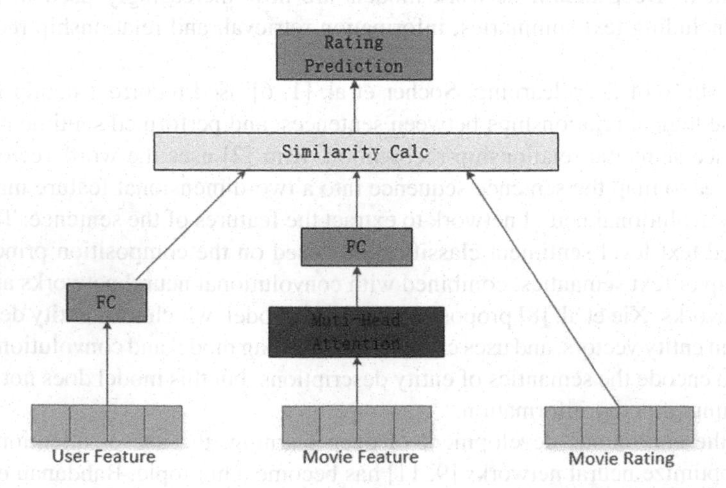

Fig. 1. User preference learning model architecture diagram with attention mechanism.

3.1 Questions Raised

In the aspect of film recommendation, the user's rating of the film as the basis for recommendation [14, 15] is the main processing method of the traditional collaborative filtering recommendation algorithm. A user who has seen the same or similar movie is generated by generating a preference model for the user's history and then filtering through the system. We recommend a movie from other user history to a user who needs to be recommended to generate a recommendation list for that user.

The traditional recommendation algorithm often only uses the scoring data [16], and the user and other various feature data of the movie, such as user age, job or gender, as well as the introduction, type or label of the movie [17] are not able to make full use of it. The result of this is a lack of fine-grained description of user behavior preferences, and in actual recommendations, accurate analysis will greatly improve recommendation accuracy.

Deep learning can not only learn nonlinear multi-level abstract feature representation, but also acquire features that are often dense and low-dimensional [18], which is not available in the traditional collaborative filtering recommendation model. We capture key information by inputting text data into a convolutional neural network and extracting features [19]. At the same time, using the attention mechanism to automatically learn the hidden relationship between the user and the movie features, and combined with the relationship between the film scores, can more effectively reflect the behavior preferences of different users, and thus improve the accuracy of the recommendation system.

3.2 Feature Extraction

The user preference learning model that integrates the attention mechanism not only captures the key information of users and movies, but also considers the interdependence between users and movies, and realizes the division and mining of effective information between users and movies. In order to further capture the relevant information between users and movies, this paper also introduces the retrofit model of attention mechanism, multi-head attention mechanism, and integrates CNN-BLSTM model to solve this problem more effectively.

Feature Extraction Based on CNN-BLSTM. We use the brief information of the movie as a sequence of sentences, and then use the CNN-BLSTM model to divide the brief information of a single movie into multiple sentences. In this way, compared to using an entire statement as input, the feature representation can be extracted sequentially through the convolutional layer neural network, and then these sentence features are sequentially integrated using LSTM [17, 18], and then the entire sentence feature representation is constructed, so that we can get the key information of the text more accurately. Figure 2 is a state diagram of the CNN-BLSTM model.

The first layer is the input layer. We use the input sequence of the movie represented by matrix $M_i^d = [m_1 \ldots m_d]$. The matrix $U_i^d = [u_1 \ldots u_d]$ represents the user's input matrix, and then the movie input sequence is used as the input of the model. The user's input matrix can be directly assigned to the convolutional neural network to extract features.

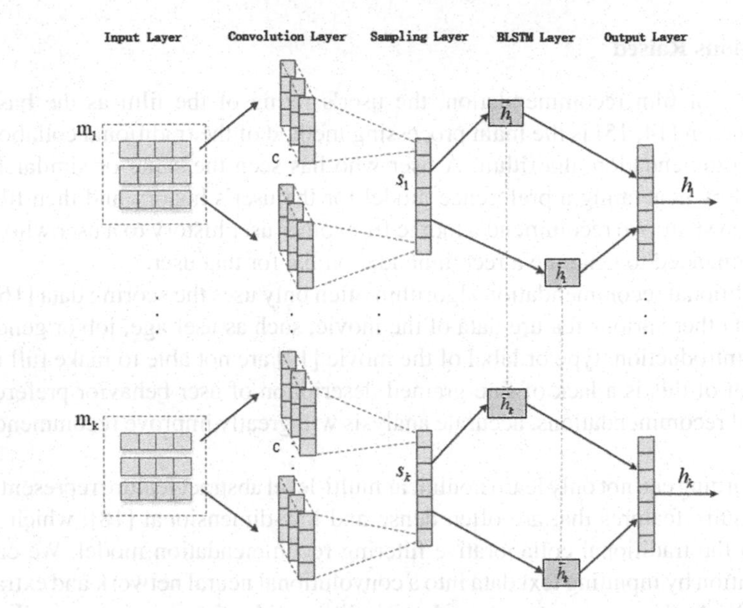

Fig. 2. CNN-BLSTM model diagram.

The second layer is the convolution layer. Drawing on the structure of the CNN model proposed by Kim [2, 19, 31], multiple convolution filters are used to extract multiple sets of local feature maps in the convolutional layer. Given the input matrix S_{ij}, for each row vector m in the matrix, a convolution operation is performed using a filter with a window size of k. The result of the convolution mapping can be expressed as:

$$y_i = \mathcal{F}(W_c \otimes m_{i:i+k-1} + b_c) \tag{1}$$

Where y_i is the *ith* element of the feature map, W_c is the coefficient matrix, b_c is the bias term, and $m_{i:i+k-1}$ is the local word window composed of k words. When the word window is stepped from $m_{i:k}$ to $m_{n-k+1:n}$, and we can get a feature map:

$$c = \{y_1, \ldots, y_{n+k-1}\} \tag{2}$$

The third layer is the sampling layer. The most representative features in each feature map are extracted at the sampling layer to obtain a feature representation at the sentence level. At the sampling layer, the feature map is sampled by the max-over-time pooling method proposed by Gollobert [20], and obtaining the feature values $\hat{c} = \max\{c\}$. The convolutional layer makes one of the filter structures that extract a local feature by the size of the window. We use the structure of h kinds of filters, and take into account the information between contexts by extracting m feature maps in each filter. A feature map obtained by extracting features from all types of filters, after the maximum pooling operation of the sampling layer, obtain the feature value s of length $h \times m$ as the output:

$$s = \left\{\hat{c}_{1,h_1}, \ldots, \hat{c}_{m,h_1}, \ldots, \hat{c}_{1,h_k}, \ldots, \hat{c}_{l,h_k}, \ldots\right\} \tag{3}$$

Where c is the *lth* feature value ($1 \le l \le m$) produced by the filter of the *kth* type ($1 \le k \le h$).

The fourth layer is the LSTM layer. The main purpose of this layer is to control the historical information retained by each LSTM unit and to memorize the currently entered information [18, 21, 22, 30], retain important features and discard unimportant features. This layer mainly contains input gates, forgetting gates, cell states and output gates.

$$i_t = \sigma(W_i s_t + W_i h_{t-1} + b_i) \tag{4}$$

$$f_t = \sigma\left(W_f s_t + W_f s_{t-1} + b_f\right) \tag{5}$$

$$g_t = \tanh\left(W_g s_t + W_g s_{t-1} + b_g\right) \tag{6}$$

$$o_t = \sigma(W_o s_t + W_o s_{t-1} + b_o) \tag{7}$$

Contains the previous cell state and based on current input and last hidden state information generated new information $c_t = i_t g_t + f_t c_{t-1}$ (The initial memory unit c_0 is marked as 0). Finally, the current hidden state of the output is obtained by multiplying the current cell state by the weight matrix of the outputs:

$$h_t = o_t \tanh(c_t) \tag{8}$$

Therefore, for a long sentence consisting of K clause sequences, the sentence feature vector $\mathrm{h}_1, h_2, \ldots, h_K$ is obtained by CNN and bidirectional LSTM.

Multi-head Attention. At present, with the deepening of deep learning research, the attention mechanism has been widely applied to various tasks of natural language processing based on deep learning [23]. The method is based on the following assumptions: the scores between different user relationships and the degree of relevance between movies are different; for important relationships or ratings, more attention is assigned, while others are less focused [24, 25]. The key is how to independently distribute attention without accepting other information [26].

Given a *Query* in the *Target*, by calculating the correlation between it and the *key* of each data pair in the *Sourse*, the weight coefficients of the Key and Value are obtained. We only need to weight the weights separately to the *value*. Can finally calculate the value of Attention.

$$\text{Attention}(Query, Source) = \sum\nolimits_{i=1}^{l} Similarity(Query, Key_i) \cdot Value_i \tag{9}$$

In order to better learn global dependence information from internal text, we used the multi-head attention mechanism proposed by Vaswani et al. [27]. The multi-head attention mechanism takes parallel calculations of multiple scaled dot products, rather than just one calculation. Then, the independent attention calculation units are spliced together, and finally converted into a dimensional output of a desired size by a linear unit.

As shown in Fig. 3. We do n linear mapping of Q, K and V matrices and learn different linear projection matrices $d_n \times d_q$, $d_n \times d_k$ and $d_n \times d_v$. Then the resulting projection

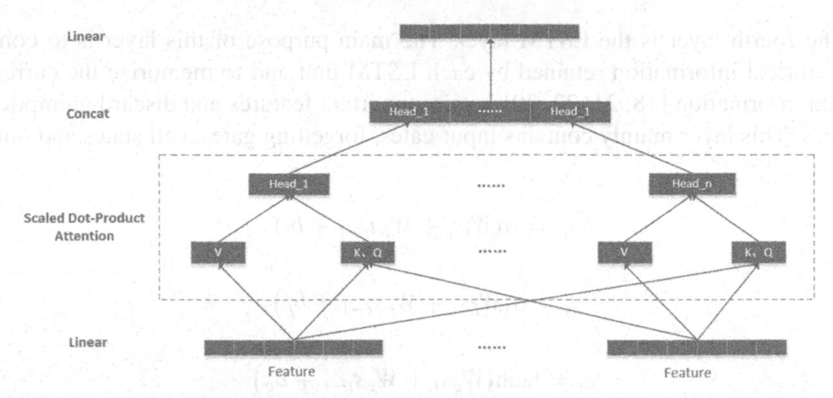

Fig. 3. Multi-head attention model.

matrix will be executed separately "Scaled Dot-Product Attention" The self-attention mechanism is essentially $X = Query = Key = Value$, which means that the Attention is used to find the interdependence within the sequence inside the sequence. The internal self-attention mechanism can solve the problem of weakening the dependence caused by the text being too long. Finally, the matrix of the output $d_n \times d_k$, and connect these values and project again, we can get the final value.

We map Q, K, and V to subspaces of lower dimensions and then perform Attention calculations in different subspaces. A lower subspace dimension reduces the amount of computation, which facilitates parallelization and captures features that represent different subspaces at different locations. This is compared to directly using the linearly mapped Q and K dot product as weighting coefficients, and then weighting and summing V. The advantage is that there will not be a situation where the dot product is too large, and there will be no problem that the gradient is too small. We scale the dot product by $\frac{1}{\sqrt{d_k}}$ to get:

$$head_i = \text{Attention}\left(QW_i^Q, KW_i^K, VW_i^V\right) = softmax\left(\frac{QW_i^Q\left(KW_i^K\right)^T}{\sqrt{d_k}}\right)VW_i^V \quad (10)$$

Finally, splicing the results, we can get the output feature value s:

$$s = MultiHead(Q, K, V) = Concat(head_1, \ldots, head_n)W^O \quad (11)$$

3.3 Recommendation Table Generation

This section describes the process of generating the pre-K recommendation tables. We extract the user feature representation through the convolutional neural network, so that we can get the user's potential feature set S_u. Then use the CNN-BLSTM plus the multi-head self-attention mechanism to calculate the set of potential features of the movie S_m. Next, the movie scoring matrix is reduced in dimension using the PCA method [28] to obtain a movie score set S_r.

First, calculate the similarity between the user preference feature and the unrated item $Similarity_1$, and select the top K term as the candidate list C_{k1}, and get the following formula:

$$Similarity_1 = \frac{\left(S_u^i\right)^T S_m^i}{\left\|S_u^i\right\| \cdot \left\|S_m^i\right\|} \tag{12}$$

Then, the similarity between the unrated items and the user's preferences is calculated in turn $Similarity_2$, and the top K items are also selected as the candidate list C_{k2}, and the similarity calculation formula is obtained:

$$Similarity_2 = \left(\frac{\left(S_m^i\right)^T S_m^j}{\left\|S_m^i\right\| \cdot \left\|S_m^j\right\|}\right) + \left(\frac{\left(S_r^i\right)^T S_r^j}{\left\|S_r^i\right\| \cdot \left\|S_r^j\right\|}\right) \tag{13}$$

Among them, S_r^i and S_m^i together constitute the potential representation of project i, S_r^j and S_m^j together constitute the potential representation of project j.

By merging the candidate lists C_{k1} and C_{k2}, the recommendation list C_k and the user relevance *Similarity* can be calculated by the user preference feature $Similarity_1$ and the user preferencc average project $\overline{Similarity_2}$. Assuming that the user has q preference items, the similarity can be calculated:

$$\overline{Similarity_2} = \frac{1}{q}\sum\nolimits_{i=1}^{q} Similarity_2 \tag{14}$$

And then, get uscr relevance:

$$Similarity = \varepsilon \cdot \left(\overline{Similarity_2}\right) + (1 - \varepsilon) \cdot Similarity_1 \tag{15}$$

Where ε is an adjustable parameter. The last calculated *Similarity* set, select the top K item as the recommended item, and recommend it to the user.

4 Experiments and Analysis

4.1 Dataset and Preprocessing

This article uses MovieLens as the experimental data set and enhances the MovieLens data set. There are several different versions of MovieLens, and we mainly select 1M and 10M data sets. We spliced the ID field in the movies.dat table in the 1M dataset with the 10M dataset in the tags.dat table, which is the newly tagged field in the movies.dat table in the 1M dataset. Add a movie profile field to the movie in the movies.dat table to make it easier to grab features from more dimensions in the movie collection.

Since the length and type of the original data are inconsistent, the data needs to be processed before the fields are entered into the embedded layer.

user.dat table:

- gender field: In the gender field, convert 'F' and 'M' to 0 and 1.
- age field: Change to a continuous number for each age group (one age range every ten years old).

movies.dat table:

- type field: First, create a dictionary of text to numbers, because a movie can be of multiple types, so we can convert the categories in type into strings, and store it in a dictionary of numbers, finally, convert the type field of each movie into a list of numbers.
- label field: Same as the type field. The description in the label is converted into a list of numbers.
- introduction field: Unsupervised word vector learning with word2vec [29], the learned word vector is stored in the vocabulary. The word is used as the basic unit of the sentence, and the word is expressed as the corresponding word vector form.

4.2 Experimental Parameters

When training the model, we selected 80% of the MovieLens data set as the training set and the remaining 20% as the test set. In the experiment, we set the learning rate to 0.0001. In the training phase, the filter window size of the convolutional neural network is set to 3, 4, and 5 respectively. To prevent over-fitting, add dropout to the neural network of the text processing module, and dropout is set 0.4. The iteration training number epoch is set to 10 and the mini-batch size is set to 128.

4.3 Experimental Results and Analysis

The experiment uses F1, MAP and NDCG to evaluate user-based collaborative filtering algorithm (UBCF), project-based collaborative filtering algorithm (IBCF), CNN-AT, bidirectional LSTM, deep learning model without multi-head attention, and a deep learning model with the multi-head attention mechanism. A total of six models were used to test the error. The results of the comparative test are shown in Table 1.

Table 1. Recommended performance comparison table

Method	@k								
	5			10			20		
	MRR	MAP	NDCG	MRR	MAP	NDCG	MRR	MAP	NDCG
UBCF	0.4358	0.5172	0.5261	0.4826	0.5738	0.5693	0.5176	0.6183	0.6137
IBCF	0.4571	0.5249	0.5412	0.5107	0.5934	0.5831	0.5318	0.6345	0.6188
CNN-AT	0.5261	0.6017	0.6372	0.5761	0.6402	0.6914	0.6033	0.6619	0.7083
BLSTM	0.5118	0.5977	0.6356	0.5704	0.6387	0.6806	0.5949	0.6581	0.7039
CNN-LSTM	0.5407	0.6284	0.6571	0.5953	0.6596	0.7027	0.6218	0.6817	0.7272
CNN-BLSTM-MHA	**0.5877**	**0.6592**	**0.6844**	**0.6376**	**0.7029**	**0.7108**	**0.6591**	**0.7282**	**0.7443**

As can be seen from Table 1, the overall performance of UBCF and IBCF is poor, mainly due to the lack of understanding of the features. The CNN-AT and the BLSTM model have relatively good effects, and both of them can be well in the entity. The CNN-LSTM model combines the advantages of the two models, so it is slightly better than the general deep learning model. Finally, the CNN-BLSTM model with the multi-head attention mechanism is better than the score prediction results produced by the former. In addition, we also compare the different k values, and found that with the increase of the recommended items, there is a better recommendation effect, and when the recommended items are increased to a certain number, the improvement of the recommendation effect is relatively small.

5 Conclusion and Expectation

In this paper, we propose a user preference learning model that incorporates attention mechanisms to recommend movies to users. We try to use the combination of convolutional neural network, BLSTM and attention mechanism to process user data and movie data, extract features from it, and finally combine movie scores to jointly generate recommendation tables. In the study of attention mechanisms, we also tried to use the multi-head attention mechanism to process and capture features of different subspaces in different locations in parallel, and there is no limit on distance dependence. A comprehensive experimental study demonstrates the effectiveness of our proposed model in film recommendation tasks.

But because our model only considers getting valid information from users and in some dimensions of the movie, it does not consider capturing features from more dimensions, such as movie posters or movie plots. Therefore, in the future we may consider improving our model, considering different features and different models to obtain more effective information in different dimensions, and finally combining to better understand user preferences. Thus providing users with higher quality and more precise recommendations. We will also try to recommend the application of the model in other areas.

Acknowledgement. The authors gratefully acknowledge support from National Key R&D Program of China (No. 2018YFC0831800), National Natural Science Foundation of China (No. 61872134), Natural Science Foundation of Hunan Province (No. 2018JJ2062), Science and technology development center of the Ministry of Education, and the 2011 Collaborative Innovation Center for Development and Utilization of Finance and Economics Big Data Property, Universities of Hunan Province.

References

1. Socher, R., Huval., B., Manning, C.D., Ng, A.Y.: Semantic compositionality through recursive matrix-vector space. In: Conference on Empirical Methods in Natural Language Processing (2012)
2. Kim, Y.: Convolutional neural networks for sentence classification. In: Proceedings of the Conference on Empirical Methods in Natural Language Processing, pp. 1746–1751 (2014)

3. Su, X., Khoshgoftaar, T.M.: A survey of collaborative filtering techniques. Adv. Artif. Intell. 1–19 (2009)
4. Zhang, Y., Zincir-Heywood, N., Milios, E.: World wide web site summarization. Web Intell. Agent Syst. Int. J. **2**, 39–53 (2004)
5. Jones, S., Staveley, M.S.: Phrasier: a system for interactive document retrieval using keyphrases. In: Proceedings of the 22nd Annual International ACM SIGIR Conference on Research and Development in Information Retrieval, pp. 160–167. ACM (1999)
6. Socher, R., Perelygin, A., Wu, J.Y., et al.: Recursive deep models for semantic compositionality over a sentiment treebank. In: Proceedings of Conference on Empirical Methods in Natural Language Processing, pp. 1631–1642 (2013)
7. Tang, D., Qin, B., Wei, F., et al.: A joint segmentation and classification framework for sentence level sentiment classification. IEEE/ACM Trans. Audio Speech Lang. Process. **23**, 1750–1761 (2015)
8. Xie, R., Liu, Z., Sun, M.: Representation learning of knowledge graphs with hierarchical types, pp. 2659–2665 (2016)
9. Jacob, T.I.: A neural model for straight line detection in the human visual system. Dissert. Theses – Gradworks (2014)
10. Bahdanau, D., Cho, K., Bengio, Y.: Neural machine translation by jointly learning to align and translate. Comput. Sci. (2014)
11. Ramirez-Moreno, D.F., Schwartz, O., Ramirez-Villegas, J.F.: A saliency-based bottom-up visual attention model for dynamic scenes analysis. Biol. Cybern. **107**, 141–160 (2013)
12. Luong, M.-T., Pham, H., Manning, C.D.: Effective approaches to attention-based neural machine translation (2015)
13. Xu, J., Chen, K., Qiu, X., et al.: Knowledge graph representation with jointly structural and textual encoding. 1318–1324 (2017)
14. Ding, Y., Li, X.: Time weight collaborative filtering. In: Proceedings of the 2005 ACM CIKM International Conference on Information and Knowledge Management, Bremen, Germany (2005)
15. Su, P., Ye, H.: [IEEE 2009 International Joint Conference on Artificial Intelligence (JCAI) - Hainan Island, China (2009.04.25–2009.04.26)] International Joint Conference on Artificial Intelligence - An Item Based Collaborative Filtering Recommendation Algorithm Usin, pp. 308–311 (2009)
16. Schafer, J.B., K., Konstan, J., Riedi, J.: [ACM Press the 1st ACM conference - Denver, Colorado, United States (1999.11.03–1999.11.05)] Proceedings of the 1st ACM conference on Electronic commerce, - EC 1999 - Recommender systems in e-commerce, pp. 158–166 (1999)
17. Cho, K., Van Merrienboer, B., Gulcehre, C., et al.: Learning phrase representations using RNN encoder-decoder for statistical machine translation (2014)
18. Cheng, H.T., Koc, L., Harmsen, J., et al.: Wide & deep learning for recommender systems. In: Proceedings of the 1st Workshop on Deep Learning for Recommender Systems, pp. 7–10. ACM (2016)
19. Zhu, J., Shan, Y., Mao, J., Yu, D., Rahmanian, H., Zhang, Y.: Deep embedding forest: Forest-based serving with deep embedding features. In: Proceedings of the ACM SIGKDD International Conference on Knowledge Discovery and Data Mining, Halifax, NS, USA, pp. 13–17 (2017)
20. Collobert, R., Weston, J., Bottou, L., et al.: Natural language processing (almost) from scratch. J. Mach. Learn. Res. **12**, 2493–2537 (2011)
21. Liu, Y., Sun, C., Lin, L., et al.: Learning natural language inference using bidirectional lstm model and inner-attention (2016)
22. Huang, C., Kuo, P.-H.: A deep CNN-LSTM model for particulate matter (PM2.5) forecasting in smart cities. Sensors **18**, 2220 (2018)

23. Wang, X., Yu, L., Ren, K., et al.: Dynamic attention deep model for article recommendation by learning human editors' demonstration (2017)
24. Vaswani, A., et al.: Attention is all you need. arXiv preprint arXiv:1706.03762 (2017)
25. Zhai, S., Chang, K.-H., Zhang, R., Zhang, Z.M.: Deepintent: learning attentions for online advertising with recurrent neural networks. In: Proceedings of the 22nd ACM SIGKDD International Conference on Knowledge Discovery and Data Mining, pp. 1295–1304 (2016)
26. Zhou, C., et al.: ATRank: an attention based user behavior modeling framework for recommendation. In: AAAI, pp. 4564–4571 (2018)
27. Vaswani, A., et al.: Attention is all you need. CoRR abs/1706.03762 (2017)
28. Dubuisson-Jolly, M.P., Gupta, A.: Color and texture fusion: application to aerial image segmentation and gis updating. Image Vis. Comput. **18**, 823–832 (2000)
29. Mikolov, T., Sutskever, I., Chen, K., et al.: Distributed representations of words and phrases and their compositionality. In: Advances in Neural Information Processing Systems, pp. 3111–3119 (2013)
30. Zhaowei, Q., Cao, B., Wang, X., Li, F., Xu, P., Zhang, L.: Feedback LSTM network based on attention for image description generator. Comput. Mater. Continua **59**(2), 575–589 (2019)
31. Feng, X., Zhang, X., Xin, Z., Yang, A.: Investigation on the Chinese text sentiment analysis based on convolutional neural networks in deep learning. Comput. Mater. Continua **58**(3), 697–709 (2019)
32. Wang, G., Liu, M.: Dynamic trust model based on service recommendation in big data. Comput. Mater. Continua **58**(3), 845–857 (2019)

Traffic Anomaly Detection for Data Communication Networks

Xiaoxiao Tang[1(✉)], Wencui Li[2], Jing Shen[2], Feng Qi[1], and Shaoyong Guo[1]

[1] Beijing University of Post and Telecommunications, Beijing 100876, China
13269900125@163.com

[2] Information and Communication Company, State Grid Henan Electric Power Company, Zhengzhou 450052, China

Abstract. The detection efficiency of the traditional data communication network traffic anomaly detection algorithm is low. And it is impossible to guarantee the accuracy of traffic detection in actual applications. The detection algorithm involves too many dimensions, and it is difficult to explore the optimal solution even if it takes a lot of time. In view of the above problems, this paper proposes an improved network traffic anomaly detection algorithm. The algorithm inherits the algorithm idea of combining the weak classifiers in the classical GBDT (Gradient Boosting Decision Tree) into the final strong classifier. The algorithm equilibrium weights are assigned to the weak classifiers in the iteration to balance the contribution of the weak classifier to the final classification model. The algorithm combines Bayesian optimization algorithm to achieve the purpose of automatically exploring the optimal super-parameter combination. The simulation results show that the proposed algorithm has an effective improvement in detection efficiency compared with the traditional traffic detection algorithm.

Keywords: Traffic anomaly detection · LightGBM algorithm · Bayesian optimization · Gradient descent

1 Introduction

With the rapid development of computer technology and network construction, the Internet plays an increasingly important and important role in people's life and social development. On August 20, 2018, China Internet Network Information Center (CNNIC) released the 42nd "Statistical Report on the Development of China's Internet Network" in Beijing. As of June 30, 2018, the number of Internet users in China reached 802 million, and the Internet penetration rate was 57.7%. [1] Along with the rapid development of networking, more and more important applications have emerged. Power communication, mobile payment, e-commerce, entertainment short video, etc., the new Internet is constantly influencing and enriching our lives and entertainment. However, Internet technology, like other application technologies, is also a double-edged sword. It also brings us severe challenges while affecting social development. Some lawless elements use the Internet and computer system vulnerabilities to conduct network attacks

X. Sun et al. (Eds.): ICAIS 2020, LNCS 12240, pp. 440–450, 2020.
https://doi.org/10.1007/978-3-030-57881-7_39

on individuals and corporate Internet communications for personal gain. The National Internet Emergency Center Internet Security Threat Report shows that in November 2018, the number of terminals infected with network viruses was more than 840,000; the number of websites that were tampered with was 1,357, of which 68 were tampering with government websites; The number of back-door websites is 2,513, of which 45 are government websites; the number of counterfeit pages for domestic websites is 6,469; and the National Information Security Vulnerability Sharing Platform (CNVD) collects 810 security vulnerabilities, including 308 high-risk vulnerabilities. There are 700 vulnerabilities that can be exploited to implement remote attacks [2]. Summarizing the characteristics of the Internet today, we can conclude that it is huge, complex, and risks and benefits coexist. Network traffic contains a lot of information, and network traffic anomaly is a manifestation of network space attacks. Therefore, how to quickly and accurately detect network traffic anomalies becomes an important issue in network technology development.

Research on network traffic anomaly detection has been ongoing at home and abroad. Hua Huiyou proposed a network traffic anomaly detection algorithm that combines KNN and K-means. The process of decomposing traffic anomaly detection is two steps: offline preprocessing and online classification. This algorithm improves the efficiency and accuracy of traditional KNN to some extent. The naive Bayesian algorithm can be easily applied to network traffic anomaly detection. Han Xiaoyan proposed a naive Bayesian optimization algorithm with weighted values to improve the accuracy of the traditional naive Bayesian algorithm. Rao xian and so on use SVM to do anomaly detection system, mainly composed of data preprocessor, SVM classifier and decision mechanism, and pass the classification result to decision mechanism to make abnormal judgment. The above algorithms all propose improved schemes on their respective basic algorithms. Although the efficiency of the traditional network traffic anomaly detection algorithm can be improved to some extent, there are certain limitations, and the efficiency needs to be further improved. The KNN and K-means algorithms are relatively old algorithms, and the poor interpretability and lazy inertia algorithms are obvious disadvantages. The theoretical basis of the naive Bayesian algorithm is based on the assumption that the variables do not affect each other independently. It is impossible to exist in real-world applications, which greatly limits the detection efficiency of naive Bayesian algorithm in network traffic anomaly detection; SVM algorithm is difficult to implement for large-scale training samples, if the amount of data is large, SVM The training time will be longer, sensitive to the selection of parameters and kernel functions, but generally rely on manpower and a large amount of time, and it is difficult to choose a better combination of parameters, the efficiency of algorithm detection needs to be improved.

Aiming at the above problems, this paper proposes a network traffic anomaly detection algorithm based on LightGBM algorithm. The weak classifiers in each iteration are given balance weights to balance the contribution of the weak classifier to the final classification model. And the Bayesian optimization algorithm is used to automatically explore the optimal super-parameter combination, which further improves the efficiency and accuracy of the algorithm's network traffic anomaly detection. LightGBM was released by the Microsoft DMTK team in January 2017. It is a gradient boosting framework that uses a learning algorithm-based decision tree. It is a greatly improved GBDT algorithm that is fast, distributed, and high performance [6]. Bayesian optimization is a very effective global optimization algorithm, the goal is to find the global optimal solution. This paper improves the detection efficiency and accuracy of network traffic anomalies by integrating Bayesian optimization and LightGBM [7].

2 Algorithm Model

2.1 Algorithm Model Framework

This section gives a brief introduction to the algorithm model framework of the algorithm. In the data traffic network network traffic anomaly detection, our goal is to design and implement algorithms with high efficiency and high accuracy. The algorithms such as the naive Bayes algorithm and the KNN-KMEANS fusion algorithm described above can achieve the high efficiency and high accuracy of the abnormality detection of the network traffic of the data communication network to a certain extent. However, with the continuous complication of the network environment and the continuous improvement of attack technologies, we need more efficient and accurate algorithms. In this chapter, we will introduce an efficient, high-accuracy algorithm implemented by the fusion of LightGBM algorithm and Bayesian optimization algorithm.

The LightGBM algorithm inherits from the classic GBDT algorithm and is the framework for implementing the GBDT algorithm. The main idea is to generate a weak classifier for each iteration through multiple iterations, and each classifier trains the residuals of the previous round of classifiers. The requirements for weak classifiers are generally simple enough and are low variance and high bias because the training process is to continuously improve the accuracy of the final classifier by reducing the bias. Each round of weak classifiers causes the model loss function to fall along the gradient direction, so that the loss function is as small as possible, and the negative gradient direction is as shown in the following figure [3].

As mentioned above, the algorithm generates a weak classifier at each iteration. The weak classifier's prediction of the result is highly biased, and the contribution to the correct classification of the entire strong classifier when combined into a strong classifier is also not variable. If it is simply a simple superposition operation on the weak classifier, it is not able to reflect the difference that the weak classifier contributes different values to the correct classification. The formula for the algorithm to obtain a strong classifier is as follows:

$$F(\mathrm{x}) = \sum_{t=0}^{T} f_t(x) \tag{1}$$

The final strong classifier does not consider the contribution of each weak classifier, and there is a certain degree of limitation on the final strong classifier classification accuracy. Therefore, the algorithm proposes to give each weak classifier a balance factor w according to the contribution of each weak classifier to the strong classifier, and the product of the weak classifier and the balance factor is superimposed to generate the final strong classifier model. The algorithm model is shown in the following figure (Fig. 1):

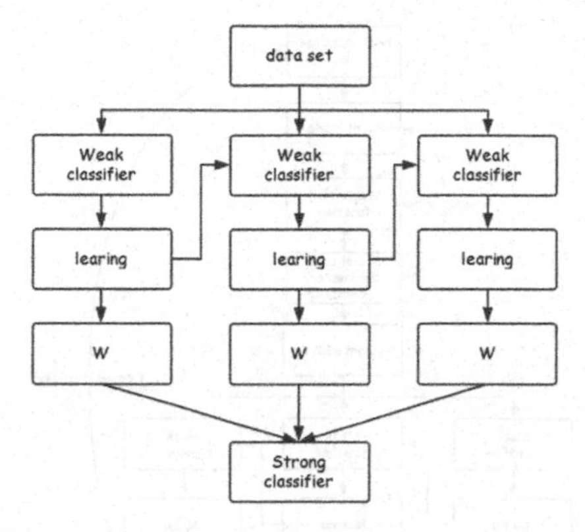

Fig. 1. Algorithm strong classifier mode figure

Get the algorithm model strong classifier formula as follows:

$$F(\mathrm{x}) = \sum_{t=0}^{T} w * f_t(x) \tag{2}$$

The LigthGBM algorithm is a collection of multiple fine-grained algorithms, which increases the complexity of the algorithm, making it difficult to manually determine the specific input parameters of the fine-grained algorithm in the set. Bayesian optimization algorithm is a very good global optimization algorithm, its goal is to find the global optimal parameter solution. Therefore, by combining the Bayesian optimization algorithm and the LigthGBM algorithm, it can solve the shortcomings of traditional algorithms such as GBDT in the abnormal traffic detection of data communication networks [4].

The overall framework of the algorithm model presented in this article is shown below (Fig. 2):

The overall framework of the algorithm model is described as follows:

1) Initializing the data set D according to the selected feature value and the collected prior data, as the input data and the training set of the entire algorithm.
2) Iteration N times.

   a. The Bayesian optimization module determines the super-parameter combination X generated by each iteration and serves as an input to the LightGBM module.
   b. The LightGBM module is iterated T times according to the input super- parameter combination X. The weights are used to balance the contribution of the weak classifier to the strong classification, and the model residuals are reduced along the direction of the gradient descent. The weak classifiers of each sub-iterative are linearly superimposed to obtain the strong classifier model M.
   c. According to the classification model M, calculate new data update into the data set D;

3) Repeat the iterated with the new data set D until the final result is obtained.

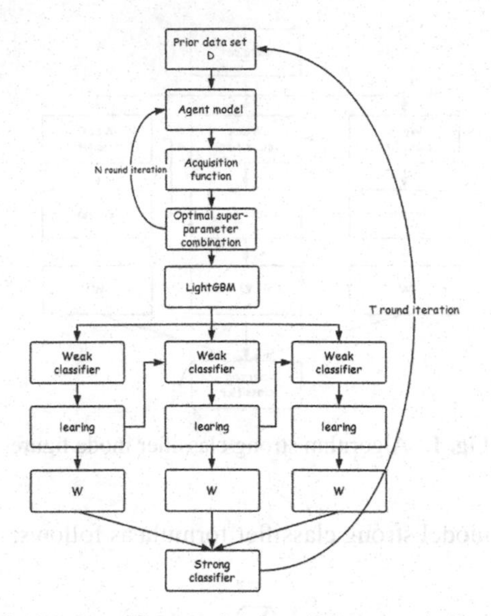

Fig. 2. Algorithm model overall framework figure

2.2 Algorithm Detailed Steps

According to the content in the previous section, the detailed steps of the network traffic anomaly detection algorithm proposed in this paper are as follows:

1) Initialize data set D based on the selected feature values and the a priori data collected.

$$D = (x_1, y_1), \ldots, (x_n, y_n) \quad (3)$$

2) The iterative model is then iteratively calculated from data set D and hypothetical model M:

$$p(y|x, D) < -FITMODEL(M, D) \quad (4)$$

Where M is the Gaussian distribution of the data. Calculated by the Gaussian process

$$p(y \mid x, D) = N(y \mid \hat{\mu}, \hat{\sigma}^2) \quad (5)$$

According to (5), the agent model is as follows.

$$\text{Agent model} = \text{S(X,P(y | X,D))} \quad (6)$$

3) Then, according to the agent model and the acquisition function, the optimal super-parameter combination X is selected:

$$X = (x_1 \ldots x_i) < - \arg\max_{x \in X} S(X, p(y|X, D)) \quad (7)$$

The acquisition function here is a trade-off between distribution and promotion. The algorithm chooses the UCB strategy as the acquisition function. Substituting the obtained optimal super- parameter combination X into the following LightGBM module. Complete the steps of inputting the training sample H, the maximum number of iterations T, the loss function L, and outputting the strong learner f(x).

4) According to training samples H

$$H = \{(x_1, y_1), (x_2, y_2), \ldots, (x_m, y_m)\} \tag{8}$$

Initialize the weak classifier

$$f_0(x) = \arg\min_c \sum_{i=1}^{m} L(y_i, c) \tag{9}$$

5) T iterations:

a. Calculate a negative gradient for the sample
i = 1, 2, 3, …, m

$$r_{ti} = -\left[\frac{\partial L(y_i, f(x_i))}{\partial f(x_i)}\right]_{f(x)=f_{t-1}(x)} \tag{10}$$

t represents the t-th iteration.

b. Use

$$(x_i, r_{ti})(i = 1, 2, \ldots, m) \tag{11}$$

fit a CART regression tree to get the t-th regression tree, and the corresponding leaf node area is $R_{tj}, j = 1, 2, \ldots, J$. Where J is the number of leaf nodes in the decision tree.

c. Calculate the best fit value for the leaf node area.

$$c_{tj} = \arg\min_c \sum_{x_i \in R_{tj}} L(y_i, f_{t-1}(x_i) + c) \tag{12}$$

d. Add a balance factor w to each weak classifier, with an initial value of 1, optimized in successive iterations. Update the strong learner.

$$f_t(x) = f_{t-1}(x) + w * \sum_{j=1}^{J} c_{tj} I(x \in R_{tj}) \tag{13}$$

Build an output strong learner and get the model.

$$f(x) = f_T(x) = f_0(x) + \sum_{t=1}^{T} \sum_{j=1}^{J} c_{tj} I(x \in R_{tj}) \tag{14}$$

6) According to the model, new results are obtained and the data set D (1) is updated. After repeated iterations T times, the final result is obtained.

3 Experiment and Simulation

3.1 Data Sources

The experimental data source is the campus network node. The data of the full-day traffic is collected by the lump function library libpcap library for the daily traffic data of a certain day. Because the campus network is relatively safe, the 3.21 million data collected on the day is preprocessed and noise is removed, and the simulated abnormal traffic is injected into the background traffic for performance testing. The experimental simulation is to generate abnormal data, which is used as a data set for algorithm simulation after injecting the detection data set.

3.2 Data Preprocessing

In this experiment, the original data is divided into Normal and Attack according to whether it is attack data, and the Attack type data is abnormal data generated by various attack means. After the original data is obtained, part of the data of the original data is randomly selected as the training set and the test set in the simulation experiment data set. Among them, the training set is 150,000 pieces of data, and the test set is 50,000 pieces of data. In the training set and test set data composition, Normal is 60%, and Attack data is 40% of the total.

3.3 Feature Selection

For the traffic anomaly detection method of the data communication network, the feature values and their meanings are selected as follows.

Count: The number of connections with the same destination address as the current connection per unit time.
Serror_rate: The number of connections with a SYN error.
Rerror_rate: The number of connections with a REJ error.
Same_srv_rate: The number of connections to establish the same service.
Diff_srv_rate: The number of connections to establish different services.
Srv_count: The number of connections per unit time that are the same as the current connection service.
Srv_serror_rate: The number of connections with a SYN error.
Srv_rerror_rate: The number of connections with REJ errors.
Srv_diff_host_rate: The number of times to connect to different hosts.

3.4 Simulation Experiment Design

- Algorithm evaluation standard

  In order to be able to visually display the abnormal flow detection results of the experiment, the experiment defines the following evaluation criteria.
  TP (True Positive): Classifies normal traffic as normal traffic.

FP (False Positive): Classifies abnormal traffic as normal traffic.
TN (True Negative): classifies abnormal traffic as abnormal traffic.
FN (False Negative): Classifies normal traffic as abnormal traffic.
AR (Accuracy Rate): The proportion of the sample with the correct classification to the total sample.
Accuracy Rate = TP + TN/(TP + FP + TN + FN)
ER (Error Rate): The proportion of samples with incorrect classifications to all samples.
Error Rate = FP + FN/(TP + FP + TN + FN)
R (Recall): The recall rate is correctly classified as the ratio of normal traffic to the normal flow of the sample.
Recall = TP/(TP + FN)
P (Precision): Accuracy rate, the actual normal flow ratio in the sample classified as normal flow.
Precision = TP/(TP + FP)
F1: The mean of the recall and accuracy.

$$F1 = \frac{2 * P * R}{P + R} = \frac{2 * TP}{2 * TP + FN + FP} \tag{15}$$

ROC (Receiver Operating Characteristic) Curve: A curve with a false positive rate (FP_rate) and a false negative rate (TP_rate) as the axis (Fig. 3).

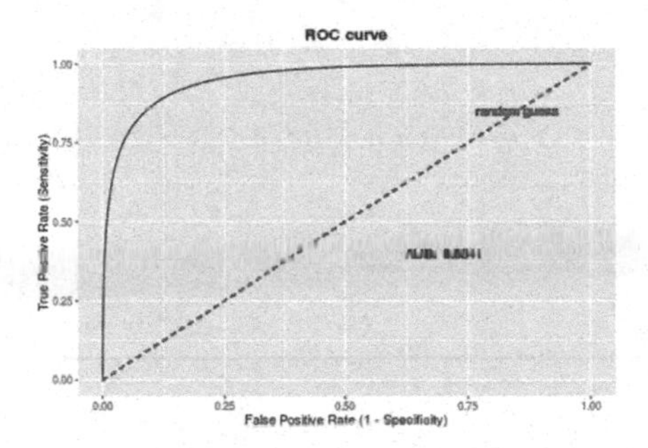

Fig. 3. ROC curve figure

When the completion curve is drawn, there will be a qualitative analysis of the model. If you want to quantify the model, you need to introduce a new concept, which is the area of AUC (Area under roc Curve). This concept is actually very simple. It refers to the area under the ROC curve, and the calculation of the AUC value only needs to be integrated along the horizontal axis of the ROC. In the real scene, the ROC curve is generally above the line y = x, so the value of AUC is generally between 0.5 and 1 The larger the value of AUC, the better the performance of the model.

- Experiment process
  This paper uses MATLAB to carry out simulation experiments for the experimental platform. Based on the experimental training set, test set data, and defined algorithm evaluation criteria obtained above, the experimental design is as follows.

The statistical unit time range of the continuous eigenvalues of the flow anomaly detection algorithm is uncertain, and a comparative experiment needs to be designed, and the optimal time interval is selected according to the experimental results. Since the flow anomaly detection requires high real-time performance, the experimental set time window is combined for 2 s, 4 s, 6 s, 8 s, 10 s, and 12 s for experiments. The simulation results of the LightGBM fusion Bayesian optimized network traffic anomaly detection algorithm, the weighted simple Bayesian algorithm, and the KNN-KMEANS fusion algorithm mentioned in this paper are compared under the same conditions. After training the model with the training set, verify it with the test set. The efficiency of the algorithm on the evaluation criteria of the above algorithm is obtained, and the efficiency of the algorithm is analyzed.

3.5 Experimental Results and Analysis

According to the above experiment, the experimental results are as follows (Fig. 4):

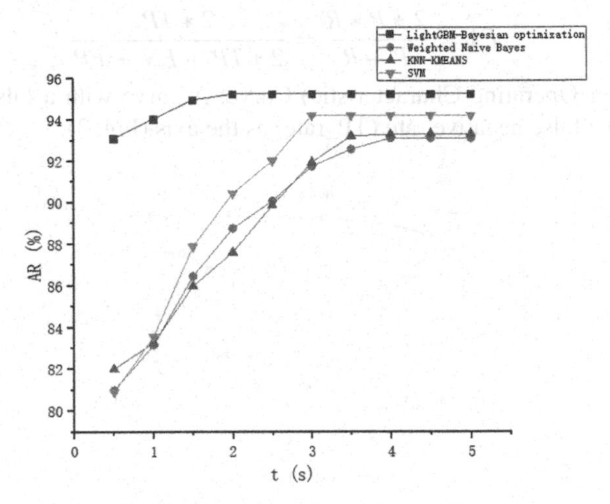

Fig. 4. Algorithm AR comparison figure

When the algorithm in this paper is characterized by statistical information in time units, it performs better in AR accuracy than traditional network traffic anomaly detection algorithms, and the time unit reaches the best in about 2 s, which is also shorter than traditional algorithms. This means that the performance of the algorithm in this paper is better than traditional algorithms (Fig. 5).

As shown in the above figure, regarding the F1 value: the algorithm in this paper reaches the best 96.85% in about 2.5 s; the weighted Naive Bayes algorithm reaches the best 95.28% in 4 s; the KNN-KMEANS fusion algorithm reaches the best 96.10 in 3.5 s %; SVM algorithm reaches the best 96.35% in about 3 s (Fig. 6).

As shown in the above figure, regarding the AUG value of the ROC curve: the algorithm in this paper reaches the best 94.47% in about 2 s; the weighted naive Bayes

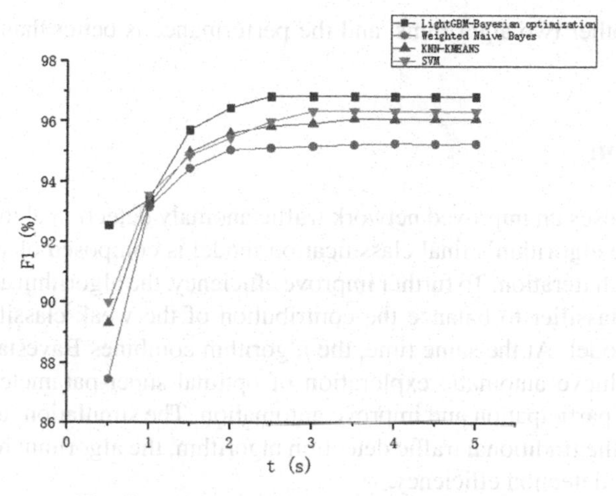

Fig. 5. Algorithm F1 value comparison figure

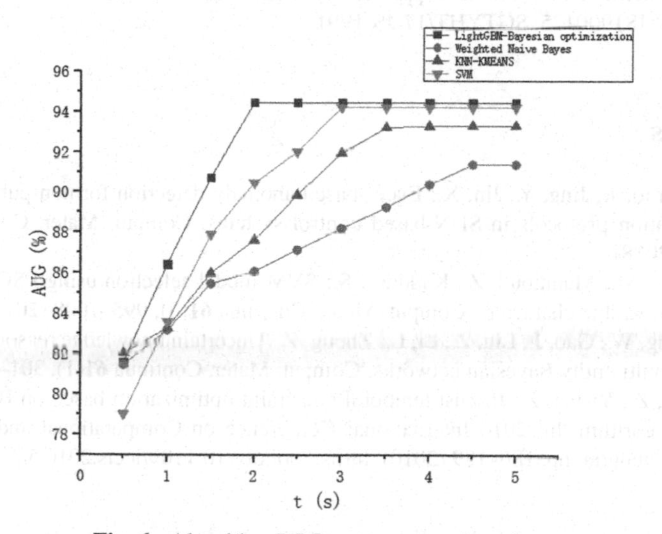

Fig. 6. Algorithm ROC curve comparison figure

algorithm reaches the best 91.38% in 4.5 s; the KNN-KMEANS fusion algorithm reaches the best in 4 s 93.29%; SVM algorithm reaches the best 94.21% in about 3 s.

From the above experiments, the algorithm proposed in this paper reaches the highest peak of each evaluation criterion in about 4 s, the KNN-KMEANS fusion algorithm is around 6 s, and the weighted naive Bayesian algorithm is around 8 s. Therefore, in terms of the real-time nature of network traffic anomaly detection, the algorithm described in this paper performs better than the traditional algorithms.

In the evaluation indexes such as algorithm accuracy, F1 value and ROC curve, it can be seen from the figure that the algorithm described in this paper is higher in peak value than the other two algorithms, and the performance is better than the traditional algorithms.

4 Conclusion

This paper proposes an improved network traffic anomaly detection algorithm based on LightGBM. The algorithm's final classification model is composed of weak classifiers generated by each iteration. To further improve efficiency, the algorithm assigns a weight to each weak classifier to balance the contribution of the weak classifier to the final classification model. At the same time, the algorithm combines Bayesian optimization algorithm to achieve automatic exploration of optimal super-parameter combination, reduce artificial participation and improve automation. The simulation results show that compared with the traditional traffic detection algorithm, the algorithm has a significant improvement in detection efficiency.

Acknowledgement. This work is supported by Henan Electric Power Technology Project (SGHAXT00JSJS1900125, SGTYHT/17-JS-199).

References

1. Wan, M., Yao, J., Jing, Y., Jin, X.: Event-based anomaly detection for non-public industrial communication protocols in SDN-based control systems. Comput. Mater. Continua **55**(3), 447–463 (2018)
2. El Mamoun, M., Mahmoud, Z., Kaddour, S.: SVM model selection using PSO for learning handwritten Arabic characters. Comput. Mater. Continua **61**(3), 995–1008 (2019)
3. Li, D., Hong, W., Gao, J., Liu, Z., Li, L., Zheng, Z.: Uncertain knowledge reasoning based on the fuzzy multi entity Bayesian networks. Comput. Mater. Continua **61**(1), 301–321 (2019)
4. Yan-guang, Z., Yi-fan, Z.: Robust temporal constraint optimization based on Bayesian optimization algorithm. In: 2010 International Conference on Computational and Information Sciences, Chengdu, pp. 186–189 (2010). https://doi.org/10.1109/iccis.2010.50

Video Action Recognition Based on Hybrid Convolutional Network

Yanyan Song, Li Tan(✉), Lina Zhou, Xinyue Lv, and Zihao Ma

Beijing Technology and Business University, Beijing, China
tanli@th.btbu.edu.cn

Abstract. Aiming at the problem of unbalanced distribution of spatio-temporal information in video images, this paper proposes a 2D/3D hybrid convolutional network that introduces attention mechanism, which fully captures video space information and dynamic motion information, and better reveals motion features. With the help of the dual-stream convolutional network structure, we built 2D convolution and 3D convolution parallel neural networks. In the 2D convolutional neural network, the residual structure and the LSTM network model are used to focus on the spatial feature information of the video behavior. Secondly, the 3D convolutional neural network constructed by Inception structure is used to extract the spatiotemporal feature information of video behavior. On the basis of the two high-level semantics extracted, the attention mechanism is introduced to fuse the features. Finally, the obtained significant feature vector is used for video behavior recognition. Compared with other network models on the UCF101 and HMDB51 datasets, it can be seen from the results that the proposed 2D/3D hybrid convolutional network has good recognition performance and robustness.

Keywords: Action recognition · Mixed convolution · Attention mechanism · Deep learning · Video understanding

1 Introduction

With the development of information technology, multimedia on the Internet is growing rapidly, and the number of shared videos is increasing. Video-based behavior recognition plays a key role in various fields, such as public intelligence monitoring, human-computer interaction, and video search. With the deep application of deep learning in images, more and more researchers have applied deep learning methods to video processing. Video data is a dynamic and continuous set of information compared to images. Therefore, in the data processing process, not only the information of the video space dimension but also the depth information in the video time dimension should be considered.

But video-based behavior recognition is a challenging issue. In the video, it is still difficult for the machine to recognize a person's simple actions due to problems such as posture, angle, illumination, and occlusion. To improve the effect of behavior recognition, it is necessary to extract the effective features of the video as much as possible. Based on the 2D convolutional neural network, remarkable achievements have been

X. Sun et al. (Eds.): ICAIS 2020, LNCS 12240, pp. 451–462, 2020.
https://doi.org/10.1007/978-3-030-57881-7_40

made in two-dimensional signal processing such as images. However, when the 2D convolutional neural network performs feature extraction on video data, the limitation of the 2D convolution method leads to ignoring the time information of the third dimension of the video during the convolution process. In theory, the 3D convolutional neural network can use the 3D convolution kernel to extract the spatio-temporal hybrid features of video data, but in practice, the 3D convolutional neural network does not show superior performance. Compared with the 2D convolutional neural network, the parameters of the 3D convolutional neural network are too large, and it is difficult to train the deep convolutional neural network, which increases the complexity of network optimization and the consumption of computing resources.

In order to capture the visual space information and dynamic motion information of video images, this paper proposes a 3D/2D hybrid convolutional neural network, which uses 3D convolution to extract the spatiotemporal features of video behavior, and another part uses 2D convolution to focus on extracting appearance information. Finally, the two high-level semantic information are integrated. When constructing a 3D/2D hybrid convolutional neural network, the 2D convolution is used to compensate the depth of the feature map while limiting the number of 3D convolutional layers, which makes it possible for the network to achieve better performance with less spatiotemporal fusion.

2 Related Work

Video behavior recognition is a core issue in computer vision. With the application of high-performance deep convolutional neural networks for image recognition tasks, many work has designed effective deep convolutional neural networks for behavior recognition [1–8, 14–22, 24, 26]. For example, the classical dual-stream convolution network proposed by Simonyan K et al. [1]. Video can naturally be broken down into spatial and temporal components. They designed the video recognition architecture and divided it into two streams. Each stream is implemented using ConvNet, and its softmax scores are combined through late fusion. The mainstream ConvNet framework relies on dense time sampling with predefined sampling intervals by focusing on appearance and short-term motion, thus lacking the ability to integrate remote time structures. To solve this problem, Wang L et al. proposed Time Segment Network (TSN) [2], a novel framework for motion recognition based on video, based on the idea of remote time structure modeling. It combines sparse time sampling strategies and video-level supervision to achieve efficient and effective learning using the entire action. Based on the time segment network, Zhou B et al. proposed an efficient and interpretable network model, Time Series Network (TRN) [3]. The network model can learn and infer the timing dependence of frames on multiple scales in video. The proposed time-series network sparsely samples the frames and then learns their causality to achieve efficient capture of timing relationships over multiple time scales.

Compared to 2DConvNet, 3DConvNet is able to better model temporal information through 3D convolution kernel and 3D pooling operations. Therefore, Tran D et al. proposed a simple and effective method for temporal and spatial feature learning using a deep three-dimensional convolutional network (C3D) [5] trained on large-scale surveillance video datasets, demonstrating that C3D networks can Simultaneously simulate appearance and motion information. Since the 3D convolutional neural network has more parameters than 2DConvNets, and the training video architecture requires additional large tag data sets, several variants of the proposed centralized 3D convolutional neural network fail to utilize long-range time information, thus limiting the performance of these architectures. Diba A et al. proposed a novel deep space-time feature extractor network (TTL) [6] that models the variable-time 3D convolution kernel depth over a short and long time horizon, and On this basis, the DenseNet architecture was extended, replacing the standard transition layer in the DenseNet architecture with TTL. Their proposed network architecture (T3D) [6] captures short-term, medium-term and long-term behavioral appearance and time information intensively and efficiently. Shuiwang Ji et al. developed a 3D convolutional neural network architecture based on 3D convolution feature extractor [8]. The CNN architecture generates multiple information channels from adjacent video frames and performs convolutional kernel sub-sampling in each channel to obtain a final feature representation by combining information from all channels.

In summary, it can be found that the idea of both the 2D convolutional neural network architecture and the 3D convolutional neural network architecture design is to make full use of the depth information of the video on the basis of extracting the appearance characteristics of the video behavior. Using the motion information provided by the optical flow frame or studying the timing relationship of the input frame is to some extent compensate for the video time dimension feature of the 2D convolutional architecture loss, while the 3D convolution architecture enhances the 3D extraction of spatio-temporal features under the limited convolutional layer. Therefore, we designed a 3D/2D hybrid convolutional neural network for the advantages and disadvantages of 2D convolution and 3D convolution.

3 Methods

In this section, we first introduce the built 2D convolution network and 3D convolution network. After that, we will introduce in detail the simple and effective 3D/2D mixed convolution behavior recognition model we provide for motion recognition in video.

3.1 2D Convolutional Neural Network

Res-RNN Model. In the 2D convolutional neural network part, the Res-RNN model is used to extract video behavior feature information. Firstly, the residual structure is used to build the CNN network to extract the appearance characteristics of the video behavior. On the basis of CNN, the RNN network is constructed to extract the depth features of the video behavior. The model structure is shown in the Fig. 1.

The video data contains depth information compared to the image data, and the 2D convolutional neural network focuses on extracting the appearance characteristics of a certain behavioral video frame. In the built CNN network, the residual network structure [9] is adopted to enhance the extraction capability of network features. Because residual learning is a good solution to the degradation of deep networks compared to other network structures. In the common convolutional neural network, as the number of network layers increases, the network performance does not increase proportionally,

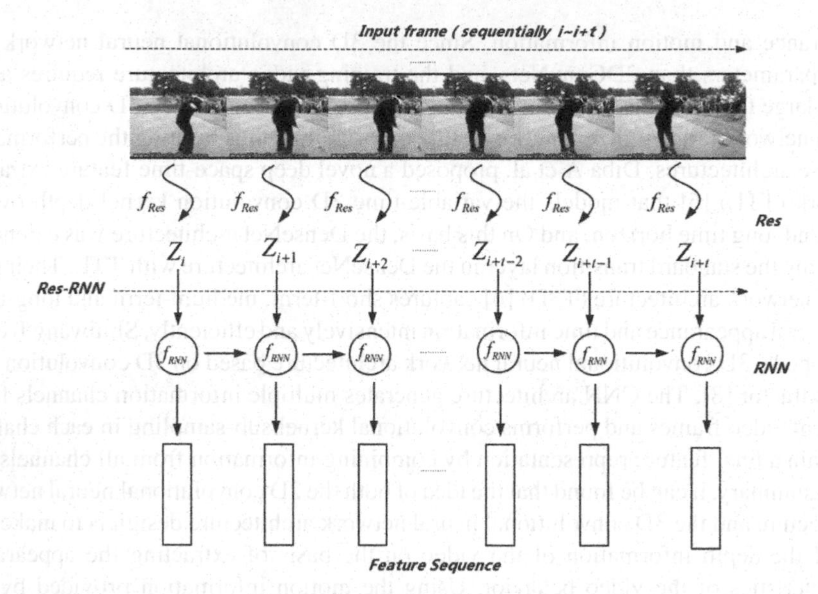

Fig. 1. Res-RNN model structure

but instead the gradient dispersion or gradient explosion problem occurs. The residual network structure utilizes the "shortcut connection" connection method to complete at least some of the redundant layer's identity mapping on the basis of the optimized network hierarchy, ensuring that the input and output of the identity layer are identical, thereby solving the problem of degraded phenomena that caused by an increase in network depth.

In order to compensate for the effect of 2D convolutional neural network on behavioral video data extraction features, and because the video input frame is time series, the long-term short-term memory model in the cyclic network [10] is used to construct the Res-RNN network. The LSTM passes the self-loop to generate a path for the gradient to continue to flow for a long time, and the weight of the self-loop is not fixed, and the accumulated time scale can be dynamically changed by the input sequence. In addition to the external RNN cycle, the LSTM also has an internal "LSTM cell" cycle (self-loop), from which it appears that the LSTM does not simply apply an element-by-element nonlinearity to the affine transformation of the input and loop elements, so LSTM is easier to learn for long-term dependencies than a simple loop architecture. It captures long-term motion information of video input frames. The structure of a certain cell network is shown in the figure below (Fig. 2).

For the input data of the behavioral video, by capturing the characteristics of the input frame at different times, the information of the previous frame of the video is used to understand the information of the current frame, and to find its dependence, thereby extracting the motion information of the behavioral video. The LSTM flows by constructing some gate control information. The important components are shown in the figure above, including the forgetting gate f, the external input gate g, the output gate q and the state unit s. Suppose that a cell unit i in the LSTM cyclic network has an input

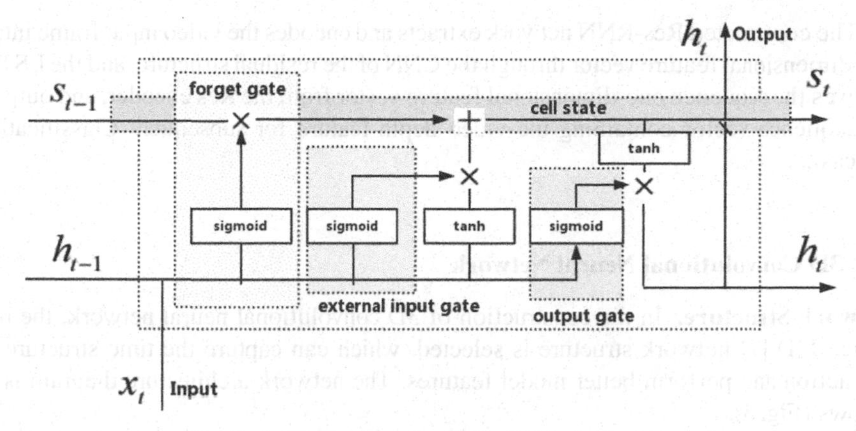

Fig. 2. LSTM cycle network "cell" block diagram

vector of $x_i^{(t)}$ at time t and an output vector of $h_i^{(t)}$. First, the weight of the self-loop is determined by the forgetting gate $f_i^{(t)}$. The weight is set by the sigmoid unit to a value between 0 and 1, where 1 represents complete reservation and 0 represents complete forgetting. The calculation formula is as follows,

$$f_i^{(t)} = sigmoid(b_i^f + \sum_j U_{i,j}^f x_j^{(t)} + \sum_j W_{i,j}^f h_j^{(t-1)}) \tag{1}$$

Where b^f, U^f and W^f are the cyclic weights of the offset, input weight, and forgetting gate, respectively. Then the external input gate $g_i^{(t)}$ determines the information to be stored in the cell, which is calculated as follows,

$$g_i^{(t)} = sigmoid(b_i^g + \sum_j U_{i,j}^g x_j^{(t)} + \sum_j W_{i,j}^g h_j^{(t-1)}) \tag{2}$$

Where b^g, U^g and W^g are the cyclic weights of the offset, input weight and forgetting gate, respectively. After that, the status unit $s_i^{(t)}$ of the LSTM cell is updated and calculated as follows,

$$s_i^{(t)} = f_i^{(t)} s_i^{(t-1)} + g_i^{(t)} \tanh(b_i + \sum_j U_{i,j} x_j^{(t)} + \sum_j W_{i,j} h_j^{(t-1)}) \tag{3}$$

Among them, b, U and W are the offset weights of the LSTM cells, the input weights and the forgetting gates. Finally, the output of the cell unit is calculated by the output gate $q_i^{(t)}$ and the state unit $s_i^{(t)}$, which is calculated as follows,

$$q_i^{(t)} = sigmoid(b_i^o + \sum_j U_{i,j}^o x_j^{(t)} + \sum_j W_{i,j}^o h_j^{(t-1)}) \tag{4}$$

$$h_i^{(t)} = \tanh(s_i^{(t)}) q_i^{(t)} \tag{5}$$

Where b^o, U^o and W^o are the cyclic weights of the offset, input weight and forgetting gate, respectively.

The constructed Res-RNN network extracts and encodes the video input frame into a one-dimensional feature vector through the CNN of the residual structure, and the LSTM receives the sequence one-dimensional feature vector from the Res encoder, and outputs the sequence vector containing the video depth feature for subsequent Classification forecast.

3.2 3D Convolutional Neural Network

Network Structure. In the construction of 3D convolutional neural network, the reference I3D [7] network structure is selected, which can capture the time structure of fine action and perform better model features. The network architecture diagram is as follows (Fig. 3),

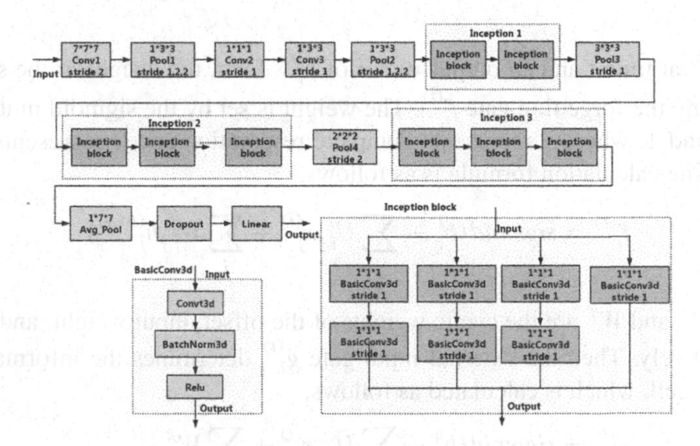

Fig. 3. 3D convolution network architecture

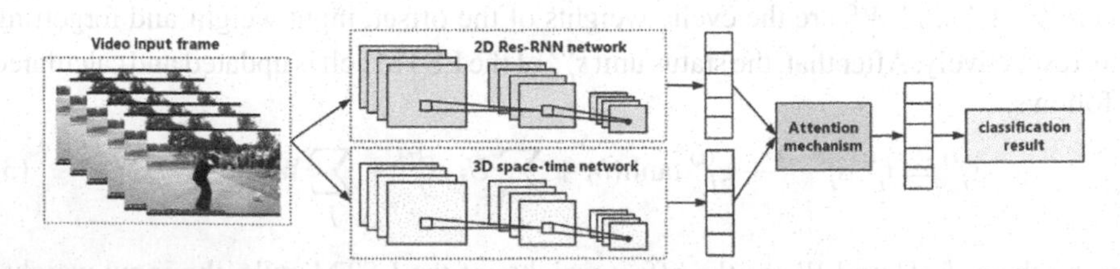

Fig. 4. Mixed convolution network behavior recognition model

3.3 Hybrid Convolutional Network

Action Recognition Model. In order to improve the accuracy of behavior recognition, the attention mechanism [11, 25] is used to fuse the features extracted by the 2D convolutional neural network and the 3D convolutional neural network. The hybrid convolutional network will learn the sequence of features with time information and convolve the input video frames layer by layer. Based on the advantages of 2D convolution and 3D convolution, the attention mechanism is used to add weights to the behavioral features extracted by the model. And the feature vectors with significant attention weights are output for the classification of video behavior. The video behavior recognition model of the 2D/3D hybrid convolutional network based on attention mechanism proposed in this paper is shown in the figure below (Fig. 4).

First, the video data is segmented into video frames using the ffmpeg tool. Using the obtained image frame data as input data of the constructed hybrid convolutional network model, feature extraction is performed separately by layer-by-layer convolution to generate two high-level semantic feature sequences. After that, the high-level semantic feature sequences obtained by convolving 2D and 3D are input into the attention mechanism. By calculating the significant attention, the corresponding weights are added to the input feature sequence, and the output obtains a new feature sequence with attention weights for the recognition of the video behavior. The attention mechanism is applied in the network, and the video frame feature vectors obtained by 2D and 3D convolution are weighted, which can enhance the saliency of the feature sequence, continuously train learning, reduce losses, and improve the accuracy of behavior recognition.

Attention Mechanism. The human brain can intentionally or unintentionally select a small amount of useful information from a large amount of input information to focus on and ignore other information. Inspired by this, the attention mechanism is introduced to select important information for the neural network to calculate. Assume that N information, $I = [x_1, \ldots\ldots, x_n]$, and question q (the object of interest, which may be itself) are input, and some task-related information is selected from input I and input to the neural network for learning. First, calculate the attention distribution α_i, that is, the probability of selecting the $i{-}th$ information, and the calculation formula is as follows.

$$\alpha_i = \mathrm{softmax}(score(x_i, q)) = \frac{\exp(score(x_i, q))}{\sum_{j=1}^{N} \exp(score(x_j, q))} \tag{6}$$

Where $score(x_i, q)$ is a scoring function, here we define the $score$ function as,

$$score(x_i) = sigmoid(W^T x_i + b) \tag{7}$$

On the basis of calculating the attention distribution, the soft attention mechanism is used to calculate the attention, and all the information is weighted and summed, and the input information is encoded as,

$$attn(I, q) = \sum_{i=1}^{N} \alpha_i x_i \tag{8}$$

Attention is essentially to integrate the information of the feature vector, reduce the computational complexity, obtain significant features, and improve the feature representation ability of the network model.

4 Experiment

4.1 Experiment Environment

The experimental platform is Dell server PowerEdge R430, operating system: Ubuntu 14.04, CPU: Intel (R) Core i3 3220, memory: 64 GB, GPU: NVIDIA Tesla K40 m × 2, video memory: 12 GB × 2.

4.2 Experimental Parameter Setting

Data Set. The data sets selected for the experiment were the UCF101 data set [12] and the HMDB51 data set [23]. The UCF101 data set has a total of 13320 video segments, and the number of categories is 101. It has the greatest diversity in motion. Some of its behavioral categories are shown on the left in Fig. 5. The HMDB51 dataset contains 6849 clips and is divided into 51 action categories. Each category contains at least 101 clips, and some action categories are shown on the right in Fig. 5.

Fig. 5. UCF101 data set (left) and HMDB51 data set (right)

Experimental Parameters. The behavioral feature extraction process uses the Pytorch framework, and the model used is the 2D/3D hybrid convolutional network built above. In the training, the batch-size is set to 4, 8, 16, and 28 frames respectively for experimental comparison. At the same time, experiments are carried out on the input order of different frames to see if it affects the accuracy of behavior recognition. The BatchNormalization layer is set in the network to perform batch normalization to solve the problem of changing the distribution of data in the middle layer during the training process. In addition, consider the model without adding the BatchNormalization layer, and compare experiments with other networks to observe the performance of the built 2D/3D hybrid convolution network. The optimization method selects the Adam method and the learning rate is set to 0.001.

4.3 Experimental Result

In order to verify the effectiveness of the 2D/3D hybrid convolutional network that introduces the attention mechanism, the experimental comparison with the basic network I3D is carried out from the number of input frames, the order of input frames and whether or not the BN layer is included. The 2D/3D hybrid convolution network we built can take advantage of the network under different convolution kernels. End-to-end training is performed by taking a well-framed RGB video sequence as input. In order to verify the performance of the hybrid network, we used the UCF101 public data set for behavior recognition to test. The results are shown in Table 1 below.

Table 1. Comparison of experimental results on the UCF101 data set

Method	Dim	BN	Input	UCF101				Speed
				4f (%)	8f (%)	16f (%)	28f (%)	Average (fps)
I3D [7]	3d	Y	Order	72.46	71.35	75.11	79.39	8.00
		N	Order	46.57	37.23	42.58	58.79	7.07
		Y	Random	74.23	70.63	74.74	73.15	6.88
		N	Random	44.26	36.93	48.34	50.48	6.95
Our	3d/2d	Y	Order	**83.51**	**82.88**	**80.72**	**80.69**	**19.18**
		N	Order	73.42	72.31	72.67	71.50	22.82
		Y	Random	82.37	80.78	78.86	77.21	20.73
		N	Random	73.12	74.71	70.48	66.09	22.75

It can be seen from the results in the table that the feature extraction by inputting 4, 8, 16, 28 different frames to the convolutional neural network has a certain influence on the video behavior recognition result. It can be seen from the data in the table that for the I3D network model, the influence of the input of different frame data on the accuracy of behavior recognition is relatively large. For the proposed hybrid convolutional network, the input of different frame data has less influence on the accuracy of behavior recognition.

The partial data in the table is compared and displayed in a visual form. The result is shown in Fig. 6 and Fig. 7 below.

For the order of input frames is ordered or unordered, we can see from Fig. 6 that when the number of network input frames is large, the sequential features extracted by ordered input can improve the accuracy of behavior recognition. Secondly, as can be seen from Fig. 7, when the I3D network model removes the BN layer, the network performance is seriously degraded, and the proposed hybrid convolutional network is relatively robust. And the network model processes data at a fast speed, so that it can be seen that the hybrid convolution network model with attention mechanism has better performance advantages.

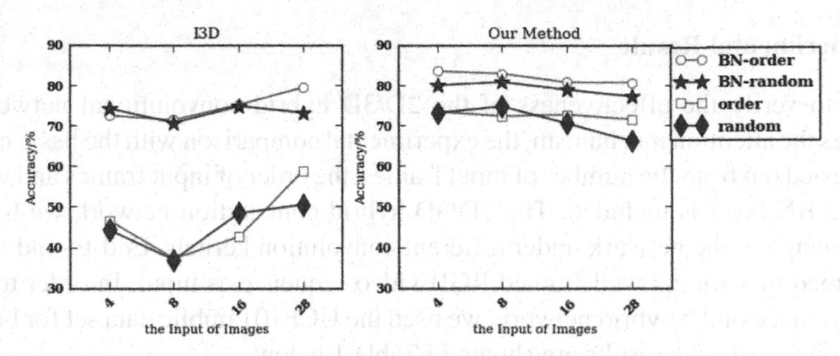

Fig. 6. Comparison of ordered or random input results

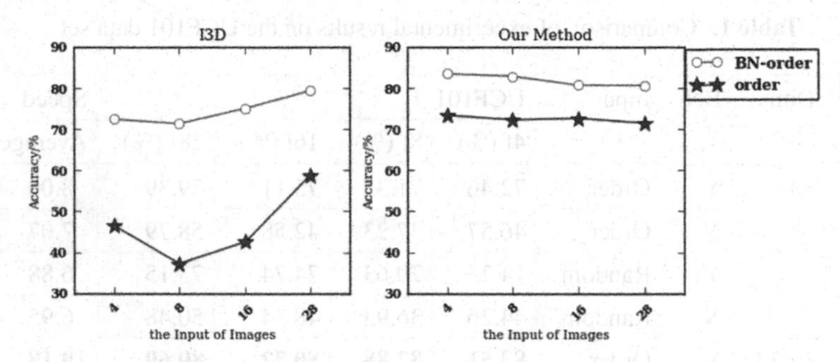

Fig. 7. Comparison of results of adding BN layers

To further understand the performance of the hybrid convolutional network model that introduces the attention mechanism, we further compare the performance of the UCF101 dataset and the HMDB51 dataset with other network models. The results are shown in Table 2.

Table 2. Comparison results with other network models on UCF101 and HMDB51

Method	Dim	UCF101	HMDB51
Two-Stream [1]	2D	78.85%	——
CRNN	2D	67.63%	34.04%
I3D [7]	3D	79.39%	28.48%
3D-ResNet [13]	3D	67.60%	25.67%
Our	3D/2D	**83.51%**	**45.71%**

5 Conclusion

A 2D/3D hybrid convolutional network with attention mechanism is introduced to extract the spatio-temporal features of video behavior through 2D/3D parallel convolutional networks, and generate two different high-level semantic information. Then, the attentional mechanism is used to apply different weights to the features to obtain a distinctive feature vector for the behavior recognition of the video. The experimental results show that the proposed methods have an accuracy of 83.51% and 45.71% on the UCF101 and HMDB101 data sets respectively. Compared with other network models, they have better recognition performance and robustness, which proves the effectiveness of the method.

Acknowledgments. This research was funded by Beijing Natural Science Foundation-Haidian Primitive Innovation Joint Fund grant number (L182007) and the National Natural Science Foundation of China grant number (61702020).

References

1. Simonyan, K., Zisserman, A.: Two-stream convolutional networks for action recognition in videos. In: Computer Vision and Pattern Recognition (2014)
2. Wang, L., et al.: Temporal segment networks: towards good practices for deep action recognition. In: Leibe, B., Matas, J., Sebe, N., Welling, M. (eds.) ECCV 2016. LNCS, vol. 9912, pp. 20–36. Springer, Cham (2016). https://doi.org/10.1007/978-3-319-46484-8_2
3. Zhou, B., Andonian, A., Oliva, A., Torralba, A.: Temporal relational reasoning in videos. In: Ferrari, V., Hebert, M., Sminchisescu, C., Weiss, Y. (eds.) ECCV 2018. LNCS, vol. 11205, pp. 831–846. Springer, Cham (2018). https://doi.org/10.1007/978-3-030-01246-5_49
4. Zhu, Y., Lan, Z., Newsam, S., Hauptmann, A.: Hidden two-stream convolutional networks for action recognition. In: Jawahar, C.V., Li, H., Mori, G., Schindler, K. (eds.) ACCV 2018. LNCS, vol. 11363, pp. 363–378. Springer, Cham (2019). https://doi.org/10.1007/978-3-030-20893-6_23
5. Tran, D., Bourdev, L., Fergus, R., et al.: Learning spatiotemporal features with 3D convolutional networks. In: IEEE International Conference on Computer Vision, Santiago, pp. 4489–4497. IEEE Computer Society (2015)
6. Diba, A., Fayyaz, M., Sharma, V., et al.: Temporal 3D ConvNets: New Architecture and Transfer Learning for Video Classification. Computing Research Repository (2017)
7. Carreira, J., Zisserman, A.: Quo Vadis, action recognition? A new model and the kinetics dataset. In: IEEE Conference on Computer Vision and Pattern Recognition, Honolulu, pp. 4724–4733. IEEE Computer Society (2017)
8. Ji, S., Xu, W., Yang, M., Yu, K.: 3D convolutional neural networks for human action recognition. IEEE Trans. Pattern Anal. Mach. Intell. **35**(1), 221–231 (2013)
9. He, K., Zhang, X., Ren, S., et al.: Deep residual learning for image recognition. CoRR abs/1512.03385 (2015)
10. Hochreiter, S., Schmidhuber, J.: Long short-term memory. Neural Comput. **9**(8), 1735–1780 (1997)
11. Vaswani, A., et al.: Attention is all you need. CoRR abs/1706.03762 (2017)
12. Soomro, K., Zamir, A.R., Shah, M.: UCF101: A Dataset of 101 Human Actions Classes From Videos in The Wild. CoRR abs/1212.0402 (2012)

13. Hara, K., Kataoka, H., Satoh, Y.: Can spatiotemporal 3D CNNs retrace the history of 2D CNNs and ImageNet? In: IEEE Conference on Computer Vision and Pattern Recognition, Salt Lake City, pp. 6546–6555. IEEE Computer Society (2018)
14. Tran, D., Wang, H., Torresani, L., et al.: A closer look at spatiotemporal convolutions for action recognition. In: IEEE Conference on Computer Vision and Pattern Recognition, Salt Lake City, pp. 6450–6459. IEEE Computer Society (2018)
15. Liu, Z., Hu, H., Zhang, J.: Spatiotemporal fusion networks for video action recognition. Neural Process. Lett. **50**(2), 1877–1890 (2019). https://doi.org/10.1007/s11063-018-09972-6
16. Li, Q., Qiu, Z., Yao, T., et al.: Action recognition by learning deep multi-granular spatio-temporal video representation. In: International Conference on Multimedia Retrieval, pp. 159–166. ACM, New York (2016)
17. Zhou, Y., Sun, X., Zha, Z.-J., Zeng, W.: MiCT: mixed 3D/2D convolutional tube for human action recognition. In: IEEE Conference on Computer Vision and Pattern Recognition, Salt Lake City, pp. 449–458. IEEE Computer Society (2018)
18. Yu, W., Wei, Y., Li, L.: Human behavior recognition model based on multi-model fusion. Comput. Eng. Des. **40**(10), 3030–3036 (2019)
19. Zeng, M., Zheng, Z., Luo, S.: Dual-convolution human behavior recognition combined with LSTM. Modern Electron. Tech. **42**(19), 37–40 (2019)
20. Ma, C., Mao, Z., Cui, J., Yi, W.: Behavior recognition based on deep LSTM and dual stream convergence network. Comput. Eng. Des. **40**(09), 2631–2637 (2019)
21. Ma, L., Yu, W., Zhu, Y., Wang, C., Wang, P.: Recognition of fall behavior based on deep learning. Comput. Sci. **46**(09), 106–112 (2019)
22. Rodríguez-Moreno, I., Martínez-Otzeta, J.M., Sierra, B., Rodriguez, I., Jauregi, E.: Video activity recognition: state-of-the-art. Sensors (Basel, Switzerland) **19**(14), 3160 (2019)
23. Wishart, D.S., Tzur, D., et al.: HMDB: the human metabolome database. Nucleic Acids Res. **35**, 521–526 (2007)
24. Shah, S.M.S., Malik, T.A., Khatoon, R., Hassan, S.S., Shah, F.A.: Human behavior classification using geometrical features of skeleton and support vector machines. Comput. Mater. Continua **61**(2), 535–553 (2019)
25. Yang, K., Wang, Y., Zhang, W., Yao, J., Le, Y.: Keyphrase generation based on self-attention mechanism. Comput. Mater. Continua **61**(2), 569–581 (2019)
26. Song, W., Yu, J., Zhao, X., Wang, A.: Research on action recognition and content analysis in videos based on DNN and MLN. Comput. Mater. Continua **61**(3), 1189–1204 (2019)

Identification of Botany Terminology Based on Bert-BLSTM-CRF

Aziguli Wulamu[1,2], Ning Chen[3], Lijia Yang[1,2], Li Wang[1,4(✉)], and Jiaxing Shi[1,2]

[1] School of Computer and Communication Engineering, University of Science and Technology, Beijing, Beijing 100083, China
wl3927@126.com

[2] Beijing Key Laboratory of Knowledge Engineering for Materials Science, Beijing 100083, China

[3] XINJIANG Research Institute of Building Sciences (Co. Ltd.), Beijing 100083, China

[4] School of Automation and Electrical Engineering, University of Science and Technology, Beijing, Beijing 100083, China

Abstract. Based on Baidu encyclopedia and CNKI botanical classification standards, we classify botanical terms into five categories in this paper: plant morphology and classification, plant ecology and geography, phytochemistry, plant heredity, plant technology and methods. In view of the difficulties in terms identification, we use a deep learning entity recognition model based on Bert-BLSTM-CRF to identify terms from large-scale corpus. In this paper, the Bert-BLSTM-CRF model is used to identify the professional terms in the field of botany, with an accuracy rate of 89.58%, a recall rate of 87.92% and an F1 value of 89.25%, indicating that the model could be effectively applied to the task of identifying the professional terms in botany. Based on the existing corpus, a comparative experiment is carried out in this paper, and the experimental results show that the model improves the recognition effect of professional terms in the field of botany.

Keywords: Deep learning · Term recognition · Bert

1 Introduction

Terminology is an important part of scientific and technological resources and conveys rich and representative domain knowledge [1]. The modern Chinese dictionary interprets professional terms as specialized terms of specialized subjects. The terms can correctly mark things, phenomena, relations and processes in the professional field and have the systematic and professional characteristics [2]. Through the study of professional terminology, people can grasp the essence of technology in a professional field. Through the understanding of the professional terminology, people can find out the research field and research direction of the literature. International academic exchange also depends on the unification and standardization of professional terms. Therefore, each country attaches great importance to the management of professional terms, set up a professional body, introduced the corresponding regulations. There are several difficulties in the

X. Sun et al. (Eds.): ICAIS 2020, LNCS 12240, pp. 463–474, 2020.
https://doi.org/10.1007/978-3-030-57881-7_41

identification of the entity. First, the knowledge system in the field of botany is complex, the number of professional terms is large, and the composition is complex, so it is impossible to define appropriate rules for named entity recognition. Second, most of the professional terms belong to unregistered words or phrases, such as "棒曲霉毒素", "拐芹", "抗褐变" and so on, so it is difficult to use Chinese word segmentation tools to segment words in the text. Third, the entity boundary is fuzzy, which is not suitable for artificial feature extraction, such as "多药耐药", "蜡梅碱" and so on. Fourthly, there is a mixture of Chinese and English, such as "NUC 基因".

According to the above problem, we use the Bert-BLSTM-CRF entity recognition model that combines pre-trained Bert language model, bidirectional Long Short-Term Memory network (BLSTM) and conditional random field (CRF) to identify the terminology entities in the field of botany.

2 Model Structure

The model is divided into Bert layer, depth feature extraction layer and output layer. Bert layer consists of Bert model. The Bert model adopts multi-layer two-way Transformer encoder structure to solve the problem of one-way information flow, enhance the semantic representation of words, and extract the features of sentences. The Bert model is pre-trained in massive corpus, which can improve the problem of insufficient corpus. The depth feature extraction layer is used to extract the features of sentences in depth. It is composed of LSTM in positive and negative directions. BLSTM is able to take advantage of state values in both directions and perform well on sequential processing tasks. The output layer is used for sequence annotation of sentences, which is composed of CRF model. CRF model can focus on the correlation between tags, realize the joint modeling of long sentences, and output the optimal annotation results (Fig. 1).

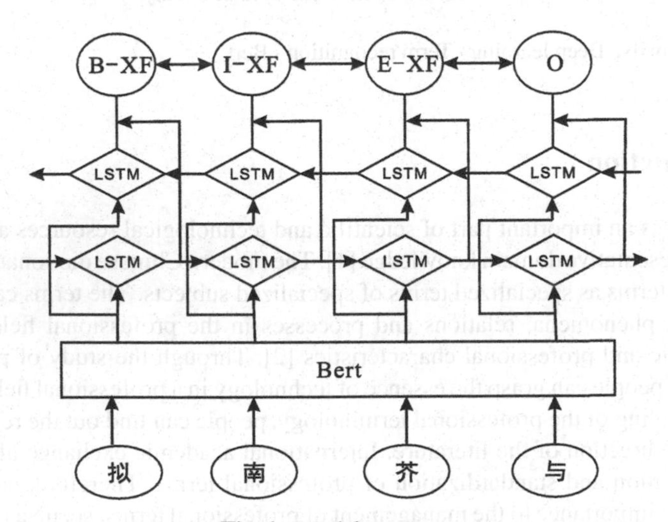

Fig. 1. Model structure.

2.1 Bert Layer

Bert model is a pre-training language model, which is trained by Google on a large amount of corpus [3]. This model is based on the multi-layer and bidirectional Transformer encoding structure. The input of Bert model can be a single sentence or a pair of sentences, and the summation of word embedding, sentence embedding and position embedding is used as the input representation of Bert model. This input representation can take full advantage of the information on the left and right sides to enhance the semantic representation of the word. The model uses the training method of masking the language model and predicting the next sentence to learn the context representation deeply and solve the one-way problem of traditional language model [4]. Bert model can be used in two ways: feature-based approach and fine-tuning-based approach. The former refers to fix the parameters of the Bert model, and the representation generated by the Bert model is directly applied to the model of down-stream tasks as an additional feature. The latter refers to fine-tuning the parameters of the Bert model according to a specific downstream task, and jointly training all parameters of the Bert model and the model of the specific downstream task.

The Masked Language Model. The masking language model borrows from the cloze idea, which is to randomly mask a certain proportion of words and let the model predict those words according to the context. Although this approach can obtain a pre-training language model of the learning context, it still has a disadvantage that the covered words never appear during the fine-tuning, resulting in a mismatch between the pre-training and fine-tuning. In response to this shortcoming, the Bert model randomly selected 15% of the words in a training sequence and performed the following steps for this.

Replace the selected word with [MASK] 80% of the time.
Replace the selected word with a random word with a 10% chance.
Keep the selected word unchanged with a 10% chance.

How to Predict the Next Sentence. In order to make the model understand the relationship between sentences, this training method can be regarded as a dichotomy task, that is, given a sentence A, choose the next sentence of A with 50% probability from the corpus, select any sentence in the corpus with 50% probability, and then ask the model to predict whether the sentence is the next sentence of A. This training method is very practical for intelligent question answering and reasoning tasks.

2.2 LSTM Layer

Compared with ordinary networks, the cyclic neural network (RNN) can use historical information to judge the current output. However, as the sequence grows longer, the gradient disappears in the calculation process of the cyclic neural network, which results in the failure of updating the model parameters. Hochreiter et al. proposed a short and long term memory neural network (LSTM) in 1997 to solve the gradient disappearance problem of RNN [5]. Since it was proposed, it has been widely used in many tasks, such as network security [6], anomaly detection [7], Hashtag Recommendation [8] and so on.

LSTM consists of input gate x_t, output gate o_t, forgetting gate f_t, and storage cell c_t. In the process of learning and training, LSTM adjusts the parameters in the door through the combination of several doors, and then adjusts the degree of information attenuation, update and retention. The gate structure consists of a Sigmoid activation function and a paired multiplication operation that outputs a number between 0 and 1. 0 indicates that the door is closed and information is forbidden to pass. 1 means the door is open to allow information to pass. The structure flow of LSTM is as follows (Fig. 2):

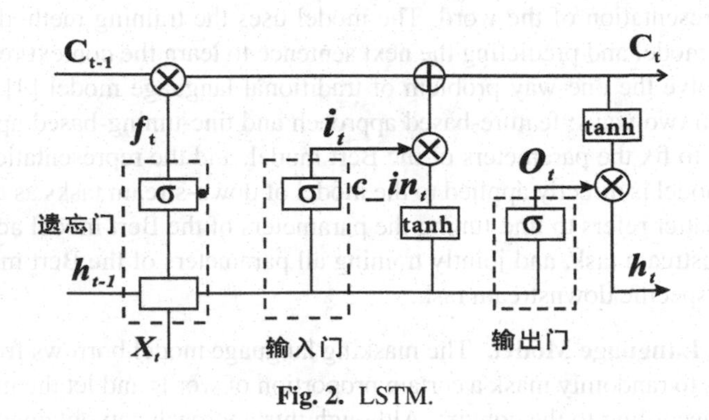

Fig. 2. LSTM.

The forgetting formula is as follows:

$$f_t = \sigma(W_f[h_{t-1}, x_t] + b_f) \quad (1)$$

The forgetting gate f_t determines what information to filter from the cell state based on the current input x_t and the previous output h_{t-1}. The formula $[h_{t-1}, x_t]$ represents the splicing of the previous output h_{t-1} and the current input vector of x_t.

The formula of internal update is as follows:

$$i_t = \sigma(W_i[h_{t-1}, x_t] + b_i) \quad (2)$$

$$c\_in_t = \tanh(W_c[h_{t-1}, x_t] + b_c) \quad (3)$$

$$c_t = f_t * c_{t-1} + i_t * c\_in_t \quad (4)$$

The input gate is i_t, which measures the importance of the current input through sigma $\sigma(\cdot)$ and decides what new information to store in the cell state. c_in_t is the candidate new information which is generated through a tanh layer. In the process of internal update, the two parts are combined to update the cell state c_t.

The output formula is as follows:

$$o_t = \sigma(W_o[h_{t-1}, x_t] + b_o) \quad (5)$$

$$h_t = o_t * \tanh(c_t) \quad (6)$$

The output gate o_t determines the output degree of the cell state of the model, and multiplies the current cell state which is processed by the tanh layer to obtain the output h_t of the hidden layer. LSTM only focuses on the previous information of the sequence and ignores the role of the latter information, so scholars proposed the bidirectional long and short term neural network model (BLSTM). The structure is capable of connecting the output of the previous and the following two directions as the current moment to improve the performance of the overall model [9].

2.3 CRF Layer

Through the feature extraction of the previous layer, the depth implicit feature representation of the text is obtained. Although LSTM can consider the long-term context information, the output results are independent of each other. If annotated directly through a softmax layer or maximum likelihood function, the relevance between tags is not concerned. In this layer, CRF layer is used to obtain binding rules between tags from the training data, to sequence annotate sentences, reduce the probability of occurrence of unreasonable sequences, and obtain the global optimal annotation result.

For a sequence $X = (x_1, x_2, x_3, \ldots, x_n)$, X is input to the model for training, then a corresponding tag sequence definition $Y = (y_1, y_2, y_3, \ldots, y_n)$ is output, CRF layer defines the total score of the prediction tag sequence:

$$score(X, Y) = \sum_{i=1}^{n} P_{i,y_i} + \sum_{i=0}^{n} A_{y_i,y_{i+1}} \tag{7}$$

$A_{y_i,y_{i+1}}$ is the score of transferring from y_i to y_{i+1}, and its size is $(k+2)*(k+2)$, P_{i,y_i} is the probability of outputting y_i at the ith position of BLSTM. The score is related to the output of LSTM and the output of CRF, so the score of each position is determined by the P_i output by BiLSTM and the transfer matrix A of CRF, and the score of the whole sequence is equal to the sum of the scores of each position.

For each sequence $X = (x_1, x_2, x_3, \ldots, x_n)$, the scores $s(X, Y)$ of all possible label sequence Y are calculated by using the Viterbi algorithm, and normalized by adding a Softmax layer. The score of the final sequence Y is:

$$p(Y|X) = \frac{\exp(score(X, Y))}{\sum_{\overline{Y} \in Y_w} \exp(score(X, \overline{Y}))} \tag{8}$$

In the process of model training, in order to maximize the probability of outputting the real annotated sequence, the loss function is defined as the logarithm of the probability $p(Y|X)$ of the correct tag sequence, and the maximum likelihood estimation method is adopted to estimate the parameters.

$$\log(p(Y|X)) = score(X, Y) - \log(\sum_{\overline{Y} \in Y_w} \exp(score(X, \overline{Y}))) \tag{9}$$

The stochastic gradient descent method was introduced for optimization, and the optimal marker sequence was finally obtained as follows:

$$Y^* = \operatorname{argmax}_{\overline{Y} \in Y_w}(score(X, \overline{Y})) \tag{10}$$

3 Experimental Details

3.1 Feature Extraction

Based on the classification of botany field by Baidu academic and CNKI, we divide the specialized terms of botany into five categories in this paper: plant morphology and classification, plant ecology and geography, phytochemistry, plant physiology and heredity, plant technology and methods. Plant morphology and classification include the internal and external morphological structures of plants, such as cells, tissues, organs, systems, etc., as well as the species classification names and classification methods of plants, such as algae, ferns, gymnosperms, double nomenclature, flora, etc. Plant ecology and geography include the relationship between plants, such as individual ecology, population ecology and group ecology, as well as the relationship between plants and the environment, such as soil, sunlight, air and disasters. Phytochemistry includes the chemicals secreted by plants themselves, the chemicals needed for plant growth and the chemical reagents used in experiments. Plant physiology and heredity include plant growth and development, metabolism, energy transformation and other mechanisms such as photosynthesis, plant respiration, stress resistance, as well as plant heredity and variation rules such as cell cycle, gene expression, protein synthesis and so on. Plant techniques and methods include techniques, methods, equipment, measures, etc. used in production and research (Table 1).

Table 1. Mark rules.

Class	Single character term	Multi-character term		
	Tag	Start	Middle	End
XF	S-XF	B-XF	I-XF	E-XF
SD	S-SD	B-SD	I-SD	E-SD
HX	S-HX	B-HX	I-HX	E-HX
SY	S-SY	B-SY	I-SY	E-SY
JF	S-JF	B-JF	I-JF	E-JF
Non-physical markup	O	O	O	O

In this paper, BIOES annotation model is adopted to annotate corpus, where B represents the beginning character of professional term entity, I represents the middle part of professional term entity, E represents the end character of professional term entity, S represents single-character professional term entity, and O represents non-professional term entity. we use XF to represent plant morphology and classification, SD to represent plant ecology and geography, HX to represent plant chemistry, SY to represent plant physiology and heredity, and JF to represent plant technology and methods. During testing, the prediction of an entity is correct only if the boundary and type of the entity are completely correct.

3.2 Experimental Process

The experiment is based on 7,492 reports of scientific research projects and 28,647 abstracts of research results. Since the collected project report is in picture format, we use an OCR technology based on CNN+LSTM+CTC to convert the project report in picture format into TXT file format. The project summaries and works on a series of data preprocessing report text, including deleting items in English in the data, to get rid of the noise data in text such as statistics, statistical figure used to display the results of visual content, deleting the header, footer, redundant symbols, such as documentary reference number to have mutual reference part of the corpus, which needs to be screened and deleted. In this paper, we break the corpus into words.

To improve the efficiency of annotation, the article builds a dictionary of terms in the field of botany, based on the Botanical Thesaurus of Sogou Input Method and the Biology Dictionary, using the keywords in papers of Baidu Academic in the field of botany as the supplementary. And we adopt the matching method of combining dictionary and rules to mark the corpus automatically by programming. Finally, the marked corpus are filtered manually. In this paper, a total of 31916 samples are marked. 60% of which are used as the training set for iterative generation model, 20% are used as the verification set to select the optimal training model, and 20% are used as the test set to test the identification effect of the model (Table 2).

Table 2. Data partitioning.

Data size	Lable size	XF	JF	HX	SY	SD
Training size	19150	5896	4471	3528	2906	2349
Validation size	6311	1865	1486	1209	968	783
Test size	6455	1813	1463	1352	1017	810

3.3 Evaluation Indicators

In this paper, accuracy (P), recall (R) and F1 values were used as evaluation indicators.

$$P = \frac{The\ correct\ number\ of\ entities\ identified}{Number\ of\ entities\ identified} \times 100\% \tag{11}$$

$$R = \frac{The\ correct\ number\ of\ entities\ identified}{The\ number\ of\ entities\ that\ should\ be\ identified} \times 100\% \tag{12}$$

$$F1 = \frac{2PR}{P+R} \times 100\% \tag{13}$$

Accuracy (P) is the ratio of the correct number of recognized entities to the number of recognized entities, reflecting the ability of the model to correctly identify entities; Recall rate (R) is the ratio of the correct number of identified entities to the number of entities that should be identified, reflecting the ability of the model to identify entities. The accuracy and recall rate reflect the performance of the model with different emphases, and individual indexes cannot reflect the overall performance of the model. Therefore, F1 value is added in this paper as a comprehensive indicator to balance the effects of accuracy and recall rate.

4 Experimental Realization

4.1 Experimental Environment Settings

In this experiment, under the Linux operating system, Tensorflow and Keras deep learning framework were used to construct the Bert-BLSTM-CRF entity recognition model, and two gpus were used (Table 3).

Table 3. Experimental environment.

Operating system	CPU model	Memory	Graphics card	Development environment
Linux	Intel(R) Xeon(R) CPU E5-2683 v3 @ 2.00 GHz	128G	Tesla P4	Keras Tensorflow

4.2 Parameter Setting and Experimental Results

In the model, the version of the Bert model selected in this paper is Bert-Base, which has 12 heads of attention mechanism, 12 hidden layers and 768 output dimensions of hidden layers. It adopts cascade mode to connect with the next layer. In this paper, we adopt the fine-tuning-based training method to jointly fine-tune all parameters of the whole model including Bert.

In this paper, we set the following types of parameters. Vocab_file is the vocabulary of the Bert model. We use the vocab.txt vocabulary provided by the official default. Bert_config_file is the pre-trained configuration file information of the Bert model. We use the default bert_config. Json. max_seq_length is the maximum sequence length for segmenting the corpus in the field of botany, and it is set to 128. For Bert, data that does not reach this length will be padded. If the length of data is longer than this length, the data will be truncated, which causes loss of information. Hidden units number is the number of neurons in the Hidden layer of BLSTM, which is set as 100. The model is optimized by back propagation algorithm. In order to improve training efficiency, Adam optimizer is adopted in this paper to optimize training, and the initial Learning rate is set as 5e−5, and the number of iterations (num_train_epochs) of training data is set as 20. For each epoch, the model processed the whole data in batches during the training process. In this paper, the batch_size of each batch is set to 64. In addition, a gradient threshold (Clip gradients) is introduced and set to 5 to solve the gradient explosion problem. In order to prevent over-fitting in the training process, we introduce the regularization mechanism (dropout) and set the Dropout_rate to 0.5, so as to improve the generalization ability of the model. In addition, we also set three optional parameters of Bert model, which all default to false and are set to true. They are do_train which decides whether to do fine tuning, do_eval which decides whether to run a validation set and do_predict which decides whether to predict (Table 4).

Table 4. Parameter settings.

Parameter	Value
max_seq_length	128
Hidden units number	100
Clip gradients	5
Dropout_rate	0.5
learning_rate	5e−5
train_batch_size	64
num_train_epochs	20
do_train	True
do_predict	True
do_eval	True

According to the experimental results and annotated corpus, the effect is affected by the amount of annotation and the degree of entity distinction. Plant morphology and classification is usually the object of research in scientific research projects or literatures. It appears most frequently throughout the whole paper and usually has the name feature words about organs and species, so the number of labels is the largest and the identification effect is the best. Plant technology and methods also appear frequently in the text, with a large number of them, often as verb phrases, with a large amount of annotation and a high entity classification. The number of plant ecology and geographical labeling is relatively small, but the entity differentiation degree is relatively high, the phenomenon of Chinese-English mixing is rarely seen, and symbols are rarely included. Therefore, plant ecology and geography, plant technology and methods have relatively ideal recognition effect. The plant physiological and genetic entities and phytochemical entities are often labeled in Chinese and English, such as “bHLH 转录因子”, and there are nesting and conflicting phenomena, so the boundary is fuzzy, and the recognition effect is poor (Table 5).

4.3 Verification and Comparison

Effect Comparison of Different Training Methods. In this paper, the Bert-BLSTM-CRF model is adopted to identify botanical term entities from large-scale corpus, and

Table 5. Terminology entity recognition results.

Species	Precision (%)	Recall (%)	F1 (%)
XF	92.85	90.13	91.46
JF	91.72	87.05	89.32
HX	86.94	87.76	87.35
SY	87.36	87.17	87.76
SD	90.18	89.57	89.87
The mean model	89.58	87.92	89.25

relevant experiments were conducted to verify the performance of the model in the task of botanical term recognition.

The use of Bert model is divided into the fine-tuning-based method and the feature-based method. Therefore, the training method of the Bert-BLSTM-CRF entity recognition model is also divided into two types. One is to fix the parameters of Bert model and only train the BLSTM-CRF model. The other is to train the entire Bert-BLSTM-CRF model. In this paper, we make a comparative experiment on the training process and results of the fine-tuning-based training method and the feature-based training method to verify the effectiveness of different training method (Fig. 3).

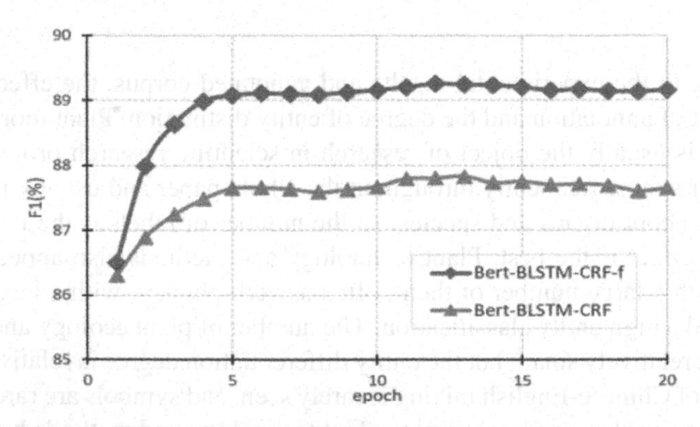

Fig. 3. Results of different training methods.

In the experiment, Bert-BLSTM-CRF represents the parameters of Bert model are fixed, and only the BLSTM-CRF model is trained. Bert-BLSTM-CRF-f represents that parameters of the entire model are trained. Both of the two training methods reach the maximum value when the 12th epoch is trained, the F1 value of Bert-BLSTM-CRF-f is 89.25%, and the F1 value of Bert-BLSTM-CRF is 87.81%. The experimental results show that for the entity recognition task in this paper, the model obtained by the fine-tuning-based training method is the best. This is because there are many model parameters, the model fits well. In addition, the total training time of the two training methods was statistically analyzed. A round of training with Bert-BLSTM-CRF took 537 s, and a round of training with Bert-BLSTM-CRF-f took 3168 s, which indicates that the fine-tuning-based training method needs more time, because the fine-tuning-based training method needs to train more parameters.

Effect Comparison of Different Models. In terms of the same corpus and entity classification, we compare the Bert-BLSTM-CRF model with the classic word2vec-BLSTM-CRF model, Bert-CRF, Lattice-LSTM-CRF and other models to verify the performance of the model in terminology entity recognition tasks in the field of botany.

The Word2Vec tool was used to generate word vectors as the input of BLSTM-CRF model in Experiment 1. In Experiment 2, a fine-tuning-based training method was used for Bert-CRF. In Experiment 3, the Bert model parameters were fixed. Word vectors were generated by the Bert and input to the BLSTM-CRF model for entity recognition. In Experiment 4, the fine-tuning-based training method was used to jointly fine-tune the parameters of the entire Bert-BLSTM-CRF model.

As shown in Table 6, the result of experiment 3 is better than that of experiment 1, indicating that the word vectors generated by Bert model can enhance the semantic representation of words. The result of experiment 4 is better than that of experiment 3, indicating that the model obtained by fine-tuning-based training is better than that obtained by feature-based training. The result of experiment 4 is better than that of experiment 2, which shows that BLSTM further extracts the sentence features on the basis of the Bert model, and also indicates that the fine-tuned Bert model can be combined with BLSTM to create a more advanced model. The result of experiment 2 is superior to that of experiment 3, which indicates that using the feature-based method to train the Bert-BLSTM-CRF model is inferior to the fine-tuning-based training approach of the Bert-CRF model, and further indicates that the feature extraction effect of the Bert model is superior to that of the BLSTM model in entity recognition tasks

Table 6. Effect of different models.

	Model	Precision	Recall	F1	Time
1	BLSTM-CRF	85.28%	84.63%	86.33%	528 s
2	Bert-CRF	88.75%	88.19%	88.47%	264 s
3	Bert-BLSTM-CRF	87.24%	88.39%	87.81%	537 s
4	Bert-BLSTM-CRF-f	89.58%	87.92%	89.25%	3168 s

For different models, we recorded the time required for each model to complete an epoch in the training process. It took the longest to train an epoch in experiment 4, indicating that although the Bert-BLSTM-CRF-f model based on the fine-tuning training method has better effect, its training efficiency is low. And because it would occupy a large amount of memory, it has higher requirements on hardware. It took less time to train an epoch in Experiment 2 than in Experiment 3, indicating that using the feature-based method to train the Bert-BLSTM-CRF model is not as good as Bert-CRF in training efficiency and results. In this paper, we conclude experience of using Bert model for entity recognition, that is when the hardware conditions are relatively ideal, the Bert-BLSTM-CRF model based on fine-tuning training method can achieve the better results, otherwise, the Bert-CRF model can improve the effect and training efficiency.

To sum up, we use the Bert-BLSTM-CRF entity récognition model and adopt the fine-tuning-based method for training in this paper, which performs well in botanical terms recognition.

5 Conclusion

In this paper, we firstly divide the specialized term entities in the field of botany into five categories: plant morphology and classification, plant ecology and geography, phytochemistry, plant heredity, plant technology and methods. Then we use the Bert-BLSTM-CRF model to identify the terminology entities from the large-scale corpus in the field of botany. We also conduct the comparative experiment about two kinds of training methods, which are the fine-tuning-based method and the feature-based method. The experiment proves that the model using the training way based on fine-tuning can obtain better entity recognition effect; Then we carry out comparative experiments on different entity recognition models, which proves that the Bert-BLSTM-CRF model has better performance in botanical terms recognition tasks.

References

1. Kageura, K., Umino, B.: Methods of automatic term recognition: A review. Terminology. Int. J. Theoret. Appl. Issues Special. Commun. **3**(2), 259–289 (1996)
2. Ananiadou, S.:A methodology for automatic term recognition. In: COLING 1994 Volume 2: The 15th International Conference on Computational Linguistics. (1994)
3. Devlin, J., Chang, M.W., Lee, K., et al.: BERT: pre-training of deep bidirectional transformers for language understanding (2018)
4. Vaswani, A., Shazeer, N., Parmar, N., et al.: Attention is all you need. In: Advances in Neural Information Processing Systems, pp. 5998–6008 (2017)
5. Sundermeyer, M., Ralf Schlüter, N.H.: LSTM Neural Networks for Language Modeling. Interspeech (2012)
6. Ding, L., Li, L., Han, J., Fan, Y., Donghui, H.: Detecting domain generation algorithms with bi-LSTM. Comput. Mater. Continua **61**(3), 1285–1303 (2019)
7. Zhu, H., et al.: Long short term memory networks based anomaly detection for KPIs. Comput. Mater. Continua **61**(2), 829–847 (2019)
8. Shen, Y., et al.: Hashtag recommendation using LSTM networks with self-attention. Comput. Mater. Continua **61**(3), 1261–1269 (2019)
9. Chen, T., Xu, R., He, Y., Wang, X.: Improving sentiment analysis via sentence type classification using BiLSTM-CRF and CNN. Expert Syst. Appl. **72**, 221–230 (2017)

Correlation Analysis of Chinese Pork Concept Stocks Based on Big Data

Yujiao Liu[1], Lin He[1](✉), Duohui Li[1], Xiaozhao Luo[1], Guo Peng[1,2], Xiaoping Fan[2], and Guang Sun[1]

[1] Hunan University of Finance and Economics, Changsha 410205, China
1781139005@qq.com
[2] University Malaysia Sabah, Kota Kinabalu, Malaysia

Abstract. This article conducts an empirical study on the correlation between Chinese pork price and the fluctuation of the pork concept index. The first is to use Tushare financial data interface, crawler tools and other technologies to obtain initial data, then use machine learning SVM sentiment analysis to convert text data into structured data to pre-process and standardize the data, which is beneficial to SPSS.24 for correlation and multiple linear regression analysis. Finally came to the following conclusions: Firstly, the Chinese pork price has a significant positive correlation at the level of 0.01 with the pork concept index, and upstream and midstream companies in the pork industry chain are more affected by changes in pork prices. Therefore, investors can focus on pork price changes to guide investment decisions. Secondly, from the long-term analysis, investor sentiment has little effect on the stock price of pork stocks. Thirdly, weak correlation between Chinese macroeconomic factors and Chinese pork stock price.

Keywords: Big data technology · Correlation analysis · SVM sentiment analysis · Pork sector index

1 Introduction

Recently, there has been a lot of news about the rise of Chinese pork prices. Pei Wu [1] used the ARIMA-GM-RBF combination model to predict Chinese pork prices, and concluded that Chinese pork prices in 2019 will continue the upward trend at the end of 2018, it continued to rise from 13.55 yuan/kg to 22.52 yuan/kg. And Xiuhong Cui [2, 3] also confirmed through actual data that Chinese pork prices have been running at a high level since March 2019. According to information from the Ministry of Agriculture and Rural Affairs of China, experts predict that pork prices may increase by more than 70% year-on-year in the second half of 2019, The China Agricultural Outlook Report (2019–2028) released on April 20, 2019 shows that the high price of pig prices in this round of price cycle is expected in the first half of 2020. At the same time, pork concept stocks also set off a new round of climax. Ke Liu [4] pointed out that since this year, the pork concept index has risen by 75%, even exceeding 70% increase in industrial cannabis concept, and the leading company New Wufeng has more than quadrupled. China's stock

X. Sun et al. (Eds.): ICAIS 2020, LNCS 12240, pp. 475–486, 2020.
https://doi.org/10.1007/978-3-030-57881-7_42

market data on July 17, 2019 directly shows that pork concept stocks are leading the rise. July 11–17, in the five trading days, a total of 333 transactions occurred in Shanghai and Shenzhen, with a transaction value of 4.398 billion yuan. From the perspective of the single transaction amount, the largest amount is Muyuan, with a transaction value of 281 million yuan and a premium of 0.16% [5]. Soochow Securities Research report believes that with the recent rise in pork prices, pork concept stocks have significant returns. In particular, the research report pointed out that back to the current round of the pig cycle, combining inventory, pork prices and pork concept stock performance, May 2018 to April 2019 is divided into the first stage, with the impact of classical swine fever, the excess income at this stage reached 63%; beginning in April 2019, with the rise in pork prices, pork concept stocks started and entered the second stage. Excess returns have reached 6%. Based on the expected rise in pork prices in the future, we are focusing on investment opportunities for pork concept stocks at this stage. This shows that rising pork prices have a significant impact on pork concept stocks. Therefore, the author uses big data technology to analyze the correlation between pork price and the price pork concept stocks.

By combing Chinese and foreign literature on the pork concept stock, we found there are many literatures on the fluctuation of pork stock price. Shuangni Zhang, Shuanglan Zhang [6] found that the CPI index is highly consistent with the periodicity and trend of the pork concept index, and the CPI index has a positive transmission effect on the market of the price of the pork concept stock through VAR model. Zhihong Sun, Xinsheng Lu [7] used VAR model to verify that there is a statistically significant correlation and a long-term equilibrium relationship between the series of changes in the prices of animal products in China and the changes in stock prices of listed companies in animal husbandry in China, and concluded that the price changes of forage and animal products in China are one of the reasons that cause the stock price of listed companies in animal husbandry to fluctuate through Granger causality test. Shaoying Hong, Qinglong Zhang [8] used the ARMA-GARCH combination model to analyze and predict the volatility of the pork concept index. Ke Liu [4] believes that the rise in pork prices has little effect on the performance of listed agricultural companies, resulting in the history of each A-share fried pork cycle ending in a bleak end. It can be seen that scholars have relatively perfect research on the volatility and influencing factors of the pork concept index. But there are still the following deficiencies: The first is there is less data. The number of pork concept stocks selected is small, and the experimental data is only traditional statistical data, and does not include semi-structured and unstructured data; the second his empirical research that does not directly involve the correlation between pork prices and pork conceptual indexes, and the relevant investment recommendations are not formed.

With the popularity of Internet technology and the arrival of the era of big data, a large amount of structured and unstructured data need to be visually analyzed. Providing information support for improving the division of the industrial of pork concept stocks through big data mining technology and combined with the characteristics [9]. At the same time, the era of big data is also an era driven by user experience or consumer [10]. Using a crawler program to crawl a large amount of online comment data and preprocess it can find the data to reflect investor sentiment faster. Therefore, the author uses big data technology to conduct empirical research to make the experimental results more accurate and objective, and to enrich Chinese research on the price of pork concept stock changes.

2 Correlation Analysis

We study the correlation between the fluctuation of the pork price index and the pork concept stock price index, and in order to improve the goodness of fit of the regression equation, several variables are added for modeling. It means a correlation analysis is performed on the correlation between multiple variables and the price of the pork concept stock and the variables are filtered to generate a model for prediction. Therefore, the author mainly uses correlation coefficients and multiple linear regression methods.

2.1 Correlation Coefficient

In the correlation analysis, the variances Lxx, Lyy respectively reflect the degree of variation between the two variables X and Y. The calculation formulas are:

$$l_{xx} = \frac{\sum_{i=1}^{n} (x_i - \bar{x})^2}{n-1} \tag{1}$$

$$l_{yy} = \frac{\sum_{i=1}^{n} (y_i - \bar{y})^2}{n-1} \tag{2}$$

$$l_{xy} = \frac{\sum_{i=1}^{n} (x - \bar{x})(y - \bar{y})}{n-1} \tag{3}$$

$$r = \frac{l_{xy}}{\sqrt{l_{xx} l_{yy}}} \tag{4}$$

The value of the correlation coefficient ranges from 1 to −1. 1 means that the two variables are completely positively correlated, −1 means that the two variables are completely negatively correlated, 0 means the two variables are uncorrelated, the sign of r indicates the direction of the correlation.

The degree of linear correlation between variables can be divided into the following 4 cases according to experience (Table 1):

Table 1. The degree of linear correlation between variables

$\|r\|$	Relevance
When $\|r\| \geq 0.8$	High correlation
When $0.5 \leq \|r\| < 0.8$	Moderate correlation
When $0.3 \leq \|r\| < 0.5$	Low correlation
When $\|r\| < 0.3$	Extremely weak, can be considered irrelevant

2.2 Multiple Linear Regression

Linear regression model:

$$\begin{bmatrix} Y_1 \\ Y_2 \\ \dots \\ Y_n \end{bmatrix} = \begin{bmatrix} 1 & X_{11} & X_{21} & \dots & X_{n1} \\ 1 & X_{12} & X_{22} & \dots & X_{n2} \\ \dots & \dots & \dots & \dots & \dots \\ 1 & X_{1n} & X_{2n} & \dots & X_{kn} \end{bmatrix} \begin{bmatrix} b_0 \\ b_1 \\ b_2 \\ \dots \\ b_k \end{bmatrix} + \begin{bmatrix} u_0 \\ u_1 \\ u_2 \\ \dots \\ u_k \end{bmatrix} \tag{5}$$

Recorded as

$$\mathrm{Y = XB + U} \tag{6}$$

The key is to calculate the corresponding parameter matrix B. The least squares criterion is still used, the optimal parameter satisfies the following formula (7) Smallest.

$$\sum u_i^2 = \sum \left(Y_i - \left(\hat{b}_0 + \hat{b}_1 X_1 + \hat{b}_2 X_2 + \dots + \hat{b}_n X_n\right)\right)^2 \tag{7}$$

Similarly, according to the first-order condition, we can have the optimal parameters $\hat{B}$ to satisfy

$$\begin{cases} \dfrac{\partial \sum \left(Y_i - \left(\hat{b}_0 + \hat{b}_1 X_1 + \hat{b}_2 X_2 + \dots + \hat{b}_k X_k\right)\right)^2}{\partial b_1} = 0 \\ \dfrac{\partial \sum \left(Y_i - \left(\hat{b}_0 + \hat{b}_1 X_1 + \hat{b}_2 X_2 + \dots + \hat{b}_k X_k\right)\right)^2}{\partial b_2} = 0 \\ \vdots \\ \vdots \\ \dfrac{\partial \sum \left(Y_i - \left(\hat{b}_0 + \hat{b}_1 X_1 + \hat{b}_2 X_2 + \dots + \hat{b}_k X_k\right)\right)^2}{\partial b_k} = 0 \end{cases} \tag{8}$$

Can be obtained by the above formula

$$\hat{B} = \left(X'X\right)^{-1} X'Y \tag{9}$$

X' is the transpose of matrix X. Multiple linear regression equations can be solved through the above steps. However, because the dependent variable Y can use any independent variable X to calculate the corresponding optimal parameters, and can use any number of independent variables to calculate the optimal parameter matrix, it is necessary to determine the reliability of the model. It means the independent variable we choose must satisfy its interpretation of the dependent variable as high as possible, it also means the denser the points near the regression line in the drawn image [11]. Introducing the Multivariate Decision Coefficient R^2. The calculation formula is shown below (10):

$$R^2 = \frac{b_2 \sum y_t x_{2t} + b_3 \sum y_t x_{3t} + \dots + b_k \sum y_t x_{kt}}{\sum y_t^2} \tag{10}$$

When the goodness of fit R^2 approaches 1, it shows that the actual data are all around the edge of the regression line, that is, the model is more efficient.

3 Empirical Research

We selected the sample data selection, then obtained the data source through the Tushare API; used the existing positive emotional corpus and negative emotional corpus and machine learning technology SVM tools to train the sentiment analysis model, transform the text data into numerical data, and preprocessed other data; filtered the data and used SPSS.24 for data standardization. Finally, the research variables are subjected to correlation analysis and multiple linear regression analysis, a regression model is established, the experimental results are analyzed, and investment recommendations are proposed.

3.1 Sample Selection and Data Source

This paper selects 18 pork concept stocks that have suffered significant fluctuation in pork prices in August 2019. By deleting companies that have not issued shares in previous years, 15 shares are retained. We classify the stocks according to company's the total number of links involved in the pork industry chain. The specific classification is shown in Table 2.

Table 2. Sample stock and classification

Category	Participate in the pork industry chain	Sample stock
Primary pork concept stock	Upstream	DabeiNong, Techbank, Haid, Luoniushan, Jingxinnong
Intermediate pork concept stock	Upstream and midstream or midstream and downstream	ShunxinAgriculture, Delisi, Muyuan, Yisheng
Senior pork concept stock	Integrated industry chain	Tangrenshen, Shineway, Tecon, NewWellful, New Hope, Zhengbang

Because the production and operation of animal husbandry is affected by natural factors and market factors, meat and meat products inevitably show instability in supply and demand in the market, and their prices are likely to fluctuate, and the risks caused by price fluctuations are directly related the problems of cost and income of animal husbandry enterprises, thereby affecting corporate stock price [7]. It can be seen that pork concept stocks not only have the characteristics of the stock market, but also have their own industry characteristics, while being restricted by the macroeconomic environment. We chose macro, meso and micro indicators to construct the indicator system. The first is the construction of a meso-indicator system. The incentive for the collective increase in pork concept stocks in 2019 was caused by the soaring pork prices caused by the decline in live pig stocks. We choose pork price and pig inventory as the main explanatory variables. The pork price data comes from China Animal Husbandry Information Network, and the live pig inventory data comes from the database of the Ministry of Agriculture and Rural Affairs. Then establish a micro-indicator system. The above investor sentiment indicators are objective manifestations of behavior under certain emotions, and investor sentiment has strong subjective characteristics, the text information of Weibo, Forums and Stock Bars more intuitively reflects investor sentiment [12–14]. Comprehensive consideration, the four indicators of investor sentiment score, individual stock turnover rate, individual stock transaction growth rate, and trading volume were selected as the secondary explanatory variables. The stock turnover rate data is derived from the turnover rate APP database. The original text data of investor sentiment is based on the network online platforms such as news sites, social media, and online message boards, text data is extracted through text mining. The growth rate of individual stock transactions and the trading volume are obtained through Tushare financial data interface program. Finally, a macro indicator system is established as a control variable to enhance the validity of the experimental results. Research by many scholars in China and abroad shows that the macroeconomic changes are systemic risks and affect the whole stock market. Among them, the consumer price index, the producer price index of industrial producers, and the exchange rate of RMB against the US dollar have significant effects on the stock prices [15, 16]. The CPP and PPI data are obtained through the Tushare financial data excuse program, and EX data comes from the China Foreign Exchange Trading Center website. All variables in this paper are shown in Table 3 below.

3.2 Data Processing

In the processing of price data of pork concept stocks, we use the closing price of the month before the suspension period to complete the data information. Then calculates the monthly average closing price of each stocks as the monthly average stock price, recorded as pps:

$$\text{pps} = \frac{\sum P_0}{n} \tag{11}$$

P_0 is the daily closing price of the stock, n is the sum of days of holding the number of shares. The calculation formula is as follows:

$$\text{PPS} = \frac{\sum PPS}{N} \tag{12}$$

The N is the number of samples in the group. In the analysis of investor sentiment, we used text analysis tools such as Web crawling tools, SVM, jieba. Web crawlers are written in the python programming language. The crawler tool first crawls the comment link on the dynamic page of users such as Sina Finance, and then uses the crawled link, collects all the messages and comment information, and finally stores the crawled text data into the csv file. After completing the collection of all the review information, we first use the existing positive emotional corpus and negative emotional corpus and machine learning technology SVM tools to train the sentiment analysis model, and then use jieba word segmentation technology to get a list of vocabulary for each comment, and finally use the emotion analysis model processes the list of words for each comment to get the emotional bias of each comment. The definition of emotional orientation in

Table 3. Variable definition table

Variable code	Variable (Metrics)	Description
PPS'	Primary concept stock price	The average number of monthly opening and closing prices for primary concept stocks
PPS''	Intermediate concept stock price	The average number of monthly opening and closing prices for intermediate concept stocks
PPS'''	Senior concept stock price	The average number of monthly opening and closing prices for intermediate concept stocks
PP	Pork price index	The average monthly price of pork bones in the Chinese market
ISS	Investor sentiment score	Sina Finance's sentiment analysis score (monthly average)
TR	Individual turnover rate	Ratio of the number of shares sold at the end of the month to the total number of shares at the end of the month
AGR	Individual stock turnover growth rateT	The ratio of the difference between the transaction amount at the end of this month and the transaction amount at the end of the month
VOL	Trading volume	The transaction volume at the end of the month reflects the relationship between stock supply and demand, reflecting the enthusiasm of investors and market activity
PBS	Live pig stocks	The quantity of live hogs in Chinese market
CPI	Consumer Price Index	(The value of a set of fixed goods at the current price divided by the value of a fixed set of goods at the base price) × 100%
PPI	Industrial producer ex-factory price index	The relative number of ex-factory prices and purchase prices of industrial production products changed during a certain period of time
EX	RMB against the US dollar exchange rate	Final exchange rate

the text is the number 5 indicates a positive emotion, the number 3 indicates a pertinent emotion, and the number 0 indicates a negative emotion, the larger the value, the more positive the emotion.

In the processing of the investor sentiment index, firstly, directly obtain the volume and transaction amount of 15 stocks through Tushare. Then use the Excel formula to calculate the growth rate of individual stocks by using the ratio of the difference between the transaction amount at the end of this month and the transaction amount at the end of the previous month to the transaction amount at the end of last month. Since the stocks lacked the data the end of 2013 and the growth rate of the transaction volume in January 2014 could not be obtained, the missing point value was replaced by the SPSS system.

Due to the difference scales of each dependent variable X, the number of original indicator variables greatly, which has a great impact on the comprehensive indicator Y. In order to eliminate the impact and reduce the quantitative difference between the original variables, we performed dimensionless processing [14] to standardize the obtained raw data and processed data.

3.3 Experiment Procedure

Correlation analysis was performed on the following outcome variables of the sample. The results are shown in Table 4:

Table 4. Primary, Intermediate, Advanced Correlation Table

Model	PP	PBS	CPI	PPI	EX	ISS	VOL	TR	AGR
PPS′	.425**	−.827**	.211	.107	.524**	−.034	.753**	.541**	.102
PPS″	.480**	−.905**	.293*	.073	.494**	−.078	.802**	.728**	−.043
PPS‴	.391**	−.826**	.365**	.077	.417**	−.005	.259*	.262*	−.248*

* Significant at the 0.05 level
**Significant at the 0.01 level

PP and the three types of PPS all have a low degree of positive correlation. In contrast, PP and PPS″ have the highest correlation. As the price of pork rises, the stock price of pork stocks increases. PBS and the three types of PPS have a very significant negative correlation and are highly correlated, and the correlation with PPS″ is as high as 90.5%. As the live pigs inventory increased, the stock price of pork stocks fell. From the analysis of the micro-indicator system, both TR and VOL have a moderate positive correlation with PPS′, PPS″, and have a very weak positive correlation with PPS‴. The higher the turnover rate and trading volume, the stronger the willingness of people to buy, and the greater the stock price of pork stocks. There is only a very weak negative correlation between AGR and PPS‴, and it can be considered that AGR is not related to the three types of PPS. ISS is not related to the three types of PPS.

We can see there is no correlation between CPI and PPS', CPI has a very weak positive correlation with PPS'', and has a low positive correlation with PPS'''. EX and three types of PPS have moderate positive correlation. PPI is not related to the three types of PPS.

In order to further measure the correlation between different types of PPS and explanatory variables, and to avoid the problem of collinearity, we perform stepwise multiple regression analysis. According to Tables 4, we get 5 factors related to the price of primary pork concept stock which are PP, PBS, EX, VOL', TR', and then put them into the model of price of primary pork concept stock. the determination coefficient R-square of the equation is 0.810, the goodness of fit was adjusted to 0.795, which is about 79.5% of the dependent variable changes can be explained by this model. The R values of the stock price of primary pork is close to 1, and the adjusted R-square is significantly larger than 0.25, indicating that the goodness of fitting of the model of the stock price of primary pork is good and a multivariate linear relationship is established. The adjusted R-square takes into account the interaction between the explanatory variables and is more rigorous than before the R-square, usually the R-square is greater than 0.25 which is considered to be good fit. In the same way, we can get 4 factors related to the price of intermediate pork concept stock price which are PP, PBS, VOL'', TR'', and then put them into the intermediate pork concept stock price model, the goodness of fit was adjusted to 0.933, the adjusted R-square of the intermediate pork stock price model is closest to 1, which means that the fitting effect is the best. In the same way, we can get 5 factors related to the price of advanced pork concept stock which are PP, PBS, CPI, TR''', and then put them into the advanced pork concept stock price model, the goodness of fit was adjusted to 0.795, it shows that multiple linear regression is valid. The results are shown in the following table (Table 5):

Table 5. Multiple regression analysis summary table

Model	Regression equation	F Value	R Square
Primary	PPS'=9.674E-16+0.201PP-0.503PBS+0.119EX+0.193VOL'+0.187TR'	53.748	.795
	1.000 .004 .000 .072 .034 .016	.000a	
Intermediate	PPS" =2.870E-15+0.234PP-0.609PBS+0.140VOL"+0.222TR"	236.524	.933
	1.000 .000 .000 .020 .000	.000b	
Advanced	PPS'"=-0.006+0.156PP-0.729PBS+0.126CPI+0.242TR'"	53.097	.795
	.909 .009 .000 .034 .000	.000c	

Comparing and analyzing the four curves in Fig. 1, we can see the pork price and the price of the three types of pork concept stock are highly consistent in form. And there is a strong correlation.

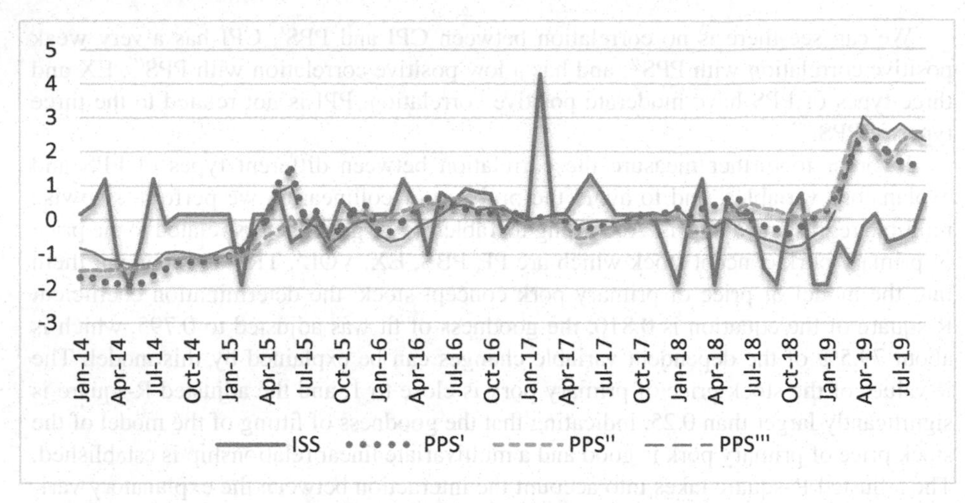

Fig. 1. Pork price and pork concept stock line chart

4 Conclusion

According to the time series chart of pork price index and pork concept index, it can be found that the strong cyclicality of pork, no matter its price or concept stock price, both has a significant cyclical characteristic, about 3 years a cycle. This is because there are many retail investors in China's stock market, and market investors are generally more sensitive than the actual market, which will amplify short-term fluctuations, investors' estimates of pork concept stocks through sudden outbreaks, national policies, and macroeconomics, these will make pork concept stock prices reflect market conditions faster than pork prices, therefore, the trend of pork concept stock index is consistent with pork price index. There is a high correlation between the hog inventory and the three types of pork concept stock prices, and there is a negative correlation. Due to factors such as African swine fever, the pig cycle, and the low sentiment of farmers, so the stock of pigs in China has decreased significantly, which has led to a rise in pork prices and a rise in the stock price of pork concept stocks. It can be seen that the quantity of live pig stocks is one of the reasons for the rise in pork prices, compared with pork prices, it can directly affect the stock price of pork concept stocks, so it contributes more to the price of pork concept stocks. To sum up, investors can pay attention to China's hog industry news, invest in pork concept stocks by comprehensively grasping the two indicators of pork price and hog inventory, combined with actual conditions.

Going to hold pork concept stocks, investor will not be too affected by TR and VOL. ISS indicator has no significant relationship with the model of the stock price of pork concept stock. On the on hand, it is limited to research breadth and technical depth. The research in this paper is only based on traditional machine learning models for text sentiment analysis, compared to other deep learning model that process data, such as convolutional neural networks, it is not deep enough and can not perform hierarchical extraction of text features. Thus failing to describe the hidden information of the input data. On the other hand, the sample data obtained is insufficient or not typical. It does not fully consider the discussion of pork prices on other social platforms and cannot fully reflect investors' expectations of the market [17]. In summary, investor sentiment has no effect on the model of pork concept stock.

Combining theory with practice, starting from various investor sentiment indicators, investors can pay attention to which type of pork stock price model, and then consider trading active pork concept stocks based on the TR and VOL indicators, and at the same time, there is an option to buy the stocks based on the discussion of the land stock forum when purchasing pork concept stocks. Combining theory with practice, starting from various investor sentiment indicators, investors can pay attention to which type of pork stock price model, and then consider trading active pork concept stocks based on the TR and VOL indicators, and at the same time, there is an option to buy the stocks based on the discussion of the land stock forum when purchasing pork concept stocks.

CPI has a positive effect on pork stock price growth. Commodity price changes have an important impact on the stock market. Specifically, when prices rise, stock prices rise; when prices fall, stock prices also fall. According to the China National Bureau of Statistics' interpretation of the CPI data for the first three quarters of 2019, pork prices rose 21.3% in the first three quarters, affecting about 0.49% points of the CPI increase, which accounted for about the total CPI increase 20%. It can be seen that the rise in pork prices has increased the increase in CPI growth. The CPI rises, the commodity market shows prosperity, when the commodity market rises and then declines, the best time to invest in stocks is formed, which causes the stock price to rise. At the same time, the rise in CPI will trigger structural adjustments in the stock market. The pork sector which are encouraged and protected by national policies will enter a new growth cycle, and market investors may amplify short-term sentiment and believe that pork prices will continue to rise. Based on the analysis of the company's fundamentals, the larger the scale and the more complete the industrial chain of the pork companies, the greater the investor's confidence in its market, and the more significantly it will be affected by CPI. Since PPI mainly focuses on the prices of industry, mining, raw materials, and semi-finished products, and has nothing to do with pork prices, so PPI has no significant impact on the price of pork concept stocks. The price of RMB against the US dollar has a moderate correlation to the Chines pork concept stock price. Due to the combined effects of African swine fever and the swine cycle, China's domestic pork supply is inadequate and it is necessary to increase a large amount of imported pork, and the rising exchange rate is beneficial to pork imports. Since China's pork supply is sufficient, it will have little impact on the profits of domestic pork companies, but the capital market is generally more sensitive to macroeconomic responses, investors may amplify short-term emotions and believe that pork prices continue to rise, and then, increase the holding of pork concept stocks, resulting pork concept stock price to price. From the perspective of pork supply, exchange rate changes have a greater impact on upstream companies in the pork industry, it means the exchange rate has a greater impact on stock price of primary concept pork than intermediate and advanced pork concept stock price.

Acknowledgments. This research work is implemented at the 2011 Collaborative Innovation Center for Development and Utilization of Finance and Economics Big Data Property, Universities of Hunan Province; Hunan Provincial Key Laboratory of Big Data Science and Technology, Finance and Economics; Key Laboratory of Information Technology and Security, Hunan Provincial Higher Education. This research is funded by the Open Foundation for the University

Innovation Platform in the Hunan Province, grant number 18K103; Open project, grant number 20181901CRP03, 20181901CRP04, 20181901CRP05; Hunan Provincial Education Science 13th Five-Year Plan (Grant No. XJK016BXX001), Social Science Foundation of Hunan Province (Grant No. 17YBA049).

References

1. Pei, W., Li, Z.: Prediction research on chinese pork price forecasting—analysis based on ARIMA-GM-RBF combination model. Price Theory Pract. (1), 19 (2019)
2. Cui, X., Zhang, J.: Significant increase in pork prices Egg prices remain low—review of Beijing agricultural products wholesale price index in March 2019. Price Theory Practice (3), 42 (2019)
3. Cui, X., Zhang, J.: Pork prices remain high Grain and oil prices continue to stabilize—review of Beijing agricultural products wholesale price index in May 2019. Price Theory Pract. (5), 43 (2019)
4. Ran Lin, J.: Laser: connected transactions support performance growth. Stock Market Dyn. Anal. (28), 14 (2019)
5. Liu, K.: Pork concept stocks "fly to heaven" need to think coldly after crazy. Financial investment newspaper, 2019-04-25 (001)
6. Zhang, S., Zhang, S.: Study on the impact of CPI on the conceptual stock market value of pork based on VAR model. Mark. Res. (02), 24–26 (2019)
7. Sun, Z., Lu, X.: An empirical study on the relationship between the stock price of agricultural listed companies and the prices of related agricultural products—an empirical study on the price of forage and livestock products and the stock price of listed companies in livestock. Econ. Surv. **02**, 118–122 (2011)
8. Hong, S., Zhang, Q.: Analysis and forecast of pork sector index based on ARMA-GARCH model. China Prices (11), 23 (2019)
9. Guo, F., et al.: Research on the law of garlic price based on big data. Comput. Mater. Continua **58**(3), 795–808 (2019)
10. Wu, C., Lee, V., McMurtrey, M.E.: Knowledge composition and its influence on new product development performance in the big data environment. Comput. Mater. Continua **60**(1), 365–378 (2019)
11. Zheng, Z.: Analysis of CPI influencing factors in Liyang City based on multiple regression model and stepwise analysis. J. Puyang Voc. Tech. College **30**(01), 80–84 (2017)
12. Zhou, H., Tu, Y., Li, P., Chen, J.: An Empirical Study of the Impact of Black Swan Events on Stock Price from the Perspective of Investor Emotion
13. Yin, H., Wu, X.: Investor's daily mood, excess rate of return and market liquidity—the correlation study based on DCC-GARCH model. J. Beijing Inst. Technol. (Soc. Sci. Edn.) **21**(05), 76–87+114 (2019)
14. Dai, D., Lan, Y.: Review of research on investor sentiment and stock price based on web text. Wuhan Finance **01**, 41–45 (2019)
15. Fan, Y.: An empirical study of factors affecting stock returns based on multiple linear regression model. Nat. Circ. Econ. **06**, 101–103 (2019)
16. Liu, L.: An empirical analysis of the influence of macro factors on stock price. Econ. Res. Guide **10**, 95–96 (2014)
17. Feng, X., Zhang, X., Xin, Z., Yang, A.: Investigation on the Chinese text sentiment analysis based on convolutional neural networks in deep learning. Comput. Mater. Continua **58**(3), 697–709 (2019)
18. Chang, J., Yan, R., Wang, L.: Research on changes in investors' sentiment in the fund market and their impacts-based on the statistical perspective of Sina finance fund review. China Dev. **18**(01), 28–34 (2018)

Visual SLAM Location Methods Based on Complex Scenes: A Review

Hanxiao Zhang[1] and Jiansheng Peng[1,2](✉)

[1] School of Electrical and Information Engineering, Guangxi University of Science and Technology, Liuzhou 545000, Guangxi, China
sheng120410@163.com

[2] School of Physics and Mechanical and Electronic Engineering, Hechi University, Yizhou 546300, Guangxi, China

Abstract. In recent years, positioning for simple static scenes has been unable to meet the requirements of people's production and life. People want to achieve accurate positioning in practical scenarios such as airports, exhibition halls and stations. Therefore, the research on visual SLAM positioning in complex dynamic scenes is increasing day by day. This article reviews the research results of SLAM positioning methods and visual SLAM positioning methods for complex scenes in recent years. Firstly, the development process of laser SLAM, visual SLAM, semantic SLAM and multi-sensor fusion is introduced, but the focus is on visual SLAM. Secondly, the paper summarizes the methods of moving object detection and visual SLAM localization in complex dynamic scenes. Then the paper describes the development of deep learning and multi-sensor fusion in visual SLAM positioning based on complex dynamic scenes. Finally, the shortcomings of visual SLAM positioning methods based on complex scenes are summarized and the research prospects are prospected.

Keywords: Complex scene · Visual SLAM · Positioning

1 Introduction

Indoor positioning and navigation technology has always been the key technology for intelligent service robots. As the most important link for robots to achieve autonomous navigation, positioning has always been the focus of scientists' research. The current positioning technology is mainly realized through a variety of sensor information sensing environments such as laser radar, code disks, cameras and ultrasonic sensors. However, the positioning effect is often related to the complexity of the indoor environment and the cost of the sensor. Most of the visible on the market is based on laser to achieve positioning and navigation, but the cost of laser radar is high and the data is single. In comparison, vision-based positioning technology has more development potential than laser radar. Visual positioning and navigation technology has developed rapidly in recent years, with the advantages of low cost, rich information, and strong scalability. Visual positioning navigation is a popular direction of current positioning navigation technology.

X. Sun et al. (Eds.): ICAIS 2020, LNCS 12240, pp. 487–498, 2020.
https://doi.org/10.1007/978-3-030-57881-7_43

The current visual SLAM positioning and navigation technology is unstable and is more commonly used in simple scenes. The application scenarios of indoor positioning technology include airports, restaurants, hotels, exhibitions, and home environments. The application scenarios of indoor positioning technology are all complex dynamic scenarios. The study of simple scene positioning is not enough to meet the requirements of production and life, so the positioning of complex scenes is more practical and has become the focus of research in recent years. The core problem of the existing visual positioning technology is that it has poor robustness to complex dynamic scenes, and it is easy to lose positioning. How to achieve high robust positioning in complex dynamic scenes is the key issue of visual positioning technology. In order to improve the recognition accuracy, many scholars have combined deep learning with visual SLAM of complex scenes, which has improved the recognition accuracy of images. However, it often takes more time to process the data of complex scenes after adding deep learning, resulting in poor real-time positioning of complex scenes and poor positioning results, which makes them widely used in complex scenes with slow backgrounds.

The second chapter of this paper summarizes the current research status of SLAM positioning technology. The second chapter mainly introduces positioning methods from laser SLAM, visual SLAM, semantic SLAM and multi-sensor fusion, but mainly focuses on visual SLAM positioning research. The third chapter summarizes the research on moving object detection methods and visual SLAM positioning methods based on complex dynamic scenes. The purpose of moving object detection is to eliminate the impact of dynamic obstacles in complex environments and improve the positioning accuracy of robots in complex environments. The fourth chapter introduces the development direction of visual SLAM positioning based on complex dynamic environment around the future of deep learning. Finally, the shortcomings of the existing methods are further summarized and prospects are discussed.

2 Research Status of SLAM Positioning Technology

According to the different sensors carried by the robot, SLAM is mainly divided into laser SLAM and visual SLAM. Semantic information extracted based on visual information is the key to intelligent robots to perform advanced tasks, so semantic SLAM is derived from the latest research directions. Researchers fuse these different sensors, and a multi-sensor fusion method has emerged. This chapter will focus on four areas: laser SLAM, visual SLAM, semantic SLAM and multi-sensor fusion, but we mainly focus on visual SLAM.

2.1 Laser SLAM

After SLAM was proposed, it has gone through several research stages. The early laser SLAM used extended Kalman filter [1], and then estimated the pose of the robot through the information provided by the odometer and laser, but the effect was not good. For some strong nonlinear systems, this method will bring more The truncation error makes it impossible to accurately locate and map. Then the particle filter-based laser SLAM gradually became the mainstream, and it was widely used in non-linear systems, and it is still a popular method until today [2–4]. However, particle filtering consumes memory and is not suitable for detecting large scenes. Grissett et al. [5] proposed an improved particle filtering method and proposed an adaptive resampling method to reduce the computational load of particle filtering. Then someone proposed a method of graph optimization instead of filtering [6–10]. It is generally considered that with the same amount of calculation, nonlinear optimization can achieve better results than Kalman filtering [11]. One of the major problems with the above method is the laser closed-loop detection problem. For the difficult closed-loop phenomenon caused by accumulated errors, Google proposed a cartographer [12], a solution that uses laser data and maps to identify closed-loops. This scheme can guarantee the accuracy of the map structure, but this scheme only relies on lasers and the closed loop is prone to errors.

2.2 Visual SLAM

Visual SLAM has become a research hotspot of SLAM due to low cost of sensors and rich image information. However, visual SLAM is more complicated than laser SLAM. Visual SLAM extracts feature points to achieve positioning and mapping. At first, Kalman filter [13] and particle filter [14] were used to extract feature points for matching to calculate the camera pose. The feature points are extracted to match the camera pose. The commonly used image features are SIFT [15], SURF [16] and ORB [17], etc. However, processing all data directly will cause excessive calculation. In order to achieve better real-time performance, the subsequent visual SLAM adopts the form of multi-threading. SLAM is divided into two parts: front-end and back-end. Nonlinear optimization method is used [6] and ORB feature is used to improve real-time performance and use the bag of words model Implement closed-loop detection [7].

Visual SLAM can also be combined with direct methods to achieve positioning and mapping. Compared with extracting feature points, the direct method is more practical for weakly textured scenes. Stuhmer et al. [8] proposed pixel calibration of dense images to construct dense three-dimensional maps. Engel et al. [18, 19] proposed dense image registration method to calculate camera pose and LSD-SLAM. LSD-SLAM is a successful example of direct method in semi-dense monocular SLAM. Subsequent researchers have successively proposed Kinect-based fusion methods [20], direct RGB-D SLAM method [21] and multi-camera realization of accurate positioning [22], etc. to perform positioning mapping work.

2.3 Semantic SLAM

Semantic SLAM has only been developed in recent years in order to make robots understand the environment closer to humans. Bao et al. [23] used image semantics to improve the robot's ability to recognize objects in the environment. Salas-Moreno et al. [24] generated dense, accurate problem-level semantic maps in order to be able to identify pre-modeled objects. Vineet et al. [25] for the first time realized a system capable of simultaneous semantic segmentation and mapping. Bowman et al. [26] and Schönberger et al. [27] implemented SLAM positioning using semantic information. The RGB-D SLAM system proposed by Rünz et al. [28] can realize scene multi-target recognition.

2.4 Multi-sensor Fusion

In terms of multi-sensor fusion, there are currently various sensors such as vision sensors, lidar, and inertial measurement unit (IMU). Currently, the mainstream approach is to fuse laser radar with vision sensors and IMU with vision sensors. Zhang Jie and Zhou Jun [29] proposed a SLAM method combining lidar and vision, which uses laser map navigation and visual map to restore the detection scene. Qin et al. [30] proposed a tight coupling scheme between vision and IMU, combining the visual structure and the residuals of the IMU structure to form a joint optimization problem. Yin Lei et al. [31, 32] proposed a method of positioning and mapping combining laser and vision. Combining laser data and image information effectively improved the accuracy of positioning and mapping.

3 Visual SLAM Positioning Method Based on Complex Scene

3.1 Moving Object Detection

Moving object detection refers to the process of detecting moving objects from complex background images. Usually, the common methods of detecting moving objects are divided into the following categories: frame-based method, background-based method and light flow-based method, etc.

Frame Difference Method. This method detects moving objects by comparing the gray value of the pixel corresponding to the current frame and the adjacent frame in the video image to find the difference. Cheng et al. [33] proposed a difference method for three consecutive frames. The basic principle is to use the difference between two adjacent frames of three consecutive frames to obtain the result. This method can effectively remove the background due to motion occlusion. The detection error caused. Min et al. [34] are prone to some loopholes in the frame difference method. The background model method is difficult to establish a model. The improved frame difference method of motion history image is combined with the background difference method based on improved Gaussian mixture model to perform moving targets detection.

Background Difference Method. The video image is divided into foreground and background images. And the current frame and the background frame are subjected to difference operation. Stauffer et al. [35] further proposed a mixed Gaussian model in order to improve the detection capability of the single Gaussian model in complex environments. Zhou Jianying et al. [36] to solve the problem that the parameters of the traditional gaussian mixture model converge slowly and it is difficult to adapt to the real-time change of the real background in the scene with time. And the traditional method leads to the increase of error detection rate of moving target. They proposed a moving target detection method based on gaussian mixture model.

Optical Flow Method. The optical flow method extracts the velocity field from the image by using the change of the gray scale information of the image, and derives the motion parameter and the object structure of the moving object due to the existence of the constraint condition.

3.2 Research on Visual SLAM Location Method Based on Complex Scenes

In the study of visual SLAM based on complex dynamic scenes, the SLAMIDE algorithm [37] uses the expectation maximization algorithm to update the feature point motion model in the scene. After updating the motion model, the dynamic object is introduced into SLAM through a reversible model selection mechanism. The disadvantage of this algorithm is that the continuous introduction of dynamic map points in the map will increase memory consumption and reduce the search speed of map points. Tan et al. [38] proposed an adaptive RANSAC algorithm based on existing knowledge in order that the system can run stably in a dynamic environment. Tan et al. Iteratively screens static feature points in the image and removes external points introduced by dynamic objects, and according to the static feature points in the scene, obtains a more uniform distribution, a larger number, and a more consistent camera motion model. The algorithm adaptively models the dynamic environment and can effectively detect and handle appearance or structural changes. Oh [39] proposed a SLAM method based on dynamic extended Kalman filter (EKF) in order to solve the robot pose estimation and environment mapping errors caused by the change of landmark position. This algorithm divides SLAM into traditional static SLAM part and single dynamic SLAM part to reduce the impact of dynamic environment on SLAM and improve the positioning accuracy of robots in dynamic landmarks. Newcombe et al. [40] proposed that the Dynamic Fusion algorithm can reconstruct the dynamic changes of the environment and is suitable for various moving objects and scenes. However, the algorithm can only be used in smaller environments and requires the GPU to complete the composition, and its deployment is not easy. Kumar [41] proposed an online spatiotemporal joint model in order to obtain motion information. This model is used to estimate joint structure. The model predicts the future motion of joint objects by integrating spatial and temporal structures and adds it to the SLAM algorithm. This model enables the algorithm to include the motion of dynamic objects in the real-time SLAM framework, improving detection accuracy. Sun [42] proposed a motion culling method based on RGB-D data and integrated it into RGB-D SLAM. Sun first performs rough detection of moving targets through image difference, then uses particle filtering to track motion, and finally determines moving objects by estimating the maximum posterior probability on depth images. Experimental results show that this method effectively improves all the performance of RGB-D SLAM, but there are still some defects. For example, when the parallax between consecutive frames is large, the homography estimation will decrease and when the moving object becomes stationary, the tracking will fail. Sun [43] proposed a method of removing moving objects in order to effectively eliminate moving objects and assist the scene modeling algorithm to establish a clear scene model. This method mainly relies on optical flow method and depth information to achieve moving object culling in RGB-D SLAM. The disadvantage of this method is that the camera is assumed to be static during scene modeling. Barsan et al. [44] proposed a three-dimensional large-scale dynamic urban environment density mapping algorithm. The algorithm uses instance-perceived semantic segmentation and sparse scene flow to divide the object into background and movement and reconstruct the background and moving objects through the depth information calculated by visual odometry and stereo vision. The algorithm outputs high-quality static backgrounds and dense model dynamic objects of backgrounds, which prevents the impact of dynamic scenes on positioning. In order to improve the performance of RGB-D SLAM in high dynamic scenes, Yu Xiang [45] proposed to build and update the motion foreground model by learning. This model can implement moving object culling and apply it to visual SLAM, but this method still has some limitations. First, the parallax between consecutive frames is required to be small. Second, requiring static objects to dominate the detection environment reduces performance in low dynamic environments. Finally, this method is very time-consuming and has poor real-time performance. Berta et al. [46] proposed a dynamic object detection system DynaSLAM based on ORB-SLAM2 in order to track and reuse scene maps more accurately. DynaSLAM can detect moving objects through multi-view geometry method and deep learning method. DynaSLAM uses MASK-RCNN [47] for instance segmentation and segmentation of objects with mobility. In highly dynamic situations, DynaSLAM is superior to standard visual SLAM in accuracy, and it can also estimate the map of the static part of the scene. However, the performance of DynaSLAM on low dynamic sequences is not obvious. Yu et al. [48] proposed DS-SLAM in order to allow the robot to complete advanced tasks and reduce the impact of dynamic objects on pose estimation. The algorithm combines semantic segmentation network with optical flow method and provides semantic representation of octree maps. The algorithm focuses on reducing the impact of dynamic objects in vision-based SLAM and improving the robot's positioning and mapping performance in complex dynamic environments. The algorithm improves the robustness and accuracy of positioning and mapping. Xu [49] combined instance segmentation into RGB-D SLAM to provide stable motion estimation. However, this combination runs at a slower speed, which can only run at a speed of 2–3 Hz and does not include instance segmentation. Zhenlong Du et al. [50] proposes an improved SLAM algorithm, which mainly improves the real-time performance of classical SLAM algorithm, applies KDtree for efficient organizing feature points, and accelerates the feature points correspondence building. Moreover, the background map reconstruction thread is optimized, the SLAM parallel computation ability is increased. The improved SLAM algorithm holds better real-time performance than the classical SLAM. Xianyu Wu et al. [51] in order to solve the natural scene image has more interference and complexity than text. They propose a new text detection and recognition method based on depth convolution neural network is proposed for natural scene image. In text detection, this method obtains high-level visual features from the bottom pixels by ResNet network, and extracts the context features from character sequences by BLSTM layer, then introduce to the idea of faster R-CNN vertical anchor point to find the bounding box of the detected text, which effectively improves the effect of text object detection. The method can replace the artificially defined features with automatic learning and context-based features. It improves the efficiency and accuracy of recognition. Zhe Liu et al. [52] In order to avoid most image segmentation methods based on clustering algorithm, single objective function is used to realize image segmentation. An image segmentation method based on multi-objective particle swarm optimization (PSO) clustering algorithm is proposed. The unsupervised algorithm not only provides a new similarity calculation method based on electromagnetic force, but also obtains the appropriate number of clusters determined by scale space theory. Zhang Jinfeng et al. [53] proposed a SLAM method based on visual features in dynamic scenes in order to reduce the impact of dynamic objects on tracking and positioning. This algorithm introduces a deep learning-based object detection algorithm into the classic ORB_SLAM2 method. The algorithm divides feature points into latent dynamic features and non-latent dynamic features. The algorithm will calculate the motion model based on the non-potential dynamic feature points and then filter out the static feature points in the scene for pose tracking. The algorithm uses static feature points in non-latent dynamic features for mapping. Compared with ORB_SLAM2, the performance of the system has been significantly improved and the running speed of the system can meet the real-time requirements. Gao Chengqiang et al. [54] proposed a semi-direct RGB-D SLAM algorithm for indoor dynamic environment. The algorithm uses a sparse image alignment algorithm to make a preliminary estimation of the camera pose. The algorithm uses the pose estimation of the visual odometer to compensate the motion of the image. The algorithm establishes a Gaussian model based on real-time update of image blocks. The algorithm divides the moving target in the image according to the variance to eliminate the local map points projected on the moving area of the image. The algorithm completes the real-time update of the map in a dynamic environment and improves the camera pose accuracy.

In order to better reflect the improvement results of different algorithms, this article selects some algorithms for comparison. The comparison results are shown in Table 1. Table 1 compares the absolute path error (ATE) of ORB-SLAM2, DS-SLAM [48] and Zhang Jinfeng's improved dynamic scene SLAM method [53]. Both DS-SLAM and Zhang Jinfeng's improved dynamic scene SLAM methods are improvements based on the original system ORB-SLAM2. It can be seen from Table 1 that the improved dynamic scene SLAM method proposed by DS-SLAM and Zhang Jinfeng has improved performance compared to the original system ORB-SLAM2.

Table 1. Absolute path error (ATE) comparison

Data		Walking_static	Walking_rpy	Walking_halfsphere
ORB-SLAM2	RMSE	0.3900	0.8705	0.4863
	Mean	0.3554	0.7425	0.4272
	S.D.	0.1602	0.4520	0.2290
DS-SLAM	RMSE	0.0081	0.4442	0.0303
	Mean	0.0073	0.3768	0.0258
	S.D.	0.0036	0.2350	0.0159
Zhang Jinfeng's improved method	RMSE	0.0088	0.0595	0.0394
	Mean	0.0079	0.0427	0.0326
	S.D.	0.0040	0.0362	0.0220

Visual SLAM started late and most of the research on visual SLAM positioning in complex dynamic scenes has only begun in recent years. Research on visual SLAM in complex scenes is relatively shallow and only applicable to simple dynamic scenes. The most common optical flow method often assumes that the largest connected area in the picture is a static background, but the opposite situation may exist in the actual scene. Using deep learning to consider semantic information into dynamic scene vision SLAM also has the problem of large amount of calculation and cannot be real-time. Moreover, the deep segmentation of the object semantics in the image through deep neural network can only obtain its own information, but cannot directly determine the movement.

4 Direction of Development

The following discussion will help to advance the development of visual SLAM positioning methods based on complex environments:

(1) Visual SLAM positioning and deep learning

Aiming at the impact of dynamic objects on the robot's current positioning in complex dynamic scenes, a lightweight dynamic semantic segmentation network architecture with real-time performance can be used. This architecture can optimize existing semantic segmentation network models and reduce the number of network model layers. Specific training samples are given for specific scenarios, for example, for airports, the environment contains a large number of people, luggage, walkers, and so on. The architecture specifically trains network models for these objects and extracts dynamic semantic information from images. The architecture redesigns a learning inference model that combines semantically segmented images with the original images for the detection and elimination of dynamic objects. The architecture reduces the variety of semantics through specialized training samples to reduce network complexity and achieve lightweight real-time semantic segmentation. At the same time, in the sample processing stage, artificially designed mobility tags are added for classified semantics. The tag can distinguish whether the object has mobility or not, and can distinguish whether the object is easily moved.

(2) Multi-sensor fusion positioning

Multiple sensor fusion positioning is also the focus of research. The laser sensor has high detection accuracy, but it is difficult to achieve a closed loop. The difficulty of closed loop greatly affects the establishment of maps by robots in indoor environments. Vision sensors are currently widely used and are easy to implement closed loops for maps. Can be visual sensor is detection accuracy is not enough. The inertial measurement unit (IMU) has large drift errors and accumulated errors. Semantic SLAM has only been developed in recent years and is mostly used in complex dynamic environments. But semantic SLAM detection accuracy is not enough to meet the requirements of production and life. At present, there are many researches on the combination of semantic segmentation and visual SLAM, but the current research is insufficient to meet people's positioning requirements for indoor robots in complex dynamic environments. Multi-sensor fusion also requires researchers to continuously improve their methods to improve robustness and real-time performance.

5 Conclusion

Visual SLAM positioning based on complex scenes has made great progress, driven by the continuous development of deep learning and computer vision. However, the visual SLAM based on complex scenes still fails to meet the requirements of people's production and life in terms of real-time, robustness and scalability. At present, the combination of SLAM and deep learning has improved the object detection of complex scenes to a certain extent. The combination of SLAM and deep learning also has an important impact on the rapid and accurate generation of high-level semantics and the construction of robot knowledge bases. However, after SLAM combined with deep learning, the detection and recognition time of complex scenes increases significantly, which makes it difficult to meet the system real-time requirements. And the training parameters greatly depend on the experience of adjusting the parameters. The detection result also greatly depends on the similarity of the current field application scenario. Therefore, the visual SLAM of complex scenes in the future needs to rely on the further development and integration of target recognition, semantic segmentation, and deep learning. And how to apply deep learning to the entire SLAM system instead of just positioning, closed-loop modules, etc. is still a huge challenge.

Acknowledgement. This work is supported by the National Natural Science Foundation of China (Grant No. 61640305). This research was financially supported by the project of Thousands outstanding young teachers' training in higher education institutions of Guangxi, The Young and Middle-aged Teachers Research Fundamental Ability Enhancement of Guangxi University (ID: 2019KY0621), Natural Science Foundation of Guangxi Province (No. 2018GXNSFAA281164). Guangxi Colleges and Universities Key Laboratory Breeding Base of System Control and Information Processing, Hechi University research project start-up funds (XJ2015KQ004), Supported by Colleges and Universities Key Laboratory of Intelligent Integrated Automation (GXZDSY2016-04), Hechi City Science and Technology Project (1694-3-2), Research on multi robot cooperative system based on artificial fish swarm algorithm (2017CFC811).

References

1. Smith, R., Self, M., Cheeseman, P.: Estimating uncertain spatial relationships in robotics. In: Proceedings of IEEE International Conference on Robotics and Automation, Raleigh, NC, USA, 31 March–3 April 1987
2. Sileshi, B.G., Oliver, J., Toledo, R., Goncalves, J., Costa, P.: On the behaviour of low cost laser scanners in HW/SW particle filter SLAM applications. Robot. Auton. Syst. **80**(C), 11–23 (2016)
3. Thallas, A., Tsardoulias, E., Petrou, L.: Particle filter—scan matching hybrid SLAM employing topological information. In: 24th Mediterranean Conference on Control and Automation, Athens, Greece, 21–24 June 2016
4. Song, W., Yang, Y., Fu, M., Kornhauser, A., Wang, M.: Critical rays self-adaptive particle filtering SLAM. J. Intell. Robot. Syst. **92**(1), 107–124 (2017). https://doi.org/10.1007/s10846-017-0742-z
5. Grisetti, G., Stachniss, C., Burgard, W.: Improved techniques for grid mapping with Rao-Blackwellized particle filters. IEEE Trans. Rob. **23**(1), 34–46 (2007)

6. Klein, G., Murray, D.: Parallel tracking and mapping for small AR workspaces. In: 6th IEEE and ACM International Symposium on Mixed and Augmented Reality, Nara, Japan, 13–16 November 2007
7. Mur-Artal, R., Montiel, J.M., Tardós, J.D.: ORB-SLAM: a versatile and accurate monocular SLAM system. IEEE Trans. Rob. **31**(5), 1147–1163 (2015)
8. Stühmer, J., Gumhold, S., Cremers, D.: Real-time dense geometry from a handheld camera. In: Goesele, M., Roth, S., Kuijper, A., Schiele, B., Schindler, K. (eds.) DAGM 2010. LNCS, vol. 6376, pp. 11–20. Springer, Heidelberg (2010). https://doi.org/10.1007/978-3-642-15986-2_2
9. Konolige, K., Grisetti, G., Kümmerle, R.: Efficient sparse pose adjustment for 2D mapping. In: International Conference on Intelligent Robots and Systems, pp. 22–29 (2010)
10. Engel, J., Sturm, J., Cremers, D.: Semi-dense visual odometry for a monocular camera. In: IEEE International Conference on Computer Vision, Sydney, NSW, Australia, 1–8 December 2013
11. Strasdat, H., Montiel, J.M.M., Davison, A.J.: Visual SLAM: why filter? Image Vis. Comput. **30**(2), 65–77 (2012)
12. Hess, W., Kohler, D., Rapp, H., Andor, D.: Real-time loop closure in 2D LIDAR SLAM. In: IEEE International Conference on Robotics and Automation, Stockholm, Sweden, 16–21 May 2016
13. Davison, A.J., Reid, I.D., Molton, N.D., Stasse, O.: MonoSLAM: real-time single camera SLAM. IEEE Trans. Pattern Anal. Mach. Intell. **29**(6), 1052 (2007)
14. Sim, R., Elinas, P., Griffin, M.: Vision-based SLAM using the Rao-Blackwellised particle filter. In: IJCAI Workshop on Reasoning with Uncertainty in Robotics, vol. 9(4), pp. 500–509 (2005)
15. Lowe, D.G.: Distinctive image features from scale-invariant keypoints. Int. J. Comput. Vision **60**(2), 91–110 (2004). https://doi.org/10.1023/B:VISI.0000029664.99615.94
16. Bay, H., Tuytelaars, T., Van Gool, L.: SURF: speeded up robust features. In: Leonardis, A., Bischof, H., Pinz, A. (eds.) ECCV 2006. LNCS, vol. 3951, pp. 404–417. Springer, Heidelberg (2006). https://doi.org/10.1007/11744023_32
17. Rublee, E., Rabaud, V., Konolige, K., Bradski, G.: ORB: an efficient alternative to SIFT or SURF, vol. 58, no. 11, pp. 2564–2571 (2011)
18. Engel, J., Sturm, J., Cremers, D.: Semi-dense visual odometry for a monocular camera. In: IEEE International Conference on Computer Vision, pp. 1449–1456 (2013)
19. Engel, J., Schöps, T., Cremers, D.: LSD-SLAM: large-scale direct monocular SLAM. In: Fleet, D., Pajdla, T., Schiele, B., Tuytelaars, T. (eds.) ECCV 2014. LNCS, vol. 8690, pp. 834–849. Springer, Cham (2014). https://doi.org/10.1007/978-3-319-10605-2_54
20. Newcombe, R.A., Izadi, S., Hilliges, O., et al.: KinectFusion: real-time dense surface mapping and tracking. In: IEEE International Symposium on Mixed and Augmented Reality, pp. 127–136 (2011)
21. Kerl, C., Sturm, J., Cremers, D.: Dense visual SLAM for RGB-D cameras. In: International Conference on Intelligent Robots and Systems, pp. 2100–2106 (2014)
22. Yang, S., Scherer, S.A., Yi, X., et al.: Multi-camera visual SLAM for autonomous navigation of micro aerial vehicles. Robot. Auton. Syst. **93**(1), 116–134 (2017)
23. Bao, S.Y., Bagra, M., Chao, Y.W., et al.: Semantic structure from motion with points, regions, and objects. In: Proceedings of IEEE Conference on Computer Vision and Pattern Recognition, pp. 2703–2710 (2012)
24. Salas-Moreno, R.F., Newcombe, R.A., Strasdat, H., et al.: SLAM++: simultaneous localisation and mapping at the level of objects. In: Proceedings of IEEE Conference on Computer Vision and Pattern Recognition, pp. 1352–1359 (2013)

25. Vineet, V., Miksik, O., Lidegaard, M., et al.: Incremental dense semantic stereo fusion for large- scale semantic scene reconstruction. In: Proceedings of IEEE International Conference on Robotics and Automation, pp. 75–82 (2015)
26. Bowman, S.L., Atanasov, N., Daniilidis, K., et al.: Probabilistic data association for semantic slam. In: Proceedings of IEEE International Conference on Robotics and Automation, pp. 1722–1729 (2017)
27. Schönberger, J.L., Pollefeys, M., Geiger, A., et al.: Semantic visual localization. In: Proceedings of IEEE Conference on Computer Vision and Pattern Recognition, pp. 1–21 (2008)
28. Rünz, M., Agapito, L.: MaskFusion: real-time recognition, tracking and reconstruction of multiple moving objects. In: Proceedings of IEEE International Symposium on Mixed and Augmented Reality, pp. 1–10 (2018)
29. Jie, Z., Jun, Z.: A study on laser and visual mapping of mobile robots with improved ICP algorithm. Mech. Electr. Eng. (12) 1480–1484 (2017)
30. Qin, T., Li, P., Shen, S.: VINS-Mono: a robust and versatile monocular visual-inertial state estimator. IEEE Trans. Rob. **3**(4), 1–17 (2017)
31. Lei, Y.: Research on simultaneous positioning and mapping of indoor robots. Guangxi University of Science and Technology (2019)
32. Yin, L., Peng, J., Jiang, G., Ou, Y.: Research on synchronous positioning and mapping of low-cost laser and vision. Integr. Tech. (2) (2019)
33. Cheng, Y.H., Wang, J.: A motion image detection method based on the inter-frame difference method. Appl. Mech. Mater. **490–491**, 1283–1286 (2014)
34. Min, H., Shu, H., Liu, Q., Xia, Y., Gang, C.: Moving object detection method based on NMI features motion detection frame difference. Adv. Sci. Lett. **6**(1), 477–480 (2012)
35. Stauffer, C., Grimson, E.: Learning patterns of activity using real-time tracking. IEEE Trans. Pattern Anal. Mach. Intell. **22**(8), 747–757 (2000)
36. Zhou, J., Wu, X., Zhang, C., Lu, W.: A moving object detection method based on hybrid Gaussian model of sliding window. J. Electron. Inf. Tech. **35**(07), 1650–1656 (2013)
37. Hahnel, D., Triebel, R., Burgard, W., et al.: Map building with mobile robots in dynamic environments. In: IEEE International Conference on Robotics and Automation, pp. 1557–1563. IEEE, Piscataway (2003)
38. Tan, W., Liu, H.M., Dong, Z.L., et al.: Robust monocular SLAM in dynamic environments. In: 12th IEEE/ACM International Symposium on Mixed and Augmented Reality, pp. 209–218. IEEE Piscataway (2013)
39. Oh, S., Hahn, M., Kim, J.: Dynamic EKF-based SLAM for autonomous mobile convergence platforms. Multimedia Tools Appl. **74**(16), 6413–6430 (2014). https://doi.org/10.1007/s11042-014-2093-0
40. Newcombe, R.A., Fox, D., Seitz, S.M.: DynamicFusion: reconstruction and tracking of non-rigid scenes in real-time. In: IEEE Conference on Computer Vision and Pattern Recognition, pp. 343–352. IEEE, Piscataway (2015)
41. Kumar, S., Dhiman, V., Ganesh, M.R., Corso, J.: Spatiotemporal Articulated Models for Dynamic SLAM (2016)
42. Sun, Y., Liu, M., Meng, Q.H.: Improving RGB-D SLAM in dynamic environments: a motion removal approach. Robot. Auton. Syst. **89**, 110–122 (2017)
43. Sun, Y., Liu, M., Meng, Q.H.: Invisibility: a moving-object removal approach for dynamic scene modelling using RGB-D camera. In: IEEE International Conference on Robotics and Biomimetics (ROBIO), Macau, China, 5–8 December 2017
44. Barsan, I.A., Liu, P., Pollefeys, M., Geiger, A.: Robust dense mapping for large-scale dynamic environments. In: International Conference on Robotics and Automation (ICRA), pp. 7510–7517 (2018)

45. Sun, Y., Liu, M., Meng, M.: Motion removal for reliable RGB-D SLAM in dynamic environments. Robot. Auton. Syst. **108**, 115–128 (2018)
46. Berta, B., Facil, J.M., Javier, C., Jose, N.: DynaSLAM: tracking, mapping and inpainting in dynamic scenes. IEEE Robot. Autom. Lett. **3**(4), 4076–4083 (2018)
47. He, K., Gkioxari, G., Dollar, P., Girshick, R.: Mask R-CNN. IEEE Trans. Pattern Anal. Mach. Intell. (2018)
48. Yu, C. et al.: DS-SLAM: a semantic visual SLAM towards dynamic environments. In: IEEE/RSJ International Conference on Intelligent Robots and Systems (IROS), pp. 1168–1174 (2018)
49. Xu, B., Li, W., Tzoumanikas, D., Bloesch, M., Davison, A., Leutenegger, S.: MID-Fusion: octree-based object-level multi-instance dynamic SLAM. In: International Conference on Robotics and Automation, Montreal, Canada, 20–24 May 2019
50. Du, Z., Ma, Y., Li, X., Lu, H.: Fast scene reconstruction based on improved SLAM. Comput. Mater. Continua **61**(1), 243–254 (2019)
51. Wu, X., Luo, C., Zhang, Q., Zhou, J., Yang, H., Li, Y.: Text detection and recognition for natural scene images using deep convolutional neural networks. Comput. Mater. Continua **61**(1), 289–300 (2019)
52. Liu, Z., Xiang, B., Song, Y., Lu, H., Liu, Q.: An improved unsupervised image segmentation method based on multi-objective particle, swarm optimization clustering algorithm. Comput. Mater. Continua **58**(2), 451–461 (2019)
53. Zhang, J., Shi, C., Wang, Y.: SLAM method based on visual features in dynamic scenes. Computer program, pp. 1–8, 04 November 2019.http://kns.cnki.net/kcms/Detail/31.1289.tp.20191025.1559.006.html
54. Gao, C., Zhang, Y., Wang, X., Deng, Y., Jiang, H.: Semi-direct method RGB-D SLAM algorithm for indoor dynamic environment. Robot **41**(03), 372–383 (2019)

Improvement of Co-training Based Recommender System with Machine Learning

Wenpan Tan, Yong He(✉), and Bing Zhu

School of Computer Science and Engineer, Hunan University of Science and Technology, Xiangtan, China
wen4p@foxmail.com, yonghe@hnust.edu.cn

Abstract. Recommender system (RS) is an application of information filtering to recommend information or items that users like according their data. Collaborative filtering algorithm is a famous algorithm in recommender system. However, the proportion of users willing to rate is very small, which causes data sparsity and cold start problem. In this paper, machine learning (ML) technology is integrated to an existed recommendation algorithm to get a more accurate predictions during filling in the user-item matrix, which can enrich data and then alleviate data sparsity and cold start problems. Follow that a new reliable confidence measure is applied in the process of enrich data to filter noisy. The experimental result of real data shows that the accuracy of the modified algorithm is better than the original.

Keywords: Recommender systems · Machine learning · Parameter optimization · Confidence

1 Introduction

With the development of website, the widespread access to the Web cause an information overload, which has brought potential problems to internet users [1]. RS is an application of information filtering, it recommend information or items (such as movies, TV programs, music, books, news, pictures, web pages and medical) users may like [1, 2, 4]. RS does not require the users to provide explicit requirements and analyzes the user's historical behavior data to model the user's interests, thereby recommend information that can satisfy their interests and needs [2–4].

Collaborative filtering (CF) algorithm is the most used technology of RS. CF algorithms mainly aggregate feedback for items from different users and use the similarities between items and items (item-based) or between users and users (user-based) to provide recommendations to a target user [2, 3]. However, there are some limitation in CF, such as data sparsity and cold start problem [2, 5]. In the real application, the number of user and item are very large, even if all the user-items interactions are recorded, the user-items interaction matrix is still very sparse, it's not enough to train the model accurately [2, 5, 6]. The cold start problem is caused by the fact that when the new user (item) appears, RS does not have any historical data about it, which makes it difficult to produce accurate recommendations. These problems are usually solved by a hybrid approach [2, 5].

X. Sun et al. (Eds.): ICAIS 2020, LNCS 12240, pp. 499–509, 2020.
https://doi.org/10.1007/978-3-030-57881-7_44

However, they narrowed down the applicable algorithms to a few that could use them and eliminated many users who do not have the required data [12].

There are lots of research for data sparsity and cold start problem in RS. Enriching matrix as preprocessing step can reduce these problem to some extent. But, it may generate noisy and consumes extra memory. For these problems are usually solved by an external source of data (such as social network relationship information) [2, 5, 7, 8]. Another alternative is to use semi-supervised learning (SSL) techniques, which is an algorithm to understand how the combination of labeled (exist ratings) and unlabeled data (missing ratings) can change the learning [19]. Basically, it exploits relationships between labeled and unlabeled data to make more accurate predictions, as opposed to using solely labeled data for training, as in the traditional supervised learning setting [12]. Co-training technique belong to SSL, it can be used to tackle traditional recommendation problems such as sparsity and cold start by using different recommendation algorithms to enriching the datasets as a pre-processing step for the RS [12], but it's more sensitive to noise. For this problem, confidence measure in RS is define as a measure to distinguish individual predictions that can be trusted from those that cannot [11]. Enrich the user-item matrix as the data preprocessing of RS, and use confidence to filter noise data, it can reduce the data sparsity and cold start problem [12].

In this paper, our contributions are the following: 1) using K-NN variability confidence in a co-training based recommender system (ECoRec) for reduce noisy; 2) using machine learning to determine two priori parameters in the CF in the ECoRec for get more accuracy predictions and convenient the parameters selection problem. Our approach uses two baseline algorithms: User-based K-NN algorithm (User-KNN) and Item-based K-NN algorithm (Item-KNN) [2, 3]. We compared our method with the well-known collaborative filtering algorithms and ECoRec. The experiment use three real datasets, and the results show that the results obtained by this method are better than the predictions of the separate recommendation algorithm.

This paper is structured as follows: The Sect. 2 we overview the machine learning application in RS. The Sect. 3 is introduces the recommendation algorithm, and then introduces our method in Sect. 4. The results are show in the Sect. 5. Finally, in Sect. 6, we present our conclusion and future work.

2 Related Work

The most popular to solve the data sparsity and cold start problem is to fill the missing ratings in the user-item matrix with default values. However, this method may generate noisy, consumes extra time and memory and lacks personalization [3, 5]. We need some more sophisticated techniques to solve these problems. Next we will introduce some applications of machine learning and confidence measure in the RS.

The characteristics of RS, such as data sparsity, the matrices sparsity is reach to 99%. Hence, using ML to reduce spares and cold star is still a challenging work. Blum A proposed an Item-KNN method in recommender system to reduce data sparsity and cold start [9]. It classifies items by K-Nearest-Neighbor (KNN) algorithm and provides recommendations for new users by using other item categories. Meanwhile, for reduce the data sparsity problem, it only focuses on this item category when making recommendations. But, this approach has not been proven in large data. Another approach is co-training which is belong to semi-supervised learning in ML. Zhang proposed a co-training method for stationary, batch-based recommender systems, using items' metadata to enrich the traditional training model [19]. A co-training based recommender system named ECoRec was proposed [13], it learn from different views of data, predicts missing data and integrates them to enrich data. This approach can reduces data sparseness and cold start problem, but need many priori parameters.

Confidence in RS can be defined as the system's trustworthiness of its recommendations or predictions [3]. Bobadilla proposed a method to verify the reliability of confidence, and verifies the reliability of K-NN variability confidence measure [10]. In Da Costa's work [12], confidence is use as the weight of predictions for fill the most confident ratings into the user-item matrix. This method can improve the accuracy of prediction and solve the problem of cold start and data sparseness, but it's more sensitive to noise.

Our proposed approach is use ML technique to determine two parameters in CF algorithms which used in ECoRec without any prior knowledge. In addition, we use a most reliability confidence measure in ECoRec.

3 Collaborative Filtering

This section introduces concepts related to our method. Before describing our approach, we introduce the collaborative filtering algorithms that will be appear later on.

Firstly, users and items are indicate as u and i respectively; the user u rate item i in the data are marked as r_{ui}r_{{ui}} . The recommendation algorithm concludes that the user's predicted ratings and confidence values for the item are expressed separately $\hat{r}_{ui}$ and C_{ui} separately.

3.1 User Based K-Nearest Neighbors (User-KNN)

Firstly, the recommendation algorithm used in our method is the more famous User-KNN recommendation algorithm [2, 3, 18]. This method mainly uses the ratings of similar users to predict the unknown ratings.

Similar users employ the description user by calculating the distance between the features vectors. The similarity measure between users can be based on a Pearson correlation coefficient or a cosine similarity. We identify the set of k users that are most similar to u, and select the subset of those users that have rated to item i, we symbol this user set as $S_k(u, i)$. The final prediction rating for unrated pair (u, i) is:

$$\hat{r}_{ui} = b_{ui} + \frac{\sum_{v \in S_k(u,i)} S_{uv}(r_{vi} - b_{vi})}{\sum_{v \in S_{k(u,i)}} S_{uv}} \tag{1}$$

Where b_{ui} is based on a user-supplied baseline assessment, derived from the neighborhood-based model proposed in [15], which is defined as:

$$b_{ui} = \mu + b_u + b_i \tag{2}$$

Where μ is the mean of all ratings, and b_u and b_i are estimates based on the ratings of user u and item i, respectively. The calculation method is as follows:

$$b_i = \frac{\sum_{u:(u,i) \in D} (r_{ui} + \mu - b_u)}{\lambda_1 + |\{u : (u, i) \in D\}|} \tag{3}$$

$$b_u = \frac{\sum_{u:(u,i) \in D} (r_{ui} + \mu - b_i)}{\lambda_2 + |\{u : (u, i) \in D\}|} \tag{4}$$

Where λ_1 and λ_2 are constants and generally set to [5, 25] according paper [3], b_{ui} is to capture the trend of higher/lower ratings given by certain users/items than other users/items, and D is the set of all know user-item pair's ratings.

3.2 Item Based K-Nearest Neighbors (Item-KNN)

The second recommendation algorithm used in this method is item-based K-NN collaborative filtering [2, 3, 15, 18]. Item-K-NN algorithm searches for similar items by similarity measure, and uses existing ratings in similar items to predict the unknown ratings. The final predicted rating is the average value of rating of similar items. The formula is as follows:

$$\hat{r}_{ui} = b_{ui} + \frac{\sum_{j \in S_k(i,u)} S_{ij} (r_{uj} - b_{uj})}{\sum_{j \in S_k(i,u)} S_{ij}} \tag{5}$$

Where b_{ui} and b_{uj} are same as formula (1).

In this paper, we use ML technique to determine two parameters in CF, which used in a co-training based recommender system (ECoRec), and use a new confidence criterion in ECoRec. We introduce more details of our method in the next section.

4 Proposed Work

In this section, it's split into two parts: 1) Review the ECoRec [13], include the improvements we made; 2) Introduce our method for determine priori parameters in the collaborative filtering algorithm. We review ECoRec in follow steps: data pre-processing, co-training step, ensemble step and recommendation task.

4.1 Data Pre-processing

The original data set is split into set of labeled (T_L^η) and unlabeled set (T_{UL}^η) to jointly train the two recommendation algorithms ($\eta = 1, 2$). These sets are the same as the two recommendation algorithms. The original labeled set consists of all existing ratings. Since the user-item matrix is sparse, there are a large number of unobserved ratings. We have not use all unrated user-item pairs as the initial unlabeled set. Instead, we selected unlabeled ratings randomly for each user. For example, if $X = 10$, each user in the original user-item matrix randomly selects 10 unrated items, includes them in the unlabeled set (T_{UL}^η). During co-training, the recommendation algorithm is to jointly learn and predict the unlabeled user-item pair, this is same as original method [13] (Fig. 1).

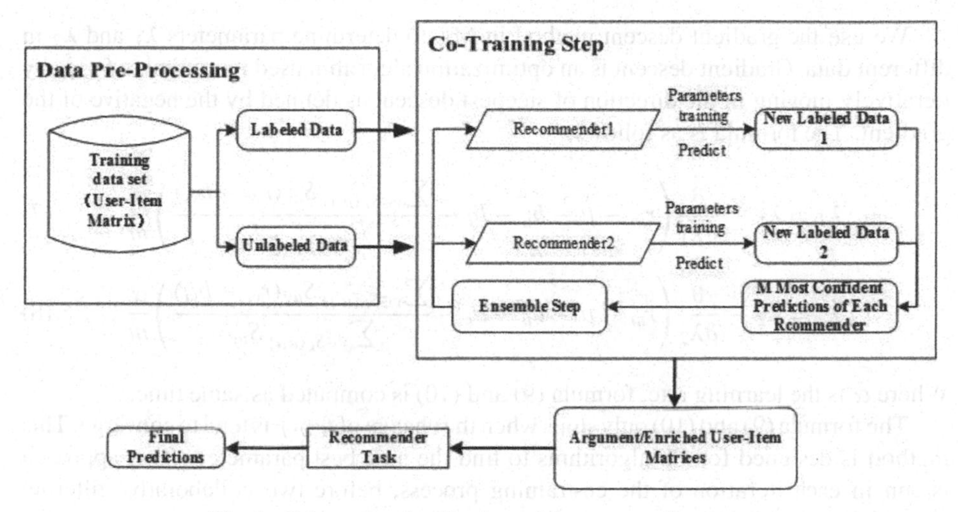

Fig. 1. Schematic visualization of the proposed system.

4.2 Co-training Step

The co-training process involves using the same data for two different recommendation algorithms to encourage them to use each other the most confident predictions. This method can use any two different recommendation algorithms, provided they represent two different data views. In this paper, we use User-KNN and Item-KNN (introduced in Sects. 3.1 and 3.2 respectively) as two different views of the same data. The first view focuses on the similarity between users in the data, while the second view focuses on the similarity between items in the data.

For each iteration of the algorithm, each recommendation algorithm $\eta(\eta = 1, 2)$ is responsible for predict unlabeled set after parameter training according to its labeled set T_{UL}^{η} (show in follow). Then, the confidence of each prediction is calculated, choose the M most confident prediction, and delete it from the unlabeled data set of the corresponding recommendation algorithm. Once this process is performed on both recommendation algorithms, the algorithm updates the label set of each recommendation algorithm, and the most confident predictions are added to the labeled data set of another recommendation algorithm, so that each model can learn from the new labeled data. The algorithm stops only when all unlabeled data of the two algorithms are labeled.

Parameters Determine

We have introduced CF, which have two parameters λ_1 and λ_2 in formula (1) and (5). These are affect the prediction accuracy of CF. Traditional methods to determine two parameters are based on the prior experience. For get better parameters, we use ML to determine parameters λ_1 and λ_2. We build the cost function by least square method, as shown in formula:

$$fcost = min \sum_{(u,i)\in R} \frac{1}{2m} \left(r_{ui} - \hat{r}_{ui} \right)^2 \tag{6}$$

Where R is the existing ratings set and m is the cardinality of R.

We use the gradient descent method in ML to determine parameters λ_1 and λ_2 in different data. Gradient descent is an optimization algorithm used to minimize f_{COST} by iteratively moving in the direction of steepest descent as defined by the negative of the gradient. The formula is as follows:

$$\lambda_1 = \lambda_1 - \frac{\partial}{\partial \lambda_1}\left(r_{ui} - \mu - b_u - b_i - \frac{\sum_{v \in S_k(u,i)} S_{uv}(r_{vi} - b_{vi})}{\sum_{v \in S_k(u,i)} S_{uv}} \right)\frac{\alpha}{m} \quad (7)$$

$$\lambda_2 = \lambda_2 - \frac{\partial}{\partial \lambda_2}\left(r_{ui} - \mu - b_u - b_i - \frac{\sum_{v \in S_k(u,i)} S_{uv}(r_{vi} - b_{vi})}{\sum_{v \in S_k(u,i)} S_{uv}} \right)\frac{\alpha}{m} \quad (8)$$

Where α is the learning rate, formula (9) and (10) is computed as same time.

The formula (9) and (10) only stops when the change of f_{COST} is tend to converge. This method is designed for CF algorithms to find the near best parameters. This approach is run in each iteration of the co-training process, before two collaborative filtering algorithms is predicted.

Confidence Measure

Recommender system should provide confidence measure related to accurate prediction to differentiate the credibility [11]. Recommender system need simple, credible and universally reliable quality confidence measure. And Bobadilla verifies the reliability of K-NN variability confidence measure [10], which using in our method for calculate confidence of each prediction:

$$C_{ui} = \frac{|S_k(u, i)|}{\sum_{v \in S_k(u,i)} |r_{vi} - \bar{v}_{ui}|} \quad (9)$$

$$\bar{v}_{ui} = \frac{\sum_{v \in S_k(u,i)} r_{vi}}{|S_k(u, i)|} \quad (10)$$

4.3 Ensemble Step

In each iteration, we will get two different predictions. In order to unify the predictions of the two different recommendation algorithms, we use the weighted average method to unify their predictions by formula (11) [13]. Our integration part is carried out in each iteration. That is, in one iteration, only the same prediction (the same user-item pair) with the most confident of the two recommendation algorithms is ensemble.

$$\hat{r}_{ui}^{E} = \frac{\sum_{\eta=(1,2)} C_{ui}^{\eta} \hat{r}_{ui}^{\eta}}{\sum_{\eta=(1,2)} C_{ui}^{\eta}} \quad (11)$$

Where $\hat{r}_{ui}^{\eta}$ is prediction which generate by algorithm η, C_{ui}^{η} is the confidence of prediction $\hat{r}_{ui}^{\eta}$.

4.4 Recommendation Task

The output of our method is three enriched sets T_L^2, T_L^2 and T_L^E, which can then be used to enrich the original user-item matrix. New recommender can use the resulting enriched matrix to make rating prediction for user-item pairs (unrated) that are still missing from these matrices, or for new users and items that have not appeared.

5 Evaluation

In our experiments, we applied a co-training process with User-KNN and Item-KNN as the recommendation algorithm (see Sect. 3), and Cosine correlation as a measure of proximity the parameter M is set 0.1. To evaluate our proposed method, we compare it to the independent results of the User-KNN and the Item-KNN as baselines. In addition, we compared it with the original method.

5.1 Datasets

- **Book Crossing**: was developed in [20], contains 278,858 users providing 1,149,780 ratings (explicit/implicit) about 271,379 books. We re-scaled the original ratings (0 to 10) into a new range (0.5 to 5), with 102,963 interactions made by 6,628 users to 7,164 books [13].
- **MovieLens 2 k**: Includes 855,599 ratings (0.5 to 5) with 2,113 users and 10,197 movies.
- **Jester**: A total of 13,500 users rated 1810455 of 100 jokes. In our experiments, we use only one sample of the original data set and selected random users according to the recommendations in [14]. The result was 5,000 users interacting with 100 projects for 360,917 interactions. The original rating (−10 to 10) has also been re-scaled to 1 to 5.

5.2 Experimental Setup and Methodology

We get three different enrich data (T_L^1, T_L^2 and T_L^E) after our method. The comparison results include two baseline algorithms, our method and ECoRec. For the parameter X, we select 20, 30 and 40 as the experimental parameters. Although in our method a most reliability confidence measure can filtering unreliable predictions, but, less noise data is still added to the matrix. For minimize noise, parameter X should not be set too large. We set two parameters in the ECoRec and baseline as 10 and 15 respectively.

For test the predictive power of different methods, we use the root mean square error (RMSE) for ever predicted ratings $\hat{r}_{ui}$, which is exist ratings r_{ui} can found in test dataset:

$$\text{RMSE} = \frac{1}{|U|}\sum\nolimits_{u \in U}\sqrt{\frac{1}{|O_u|}\sum_{i \in O_u}\left(\hat{r}_{ui} - r_{ui}\right)^2} \quad (6)$$

Where U is a set of all users, O_u is a set of all items rated by user u.

In order to improve the reliability of the final test results, we divided the data into 10 equal parts for cross-validation, and the final result was taken as the average value after multiple tests.

5.3 Result

Table 1 compares the results of User-KNN and Item-KNN, it's include applied to the original training data (T_L) and three enriched datasets T_L^1, T_L^2 and T_L^E, for different parameters X (new ratings/users). Notice, bold typeset indicates the best performance in each dataset.

Table 1. Comparison between our method and standalone KNN-based recommenders in terms of RMSE

Recommender	User-KNN				Item-KNN			
User-item matrix	T_L	T_L^1	T_L^2	T_L^E	T_L	T_L^1	T_L^2	T_L^E
Book crossing								
20 new ratings	0.9635	0.9153	0.9351	0.8934	0.8705	0.8220	0.8204	0.8201
30 new ratings		0.9234	0.9144	0.9531		0.8049	0.8021	0.7937
40 new ratings		0.9101	0.9202	0.8233		0.7953	0.7942	**0.7731**
MovieLens 2k								
20 new ratings	0.8014	0.7957	0.8001	0.7950	0.7751	0.7726	0.7783	0.7619
30 new ratings		0.7963	0.7932	0.7903		0.7673	0.7767	0.7601
40 new ratings		0.7864	0.7903	0.7832		0.7632	0.7726	**0.7573**
Jester								
20 new ratings	0.9643	0.9363	0.9421	0.9329	0.9342	0.9297	0.9361	0.9250
30 new ratings		0.9357	0.9414	0.9322		0.9287	0.9322	0.9213
40 new ratings		0.9346	0.9420	0.9243		0.9253	0.9301	**0.9177**

Table 2 is analogous, but it is specifically designed to evaluate cold start problems. It reports an RMSE calculated only for users with number of ratings below 20 in the training dataset. Table 2 shows the cold start problem can be reduced by our method. Compare to the baselines, our method.

Figure 2 shows the results of comparing our method and ECoRec. The best enrich matrix and recommendation algorithm for each data set in Table 1, and the best result obtained from ECoRec. The best result of all three data sets, our method yielded better results than ECoRec in terms of overall RMSE.

The experiments have shown that using our method to enrich the user-item matrix, we get better results than using original dataset to train standalone recommenders. The reason is that sparsity and cold start in the original datasets can be reduced by enriching unobserved entries with most confident predictions obtained from different recommendation algorithms.

Our improved ECoRec approach can successfully increase the number of ratings, resulting in a reduction in RMSE in User-KNN and Item-KNN. The results of Tables 1, 2 and Fig. 2 show that our method has some improvement in different fields of data.

Table 2. KNN-based recommenders with our method data enrichment in terms of RMSE: cold-start scenario.

Recommender	User-KNN				Item-KNN			
User-item matrix	T_L	T_L^1	T_L^2	T_L^E	T_L	T_L^1	T_L^2	T_L^E
Book crossing (2,675 new users)								
20 new ratings	1.2427	1.1003	1.1341	1.023	1.0926	0.9721	0.9932	0.9738
30 new ratings		0.8973	0.9032	0.8528		0.8221	0.8392	0.8002
40 new ratings		0.8652	0.8831	0.8023		**0.8021**	0.8521	0.8323
MovieLens 2k (113 new users)								
20 new ratings	1.0301	1.0201	1.1016	1.0032	0.9841	0.9811	0.9883	0.9803
30 new ratings		0.9832	0.9876	0.9632		0.9523	0.9632	0.9552
40 new ratings		0.9532	0.9732	0.9421		0.9472	0.9534	**0.9362**
Jester (51 new users)								
20 new ratings	1.0031	0.9832	1.0058	1.0103	1.0204	0.9923	1.0076	1.0023
30 new ratings		0.9726	0.9934	0.9682		0.9762	0.9804	0.9622
40 new ratings		0.9688	0.9882	0.9662		0.9542	0.9773	**0.9422**

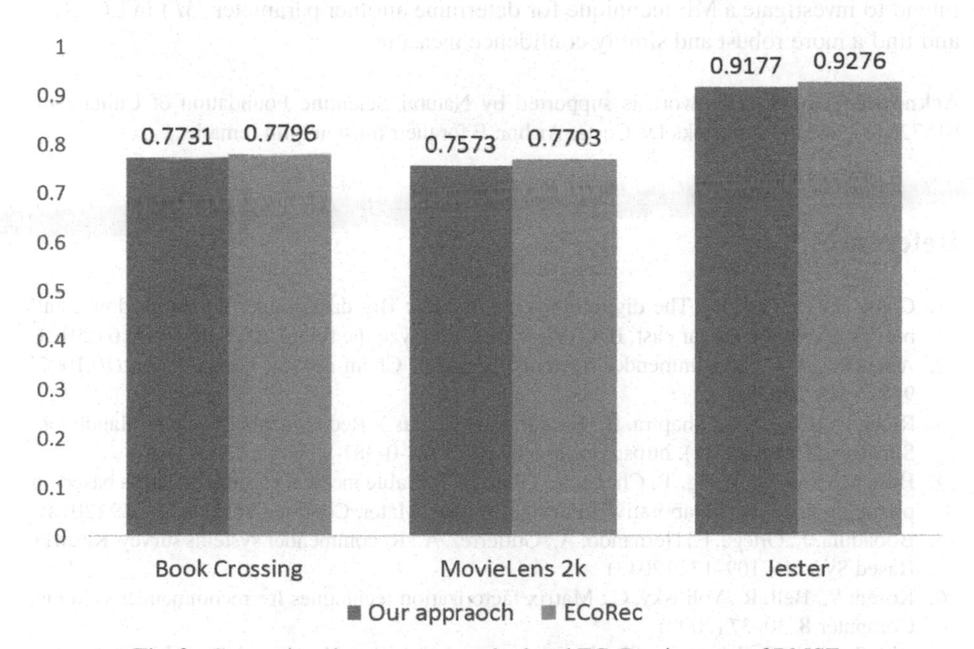

Fig. 2. Comparison between our method and ECoRec in terms of RMSE.

Improvements in RMSE indicate that the increased ratings accurate due to three factors: 1) After enrich the user-item matrix with our method, the prediction accuracy be improved and the spares and cold start problem can be reduced; 2) we use ML to determine parameters in CF which used in ECoRec for improve the accuracy; 3) we use a most reliability confidence measure in ECoRec to filter noisy.

The number of new ratings per user affects performance in a reasonably predictable way. In general, increasing will tend to improve performance, but only to a certain extent depending on the data, beyond which performance may stabilize or begin to decline. Performance dropped in most datasets when $X > 40$, this is because of that samples become less reliable.

6 Conclusion and Future Work

In this paper, two improvements, machine learning methods to determine a priori parameters and a new confidence computation method to filtering noisy, are made to the ECoRec. A variety of (two or more) different recommendation algorithms can be implemented in ECoRec. The main advantage of our method is to reduce the data sparsity and cold start by enrich the original matrix and use the ML technique to determine a priori parameters without any prior knowledge in ECoRec. The result of all datasets shows that the accuracy of our approach is better than ECoRec.

As future work, we address the low efficiency of the proposed system for processing large data sets, and aim to how to incorporate different types of datasets. Finally, we intend to investigate a ML technique for determine another parameter (M) in ECoRec, and find a more robust and simply confidence measure.

Acknowledgement. This work is supported by Natural Scientific Foundation of China (No. 61572013), and author thanks Da Costa, Arthur. F for their most helpful remarks.

References

1. Gantz, J., Reinsel, D.: The digital universe in 2020: Big data, bigger digital shadows, and biggest growth in the far east. IDC iView: IDC Analyze the future 2007(2012) 1–16 (2012)
2. Aggarwal, C.C.: Recommender Systems. Springer, Cham (2016). https://doi.org/10.1007/978-3-319-29659-3
3. Ricci, F., Rokach, L., Shapira, B., Kantor, Paul B. (eds.): Recommender Systems Handbook. Springer, Boston (2011). https://doi.org/10.1007/978-0-387-85820-3
4. Hou, M., Wei, R., Wang, T., Cheng, Y., Qian, B.: Reliable medical recommendation based on privacy-preserving collaborative filtering. Comput. Mater. Continua **56**(1), 137–149 (2018)
5. Bobadilla, J., Ortega, F., Hernando, A., Gutiérrez, A.: Recommender systems survey. Knowl.-Based Syst. **46**, 109–132 (2013)
6. Koren, Y., Bell, R., Volinsky, C.: Matrix factorization techniques for recommender systems. Computer **8**, 30–37 (2009)
7. Bin, S., et al.: Collaborative filtering recommendation algorithm based on multi-relationship social network. Comput. Mater. Continua **60**(2), 659–674 (2019)
8. Liu, G., Meng, K., Ding, J., Nees, J.P., Guo, H., Zhang, X.: An entity-association-based matrix factorization recommendation algorithm. Comput. Mater. Continua **58**(1), 101–120 (2019)

9. Blum, A., Mitchell, T.: Combining labeled and unlabeled data with co-training. In: Proceedings of the Eleventh Annual Conference on Computational Learning Theory (COLT 1998), pp. 92–100. ACM, New York (1998)
10. Bobadilla, J., Gutiérrez, A., Ortega, F., Zhu, B.: Reliability quality measures for recommender systems. Inf. Sci. **442–443**, 145–157 (2018)
11. Mazurowski, M.A.: Estimating confidence of individual rating predictions in collaborative filtering recommender systems. Expert Syst. Appl. **40**(10), 3847–3857 (2013)
12. Da Costa, A.F., Manzato, M.G., Campello, R.J.G.B.: CoRec: a co-training approach for recommender systems. In: SAC 2018. ACM, New York (2018)
13. Da Costa, A.F., Manzato, M.G., Campello, R.J.G.B.: Boosting collaborative filtering with an ensemble of co-trained recommenders. Expert Syst. Appl. **115**, 427–441 (2019)
14. Goldberg, K., Roeder, T., Gupta, D., Perkins, C.: Eigentaste: a constant time collaborative filtering algorithm. Inf. Retrieval **4**(2), 133–151 (2001)
15. Koren, Y.: Factorization meets the neighborhood: a multifaceted collaborative filtering model. In: Proceedings of the 14th ACM SIGKDD International Conference on Knowledge Discovery and Data Mining, pp. 426–434. ACM, New York (2008)
16. Su, X., Khoshgoftaar, T.M., Zhu, X., Greiner, R.: Imputation-boosted collaborative filtering using machine learning classifiers. In: Proceedings of the 2008 ACM Symposium on Applied Computing, pp. 1–2. ACM, New York (2008)
17. Sun, S.: A survey of multi-view machine learning. Neural Comput. Appl. **23**(7–8), 2031–2038 (2013). https://doi.org/10.1007/s00521-013-1362-6
18. Wang, J., De Vries, A.P., Reinders, M.J.T.: Unifying user-based and item-based collaborative filtering approaches by similarity fusion. In: Proceedings of the 29th Annual International ACM SIGIR Conference on Research and Development in Information Retrieval, pp. 501–508. ACM, New York (2006)
19. Zhang, M., Tang, J., Zhang, X., Xue, X.: Addressing cold start in recommender systems: a semi-supervised co-training algorithm. In: Proceedings of the 37th International ACM SIGIR Conference on Research & Development in Information Retrieval, pp. 73–82. ACM, New York (2014)
20. Ziegler, C.-N., McNee, S.M., Konstan, J.A., Lausen, G.: Improving recommendation lists through topic diversification. In: Proceedings of the 14th International Conference on World Wide Web, pp. 22–32. ACM, New York (2005)

Privacy Security Classification (PSC) Model for the Attributes of Social Network Users

Yao Xiao[(✉)] and Junguo Liao

School of Computer Science and Engineering, Hunan University of Science and Technology, Hunan, China

xiaoyao6812@gmail.com

Abstract. With the development of technology and the increasing popularity of smart devices, more and more people use social network sites. When users express their opinions on the Internet, their personal privacy may be inadvertently exposed, which make the privacy issue more obvious. At present, many existing studies consider methods to encrypt or calculate privacy ratings without considering whether this is unreasonable for social networks. Meanwhile, these studies ignore the limitation of memory resources. In order to protect the privacy of sensitive information with limited resources, classification of privacy information is particularly important. In this paper, we discuss a privacy security classification model for the attributes of online social network users. Through this method, the privacy risk degree of user attributes on social network can be clearly understood. In addition, we introduce visibility innovatively into the privacy security classification model and provide a reference for future research.

Keywords: Social network · Privacy security classification · Privacy risk · Visibility

1 Introduction

With the rapid development and popularization of mobile intelligent terminal devices such as notebook computers, the mobile internet has developed rapidly and surpassed the traditional fixed Internet in just a few years. According to CNNIC's "42nd Statistical Report on Internet Development in China" [1], as of June 2018, the number of mobile Internet users in China reached 788 million. In the first half of the year, 35.09 million new mobile Internet users were added, an increase of 4.7% from the end of 2017. These mobile devices expand the traditional network service platform and bring a lot of convenience. The premise of using these services is to allow the service provider to collect personally data.

In the process of data collection and processing, the data should be protected according to the type and importance of the data. If the data is stored improperly or not identified, it is easy to lose and leak important data, especially personal information. At present, there is no mature data classification and grading standard in China. Each application field often only relies on data resource catalogs for data differentiation. For different types

X. Sun et al. (Eds.): ICAIS 2020, LNCS 12240, pp. 510–522, 2020.
https://doi.org/10.1007/978-3-030-57881-7_45

of data and different security protection levels, we still can't specify different security protection requirements, and lack of data classification and grade protection mechanism. For online social network, due to their nature, the site data flow is extremely large. If operators implement the same level of privacy protection for data security, it is easy to cause great waste of resources. In that case, it is a big challenge that how to make rational use of limited resources in the privacy protection of social networks.

In recent years, online social networks (OSNs) have been greatly expanded and become an indispensable part of human life. People can chat and share news, photos, video and other resources through OSNs. In this environment, many researchers are concerned with social identity links across multiple OSNs [2–4]. In this trend, malicious attackers can integrate a complete online role through configuration files across multiple ONSs, which may cause serious harm to individuals. At present, there is no perfect solution to this problem because users have varied requirements for privacy protection that depend on the context. In our study, we found that social networking sites protect their privacy through profiles. Therefore, using profile information is a good way to measure personal privacy. We must determine a method to quantify the privacy risk of social network users' attributes to protect privacy.

For social network user's attribute, we measure the privacy risk as a combination of the privacy risk of user profile attribute, for example, the user's name, email, hometown, phone number, etc. The contribution of each profile attribute to the privacy risk depends on the sensitivity of the attribute and the visibility it gets due to user's privacy settings and user's position in the network.

Attribute sensitivity and visibility are important factors that affect the security risks of users' private information. The sensitivity of attributes increases, so does the risk of disclosure of personal information. The more it spreads, the higher the attribute's visibility, the greater the risk of privacy. According to the importance of personal information attributes and the degree of spread in the network, the classification and grade protection method can be adopted for personal information. It can be avoiding the disadvantages of one-sided protection caused by "one size fit all", and carry out comprehensive and targeted protection of information.

The next section discusses related works. Section 3 and Sect. 4 specify the design and the mathematical formulation for calculating the attribute sensitivity and visibility. Sections 5 specify the design for privacy classification model. Section 6 presents the experimental results of the model using real social network data. The final section presents conclusions and future direction for this work.

2 Related Work

With the unrestrained success of online social networks, there has been increasing research interests about privacy protection methods for individuals that participate in them. Most of the research main focuses on three strategies: (1) use data anonymity and generalization technology to protect the identity of social network users; (2) design privacy protection mechanism involving access control rules and policy defense; (3) measure the privacy level of users and provide practical methods for users to assess the actual risks.

In most cases, disclosure of information online is the result of voluntary activity, which undoubtedly leads to many privacy issues. In the realm of social networking privacy, there is a common view that users should care about their privacy when interacting with other social network users. As a result, the problem of measuring and improving risk perception is becoming more and more popular among researchers. For example, Cetto [5] provided an online game that allows Facebook users to test their knowledge of the actual sharing of personal items and provide them with suggestions for improving their privacy settings. Fang and LeFevre [6] also used the active learning methods to estimate user risk, because user interaction was rarely required in the methods. The privacy metric can help people to measure the degree of privacy enjoyed by users in the system and the level of protection provided by privacy enhancement technology [7], which has attracted some interest in important research. For example, Liu and Terzi [8] proposed a framework for calculating privacy scores that measure the potential privacy risks of users. This score increases with the sensitivity of the information item and its visibility, and they use the project response theory [9] as the theoretical basis for the mathematical formula of the score. Similarly, Wang [10] use a privacy index to measure user privacy exposure in social networks. However, unlike Liu and Terzi, this index requires a predefined sensitivity value for the user's project and requires the availability of user privacy settings. Gao [11] proposed a classification and grade method for applying personal information to big data, but their method only starts from the frequency of data flow and the number of users, and does not study the data itself and the privacy attitude of users. Lou [12] obtained the classification result of medical data by the k-means clustering, without considering the spread of data in the population. In addition, some studies from the social network itself, discussed the user information in the network transmission nature, and gave substantial protection scheme [13–15].

The work is inspired by the classification and grade method defined by Gao [11], for which we have greatly improved the measurement method in [11] to adapt it to scenarios with multiple social networks to reflect how OSNs are currently used by users. In addition, we intuitively added discussion and use the attribute visibility to make the measurement of privacy risk more accurate.

In this paper, our main contributions are as follows:

- Based on the research in the privacy attributes of social network users, we propose a method to calculate the sensitivity and visibility of user attributes;
- We proposed a user privacy security classification model (PSC);
- We report the classification results for real network user raw data.

3 Attribute Sensitivity

3.1 Definition of Sensitivity

Sensitivity shows the risk associated with the attributes of the user. When the sensitivity of an attribute increases, the risk posed by information disclosure of the individuals also increases. It depends on the nature of the attribute [8]. Some attributes are, by nature, more sensitive than others. For example, the mobile-phone number is considered more sensitive than employer for the same privacy level.

3.2 Sensitivity Calculation Method

Privacy leakage is ubiquitous in social network user information, past research has identified inherent privacy risks. Even if users' activities in social networks are known, measuring their inherent privacy risks is not easy. The sensitivity calculation method is based on the privacy score defined by Liu and Terzi. Each user i can disclose information related to a set of n attribute $T = \{t1, t2, \ldots, tn\}$, such as religion, workplace, political opinion, health status, place of birth, gender, age, vacation.

Here, we assume that all n users specify their privacy settings for the same N attribute. These settings are stored in an $N \times n$ response matrix R. The value of the matrix is represented by $r(i,j)$, $r(i,j) \in \{0, 1, 2, \ldots, \ell\}$, $r(i,j) = 0$ means that the attribute j of the user i refuses to be viewed by anyone, $r(i,j) = 1$ means that the friend of the user i can view the attribute j, $r(i,j) = 2$ means that the friends of friends of user i can view the attribute j, and so on. An example of item response matrix R for four users and three attributes is shown in Table 1:

Table 1. An example of item response matrix

User	Attribute		
	Job	Age	Name
User 1	2	2	1
User 2	3	2	3
User 3	3	3	2
User 4	4	2	4

The response matrix R represents the user's attitude towards related attributes, so it can be used to calculate the sensitivity of the user's attributes. Liu and Terzi [8] used a mathematical model based on polynomial response theory to calculate sensitivity. Unlike Liu and Terzi, we adopt a simpler and more efficient method, which takes less time to calculate. In the method, for any value of $r(i,j) = h = \{0, 1, 2, \ldots, \ell - 1\}$, the sensitivity is calculated as follows:

$$\beta_{jh} = \frac{1}{2}\left(\frac{n - \sum_{i=1}^{n} 1_{(r(i,j) \geq h)}}{n} + \frac{n - \sum_{i=1}^{n} 1_{(r(i,j) \geq h+1)}}{n}\right) \quad (1)$$

Where 1_A is the indication function that returns 1 when the condition A is true (0 otherwise).

When $h = \ell$, the sensitivity value of attribute j is expressed as

$$\beta_{j\ell} = \frac{n - \sum_{i=1}^{n} 1_{(r(i,j) \geq \ell)}}{n} \quad (2)$$

For the above formula, the more users adopt the privacy level of at least h for attribute j, the less sensitivity is with regard to level h. In the paper, as shown in the following formula (3), we define the sensitivity of user attribute j is the mean of all privacy levels sensitivity of the attribute.

$$\beta_j = \frac{\sum_{h=0}^{\ell} \beta_{jh}}{\ell + 1} \tag{3}$$

4 Attribute Visibility

4.1 Definition of Visibility

The visibility of attribute depends on the spread of the attribute in the Internet. The more it spreads, the higher the attribute's visibility [8].

4.2 Visibility Calculation Method

In the privacy protection of sensitive information, it is not advisable to focus only on information sensitivity and ignore visibility. We assume that the identity information is very sensitive and the visibility is very high. The information may be having a significant privacy risk because the information can be easily accessed by an attacker, and may be posed a great threat to the information owner. Compared to other information with high sensitivity and low visibility, the identity information needs stronger privacy protection.

The visibility of attributes is derived from Liu and Terzi [8], and it depends on the spread of the attribute in the Internet. Based on their work, we added a fine-grained method to obtain more accurate results. In this paper, we calculate the information accessibility δ, the difficulty ε of information data extraction, the data reliability γ, and the privacy attitude ω for each attribute on each platform. These values as input and we use the fuzzy c-means clustering algorithm to obtain the visibility value.

4.3 Accessibility

In general, OSN operators allow users to set accessibility for each attribute in their profile. Accessibility represents how many people can access attribute content. According to OSN Settings, we define four different levels of accessibility: 1 means that content is accessed only by the information owner; 2 means that close friends can access relevant content; 3 represents that the content can be access by specific groups of people, such as teachers; and 4 represents public available content.

Due to our research is based on multiple platforms, we calculate each attribute separately for each platform. Moreover, we consider that not all users fill in the same content for one attribute on each platform. Therefore, we statistics the user profile information, as shown in Table 2, 1–4 represents the accessibility of the attribute and A–Z represents the content of the attribute. In addition, the same letter in the same attribute of different users does not mean that the content is the same, 0 means that the user does not disclose any attribute information.

Table 2. Attribute content and accessibility

Attribute	Platform			
	Platform 1	Platform2	······	Platform s
Attribute 1	1A	3A	······	4B
Attribute 2	2A	0	······	2B
Attribute 3	1A	3B	······	2A
Attribute 4	······	······	······	······
Attribute 5	3A	2A	······	3A

Through Table 2, we can use following formula (4) to calculate the accessibility for attribute.

$$\delta_n = \frac{\sum_{f=1}^{s} \delta_n^f}{s} \tag{4}$$

Where δ_n^f is the accessibility of attribute n on the f th platform, and we sum all the values of δ_n^f and average it to obtain the δ_n that represents the total accessibility of attribute n on social network platforms with the number of s.

Here, we assume that the content of the user attribute n is (2A, 2A, 3B, 4A), accessibility δ_n of the attribute n is $(2 + 3 + 4)/3 = 3$. Repeated items are removed because they do not provide additional privacy losses. It is worth noting that values of 0 is excluded to address cases like (0, 0, 4C, 0), where if we do not remove 0, the result is 2, which is unreasonable because this attribute has been publicity.

4.4 Extraction Difficulty

An important factor in measure privacy risk is the difficulty of extracting privacy information from data in different formats. It is much easier to extract attributes from structured data than from unstructured data [16].

Extracting the attribute content from the user profile is obviously much easier, but some attribute content may be not provided. On this case, the information has to be inferred from video and other approaches. It is more difficult for a person's religious belief to be judged from the picture than from the user's personal summary. According to the difficulty of data extraction, we divide it into three levels: 1 is difficult, 2 is relatively easier, and 3 is easy. The specific definition is as follows: 1 represents content obtained from a picture or other approaches, 2 represents content acquired from a text message, and 3 represents content directly acquired from user profile information. This approach is taken because not every platform provides all the attributes content in the profile.

In order to get the difficulty of data extraction, we consider that different platforms have different difficulty in data extraction, the calculation function of difficulty of data extraction given by:

$$\varepsilon_n = \frac{\sum_{f=1}^{s} \varepsilon_n^f}{s} \quad (5)$$

Where ε_n^f refers to the difficulty of extracting the attribute n on the fth platform, and ε_n is the total extraction difficulty of attribute n on social network platforms with the number of s.

4.5 Reliability

Reliability is a criterion that can provide with what confidence a particular attribute has been disclosed in one or multiple sources [16]. In this case, for each attribute of the user, we consider the overall reliability of data disclosure for each attribute of the users to consider it in total visibility calculation. The reliability of sensitive information increases with the number of validation resources. As show in formula (6), we use the sigmoid function to measure the reliability of the data.

$$\gamma_n = \frac{2}{1 + e^{-q_n}} - 1 \quad (6)$$

Where q_n is the maximum number of occurrences same content in the platform for attribute n.

4.6 Privacy Attitude

Privacy attitude refers to the user's understanding of the privacy options available on social networking sites [17]. Users with a higher level of privacy attitude often hide sensitive information or fill in false attribute content to confuse the attacker. Our approach for calculating user privacy attitudes is measured by the number of platforms on which users fill with the same content, and it is theoretically possible. For example, the content of the user attribute n is (A, A, B, A), in which at least one of content is mendacious. The calculation functions of privacy attitude given by:

$$\omega_n = 2 - \frac{2}{1 + e^{-0.5q_n}} \quad (7)$$

This function is designed because it is a 3-parameter logical model (3PL), which is a variant of the IRF. IRF provides the probability that a person with a given ability level will answer correctly. People with lower ability have less chance of answering correctly, while persons with high ability are very likely to answer correctly [18]. Therefore, this method can well express human attitude of privacy.

4.7 Visibility

Through the abovementioned work, extraction difficulty, accessibility, reliability and privacy attitude can be quantified. In order to calculate the visibility, we refer to the semi-suppressive fuzzy c-means clustering algorithm [19], which is a clustering algorithm based on FCM (fuzzy c-means algorithm). In this paper, we changed the iteration of the algorithm. If the final step determines that the iteration has not ended, then the new cluster center is restored to the preset cluster center. This clustering method was chosen because it requires only a small number of training samples to get enough results, especially considering the openness of the large sample set and the specificity of the data. Moreover, the algorithm can manually specify the cluster center, which is easy to implement in the research. According to common sense, we can establish a great clustering center similar to the difference between (1A, 1B, 1C, 1D, 1E) and (4A, 4A, 4A, 4A, 4A). When we specify a cluster center, we can use the cluster center of the aggregated sample to determine the visibility of the sample.

In the sample space, $X = \{x_j | j = 1, 2, \ldots, n; x_j \in R^p\}$ is an n-dimensional vector, and $v_i (i = 1, 2, \ldots, c)$ represents clustering of cluster i Center, u_{ij} is the membership degree of instance x_j in cluster i, which satisfies the following constraints:

$$\sum\nolimits_{i=1}^{c} u_{ij} = 1; j = 1, 2, \cdots, n \tag{8}$$

$$u_{ij} \geq 0; i = 1, 2, \cdots, c; j = 1, 2, \cdots, n \tag{9}$$

$$0 \leq \sum\nolimits_{j=1}^{n} u_{ij} < n; i = 1, 2, \cdots, c \tag{10}$$

$$v_i = \frac{\sum_{i=1}^{n} u_{ij}^{m} x_j}{\sum_{j=1}^{n} u_{ij}^{m}} \tag{11}$$

The specific algorithm process is as follows:

(1) Initialize the cluster center $v_i^{(0)}$, α is the suppression factor, β is the suppression threshold, m is the prime index factor, the error threshold is $\varepsilon > 0$, the maximum number of iterations is K, and set the number of iterations $k = 0$.

(2) Calculate $U^{(k)} = \left[u_{ij}^{(k)}\right]$ by the formula $u_{ij} = \frac{\left(\frac{1}{\|x_j - v_i^2\|}\right)^{\frac{1}{m-1}}}{\sum_{k=1}^{c}\left(\frac{1}{\|x_j - v_k^2\|}\right)^{\frac{1}{m-1}}}$.

(3) Update the fuzzy classification matrix according to the following formula (12) and formula (13) to obtain $U^{(k)'}$.

$$u_{rj} = \begin{cases} \alpha u_{rj} + (1 - \alpha), u_{rj} > \beta; \\ u_{rj}, u_{rj} \leq \beta; \end{cases} \tag{12}$$

$$u_{ij} = \begin{cases} \alpha u_{ij}, u_{ij} > \beta; \\ u_{ij}, u_{ij} \leq \beta; \end{cases} (i = 1, 2, \cdots, c; i \neq r) \tag{13}$$

(4) According to $v_i = \frac{\sum_{j=1}^{n} u_{ij}^{m} x_j}{\sum_{j=1}^{n} u_{ij}^{m}}$ and $U^{(k)'}$ calculate the new cluster center.

(5) If $\left(\sum_{i=1}^{c} \left\| v_i^{(k+1)} \right\|^2\right)^{\frac{1}{2}} < \varepsilon$ or $k < K$, the iteration ends, otherwise $k = k + 1$, $v_i = v_i^{(0)}$ return to step 2 to restart.

We use accessibility, reliability, extraction difficulty and privacy attitude as four-dimensional sample input; then we can obtain visibility through the data result after clustering method. Use the collected data through multiple experiments, the better model parameters are as follows: suppression factor was 0.5, suppression threshold was 0.5, the prime index factor $m = 2$, error threshold $\varepsilon = 0.01$, the maximum number of iterations $K = 20$. We used the above parameters in the subsequent comparative analysis and set the outcome to six categories.

5 Privacy Security Classification (PSC)

The security level of user attributes not only depends on the degree of damage to the privacy information itself, but also depends on the scope of the people involved in the privacy disclosure and the degree of privacy confidentiality [16]. This is the key to classify privacy information. Through the above method, we can calculate the sensitivity and visibility of user attribute respectively. In PSC, we ranked the attribute sensitivity value from high to low. The first 25% was defined as high sensitivity attribute; the first 50%–25% as medium sensitivity attribute; and the rest as low sensitivity attribute. Similarly, the value of visibility is five or six was defined as high visibility attribute; the value of visibility is three or four was defined as medium visibility attribute; and the value of visibility is one or two was defined as low visibility attribute. The Table 3 gives the user attribute privacy risk level determination rule.

Table 3. Privacy risk determination rule of PASC

Sensitivity	Visibility		
	Low	Medium	High
Low	Low	Low	Medium
Medium	Low	Medium	High
High	Medium	High	High

High-risk privacy: High-risk privacy refers to privacy in which the attribute sensitivity and visibility are at least one high and the other cannot be low.

Medium-risk privacy: Medium-risk privacy refers to privacy in which the attribute sensitivity and visibility are both medium, or one is low, the other one is high.

Low-risk privacy: Low-risk privacy refers to privacy in which the attribute sensitivity and visibility are at least one low and the other cannot be high.

6 Experiment

In the sensitivity study, we use Li's questionnaire data as experiment date [17]. The questionnaire surveyed Internet users' opinion about user information leakage in the popular social network platforms, and divided opinion into five levels from extremely worried to indifferent. Restricted by user profiles, we selected 11 user profile attributes such as birthday and email as research attributes, and used the above method to calculate the sensitivity of user information. Base on PSC, we obtained the sensitivity level and value of the attribute sensitivity. The results shown in Table 4.

Table 4. Sensitivity level and value of the attribute sensitivity

Attribute	Sensitivity	Level
Phone number	0.530	High
E-mail	0.370	Medium
Address	0.423	High
Birth date	0.330	Medium
Hometown	0.304	Low
Current town	0.434	High
Job details	0.286	Low
Relationship status	0.266	Low
Interests	0.223	Low
Photo	0.390	Medium
Education	0.258	Low

Through the Tablc 4, we can see that phone number, address and current town are high sensitivity attribute; e-mail, birth date and photo are medium sensitivity attribute; hometown, job details, relationship status, interests and education are low sensitivity attribute.

In the visibility study, we use the collected dataset to measurement of attribute visibility. The QQ, Weibo, WeChat, and Vibrato are the popular social platforms in China. In our experiment, we investigated 286 people in the school, including graduate students and their friends. Respondents were asked to truthfully record their personal information in these four platforms include login account and password after getting permission and make sure not to reveal true information. In the end, we get data into 2 groups, 140 user data as sample to train clustering algorithm, other data through clustering algorithm to obtain visibility result. In this paper, we sum all the visibility values of an attribute and average it to obtain the visibility. Base on PSC, shown in Table 5, we obtained the sensitivity level and value of the attribute sensitivity.

The Table 6 is the PSC classification result. It shows that phone number, address, current town have high privacy risks; e-mail, birth date and other attributes have medium privacy risk; interests, hometown have low privacy risk. For high privacy, even if these

Table 5. Visibility level and value of the attribute visibility

Attribute	Visibility	Level
Phone number	4	Medium
E-mail	3	Medium
Address	3	Medium
Birth date	3	Medium
Hometown	4	Medium
Current town	5	High
Job details	5	High
Relationship status	2	Low
Interests	4	Medium
Photo	3	Medium
Education	5	High

attributes of information are important to the information owner, the information still is used to exchange for better services, and it may be caused information leakage. The leakage of this information can allow an attacker to directly identity personal information, and personal security is seriously threatened. For medium privacy, these attributes are widely spread in social networks, but it is not particularly sensitive to users. Social networking sites are a platform for user communication. For users of social networking sites, the interest is the least important information and they are willing to share it with their friends.

Table 6. PAC classification results

High-risk privacy	Medium-risk privacy	Low-risk privacy
Phone number	E-mail, Birth date, Photo	Hometown
Address	Job details	Relationship status
Current town	Educations	Interests

7 Conclusions and Future Work

With the popularity of social networks, a large number of users are willing to share personal information in social network. Social network users usually have accounts on multiple social platforms. Once information is leaked on one platform, users will also have great privacy risks on other platforms. The privacy protection of user information across multiple platforms becomes increasingly important. Common privacy protection method is "one size fits all", all the information is encrypted in the same way to achieve the effect of privacy protection, and this method may be caused waste of storage resources. Therefore, graded protection of privacy has become particularly important. Before that, there is no doubt that we need to classify social network users' privacy information. In this paper, we simplify the calculation method of sensitivity and discussed four factors that affect visibility, such as the accessibility, difficulty of data extraction, reliability and privacy attitude. Then, we use the simplified semi-suppressed fuzzy c-means clustering algorithm to calculate the visibility. Finally, we propose a privacy security classification model, and creatively take information visibility as a factor that can affect information security. Through the privacy security classification model, we have implemented privacy classification of user information on social network.

There are some deficiencies in the research. It mainly focuses on user profile information classification, and there is a lack of consistency and rationality test of classification results. In the future, we would like to study the privacy protection methods of different security levels of information.

References

1. Statistical Report on the Development of China's Internet. www.cnnic.net.cn/hlwfzyj/hlwxzbg/hlwtjbg/201808/t20180820_70488.htm. Accessed 20 Aug 2018
2. Zhang, Y., et al.: CosNet: connecting heterogeneous social networks with local and global consistency. In: Proceedings of the 21st ACM SIGKDD International Conference on Knowledge Discovery and Data Mining, pp. 1485–1494. ACM, Sydney (2015)
3. Nie, Y., et al.: Identifying users across social networks based on dynamic core interests. Neurocomputing **2**(10), 107–115 (2016)
4. Shu, K., et al.: User identity linkage across online social networks: a review. ACM SIGKDD Explor. Newslett. **18**(2), 5–17 (2017)
5. Cetto, A., et al.: Friend inspector: a serious game to enhance privacy awareness in social networks. arXiv preprint arXiv:1402.5878 (2014)
6. Fang, L., LeFevre, K.: Privacy wizards for social networking sites. In: Proceedings of the 19th International Conference on World Wide Web, pp. 351–360. ACM, New York (2010)
7. Vuokko, N., Terzi, E.: Reconstructing randomized social networks. In: Proceedings of the 2010 SIAM International Conference on Data Mining, pp. 49–59. SIAM (2010)
8. Liu, K., Terzi, E.: A framework for computing the privacy scores of users in online social networks. ACM Trans. Knowl. Discov. Data **5**(1), 6 (2010)
9. Nering, M.L., Ostini, R.: Handbook of Polytomous Item Response Theory Models. Taylor & Francis, Routledge (2011)
10. Wang, Y., Nepali, R.K., Nikolai, J.: Social network privacy measurement and simulation. In: International Conference on Computing, Networking and Communications, pp. 802–806. IEEE, Honolulu (2014)
11. Gao, L., Li, C.S., et al.: Research on hierarchical protection of personal information in big data application. Inf. Secur. Res. **5**(5), 394–399 (2019)
12. Lou, P., Liu, L., et al.: Classification and grading of medical data based on questionnaires. Chin. J. Med. Inf. **27**(6), 22–27 (2018)
13. Sun, G., et al.: Research on public opinion propagation model in social network based on blockchain. Comput. Mater. Continua **60**(3), 1015–1027 (2019)
14. Zhu, D., et al.: MMLUP: multi-source & multi-task learning for user profiles in social network. Comput. Mater. Continua **61**(3), 1105–1115 (2019)

15. Wang, P., Wang, Z., Chen, T., Ma, Q.: Personalized privacy protecting model in mobile social network. Comput. Mater. Continua **59**(2), 533–546 (2019)
16. Aghasian, E., Garg, S., Gao, L., et al.: Scoring users' privacy disclosure across multiple online social networks. IEEE Access **5**, 13118–13130 (2017)
17. Li, X., Yang, Y., Chen, Y., et al.: A privacy measurement framework for multiple online social networks against social identity linkage. Appl. Sci. **8**(10), 1790 (2018)
18. Bock, R.D., Aitkin, M.: Marginal maximum likelihood estimation of item parameters: Application of an EM algorithm. Psychometrika **46**(4), 443–459 (1981)
19. Zhao, Q.H., Ha, M.H., Peng, G.B., et al.: Support vector machine based on half-suppressed fuzzy c-means clustering. In: International Conference on Machine Learning and Cybernetics, vol. 2, pp. 1236–1240. IEEE, Hebei (2009)

Clustering Analysis of Extreme Temperature Based on K-means Algorithm

Wu Zeng, YingXiang Jiang(✉), ZhanXiong Huo, and Kun Hu

Department of Electrical and Electronic Engineering, Wuhan Polytechnic University, Hubei 430023, China
m18290018106@163.com

Abstract. The research is based on the 2 GB extreme climates data recorded from 825 meteorological stations in mainland cities of China which was published online by the National Climate Center of the China Meteorological Administration. It analyzes the changes of extreme temperature in different time in the country and the spatial distribution of extreme temperature. The k_means clustering algorithm in data mining is used to study the regional aggregation of extreme temperature time across the country, and the experimental results are visually displayed and introduced. The analysis results show that the frequency and area of extreme high temperature events are increasing over time throughout the country. Extreme temperature events not only have regional aggregation, but also the occurrence of extreme temperature events with regional mobility. In this paper, the Apriori Algorithm of mining association rules in data mining technology is used to study the regional association pattern of extreme climate events in small areas. On this basis, the K-means Clustering Algorithm is used to identify and express the regional aggregation of time and space data in the extreme temperature events nationwide, which provides a new idea and method for the study of extreme climate events.

Keywords: Extreme temperature events · Data mining · Clustering algorithm

1 Introduction

Since the second half of the 20th century, global warming and the increase in extreme weather and climate events have had a huge impact on the economies and societies of countries around the world. As a low-probability event, extreme climate events do not yet have very uniform definition standards. However, from the perspective of various defined standards, extreme climate events are extreme conditions in climate events and have strong destructive and influential characteristics. Extreme weather events often lead to the occurrence of various meteorological disasters, which not only cause huge casualties and economic losses to people, but also have different degrees of influence on the environment, society and agriculture in different fields. With the rapid development of atmospheric exploration, a large amount of meteorological data has been generated, and the daily data growth of meteorological departments is very large. At present, the

X. Sun et al. (Eds.): ICAIS 2020, LNCS 12240, pp. 523–533, 2020.
https://doi.org/10.1007/978-3-030-57881-7_46

new meteorological data in China reaches PB order of magnitude every year, which has increased thousands of times from the 1990s and is still growing rapidly. This data is huge and includes different data types. Experts (Shen 2014) said that meteorological data has the common characteristics of "big data", that is, the characteristics of huge amounts of data, rapid data growth, and diverse data types, and so on (Sui et al. 2016). The efficient organization and full use of massive data have provided solid support for the advancement and development of science and technology in meteorology and other fields.

In the past, statistical methods were commonly used to study and accumulate meteorological data. Using the methods of anomaly, sliding average, least squares, and linear regression (Zhai and Pan 2003; Zhou and Ren 2010), it was concluded that the average maximum temperature of our country was slight increased. The lowest temperature was significantly warmer, and the daily difference was significantly smaller. The linear changed trend of the lowest and highest temperature showed a more consistent decadal variation, but it reflected a very obvious trend of asymmetry. Statistical methods usually need to convert these data statistics into the patterns we need, and then bring the results into the model through artificial experience and simulations and establish a digital model. Then we can know how many days of precipitation in the country and where the distribution. We know how many places in the country experience extreme weather events and how the trend is compared with the past. From that we can get the climate distributions with accumulated data for many years.

With the rapid growth of meteorological data, traditional statistical methods need to gradually analyze each type of data step by step when dealing with a large amount of data, and the efficiency is slow. The K-means algorithm in the clustering algorithm (Yang et al. 2019; Sun et al. 2013) can model the climate data, cluster iteratively the spatial attribute data of China's extremely high-temperature data and extend the data volume of the small-scale region to the data of the large-scale region. As a result, all spatial sites are clustered and classified, and then the clustered spatial sites are visualized on the map of China (Sui et al. 2016). It is used to identify and express the regional concentration of spatial and temporal data in extreme temperature events across the country, which has the advantage of strong portability. This provides a new idea and method for the study of extreme climate events.

2 Research Status of Clustering in Weather Processing

With the development of the database and data mining technologies, cluster analysis has become more and more widely used in meteorology.

Chinese scholars have made many achievements in spatiotemporal analysis of extreme temperature events. In the study of temporal and spatial variation characteristics of local areas, there are spatial-temporal changes in the temperature in the Yangtze River valley (Wang and Yin 2015), the Huaihe River valley (She et al. 2011), and Tibet (Wang et al. 2004) and other areas during the last 50 years. In the analysis of independent stations, there are analyses of the characteristics of extreme temperature changes in Shanghai and the impact on urbanization processes. In addition, there are studies on the clustering laws of extreme temperature events, defining K-order nearest-neighbor clustering point extraction algorithms and analyzing the spatial distribution characteristics of extreme temperature and precipitation (Jin et al. 2010). Professor Yang Bingru, School of Information Engineering, University of Science Technology Beijing, established a brand new multi-level hierarchy in the sub-item of the project "Data Mining and Knowledge Discovery in Agricultural Meteorological Forecasts" of the National Meteorological Administration "Study on Environmental Meteorological Information System in Big Cities". The intelligent meteorological data knowledge discovery system architecture and the close integration of the application background, proposes and implements a new mechanism for knowledge information processing of complex data types. Dr. Ping Yang used K-order nearest neighbor distance extraction algorithm to study the regional crowding of extreme temperature events (Yang 2009). K-order nearest neighbor is an algorithm based on analog learning, and it can be used for both clustering and classification. In the research of meteorological data, the decision tree or decision table can be used to process the data to obtain meteorological rules by using cluster analysis method and data mining technology, which can help to further improve data processing quality for professional meteorological services, website construction and related software development.

3 Data and K-means Algorithm

3.1 Data Preparation and Processing

The temperature data used in this paper is in mainland China published by the National Climate Center of the China Meteorological Administration from 1951 to 2010. The data of each meteorological station was used to select the daily maximum (low) temperature to build the weather data set in 52 years. Removing the data which has l large number of missing or obvious errors from the stations, counting the time interval and picking up the stations that will be used. At the same time, design the R program. Using the data from 1971 to 2000 as the reference period data, the extreme temperature event threshold was calculated and the original transaction set of data mining was obtained.

The frequency of national extreme weather events after data processing is shown in Fig. 1 and Fig. 2. The trend of the frequency of extreme weather events (days above the threshold) for each site during 1958–2010 was estimated from a linear trend.

At this point, the data pre-processing process is completed, and related issues can be studied next.

3.2 Algorithm

The Apriori algorithm of mining association rules in data mining technology is used to study the regional association pattern of small-scale extreme temperature events. On this basis, K-means clustering algorithm is used to identify and represent the regional aggregation and regional mobility in the national extreme temperature events spatiotemporal data set. The following is a preliminary introduction to two algorithms:

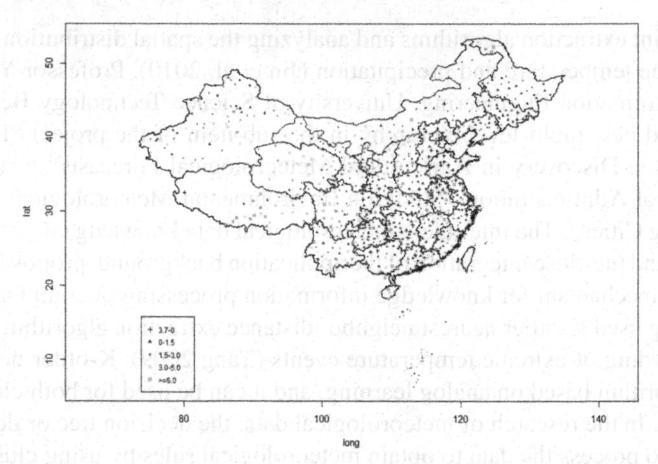

Fig. 1. Spatial distribution of frequency trends of extremely high-temperature in China from 1958 to 2010. The abscissa is longitude, and the ordinate is latitude

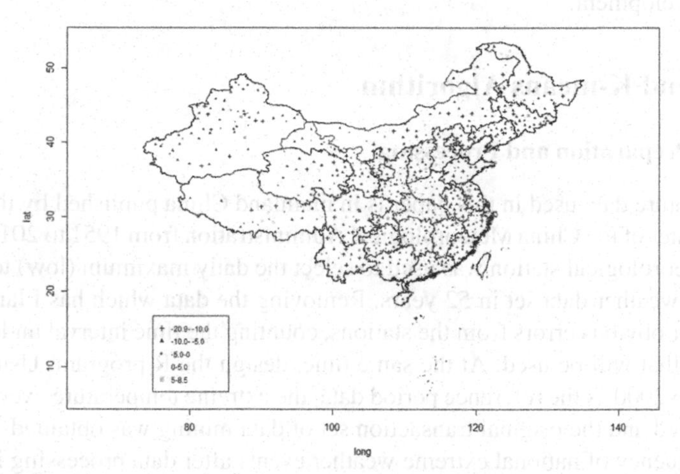

Fig. 2. Spatial distribution of frequency trends of extremely low temperatures in China from 1958 to 2010. The abscissa is longitude, and the ordinate is latitude

Apriori Algorithm

The essence of Apriori algorithm uses candidate item set to find frequent item set. Apriori algorithm is one of the most influential algorithms for mining frequent itemsets of Boolean association rules. Therefore, Apriori algorithm is chosen to mine spatial association rules in a small range.

The name of the algorithm is based on the fact that the algorithm uses prior knowledge of the properties of frequent itemsets, as we will see. Apriori uses an iterative method called layer by layer search. The k-term set is used to explore the (K + 1) - term set. First, we find the set of frequent 1-term sets. This set is recorded as L1. L1 is used to find the set L2 of frequent 2-itemsets, and L2 is used to find L3, so on until frequent k-itemsets cannot be found. A database scan is required to find each LK. Its core is a recursive algorithm based on two-stage frequency set. The association rules belong to single dimension, single level and Boolean association rules. In this case, all the item sets with more support than the minimum support are called frequent item sets, or frequency sets for short.

The basic idea of the algorithm is to find out all the frequency sets, and the frequency of these item sets is at least the same as the predefined minimum support. Then, strong association rules are generated by frequency set, which must satisfy the minimum support degree and the minimum credibility. Then use the frequency set found in step 1 to generate the expected rules, and generate all the rules that only contain the items of the set. There is only one rule on the right side of each rule. Here is the definition of the rule in. Once these rules are generated, only those rules that are greater than the minimum confidence level given by the user will be left behind. In order to generate all frequency sets, the recursive method is used.

The key of Apriori algorithm is to find LK with LK − 1, which is composed of the following two steps:

(1) Linking and pruning: Connection step: to find LK, connect LK − 1 with itself to generate a set of candidate k-term sets. The set of the candidate set is recorded as CK. Let l1 and l2 be the set of terms in LK − 1. The mark li [J] indicates the j-th term of Li. For ease of calculation, it is assumed that items in a transaction or item set are ordered in dictionary order. The link LK − 1 is executed; where the elements of LK − 1 are connectable if their first (K − 2) items are the same; that is, the elements l1 and l2 of LK − 1 are connectable if $(l_{1[1]} = l_{2[1]}) \wedge (l_{1[2]} = l_{2[2]}) \wedge \ldots \wedge (l_{1[k-2]} = l_{2[k-2]}) \wedge (l_{1[k-1]} = l_{2[k-1]})$. The condition $\left(l_{1[k-1]} = l_{2[k-1]}\right)$ is simply that there is no repetition. The resulting set of items generated by connecting L1 and L2 is $l_{1[1]}\, l_{2[1]} \cdots l_{1[k-1]}\, l_{2[k-1]}$.
(2) Pruning step: CK is a superset of LK; that is, its members may or may not be frequent, but all frequent k-term sets are included in CK. Scan the database to determine the count of each candidate in the CK, so as to determine the LK (that is, according to the definition, all candidates whose count value is not less than the minimum support count are frequent, so they belong to LK). However, CK may be very large, so the amount of calculation involved is very large. To compress CK, Apriori property can be used in the following way: any non frequent (k − 1) - item set cannot be a subset of frequent k-item set. Therefore, if the (k − 1) - subset of a candidate k-term set is not in LK − 1, then the candidate can not be frequent, so it can be deleted by CK. This subset test can be done quickly using a hash tree of all frequent itemsets.

In data mining, the key problem lies in the selection of support and confidence threshold. Support minsup is the percentage of tuples (or transactions) related to tasks whose patterns are true, which is used to measure the potential usefulness of the patterns; confidence minconf is used to measure the accuracy or reliability of the discovered patterns; meanwhile, association rules that meet the user-defined minimum confidence threshold and minimum support threshold are called strong association rules, which are considered to be interesting. The setting of threshold will directly affect the quantity and quality of mining results. Appropriate threshold can ensure the accuracy, interest and scientificity of association rules and the quality of algorithm mining results. Therefore, it is necessary to set the threshold reasonably according to the internal characteristics of data sets, generally through experiments or thresholds.

The K-means Algorithm
The K-means algorithm is a clustering method based on the nearest neighbor method (Xiong et al. 2011). It is suitable for mining large data sets, and it selects the related ones to attributes from multiple dimensions, removes the irrelevant dimensions, clusters them along relevant dimensions to cluster high-dimensional data. K-means algorithm is an indirect clustering method based on the similarity measure between samples (Tan et al. 2019), which belongs to unsupervised learning method (Tian and Wen 2011). This algorithm takes k as a parameter, and divides n objects into k clusters, so that the clusters have higher similarity, and the similarity between clusters is lower.

Firstly, this algorithm selects k objects in randomly, each of which represents the centroid of a cluster. For each of the remaining objects, it is assigned to the most similar cluster based on the distance between the object and each cluster's centroid. Then, calculate the new centroid for each cluster. Repeat the above process until the criterion function converges. The k-means algorithm is a typical dynamic clustering algorithm with the function of points iteration modification (Maamar and Benahmed 2019; Jiang and Li 2006). The main purpose is to use the sum of errors squares as a standard function to modify the class center point by point. After a pixel sample is assigned to a group of classes according to a certain principle, the mean value of this group should be recalculated, and the next average image element clustering should be performed with the new average value as the focal point; Modifying the class center batch by batch: After all the pixel samples are classified according to the class center of a certain group, the mean value of each class is calculated and modified as the cohesive center point of the next class.

The specific steps of K-means clustering algorithm are as follows:

(1) Considering that the spatial distance between samples represents the similarity between samples, Euclidean distance is often used to calculate this characteristic, and sum squared error (SSE) is used as the objective function. The minimum SSE is taken as the initial clustering center (Wang et al. 2012), and the k-mean is obtained from the data set.
In other words, we calculate the error for each data point, that is, its Euclidean distance to the nearest centroid, and then calculate the sum of squared errors. Given two different clusters resulting from two runs of the K-means, we select the one with the smaller squared error, because this indicates that the centroid of the cluster can better represent the cluster center (Zhai et al. 2014). SSE is defined formally as follows:

$$SSE = \sum_{i=1}^{k} \sum_{x \in C_i} dist(C_i, x)^2 \tag{1.1}$$

Among them, dist is the Euclidean distance (L_2) between two objects in Euclidean space, C_i is the ith cluster, and x is the point in C_i. The centroid of the i-th cluster is defined by formula (1.2).

$$C_i = \frac{1}{m_i}\sum\nolimits_{x \in C_i} X \tag{1.2}$$

(2) Assign each object to the aggregation that is most similar to it. Each aggregation is represented by the mean of all the objects in it. The "most similar" means that the distance is the smallest. For each point V_i, find a centroid c_i, minimize the distance d (V_i, c_i) between them, and assign V_i to the i-th group;
(3) After all points are assigned to the corresponding group, the centroid c_i of each group is recalculated;
(4) Perform steps (2) and (3) cyclically until the division of the data no longer changes.

The algorithm has good scalability, and its computational complexity is $O(n^{kt})$, where t is the number of cycles.

4 Experimental Results and Analysis

Due to the large time span of the experimental use of the data, the experimental results only selected the clustering results from 1960–1969 and 2000–2010 for comparative analysis. In the study of extreme climate, the percentile definition method is often used, and 95% is selected as the threshold of extreme climate in this paper (Justin and Scott 2016).

Figure 3 shows that K-means clustering clustered the extremely high-temperature data from 1960 to 1969 into eight categories, where each category was identified by color.

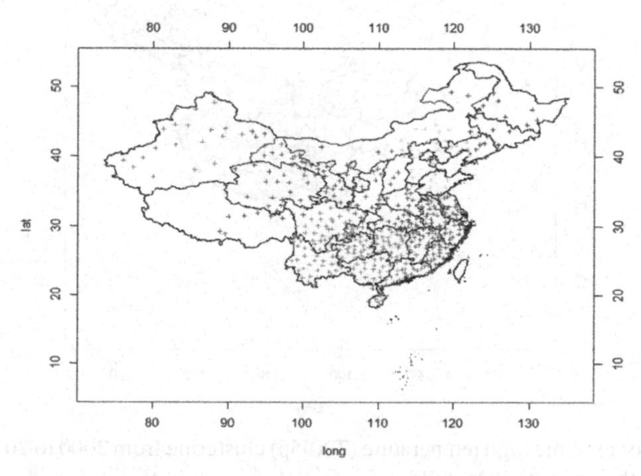

Fig. 3. The clustering of extreme high temperature (TX95p) space site in China from 1960 to 1969. The abscissa is longitude, and the ordinate is latitude

It shows that extreme high-temperature events do have regional aggregation. And some categories contain only one province, some clusters contain multiple provinces. Take the purple region on the map as an example. The region includes five regions including Hunan, eastern Guizhou, Guangxi, Guangdong, and Hainan. This shows that when extreme temperature events occur in Hunan, China is affected by monsoons and currents, the high temperature in the air will shift or spread with the flow. Therefore, the possibility of extreme high-temperature events in the four provinces of eastern Guizhou, Guangxi, Guangdong and Hainan are extremely high. Therefore, all departments in these areas should be vigilant, take measures to prevent disasters, and remind the general public to reduce outdoor activities and do cooling work to reduce the casualties caused by high temperatures. It can also remind forestry workers to be vigilant from forestry occurring fires. Extreme weather events often lead to the occurrence of various meteorological disasters, resulting in huge economic losses and casualties. And the occurrence of extreme climate events is regional and mass (Yang et al. 2010). If we can know in advance which regions are extremely related to extreme weather events, or if there is a strong clustering of extreme weather events in these regions, we can make protective measures in advance. Reducing various human and social losses caused by extreme weather events. Therefore, it is very important to study the regional correlation of extreme weather events (Wu 2016).

From the extreme high temperature (TX95p) clustering space sitemap in Fig. 4, it can be seen that a total of 8 clusters have been generated. As in Fig. 3, it is divided into eight clusters. The difference is that Fig. 3 is a cluster analysis of extreme high-temperature events in the 1960s. Figure 4 is a cluster analysis of extreme high-temperature events in the twentieth century. Compare two figures:The extreme high-temperature events that

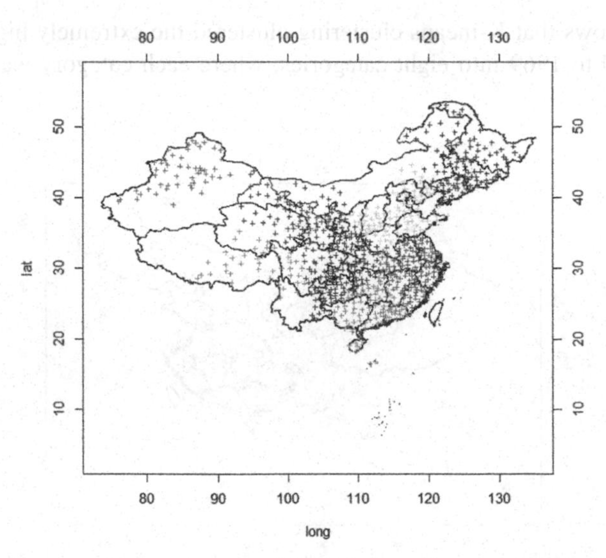

Fig. 4. Sitemap of extreme high temperature (TX95p) clustering from 2000 to 2010. The abscissa is longitude, and the ordinate is latitude

occurred in the 1800s have greatly increased the frequency and range of extreme high-temperature events that occurred in the 20th century, and the extreme high-temperature events in the country have shown an increasing trend.

As shown in the above two figures, in the early 20th century, the extremely high-temperature events mainly occurred in the southern region, mainly including Guangdong, Guangxi, Guizhou, Chongqing, Fujian and Anhui. In the aggregation area of cluster events in the twentieth century, Shandong, Hebei, Liaoning, and Heilongjiang were also included. It shows that extreme temperature events not only have regional aggregation but also regional mobility.

The interdecadal populations obtained in the literature (Wei 2006) have obvious regional and asynchronicity, and the concordance of the trend of consistency in adjacent regions is consistent. In 2016, the strongest hot weather occurred in 2016, including the middle and lower reaches of the Yangtze River and most of the southern part of the Yangtze River. The Central Meteorological Observatory continued to raise the high-temperature orange warning. Shanghai, Hubei, Chongqing and other places have successively sounded the first high-temperature red warning in the summer of 2016 (Cui et al. 2009), and the high-temperature "barbecue" model was opened in many places across the country. Chongqing citizens reported on the "news" of collective watering in the park due to hot weather on TV stations and on the Internet. This phenomenon can be explained by the red region in Fig. 4. The red clustered regions include Chongqing, Hubei, Guizhou, Hunan and other regions, indicating that there is a regional agglomeration of extreme high-temperature events in these regions, and extremely high-temperature events may occur in succession. The successive high-temperature events in Chongqing and Hubei in 2016 confirmed the correctness and practicality of this clustering study. Therefore, we should pay close attention to the occurrence and occurrence of extreme temperature events in order to prepare for follow-up work, prevent the occurrence of extreme events in advance, and reduce their various adverse effects on humans and society and disaster losses.

5 Summary

This paper uses the K-means clustering algorithm to analyze and study the national extreme climate data. Through cluster analysis, it is found that with the increase of time, the frequency and area of extreme high-temperature events are increasing, which is consistent with the actual national extreme temperature change trend. And found that extreme temperature events not only have local aggregation, but also there is liquidity. Using k-means algorithm not only excavates the internal relationship of climate data, but also effectively avoids the low efficiency and slow processing of traditional methods. Therefore, clustering algorithm is very important for the study of extreme climate events.

References

Yang, Y., Zhou, D., Yang, X.: A multi-feature weighting based K-means algorithm for MOOC learner classification. Comput. Mater. Continua **59**(2), 625–633 (2019)

Maamar, A., Benahmed, K.: A hybrid model for anomalies detection in AMI system combining K-means clustering and deep neural network. Comput. Mater. Continua **60**(1), 15–39 (2019)
Tan, L., Li, C., Xia, J., Cao, J.: Application of self-organizing feature map neural network based on K-means clustering in network intrusion detection. Comput. Mater. Continua **61**(1), 275–288 (2019)
Shen, W.H.: Analysis of "big data application" of meteorological data—discussion on the applicability of thinking change in "the era of big data". China Inf. Technol. **11**, 20–31 (2014)
Sui, W.H., Pei, S.Q., Zhang, F.: Application and visualization of media meteorological service products in the context of big data. In: The 33rd Annual Meeting of China Meteorological Society (2016)
Zhai, P.M., Pan, X.H.: Changes in temperature and precipitation extreme events in northern China in the past 50 years. Geogr. J. **58**, 1–10 (2003)
Zhou, Y.Q., Ren, G.Y.: Analysis of changes in extreme temperature events in mainland China from 1956 to 2008. Clim. Environ. Stud. **15**, 405–417 (2010)
Sun, Z.J., Liang, Y.Q., Fan, J.C., Ma, Y.K., Liang, T.Y.: Research and application of k-value optimization of K-means algorithm. Comput. Dig. Eng. **41**(11), 1713–1716 (2013). https://doi.org/10.3969/j.issn.1672-9722.2013.11.001
Wang, M., Yin, S.Y.: Spatial and temporal variation characteristics of extreme precipitation in the middle and lower reaches of the Yangtze river near 52a. Resour. Environ. Yangtze River Basin **24**(7), 1221–1229 (2015)
She, D.X., Xia, J., Zhang, Y.Y., Du, H.: Spatial and temporal variation and statistical characteristics of extreme precipitation over the Huaihe river basin in the past 50 years. Acta Geodeica Sin. **66**(9), 1200–1210 (2011)
Wang, Y., Li, X., Miao, Q.L.: A study on the characteristics of temperature changes over the last 50 years on the Qinghai-Tibet plateau. Geogr. Arid Reg. (Chin. Vers.) **1**, 42–47 (2004)
Jin, D., Liu, J., Jia, Z.X., Liu, D.Y.: Data clustering algorithm based on k-nearest neighbor network. Pattern Recog. Artif. Intell. **23**(4), 546–551 (2010)
Yang, P.: China nearly 40 years of extreme temperature and extreme precipitation events clusters research. Doctoral dissertation, Lan Zhou University (2009)
Xiong, Z.Y., Chen, R.T., Zhang, Y.F.: An effective method for initialization of clustering centers in K-means clustering. Comput. Appl. Res. (2011)
Tian, S.P., Wen, L.W.: Automatic acquisition algorithm of parameter value K based on dynamic K-means. Comput. Eng. Des. **32**(1), 273–274 (2011)
Han, S.W., Seo, J.Y., Kim, D.-Y., Kim, S.H., Lee, H.M.: Development of cloud based air pollution information system using visualization. Comput. Mater. Continua **59**(3), 697–711 (2019)
Jiang, S.Y., Li, T.: An improved k-means clustering algorithm. Comput. Eng. Sci. **28**(11), 56–59 (2006)
Wang, Q., Wang, C., Feng, Z.Y., et al.: Overview of K-means clustering algorithm. Electron. Des. Eng. (2012)
Zhai, D.H., Jiang, Y.G., et al.: K-means text clustering algorithm based on the selection of initial clustering center of maximum distance. Comput. Appl. Res. (2014)
Justin, T.S., Scott, M.R.: Projecting changes in regional temperature and precipitation extremes in the United States. Weather Clim. Extremes **11**, 28–40 (2016). https://doi.org/10.1016/j.wace.2015.09.004
Yang, P., Hou, W., Guo, L.F.: Study on the characteristics of group extreme events in China. Clim. Environ. Res. **15**(4), 365–370 (2010)
Wu, Y.J.: Temporal and spatial distribution characteristics and future prediction of extreme precipitation in China under the background of climate change. Shanghai Normal University (2016)

Wei, F.Y.: Progress in research on methods of diagnosis and prediction of climate statistics–commemoration of the 50th anniversary of Chinese academy of meteorological sciences. Acta Appl. Meteorol. **17**(6), 736–742 (2006)

Cui, L., Shi, J., Zhou, W.: Characteristics of extreme temperature change in Shanghai and its response to urbanization. Geosciences **29**(1), 93–97 (2009)

Network Topology Boundary Routing IP Identification for IP Geolocation

Fuxiang Yuan[1], Fenlin Liu[1], Rui Xu[2,3](✉), Yan Liu[1], and Xiangyang Luo[1]

[1] PLA Strategic Support Force Information Engineering University, Zhengzhou 450001, China
[2] Cyberspace Security Key Laboratory of Sichuan Province, Chengdu 610000, China
jasmine_x@163.com
[3] China Electronic Technology Cyber Security Co., Ltd., Chengdu 610000, China

Abstract. The only few existing methods for network topology boundary routing IP identification are often based on a single network characteristic, with poor applicability, low accuracy, and difficult to meet the IP geolocation requirements. This paper proposes a network topology boundary routing IP identification algorithm for IP geolocation. The proposed algorithm is no longer limited to a specific delay distribution used in D-Based, but identifies the boundary routing IP according to the difference between each single-hop delay in the path and the city delay threshold; more different from traditional methods, when the delay difference characteristic is not obvious, the proposed algorithm combines the differences of routing IP hostnames in different cities to further identify the boundary routing IP and the proportion of identifiable paths is increased. Experiments are carried out using the probing results of millions of landmarks in 10 target cities in China and the US. Results show that compared with D-Based and P-Based, the ratio of identifiable paths is increased by about 103.0% and 69.5%; the number of nodes in the target city is about 1.7 and 1.9 times, and the accuracy of the node location is improved by about 48.3% and 16.9%; using the topology obtained by the proposed algorithm, the success rate of three classic street-level IP geolocation methods such as SLG, LENCR, and Geo-RMP is increased by 75.8%, 61.2%, and 71.7%.

Keywords: Boundary IP · IP geolocation · Network topology · Network measurement

1 Introduction

Along with the development of the Internet, a large number of physical devices configured with IPs access to the network, such as terminal hosts, routers, etc. [1–3]. Accurately geolocate IPs of these network entities is very important for detecting suspicious nodes, tracking sensitive users, and maintaining cyberspace security [4–7]. Existing IP geolocation methods such as SLG [8], LENCR [9], and Geo-RMP [10] often rely on the links between target IPs, landmarks, and routing IPs. Accurate identification of the network topology boundary routing IPs (hereafter boundary IP) of a target city is a key part of these geolocation methods. Therefore, it is of great significance to carry out research on

X. Sun et al. (Eds.): ICAIS 2020, LNCS 12240, pp. 534–544, 2020.
https://doi.org/10.1007/978-3-030-57881-7_47

boundary IP identification for extracting links belonging to the target city from probing paths across cities and accurately geolocating target IPs [11, 12].

Although in the existing work, the concept or definition of the boundary IP has not been explicitly mentioned [13], in view of the regional attribute of the network, when a packet is forwarded from the source IP to the destination IP, it must go through some routing IPs responsible for forwarding data to the specific city [12]. And all routing IPs from these IPs in the path tend to belong to the target city. These IPs, which indicate that probing paths enter the target city, constitute the boundary of the target city network topology, which are the boundary IP referred to herein.

Currently, there are few published methods for boundary IP identification, such as D-Based [11], P-Based [12]. D-Based statistically analyzes the single-hop delay in probing paths and finds that delays exhibit a "low-high low" distribution. Based on this distribution characteristic, boundary IPs are found and paths are divided. When single-hop delays meet the above-mentioned characteristic, the method can obtain the boundary IP, but after analyzing a large amount of data, it is found that when single-hop delays have no obvious distribution characteristic, it is impossible to determine which hop in the path is the boundary IP with this method. P-Based identifies the boundary IP based on the statistics of probing paths of landmarks. P-Based probes some landmarks in the target city. For each landmark, the method extracts each routing IP from the source on the path in turn, and check whether other landmarks whose paths pass the routing IP are in the same city. If landmarks belong to the same city, the routing IP is considered to be the boundary IP of the corresponding city, and the above analysis of the next hop in the path is continued. Through analysis, it is found that IPs that passed by probing paths of landmarks belonging to the same city are not necessarily boundary IPs of the city. These IPs may be the next hops corresponding to the boundary IPs. Therefore, IPs obtained by the P-Based method are not all boundary IPs. It is difficult to accurately divide paths according to these IPs, and it is difficult to meet the IP geolocation requirements.

Aiming at these problems of the above methods, this paper proposes a network topology routing IP identification algorithm for IP geolocation. Through the proposed algorithm, boundary IPs of the target city can be more accurately identified. The main contributions are as follows:

Firstly, a network topology routing IP identification algorithm for IP geolocation is proposed. The algorithm identifies boundary IPs based on two characteristics that are the difference in single-hop delays in paths and the difference in router hostnames in different cities.

Secondly, boundary IP identification based on single-hop delay difference is designed. In the case that the difference between two single-hop delays is obvious, the boundary IP is identified by comparing each single-hop delay with the target city delay threshold.

Thirdly, boundary IP identification based on hostname differences is designed. For paths that the difference between single-hop delays is not obvious, the composition of IP hostname per hop in the path is analyzed, and the boundary IP is further identified based on the difference of the hostname strings.

The rest of the paper is organized as follows: In Sect. 2, the boundary IP and the importance of its identification are introduced; the main steps of the proposed algorithm are given in Sect. 3; In Sect. 4, the experimental setup is given, and the proposed algorithm and two existing typical methods are tested and compared in the effect of boundary IP identification and the application of target IP geolocation; in Sect. 5 the full text is summarized.

2 Problem Description

Nodes inside the city are interconnected to form a network. In order to ensure nodes of different networks can communicate with each other, networks of adjacent cities are interconnected and multiple networks constitute larger networks [14–16]. Since network devices such as routers (except backbone routers) and terminal hosts often belong to a certain city, the boundaries between networks in adjacent cities are formed [17, 18]. IPs of the target city is probed from the vantage points (hereafter vps) deployed in other cities, and paths through which packets are forwarded by the source IP to the destination IP can be obtained [19–22]. The intermediate routing IPs that divide paths into the city external path segments and internal path segments together constitute the boundary IPs of the city network topology.

In order to obtain the network topology of the target city, understand the characteristics of the network route and structure, and then use these characteristics to geolocate target IPs, it is important to obtain as many accurate nodes and links as possible. Generally, existing researches often deploy a large number of vps from inside and outside the target city, and perform multi-source probing of IPs in the city to obtain a large number of paths. Each hop in the path of the target IP obtained from the internal vp of the city may be located inside the city. These nodes and links can be directly used for analysis of network characteristics. But the scale of the data is not enough to understand the characteristics of a city network. It is still necessary to supplement paths of target IPs from the external vps of the city. However, paths obtained from the external vps cannot be directly used. This is because although paths can be theoretically divided into the city internal path segments and the external path segments, the actual paths obtained does not include the flag information that can distinguish the two segments, and that is, boundary IPs of the target city. After obtaining these paths, it is impossible to directly extract the path segments belonging to the target city.

Figure 1 shows the comparison of the actual path and probing path when the vp is outside the target city B. Figure 1(a) is a complete path from the vp through the routing node inside city A, the backbone network routing node, the routing node inside city B to the target IP, a total of 10 hops; Fig. 1(b) shows the probing path corresponding to Fig. 1(a). Each row represents one hop information. The three fields in each row are separated by spaces. The first field indicates the hop count, the second field indicates the IP, and the third field indicates the delay between the current hop and the previous hop. It can be seen from Fig. 1(a) that the path can be divided into a segment inside A, a backbone segment, and a segment inside B. However, from the actual probing path shown in Fig. 1(b), the segment and corresponding nodes and links inside B cannot be directly obtained. As a result, the topology of B cannot be constructed. However, these nodes, links, and topology are the basic data of topology-based street-level IP geolocation algorithms such as SLG, LENCR, and Geo-RMP. Therefore, accurately identifying boundary IP (such as the 7th hop in Fig. 1(b)) of the target city is very important for obtaining routing nodes and links, analyzing the network characteristics, and geolocating target IPs.

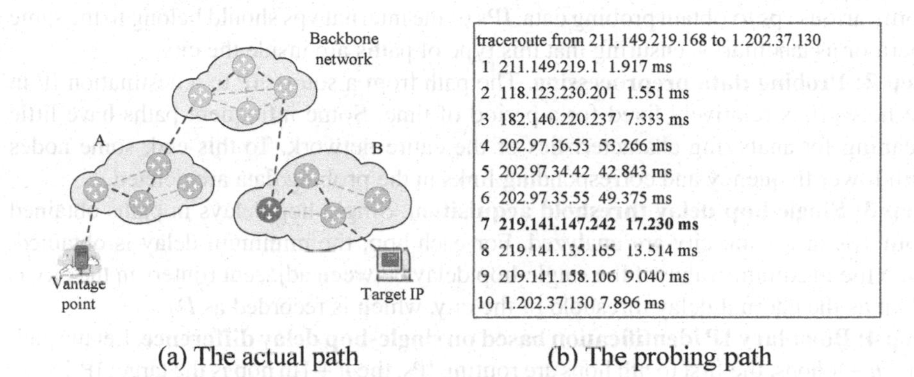

Fig. 1. Comparison of the actual path and the probing path

3 The Proposed Algorithm

The main steps of the proposed algorithm are as follows, and its schematic diagram is shown in Fig. 2.

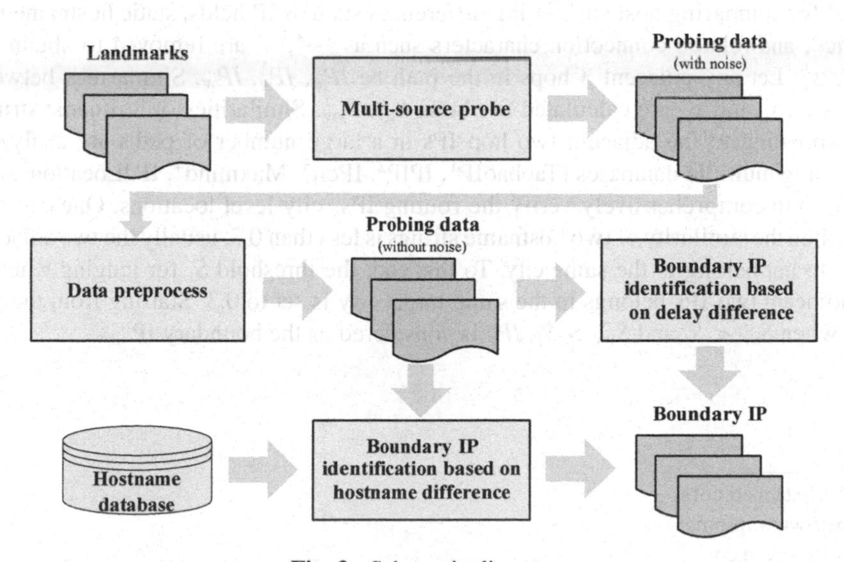

Fig. 2. Schematic diagram

Input: landmarks of a target city
Output: boundary IPs of the target city
Step 1: Target city landmarks probing. A large number of vps are deployed in and out of the target city. TCP, UDP, and ICMP protocols are used to repeatedly probe landmarks from various vps to obtain probing data. IPs of the internal vps should belong to the same operator as landmarks, ensuring that this type of paths are inside the city.
Step 2: Probing data preprocessing. The path from a source IP to a destination IP in the network is relatively fixed for a period of time. Some infrequent paths have little meaning for analyzing characteristics of the entire network. To this end, some nodes with lower frequency and corresponding links in the probing data are deleted.
Step 3: Single-hop delay threshold acquisition. Single-hop delays in paths obtained from vps inside the city are analyzed. For each hop, the minimum delay is obtained. Then the maximum value of the single-hop delay between adjacent routers in the city is taken as the internal delay threshold of the city, which is recorded as D_T.
Step 4: Boundary IP identification based on single-hop delay difference. Let the path P be $n+1$ hops, the first to nth hops are routing IPs, the $n+1$th hop is the target IP. Delay between the ith hop and the $i+1$ hop is T_i $(1 \leq i \leq n)$. Each hop is identified one by one from the back to the front of the path. If $\exists IP_i$, $T_i < D_T$, $T_{i-1} > D_T$, this indicates that IP_i is far from IP_{i-1} and is closer to IP_{i+1} which is the internal routing IP of the city. At this time, IP_i is considered to be the internal routing IP of the city and is the boundary IP. For a path that the differences between single-hop delays and the threshold are not obvious, perform Step 5.
Step 5: Boundary IP identification based on hostname difference. Let path Q have a total of m hops and intermediate routing IP to form the set V. For $IP_x \in V$, the corresponding hostname s_x is obtained and the characters included in s_x that are not useful for comparing hostname string differences such as IP fields, static hostname flag "static", and related connection characters such as '–', '.' are removed to obtain the string s'_x. Let any adjacent 3 hops in the path be IP_u, IP_v, IP_w. Similarities between s'_u and s'_v, s'_v and s'_w are calculated to obtain S_{uv}, S_{vw}. Similarities of hostname strings corresponding to the adjacent two hop IPs in a large number of paths are analyzed. And using multiple databases (TaobaoIP[1], IPIP[2], IPcn[3], Maxmind[4], IP2Location[5], and Hostip[6]) to comprehensively verify the routing IPs' city-level locations. One can find that when the similarity of two hostname strings is less than 0.3, usually the two adjacent hops do not belong to the same city. To this end, the threshold S_t for judging whether the adjacent two IPs belongs to the same target city is set to 0.3. Starting from the last hop, when $S_{uv} < S_t$ and $S_{vw} > S_t$, IP_v is considered as the boundary IP.

[1] http://ip.taobao.com.
[2] https://www.ipip.net.
[3] http://www.ip.cn.
[4] http://www.maxmind.com.
[5] http://www.ip2location.com.
[6] http://www.hostip.info.

4 Experimental Results

In order to verify the effect of the proposed algorithm, this section performs tests for boundary IP identification, topology acquisition and IP geolocation of multiple target cities.

4.1 Experiment Setting

10 cities such as Beijing, Shanghai, Guangzhou, Zhengzhou, Hangzhou in China, as well as New York, Chicago, Atlanta, Washington, and Miami in the US are selected as target cities. First of all, using the landmark mining methods based on Web [23], Internet Forum [24], navigation pages [25, 26], and multi-database fusion, a large number of city-level landmarks in the above cities are obtained as probing targets. Then, vps are deployed inside the above 10 cities. From vps inside and outside each target city, landmarks are probed for about 110 days, and each landmark is probed for 50 times. Hostnames corresponding to routing IPs are mainly obtained from RAPID7[7]. The experimental parameters are shown in Table 1.

Table 1. Experiment setting.

Country	Target cities and number of landmarks	Probing times	Period (day)	Hostname source
CN	Beijing: 232458 Shanghai: 171637 Guangzhou: 119250 Zhengzhou: 32859 Hangzhou: 116932	50	110	RAPID7
US	New York: 122470 Chicago: 192005 Atlanta: 416322 Washington: 179758 Miami: 62068			

4.2 Boundary IP Identification Tests

The proposed algorithm performs boundary IP identification for 10 target cities. For different cities, Fig. 3 shows the proportion of paths that the algorithm can identify.

It can be seen from Fig. 3 that the average proportion of paths for successful identification in five target cities in China is about 78.3%, and that is about 78.8% in the US. Among the five cities in China, the proportion of paths based on delay difference identification is higher, with an average of about 48.1%; paths with no significant difference in single-hop delay is about 30.2% and these paths can be processed based on the

[7] https://opendata.rapid7.com.

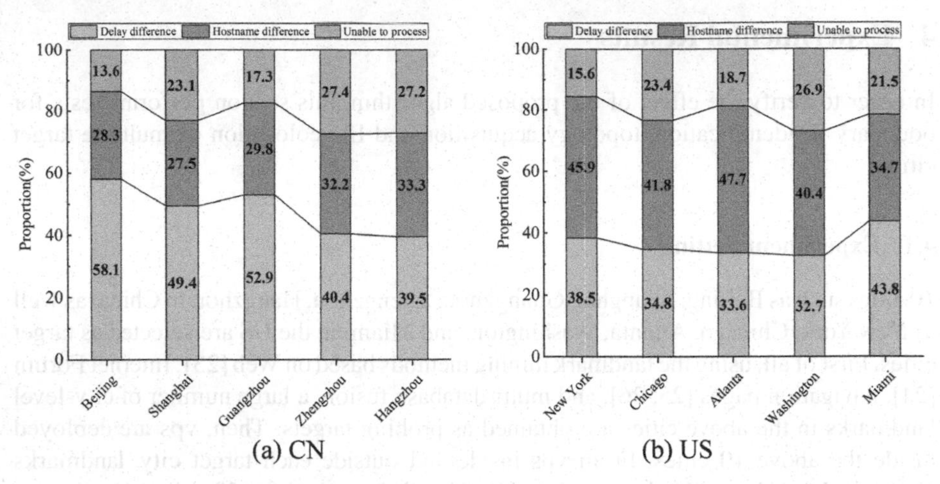

Fig. 3. Proportion of paths that can be identified

difference of hostnames; 21.7% of the paths cannot be processed. In the US these values are 36.7%, 42.1%, and 21.2%, respectively. The differences between the networks of China and the US lead to different proportions of paths identified based on different characteristics. However, by combining the delay difference with the hostname difference, the proportion of paths identified by the proposed algorithm is high.

4.3 Comparison with Typical Methods

As mentioned above, D-Based [11] and P-Based [12] can also identify boundary IPs of a target city, divide paths, and acquire network topology. So different methods are tested with data from these 10 cities in this section. Combined with the results in Sect. 4.2, the results obtained by different methods are compared in terms of the proportion of paths can be identified, the number of routing IPs obtained, and the city-level location accuracy of IPs. Table 2 shows the comparison of results obtained by different methods in 10 cities.

As can be seen from Table 2, in China the average proportions of paths identified by D-Based, P-Based and the proposed algorithm are about 44.0%, 53.0%, and 78.3%, respectively; these proportions are 33.4%, 39.6%, and 78.8% in the US. The effect of the proposed algorithm remains stable, while D-Based and P-Based become significantly worse. This is mainly because networks in China have hierarchical structure, the difference characteristics of single-hop delays are obvious, and multiple paths often pass through the same routing IPs. D-Based and P-Based are overly dependent on these characteristics. But these characteristics are not obvious in the US, so the poor effect of the two methods is inevitable. However, the proposed algorithm identifies boundary IPs based on the single-hop delay difference and hostname difference. Most of the paths obtained can always satisfy one or both of the two characteristics. Therefore, the proposed algorithm can always obtain relatively good results for cities of different countries.

Due to the limited proportion of paths that D-Based and P-Based can identify, the number of routing IPs and links between IPs in each city is small. The average number

Table 2. Results comparison.

City	Comparison of different methods								
	Proportion of paths can be identified (%)			Number of routing IPs			City-level location accuracy (%)		
	[11]	[12]	Proposed	[11]	[12]	Proposed	[11]	[12]	Proposed
Beijing	52.0	66.5	**86.4**	50263	43669	**69842**	71.2	80.4	**91.5**
Shanghai	43.3	53.2	**76.9**	49338	40256	**65338**	66.4	72.1	**93.6**
Guangzhou	49.6	43.5	**82.7**	34112	23171	**58521**	72.1	78.8	**92.2**
Zhengzhou	38.2	52.1	**72.6**	25113	10866	**46125**	63.3	80.1	**88.6**
Hangzhou	36.7	49.8	**72.8**	18695	15325	**24038**	75.6	82.5	**93.3**
New York	33.3	40.2	**84.4**	31225	36335	**63396**	50.8	75.6	**90.0**
Chicago	31.7	41.8	**76.6**	27251	29356	**60526**	55.7	69.9	**86.7**
Atlanta	32.1	39.3	**81.3**	24401	25631	**49155**	52.1	73.1	**94.1**
Washington	30.0	36.7	**73.1**	20556	19881	**38897**	46.6	81.7	**90.8**
Miami	39.9	40.2	**78.5**	19325	18336	**36615**	60.0	84.3	**89.4**

of IPs obtained by the proposed algorithm in China is 1.5 and 2.0 times than that of D-Based and P-Based respectively, and 2.0 and 1.9 times in the US. Multi-database voting method is used to verify the city-level locations of the acquired IPs. The verification results show that the accuracy of the city-level locations of IPs obtained by the proposed algorithm is about 91%, which is significantly higher than that of D-Based and P-Based.

4.4 IP Geolocation Tests

In order to further verify the effect of the proposed algorithm, D-Based, P-Based and the proposed algorithm are compared in the application of target IP geolocation in this section.

Based on topology obtained by D-Based, P-Based and the proposed algorithm in four cities in China and the US, three typical street-level IP geolocation methods such as SLG, LENCR and Geo-RMP are used for geolocation test. Target IPs are mainly composed of street-level landmarks and each city takes 10,000 targets.

As can be seen from Fig. 4(a)–(d), no matter in China or in the US, based on the network topology obtained by the proposed algorithm, the three geolocation algorithms have the highest success rate, reaching 82.8%, 87.7%, and 84.1%. However, based on the data obtained by D-Based and P-Based, the highest success rates are 54.3%, 59.2%, 67.6%, and 61.2%, 61.1%, and 69.4%. This is mainly because the proportion of paths identified by the algorithm in this paper is high, so the topology acquired contains more IPs and links. Moreover, the accuracy of boundary IP identification by the algorithm is high, and the location of the acquired IPs and links are highly accurate. This will greatly improve the possibility of different geolocation algorithms successfully finding the connection relationship between landmarks, target IPs and intermediate routers, thereby improving the geolocation success rate. The above results illustrate the good effect of the proposed algorithm on boundary IP identification.

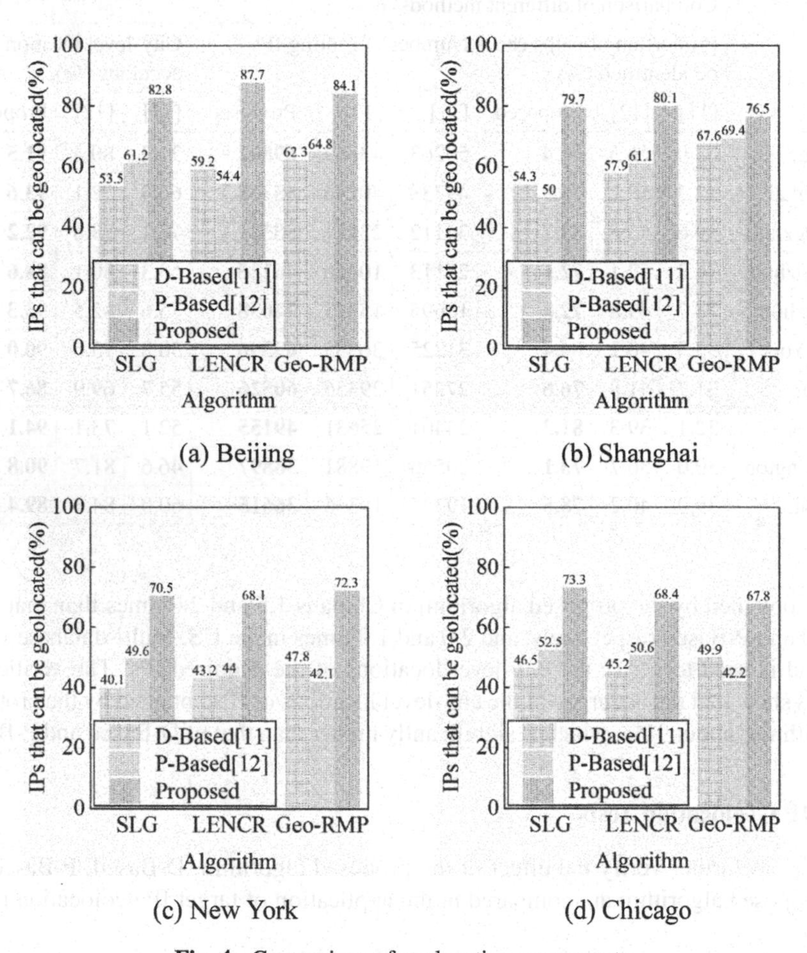

Fig. 4. Comparison of geolocation success rate

5 Conclusion

This paper proposes a network topology boundary routing IP identification algorithm for IP geolocation. Based on the single-hop delay analysis in the existing research, the proposed algorithm combines the difference between single-hop delays and the delay threshold, and the difference of hostnames to identify boundary IPs. Experiments are carried out on the proposed algorithm using the probing data of landmarks from different cities. Results show that the proposed algorithm can identify the boundary IP more accurately than the existing typical methods. And the topology obtained based on the proposed algorithm includes a large number of IPs with high accuracy, which can provide reliable support for street-level IP geolocation.

References

1. Nur, A.Y., Tozal, M.E.: Cross-AS (X-AS) internet topology mapping. Comput. Netw. **132**, 53–67 (2018)
2. Huang, J., Duan, Q., Xing, C.C.: Topology control for building a large-scale and energy-efficient internet of things. IEEE Wirel. Commun. **24**(1), 67–73 (2017)
3. Li, X., Xi, J., Cai, Z.: Analyzing the structure and connectivity of continent-level internet topology. Comput. Mater. Continua **59**(3), 955–964 (2019)
4. Wang, Z., Li, H., Li, Q.: Towards IP geolocation with intermediate routers based on topology discovery. Cybersecurity **2**(1), 1–13 (2019)
5. Yuan, F., Liu, F., Huang, D.: A high completeness PoP partition algorithm for IP geolocation. IEEE Access **7**, 28340–28355 (2019)
6. Kardes, H., Gunes, M.H., Sarac, K.: Graph based induction of unresponsive routers in internet topologies. Comput. Netw. **81**, 178–200 (2015)
7. Afzal, B., Anwar, G., Muhammad, A.Q.: Smart security framework for educational institutions using internet of things (IoT). Comput. Mater. Continua **61**(1), 81–101 (2019)
8. Yong, W., Daniel, B., Marcel, F.: Towards street-level client-independent IP geolocation. In: 8th USENIX Symposium on Networked Systems Design and Implementation, Boston, USA, pp. 365–379. USENIX (2011)
9. Chen, J., Liu, F., Shi, Y.: Towards IP location estimation using the nearest common router. J. Internet Technol. **19**(7), 2097–2110 (2018)
10. Zhao, F., Xu, R., Li, R.: Street-level geolocation based on router multilevel partitioning. IEEE Access **7**, 59237–59248 (2019)
11. Liu, S., Liu, F., Zhao, F.: IP city-level geolocation based on the pop-level network topology analysis. In: 6th International Conference on Information Communication and Management, Hatfield, UK, pp. 109–114. IEEE (2016)
12. Zhao, F., Luo, X., Gan, Y.: IP geolocation based on identification routers and local delay distribution similarity. Concurr. Comput. Pract. Exp. **31** e4722 (2018)
13. Luckie, M., Dhamdhere, A., Huffaker, B.: Bdrmap: inference of borders between IP networks. In: the 2016 Internet Measurement Conference, Santa Monica, California, USA, pp. 381–396. ACM (2016)
14. Marder, A., Smith, J. M.: MAP-IT: multipass accurate passive inferences from traceroute. In: the 2016 Internet Measurement Conference, Santa Monica, California, USA, pp. 397–411. ACM (2016)
15. Baskaran, S.B.M., Raja, G.: Blind key distribution mechanism to secure wireless metropolitan area network. CSI Trans. ICT **4**(2–4), 157–163 (2016)
16. Zeng, F., Ren, Y., Deng, X.: Cost-effective edge server placement in wireless metropolitan area networks. Sensors **19**(1), 32 (2019)
17. Ciavarrini, G., Greco, M.S., Vecchio, A.: Geolocation of internet hosts: accuracy limits through Cramér-Rao lower bound. Comput. Netw. **135**, 70–80 (2018)
18. Qazi, S., Kadri, M.B.: Revisiting constraint based geolocation: improving accuracy through removal of outliers. Int. Arab J. Inf. Technol. **15**(2), 232–239 (2018)
19. Khanna, N.: Mitigation of collaborative blackhole attack using traceroute mechanism with enhancement in AODV routing protocol. Int. J. Future Gener. Commun. Netw. **9**(1), 157–166 (2016)
20. Belghachi, M., Debab, N.: An efficient greedy traffic aware routing scheme for internet of vehicles. Comput. Mater. Continua **60**(3), 959–972 (2019)
21. Fontugne, R., Pelsser, C., Aben, E.: Pinpointing delay and forwarding anomalies using large-scale traceroute measurements. In: The 2017 Internet Measurement Conference, London, UK, pp. 15–28. ACM (2017)

22. Gansner, E.R., Krishnamurthy, B., Willinger, W.: Demo abstract: towards extracting semantics by visualizing large traceroute datasets. Computing **96**(1), 81–83 (2014)
23. Guo, C., Liu, Y., Shen, W.: Mining the web and the internet for accurate IP address geolocations. In: International Conference on Computer Communications, Rio de Janeiro, Brazil, pp. 2841–2845. IEEE (2009)
24. Zhu, G., Luo, X., Liu, F.: An algorithm of city-level landmark mining based on internet forum. In: 18th International Conference on Network-Based Information Systems, Taipei, Taiwan, CN, pp. 294–301. IEEE (2015)
25. Ma, T., Liu, F., Luo, X.: An algorithm of street-level landmark obtaining based on yellow pages. J. Internet Technol. **20**(5), 1415–1428 (2019)
26. Jiang, H., Liu, Y., Matthews, J. N.: IP Geolocation estimation using neural networks with stable landmarks. In: International Conference on Computer Communications, San Francisco, CA, USA, pp. 170–175. IEEE (2016)

Machine Vision Based Novel Scheme for Largely, Reducing Printing Errors in Medical Package

Bin Ma[1(✉)], Qi Li[1,2(✉)], Xiaoyu Wang[1,2(✉)], Chunpeng Wang[1], and Yunqing Shi[3]

[1] School of Cyber Security, Qilu University of Technology, Shandong 250353, China
sddxmb@126.com, qluliqi@163.com, qluwxy@163.com

[2] School of Information Science and Technology, Dalian Martime University, Dalian 116026, China

[3] Department of Electrical and Computer Engineering, New Jersey Institute of Technology, Newark, NJ 07102, USA

Abstract. Aiming at the problem of misprint or obscure of production date, batch number and the validity on the medicine package, a delay-based misplaced difference scheme for medical information detection is put forward. To be specific, medical images with delay in packaging are acquired by a vision detection system, based on which, character images are obtained through misplaced subtraction operation; and a convolution kernel is designed based on gray value distribution of the character images for multi-step convolution to remove speckle noise. Then a specific operation with corrosion and dilation is further utilized to remove speckle noise and enhance the target character area. In the end, the modified weighted median filter is adopted for noise inhibition to further improve the recognition accuracy. Experimental results show that when the double image overlap η is 80% and threshold λ (the percentage of non-zero gray values in each convolution block) is 60%, the scheme's recognition can be as accurate as 97.8% and detection speed can reach up to 0.1373 s/image, the detection precision and efficiency can satisfy medical information recognition requirements in medical package.

Keywords: Delay-based misplaced difference scheme · Convolution kernel · Modified weighted median filter · Double image overlap

1 Introduction

In medical package industry, the information of medicine production date, batch number and validity are essential. Once wrongly identified medicines are circulated on market, they can badly threaten physical health and even life and cause medical accidents [1–3]. Therefore, the significance of medicine information detection is self-evident. At the moment, most medical manufacturers adopt a manual quality inspection method which is susceptible to subjective factors and has a high error rate. As a result, pharmaceutical information detection systems based on machine vision have aroused the attention of pharmaceutical manufacturers [4–6]. However, the precision and efficiency of machine vision detection systems are restricted by noise made by industrial cameras for real-time captured image and poor circumstances of production lines, thus the

X. Sun et al. (Eds.): ICAIS 2020, LNCS 12240, pp. 545–556, 2020.
https://doi.org/10.1007/978-3-030-57881-7_48

effective preprocessing for the medical images collected in real-time is the premise to improve the inspection accuracy.

In the wake of faster computer operation and rapid growth of artificial intelligence (AI) technology [7–9], a tremendous amount of progress has been made in defect detection and character recognition by machine vision technology [10, 11]. Min et al. put forward a railway track defect detection system based on machine vision and expand the defects of rail image to improve detection accuracy [12]. Shi et al. raised a method of defect detection in commercial note printing, which was used in segmentation for digital area to improve the detection efficiency [13]. Wang et al. presented a real-time defect detection algorithm based on machine vision, and the algorithm's detection precision can meet production line requirements completely [14]. Iyshwerya et al. proposed a welding chip defect detection algorithm based on machine vision, and the algorithm's detection accuracy can be as high as 96.31% [15]. Huang et al. offered a nopre-processing Tibetan character feature extraction method based on wavelet transform and gradient direction, and the algorithm improved recognition precision, whose recognition rate can be as high as 95.17% and recognition efficiency can reach up to 82.5 ms/piece [16]. Li et al. provided an intelligent vision detection system for surface defects, wherein a new automatic threshold segmentation method–proportion emphasized maximum entropy (PEME) threshold algorithm [17].

Currently, most character-oriented machine vision technologies restrain noise of image by mean filtering [18, 19], median filtering [20, 21] and other algorithms to improve precision of recognition, whereas such algorithms for static image de-noising does not perform well in medical package image. Typically, images acquired in real time from industrial production lines show the following features: (1) Charge Coupled Device (CCD) industrial cameras have their own inherent noise [22]. (2) The towing phenomenon in the real-time image captured by CCD industrial camera. (3) Different light intensities cause a toll on image quality [23]. (4) Dust and other types of dirt may adhere to the surface of medical packages at random and make stochastic noise [24].

In order to solve the above problems, a delay-based misplaced image difference scheme is put forward in this paper, pursuant to structural characteristics and character recognition scenes in medical package image. Subtraction operation is done on the image by the delay effect in medical package manufacturing to and eliminate inherent noise of industrial cameras remove a large proportion of Gaussian noise in the image.

The remainder of this paper is organized as follows. Section 2 gives an introduction to the related work in this paper. Section 3 gives the selection results of the optimal parameters, and gives the final detection results. Section 4 provides a summary of this paper.

2 Related Work

2.1 The Proposed Method

(A) Delay-Based Misplaced Addition Scheme

Medical package images can be de-noised by superposition of two images with delay. To be specific, as noise is a random occurrence in the background image, original image and delayed image are superimposed, a product factor ☆ is designed to eliminate the background noise, and the coverage of target characters in space is increased to enhance the accuracy Δ of character recognition. Through the above analysis, after medical package images at *T1* moment and *T2* moment are superposed, gray values in the background area further rise but in the character area vary a little, hence a suitable designed product factor can effectively segment target characters from the background area in a medical package image.

$$S_{T1} = G_{T1} + F_{T1} \tag{1}$$

$$S_{T2} = G_{T2} + F_{T2} \tag{2}$$

$$S = \partial(S_{T1} + S_{T2}) = \partial((G_{T1} + G_{T2}) + (F_{T1} + F_{T2})) \tag{3}$$

Herein ☆ is a product factor whose value range is [0.1, 0.9], G_{T1} and G_{T2} are target characters areas, F_{T1} and F_{T2} are background areas. The image superposition scheme can be used in a medical package image with a simple background, wherein a product factor ☆ is designed to implement extraction and recognition of target characters in the medical package image. However, as for target characters on medical package images with a complicated background, superposed textures from the background image may have the same gray values distribution as superposed characters, hence accurate recognition of target characters on medical package image with a complicated background can hardly be implemented in this superposition difference method. To solve this problem, a delay-based misplaced subtraction scheme is designed in this paper.

(B) Delay-Based Misplaced Subtraction Scheme

As medical package images are detected, the accuracy of detection results may be lowered on account of random disturbance in production lines (such as stains and scratches) and inherent electronic noise of industrial cameras. As shown in Fig. 1.

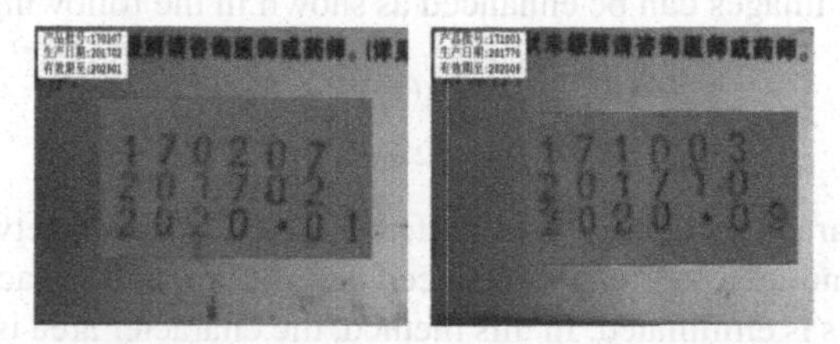

Fig. 1. Wrong recognition results caused by inherent noise of industrial cameras. Left: the detection result with "0" is recognized as "9" under poor image quality. Right: the detection result with "1" is recognized as "7" under good image quality. This example shows that the influence of camera inherent noise on recognition results is difficult to eliminate by traditional de-noising algorithms.

In order to eliminate the impact of industrial cameras inherent noise on detection results, a printing delay time *T* is designed first of all in this paper, under which in a particular trigger delay, two images (original image and delayed image) are subtracted (size of each image is normalized to 630 × 380) to get a difference image. Given that the original image is S_1 and image after delay at the T moment is S_2, a difference image S_d can be obtained by the difference algorithm with delay:

$$S_d = S_2 - S_1 \tag{4}$$

Whereas the disturbing signals created by industrial cameras and random noise in an industrial production environment are similar to white noise, noise at *T1* moment is basically similar to that at *T2* moment, indicating that noise signals of the original image can be eliminated while edge information of the target characters can be reserved by the delay-based misplaced subtraction method. T As the gray values of the background area in the image are the majority, the gray values of the background area are changed to 0 after subtraction operation and random noise in the medical package image is eliminated; the character area in the image captured under particular trigger delay moves rightward a little as a whole, whose gray values after subtraction operation is solved turn to:

(1) The right side of characters area can be regarded as $|A - B|$;
(2) The left side of characters area can be regarded as $|B - A|$;
(3) The overlapping part of characters area can be regarded as $|A - A|$, it also can be seen as the filling of gray values.

In the above description, A refers to the character area and B refers to the background area. Assuming that the character area of image moves rightward a little as whole in the wake of delay, the subtraction operation means that subtraction of medical package images at *T1* moment and *T2* moment, hence the left part of characters is obtained by the background area subtracted by the character area while the overlap part is acquired by the background area subtracted by the character area. Since the processed gray values can be positive or negative, whose value is lower than the subtraction threshold are set as 0 while ones whose value is higher than the subtraction threshold are set as $C(x, y)$. In Eq. (8), H refers to the gray threshold range, which is [0, 10]. As a result, target characters in original medical images can be enhanced as shown in the following Eq. (8):

$$S(x, y) = \begin{cases} 0, & P_2(x, y) - P_1(x, y) < H \\ C(x, y), & P_2(x, y) - P_1(x, y) \geq H \end{cases} \tag{5}$$

In this case, the character area can be extracted while the contrast between the character area and background area is greatly enhanced and the impact of background noise on the detection results is eliminated. In this method, the character area is extracted and the impact of inherent noise made by industrial camera on the detection results is minimized. Figure 2 shows a subtraction done on two random pharmaceutical images, wherein the gray histogram indicates that the gray values of the largest part of the processed image turn 0 and only the character area's gray values are higher than 0 distributed. To solve hollow strokes and discontinuity caused by subtraction operation, the multi-step convolution operation is adopted to optimize the difference image in this paper.

In order to compensate middle gray values among target characters, a convolution kernel for compensating gray values in the target area is designed pursuant to stroke

Fig. 2. Resulting image after subtraction operation under particular trigger delay.

features of target characters, it can be used to enhance the character area after delay based misplaced subtraction operation and improve the recognition capacity. According to the statistics, the image after subtraction operation has 80957 pixels whose gray value is 0 (33.3% of total pixels) and 231165 pixels whose gray value is in the 0–5 range (95.2% of total pixels), indicating that the processed image still has much speckle noise. In order to eliminate the impact of such noise points, a convolution kernel for compensating the target area is designed in this paper to convolute the image to enhance the character area while setting each gray value in the 0–5 range as 0 and raising the contrast between the character area and background area. To solve hollow strokes and discontinuity caused by subtraction operation, the multi-step convolution operation is adopted to optimize the difference image in this paper. The steps are as follows:

(1) The difference image is divided into blocks according to the size of convolution kernel under the floor principle. (The character area is in the center of the image, and there is almost no influence on the periphery of the image).
(2) As the convolution kernel is moving, its step size agrees with its width, hence the difference image is filtered for optimization as it moves by step size from the upper left corner to the lower right corner.
(3) A convolution threshold λ is set to process each image block traversed by the convolution kernel and if the result of operation between an image block and the convolution kernel is lower than the threshold λ, the image block will be recognized as part of the background area and all the gray values herein will be set as 0; otherwise all the gray values in the image block shall be normalized and filled, i.e.,

$$X = \sum_{m-1}^{N} A_m X_m \tag{6}$$

In Eq. (6), A refers to the convolution kernel, X refers to the image block traversed by the convolution kernel, m represents the size of the convolution kernel and N represents the size of medical package image. And the threshold λ refers to the percentage of non-zero gray values in each convolution block, i.e.

$$\lambda = \frac{C}{T} \tag{7}$$

In Eq. (7), wherein C represents the count of non-zero gray values in a convolution block and T represents the size of the convolution kernel.
(4) The filtered image is compared with the original character image (standard image) to work out SSIM, based on which, the best threshold λ and the optimal convolution kernel are selected.

In the process of multi-step convolution operation, the edges of target character area are prone to burrs and break lines, as shown in Fig. 3. To further remove speckle noise and enhance the target character area, a specific operation with corrosion and dilation is adopted in this paper. In order to solve the influence of interference items, the corrosion operation is firstly adopted to the character area. According to the position property of speckle noise, a special structural element is designed to eliminate the isolated small holes. In the paper, a large number of experiments have been carried out to select the optimal structural elements of corrosion and dilation

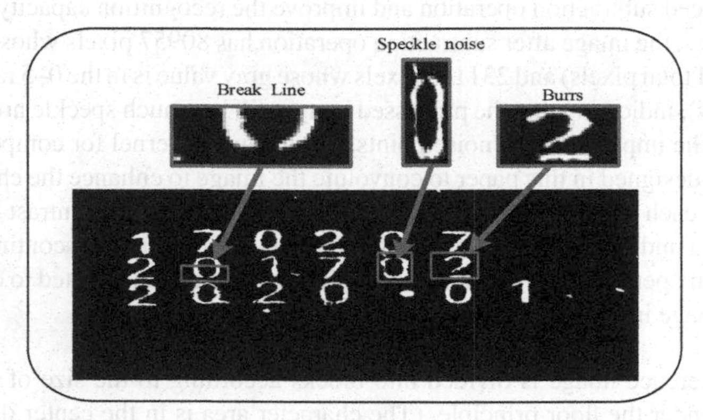

Fig. 3. (a): The burrs, break lines and speckle noise existed in images after the multi-step convolution operation.

3 Experimental Design, Analysis and Results

Considering the characteristics of the time-delay caused misplaced image subtraction method, three different character recognition methods are presented in our paper to demonstrate the feasibility of the proposed scheme. They are characters recognition on original image, characters recognition on addition image and characters recognition on subtraction image. The flowchart of the three schemes are shown in Fig. 4.

In order to evaluate the performance of the proposed method in this paper, the SSIM (structural similarity index measure) is employed to measure the quality of processed images, so as to obtain the optimal parameters. The simplified form of SSIM is as follows.

$$SSIM(x, y) = \frac{(2u_x u_y + C_1)(2\sigma_{xy} + C_2)}{(u_x^2 + u_y^2 + C_1)(\sigma_x^2 + \sigma_y^2 + C_2)} \quad (8)$$

Where, x and y are image blocks in the original image and processed image respectively, which take the same position and correspond to each other. The image block x is in the size of $M \times N$, and the image is in the size of 630×380 in the experiment. The luminance similarity of images is measured by the average gray value u_x and u_y; the contrast similarity of images is estimated by deviation information σ_x and σ_y; the structural similarity σ_{xy} is measured by correlation between the image block x and y; C_1 and C_2 are constants set, which are calculated with $C_1 = (K_1 L)^2$, $C_2 = (K_2 L)^2$ separately, wherein, $K_1 = 0.01$ and $K_2 = 0.03$ in the experiment.

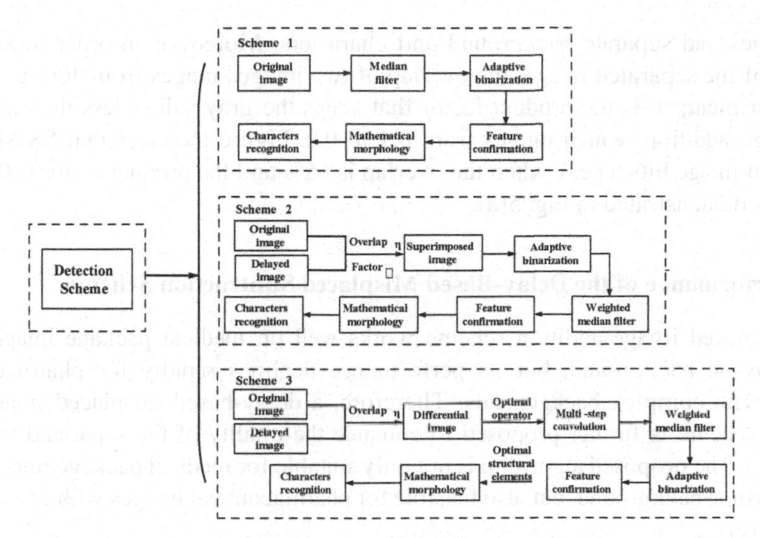

Fig. 4. The flowchart of three detection schemes in this paper

3.1 Performance of the Delay-Based Misplaced Addition Scheme

The delay-based misplaced addition scheme is more suitable for image with monochrome background. As the gray values of character pixels differ widely to those of background, the characters can then be separated from the background with a proper threshold. In this way, the noise in the background is eliminated and the quality of the separated image can be modulated. In the experiments, we choose the average value of the whole image pixels as the threshold and then a modified weighted median filter is employed to enhance the quality of the separated image. Then, 1000 medical package images with different backgrounds are employed to test performance of the misplaced addition algorithm. Here, SSIM is adopted to verify the simulation of the separated image with the original one. The results is demonstrated in Fig. 5(a).

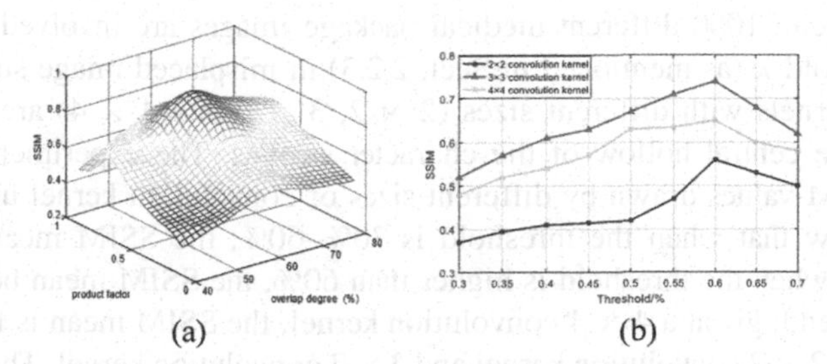

Fig. 5. (a): The changing trend of SSIM value under different overlap degrees and different product factors. (b): The averaging SSIM values obtained by different sizes of convolution kernel under different thresholds

As analyzed from Fig. 5(a), the misplaced addition scheme can be applied to medical package images with a single background to efficiently eliminate background noise in the images and separate background and characters. Moreover, in order to keep the quality of the separated image, the overlap of two images ranges from 40% to 80% in the experiment; ☆ is the product factor that keeps the gray values less than 255 after the image addition, which ranges from 0.1 to 0.9. Figure indicates that SSIM of the separated image hits a peak when the overlap is 65% and the product factor is 0.6. The results is demonstrated in Fig. 5(a).

3.2 Performance of the Delay-Based Misplaced Subtraction Scheme

The misplaced image addition scheme works well on medical package images with monochrome background, but its performance declines rapidly for pharmaceutical images with complex backgrounds. Therefore, a delay-based misplaced image subtraction scheme is further proposed to enhance the quality of the separated image in this paper. The proposed algorithm is not only suitable for medical package images with monochrome background, but also feasible for pharmaceutical images with complicated texture well.

(1) Effects of Different Overlap Coefficients on Experimental Results

As the overlap degree of two object images would highly influence the performance of misplaced image subtraction scheme, it should be carefully modulated in the experiment to achieve high quality separated image. Most of the character pixels would be eliminated if the overlap degree to large, and the characters can't be recognized, contrarily, if the overlapping degree is too small, character doubling would happen and the background still can't be removed. Experimental results show: when the overlap degree through delay-based misplaced difference subtraction is 80%, the character area can be extracted in its entirety.

(2) Effects of Convolution Kernels Different in Size and Type on Experimental Results

In the experiment, 1000 different medical package images are involved to determine the best threshold λ (as mentioned in Sect. 2.2.3) in misplaced image subtraction, the convolution kernels with different sizes (2×2, 3×3 and 4×4) are employed to compensate the central hollow of the character strokes. The experimental results of averaging SSIM values drawn by different sizes of convolution kernel under different thresholds show that when the threshold is 30%–60%, the SSIM mean presents an upward trend; when the threshold is higher than 60%, the SSIM mean begins to show a downward trend; given a 4×4 convolution kernel, the SSIM mean is far lower than ones drawn by 2×2 convolution kernel and 3×3 convolution kernel. Therefore, given that the threshold is 60%, all types of 3×3 convolution kernel s are tested in this paper, among which, the convolution kernel $[1, 0, 1; 1, 1, 1; 1, 0, 1]$ gets the highest SSIM value that is 0.8863; as the image processed by this convolution kernel is inflicted by a small amount of noise, different median filtering windows are used to process and recognize the image.

(3) Effects of Different Structural Elements During Corrosion and Dilation

In the experiment, mathematical morphology is employed to the processed image after multi-step convolution. Different structural elements for mathematical morphology are designed to enhance the processed image. In the progress of morphological operation, the structural elements with size of 2×2, 3×3, 4×4 and 5×5 are employed to eliminate the isolated small holes and enhance the target characters. Then SSIM is adopted to evaluate the processed image quality by comparing the convolutional processed image. Because too many types of structural elements are tested, therefore, some representative experimental results are chosen for display in this paper, and we came to a conclusion based on the SSIM obtained in the experiment. The conclusion is: in the progress of corrosion, the best SSIM and the highest image fidelity arc achieved when the element matrix size is 3×3. The SSIM decreases with other large or smaller element matrix, such as the size of 2×2, 4×4 and 5×5, even 7×7. However, in the progress of dilation, the best SSIM and the highest image fidelity are achieved when the element matrix size is 5×5. The specific experimental results are shown below (Fig. 6).

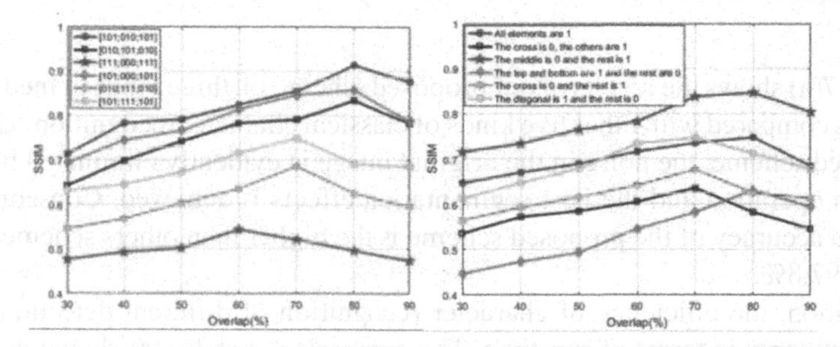

Fig. 6. Some representative SSIM values obtained by corrosion and dilation

According to the character pattern matching, the character area is enhanced and the isolated small holes and noise points are removed. According to the experimental results, the optimal element matrix of corrosion is [1, 0, 1; 1, 0, 1; 1, 0, 1] and the optimal element of dilation is [1, 0, 1, 0, 1; 0, 1, 1, 1, 0; 0, 0, 1, 0, 0; 0, 1, 1, 1, 0; 1, 0, 1, 0, 1]; their structures are shown in Fig. 3(b).

3.3 Experiments Evaluation of Medical Package Images

To further verify the performance of the proposed scheme, some typical medical package images are employed for character recognition test. In the experiment, the image overlapping threshold is set to 80%, the threshold λ (the percentage of non-zero gray values in each convolution block) is set to 60% and the optimal convolution kernel is [1, 0, 1; 1, 1, 1; 1, 0, 1]; the structural elements for corrosion and dilation are [1, 0, 1; 1, 0, 1; 1 0, 1] and [1, 0, 1, 0, 1; 0, 1, 1, 1, 0; 0, 0, 1, 0, 0; 0, 1, 1, 1, 0; 1, 0, 1, 0, 1] separately. 1000 medical package images (including paper package, plastic package and aluminum package) are obtained with industrial camera in real time, the characters are recognized with the optimal coefficients consequently. The results of the proposed scheme are compared with other state-of-the-art de-nosing schemes, and they are shown in Fig. 7(a).

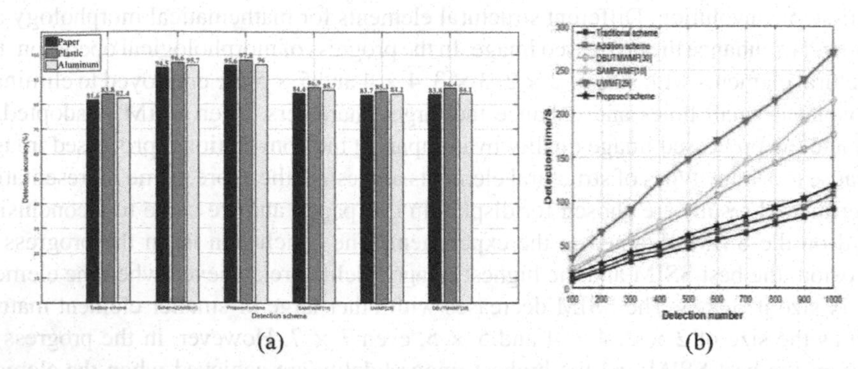

Fig. 7. (a): Characters recognition accuracy of different detection schemes for different medical package images. (b): Time comparison results of different detection schemes which use different de-noising algorithms

Figure 7(a) shows the results of our proposed scheme on three kinds of medical package images compared with other five kinds of classical character recognition schemes. In the proposed scheme, the noise in the original image is evidently eliminated by delayed subtraction operation and the best segmentation effects is achieved. Consequently the recognition accuracy of the proposed scheme is the higher than others scheme, which is as high as 97.8%.

In addition, the efficiency of character recognition of different detection schemes are also compared in terms of run-time. The comparison results are shown in Fig. 7(b). According to the experimental results, the time of character recognition is no more than 120 s on 1000 medical package images with the size of 630 × 380 for the three kinds of schemes proposed in the paper. However, the current mainstream de-noising algorithms, such as SAMFWMF (switching adaptive median and fixed weighted mean filter) [18], UWMF (Unbiased-based weighted mean filter) [29] and DBUTMWMF (Decision based unsymmetrical trimmed modified winsorized mean filter) [30], which although can greatly eliminate the impact of industrial cameras inherent noise, but the detection efficiency of those algorithms can't satisfy the industrial requirement. Although the character recognition speed of the traditional scheme takes slightly less time than other schemes, its detection accuracy is also lower than others. The detection accuracy of the proposed scheme can be approximately the same as other mainstream denoising schemes, and its efficiency is still able to satisfy the requirement of real-time character recognition.

4 Conclusion

A novel scheme to reduce printing errors during medicine package is put forward in this paper, wherein, the delay medical package images are acquired by a machine vision detection system, and the two images are subtracted one from another to achieve a timedelay difference image. The difference image is then de-noised with the optimal overlapping and convolution kernel for character extraction, then a specific mathematical morphology is adopted to obtain a visual quality enhanced image, by which, the edges of the characters are clarified. The character image without industrial cameras inherent noise are extracted through misplaced subtraction operation; the character area is enhanced by multi-step convolution, and the speckle noise and break lines are compensated by the specific structural element of corrosion and dilation. Experimental results show that the proposed scheme gains high performance than those classical schemes in terms of character recognition accuracy and efficiency, hence it can satisfy the needs of real time industrial medical package character recognition.

References

1. Sun, H., Sun, C., Liao, Y.: The detection system for pharmaceutical bottle-packaging constructed by machine vision technology. In: 2013 Third International Conference on Intelligent System Design and Engineering Applications, pp. 1423–1425. IEEE (2013)
2. Khan, S.J., Sharif, M., Raza, M., et al.: Capsule deformation in pharmaceutical industry by a non-contact metrology algorithm. J. Med. Imaging Health Inform. **5**(2), 210–215 (2015)
3. Skeppstedt, M., Kvist, M., Nilsson, G., et al.: Automatic recognition of disorders, findings, pharmaceuticals and body structures from clinical text: an annotation and machine learning study. J. Biomed. Inform. **49**, 148–158 (2014)
4. He, K., Zhang, Q., Hong, Y.: Profile monitoring based quality control method for fused deposition modeling process. J. Intell. Manuf. **30**(2), 947–958 (2018). https://doi.org/10.1007/s10845-018-1424-9
5. Zhang, H., Wang, Y., Zhou, B.: Research on foreign substance detection system for medical solution based on machine vision. Chin. J. Sci. Instr. **30**(03), 548–553 (2009)
6. He, C., Wang, Y.: Machine vision detection and recognition of quality of filling liquid medicine. J. Central South Univ. (Sci. Technol.) **40**(4), 10031007 (2009)
7. Tenhunen, H., Pahikkala, T., Nevalainen, O., et al.: Automatic detection of cereal rows by means of pattern recognition techniques. Comput. Electron. Agric. **162**, 677–688 (2019)
8. Taborri, J., Palermo, E., Rossi, S.: Automatic detection of faults in race walking: a comparative analysis of machine-learning algorithms fed with inertial sensor data. Sensors **19**(6), 1461 (2019)
9. Li, L., Luo, H., Ding, S., et al.: Performance-based fault detection and fault-tolerant control for automatic control systems. Automatica **99**, 308–316 (2019)
10. Tian, S., Bhattacharya, U., Lu, S., et al.: Multilingual scene character recognition with co-occurrence of histogram of oriented gradients. Pattern Recogn. **51**, 125–134 (2016)
11. Xiao, X., Jin, L., Yang, Y., et al.: Building fast and compact convolutional neural networks for offline handwritten Chinese character recognition. Pattern Recogn. **72**, 72–81 (2017)
12. Min, Y., Xiao, B., Dang, J., Yue, B., Cheng, T.: Real time detection system for rail surface defects based on machine vision. EURASIP J. Image Video Process. **2018**(1), 1–11 (2018). https://doi.org/10.1186/s13640-017-0241-y
13. Shi, H., Yu, W.: Research on the way of on-line quality inspection of commerce note printing. J. Wuhan Univ. Technol. **30**(5), 148–150 (2008)
14. Wang, L., Shen, Y.: Design of machine vision applications in detection of defects in highspeed bar copper. In: 2010 International Conference on E-Product E-Service and E-Entertainment, pp. 1–4. IEEE (2010)

15. Iyshwerya, K., Janani, B., Krithika, S., et al.: Defect detection algorithm for high speed inspection in machine vision. In: International Conference on Smart Structures and Systems, ICSSS 2013, pp. 103–107. IEEE (2013)
16. Huang, H., Da, P., Han, X.: Wavelet transform and gradient direction based feature extraction method for off-line handwritten Tibetan letter recognition. J. Southeast Univ. (Engl. Ed.) **30**(1), 27–31 (2014)
17. Li, Q., Ren, S.: A visual detection system for rail surface defects. IEEE Trans. Syst. Man Cybern. Part C (Appl. Rev.), **42**(6), 1531–1542 (2012)
18. Mafi, M., Rajaei, H., Cabrerizo, M., et al.: A robust edge detection approach in the presence of high impulse noise intensity through switching adaptive median and fixed weighted mean filtering. IEEE Trans. Image Process. **27**(11), 5475–5490 (2018)
19. Rakshit, S., Ghosh, A., Shankar, B.U.: Fast mean filtering technique (FMFT). Pattern Recogn. **40**(3), 890–897 (2007)
20. Gupta, G.: Algorithm for image processing using improved median filter and comparison of mean, median and improved median filter. Int. J. Soft Comput. Eng. (IJSCE) **1**(5), 304–311 (2011)
21. Li, W., Ni, R., Li, X., Zhao, Y.: Robust median filtering detection based on the difference of frequency residuals. Multimed. Tools Appl. **78**(7), 8363–8381 (2018). https://doi.org/10.1007/s11042-018-6831-6
22. Ma, B., Shi, Y.-Q.: A reversible data hiding scheme based on code division multiplexing. IEEE Trans. Inf. Forensics Secur. **11**(9), 1914–1927 (2016)
23. Wang, C., Wang, X., Xia, Z., Zhang, C.: Ternary radial harmonic Fourier moments based robust stereo image zero-watermarking algorithm. Inf. Sci. **470**, 109–120 (2019)
24. Wang, C., Wang, X., Xia, Z., Ma, B., Shi, Y.-Q.: Image description with polar harmonic fourier moments. IEEE Trans. Circuits Syst. Video Technol. (2019). https://doi.org/10.1109/TCSVT.2019.2960507
25. Fan, C., Chu, C., Šarler, B., et al.: Numerical solutions of waves-current interactions by generalized finite difference method. Eng. Anal. Boundary Elem. **100**, 150–163 (2019)
26. Singla, N.: Motion detection based on frame difference method. Int. J. Inf. Comput. Technol. **4**(15), 1559–1565 (2014)
27. Gatos, B., Pratikakis, I., Perantonis, S.: Adaptive degraded document image binarization. Pattern Recogn. **39**(3), 317–327 (2006)
28. Zana, F., Klein, J.C.: Segmentation of vessel-like patterns using mathematical morphology and curvature evaluation. IEEE Trans. Image Process. **10**(7), 1010–1019 (2001)
29. Kandemir, C., Kalyoncu, C., Toygar, Ö.: A weighted mean filter with spatial-bias elimination for impulse noise removal. Digit. Signal Proc. **46**, 164–174 (2015)
30. Vasanth, K., Manjunath, T., Raj, S.: A decision based unsymmetrical trimmed modified winsorized mean filter for the removal of high density salt and pepper noise in images and videos. Procedia Comput. Sci. **54**, 595–604 (2015)

A Mutual Trust Method for Energy Internet Agent Nodes in Untrusted Environment

Wei She[1,2,3,4], Jiansen Chen[1,4], Xianfeng He[2,3], Xiaoyu Yang[1], Xuhong Lu[1], Zhihao Gu[1], Wei Liu[1,4], and Zhao Tian[1](✉)

[1] School of Software Technology, Zhengzhou University, Zhengzhou 450000, China
tianzhao@zzu.edu.cn
[2] Yellow River Institute of Hydraulic Research, Zhengzhou, China
[3] Research Center on Levee Safety Disaster Prevention, Zhengzhou, China
[4] Cooperative Innovation Center of Internet Healthcare, Zhengzhou University, Zhengzhou 450000, China

Abstract. The rapid development of distributed energy puts forward new requirements for real-time scheduling capability. Energy agents may steal or even tamper with relevant information to maximize their own interests, which will eventually lead to malicious competition. First, in order to solve the possible problems of malicious behaviors such as price channeling and malicious quotation of energy agency nodes in the energy blockchain, this paper provide an energy blockchain consensus method based on regional alliance node credit by adding a credit mechanism in the process of block generation and verification. Second, the method consists of two parts, one of which is that the scheduling network is divided into different regional alliances by using the regional partitioning algorithm to improve the efficiency of node verification, and the other is that we propose the method of accounting credit and verification credit evaluation of energy agency node by using the credit evaluation mechanism, and weigh the cost of nodal rewards and punishments by credit mechanisms, so as to restrain the malicious behavior of the node. Finally, the evaluation and analysis of simulation experiment show that this method can restrain the malicious behavior of nodes, reduce the consumption of the process of verification and forwarding, and improve the efficiency of verification.

Keywords: Blockchain · Regional leagues · Consensus mechanism · Credit assessment

1 Introduction

In recent years, the emergence and development of blockchain technology has provided new ideas for the scheduling of energy internet and the access and consumption of clean

This work is supported in part by Key Laboratory Open Project Fund of Engineering and Technical Research Center of Embankment Safety and Disease Control of Ministry of Water Resources (2018007), National Key R&D Program of China (2018YFB1201403) and CERNET Innovation Project (NGII20180702).

X. Sun et al. (Eds.): ICAIS 2020, LNCS 12240, pp. 557–569, 2020.
https://doi.org/10.1007/978-3-030-57881-7_49

energy [1]. In the energy internet, blockchain technology can effectively support the open interconnection of multi-type energy systems and the extensive and deep participation of multiple users, and achieve the reasonable utilization of energy and efficient interconnection of requirements [2, 3]. At present, most of the research focuses on the development of blockchain technology. There is still a lack of systematic research on the information security of energy blockchain on the patterns, key technologies and development ideas of each blockchain application scenarios in the energy internet [4]. In the energy blockchain, energy agents may steal relevant information in order to maximize their own interests and lead to malicious competition, which will result in low system operation efficiency and frequent failures. The information tampered with by malicious nodes in the energy blockchain will lead to the failure of power grid facilities and cause immeasurable damage [5]. Therefore, the energy blockchain is in urgent need of a credit consensus mechanism to restrain malicious behavior.

Relevant research status: The authors [6, 7] present a proof of work (PoW) mechanism for IoT devices based on credit, which can guarantee both system security and transaction efficiency. The authors [8] provided Byzantine Fault Tolerance (DBFT) consensus algorithm based on reputation delegation and a new energy allocation mechanism to allocate limited renewable energy to electric vehicles. With the power system demand response as the application background, in order to achieve the strong adaptability of power dispatching, the characteristics of blockchain decentralization, cooperative autonomy and interconnection are applied to the independent design of the energy market in a few works [9–11]. He, Xu, Yan [12] provided a distributed consensus principle and incentive mechanism, and analyzed the different autonomy strategies of nodes. Li, Li, Hou [13] studied a consensus algorithm for Proof of Vote (POV) for consensus issues in the coalition blockchain, its core idea is to establish different security authentication for different users in the whole network. Because of the different voice rights of each user, the way of obtaining consensus is changed, and the transmission process of data verification is greatly simplified, thus improving the efficiency of consensus. The above works contribute to the trust and security issues in blockchain nodes by proposing different consensus algorithms. Although they are not applied in the energy Internet, they provide a new idea of node credit evaluation for the trust between nodes in this paper.

In order to solve the series of problems caused by various malicious ACTS of malicious agent nodes in the energy blockchain, such as the slow block generation speed, the low efficiency of new block broadcasting authentication and the information tampering, this paper provide a consensus method based on regional alliance of credit.

2 Energy Blockchain Consensus Method Based on Regional Alliance Node Credit

2.1 Regional Alliance Energy Blockchain Node

In order to minimize malicious accounting and dishonest verification behavior of malicious nodes in Energy Proxy Nodes (R_{EN}), firstly, we mapped them into blockchain network. And then we divided them into three categories according to different functions in the blockchain, such as Accounting Node, Verification Node and Light Node.

1) Accounting Node: Accounting Node is mainly responsible for collecting the records of transactions [14]. In this paper, the Accounting Node is recorded as R_{AN}.
2) Verification Node: Verification Node is mainly responsible for transaction verification [15]. And any renewal of power resource transaction data will be stored in R_{VN} in the form of a transaction. After collecting transaction information and broadcasting to the whole network, the new blocks generate locally. In this paper, Verification Node is recorded as R_{VN}.
3) Light Node: Light Node [16] is the user node which participates in the independent transaction in the energy node. In this paper, Light Node is recorded as R_{LN}.

2.2 Consensus Method

Firstly, the autonomous alliance node group of each region is obtained, which is recorded as Consortium Blockchain (CB), through the alliance region partition method and the energy routing agent node credit evaluation method. Secondly, R_{EN} is divided into R_{AN}, R_{VN} and R_{LN} according to the different functions that R_{EN} performs in the alliance. Finally, new blocks are generated at the end of the consensus process. The specific process of alliance division and credit evaluation in the whole new block generation is shown in Fig. 1.

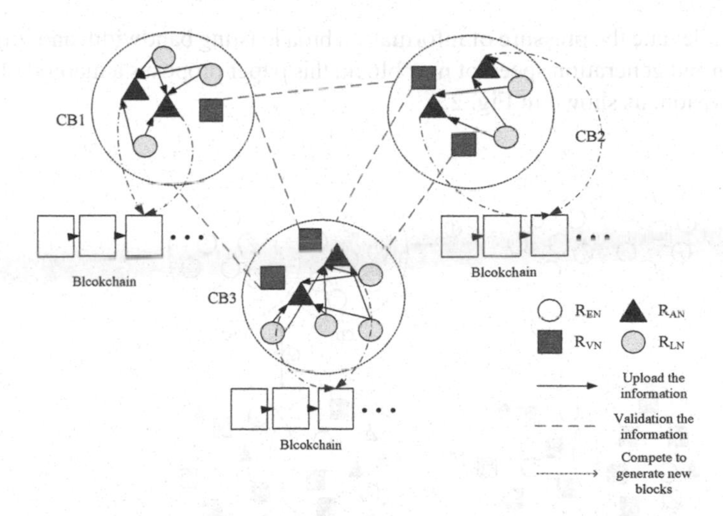

Fig. 1. League division and credit consensus process

1) R_{EN} are divided into CB_1 and CB_2 according to their own geospatial attributes. At the same time, according to the functions they need to achieve, they can be divided into three types: R_{AN}, R_{VN} and R_{LN}.
2) R_{LN} collect transaction info in the network and submit some data to R_{AN}.
3) R_{AN} collects and receives the data of power trade in the whole network constantly, and at the same time, through his own calculation, he seeks the random number with his own credit as the complex condition to solve the difficulty.

4) R_{EN} will shield malicious nodes in order to maximize the internal interests of the alliance. So R_{VN} cannot verify the new data blocks generated by R_{AN} in the same regional alliance to ensure fairness and objectivity.
5) When R^i_{AN}, a certain Accounting Node, takes the lead in obtaining the right to account, and the accumulated credit value of the transaction verification credit pool exceeds a certain proportion, the new block is certified as reasonable and effective.
6) R_{AN} and R_{VN} in the whole network synchronously connect the new data block to the previous block. And the consensus ends.

Among the three types of R_{EN}, R_{AN} and R_{VN} are in the key position of connecting each regional alliance CB to make physical energy and information data interoperable. If these two types of nodes provide illegal proxy services, they will have immeasurable impact on the grid and block chain network. In this section, the behavior of R_{AN} and R_{VN} in the block processing module is evaluated by accounting credit and verifying credit, in order to restrain the malicious behavior of R_{EN}. Because R_{LN} only participates in data query application in the whole blockchain system, and it has little influence on Energy Internet in practice, so this paper does not carry out credit evaluation for it at present.

2.3 Division of Regional Alliances

In order to alleviate the pressure of information broadcasting bandwidth and improve the verification and generation speed of new block, this paper proposes a method of regional alliance division, as shown in Fig. 2.

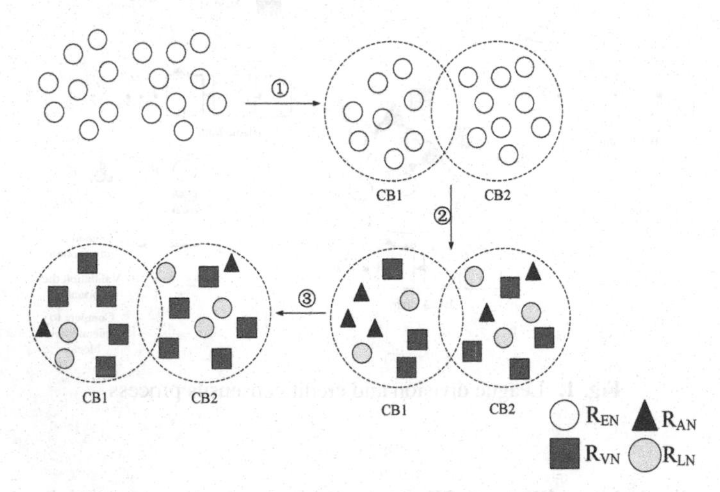

Fig. 2. Schematic diagram of league node division process

The process 1 divides R_{EN} into alliance networks of different regions according to geospatial location and management jurisdiction. And the neighboring R_{EN} form alliance node groups CB1 and CB2 of regional autonomy. When the energy block chain network undergoes the "mining" process, as shown in process 2, R_{EN} will assume different roles according to their own needs, for example R_{AN}, R_{VN} and R_{LN}. At the same time, there will be multiple R_{AN} competing for accounting rights through "mining". When a R_{AN} completes "mining", as shown in procedure 3, the only R_{AN} currently completing "mining" is determined. And it packaged data into a new block. Then the node broadcasts the new block information to the whole network and waits for the R_{VN} to verify the accuracy of the information. If the number of R_{VN} verified is larger than the validation threshold in the network, the newly generated block is considered valid. Otherwise, the new block is considered invalid and all transactions contained in the block are rolled back.

Due to the different location of different R_{EN}, the environment and resources around it are different, which makes the energy supply and consumption types different in different time periods. In this regard, In the energy market where there is a wide range of interest competition, there are deliberate information collection in the electricity sales data, power generation habits and the personalized behavior of users within their scope. Based on this, the possibility of node's real identity is speculated, which makes malicious acts to achieve privacy theft and profit-seeking misconduct occur. In order to protect the privacy of the energy blockchain, each alliance has a specific district number (District Number, DN). The domain coding of CB in alliance area is encrypted by hash algorithm and replaced every other time. DN is encrypted by hash algorithm and stored on the blockchain. New nodes connected to the grid can find the corresponding jurisdiction by forward calculation. Malicious nodes cannot deduce the exact jurisdiction code by reverse hash value, so it ensures a certain degree of anonymity. Energy block alliance formed by regional division can effectively reduce the cost of data circulation, share within the alliance, and improve the efficiency of consensus.

2.4 Accounting Credit of Energy Routing Agent Node

In order to ensure the accuracy and credibility of the energy transaction data, R_{AN} must verify each transaction data before collecting the distributed energy dispatch data of the whole network and counting it into the transaction storage pool. If the corresponding R_{AN} has malicious behavior, it will maliciously modify the data of a certain or some transactions and broadcast it, which will result in a large waste of data verification resources in the whole network. If multiple organizations form a malicious node alliance, it will have an overall structural damage to the entire energy Internet. For the malicious behavior of R_{AN}, the indicators for rating the accuracy of electrical energy transactions are:

(1) c_1: Check the integrity of power dispatching parameters, including the public key of the energy main body (*publicKey*), the address of the generation main body ($Address_g$), the address of the electricity main body ($Address_p$), the size of the transaction power (E), the dispatching unit price (P) and the transaction hash value (*TxHash*).
(2) c_2: Check the existence of $Address_g$,
(3) c_3: Check the rationality of transaction time. Determine whether the timestamp is larger than the generation time of the previous block. And avoid duplicate transactions.

(4) c_4: Check the matching between the $Address_p$ and its *publicKey*.
(5) c_5: Check the adequacy of the balance of the main account of electricity users. If the balance is lower than the single transaction price, the certification is illegal.

In order to evaluate the authentication credit of R_{AN}, the main evaluation attributes of single transaction verification behavior of R_{EN} can be expressed by formulas (1).

$$\begin{cases} Credit = \{c_s |\ s = 1, 2, 3, 4, 5\} \\ c_s = 0, False \\ c_s = 1, True \end{cases} \tag{1}$$

Where $c_s = 0$, it indicates that the first item s is unqualified. And if $c_s = 1$, it indicates that the first item s is calibrated accurately. Then the rationality of a single transaction can be expressed by formula (2).

$$T_{single}^{K,n} = c1 \ \&\ c2 \ \&\ c3 \ \&\ c4 \ \&\ c5 \tag{2}$$

In the formula, T represents the transaction accuracy; K represents the blockchain number; n represents the number of transactions in the first block n. If $T_{\sin gle}^{K,n} = 1$, it means that the firth transaction n record in the kth block in the energy blockchain is accurate. Conversely, if $T_{\sin gle}^{K,n} = 0$, the transaction record is incorrect. And the transaction is illegal. And there is a possibility of malicious tampering.

In order to indicate the accounting credit of the R_{AN}^{i}, we take the record accuracy of the previous block of R_{AN} as the criterion. As shown in Formula (3):

$$e_{AN}^{i} = \frac{\sum_{n=1}^{N^K} T_{single}^{K,n}}{N^K} \tag{3}$$

In the formula, e_{AN}^{i} is the accounting credit value of R_{AN}^{i}. And N^K is the total number of transactions in block K of the R_{AN}^{i} record.

In the whole accounting process of blockchain, R_{AN} continuously collects the information of electric energy transactions in the energy market. At the same time, it is able to obtain a qualified random number by continuously solving complex mathematical problems. In the typical blockchain application of Bitcoin, the solution process is as shown in Eq. (4):

$$\text{Hash}(R_m^i, R_s^i) <= \text{TARGET} = 2^{256-targetbits} \tag{4}$$

In the formula, Hash() is the hash algorithm. R_m^i is the merkle root of the newly generated data recorded by the first number i of R_{AN}. R_s^i is the random number of the forward solution of R_{AN}^{i} forward. TARGET is the solution target value. *targetbits* is the default of the blockchain degree of difficulty. It can be seen from Eq. (4) that since the hash function has the characteristics of positive solution and backward solution, R_{AN}^{i} needs to constantly try differently until the solution value is less than or equal to the solution value through the R_s^i hash algorithm.

In order to better restrict the accounting behavior by the credit value, the accounting credit value is added to the process of solving random numbers. As shown in Formula (5):

$$\text{Hash}(R_m^i, R_s^i) <= \text{TARGET} \times e_{AN}^i \tag{5}$$

Since the random number to be solved in the forward-direction is uniformly mapped to the interval of $0\text{–}2^{256-1}$ after hashing operation, the probability of obtaining a qualified random number in a single solution can be expressed by formula (6).

$$p_{single}^i = \frac{\text{TARGET} \times e_{AN}^i}{2^{256}} = \frac{2^{256-\text{targetbits}} \times e_{AN}^i}{2^{256}} = \kappa \times e_{AN}^i \tag{6}$$

In the formula, $p_{success}^i$ is the probability that R_{AN}^i solves the random number successfully. And κ is the probability difficulty constant.

Assuming that all R_{AN} in the energy block chain have the same computing force, that is the time of single operation hash algorithm is the same. And different R_{AN} has different degrees of difficulties in obtaining accounting right because of different e_{AN}. At the same time, the probability of R_{AN}^i competing for the right to account can be expressed by formula (7).

$$p_{success}^i = \frac{p_{\text{single}}^i}{\sum_{j=1}^{l} p_{\text{single}}^j} = \frac{e_{AN}^i}{\sum_{j=1}^{l} e_{AN}^j} \tag{7}$$

In the formula, $p_{success}^i$ is the probability that R_{AN}^i will obtain the accounting rights in a single operation. l is the number of R_{AN} in which the whole network participates in accounting at the same time.

Formula (7) shows that when the computing force of each node of the whole network is the same, the credit value of accounting in R_{AN} is positively correlated with the probability of success of "mining". Accounting credit value obtained by processing $K - 1$ blocks of each routing node can effectively reduce the generation of malicious nodes in $K - 1$ blocks. Similarly, the higher the credit value of block K, the easier the R_{AN} to get the probability of block $K + 1$. Nodes with higher credit value are more likely to win in the process of "mining" competition to obtain the right to account. Malicious nodes will not be rewarded because of their low credit value, thus curbing malicious intentions and malicious behavior.

2.5 Verification Credit of Energy Routing Agent Node

In the traditional blockchain network, each transaction data in the newly generated block needs to be broadcasted to the whole network. The new block data can be fully agreed, after most nodes verify that the node is verified. Subsequently, each blockchain node incorporates validated blocks into the local blockchain to achieve data synchronization. Verification of node information is the key link to achieve data link authentication, independent synchronization and consensus.

If R_{VN} is controlled by malicious nodes and intentionally tampers with local data, the verification information will be inaccurate. The malicious behavior of R_{VN}, such as price channeling and collective quotation, will lead to the failure of validation of correct real-time power transaction data. Thus increasing the number of validation nodes and the number of repeated validation times of the whole network, affecting the validation efficiency and causing a waste of physical resources. What's more, if there are a lot of malicious authentication nodes in the energy network, the message to be verified will not be authentic. The correct transaction data is illegally authenticated, that is, "make the truth false". Or falsely report the data to verify that it is correct, that is, "fake the truth". All these will lead to confusion in the energy blockchain, enabling malicious nodes to achieve the purpose of data tampering.

In view of the malicious behavior of R_{VN}, this paper intends to restrict it through the verification reputation of nodes. Similar to R_{AN}, the verification credit is expressed by verifying the accuracy of the new block data in the verification broadcast as shown in Formula (8):

$$e_{VN}^{i} = \frac{N_{ture}^{K}}{N^{K}} \tag{8}$$

In the formula, e_{VN}^{i} is the verification credit of R_{VN}^{i} and N_{ture}^{K} is the exact number of transactions. And N^{K} is the total number of transactions of block K, respectively.

In blockchain, different scenarios often require different consensus algorithms. The validation methods of different consensus algorithms are slightly different, but consensus can be reached only after a certain proportion of full-node validation. The method of judging the verification result based on R_{VN} credit is shown as (9):

$$e_{pool,VN}^{K} = \sum_{s=1}^{m} e_{VN}^{s} >= h \sum_{j=1}^{r} e_{VN}^{j} \tag{9}$$

In the formula, $e_{pool,VN}^{K}$ is the transaction verification credit pool of block K. m is the number of R_{VN} corresponding to the accumulation of transaction credit value. And η is the percentage of validation confirmation. And r is the number of R_{VN} participating verification in block K. It can be known from the formula that the verification accuracy requirement can be met when the verification credit accumulated in the reputation pool exceeds the accumulated standard amount of the prescribed ratio. R_{VN} with a higher credit value has a greater "speaking power" in the verification credit pool, which affects the final verification result. Conversely, R_{VN} with a smaller credit value has less influence on verification. Compared with the traditional method of determining the number of R_{VN}, the accumulation of verification credit reduces the workload of network-wide verification and confirmation.

3 Experimental Analysis

3.1 Design of Simulation Experiment

In order to verify the restrictive effect of credit evaluation mechanism on malicious behavior in energy blockchain, this paper builds a blockchain simulation experimental environment by JAVA language, and builds a P2P network on several servers with the same configuration to simulate the energy alliance blockchain scenario. Five R_{EN} were set up in the simulation experiment, and their credit scores were different. In addition, through MATLAB simulation experiments, 60 nodes are randomly simulated and their credit is randomly distributed in [1,100], which is used to simulate the relationship between the average credit of nodes and the minimum number of verification nodes needed for successful verification.

3.2 Influence of Accounting Credit on Block Output

This section presets that all participating nodes are configured consistently, that is the computing power is the same. The default value of mining difficulty is 2^{16} considering the calculation of equipment and the time of block generation.

As shown in Fig. 3: the three broken lines respectively represent the time required to generate 30 blocks under the default difficulty of "mining" when the accounting credit is 100, 90 and 80. It takes about 250 s to do the mining when the credit value is 100. It cost about 400 s when the credit value is reduced to 90. And it cost about 450 s when the credit value is reduced to 80 again. Therefore, the higher the credit value, the less time it takes to generate 30 blocks. When the credit value decreases, the more time it takes to mining successfully.

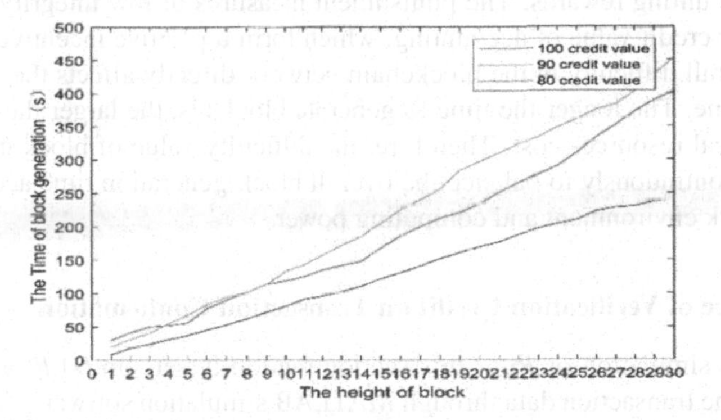

Fig. 3. The time of block generation

The three broken lines in Fig. 4 respectively show that the time for each block in the process of generating 30 blocks when the credit value is 100, 90 and 80. Then the average time of each block generated under three credit values is expressed by dotted lines. The average time of generating a block is 8.2 s, when the credit value is 100. And the average time to generate a block is 13.1 s, when the credit value is reduced to 90. While the credit value falls to 80 again, the average time to generate a block is 14.1 s. Therefore, as the credit value of R_{AN} decreases, the average time required to generate blocks increases, and the "mining" rate of R_{AN} decreases.

Under the condition that each node has the same computing power, the node's credit value of account keeping is positively correlated with the success rate of "mining".

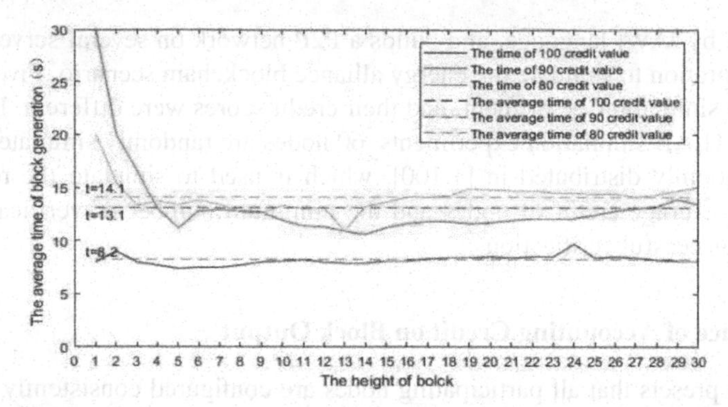

Fig. 4. The average time of block generation for different accounting credits

The higher the credit value of R_{AN} is, the greater the probability that R_{AN} obtains the accounting right of a certain block in the network, that the shorter the mining time is. With the decrease of the average credit value, the time to generate blocks increases.

The introduction of accounting credit can effectively restrict the block time of network generation. The more honest the node, the lower the cost of physical resources, faster to find the appropriate random number. Nodes with low credit value are more difficult to "mine", perform more operations, spend more time, and are also more difficult to obtain accounting rewards. The punishment measures of low integrity force R_{AN} to improve their credit value of accounting, which form a positive incentive. At the same time, the overall difficulty in the blockchain network directly affects the average block generation time. The longer the time to generate blocks is, the larger the total network loss of physical resources cost. Therefore, the difficulty value of block generation can be adjusted continuously to balance the overall block generation time according to the actual network environment and computing power.

3.3 Influence of Verification Credit on Transaction Confirmation

Based on the simulation of R_{VN}, this section randomly sets up 60 R_{VN} to verify the accuracy of the transaction data through MATLAB simulation software. On the basis of the works [13, 17], it is assumed that the average verification credit value is 90, and η is 2/3. That is 2/3 of the R_{VN} in the whole network can be verified to be valid for the new block.

We calculate the verification credit value of 60 R_{VN} according to formula (8). The average credit value of R_{VN} in the whole network can reflect the overall credit in the block verification process of current whole network. Under different verification credits, the effect of verification credit on overflow effect of verification storage pool can be reflected by solving the number of nodes required for verification and confirming and rounding upward. As shown in Fig. 5, when the average credit value of R_{VN} is 100, the number of new block verification confirmation nodes is 36. As the average credit of R_{VN} decreases, the number of R_{VN} increases. When the verification credit is lower than 90 and greater than 85, the number of R_{VN} in the entire network exceeds the preset 2/3 nodes. In this case, the verification efficiency is lower than the normal node verification efficiency. The experimental condition is that the average credit of the whole network is 90. When the average credit of R_{VN} is lower than 85, the average credit of the nodes in the whole network cannot be satisfied, which is contradictory to the condition. Therefore, when the average credit of R_{VN} is lower than 85, the verification fails and is regarded as invalid verification. And the verification needs to be carried out again, until the average credit value of R_{VN} is higher than 85. Therefore, the nodes with high credit value have higher "discourse power". When the credit environment of the whole network is good, fewer R_{VN} can be required to complete the verification, and the specific standard of the verification credit pool can be reached faster, which reduce the verification times. When the credit environment of the whole network is relatively low, more verification nodes are required, which will increase the verification time. And even a round of verification is required, until the average credit of the R_{VN} reaches at least 85.

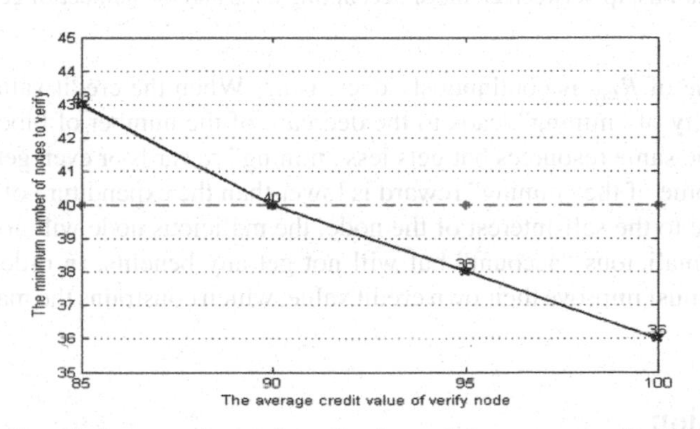

Fig. 5. The relationship between the number of nodes and the average credit of the verified nodes

3.4 Constraints of Credit Mechanisms on Nodes

This section will verify the binding effect of the reference mechanism on malicious nodes by comparing the number of "mining" of the consensus algorithm based on credit mechanism with the number of "mining" of the traditional POW consensus algorithm in a certain period of time, and comparing the income of "mining" rewards with the expenditure of "mining" resources. In this experiment, we choose the difficulty range of 26–28, and set 27 as the initial difficulty of the two algorithms, which is a relatively suitable initial value. The evaluation of "mining" based on the consensus algorithm of credit mechanism is shown in Fig. 6.

We record the relationship between the average credit value of the R_{EN} and the number of successful "mining" within 30 min based on the credit consensus mechanism of energy blockchain. When the average credit value of R_{EN} is greater than 90, the number of generated blocks is at most 69. So R_{EN} get the most "mining" rewards. However, when R_{EN} is attacked by a malicious attack, the credit value is reduced to 70. At this time, the number of blocks generated is relatively reduced, and the "mining" rewards obtained by R_{EN} are also relatively reduced. However, as the number of malicious attacks increases,

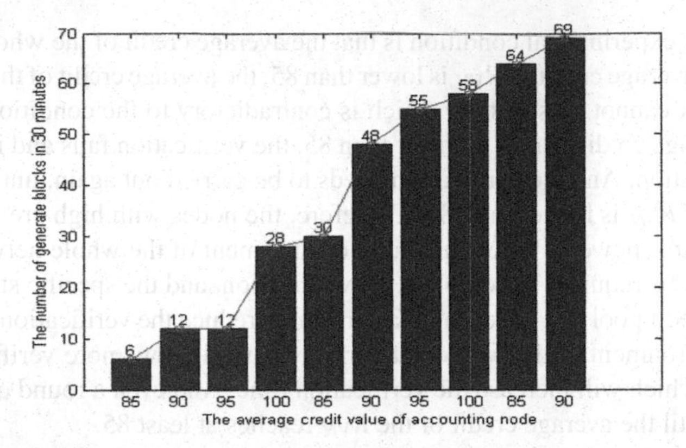

Fig. 6. The relationship between credit of accounting node and the number of generated blocks

the credit value of R_{EN} is continuously decreasing. When the credit value is less than 40, the difficulty of "mining" leads to the decrease of the number of blocks generated. It consumes the same resources but gets less "mining" rewards or even gets no rewards. When the income of the "mining" reward is lower than the expenditure of the resources consumed, due to the self-interest of the node, the malicious node will no longer waste resources for malicious "account" but will not get any benefits. In order to get more rewards, they must improve their own credit value, which constrains the malicious attack of nodes.

4 Conclusion

Firstly, this paper analyzes the problem of credit shortage of potential malicious nodes in the energy blockchain. Then we propose a regional alliance division method, and introduce the credit mechanism into the consensus process of the energy agent nodes. Meanwhile we give the credit evaluation methods of R_{AN} and R_{VN} respectively, and we also implement the process of new block generation. Finally, the simulation results show that the method can reduce the verification forwarding process, improve the verification efficiency, and reduce the malicious behavior of the energy blockchain proxy nodes in the non-trusted environment.

References

1. Mengelkamp, E., Notheisen, B., Beer, C., et al.: A blockchain-based smart grid: towards sustainable local energy markets. Comput. Sci.-Res. Dev. **33**(1–2), 207–214 (2018)
2. Pop, C., Cioara, T., Antal, M., et al.: Blockchain based decentralized management of demand response programs in smart energy grids. Sensors **18**(1), 162 (2018)
3. Imbault, F., Swiatek, M., De Beaufort, R., et al.: The green blockchain: managing decentralized energy production and consumption, pp. 1–5. IEEE (2017)
4. Fernández-Caramés, T.M., Fraga-Lamas, P.: Towards post-quantum blockchain: a review on blockchain cryptography resistant to quantum computing attacks. IEEE Access **8**, 21091–21116 (2020)

5. Song, R., Song, Y., Liu, Z., et al.: GaiaWorld: a novel blockchain system based on competitive PoS consensus mechanism. CMC-Comput. Mater. Continua **60**, 973–987 (2019)
6. Huang, J., Kong, L., Chen, G., et al.: Towards secure industrial IoT: blockchain system with credit-based consensus mechanism. IEEE Trans. Ind. Inform. **15**, 3680–3689 (2019)
7. Prieto-Castrillo, F., Kushch, S., Corchado, J.M.: Distributed sequential consensus in networks: analysis of partially connected blockchains with uncertainty. Complexity (2017)
8. Su, Z., Wang, Y., Xu, Q., et al.: A secure charging scheme for electric vehicles with smart communities in energy blockchain. IEEE Internet Things J. **6**, 4601–4613 (2018)
9. Wu, G., Zeng, B., Li, R., et al.: Research on the application of blockchain in the integrated demand response resource transaction. In: Proceedings of the CSEE, pp. 3717–3728 (2017)
10. Zeng, M., Cheng, J., Wang, Y., et al.: Primarily research for multi module cooperative autonomous mode of energy internet under blockchain framework. Proc. CSEE **37**(13), 3672–3681 (2017)
11. Li, B., Lu, C., Cao, W., et al.: A preliminary study of block chain based automated demand response system. Proc. CSEE **37**(13), 3691–3702 (2017)
12. He, Q., Xu, Y., Yan, Y., et al.: A consensus and incentive program for charging piles based on consortium blockchain. CSEE J. Power Energy Syst. **4**, 452–458 (2018)
13. Li, K., Li, H., Hou, H., Li, K, Chen, Y.: Proof of vote: a high-performance consensus protocol based on vote mechanism & consortium blockchain, pp. 466–473. IEEE (2017)
14. Deng, Z., Ren, Y., Liu, Y., et al.: Blockchain-based trusted electronic records preservation in cloud storage. Comput. Mater. Continua **58**(1), 135–151 (2019)
15. Watanabe, H., Akutsu, A., Miyazaki, Y., et al.: Blockchain generation apparatus, blockchain generation method, blockchain verification apparatus, blockchain verification method, and program. U.S. Patent Application 15/770,803[P], 21 February 2019
16. Palai, A., Vora, M., Shah, A.: Empowering light nodes in blockchains with block summarization, pp. 1–5. IEEE (2018)
17. Decker, C., Wattenhofer, R.: Information propagation in the bitcoin network, pp. 1–10. IEEE (2013)

Multilayer Perceptron Based on Joint Training for Predicting Popularity

Wei She[1,2,3], Li Xu[4], Huibo Xu[1], Xiaoqing Zhang[5], Yue Hu[1,2,4], and Zhao Tian[1](✉)

[1] School of Software, Zhengzhou University, Zhengzhou 450003, China
tianzhao@zzu.edu.cn
[2] Yellow River Institute of Hydraulic Research, Zhengzhou, China
[3] Research Center on Levee Safety Disaster Prevention, Zhengzhou, China
[4] Cooperative Innovation Center of Internet Healthcare, Zhengzhou University, Zhengzhou 450052, China
[5] Southern University of Science and Technology, Shenzhen, China

Abstract. For predictive analysis, Independent features and feature combination are of equal importance, but most models only focus on either independent features or feature combinations. In this paper, we propose a novel deep network model for predictive analysis. It incorporates two components: wide simple feed-forward neural network and MLP (multilayer perceptron) neural network. The wide simple feed-forward neural network is used to generalize to unseen feature combinations, and MLP neural network's aim to select and memorize vital independent features. The Feed-forward & MLP models are jointly trained for the Feed-forward & MLP model, in order to combine the benefits of selection, memorization and generalization. The results from the experiments show the jointly trained Neural Networks model can achieve ideal accuracy.

Keywords: MLP · Popularity prediction · Memorization · Generalization · Selection

1 Introduction

In the information era, reading and sharing news have become an indispensable part of people's entertainment lives. According to related studies of the Pew Research Center, about 40% U.S. adults used social media in 2009, and this proportion has rised to 72% in 2019. Although new social media platforms provide new opportunities for users to easily distribute their articles, they also have led to a tremendous bias in terms of popularity of the articles. A tiny percentage of the articles become popular, other articles are unknown to social media users. Hence, it would be greatly valuable if we could accurately predict the popularity of news articles ahead of its publication, for social workers (authors,

This work is supported in part by Key Laboratory Open Project Fund of Engineering and Technical Research Center of Embankment Safety and Disease Control of Ministry of Water Resources (2018007), National Key R&D Program of China (2018YFB1201403) and CERNET Innovation Project (NGII20180702).

X. Sun et al. (Eds.): ICAIS 2020, LNCS 12240, pp. 570–580, 2020.
https://doi.org/10.1007/978-3-030-57881-7_50

advertisers, etc.) and big companies such as Amazon, Baidu, Facebook and Alibaba. It also indicates that there are many business opportunities to be harnessed in the field of predictive analysis.

With our knowledge, there are few works that have used deep learning algorithm to predict popularity of news articles. In this paper, we propose a novel deep network model that analyzes the popularity of news articles prior to their publication. Assuming an deep network model, the popularity of a candidate article is first estimated using a prediction module and then according to the prediction result that the authors can make some changes in the article content and structure, in order to increase its share number. The main contributions of the paper include:

(1) In this paper, we propose a novel deep network model, which is inspired by Wide & Deep Learning for Recommender Systems [2], which called Feed-forward & MLP model. While our idea is very simple, by the combination of feed-forward neural network and MLP neural network are jointly trained for input features, and the results prove that it is a good idea. Different to conventional deep learning models, in Feed-forward & MLP model, feed-forward neural network module is used to generalize new feature combinations that have never or rarely occurred in the past, while MLP neural network module task is to select and memorize important independent features, which is always ignored by classic deep learning models.
(2) For traditional machine learning algorithms, input features must require much feature engineering effort to select appropriate features from raw data, and it is a laborious and time consuming task, but they are skilled at memorizing, which can be loosely defined as learning the frequent co-occurrence of items or features in the historical data. While conventional deep learning models are good at generalizing, which is s based on transitivity of correlation and explores new feature combinations that have never or rarely seen in the past [2]. Feed-forward & MLP models not only draw both of their strong points, but also are skilled at selecting vital independent features without feature engineering effort, which play an essential role in predictive analysis.
(3) In Highway Networks [3], the authors proposed a method that if there are different deep networks in a model that can be trained with different activation functions may help to get an efficient deep network architecture. In Feed-forward & MLP model, feed-forward neural network module and MLP neural network module belong to different deep networks, the method proposed in Highway Networks is virtually suitable to the proposed deep network model. So in our experiments, we attempt to use different activation functions to train the model, and found that some different activation function combinations can help to improve the model's performance indeed.

The remainder of the paper is organized as follows. Section 2 introduces related work. The source of the data, the number of features, data type of the feature and data preprocessing is elucidated in Sect. 3. Section 4 illustrates our deep network model, its optimization, and three comparing models. Section 5 provides the experiments, metrics are adopted, results and analysis. Section 6 concludes the paper with future work.

2 Related Work

According to Tatar et al. [4], there are two main popularity prediction approaches: those that use features only known after publication and those that do not use such features. The first approach is more common. Since the prediction task is easier, higher prediction accuracy are often achieved. The latter approach is more scarce, while a lower prediction performance might be expected, the predictions are more useful in predictive analysis. At the same time, Anitescu [5], Alhussain [6] and Maamar [7] proposed a novel neural network structure is very inspiring.

Using the second approach, Kelwin et al. [8] proposed a proactive intelligent decision support system to predict the popularity of online news with Random Forest in the Online News Popularity data set, the result showed that the Random Forest can achieve with an accuracy of 67%. The authors [9, 10] also use a Random Forest in the same dataset, the accuracy rates of 69% and 74% were obtained respectively by optimizing the Random Forest and feature selection. Petrovic et al. [11] predicted message propagation using features related with the tweet content and social features related to the author. They used a binary task to discriminate retweeted from not retweeted posts, a top F1 score of 47% was achieved when both tweet content and social features were used.

3 Dataset and Data Preprocessing

3.1 Dataset

The dataset was collected from UCI machine learning repository [8], originally acquired and preprocessed by K. Fernandes et al. The Online News Popularity (ONP) dataset with 39644 articles' textual extracted data in the last two years from Mashable website, each containing 61 attributes (58 predictive attributes, 2 non-predictive, 1 goal field). The attributes are either integer or real, and the dataset does not contain any missing values. The full feature set is mainly categorized as in Table 1 [9].

3.2 Data Preprocessing

Data preprocessing is an important step in the data mining process, which involves transforming raw data into an understandable format. In our dataset, we take the standard score method to normalize (or standardize) the data. In statistics, the standard score is the signed number of standard deviations by which the value of an observation or data point is above the mean value of what is being observed or measured. Observed values above the mean have positive standard scores, while values below the mean have negative standard scores. The standard score is a dimensionless quantity obtained by subtracting the population mean from an individual raw score and then dividing the difference by the population standard deviation. Standard scores are also called z-values, z-scores, normal scores, and standardized variables. They are most frequently used to compare an observation to a standard normal deviate, though they can be defined without assumptions of normality, which is defined as:

Where μ is the mean of the population, σ is the standard deviation of the population, and the absolute value of z represents the distance between the raw score and the

Table 1. All available features

Aspects	Features
Words	Number of words of the title/content; Average word length; Rate of unique/non-stop words of contents
Links	Number of links; Number of links to other articles in Mashable
Digital media	Number of images/videos
Publication time	Day of the week/weekend
Keywords	Number of keywords; Worst/best/average keywords (#shares); Article category
NLP	Closeness to five LDA topics; Title/Text polarity/subjectivity; Rate and polarity of positive/negative words; Absolute subjectivity/polarity level
Target	Number of shares at Mashable

population mean in units of the standard deviation, z is negative when the raw score is below the mean, positive when above.

To understand the data, a scatter plot of the share distribution was created as shown in Fig. 1, and the method we adopt from [10]. From Fig. 1, it is obvious that most of the articles situated in a band of relatively low share count values, while a few articles are outliers – popular articles.

The threshold of share number is adopted to classify identify news articles in the ONP dataset as either popular or unpopular is a key factor to impact model performance. In this paper, we used 2800 as a benchmark. In the Online News Popularity (ONP) dataset, the authors transformed the task into a binary task using a decision threshold of 1400, which keep these data in a balance proportion. The share threshold is 2800, the authors [10] used Share Genie, which is an application to identify articles in the ONP dataset as either popular or unpopular and a satisfactory indicator of share threshold for popular news articles was chosen to be in the 75th percentile of the share counts, and the value at the 75th percentile to be the threshold for popular articles, data points below the 75th percentile value, were labeled as 0 and above this as 1.

We split our dataset to 20% for testing, 20% for validation and 60% for training, so that there is no data overlap among the training, validation and test datasets. In order to obtain more general characteristics, we first took out the data shares more than 6200, and then in the middle of the 12 characteristics separated. These 12 characteristics for simple feed-forward network, and the rest used for MLP neural network. Respectively for the two data sets used Min-Max Scaler normalization, and then shuffled the data.

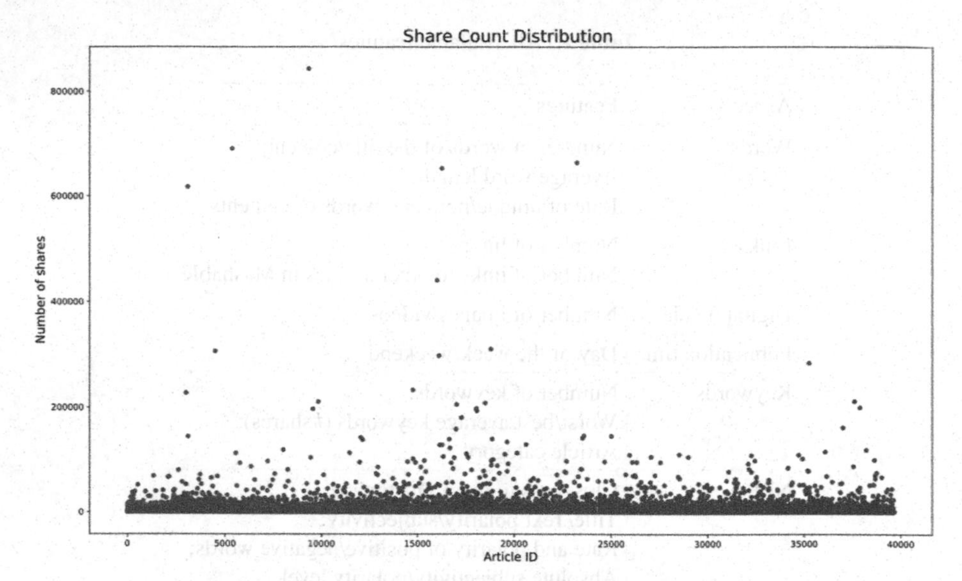

Fig. 1. Scatter plot showing the distribution of share counts of the Mashable articles.

4 Model

We first highlight the key components of our proposed deep network model: MLP neural network and feed-forward neural network in Sect. 4.1 and Sect. 4.2 respectively. Then we detail the overall the proposed deep network model in Sect. 4.3. And in Sect. 4.4, we will introduce three comparing models. The architecture of the network is shown in Fig. 2.

4.1 Wide Feed-Forward Neural Network

Feed-forward neural network refers to a broad family of artificial neural network (ANN), whose architecture consists of different interconnected layers, and was the first and simplest type of artificial neural network devised [12]. Theoretically, feed-forward neural network provided with a simple layer of hidden units is sufficient to map any function $y = f(x)$. Practically, it is often necessary provide these ANNs with at least 2 layers of hidden units. In our feed-forward neural network component, feed-forward neural network has 3 hidden layers as shown in Fig. 2 (left), and each hidden layer performs the following computation:

$$\mathrm{a}^{(l+1)} = f(\mathrm{W}^{(l)} a^{(l)} + b^{(l)}) \quad (1)$$

Where Eq. (1) is the layer number and f is the activation function. $\mathrm{a}^{(l)}$, $\mathrm{b}^{(l)}$, and $\mathrm{W}^{(l)}$ are the activation, bias, and model weights at l-th layer. In feed-forward neural network, it is usually to use gradient descent optimization algorithm to adjust the weight and the bias after a batch of data. In our model, feed-forward neural network's main work is to explore new feature combinations, which no other than deep network models are expert in.

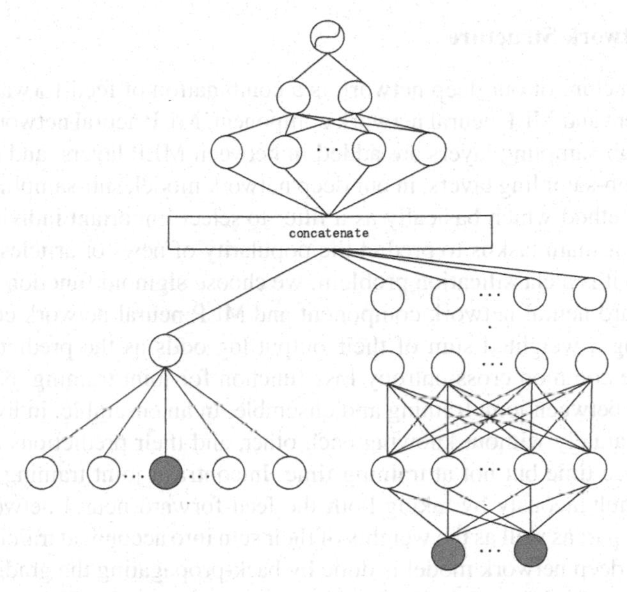

Fig. 2. The spectrum of feed-forward & MLP models.

4.2 MLP Neural Network

Convolutional neural networks (CNN) consist of alternating convolutional layers and pooling layers [10]. The convolution filter in CNN is a generalized linear model (GLM) for the underlying data patch. The authors had proved [1, 14] that replacing the GLM with a more potent nonlinear function approximate can enhance the abstraction ability of the local model, that is multilayer perceptron (MLP) convolution layers, which is equivalent to a convolution layer with 1×1 convolution kernel. The reasons we choose MLP convolutional neural network as another component as follow. First, conventional deep learning models are skilled at feature combinations' exploration and selection, but always lose sight of some important individual features. In some fields, individual features play an irreplaceable role in predicting and classifying, the prediction of online news popularity is one of them. In practice, we find that MLP layers [13] can be a pretty potent tool to select and memorize individual features with less feature engineering effort or even without feature engineering effort. Second, MLP layers can do the same work as convolution layers do, which are trained with back-propagation. Third, MLP layer can be a deep model itself, which is consistent with the spirit of feature re-use [12]. The calculation performed by MLP layers are shown as follows

$$
\begin{aligned}
f_{i,j,k_1}^{1} &= \max(w_{k_1}^{1^T} x_{i,j} + b_{k_1}, 0) \\
f_{i,j,k_n}^{n} &= \max(w_{k_n}^{n^T} x_{i,j}^{n-1} + b_{kn}, 0)
\end{aligned} \tag{2}
$$

Here n is the number of layers in MLP convolutional neural network as shown in Eq. (2) (right), more detail description about equations you can see reference [1].

4.3 Deep Network Structure

The overall structure of our deep network is a combination of feed-forward neural network component and MLP neural network component. MLP neural network is a stack of MLP layers, sub-sampling layers are added in between MLP layers, and differential to conventional sub-sampling layers, in our deep network model, sub-sampling layer adopt max pooling method which basically as a filter to select important individual features. In our study, our main task is to predict the popularity of news or articles prior to their publication, so it is a classification problem, we choose sigmoid function as classifier.

Feed-forward neural network component and MLP neural network component are combined using a weighted sum of their output log odds as the prediction, which is then fed to one common cross-entropy loss function for joint training. Note that there is a distinction between joint training and ensemble. In an ensemble, individual models are trained separately without knowing each other, and their predictions are combined only at inference time but not at training time. In contrast, joint training optimizes all parameters simultaneously by taking both the feed-forward neural network and MLP neural network part as well as the weights of their sum into account at training time. Joint training of our deep network model is done by back-propagating the gradients from the output to both the feed-forward neural network component and MLP neural network of the model simultaneously using the Adam optimizer. Before each layer we apply batch normalization and follow by an activation function, adopting the pre-activation block design [15]. Batch Normalization allows us to use much higher learning rate be less careful about initialization. We also apply Dropout [16] technique to the proposed deep network model. In our deep network, dropout technique not only is applied to prevent overfitting and over co-adapting, which is a technique for addressing this problem, but also drop some nonsignificant units randomly, which can help to discover and save important features. In our model, we apply a modified rectified linear activation function to wide feed-forward neural network and MLP neural network. By the experiments, it is actually a great improvement.

4.4 Comparison Models

In this paper, in order to compare the wide Feed-forward & MLP model performance with other models performance, we also tested 3 universal classification models: Random Forest (RF); SVM with a Radial Basis Function (RBF) kernel; K-Nearest Neighbors (KNN).

5 Experiments and Results

5.1 Experiments

We used the Scikit learn library [17] and TensorFlow [18] to implement the models we introduce above. We train from the Feed-forward & MLP model from scratch, manually set proper initialization for the learning rates and weights were initialized following the scheme introduced by He et al. [19]. We experimented with batch sizes ranging from 10 to 100, and our default batch size is 50. The training process starts from the initial weights and learning rates, and it continues until the accuracy on the training set stops improving, and then the learning rate is lowered by a scale of 2. This procedure is repeated once such that the final learning rate reaches the minimum learning rate we set. The layers of models is shown in Fig. 3.

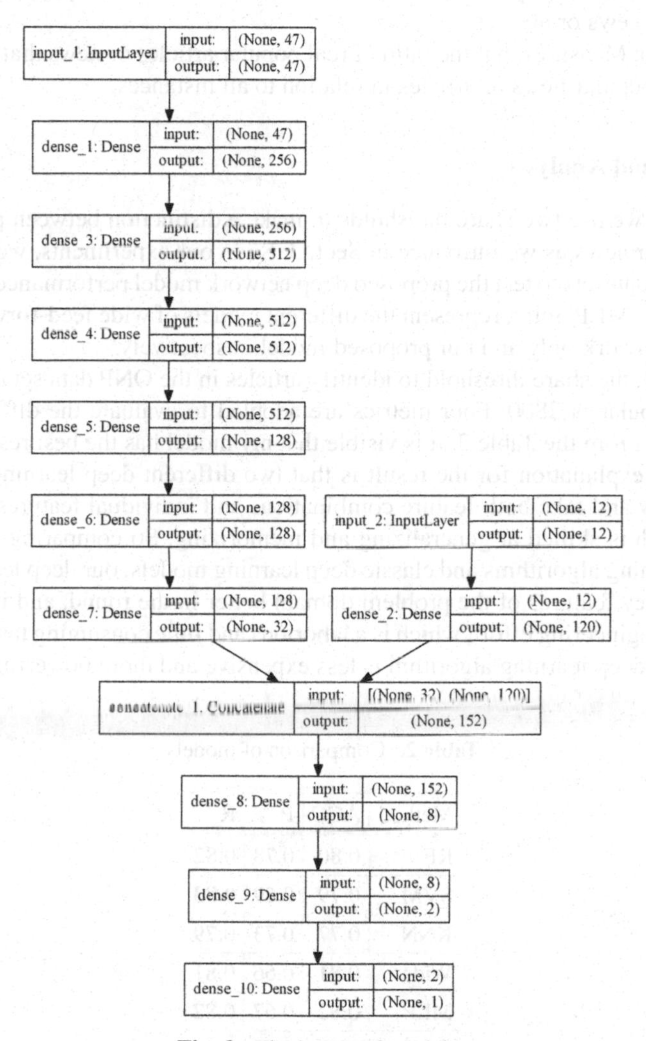

Fig. 3. The layers of models.

5.2 Evaluation Metrics

In this paper, our main research is to predict the popularity of news or articles prior to their publication. It is a classic classification model. Therefore, we choose four metrics to measure model performance: accuracy, F-measure, recall and precision.

Accuracy (ACC): It is an overall metric that measures both the correctly popular news or articles and unpopular news or articles in relation to all examples.

Precision (P): In Data Mining, precision denotes the proportion of Predicted Positive cases that are correctly Real Positives. In this paper, precision is used to measure the proportion of predicted popular news or articles that are correctly real popularity in relation to all news or articles.

Recall (R): Measures that the ratio of real popular articles or news that are correctly classified as popular news or articles in relation to all instances.

5.3 Result and Analysis

In this paper, we use two share thresholds to make a distinction between popular news and unpopular news as we introduce in Sect. 3. So in our experiments, we actually use two different datasets to test the proposed deep network model performance. WFF (wide feed-forward), MLP, Joint, represent the different models of wide feed-forward network only, MLP network only and our proposed model, respectively.

In Table 2, the share threshold to identify articles in the ONP data set as either popular or unpopular is 2800. Four metrics are adopted to evaluate the different models performance. From the Table 2, it is visible that my model has the best result out of the models. One explanation for the result is that two different deep learning models are trained jointly and take both feature combinations and individual features into consideration, which is skilled at generalizing and memorizing. So comparing to traditional machine learning algorithms and classic deep learning models, our deep learning model can capture key features of the problem domain better in the round, and it also require less feature engineering effort, which is a laborious and time consuming task. Therefore, the proposed deep learning algorithm is less expensive and more powerful.

Table 2. Comparison of models

Method	ACC	P	R
RF	0.80	0.78	0.82
SVM	0.79	0.69	**0.83**
KNN	0.77	0.73	0.79
Wide	0.81	0.66	0.81
MLP	0.82	0.67	0.82
Joint	**0.82**	**0.79**	0.82

In Fig. 4, We enlarged the pictures of the accuracy of training sets and verification sets, and we could see that the results were better at about 70 epochs. Even though we used kernel regularizer, after 100 epochs the validation accuracy sharply declined.

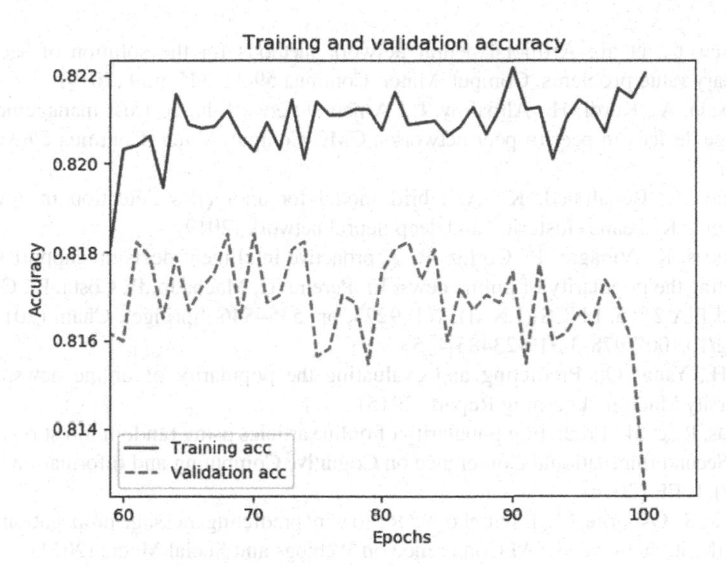

Fig. 4. Training and validation accuracy

6 Conclusion and Future Work

With the expansion of the Web, social media applications have become an essential part of people daily life. There is a growing interest in predictive analysis. For predictive analysis, memorization, selection and generalization are equally important. Feed-forward neural network models are able to automatically generalize to previously unseen feature interactions from high-level inputs. While MLP convolutional neural network as filters or choosers can effectively choose and memorize indispensable individual features. We presented the Feed-forward & MLP framework to combine the strengths of both types of model. Overall, from the experiments, the best result was achieved with an accuracy of 82%. Even though the proposed model are state-of-the-art, there are much room in our model optimization and feature selection.

In the future, we will intend to explore more advanced features related to content and topics and try to analyze relation between time and number of shares. We will also try other different deep model combinations. These approaches should improve the accuracy of prediction if combined with our current work.

References

1. Lin, M., Chen, Q., Yan, S.: Network in network. arXiv preprint arXiv:1312.4400 (2013)
2. Cheng, H.-T., et al.: Wide & deep learning for recommender systems. In: Proceedings of the 1st Workshop on Deep Learning for Recommender Systems (2016)
3. Srivastava, R.K., Greff, K., Schmidhuber, J.: Highway networks. arXiv preprint arXiv:1505.00387 (2015)
4. Tatar, A., de Amorim, M.D., Fdida, S., Antoniadis, P.: A survey on predicting the popularity of web content. J. Internet Serv. Appl. **5**(1), 1–20 (2014). https://doi.org/10.1186/s13174-014-0008-y

5. Anitescu, C., et al.: Artificial neural network methods for the solution of second order boundary value problems. Comput. Mater. Continua **59**(1), 345–359 (2019)
6. Alhussain, A., Kurdi, H., Altoaimy, L.: A neural network-based trust management system for edge devices in peer-to-peer networks. CMC-Comput. Mater. Continua **59**(3), 805–815 (2019)
7. Maamar, A., Benahmed, K.: A hybrid model for anomalies detection in AMI system combining K-means clustering and deep neural network (2019)
8. Fernandes, K., Vinagre, P., Cortez, P.: A proactive intelligent decision support system for predicting the popularity of online news. In: Pereira, F., Machado, P., Costa, E., Cardoso, A. (eds.) EPIA 2015. LNCS (LNAI), vol. 9273, pp. 535–546. Springer, Cham (2015). https://doi.org/10.1007/978-3-319-23485-4_53
9. Ren, H., Yang, Q.: Predicting and evaluating the popularity of online news. Standford University Machine Learning Report (2015)
10. Shreyas, R., et al.: Predicting popularity of online articles using random forest regression. In: 2016 Second International Conference on Cognitive Computing and Information Processing (CCIP). IEEE (2016)
11. Petrovic, S., Osborne, M., Lavrenko, V.: RT to win! predicting message propagation in Twitter. In: Fifth International AAAI Conference on Weblogs and Social Media (2011)
12. Bengio, Y., Courville, A., Vincent, P.: Representation learning: a review and new perspectives. IEEE Trans. Pattern Anal. Mach. Intell. **35**(8), 1798–1828 (2013)
13. Svozil, D., Kvasnicka, V., Pospichal, J.: Introduction to multi-layer feed-forward neural networks. Chemometr. Intell. Lab. Syst. **39**(1), 43–62 (1997)
14. LeCun, Y., et al.: Gradient-based learning applied to document recognition. Proc. IEEE **86**(11), 2278–2324 (1998)
15. He, K., Zhang, X., Ren, S., Sun, J.: Identity mappings in deep residual networks. In: Leibe, B., Matas, J., Sebe, N., Welling, M. (eds.) ECCV 2016. LNCS, vol. 9908, pp. 630–645. Springer, Cham (2016). https://doi.org/10.1007/978-3-319-46493-0_38
16. Srivastava, N., et al.: Dropout: a simple way to prevent neural networks from overfitting. J. Mach. Learn. Res. **15**(1), 1929–1958 (2014)
17. Pedregosa, F., et al.: Scikit-learn: machine learning in Python. J. Mach. Learn. Res. **12**(Oct), 2825–2830 (2011)
18. Abadi, M., et al.: Tensorflow: large-scale machine learning on heterogeneous distributed systems. arXiv preprint arXiv:1603.04467 (2016)
19. He, K., et al.: Delving deep into rectifiers: surpassing human-level performance on imagenet classification. In: Proceedings of the IEEE International Conference on Computer Vision (2015)

A Medical Blockchain Privacy Protection Model Based on Mimicry Defense

Wei Liu[1,3], Yufei Peng[2,3], Zhao Tian[1], Yang Li[2,3], and Wei She[1,3](✉)

[1] School of Software, Zhengzhou University, Zhengzhou 450000, China
wshe@zzu.edu.cn
[2] School of Information Engineering, Zhengzhou University, Zhengzhou 450000, China
[3] Collaborative Innovation Center for Internet Healthcare, Zhengzhou 450000, China

Abstract. With the advent of the medical big data era, the electronic medical records of patients will generate huge medical value and social benefits in the process of production, authorization, circulation and sharing. How to realize the efficient circulation of medical data under the premise of ensuring patients' privacy is the key problem that needs to be solved in the process of medical informationization. In order to solve the problem, this paper presents a dynamic heterogeneous medical alliance chain model named MAPM. By introducing the idea of mimicry defense, this model adds dynamic random attributes on the basis of the original isomorphic redundant blockchain consensus mechanism. The diversified workload demonstrates the consensus algorithm, improves the blockchain system's ability to withstand unknown attacks, and achieves privacy protection for patients during medical data sharing.

Keywords: Medical data · Privacy protection · Blockchain · Mimic defense

1 Introduction

Medical records recording the process of the disease occurrence, development and outcome, are valuable digital assets of patients in the medical field. It is the data foundation for treatment due to its high practical value. However, different hospitals has different data systems which lack of interoperability in the current medical information system. So, it is difficult to enjoy medical records from different sources because of the different versions of the records. In addition, data security is a very important issue owing to the sensitivity of medical data, which is vulnerable to many kinds of attacks. According to the statistics, there were 15 major medical information leakage events alone in the United States in 2017. It is conservatively estimated that about 3 million patients' information has been disclosured [1]. In September 2017, the *Legal Daily* reported that the

This work was supported in part by the National Key Research and Development Program of China under Grant 2018YFB1201403, in part by the Henan Province Foundation for University Key Project under Grant 20A520035, in part by the Henan Province Foundation for University Key Youth Teacher under Grant 2019GGJS018, and in part by the CERNET Innovation Project under Grant NGII20190707 and Grant 20180702.

X. Sun et al. (Eds.): ICAIS 2020, LNCS 12240, pp. 581–592, 2020.
https://doi.org/10.1007/978-3-030-57881-7_51

service information system of a hospital in China was hacked, which led to more than 700 million pieces of citizen information had been leaked, and more than 80 million had been trafficked [2]. Therefore, it is of great significance for the development of medical information to protect patients' privacy, realize the safe sharing and access of medical data.

As a distributed storage technology based on cryptography, blockchain provides the possibility for medical data of safe circulation and playing full value [3–5]. With the development of blockchain technology, more and more companies and research institutions have begun to try the research and application of blockchain in the field of medical health, hoping to bring huge advantages to the development of the medical field through blockchain technology. Many people believe that the medical field is the second largest blockchain landing area besides the financial field [4–9]. The privacy protection of medical data sharing is the key to realize data sharing. Tengfei Xue [4] proposed a medical sharing model combining improved DPOS and proxy re-encryption mechanism, which solved the problems of data security sharing, over-reliance on central institutions, and realized decentralized security medical data sharing. Chao Zhang [5] designed an alliance medical blockchain system based on practical Byzantine fault-tolerant algorithm (PBFT) to solve the medical problems of data sharing, tampering and leakage. Jieying Chen [6] designed a blockchain-based medical data sharing platform to achieve secure sharing and verification of medical data among patients, hospitals, and third-party organizations. MedRec [7] providing patients with a comprehensive and unchanging log, can easily access medical information across providers and treatment sites. Wenyu Xu [8] combined the homomorphic encryption and Ethereum smart contract technology to realize the function of the insurance claims. The insurance company can determine whether to settle claims even if the user's EHR plaintext and claim object ID could not be obtained. Qi Xia [9] proposed a data sharing framework based on the permission blochchain, which fully solved the access control challenges related to sensitive data stored in the cloud. All of these medical blockchain systems adopt static isomorphic redundancy model, in other words, each node holds the same algorithm and data. When the components (such as consensus algorithms, signature algorithms, etc.) that all nodes depend on are attacked, the whole system will face security threats.

In order to solve the problem, this paper proposes a privacy protection model of medical data based on blockchain and mimicry defense mechanism. The ability of the system to resist attacks is improved due to the dynamic and heterogeneous properties of the blockchain system by introducing mimicry defense technology.

2 Related Technology

2.1 Mimic Defense Mechanism

The core idea of Dynamic Heterogeneous Redundance [10] (DHR) architecture is to introduce uncertainty of the structural, so that the heterogeneous executors are dynamic and random. Mimic defense mechanism implements intrusion tolerance for specific system vulnerabilities or backdoors by building a dynamic heterogeneous redundant system architecture and operational mechanism. In the mimic defense system, the redundant controller firstly generates a scheduling policy and an arbitration policy by controlling parameters, respectively sends them to the input agent and the output agent. Then the input agent selects the corresponding heterogeneous functional equivalent according to scheduling policy to respond to the external service request. The heterogeneous functional equivalents send the result to the output agent, and the output agent decides each result according to the arbitration strategy. Finally the output agent selects a result as the response output [11]. The mimicry defense system architecture is shown in Fig. 1.

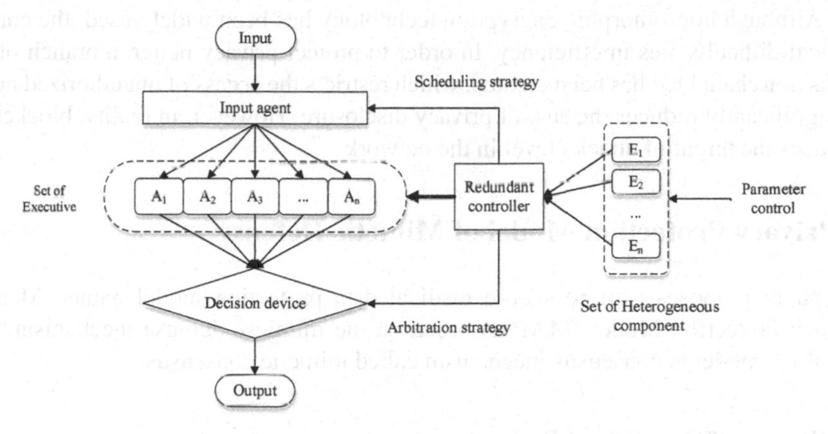

Fig. 1. Mimetic defense system architecture

2.2 The Consensus Mechanism of Blockchain

Blockchain, first appeared in Bitcoin, is an extended chain structure consisting of blocks, each of which includes header and transactions. The header includes the Merkle root, timestamp, random number, difficulty and hash value of the previous block. The Merkle tree is a binary tree structure that stores the hash of the transactions and can verify the integrity of the data. Blockchain system has the characteristics of distributed, autonomous, collective maintenance and so on. There is no centralized special node, and each equal node undertakes the process of block generation and verification in P2P network. The data in the blockchain is almost unchangeable after verifying by the whole network and maintaining by participants. Consensus mechanism is the core element of blockchain, so its security is the basis to ensure the security of the whole blockchain system. Most of the existing blockchains use single or multiple serial consensus algorithms. The security of the consensus algorithm is guaranteed under certain assumptions. For example, POW requires more than 50% honest nodes in the system [12].

2.3 Privacy Protection for Blockchain Data

Privacy includes identity privacy and transaction privacy in the blockchain. Identity privacy is the association between user identity and blockchain address. Transaction privacy is the records and the knowledge behind transaction records. Ring signature, first proposed by Rivest, Shamir and Tauman, is a simplified group signature. The signer first selects a temporary set of signers including himself. The signer only needs to use his private key and the public key of others in the signature set to generate the signature independently. The attacker cannot determine which member of the ring generates the signature, even if the private key of the ring member is obtained, the probability is not more than 1/n. Homomorphic encryption is a cryptographic technique based on the theory of computational complexity. Data can be operated without decrypting and it can be compared and retrieved in the case of encryption to get the correct conclusion. Although homomorphic encryption technology has been widely used, the current practical difficulty lies in efficiency. In order to protect privacy better, a branch of the permission chain [13] has been created, which restricts the access of unauthorized nodes and significantly reduces the risk of privacy disclosure. However, in reality, blockchain still faces the threat of attacks level in the network.

3 Privacy Protection Model of Mimetic Defense

This paper proposes a more secure medical data protection model named Medical Alliance Protection Model (MAPM) based on the mimicry defense mechanism. The core of the model is consensus mechanism called mimetic consensus.

3.1 Storage of the Medical Data

Many medical data files have a large amount of data. Cloud storage faces security risks although the low cost. Storing large-sized data directly on the blockchain is not suitable because of its relatively slow speed. So in MAPM, large-scale data are stored in public or private clouds after symmetric encryption, then the storage address and hash value of these data are linked with blockchain. That is, the off-chain storage is combined with the encrypted storage.

In order to convenient for downloading and searching data, MAPM is logically stored according to a certain level, while the specific medical data is stored in the leaf node, the root node stores the user ID, and the intermediate node stores the classification information. The index naming rules are as follows: User ID + Node sequence number. As shown in Fig. 2, the encrypted data 3, whose index seq is ID||221. The data owner creates the index when the data is generated, and at the same time generates a smart contract for finding the index, which is stored in the blockchain system. When the user meets the data access requirements, the contract code is executed to find the data index in the cloud storage server.

The data retrieval speed and search efficiency are high due to the large amount of medical data access and strong randomness. In addition, medical data is frequently updated (including insert and delete), and the chain structure is not suitable for storing medical data. Therefore, our model adopts index storage in the cloud server. The user's search behavior triggers the smart contract execution and finds the index of the data, so as to find the storage address of the data and perform the download operation.

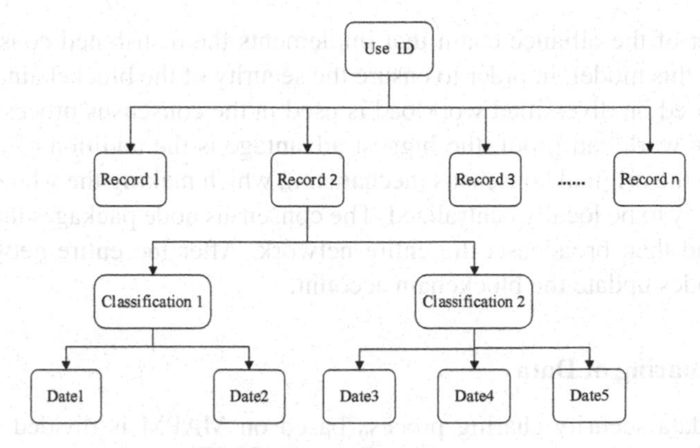

Fig. 2. Data storage structure

3.2 Model Architecture of MAPM

As shown in Fig. 3, the MAPM model includes the following parts: blockchain, cloud storage server, owner, consensus node, and user. The owner is the entity that created the electronic medical records (EHR), which can be a person (e.g., a patient) or an organization (e.g., a hospital, clinic, etc.). The user is an entity that meets certain attribute characteristics, similar as the owner, which can be a person (such as a medical practitioner) or an organization (such as a hospital, medical research institute, health insurance company, etc.).

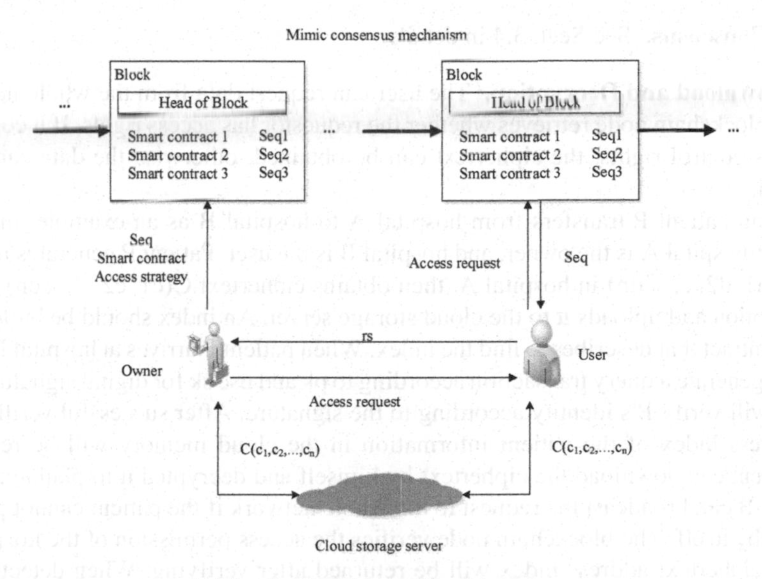

Fig. 3. Model overall architecture

Blockchain is an entity that stores indexes, all smart contracts, and access control policies. Users can search for certain EHRs in the blockchain. Consensus node, that is, the member of the alliance chain that implements the distributed consensus of the blockchain. In this model, in order to ensure the security of the blockchain, a consensus mechanism based on diversified workload is used in the consensus process. Compared with the single workload proof, the biggest advantage is the addition of dynamic and randomness to the original consensus mechanism., which making the whole blockchain network not easy to be locally centralized. The consensus node packages the transaction into blocks and then broadcasts the entire network. After the entire network node is verified, all nodes update the blockchain account.

3.3 Secure Sharing of Data

The medical data security sharing process based on MAPM is divided into the following stages: initialization, encryption and upload, mimetic consensus, download and decryption.

Initialization. The identity certificate is issued by CA, including user's attribute set $R = (r1, r2, r3, \ldots rn)$, ID, symmetric encryption private key rs, public keys pk and private keys sk.

Data Encryption and Upload. Before encrypting and uploading the data, the data owner first generates the data index, and creates a smart contract to describe how to search it. An access strategy for the data develops based on the visitor attributes. Once this is done, the data owner sends the smart contract, index, and access control policies to the blockchain.

Mimic Consensus. See Sect. 3.4 in details.

Data Download and Decryption. The user can request data from the whole network, and the blockchain node retrieves whether the requestor has access rights. If it conforms to access control rights, the ciphertext can be obtained, otherwise the data cannot be accessed.

Taking patient P transfers from hospital A to hospital B as an example, in which patient P, hospital A is the owner, and hospital B is the user. Patient P generates medical data $D(d1, d2, \ldots, dn)$ in hospital A, then obtains ciphertext $C(c1, c2, \ldots, cn)$ by key rs encryption and uploads it to the cloud storage server. An index should be build and a smart contract that describes to find the index. When patient P arrives at hospital B, he or she will generate a query transaction according to pk and use sk for digital signature. The system will verify P's identity according to the signature. After successful verification, the address index of the patient information in the cloud memory will be returned. The patient can download the ciphertext by himself and decrypted it to plaintext by rs. Hospital B can broadcast the request to the whole network if the patient cannot provide the data by itself. The blockchain node verifies the access permission of the hospital B, and the ciphertext address index will be returned after verifying. When detecting the transaction request, owner (hospital A) sends the rs to user (Hospital B) through the secure channel after verifying the identity.

Assuming the owner sets the data $D(d1, d2, \ldots, dn)$ as accessible to the hospital B in the access control policy, then the specific data request process is as follows:

Algorithm Description
1.P and A create address index of data: seq = (id | | Loc) 2.Develop access policies based on visitor attributes: StrGen(R)→acp 3.Encrypts the data with a symmetric encryption key rs: SEnc(D(d1,d2,...,dn), rs)→C(c1,c2,...,cn) 4.Encrypted data is stored in the cloud, access control policies and the indexes in the cloud are linked with blochchain. 5.Decryption of data: SDec(C(c1,c2,...,cn),rs)→D(d1,d2,...,dn).For users who do not meet the permissions, they cannot pass the authentication, so they cannot get the data index and key, and return null. 6.After B completes the above operation, the blockchain will record this operation. REQ((B→A) , D(d1,d2,...,dn)) →Blockchain

3.4 Mimic Consensus Mechanism

Consensus algorithm is an important factor to ensure the security of blockchain. The core idea of POW is that guarantee the consistency of data and consensus security among distributed nodes through computing power competition. All nodes (miners) in the system solve a SHA256 mathematical problem (mining) simultaneously based on computing power, and the node that solves the problem first gets the billing right of the next block and the reward generated by the system. However, with the emergence and advent of mining machines, the calculation speed has been greatly improved, resulting in the local centralization of the computing power. If the node gets more than 50% of the computing power, it may cause serious harm to the system.

In the mimicry consensus mechanism of MAPM, there are four evaluation standards for realizing node competition accounting rights, namely computing power, memory, storage space and the existence time of nodes in the system. The generation of a new block should simultaneously complete the POW consensus algorithm based on computing power, memory, storage space and the time of the node in the system, which is denoted as POW(1), POW(2), POW(3) and POW(4). During the consensus process, when the certain conditions are reached, the POW based on certain evaluation standard will be adjusted to another POW consensus. Figure 4 shows the mimicry consensus process.

There are four sub-chains formed by the POW algorithm in the system, which are respectively workload proof consensus POW(1) based on computation force, the consensus-based POW(2) for the memory-based workload, POW(3) workload proof consensus based on storage space, and workload proof consensus POW(4) based on the existence time of the node in the system. To achieve the goal of dynamic randomization, the consensus algorithm is adjusted by generating random seeds. As shown in Fig. 5, the system control process generates random seeds *I* according to the hash of the previous block, the address *add* and the difficulty *diff* of the previous consensus node, and the number of blocks *n* on the current blockchain. Define the judgment function as Ver_j^i $(i, j = 1, 2, 3, 4$ and $i \neq j)$. When the condition POW(i) adjusts to POW(j) is satisfied, the judgment function $\mathrm{Ver}_j^i(I) = 1$, otherwise $\mathrm{Ver}_j^i(I) = 0$.

Assuming that the r-round consensus is being carried out currently, the specific steps of the mimicry consensus are given below.

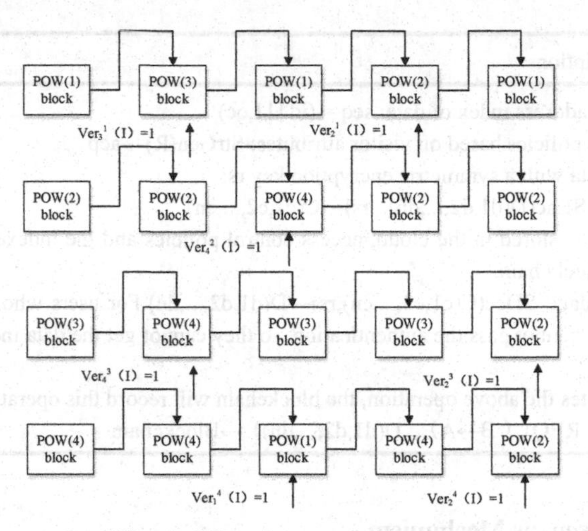

Fig. 4. Mimetic consensus process

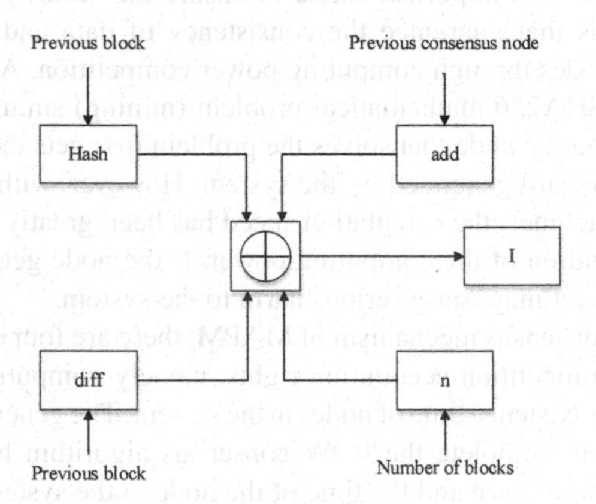

Fig. 5. Generation of random seed generation

Implementation Consensus. The blockchain system executes 4 kinds of consensus respectively to generate $block_r$ ($r = 1, 2, 3, 4$).

Block Generate. The whole network verifies the data and signature in the $block_r$. If the $block_r$ is valid and consistent, it is verified and added it into the verification node's own blockchain. If the $block_r$ results are inconsistent, the consensus will be re-executed. If a consensus node signs too many error blocks, the billing rights of new blocks may be restricted for a period of time.

Consensus Adjustment Process. When the current blockchain system meeting certain conditions, consensus algorithm will be adjusted to the corresponding other consensus algorithm. The system generates seeds randomly from the difficulty value, the address of the previous consensus node, the Hash of the previous block and the number of current nodes. If the condition that POW(i) adjusted to POW(j) is satisfied, $\mathrm{Ver}_j^i(\mathrm{I}) = 1$, otherwise $\mathrm{Ver}_j^i(\mathrm{I}) = 0$.

In the process of mimicry consensus, it is defined that the benefits of the attacker to attack POW(1), POW(2), POW(3), and POW(4) successfully respectively is I1, I2, I3 and I4, while the consumption is E1, E2, E3 and E4. If E1 + E2 + E3 + E4 < I1 + I2 + I3+ I4, the attack is successful, otherwise it is unsuccessful. In order to attack the consensus mechanism, four types of POW must be attacked at the same time, however, the workload of the nodes in the four types of POW is based on different standards. The attacker needs not only computational power, but also a large amount of memory and storage space, so the cost of the attack is greatly increased.

4 Comparison and Analysis

This section analysis the characteristics of MAPM by comparing with the existing medical data privacy protection model and methods.

Table 1 compares MAPM with other models from four aspects. The model has certain advantages in the consensus mechanism. In the traditional POW, if the computing power exceeds 51% of the whole network, the blockchain system can be controlled, such as split a new chain. The improved DPOS selects some node representatives to participate in the verification and accounting of the future transactions. Both are easy to cause local centralization or weak centralization. Comparing with MDSM and MedRec, the cost of partial centralization in MAPM is higher. PBFT algorithm does not need to mine, but under PBFT, the system has a low fault tolerance rate. If more than 1/3 of the nodes fail, the whole blockchain cannot run. In the three consensus phases, the traffic is very large, and the efficiency of PBFT consensus algorithm will be reduced in the medical system with more nodes.

Table 1. Model versus existing solution

System	Based on blockchain	Consensus mechanism	Pay	Public chain
MDSM[4]	Yes	Improved DPOS	No	No
MedRec[7]	Yes	POW	Yes	Yes
Medical Chain[5]	Yes	Improved PBFT	No	No
MAPM	Yes	Various	Yes	No

In the mimetic consensus, there is still a consensus adjustment. Assuming that the probability of satisfying the attacker's return greater than consumption is $P'_S = \mathrm{P}(\mathrm{E1} + \mathrm{E2} + \mathrm{E3} + \mathrm{E4} < \mathrm{I1} + \mathrm{I2} + \mathrm{I3} + \mathrm{I4})$, and the consensus algorithm POW(i) adjusting to POW(j) is $\mathrm{P}_j^i = \mathrm{P}\left[\mathrm{Ver}_j^i(\mathrm{I}) = 1\right]$, where i, j = 1, 2, 3, 4. The probability of attacking the mimetic consensus successfully is:

$$P_S = P'_S \prod_{i=1}^{4} \prod_{j=1}^{4} (1 - P_j^i) \tag{1}$$

For simplicity, all P_j^i are equal, then $P_S = P_S'\left(1 - P_j^i\right)^{12}$. When P_j^i is increased to a certain extent, that is, $P_j^i \approx 1$, P_S is close to 0 regardless of the value of P_S'. Therefore, from the perspective of attack success rate, the mimetic consensus mechanism is superior to the static consensus mechanism.

Medical informatization has gradually matured after years of development and the medical process has become more standardized. However, the medical industry still faces many problems and challenges. The existing medical problems and the solution of this model are shown in Table 2.

Table 2. Current problems and model response plans

Types	Problems	Solution
Privacy and security	Hacking	Dynamic heterogeneous consensus mechanism
	Data protection	Symmetric encryption storage
	Non-repudiation	Regularly anchor transactions to blockchains
Access control	Seeing previous medical data difficult	Develop access control policies based on attributes
Data storage	Data island	Open data islands between medical institutions to achieve data sharing
Abuse and fraud	Investigate	Digital signature
Participation of user	Unable to manage data	View data from various hospitals on the client platform and develop access policies to protect privacy

Although this model has certain advantages in the security of consensus algorithms, there are still weaknesses and areas for improvement. This model adopts symmetric encryption. Although it has high security, how to ensure the secure storage of keys and reasonable access control strategies in practical applications needs further consideration. In addition, how to optimize system performance, improve algorithm speed and system throughput also needs further improvement.

5 Conclusion

As a application of blockchain technology, medical blockchain has many advantages in data sharing. It is an important scenario and a direction for future research and application. Aiming at the problem that the existing workload proof mechanism based on calculation power is easy to cause partially centralized. The randomness and dynamic are added to the consensus mechanism combining with the idea of mimicry defense. Multiple workload proof mechanism are used in the process of mimicry consensus to further improve the security guarantee of medical blockchain system.

Comparison and analysis of the model show that MAPM improves the security of the consensus algorithm and reduces the possibility of successful attack by attacker. In the next research, we will optimize existing problems such as storage limit, throughput, and algorithm speed.

References

1. Liu, X., Fu, W., Li, L.: Opportunities and challenges faced by information security in the medical industry in the age of big data. China Inf. Secur., 100–102 (2018)
2. The editorial department of the journal: Inventory: major data breaches at home and abroad in 2017. China Inf. Secur. (03), 62–68 (2018)
3. Yuan, Y., Wang, F.: Development status and prospect of blockchain technology. J. Autom. **42**(4), 481–494 (2016)
4. Xue, T., Fu, Q., Wang, W., Wang, X.: Research on medical data sharing model based on blockchain. J. Autom. **43**(09), 1555–1562 (2017)
5. Zhang, C., Li, Q., Chen, Z., et al.: Medical chain: alliance medical blockchain system. J. Autom. **45**(8), 1495–1510 (2019)
6. Chen, J., et al.: A blockchain application for medical information sharing. In: 2018 IEEE International Symposium on Innovation and Entrepreneurship, TEMS-ISIE, pp. 1–7. IEEE (2018)
7. Azaria, A., Ekblaw, A., Vieira, T., et al.: MedRec: using blockchain for medical data access and permission management. In: 2016 2nd International Conference on Open and Big Data (OBD), pp. 25–30. IEEE (2016)
8. Xu, W., Wu, L., Yan, Y.: Privacy protection scheme for electronic health record based on blockchain and homomorphic encryption. Comput. Res. Dev. **55**(10), 2233–2243 (2018)
9. Xia, Q., Sifah, E., Smahi, A., et al.: BBDS: blockchain-based data sharing for electronic medical records in cloud environments. Information **8**(2), 44 (2017)
10. Xu, M., Yuan, C., Wang, Y., et al.: Mimic blockchain—blockchain security solution. J. Softw., 1681–1691 (2019)
11. Liu, Q., Lin, S., Gu, Z.: A heterogeneous functional equivalent scheduling algorithm for mimic security defense. J. Commun. **39**(07), 188–198 (2018)
12. Yuan, C.: Research on key technologies of blockchain privacy protection. Strategic Support Force Information Engineering University (2018)
13. Zhu, L., et al.: Summary of research on blockchain privacy protection. J. Comput. Res. Dev. **54**(10), 2170–2186 (2017)
14. Wang, X., Jiang, X., Li, Y.: A data access control and sharing model based on blockchain. J. Softw. (06), 1661–1669 (2019)
15. Theodouli, A., Arakliotis, S., Moschou, K., et al.: On the design of a blockchain-based system to facilitate healthcare data sharing. In: Computing and Communications/12th IEEE International Conference on Big Data Science and Engineering, TrustCom/BigDataSE, pp. 1374–1379. IEEE (2018)
16. Wang, Z., Liu, J., Zhang, Z., et al.: Fully anonymous blockchain based on aggregate signature and encrypted transactions. J. Comput. Res. Dev. **55**(10), 2185–2198 (2018)
17. Liu, A., Du, X., Wang, N., Li, S.: Blockchain technology and its research progress in the field of information security. J. Softw. **29**(07), 2092–2115 (2018)
18. Wang, J., Gao, L., Dong, A., Guo, S., Chen, H., Wei, X.: Research on data security sharing network system based on blockchain. J. Comput. Res. Dev. **54**(04), 742–749 (2017)
19. Qiao, R., Dong, S., Wei, Q., Wang, Q.: Research on dynamic data storage security mechanism based on blockchain technology. Comput. Sci. **45**(02), 57–62 (2018)

20. Jiang, H., Peng, H., Dian, S.: A design of medical information sharing model based on blockchain technology. In: IOP Conference Series Materials Science and Engineering, vol. 428 (2018)
21. Yuan, Y., Ni, X., Zeng, S., Wang, F.: Development status and prospect of blockchain consensus algorithm. J. Autom. **44**(11), 2011–2022 (2018)
22. Nguyen, G.T., Kim, K.: A survey about consensus algorithms used in blockchain. J. Inf. Process. Syst. **14**(1), 101–128 (2018)
23. Lin, Z., Li, K., Hou, H., et al.: MDFS: a mimic defense theory based architecture for distributed file system. In: 2017 IEEE International Conference on Big Data, Big Data, pp. 2670–2675. IEEE (2017)
24. Tosh, D.K., Shetty, S., Liang, X., et al.: Consensus protocols for blockchain-based data provenance: challenges and opportunities. In: 2017 IEEE 8th Annual Ubiquitous Computing, Electronics and Mobile Communication Conference, UEMCON, pp. 469–474. IEEE (2017)
25. Halpin, H., Piekarska, M.: Introduction to security and privacy on the blockchain. In: 2017 IEEE European Symposium on Security and Privacy Workshops, EuroS&PW, pp. 1–3. IEEE (2017)
26. Xia, Z., Lu, L., Qiu, T., Shim, H.J., Chen, X., Jeon, B.: A privacy-preserving image retrieval based on AC-coefficients and color histograms in cloud environment. Comput. Mater. Continua **58**(1), 27–43 (2019)
27. Huang, J., Wang, Y., Pei, B.: Research on security technology of health medical big data. Comput. Age (11), 45–48 (2018)
28. Wang, Y., Shi, H., Gu, R., Zhang, B., Wang, H.: Research on privacy protection of regional medical big data sharing platform. Inf. Comput. (Theor. Ed.) (09), 211–213 (2018)
29. Zheng, B.K., Zhu, L.H., Shen, M., et al.: Scalable and privacy-preserving data sharing based on blockchain. J. Comput. Sci. Technol. **33**(3), 557–567 (2018)
30. Sankar, L.S., Sindhu, M., Sethumadhavan, M.: Survey of consensus protocols on blockchain applications. In: 2017 4th International Conference on Advanced Computing and Communication Systems (ICACCS), pp. 1–5. IEEE (2017)
31. Jiang, X., Liu, M., Yang, C., Liu, Y., Wang, R.: A blockchain-based authentication protocol for WLAN mesh security access. Comput. Mater. Continua **58**(1), 45–59 (2019)
32. Alabdulkarim, A., Al-Rodhaan, M., Tian, Y., Al-Dhelaan, A.: A privacy-preserving algorithm for clinical decision-support systems using random forest. Comput. Mater. Continua **58**(3), 585–601 (2019)

Securing Data Communication of Internet of Things in 5G Using Network Steganography

Yixiang Fang(✉), Kai Tu, Kai Wu, Yi Peng, Junxiang Wang, and Changlong Lu

Jingdezhen Ceramic Institute, Jingdezhen 333000, China
fangyixiang@jci.edu.cn

Abstract. Data transmission security is the foundation of reliable application of the Internet of Things. With the rapid development of the Internet of Things, big data, 5G technologies in recent years, the Internet of Things has been widely used in industrial production, healthy domain, smart environment domain, national security, and other fields. The problem that monitoring data can be easily eavesdropped and tampered in wireless channel transmission has been recognized gradually. Some confidential data with sensitive information of users, such as medical data, must be protected during transmission. The traditional method cannot meet the requirement of some privacy data transmission stage. In this paper, we proposed a novel approach based on network steganography to improve the security of data transmission in 5G IoT network, which could improve the security of private data with less communication overhead.

Keywords: Confidential data security · IoT · 5G · Network steganography · AES encryption algorithm · BCH error-correcting codes

1 Introduction

The emerging of 5th generation wireless system has created high-quality conditions for the Internet of everything. With the advantages of high bandwidth, low latency and mass Internet of Things, it is widely applied in the fields of intelligent manufacturing, smart home and Internet of vehicles. In order to realize the functions such as intelligent identification, positioning, tracking, monitoring and etc., everything will be connected to the Internet by means of Internet of Things via information sensing equipment in according with the appointed protocol for information exchange and communication. Owing to wireless communication being widely used in Internet of Things devices, some threats and attacks are exacerbated. Common types of attacks include: spoofing, message altering & replaying, flooding & wormhole attacks [1]. Moreover, wireless communication is much more vulnerable and eavesdropping than cable communication. Attackers can monitor the traffic and take actions to capture and modify data packets in the process of data transmission [2]. Therefore, it will be maliciously invaded during the period of data transmission. In such circumstances, it will greatly reduce the security of the Internet of Things due to the leakage of confidential data and damage to the IoT equipment. Various communication protocols are widely used in the current industrial

X. Sun et al. (Eds.): ICAIS 2020, LNCS 12240, pp. 593–603, 2020.
https://doi.org/10.1007/978-3-030-57881-7_52

Internet of Things system, which security measures are not uniform, especially for communication encryption and identity authentication. If the communication content is not encrypted, attackers can easily intercept the communication data, and steal important industrial production data once they penetrate the Internet of Things system. Just for these reasons, the security maintenance of Internet of Things devices in the process of data transmission is particularly important.

The data of the Internet of Things needs various cryptographic algorithms to encrypt, and then cryptographic algorithms need to use various keys to participate in the operation in order to truly realize safe and effective ciphertext information storage and exchange. That is to say, the confidentiality of the key determines the security of the cryptographic system. The original cryptographic technology system could not be suitable for current Internet of Things environment directly. It is necessary to consider the characteristics of Internet of things and secret key management as a whole picture so that the secret key system could be designed much more in detail. In a word, information security and privacy protection are the most fundamental requirements for the application of secret key management in the process of data transmission among Internet of Things devices.

As to secret key management, the current secret key system is mainly divided into the following three types: (1) Key system based on PKI. It is used to realize the generation, management, storage, distribution and revocation of key and certificate based on public key cryptography; (2) Key system based on symmetric ciphers, that is, the sender and receiver of information use the same key to encrypt and decrypt data, which is suitable for encrypting large amounts of data; (3) Key system based on IBC (identity-based cryptograph) is a cryptographic technology that uses entity identity as public key. It can effectively meet the requirements of identity authentication, data security, transmission security in Internet of Things applications. The current commonly used AES encryption algorithm, which is difficult to crack with the present computing power, belongs to the traditional symmetric encryption algorithm. However, once the key is intercepted by the attacker during the period of data transmission, the whole system will be completely ruined. Moreover, in the real life enormous redundancy exists in the message source, and thus, the attacker can infer hiding information about the key from a large amount of statistical data [3]. In addition, Internet of Things devices need to use different keys when communicating with multiple parties, meanwhile, the management of large number of keys is extremely complicated. What's worse, if the device is hijacked by the attacker and the secret keys are exposed, more serious consequences may occur.

With the rapid development of the Internet of Things in recent years, various data security transmission and secret key management schemes have been proposed. Suat Ozdemir et al. summarized the secure data aggregation schemes in wireless sensor networks [4]. Shaohua Tang et al. analyze the problems existing in three key management schemes in [5]. Dae-Young Kim et al. proposed a DPN mechanism for secure data transmission in IoT services [6]. Changting Shi proposed a physical layer method with which Radio Frequency Fingerprinting could be extracted from the radio signals for IoT devices indentification [7]. Szczypiorski and k. first proposed the concept of network steganography in 2003 [8]. Khan Muhammad proposed that steganography is an important tool in data transmission security in Internet of Things applications, and network steganography is one of its major branches [9]. Józef Lubacz et al. introduced some methods of network steganography. Due to the increasing complexity of communication protocols, the need for network steganography is improving and the importance of network steganography is increasing [10]. In order to solve the security problem of data transmission in IoT devices, we must take some measures to ensure the data authenticity, freshness, replay protection, integrity, and confidentiality [11].

For the confidentiality of data transmission in IoT, an Error-correcting Retransmission STEGanography algorithm "ERSTEG" is proposed in this paper. We focus on this method to transmit the AES secret key after BCH coding. This method not only solves the problem that large number of secret keys are extremely difficult to manage after using AES encryption algorithm, but guarantees the accuracy and security of secret key during the transmission stage.

The rest of the paper is organized as below. The workflow of the proposed model based on network steganography including AES encryption algorithm and BCH error-correcting coding is demonstrated in Sect. 2. The implementation details are illustrated in Sect. 3. The paper concludes in Sect. 4.

2 Framework for the Model

2.1 AES Encryption Algorithm

The AES encryption algorithm is an iterative block cipher composed mainly of nonlinear components, linear components, and round keys [11, 12]. The AES specifics the Rijndael algorithm, in which a symmetric block cipher that can process data blocks of 128 bits, using cipher keys with lengths of 128, 192, and 256 bits [11]. If the key length is equal to 128, 192, 256, the number of encryption rounds is 10,12 and 14 respectively. Each round of data encryption includes round key xor, s-box transformation, row permutation, and mixing column.

DES has been gradually replaced by AES that becomes the mainstream encryption algorithm. Comparing with other encryption algorithms, AES algorithm has higher efficiency and less overhead with symmetrical encryption technique. Furthermore, it is very suitable to encrypt the data collected by IoT equipment.

2.2 BCH Error-Correcting Coding

BCH coding is a kind of error-correcting codes with good performance which is raised individually by Hocquenghem in 1959 and Bose and Chandha ri in1960 [13]. It has strong error correcting ability and is used to correct multiple random errors. Its performance is close to the theoretical value in short and medium code length, so it plays an important role in coding theory.

The relationship between the code length n of BCH code and the supervisory bits and error correcting ability t is as follows: For positive integers $m >= 3$ and positive integers $t < \frac{m}{2}$, There must be a primitive BCH code with a code length of $n = 2^m - 1$ and supervisory bits of $n - k <= mt$, which can correct no more than t random errors. If the code length

$$n = (2m - 1)/I \left(I > 1, \textit{and n can be divided by} \left(2^m - 1\right)\right) \tag{1}$$

it is imprimitive BCH code.

BCH coding can find the generation polynomial of codes under the condition of given error correcting ability. The encoding and decoding process is as follows:

Step-1: Select a primitive polynomial $P(x)$ according to m, that is used to compute the Galois field $GF(q^m)$.
Step-2: Generate field $GF(q^m)$ from the irreducible polynomial $P(x)$.
Step-3: Compute primitive element, according to error correct capacity t. Compute minimal polynomial $f_i(x)$
Step-4: Compute the generator polynomial

$$\mathrm{g}(x) = \mathrm{LCM}\left[f_1(x), f_2(x), f_3(x) \ldots f_{2t}(x)\right] \tag{2}$$

Step-5: Use

$$\mathrm{C}(x) = x^{\mathrm{n-k}} m(x) + Rem_{g(x)}\left[x^{\mathrm{n-k}} m(x)\right] \tag{3}$$

to encode. ($m(x)$ is information polynomial)

Decoding Process:

Step-1: Compute the concomitant formula

$$\mathrm{S} = (s_1, s_2, s_3, \ldots s_{2t}) \tag{4}$$

by the received $R(x)$.
Step-2: Compute the error location polynomial

$$\sigma(x) = \sigma_t x^t + \sigma_{t-1} x^{t-1} + \ldots + \sigma_1 x + 1 \tag{5}$$

by concomitant formula S.
Step-3: Compute the roots of $\sigma(x)$, these roots are error locations.

2.3 Network Steganography

"The best way of keeping a secret is to pretend there isn't one [14]." "Steganography is the art and science of communicating in a way which hides the existence of the communication. The goal of steganography is to hide messages inside other harmless messages in a way that does not allow any enemy to even detect that there is a second message present [15]."

Network steganography is a branch of steganography. Network steganography aims to hide the data transmission process. It uses overt network traffic as a carrier to transmit secret message. Generally, network steganography can be classified into timing method and storage method. Timing method hides secret messages in rate or throughput of network traffic. Storage method hides secret message in the unused area of data packet, such as unused bits of a protocol header or payload field of data packet [14].

Network steganography may be classified [16] into three broad groups (Fig. 1). The data packet could be modified (MP) within the network protocol header or the payload field. Different MP methods could achieve different steganographic abilities, meanwhile the difficulties of implementation and detection differ from one another. It can be selected according to the needs of various situations. The steganographic methods that modify the structure of packet streams (MS) have a high demand for synchronization between the sender and the receiver, have low steganographic capability but are more difficult to detect. Its implementation is simple but the delay may affect the transmission quality. The hybrid steganographic approaches (HB) that modify the packets both in contents and timing and ordering aspects have high steganographic capability. It is difficult to detect but brings in a loss of connection quality. In our paper, we use a hybrid method called “Retransmission steganography”. The method was proposed by Krzysztof Szczypiorski et al. [17] (Fig. 2).

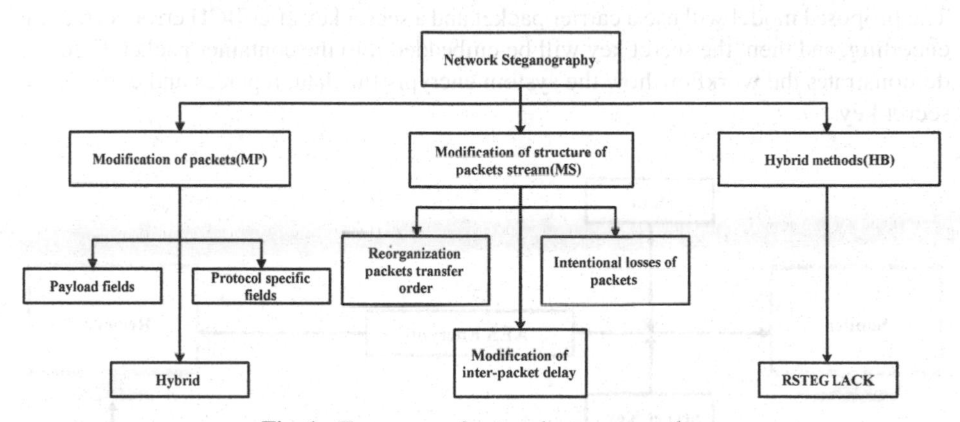

Fig. 1. Taxonomy of network steganography

TCP is a reliable network transport protocol. After receiving the data packet, an ACK signal must be sent back by the receiver, otherwise the sender will retransmit the data packet after the fixed duration.

Based on this characteristic of TCP protocol, we imagine a scenario that Bob and Alice start normal data transmission after the connection being established. Once, Bob sends a packet to Alice. However, Alice does NOT reply intentionally after receiving it. As the time is out, the data packet will be retransmitted. Bob replaces the payload with some secret messages, and then sends the packet to Alice. Alice receives the retransmitted packet and extracts steganogram.

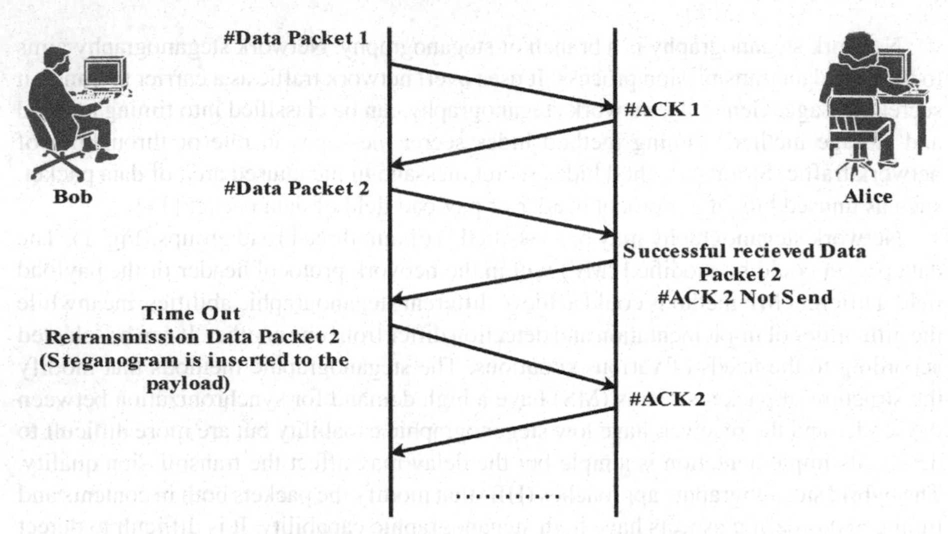

Fig. 2. Time-out based Retransmission Steganography

2.4 Workflow of Proposed Model

The proposed model will use a carrier packet and a secret key after BCH error correcting encoding, and then, the secret key will be embedded into the container packet. Figure 3 demonstrates the workflow how the system encrypts the data, replaces and extracts the secret key.

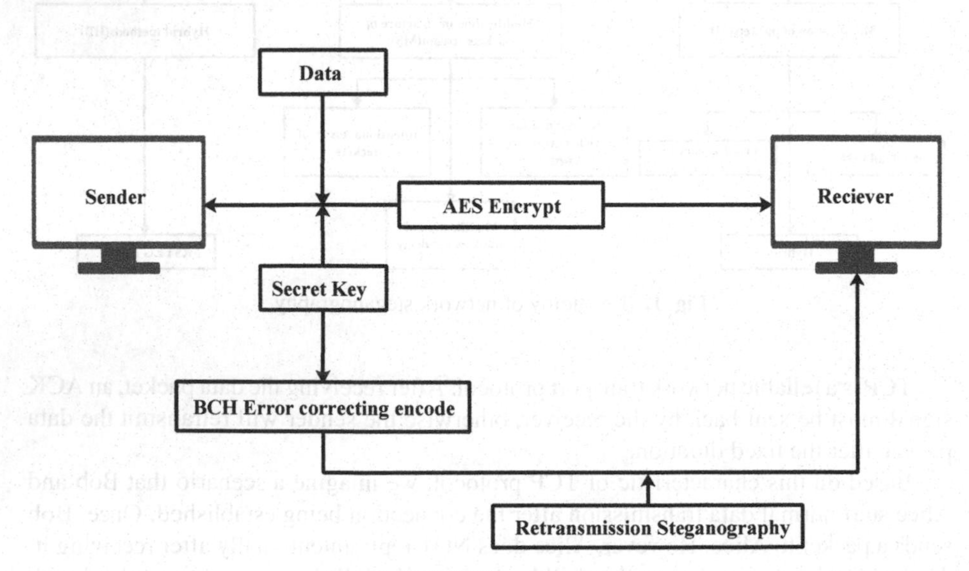

Fig. 3. ERSTEG system diagram

Step-1: Generate a secret key randomly.
Step-2: Data encryption will be done using AES encryption algorithm with a generated secret key in step-1.
Step-3: For the accuracy of the secret key after transmission, use the BCH error correcting encoding.
Step-4: Use ERSTEG to transmit the secret key after BCH error correcting encoding.
Step-5: When the receiver gets the data packet with the steganogram, extract the secret key after BCH error correcting encoding, and use the BM algorithm to decode the secret key.
Step-6: Use the secret key to decrypt the data.

3 Implementation and Analysis

In this section, the implementation details of this system are introduced, meanwhile, its practicality is discussed.

Firstly, generate a secret key randomly to encrypt the data using AES encryption algorithm. After the encryption is completed, BCH error correcting encoding is performed on the key (Fig. 4).

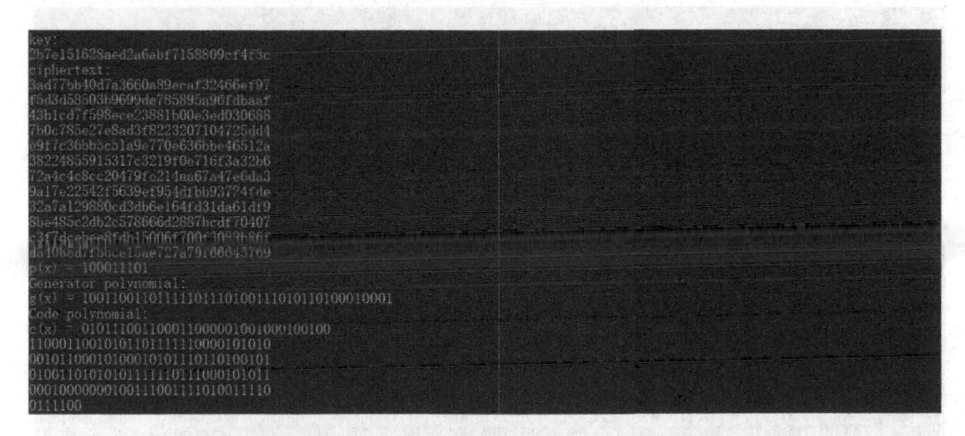

Fig. 4. Data after encryption and secret key after BCH encoding

Once the connection is established, data transmission starts immediately, and the secret key encoded by BCH error correction is transmitted by employing ERSTEG.

The retransmission steganography part of this system is implemented by adopting two Free Open Source libraries. One is WinPcap [18] library which allows applications to capture and transmit network packets bypassing the protocol stack. The library is used to capture and analyze whether there is RetransmissionID during the retransmission steganography phase. The other one is Windivert [19]. The library, which allows user-mode applications to capture, modify, drop network packets from the network stack, is applied to drop ACK packet.

The ERSTEG algorithm does not specify that when the receiver will refuse to reply. When transmitting a key using the Retransmission steganography algorithm, a RetransmissionID is required as the retransmission request. In this way, the receiver knows when to refuse, and the next received content will contain the Steganogram (that is, key). It is important to note that the RetransmissionID should not differ from other data sent during a connection, otherwise it is likely to cause suspicion [17].

When the connection is established, payload analysis routine is running on both sides. If the sender detects the RetransmissionID, it will set the Stegflag to 1 and insert the key into the payload filed of the container packet before the packet retransmission. After ERSTEG process finishes, set the Stegflag to 0. If a packet loss occurs during transmission, thereafter, the same content will be retransmitted. The receiver checks the payload field of each incoming packet and determines whether the received packet contains RetransmissionID. If the RetransmissionID is detected, the packets dropping subroutine will be carried out.

After sending the next data, ACK will be dropped so that the sender will not get ACK response. When time expiration is done on the sender side, the sender initiates to retransmit the data, however, at this time, the sender will replace some segment of the payload field with the secret key and send the data out (Fig. 5).

Fig. 5. ACK packet captured and dropped

Fig. 6. Payload field modified

Fig. 7. Secret key decoding

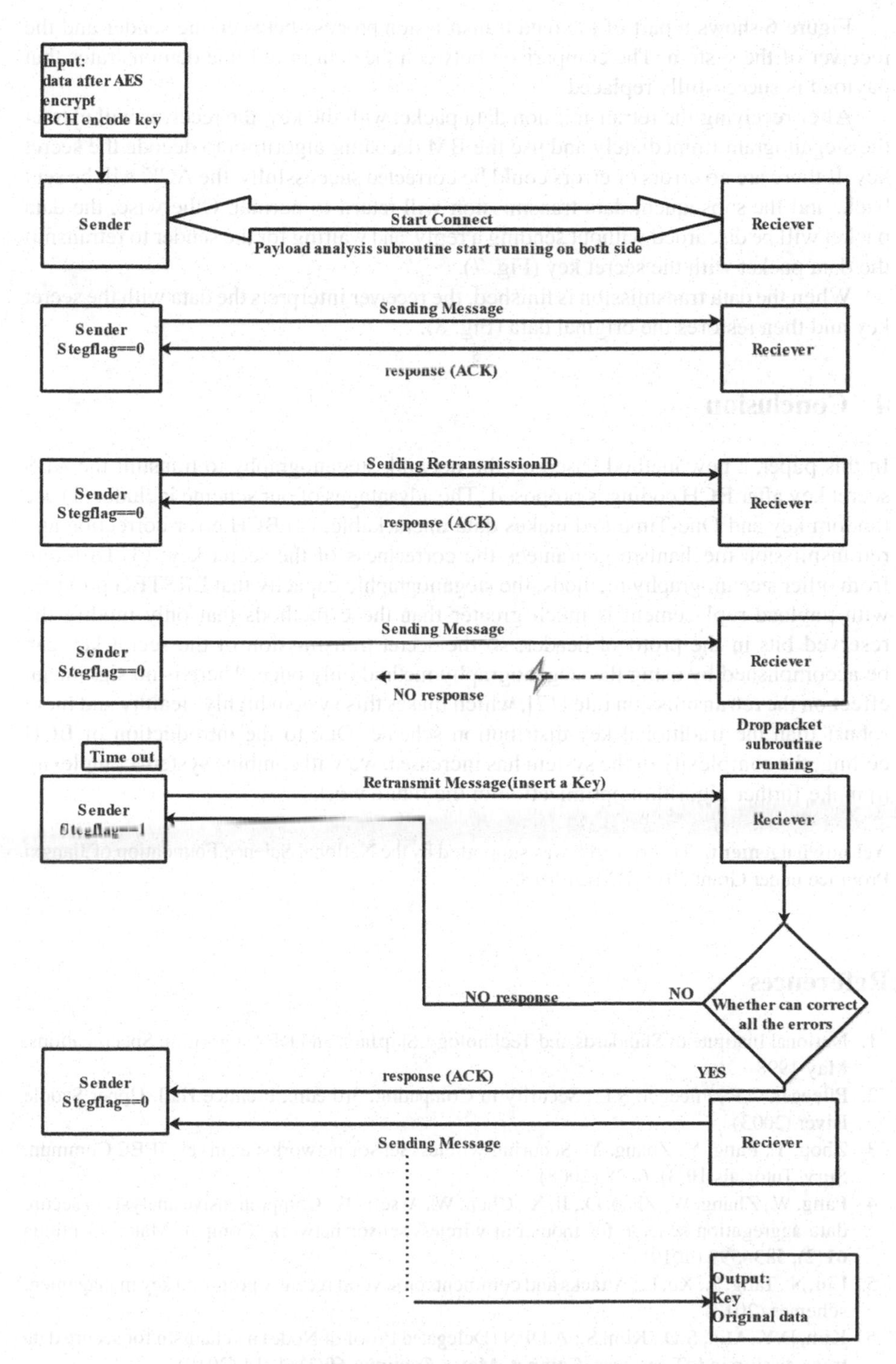

Fig. 8. ERSTEG program flow chart

Figure 6 shows a part of the data transmission process between the sender and the receiver of the system. The comparison between the data in red line demonstrates that payload is successfully replaced.

After receiving the retransmission data packet with the key, the receiver will extract the steganogram immediately and use the BM decoding algorithm to decode the secret key. If there are no errors or errors could be corrected successfully, the ACK will be sent back, and the subsequent data transmission will return to normal. Otherwise, the data packet will be discarded without sending a reply, and waiting for the sender to retransmit the data packet with the secret key (Fig. 7).

When the data transmission is finished, the receiver interprets the data with the secret key and then restores the original data (Fig. 8).

4 Conclusion

In this paper, a new method based on the network steganography to transmit the AES secret key after BCH coding is proposed. The advantages of our scheme include: (1) use random key and One-Time Pad makes data uncrackable; (2) BCH error-correction and retransmission mechanism guarantees the correctness of the secret key; (3) Different from other steganography methods, the steganographic capacity that ERSTEG provides with payload replacement is much greater than those methods that only modify the reserved bits in the protocol header, so the secret transmission of the secret key can be accomplished by using the steganography method only once. There is no significant effect on the retransmission rate [17], which makes this system highly stealthy and more robust than the traditional key distribution scheme. Due to the introduction of BCH coding, the complexity of the system has increased, we will combine system complexity to make further algorithm optimization in the future work.

Acknowledgement. This research was supported by the National Science Foundation of Jiangxi Province under Grant 20151BAB217018.

References

1. National Institute of Standards and Technology, SkipJack and KEA algorithm Specifications, May 1998
2. Pfleeger, C.P., Pfleeger, S.L.: Security in Computing, 3rd edn., Prentice Hall, Upper Saddle River (2003)
3. Zhou, Y., Fang, Y., Zhang, Y.: Securing wireless sensor networks: a survey. IEEE Commun. Surv. Tutorials **10**(3), 6–28 (2008)
4. Fang, W., Zhang, W., Zhao, Q., Ji, X., Chen, W., Assefa, B.: Comprehensive analysis of secure data aggregation scheme for industrial wireless sensor network. Comput. Mater. Continua **61**(2), 583–599 (2019)
5. Liu, N., Tang, S., Xu, L.: Attacks and comments on several recently proposed key management schemes (2013)
6. Kim, D.Y., Min, S.D., Kim, S.: A DPN (Delegated Proof of Node) mechanism for secure data transmission in IoT services, Comput. Mater. Continua **60**(1), 1–14 (2019)

7. Shi, C.: A novel ensemble learning algorithm based on D-S evidence theory for IoT security. Comput. Mater. Continua **57**(3), 635–652 (2018)
8. Szczypiorski, K.: Steganography in TCP/IP Networks. State of the Art and a Proposal of a New System–HICCUPS, Institute of Telecommunications' seminar, Warsaw University of Technology, Poland, November 2003
9. Muhammad, K.: Steganography: a secure way for transmission in wireless sensor networks, November 2015
10. Lubacz, J., Mazurczyk, W., Szczypiorski, K.: Principles and overview of network steganography, Institute of Telecommunications, Warsaw University of Technology, Warsaw, Poland (2011)
11. Trad, A., Bahattab, A.A., Othman, S.B.: Performance trade-offs of encryption algorithms for wireless sensor networks. In: 2014 World Congress on Computer Applications and Information Systems (WCCAIS). IEEE (2014)
12. Daemen, J., Rijmen, V.: The Design of Rijndael, AES–The Advanced Encryption Standard, Springer, Heidelberg (2002). https://doi.org/10.1007/978-3-662-04722-4
13. Wilson, S.G.: Digital Modulation and Coding, pp. 179–183. Publishing House of Electronic Industry, Beijing (1998)
14. Mazurczyk, W., Wendzel, S., Zander, S., et al.: Information Hiding in Communication Networks: Fundamentals, Mechanisms, Applications, and Countermeasures. Wiley-IEEE Press, Hoboken (2016)
15. Thampi, S.M.: Information hiding techniques: a tutorial review. Eprint Arxiv (2008)
16. Mazurczyk, W., Szczypiorski, K.: Steganography of VoIP Streams. In: Meersman, R., Tari, Z. (eds.) OTM 2008. LNCS, vol. 5332, pp. 1001–1018. Springer, Heidelberg (2008). https://doi.org/10.1007/978-3-540-88873-4_6
17. Mazurczyk, W., Smolarczyk, M., Szczypiorski, K.: Retransmission steganography and its detection. Soft Comput. **15**, 505–515 (2011). https://doi.org/10.1007/s00500-009-0530-1
18. https://www.winpcap.org/
19. https://reqrypt.org/windivert.html

The Use of the Multi-objective Ant Colony Optimization Algorithm in Land Consolidation Project Site Selection

Hua Wang[1](✉), Weiwei Li[1](✉), Jiqiang Niu[2](✉), and Dianfeng Liu[3](✉)

[1] School of Computer and Communication Engineering, Zhengzhou University of Light Industry, Zhengzhou, China
whuwanghua@163.com
[2] Xinyang Normal University, Xinyang, China
[3] School of Resource and Environmental Science, Wuhan University, Wuhan, China

Abstract. The present study abstracts the land consolidation project site selection (LCPSS) problem to a multi-objective spatial optimization problem. In view of the shortcomings of traditional site selection methods in coordinating multiple objectives, the present study, considering the scale-related constraint conditions and general site selection rules of land consolidation, proposes a multi-objective LCPSS model (MOLCPSSM) based on an ant colony optimization algorithm with social, economic and ecological benefits as the optimization objectives. In addition, the present study focuses on the investigation of the mapping and coding relationships between the artificial ants and vector patches and also improves the ants' spatial unit selection scheme and pheromone update mechanism. Furthermore, the present study verifies the MOLCPSSM through a case study of Jiayu County, Hubei Province, China. The results demonstrate the operability of the MOLCPSSM in solving practical land consolidation problems.

Keywords: Land consolidation · Project site selection · Ant colony optimization algorithm · Multi-objective decision-making · Spatial optimization

1 Introduction

Land consolidation project site selection (LCPSS) is a task that determines the area where a land consolidation project (LCP) is implemented within a spatial range, and the scientific reasonableness of the site selection is a precondition for the successful implementation of a LCP [1]. A number of researchers have studied the effect of natural and socio-economic conditions on the LCPSS process, including the terrain and landscape, soil quality, climate and transportation accessibility. Some studies have also quantitatively evaluated the feasibility of a LCP using methods such as the analytical hierarchy process and the Delphi method to examine the scientificity of the preliminary project site selection [2–4]. The aforementioned studies and methods have certain operability in solving specific problems arising from the LCPSS process. However, because of the

X. Sun et al. (Eds.): ICAIS 2020, LNCS 12240, pp. 604–616, 2020.
https://doi.org/10.1007/978-3-030-57881-7_53

lack of support for a unified, scientific LCPSS system that is theoretical and methodological, the selection of sites for some projects is unfounded and random. The excessive fragmentary spatial layout of a project and the relatively low land suitability constrain the realization of the scale merit of land consolidation.

In terms of its nature, the LCPSS problem can also be viewed as a type of decision-making problem of maximally optimizing multiple objectives within a given feasible region, i.e., a multi-objective spatial optimization problem [5]. In recent years, the development of artificial intelligence (AI) and multi-objective optimization [6–8] and geographical information system (GIS) technologies has provided important technical support for solving multi-objective high-dimensional spatial optimization problems. In addition, AI and GIS technologies have also been successfully used to solve such problems as spatial site selection, spatial layout optimization, allocation of land resources and land-use zoning [9–13].

Characterized by fast convergence, a high global search ability and high robustness, the ant colony optimization (ACO) algorithm solves complex optimization problems based on the positive feedback mechanism of pheromone, distributed and parallel computing, and greedy search and has been extensively used in areas such as systematic parameter optimization, classification of characteristics and optimal allocation of resources. Based on existing LCPSS studies, the present study establishes a multi-objective LCPSS model (MOLCPSSM) with vector patches of the current land use situation as the spatial division unit, based on the ACO algorithm, and verifies this model through a case study of Jiayu County, Hubei Province, China with the objective of formulating an intelligent, efficient LCPSS method.

2 Model Building

2.1 General Framework of the Model

The present study investigates the LCPSS problem based on the spatial search ability of the ACO algorithm and the concept of multi-objective optimization. The basic premise of establishing a LCPSS model using the ACO algorithm is to view the LCPSS problem as a multi-objective spatial optimization problem with vector patches, $X = \{x_1, x_2, \ldots, x_n\}$, as the decision-making variables. When the ACO algorithm is used to solve the LCPSS problem, one ant represents one candidate LCPSS scheme, and one ant colony represents one set of candidate solutions. After each ant traverses all the land-use units, some units were selected as the project implementation units based on the positive feedback mechanism of pheromone.

2.2 Optimization Objectives and Constraint Conditions

Social, economic and ecological benefits should be comprehensively considered during the LCPSS process. Therefore, the present study selects the maximum potential of newly added cultivated land, the optimum land consolidation suitability and maximum spatial compactness as the optimization objectives.

According to the Technical Specification for the Survey and Evaluation of Reserve Resources of Cultivated Land (TD/T 1007–2003), the net increase in the amount of cultivated land resulting from land consolidation can be calculated based on the following equation:

$$M_Z = (1 - R_2) \cdot (M_1 - M_2) + (R_1 - R_2) \cdot M \tag{1}$$

$$f_{EB}(s) = M_z + RS + P - FS \tag{2}$$

where M_Z represents the net increment in the area of cultivated land after land consolidation; M represents the original cultivated land area of the region to be subjected to land consolidation; R_1 represents the ridge coefficient before land consolidation of the region to be subjected to land consolidation; R_2 represents the ridge coefficient after land consolidation; M_1 represents the non-cultivated land area of the region to be subjected to land consolidation; and M_2 represents the non-cultivated land area of the region to be subjected to land consolidation that must be preserved after land consolidation. $f_{EB}(s)$ represents the potential of newly added cultivated lands; s represents the LCPSS scheme; RS represents the area of isolated rural residential areas; P represents the area of pit-ponds that can be consolidated; and FS represents the area of cultivated lands that have been converted back to forestlands and grasslands. The unit of all of the aforementioned variables is ha. The unit of all of the aforementioned variables is ha (hectare).

The evaluation of land suitability not only can provide a basis for the spatially optimal allocation of land-use resources and zoning of land-use resources for different uses but also can provide guidance for the scientific implementation of land consolidation. Based on the land consolidation suitability evaluation index system, a suitable method is selected to evaluate the land consolidation suitability of the evaluated object. In the MOLCPSSM, the variable Suit_i is used to represent the land consolidation suitability coefficient of the land-use unit Parcel_i. Thus, the overall level of suitability of the LCP area can be determined using the area-weighted evaluation method. The suitability objective can be described as follows:

$$f_{ST}(s) = \frac{\sum_{i=1}^{I} u_i \cdot Suit_i \cdot Area_i}{\sum_{i=1}^{I} u_i \cdot Area_i} \tag{3}$$

where $f_{ST}(s)$ represents the land consolidation suitability and u_i indicates whether Parcel_i is selected (if Parcel_i is selected, $u_i = 1$; otherwise, $u_i = 0$). The goal of LCPSS is to include the land units within the area whose natural and social attributes both meet the requirements of land consolidation in the LCP area as much as possible, i.e., the $f_{ST}(s)$ of the site selection scheme tends to a maximum.

LCPSS requires that the project area be concentrated and continuous as much as possible. The present study measures the objective of the degree of concentration and continuity of the LCPSS scheme using the neighborhood identity index. Because a vector patch-type solution space is used in the present study, an appropriate modification is needed when calculating the neighborhood identity index of the site selection scheme. In the MOLCPSSM, the land-use units only have two states, 0 and 1; in addition, only the neighborhood identity index of units in the first order needs to be calculated. For vector patches, the MOLCPSSM defines the neighborhood of the current statistical unit using the first-order spatial contiguity relationship (the second-order or a higher-order spatial contiguity relationship can also be defined based on the specific requirements). If vector unit Parcel$_i$ is included in the candidate project area and b_i represents the number of units within the first-order neighborhood of Parcel$_i$ that are also included in the project area, then: $b_i = \sum_{l \in c} u_l$. Where C represents the subscript set of neighborhood unit b_i and u_l indicates whether the units in the set are selected (if the units in the set are selected, $u_l = 1$; otherwise, $u_l = 0$). Therefore, the objective function of the degree of concentration and continuity of the LCPSS is expressed as follows:

$$f_{AD}(s) = \sum\nolimits_{i=i}^{I} u_i b_i \tag{4}$$

where $f_{AD}(s)$ represents the neighborhood identity index of the project area, which is used to represent the degree of concentration and continuity of the project area. The larger the value of $f_{AD}(s)$ is, the higher the degree of concentration and continuity of the project area.

When solving a multi-objective optimization problem, the MOLCPSSM assigns a different weight factor to each objective function using a simple, effective and practical linear weighting method to integrate all the objective functions. The relative weight coefficients between the objective items, w_1, w_2 and w_3, are established. Based on the aforementioned analysis, the objective function of the proposed site selection model can be expressed as follows:

$$\max\{w_1 \cdot f_{EB}(S) + w_2 \cdot f_{ST}(S) + w_3 \cdot f_{AD}(S)\} \tag{5}$$

where $Max\{\}$ represents the maximum value of the function after weighting.

Ant initialization and spatial unit selection also require consideration of some constraint conditions and site selection rules. The constraint conditions of the model primarily include the scale of the project area and the newly-increased cultivated land ratio. Based on the definition of the solution space of the model, the following scale-related constraint condition needs to be defined within any arbitrary area C_k in the scheme:

$$A_{low} \leq \sum_{i=1}^{I} v_{ik} \cdot u_i \cdot \text{Area}_i \leq A_{\text{upper}} \tag{6}$$

where v_{ik} indicates whether parcel Parcel$_i$ is included in area C_k (if parcel Parcel$_i$ is included in area C_k, $v_{ik} = 1$; otherwise, $v_{ik} = 0$); Area$_i$ represents the area of parcel Parcel$_i$ (ha); and A_{low} and A_{upper} represent the upper and lower limits of the area of area C_k (ha), respectively.

In the MOLCPSSM, the rate at which new cultivated lands are added in area C_k can be calculated using the following equation:

$$R_k = \frac{Z_k}{X_k} \tag{7}$$

where Z_k represents the potential of newly added cultivated lands in project area C_k ($Z_k = M_Z + RS + P - FS$) and X_k represents the total area of the project area. For project area C_k,the following condition must be met: $R_k \geq r$, where r represents the minimum newly-increased cultivated land ratio.

The following rules must also be followed when selecting the spatial unit for the ant system: (1) construction land and water-type land units should be excluded in advance; (2) the inclusion of sloping cultivated lands with a slope greater than 25° in the project area is forbidden; and (3) to ensure transportation accessibility, all the units within the project area must be accessible via roads that are at least above the township level.

2.3 Key Techniques to Ensure the Implementation of the MOLCPSSM

Ants are generally based on coding strings. In the vectorial land-use data, each patch is assigned a unique identification (ID) number. An ant can determine its own coding framework based on the internal codes of the land-use patch. Figure 1a shows the patches included in the project area. Figure 1b shows the formed project areas. Figure 1c shows the ant coding scheme, which includes an S × 2 matrix (solution) (S represents the number of selected land units). The first column of this matrix is composed of a unique code for each patch. The second column of this matrix is composed of the coding number of the project area in which each current patch is included (starting from 1). In this matrix, only the ID number of each land-use unit that has been visited by ants and the coding number of its corresponding project area are retained. In addition, the ant coding scheme also includes a tabu list (TabuList) that records the land-use units that have been visited by ants as well as the fitness of the ants. At the initialization stage, the hash table (solution) of each ant used to store the scheme needs to be cleared and the TabuList of each ant used to store the records of visits also needs to be cleared. After the aforementioned operations are completed, ants need to be randomly distributed into each spatial unit of land use. Only one ant is allowed to be distributed in each unit. In addition, it is also necessary to ensure that each unit adjacent to a unit in which an ant is distributed does not contain an ant.

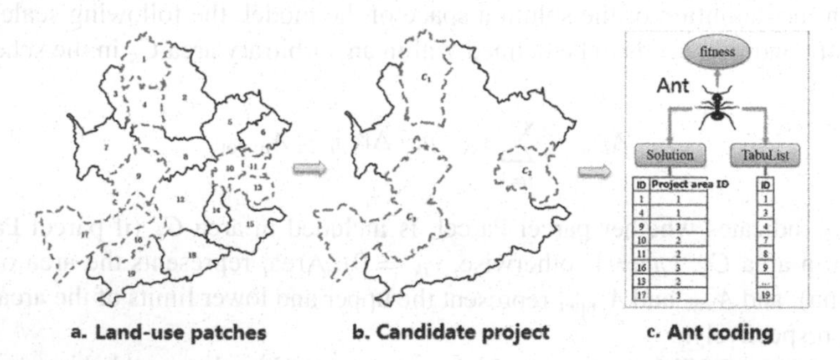

Fig. 1. Ant coding scheme.

The ants' selection of land-use units is a key step in the formulation of a LCPSS scheme. During the process in which an ant selects a land-use unit, the ant not only bases its selection on the concentration of pheromone accumulated in each unit but also determines the direction of its next turn based partly on heuristic information (this information refers to the length of each connecting side in the travelling salesman problem

(TSP)). In the MOLCPSSM, the probability for land-use unit i to be visited by ant k at time t is:

$$p_i^k(t) = \begin{cases} \frac{[\tau_i(t)]^\alpha \cdot [\eta_i(t)]^\beta}{\sum_{s \in allowed_k} [\tau_i(t)]^\alpha \cdot [\eta_i(t)]^\beta} \text{if } i \in allowed_k \\ 0, \, else \end{cases} \tag{8}$$

where α and β represent the pheromone heuristic factor and the expectation heuristic factor, respectively, which reflect the impact of the ant's pheromone and heuristic information on the ant's selection of a land-use unit that the ant is allowed to visit in its next step; and η_i(t) represents a heuristic function redesigned in the present study based on the land consolidation suitability index of units, which is used to guide the ant in selecting a land-use unit. The heuristic function of the ACO-based MOLCPSSM can be defined as:

$$\eta_i(t) = \frac{Suit_i}{\sum_S Suit_S} \tag{9}$$

where Suit_i represents the consolidation suitability index of land-use unit i.

After the probability for the selection of each land-use unit is calculated based on Eq. (8), each ant will select a unit using the roulette wheel technique. The implementation of this technique requires a randomly generated cursor. Let us assume that there are four units, a, b, c and d, and that the probability for the selection of each unit a, b, c and d is 0.1, 0.2, 0.3 and 0.4, respectively (the sum of the probabilities for the selection of each unit is 1). Then, the cumulative probability for the selection of unit a alone, units a and b alone, units a, b and c alone, and units a, b, c and d is 0.1, 0.3, 0.6 and 1, respectively. A random function is used to generate a cursor decimal between 0 and 1, which is then compared with the cumulative probability for the selection of each unit. Assume the decimal = 0.8. If the cumulative probability for the selection of the first n number of units is greater than 0.8, then the nth unit will be selected. In the aforementioned example, because the cumulative probability for the selection of the first four units is 1 (>0.8), unit d will be selected.

The pheromone update mechanism is also a core component of the ACO algorithm. When solving the TSP, the pheromone left by the mth ant on path (i, j) at time t can be expressed as:

$$\Delta\tau_{ij}^k(t) = \begin{cases} \frac{Q}{L_k}, \text{ if the } k^{th} \text{ ant passes path } ((i,j)) \text{ in the current cycle} \\ 0, \text{ else} \end{cases} \tag{10}$$

where Q represents the intensity of the pheromone and L_k represents the total length of the paths that the kth ant has visited in the current cycle.

When using the ACO algorithm to solve the LCPSS problem, the aforementioned equation must be modified. The $1/L_k$ variable in the model is replaced by the determined comprehensive objective function of the MOLCPSSM, i.e., Eq. (5) is used to modify Eq. (10). If Eq. (5) is defined as F(S), then Eq. (1) can be modified to:

$$\Delta\tau_{ij}^k(t) = \begin{cases} Q \times F(S), \text{ if the } k^{th} \text{ ant has visited spatial unit } i \\ 0, \text{ else} \end{cases} \tag{11}$$

2.4 Establishing the Main Parameters

The parameters of the MOLCPSSM proposed in the present study primarily include the scale of the ant colony (N_{ACO}), the pheromone heuristic factor (α), the expectation heuristic factor (β), the pheromone evaporation coefficient (ρ), the intensity of the pheromone (Q) and the initial pheromone level (τ_0). The parameter setting of the ACO algorithm has a significant impact on its performance. The present study establishes the parameters of the ACO algorithm based on existing studies [14] as follows: $N_{ACO} = 30$, $\alpha = 1$, $\beta = 1.5$, $\rho = 0.7$, $Q = 100$ and $\tau_0 = 0.1$.

The present study processes the multi-objective function using the linear weighted method. The weight of each objective function can, to a certain degree, reflect the preference information of the decision-maker. To investigate the effect of the weight of each objective function on the site selection results, the present study designs weight schemes (Table 1) based on the study conducted by Santé et al.

Table 1. Weight schemes for the MOLCPSSM

Weight scheme	w_1	w_2	w_3
A	0	1	0
B	0.5	0.5	0
C	0.25	0.75	0
D	0.75	0.25	0
E	0	0.5	0.5
F	0	0.75	0.25
G	0	0.25	0.75
H	0.33	0.34	0.33
I	0.5	0.25	0.25
J	0.25	0.5	0.25
K	0.25	0.25	0.5

Based on the implementation of LCPs in Hubei Province in recent years, the actual topographic conditions of Jiayu County and the investigation conducted by our research group on the land-use planning of Jiayu County, the present study sets the scale standard for each land consolidation project area to be 400–1,200 ha.

In principle, the rate at which new cultivated lands are added in a plain region is required to not be lower than 3%. However, because the current focus of land consolidation is on comprehensive consolidation and remediation and the land of Jiayu County has already been consolidated on a large scale since 2005, the requirement for the rate at which new cultivated lands are added should be lowered to some extent, and the focus should be on improving the quality of the current cultivated lands and the ecological environment. Based on the overall land-use planning of Jiayu County, the present study sets the standard rate at which new cultivated lands are added to 2%, i.e., $R_k \geq 0.02$.

3 Case Study with the Proposed MOLCPSSM

3.1 Experimental Area and Data

Jiayu county (29°48′~30°19′ N, 113°39′~114°22′ E) is chosen as the study area, which is located in the southeast of Hubei Province and on the south bank of middle Yangtze proper, with an area of 1017 km^2 (Fig. 2). The length of the county is 85 km and the width of the county is range from 5.7 to 17.9 km. For this area, the population is 3.65×10^5 and there are 8 towns.

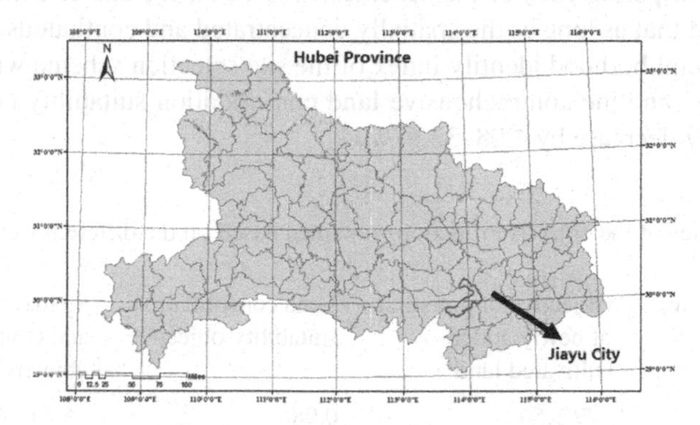

Fig. 2. Location of the study area

Multiple data sources were used in this study. The land use data was obtained from the database of the second national land survey and land use change survey of recent years, the soil quality data were obtained from the database of the agricultural land grade and evaluation (1:50000) which was provided by local bureau of land and resources. A digital elevation model (DEM of 30 m resolution, provided by the International Scientific & Technical Data Mirror Site, Computer Network Information.

Center, Chinese Academy of Sciences) was used to provide the slope and elevation information. Some vector traffic maps (1:10,000) were used to provide the spatial information such as centers of towns, railways, highways, roads, and administrative boundaries (provided by local bureau of land and resources). Some socioeconomic data such as the population, gross domestic product, fiscal revenue, et., were obtained from the Jiayu statistical yearbook (2011). The mean annual temperature and the mean annual rainfall were provided by local weather bureaus. Special plan on development of agricultural, forestry, traffic and tourism are the necessary basis for the site selection of land consolidation project, so the eleventh five-year agriculture development plan of Jiayu, Jiayu overall urban planning (2008–2020), overall planning of the development of tourism in Jiayu county were also collected.

All the spatial data were processed and computed by the spatial analysis tool in ArcGIS 10.2. To ensure consistency of the data, all the images and maps were geometrically rectified with each other and subsequently referenced to the Gauss-Kruger projection Xian_1980_3_Degree_GK_Zone_38 Projected Coordinate System.

3.2 Results and Analysis

The MOLCPSSM run 10 times using each of the weight schemes listed in Table 1 to obtain the average value of each objective function corresponding to each weight scheme (Table 2). By comparing each of the schemes B, C, D, H, I, J and K with scheme A, it can be observed that once the objective of the potential of newly added cultivated land is included in the comprehensive objective function, the number of newly added cultivated lands will increase significantly from 12.67% to 64.19%, and the land consolidation suitability coefficient will correspondingly decrease from 11.22% to 33.67%. Similarly, by comparing each of the schemes E, F, G, H, I, J and K with scheme A, it can be observed that as long as the spatially concentrated and continuous objectives are included, the neighborhood identity index of the site selection scheme will increase by 45.93–130.98%, and the comprehensive land consolidation suitability coefficient will correspondingly decrease by 4.08–33.67%.

Table 2. Values of the objective functions obtained based on the different weight schemes

Weight scheme (w_1, w_2 and w_3)	Objective of potential of newly added cultivated land	Land consolidation suitability objective	Spatial concentration and continuity objective
A (0/1/0)	761.50	0.98	5,542.42
B (0.5/0.5/0)	1,167.35	0.81	4,338.45
C (0.25/0.75/0)	924.80	0.87	5,128.69
D (0.75/0.25/0)	1,250.34	0.71	4,124.99
E (0/0.5/0.5)	681.31	0.85	10,870.33
F (0/0.75/0.25)	676.60	0.94	9,084.96
G (0/0.25/0.75)	686.69	0.73	12,801.88
H (0.33/0.34/0.33)	929.16	0.75	8,985.27
I (0.5/0.25/0.25)	1,068.75	0.68	8,088.04
J (0.25/0.5/0.25)	896.87	0.74	8,185.22
K (0.25/0.25/0.5)	857.99	0.65	9,298.12

By comparatively analyzing schemes C, B and D, it can be found that the maximum value of the potential of newly added cultivated land objective function gradually increases with increasing weight coefficient w_1 at a rate of increase of approximately 1.35, whereas the land consolidation suitability index gradually decreases to 0.72 (scheme A) with decreasing w_2. The maximum land consolidation suitability objective will guide the ants to select the land-use units with a relatively high suitability coefficient as much as possible during the iterative process of the model, and consequently, the number of newly added cultivated lands will be ignored. In comparison, the maximum potential of newly added cultivated land objective will induce the model to include the land units that can be converted to cultivated lands through consolidation in the project area as much as possible during the site selection process, thereby resulting in a decrease in the comprehensive consolidation suitability coefficient of the site selection scheme to some extent and in a significant increase in the number of newly added cultivated lands. The sum of the land area of each of the four types of land—grassland, ditch, garden and pit-pond—in each of the site selection schemes A, B, C and D is calculated. The result shows that scheme D has the largest total land area of the four types of land, followed by schemes B, C and A. This statistical result suggests, to a certain degree, that the maximum potential of newly added cultivated land objective can increase the capacity of the model to guide the ants to select the land units with relatively high consolidation potential.

By comparatively analyzing schemes A, F, E and G, it can be observed that the neighborhood identity index of the site selection scheme increases with increasing w_3, and the maximum rate of increase is approximately 2.2, whereas the land consolidation suitability coefficient decreases to 0.74 (scheme A) with decreasing w_2. The basic principle of land consolidation measures that suit the local conditions requires that land units with relatively high consolidation suitability be given preference when selecting a site for a project, which can easily result in the formation of a fragmented project area. However, after introducing the spatial concentration and continuity objective, each intelligent individual in the model will preferably select the neighboring units of the selected unit and simultaneously amalgamate some isolated units, thereby effectively increasing the neighborhood identity index of the project area and reducing the degree of fragmentation. The active role of the maximum spatial concentration and continuity objective in the optimization of the spatial layout of the LCPSS scheme can be clearly observed in Fig. 3. As the proportion of the maximum spatial concentration and continuity objective in the comprehensive function increases, the spatial layout of the site selection scheme becomes increasingly concentrated and continuous, and the number of spatial clusters also gradually decreases.

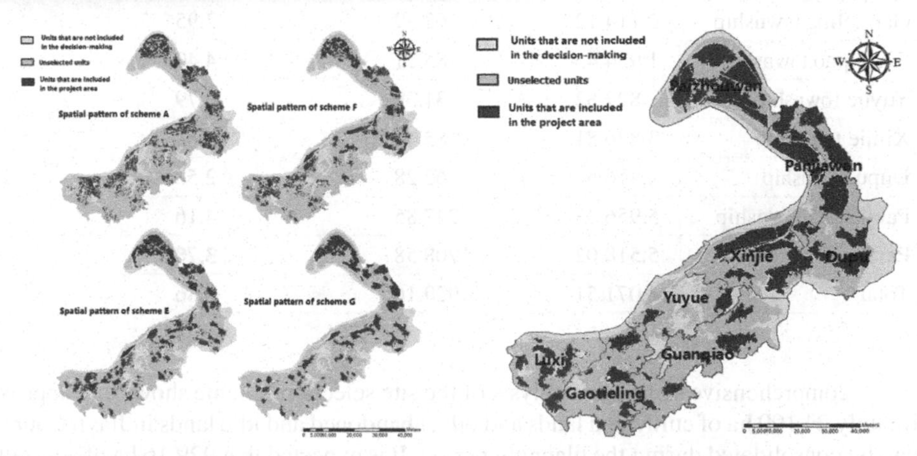

Fig. 3. Spatial patterns of site selection schemes A, E, F and G

Fig. 4. LCPSS scheme for Jiayu County (H)

Based on weight scheme H, a LCPSS scheme for Jiayu County is obtained using the MOLCPSSM (Fig. 4). Before running the model, the spatial units with a land-use type of construction, land or water were subjected to special treatment based on the site selection rules. The artificial ant colony can directly exclude these land-use units during the iterative process of the model. In the final site selection scheme, these land-use units are included in the non-consolidated area.

The scale of each project area is defined in advance as a model constraint: the scale of each project area should be 400–1,200 ha. In addition, this constraint is also treated flexibly using the penalty function method. However, because of the random selection mechanism used by the ACO algorithm, the selected land units exhibit a relatively fragmentary spatial layout. Table 3 lists the area of consolidated lands for each township and its potential value of newly added cultivated land calculated based on Eq. (2). Panjiawan Township and Paizhouwan Township have the largest areas of land that can be consolidated in Jiayu County, and their respective candidate area exceeds 5,000 ha. Panjiawan Township and Paizhouwan Township are the primary agricultural production townships in Jiayu County, and each of them has flat terrain and relatively abundant water resources. However, the agricultural land for each of these two townships is relatively small in overall scale and has a scattered local spatial layout. Yuyue Township is the political, economic and cultural center of Jiayu County and has the smallest amount of land that can be consolidated due to urban population growth and urban land expansion.

Table 3. Area of the candidate land consolidation regions in each township of Jiayu County, Hubei Province

Township name	Land consolidation area/ha	Potential newly added cultivated land area/ha	Newly cultivated land addition coefficient/%
Luxi township	1,467.35	46.02	3.14
Gaotieling township	2,114.12	62.39	2.95
Guanqiao township	1,934.43	85.21	4.40
Yuyue township	827.82	31.39	3.79
Xinjie township	3,836.81	185.44	4.83
Dupu township	2,416.53	62.28	2.58
Panjiawan township	5,956.43	247.85	4.16
Paizhouwan township	5,518.02	208.58	3.78
Total	24,071.51	929.16	3.86

A comprehensive statistical analysis of the site selection scheme shows that approximately 24,100 ha of cultivated lands and odd, abandoned and idle lands in Jiayu County can be consolidated during the planning period. It is expected that 929.16 ha of new cultivated lands can be added through comprehensive construction measures such as land flattening, merging of odd parcels and soil improvement, as well as road, shelter belt and ditch improvement. Xinjie Township, Panjiawan Township, Paizhouwan Township and Guanqiao Township should be focused on as the consolidation regions.

4 Concluding Remarks

The present study treats the LCPSS problem as a multi-objective optimal decision-making problem, establishes the mapping relationship between the spatial decision-making units and artificial ant coding and performs a spatial improvement on the ants' spatial unit selection strategy and pheromone update mechanism, based upon which the present study establishes a LCPSS model using the multi-objective ACO algorithm—the MOLCPSSM. The experimental results demonstrate that the MOLCPSSM can coordinate the three mutually conflicting optimization objectives relatively well, noticeably optimize the spatial layout of the LCP, and increase the social and ecological benefits while ensuring the economic benefits of the project. In addition, decision-makers can also obtain a project site selection scheme that meets their preference by adjusting the weight coefficient of each objective function. However, because the LCPSS problem is a complex problem involving multiple aspects, establishing an improved objective and constraint system is worth considering when using the proposed model to solve practical problems. Furthermore, traditional LCPSS uses villages/towns or small river basins as the evaluation units. The present study, however, selects small-scale vector patches of the current land use situation as the evaluation units. Therefore, further investigation is necessary to understand the effect of the difference in spatial scale on the site selection results.

Acknowledgments. This research was supported by the National Natural Science Foundation of China (No. 41601418) and Key Scientific and Technological Research Projects in Henan Province (No. 152102210356,162102210059, 172102210539). The authors would like to thank the anonymous reviewers for their suggestions and comments.

References

1. Yang, L., He, X., Yang, Q., et al.: Research on influencing factors of location for advancement project. Resour. Dev. Market **2017**(3), 289–294 (2017)
2. Ren, Y., Xu, Y., Liu, Y.: Study on spatial-temporal collocation of land reclamation based on dual self-organizing model. Acta Scientiarum Naturalium Universitatis Pekinensis **53**(2), 360–368 (2017)
3. Fan, P., Shao, H., Yang, Q., et al.: Study on land consolidation project location based on cultivated land quality evaluation: a case study of Yanjin county in Henan. Hubei Agric. Sci. **2017**(16), 3037–3041 (2017)
4. Hong, K., Liu, H., Wang, H.: The application of interval-valued intuitionistic fuzzy group decision making method in social benefit evaluation of land reconsolidation. Econ. Geogr. **35**(7),163–167 (2015)
5. Wang, H., Zhu, F.: Site selection model of land consolidation projects based on multi-objective optimization PSO. Trans. Chin. Soc. Agric. Eng. **31**(14), 255–263 (2015)
6. He, J.: Directional antenna intelligent coverage method based on traversal optimization algorithm. Comput. Mater. Continua **60**(2), 527–544 (2019)

7. Shamshirband, S., Rabczuk, T.: Parkinson's disease detection using biogeography-based optimization. Comput. Mater. Continua **61**(1), 11–26 (2019)
8. Zhe, L., Bao, X., Lu, H., Song, Y., Liu, Q.: An improved unsupervised image segmentation method based on multi-objective particle, swarm optimization clustering algorithm. Comput. Mater. Continua **58**(2), 451–461 (2019)
9. Liu, X., Li, X., Tan, Z..: Zoning farmland protection under spatial constraints by integrating remote sensing, GIS and artificial immune systems. Int. J. Geogr. Inf. Sci. **25**(11), 1829–1848 (2011)
10. Li, X., Lao, C., Liu, X.: Coupling urban cellular automata with ant colony optimization for zoning protected natural areas under a changing landscape. Int. J. Geogr. Inf. Sci. **25**(4), 575–593 (2010)
11. Eldrandaly, K.: A GEP-based spatial decision support system for multisite land use allocation. Appl. Soft Comput. **10**(3), 694–702 (2010)
12. Niu, J., Xu, F.: Establishing land use zoning model by clonal selection algorithm. Geomatics Inf. Sci. Wuhan Univ. **39**(2), 172–176 (2014)
13. Wang, H., Liu, Y., Ji, Y..: Land use zoning model based on multi-objective particle swami optimization algorithm. Trans. Chin. Soc. Agric. Eng. **28**(12), 237–244(2012)
14. Toksari, M.D.: Ant colony optimization for finding the global minimum. Appl. Math. Comput. **176**(1), 308–316 (2006)

Image Content Location Privacy Preserving in Social Network Travel Image Sharing

Wang Xiang[1(✉)], Canji Yang[1], Lingling Jiao[2], and Qingqi Pei[3]

[1] State Key Laboratory of Integrated Service Networks, Xidian University, Xi'an 710071, Shanxi, China
wangxiang@xidian.edu.cn

[2] College of Information and Communication, National University of Defense Technology, Changsha, China

[3] Shanxi Key Laboratory of Blockchain and Security Computing, Xidian University, Xi'an 710071, Shanxi, China

Abstract. It is a well-known truth that online image sharing can lead to privacy leakage. Although the privacy information may vary in forms, image contents only with private property will be concerned, while other contents that seems public but could reveal personal information are always ignored. In online image sharing, some images may contain labels or landmarks which could reveal the geographic positions where these photos were taken, in which the public content could disclose private location information. To handle such problem, we proposed a travel image location privacy protection system that aims at protecting travel images location. The key issue in this problem is which part of image content is related to the scene and to what extent. To determine such relevance, in our proposed system, several image process methods are utilized to find out potential objects and their belonging classes, and two proposed privacy strategies that respectively focus on quantity and relative position are further implemented to define initial privacy level of each classes. Finally, a privacy predict model is trained in online learning way so that it can be updated with future user feedbacks. We conducted experiment on travel image dataset that related to specific keyword, and the results have demonstrated the effectiveness of our proposed system.

Keywords: Online image sharing · Travel image location privacy · Image content privacy preserving

1 Introduction

With the ever-increasing popularity of smart devices equipped with high quality on-board cameras and online social networks, capturing and sharing photos online have become extremely convenient and popular for many people. Various social platforms like Facebook, Flickr, Tumblr and Instagram have provided image sharing service in which people can upload photos anytime they want and wherever they are. However, images contain significant amount of personal information which can reveals the owner's daily life routines and social relationships as the images can be accessed by anyone else.

X. Sun et al. (Eds.): ICAIS 2020, LNCS 12240, pp. 617–628, 2020.
https://doi.org/10.1007/978-3-030-57881-7_54

Besides, the shared images will be collected together with other format of data by some third-party corporations for user data analysis. Thus, image privacy protection has become a critical issue to be addressed.

Considering such problem, most of the social networks now offer the user access control system so that they can modify the settings to determine who has the authority to access the information and who does not. However, the access control system usually can't achieve the goal of protecting privacy information for reasons like some people do not realize the possible consequences privacy leakage may lead to, or some others being tired of the tedious step in configuration, or for lacking of enough privacy knowledge. This phenomenon appears more often when it comes to personal images sharing, due to the reason that privacy information in image is not as obvious as in text. To reduce the burdens on privacy settings and archive effective privacy preservation, it is necessary to develop automatic method to help the users protect image privacy.

With regard to location privacy, Researchers have proposed many approaches to address such problem in Location-Based-Services (LBSs). While in such condition, the privacy is just about the geography location of people's devices requested by the LBS server, image content location has become a blind spot in location privacy protection.

Generally speaking, there are two kinds of things in image that can be used to infer the actual location: text and some iconic objects in the scene. Text could explicitly indicate where the image was being taken if its content contains some texts about places or streets; as for other objects in the scene, some of which may have correlation with the scene to some degree; that is to say, people can infer where a photo was taken from some specific objects in it. However, text contents are not easy to recognize at most time in pictures, so other objects could be considered as the "keywords" to the scene. On the other hand, pictures taken in courtyard, parks or streets may be strange to most people who haven't been there, while pictures of some well-known attractions shall be recognized for most people at a glance. Therefore, we mainly consider the location privacy issue in travel images.

By assuming that there is relationship between the attraction location and some objects in the scene, we proposed a system which aims to deal with the travel images location privacy protection problem by means of finding objects which may have a correlation with the tourist attraction, and match them with different privacy level according to proposed privacy strategies, and implement different image content privacy protection method.

The rest of this paper is organized as follows. Section 2 reviews the related work briefly. Section 3 presents the system model and design goal of our proposed travel image location privacy protection system. Section 4 presents our system design details and our proposed location privacy strategy. Section 5 reports the experimental and evaluation results of our system. Finally, we make the conclusion.

2 Related Work

Most online social networks allow users to decide their own privacy settings about images they share. A lot of studies have been focused on protecting image privacy in social networks image sharing. Several works have studied how to automatically recommend privacy settings to users. By utilizing template, Bonneau et al. [1] proposed privacy setting suits specified by friends or trusted experts so that others can easily choose from. Adu-Oppong et al. [2] proposed a technique that analyzes social circle according to partition of users' friend lists. Fang et al. [3] proposed a wizard to help user grant privileges to their friends, and construct a classifier which can automatically classifies friends according to the user profile. Klemperer et al. [4] studied whether the keywords and captions with which users tag their photos can be used to help users more intuitively create and maintain access-control policies. Squicciarini et al. [5] proposed an A3P system that determines the best available privacy policy for the user's images being uploaded using the user's available history on the site. More recently, Yu et al. [6] proposed a deep learning based approach called iPrivacy which can automate the process of privacy settings during image sharing. Considering the effect of users' social behaviors, they [7] further developed a new approach to recommend fine-grained privacy settings for social image sharing.

For image content and location, Gallagher et al. [8] explored the benefit of employing user-tags along with image content to infer geo-location of Web images, and constructed global location probability maps. Qian et al. [9] built up visual word group as position descriptor, and used these as image index to find out image location by image retrieval in image dataset. These works emphasis on finding image location using visual features, which is meaningless to human vision. While ours aims at finding relevance between specific content and scene location, and protecting the images from leaking location privacy.

3 System Model and Design Goals

To solve the image location privacy problem mentioned in Sect. 1, our system design follows three main goals: key objects detection, to precisely detect the scene related objects defined in the system's initial stage from the incoming images of users; location privacy protection, to adopt some kind of information concealing methods for the protection of image content according to pre-defined privacy policy; Privacy prediction model, to train a machine learning model that can predict the privacy level of different objects and update with users' feedbacks on privacy definition.

The overall system architecture contains three main steps: (1) Pre-trained deep convolutional neural networks are used to segment image data collected from internet to get object regions, and features extracted from which are clustered so that objects can be classified into different classes, and object detection model will be trained with these data; (2) The classified objects and features are further used to train privacy prediction model on the basis of privacy strategies; (3) Image processing methods are adopted to protect the privacy region in images.

4 Travel Image Location Privacy Protection System

4.1 System Pre-preparation

As for the travel image data, images are crawled from the internet according to the keywords of tourist attraction. For location privacy in LBS, the privacy information to be protected is the geographic coordinates of people's mobile device, while in images, the privacy contents are not obvious as in LBS. Therefore, the first step to do is to segment the image into different portions so that we can find out which part may be related to the scene, e.g., objects that could indicate the geographical position of the scene. Image segmentation is an important part in image understanding, and a lot of research has been conducted on it. Traditional ways of image segmentation like threshold-based, region growing-based and edge detection-based method generally can only segment the image into foreground and background; or partition the image into different regions according to color and texture. For travel images, we intend to find objects in them while the traditional methods cannot meet with requirement. So we choose a CNN based image segmentation technique [10–12], e.g., a pre-trained convolutional network to segment the image data collected from internet which belongs to the same tourist attraction keyword. Then, different object regions are obtained.

To determine the class of these unlabeled segmented objects, we utilize deep features extracted by DCNN called VGGnet [13], and the output of layer FC7 is used as deep visual features for object regions in images. After feature extraction, these features are further clustered into different classes. In view of this unsupervised learning problem where the exact amount of classes is unknown, an unsupervised clustering algorithm called affinity propagation [14] is utilized to cluster these deep features, which do not need pre-defined number of clusters.

With potential privacy classes determined, we need to further detect the exact position of these classes in incoming travel images for the purpose of privacy protection, and object detection method is employed. With regard to object detection in image, the most advanced algorithm is also based on deep neural network, which shows great advantages on accuracy and speed than traditional algorithms like sliding-window based method with some hand-crafted features. In our location privacy protection system, we utilize a CNN based method in work [15] proposed by Ren et al., which combines a region proposal network an d a classification network, while the previous network optimizes the region selection problem and the later network takes the regions gained from the former one as input, and output the object region and its probability in images.

4.2 Privacy Strategies

After the preceding procedures, objects and its class information are obtain. Next, we need to determine the privacy level of these objects, e.g., to what degree can the object indicate its scene location. Based on such observation, we propose a privacy strategy where the relevance between objects and scene are defined as two parts: quantity and relative position.

Quantity Privacy Strategy. Generally speaking, among varies objects in a scene, some of which are just common things that can be seen in other places, while some others could be special. Accordingly, the privacy strategy is proposed for the aim of discriminating the relevance score between diverse objects and the scene. But there remains another question that in most places, the common objects are often in majority and could not help people identify where the location is. However, when traveling in attractions, people usually tend to take photos with landmark architectures or scene that has nothing special but famous; that is to say, in most time, the majority objects in photos are special things like buildings and statues rather than some ordinary decorations. For this reason, the image data collected with keywords will have the same features that can be used to infer the potential relationship between objects and scene. We define the single class relevance score of class C_i is defined as:

$$S(C_i) = \frac{\|C_i\|}{N} \tag{1}$$

where $\|C_i\|$ denotes the appearance times of class C_i in the dataset, and N is the total number of the images.

On obtaining the single class relevance score of each object class, we can roughly obtain a range of [min-value, max-value] (both value denote a relevance score) and split the range into three adjacent intervals according to a certain ratio, and let θ_1 and θ_2 be the threshold that separate the intervals. The relevance class is made by thresholding the score S:

$$y_i = \begin{cases} \text{"most relevant",} & \text{when } \theta_1 \le S_i \\ \text{"less relevant",} & \theta_2 \le S_i < \theta_1 \\ \text{"little relevant",} & S_i < \theta_2 \end{cases} \tag{2}$$

For short, we denote these three classes as “most”, “less” and “little” temporarily. However, single relevance score is not convincing enough to explain the real relevance. Considering different objects in the same image, they are not just stand there independently, rather together form an impression to people. So one objects will be affected by its co-occurrence objects, and so as its relevance score. Therefore, we define the co-occurrence score of C_i and C_j as:

$$\phi(C_i, C_j) = S(C_i, C_j) \log \frac{S(C_i, C_j)}{S(C_i) + S(C_j)} \tag{3}$$

where $S(C_i, C_j)$ is their co-occurrence probability:

$$S(C_i, C_j) = \frac{N(C_i, C_j)}{N} \tag{4}$$

Now single and co-occurrence score of each object class are obtained, the final relevance score is defined with follow rules: for class C_i, if its co-occurrence class C_j belongs to higher relevance class (like “most” to “less”) defined in Eq. (2), then their co-occurrence score should have a positive contribution in the final score of C_i; in the contrary, a lower relevance class should contribute a negative value on score of C_i; if they are in the same relevance class, do nothing. Accordingly, the final relevance score of C_i is defined as:

$$\theta(C_i) = \lambda S(C_i) + \mu \sum_{C_j \in H(C_i)} \phi(C_i, C_j) - \gamma \sum_{C_k \in L(C_i)} \phi(C_i, C_k) \tag{5}$$

where $H(C_i)$ is the set whose relevance class is higher than C_i, and $L(C_i)$ is opposite; λ, μ and γ are positive coefficients less than 1. Finally, object classes are re-classified according to the final relevance score into different relevance class.

Relative Position Privacy Strategy. Besides the quantity property, there remains another potential characteristic in travel images: photo direction, e.g., the direction people chose when they take photos of scenery or buildings. Consider some famous places, for people who have never been there, the most common way to get acquainted with is through pictures, whose contents usually present in some certain perspective, and people will become familiar with these photos sometimes even if the photo itself does not contains something special. Base on this observation, we propose a relative position privacy strategy aiming to find out specific inter-class position patterns in the travel image dataset.

Generally, there are varying amounts of objects in one image, and we only choose three-elements-pattern or four-elements-pattern which represents the sets consist of three or four objects with the highest quantity relevance score in an image, which should at least belong to "most relevant" or "less relevant" class. To modeling such position relationship, we use geometric center to represent the whole object; the center points of top four (if there are not enough qualified objects to form four-pattern, the number will be three) objects form a polygon and stored. A relative position dataset is obtained by implementing above process to each image. Meanwhile, these three-pattern and four-pattern will be classified into different group, and patterns in same group must have every recorded object matched. Groups that satisfy the quantity requirement will be reserved.

To measure the similarity of these relative position, a modified method called turning function [16] is used to describe such relationship in images. Given a polygon, the first step is to normalize the total length to 1 for the purpose of making the method invariant to scaling, which is done by dividing each edge's length by the perimeter of the polygon. The turning function of a polygon is angles as a function of distances along the polygon edges clockwise. To be specific, starting from a vertex, the first value is the angle between horizontal axis and the first edge; the succeeding value is inner angles of other vertex in order, and an example of turning function is showed in Fig. 1.

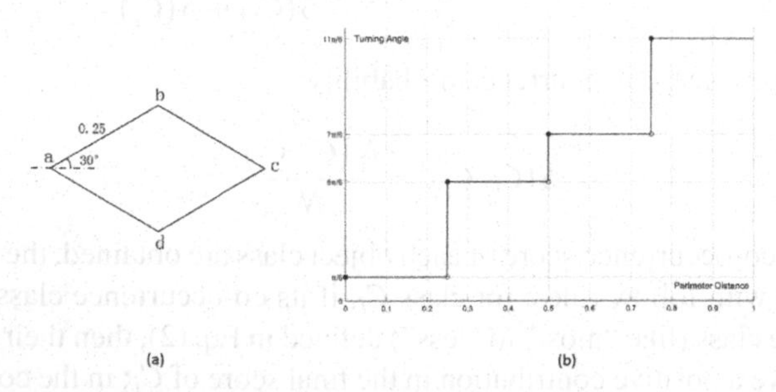

Fig. 1. (a) A simple polygon. (b) The turning function for the polygon showed in (a).

Then, the distance between shape A and B is defined as:

$$\mathrm{d(A,B,}\theta) = \int_0^l (t_A(s) - t_B(s) + \theta)^2 ds \tag{6}$$

where $t_A(s)$ and $t_B(s)$ are turning function of polygon A and B; θ denotes angle of rotation. To find out that minimizes the distance function, the partial derivative of d(A, B, θ) is set to zero and the value of θ is obtain:

$$\theta = \frac{-\int_0^l (t_A(s) - t_B(s))ds}{l} \tag{7}$$

and the distance function can be computed as:

$$\begin{aligned} d(A,B) &= \int_0^l \left(t_A(s) - t_B(s) - \frac{\int_0^l (t_A(s) - t_B(s))ds}{l} \right)^2 ds \\ &= \int_0^l (t_A(s) - t_B(s))^2 ds - \frac{1}{l}\left[\int_0^l (t_A(s) - t_B(s))ds \right]^2 \\ &= II(A,B) - \frac{1}{l} I(A,B)^2 \end{aligned} \tag{8}$$

where

$$I(A,B) = \int_0^l (t_A(s) - t_B(s))ds \tag{9}$$

And

$$II(A,B) = \int_0^l (t_A(s) - t_B(s))^2 ds \tag{10}$$

then, distance of every two patterns in the same group are computed and form an n * n distance matrix, where n represents the total number of patterns in one group and differs in different group.

In order to find out right range of patterns in each group, e.g., the rational range of majority patterns, an outlier detection algorithm called local outlier factor (LOF) [17] is used, which is able to detect the outlier in a dataset based on the distances of each two points.

Given the distance matrix, first let the k-distance (P) be the distance between pattern A and its k-th nearest neighbor. Then, reachability distance of P and Q is the maximum value of k-distance (Q) and distance between P and Q:

$$\textit{reach-dist}_k(P,Q) = \max\{\textit{k-distance}(Q), d(P,Q)\} \tag{11}$$

and the local reachable density is defined as:

$$lrd_k(P) = 1 \Bigg/ \left(\frac{\sum_{Q \in N_k(P)} \textit{reach-dist}_k(P,Q)}{|N_k(P)|} \right) \tag{12}$$

where $N_k(P)$ denotes those patterns whose distance to P equals or less than k-distance(P). Eventually, the local outlier factor is defined as:

$$LOF_k(P) = \frac{\sum_{Q \in N_k(P)} \frac{lrd(Q)}{lrd(P)}}{|N_k(P)|} = \frac{\sum_{Q \in N_k(P)} lrd(Q)}{|N_k(P)|} \Bigg/ lrd(P) \tag{13}$$

and LOF value of each pattern is computed (distances that are extremely small are not considered), and the pattern with the minimum LOF is chose as the center point, and patterns whose distance to which is within its k-distance will be regarded as privacy.

4.3 Machine Learning Model and Privacy Protection

Although the proposed privacy strategy has defined the privacy level of objects that can expose geographic location, this kind of judgment can be subjective for different people inevitably. For this reason, we need to train a privacy prediction model incrementally as user respond to prompts over time, and online learning model is utilized for it adapting to new data. For this purpose, the passive-aggressive algorithms [18] is employed, as it can be used to train an online model without using the whole training data each time, and the model will slowly forget the previous one and learning the new distribution. In our system, we choose deep features extracted from the objects as training data, and three types of relevance classes as target classes; the multi-class classifier will be trained in a one-vs-rest way.

Finally, according to different privacy level, the detected object region will be processed in different way: for most relevant class, the image that contains these classes will be marked with "not recommend", which means these images can be easily recognized with high probability, so they are not recommended for sharing; for less relevant class, we simply blur the detected region, and do nothing for little relevant class.

5 Experimental Results

To evaluate the effectiveness of our proposed method, we collected attraction images according to a certain chosen keyword Ci'en Temple which is located in Xi'an, Shaanxi Province, China from Internet with total amount of 1523 after filter. Such a dataset with pictures less than two thousand is collected mainly based on two main observations: Unlike image datasets in other researches, a large number of images with respect to some specific keywords that is related to complicated attractions always show high repeatability, let alone keywords like Eiffel Tower, London Eye which could directly indicate the landmark building; On the other hand, too many pictures can also dilute the existence of other object classes that are less in quantity, which may influence the potential possibility in privacy.

Therefore, the dataset is made under the above consideration and has plenty of cultural symbols and architectures, which makes its suitable enough for our research. In this section, we mainly discuss and evaluate different parts in our system from initialization stage to privacy protection stage, and the experimental results have showed that our proposed system has achieved well result in terms of privacy protection.

5.1 System Initialization Stage

In the initial stage, we adopt the first three main processes: image segmentation, image clustering and object detection model training. Figure 2 shows some segmentation results in the collected dataset, from which we can find that the results are shown in the form of transparent mask with different light colors. These processed images are filtered and union areas are chosen as objects, in which morphology methods include dilating and eroding are adopted to control the connected domains.

Then, these extracted objects are clustered into serval clusters according to their CNN features, and the result can be seen in Fig. 3. With the above clustering result

Fig. 2. Some results of the image segmentation part, which are further processed to get rectangular regions that represent objects.

and statistic data (quantities and relative positions of classes), the privacy score of each class can be calculated with defined privacy strategies. Then, the two thresholds that determine the relevance privacy class of each object class can be obtained, which will be as privacy labels then.

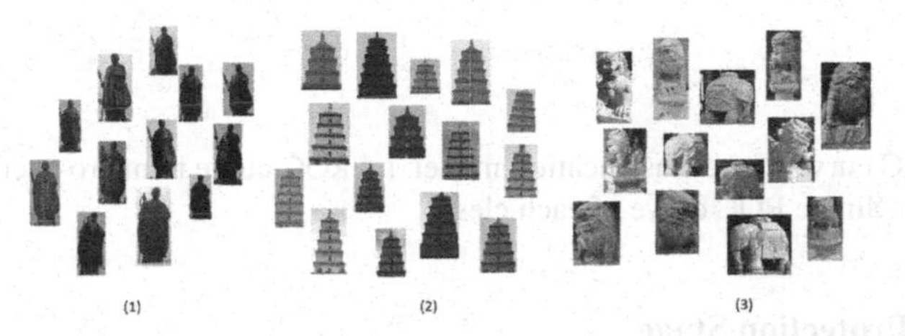

Fig. 3. Three classes in the clustering results (There are total 15 kinds of objects actually)

Meanwhile, on obtaining the labeled images with category and position information, the object detection model is trained, and some results are showed in the Fig. 4. For the evaluation of the detection model, Table 1 shows the average precision of each class and the mAP. During training period, 3 classes are abandoned for their small amount; due to characteristic of travel image datasets where each class is more unified in image form than ordinary object detection datasets, the rest classes have reached an mAP of 0.825, which is an acceptable value for object detection model.

Fig. 4. Some detection result of the trained model, and the accuracy is basically above 0.9.

To further determine the privacy level of each object in practice, a machine learning model is trained with deep features and its label (e.g. privacy relevance class), and Fig. 5 shows the ROC curve of the privacy classification model, where class 0, 1, 2 respectively represent three kind of privacy level from most relevant to little relevant. Though the total accuracy score is 0.82, the model can be updated by learning feedbacks from users.

Table 1. Detection results on the datasets, which contains the average precision of each class and the average value.

C	1	2	3	4	5	6	7	8	9	10	11	12	Avg
AP	0.899	0.786	0.909	0.780	0.640	0.883	1.000	0.818	0.746	0.755	0.980	0.702	0.825

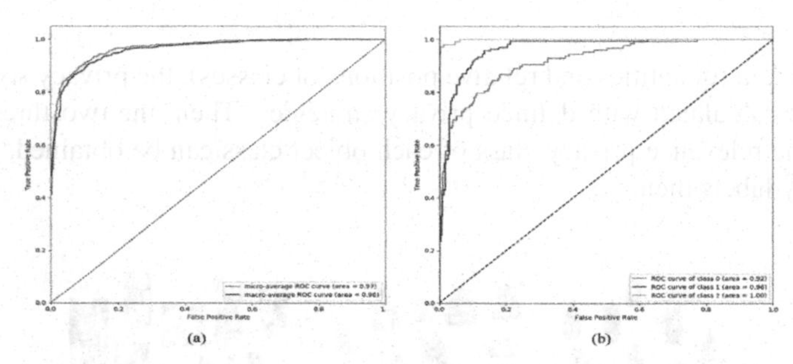

Fig. 5. The ROC curve of the classification model: (a) ROC curve in micro-average and macro-average way; (b) Single ROC curve of each class.

5.2 Privacy Protection Stage

In this stage, each part of the system is already prepared, and is able to handle new images to detect privacy objects or regions. As shown in Fig. 6, the objects belong to the first privacy level are marked in the system output (which can not be seen from the image), and objects belong to less relevant are blurred to make it difficult to recognize. And the processed regions are based on masks obtained by object detection model, and the masks are further expanded in area to some extent, for the reason of making sure the privacy region covered.

Fig. 6. Some experimental results on location privacy protection: the first row are origin images with detection mark; the second row are images after blurring.

However, blurring will significantly affect the quality of images, the results will just be as a choice and can be modified by subjective feedbacks of users, and the feedback will also be used as training data to update the classification model and make it more reasonable.

6 Conclusion

In this paper, we proposed a travel image privacy protection system which aims at solving location privacy leakage in image content. By leveraging deep CNN based methods, our system can find out objects and their classes in images; two types of privacy strategies are further implemented to determine the initial privacy level of each class; then, a privacy classification model is trained in online learning way so that it can updated with user feedbacks to make the system more reasonable. Our experimental results show that our system can effectively protect the location privacy in image.

References

1. Bonneau, J., Anderson, J., Church, L.: Privacy suites: shared privacy for social networks. In: Proceedings of the 5th Symposium on Usable Privacy and Security. ACM (2009)
2. Adu-Oppong, F., Gardiner, C.K., Kapadia, A., et al.: Social circles: tackling privacy in social networks. In: Symposium on Usable Privacy and Security (SOUPS) (2008)
3. Fang, L., LeFevre, K.: Privacy wizards for social networking sites. In: Proceedings of the 19th international conference on World Wide Web, Raleigh, North Carolina, USA, pp. 351–360. ACM (2010)
4. Klemperer, P., Liang, Y., Mazurek, M., et al.: Tag, you can see it!: using tags for access control in photo sharing. In: Proceedings of the SIGCHI Conference on Human Factors in Computing Systems, pp. 377–386. ACM (2012)
5. Squicciarini, A.C., Lin, D., Sundareswaran, S., et al.: Privacy policy inference of user-uploaded images on content sharing sites. IEEE Trans. Knowl. Data Eng. **27**(1), 193–206 (2015)
6. Yu, J., Zhang, B., Kuang, Z., et al.: iPrivacy: image privacy protection by identifying sensitive objects via deep multi-task learning. IEEE Trans. Inf. Forensics Secur. **12**(5), 1005–1016 (2017)
7. Yu, J., Kuang, Z., Zhang, B., et al.: Leveraging content sensitiveness and user trustworthiness to recommend fine-grained privacy settings for social image sharing. IEEE Trans. Inf. Forensics Secur. **13**(5), 1317–1332 (2018)
8. Gallagher, A., Joshi, D., Yu, J., et al.: Geo-location inference from image content and user tags. In: 2009 IEEE Computer Society Conference on CVPR Workshops 2009 Computer Vision and Pattern Recognition Workshops, pp. 55–62. IEEE (2009)
9. Qian, X., Zhao, Y., Han, J.: Image location estimation by salient region matching. IEEE Trans. Image Process. **24**(11), 4348–4358 (2015)
10. Cui, Q., McIntosh, S., Sun, H.: Identifying materials of photographic images and photorealistic computer generated graphics based on deep CNNs. Comput. Mater. Continua **055**(2), 229–241 (2018)
11. Meng, R., Rice, S.G., Wang, J., Sun, X.: A fusion steganographic algorithm based on faster R-CNN. Comput. Mater. Continua **55**(1), 001–016 (2018)
12. Liu, Z., Xiang, B., Yuqing Song, H.L., Liu, Q.: An improved unsupervised image segmentation method based on multi-objective particle, swarm optimization clustering algorithm. Comput. Mater. Continua **58**(2), 451–461 (2019)
13. Simonyan, K., Zisserman, A.: Very deep convolutional networks for large-scale image recognition. In: ICLR (2015)
14. Frey, B.J., Dueck, D.: Clustering by passing messages between data points. Science **315**(5814), 972–976 (2007)

15. Ren, S., He, K., Girshick, R., et al.: Faster R-CNN: towards real-time object detection with region proposal networks. In: Advances in Neural Information Processing Systems, pp. 91–99 (2015)
16. McCreath, E.: Partial matching of planar polygons under translation and rotation. In: CCCG (2008)
17. Breunig, M.M., Kriegel, H.P., Ng, R.T., et al.: LOF: identifying density-based local outliers. ACM SIGMOD Rec. **29**(2), 93–104 (2000)
18. Crammer, K., Dekel, O., Keshet, J., et al.: Online passive-aggressive algorithms. J. Mach. Learn. Res. **7**(Mar), 551–585 (2006)

A Robust Blind Watermarking Scheme for Color Images Using Quaternion Fourier Transform

Renjie Liang, Peijia Zheng(✉), Yanmei Fang, and Tingting Song

School of Data and Computer Science,
Guangdong Key Laboratory of Information Security Technology,
Sun Yat-Sen University, Guangzhou 510006, China
{liangrj5,songtt3}@mail2.sysu.edu.cn, {zhpj,fangym}@mail.sysu.edu.cn

Abstract. With the advancement of image devices, massive image data are being generated exponentially, which indirectly leads to more copyright issues over the images. Image watermarking is an important method for copyright protection. Considering that color images dominate the real-world image data, designing copyright protection techniques specially for color images are very meaningful. In this paper, we propose a robust blind watermarking scheme for color images based on quaternion Fourier transform (QFT). After converting the color image data into QFT coefficients, we embed the watermark message into the four channels in QFT domain, to improve the embedding capacity and robustness. We also conduct experiment to investigate the performance of visual quality and robustness, by comparing with some other common signal transforms. The experimental results show that our scheme achieves a satisfactory performance.

Keywords: Color image watermarking · Quaternion Fourier transform · Blind watermarking · Robust watermarking

1 Introduction

With the exponential growth of image data in the era of big data, multimedia security issues, such as tamper detection, copyright protection, source tracking, etc., have attracted more and more attentions. As an important solution, image watermarking has been extensively studied during the past three decades. Image watermarking schemes can be generally divided into spatial domain and frequency domain watermarking schemes. The spatial domain watermarking scheme has the advantage of low computation complexity but are less robust against watermarking attacks [1]. In the frequency domain watermarking scheme [2–4], the watermark message is embedded into the host image by modifying the transform coefficients of the host image. In [5], the authors proposed an image watermarking method based on block-based Discrete cosine transform (DCT) coefficient modification and the modification of DCT coefficients in the

X. Sun et al. (Eds.): ICAIS 2020, LNCS 12240, pp. 629–641, 2020.
https://doi.org/10.1007/978-3-030-57881-7_55

adjacent blocks at the same position. In [6], Zheng *et al.* presented a blind image watermarking scheme by using Walsh-Hadamard transform (WHT) in the encrypted domain. Jayashree *et al.* proposed a chaotic hybrid watermarking method with discrete wavelet transform (DWT), Z-transform and bidiagonal singular value decomposition [7]. In [8,9], Liu *et al.* presented two encrypted medical image watermarking schemes by using scale invariant feature transform-DCT and dual-tree complex wavelet transform, respectively.

Compared with DCT and DWT, Quaternion Fourier transform (QFT) can transform the four-dimensional signal into the transform domain that is also four-dimensional. Thus, QFT is able to be used to jointly process the three color channels of the color image. There are already many works on using QFT in different applications of color image processing. Grigoryan *et al.* [10] presented a simple measure to evaluate the quality of enhanced color images based on QFT. In [11], QFT was used to design robust color image watermarking algorithms. Chen *et al.* [12] proposed an color image watermarking scheme by combining QFT and least squares support vector machine, in order to enhance the robustness against geometrical distortions. However, in these existing color watermarking schemes, the watermark messages are embedded into the real channel of the QFT transform domain. Therefore, designing a color image watermarking scheme that can fully use the other three imaginary channels of the QFT transform domain is worthy of study.

In this paper, we propose a color image watermarking scheme based on QFT, which has the blind extraction property and is robust against some common watermarking attacks. We firstly split the color image into different blocks, and the apply QFT to every sub-block. Different from the previous works, we perform watermark embedding on the four components of the QFT transform coefficients. The proposed scheme has the following two advantages. 1) Our scheme produces higher perpetual quality watermarked images than the related work [13] that performs embedding only on the real channel of the QFT transform domain. 2) Compared with the methods based on WHT and DCT, our scheme is more robust against noise attack, cropping attack, and JPEG 2000 compression attack. We also conduct experiments to evaluate the performance of visual quality and robustness. The experimental results demonstrate that our performance is adequate.

The rest of paper has been organized as follows. Section 2 provides a brief introduction on quaternion and QFT. The proposed algorithm is presented in Sect. 3 and the experimental results are given in Sect. 4. Section 5 draws a conclusion of this paper.

2 Preliminaries

2.1 Quaternion

Quaternion is a number system that extends the complex numbers. It can be used in computer graphics, computer vision, signal/image processing, etc.

A quaternion Q has four components, i.e.,

$$Q = a + b\mathbf{i} + c\mathbf{j} + d\mathbf{k} \tag{1}$$

where a is the scalar coefficient of the real channel, and b, c, and d are scalar coefficients of the imaginary channels. $\mathbf{i}$, $\mathbf{j}$, and $\mathbf{k}$ are imaginary operators that satisfy the four rules: $\mathbf{i}^2 = \mathbf{j}^2 = \mathbf{k}^2 = -1$, $\mathbf{ij} = -\mathbf{ji} = \mathbf{k}$, $\mathbf{jk} = -\mathbf{kj} = \mathbf{i}$, and $\mathbf{ki} = -\mathbf{ik} = \mathbf{j}$.

2.2 Quaternion Fourier Transform

Ell *et al.* [14] proposed three types of QFTs named left-side QFT, right-side QFT, and both-side QFT. In this paper, we adopt the left-side QFT that is widely used in image processing. Suppose that the input signal is denoted by $f(x,y) = f_w(x,y) + f_i(x,y)\mathbf{i} + f_j(x,y)\mathbf{j} + f_k(x,y)\mathbf{k}$ sized of $M \times N$. The QFT of $f(x,y)$ is given as

$$F(u,v) = \frac{1}{\sqrt{MN}} \sum_{M-1}^{x=0} \sum_{N-1}^{y=0} e^{-\mu 2\pi (\frac{xu}{M} + \frac{yv}{N})} f(x,y) \tag{2}$$

where μ is a unit pure quaternion.

3 Proposed Scheme

We assume that the input image is a RGB color image $I(x,y) = I_r(x,y) + I_g(x,y) + I_b(x,y)$ sized of $M \times N$. The original watermark message is denoted by w sized of $m \times n$. We convert the color image $I(x,y)$ into a quaternion matrix I_s as $I_s(x,y) = I_r(x,y)\mathbf{i} + I_g(x,y)\mathbf{j} + I_b(x,y)\mathbf{k}$.

3.1 Watermarking Embedding

We detail our watermark embedding algorithm as follows.

Step 1: Based on the original watermark w, we generate a quaternion matrix $W_s = w + w\mathbf{i} + w\mathbf{j} + w\mathbf{k}$.

Step 2: We divide the quaternion matrix I_s into non-overlapping $m \times n$ blocks, each of which has the size of $\lfloor M/m \rfloor \times \lfloor N/n \rfloor$.

Step 3: We perform QFT on all the blocks to obtain the frequency matrices $\{I_f^{[i,j]}\}_{i,j}$. W_s is also converted into QFT domain to obtain $W_f(i,j)$. Then every element of $W_f(i,j)$ corresponds to a sub-block frequency matrix $I_f^{[i,j]}$.

Step 4: For $1 \le i \le m$ and $1 \le j \le n$, we perform watermark embedding as

$$I_{wf}^{[i,j]}(x,y) = I_f^{[i,j]}(x+1, y+1) + \alpha W_f(i,j) \tag{3}$$

where (x,y) is the embedding position and α is the watermark embedding strength.

Step 5: We perform the inverse QFT on $I_{wf}^{[i,j]}$ and obtain I_{ws}, from which we can extract the three imaginary channels as the watermarked color image I_w.

We sketch the watermarking embedding procedure in Fig. 1.

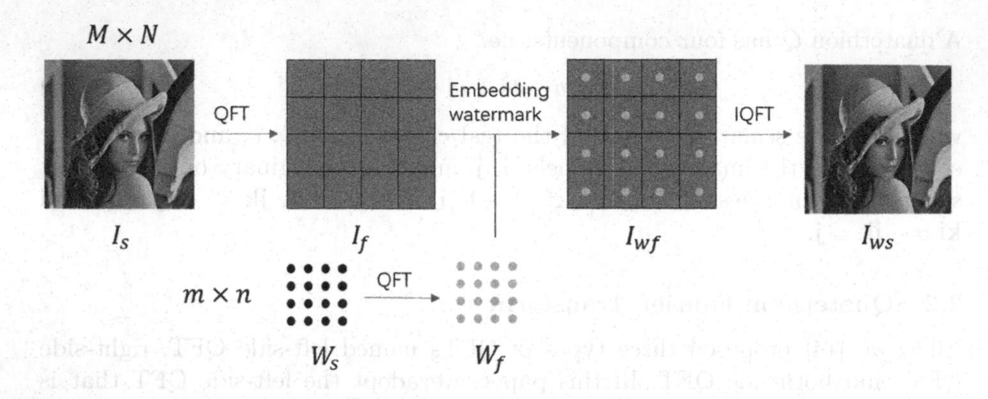

Fig. 1. The procedure of watermark embedding.

3.2 Watermark Extraction

The detailed watermark extraction procedure is described as follows. The input is a color image $I'(x, y)$ may contain the watermark message.

Step 1: We convert $I'(x, y)$ into a quaternion matrix $I'_s(x, y)$, and then divide it into non-overlapping blocks sized of $\lfloor M/m \rfloor \times \lfloor N/n \rfloor$.
Step 2: We perform QFT on every block and get $I'^{[i,j]}_f$.
Step 3: The watermark extraction is given as

$$W'_f(i, j) = (I'^{[i,j]}_f(x, y) - I'^{[i,j]}_f(x + 1, y + 1))/\alpha. \quad (4)$$

Step 4: We perform the inverse QFT on W'_f to obtain W'_s. The extracted watermark w' is then generated from the three imaginary channels of W'_s.

4 Experimental Results

In our experiments, we select 200 color images from COCO 2014 dataset [15], each of which is then resized to be $512 \times 512 \times 8$ bits. We also use some standard test images such as Lena, Baboon, Pepper, etc. We show some example images in Fig. 2. We choose a random $8 \times 8 \times 1$ matrix as the watermark message and show it in Fig. 2(i). We have implemented the proposed scheme in Matlab on a 64-bit Windows 10 PC with Intel Core i5-4210M CPU @2.60 GHz and 8 GB memory. We use three measures in our experiments, including Peak signal-to-noise ratio (PSNR), normalized correlation coefficient (CORR), and bit error ratio (BER).

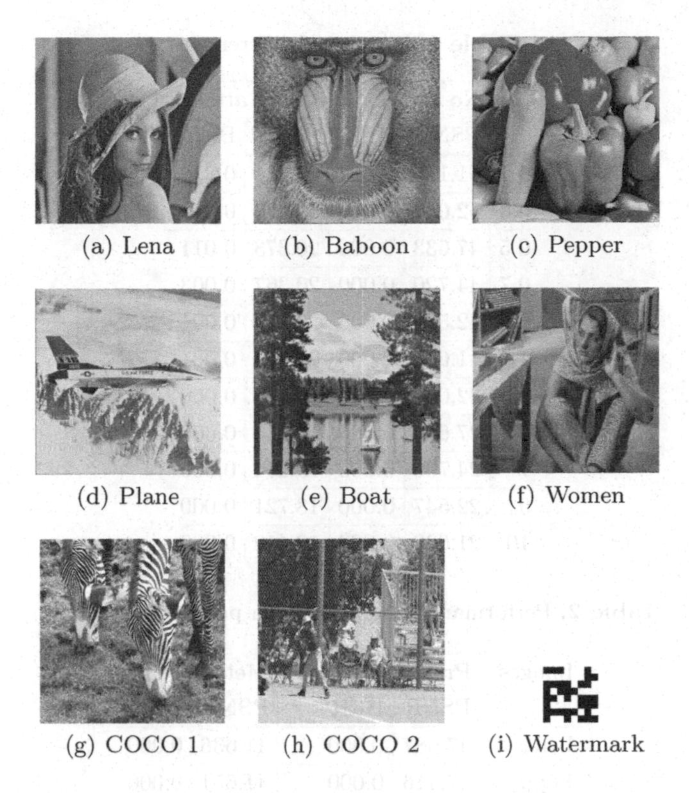

Fig. 2. Some examples of the original images. (a) Lena. (b) Baboon. (c) Peppers. (d) Plane. (e) Boat. (f) Women. (g–h) Coco 2014 dataset. (g) Watermarking

4.1 Perceptual Quality

In order to find a suitable embedding strength, we conduct experiments to test different values of α under Gaussian noise attack with the standard deviation of 0.01. The experimental results are shown in Table 1.

We also compare the perceptual quality between the proposed scheme and the method in [13]. We set the watermarking strength to be 0.5. The comparison results are shown in Table 2. We can see that our scheme can keep the watermarked images in higher perceptual quality than the compared scheme.

4.2 Robustness Under Different Watermarking Attacks

We conduct experiments to study the robustness of the proposed scheme under different types of watermarking attacks. We plot the curves of PSNR versus BER. Specifically, we consider three types of watermarking attacks as follows.

Table 1. Results of different α

α	No attacks		Under attacks	
	PSNR	BER	PSNR	BER
0.1	61.155	0.002	20.381	0.312
0.3	52.034	0.000	20.378	0.071
0.5	47.633	0.000	20.373	0.011
0.7	44.720	0.000	20.367	0.003
0.9	42.542	0.000	20.357	0.001
1	41.628	0.000	20.351	0.001
3	32.089	0.000	20.136	0.000
5	27.653	0.000	19.753	0.000
7	24.730	0.000	19.264	0.000
9	22.547	0.000	18.721	0.000
10	21.632	0.000	18.442	0.000

Table 2. Performance comparison on perceptual quality.

Images	Proposed scheme		Method [13]	
	PSNR	BER	PSNR	BER
Lena	47.687	0.000	41.636	0.000
Peppers	47.416	0.000	41.651	0.000
Baboon	47.676	0.000	41.651	0.000
Boat	47.658	0.000	41.630	0.000
Plane	47.569	0.000	41.601	0.000
Women	47.615	0.000	41.634	0.000

Noise attack. Add 1% noise into the watermarked image. Three types of noises are used, i.e., speckle noise, gaussian noise, and salt & pepper noise.

Cropping attack. Crop block of size 250 from watermarked images sized of 512. The crop positions are top left corner, the center and the bottom right corner.

JPEG 2000 compression. Perform JPEG 2000 lossy compression with three compression factors, i.e., 20, 10, and 5.

We show the noised watermarked image of two sample images under three types of noises in Fig. 3. We can see that the original watermarks can be correctly extracted from the noised watermarked images. In Fig. 4, we show the cropped watermarked images of two sample images under the three types of cropping attacks. For the cropping attacks in the center and on the bottom right corner, we can correctly extract the watermark messages. But when the cropping attack occurs on the left top corner, the extracted watermarks will have a high error rate. This is because of the QFT will gather the image energy at the left top corner. Our scheme has excellent robust performance against JPEG 2000 compression attacks. We show some example results in Fig. 5, from which we can see that all the watermarks can be correctly extracted.

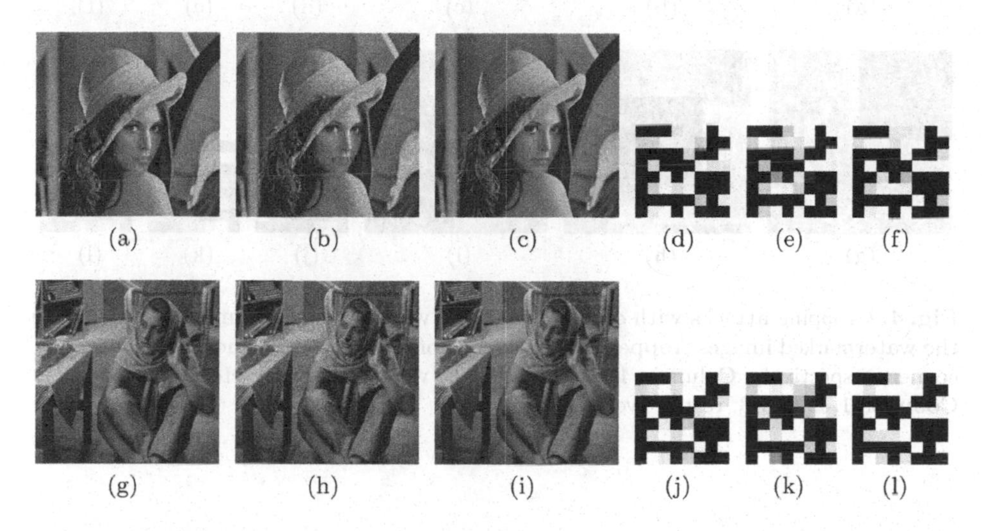

Fig. 3. Different kinds of noise attacks, where $\alpha = 1$. Columns 1, 2, and 3 are the watermarked images attacked with speckle noise, Gaussian noise, and salt & pepper noise, respectively. Columns 4, 5, and 6 are the watermarks extracted from images in Columns 1, 2, and 3 respectively.

To further investigate the robust performance of the proposed scheme, we conduct more experiments to compare our method with the schemes with other signal transformation, i.e., DCT and DWT. For all three signal transformation DCT, DWT, and QFT, we study the relationship between PSNR and BER, and the relationship between PSNR and CORR under different watermarking attacks.

We show the results on the robustness against noise attacks in Fig. 6. All the watermarked images are attached by the same noises, i.e., 1% speckled noise, 1% strength Gaussian noise, and 1% salt & pepper noise. From Fig. 6, we can see that the PSNR-BER curves of the our scheme are all below those of DCT and WHT. Thus, the proposed method with QFT has better robustness resisting to different noising attacks than those of DCT and WHT.

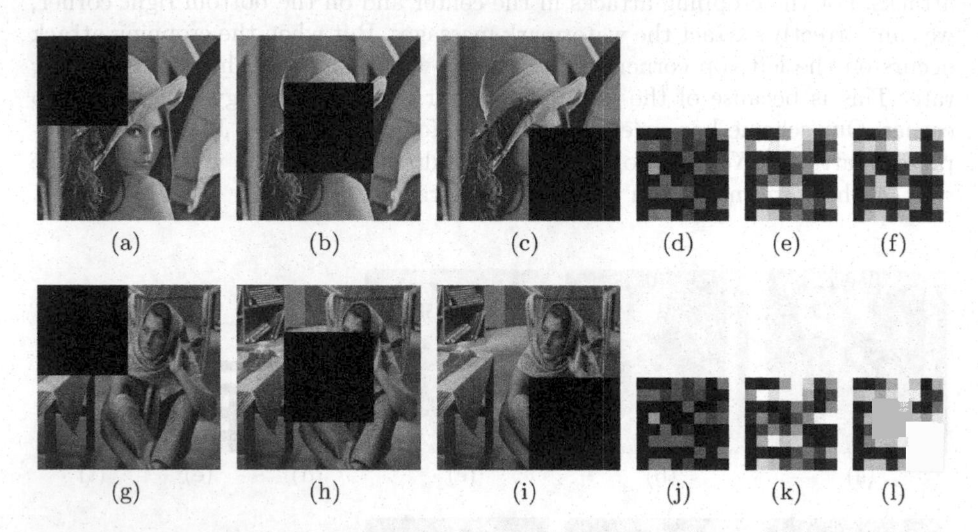

Fig. 4. Cropping attacks with different position, where $\alpha = 1$. Columns 1, 2, and 3 are the watermarked images cropped at the top left corner, the center, and the bottom right corner, respectively. Columns 4, 5, and 6 are the watermarks extracted from images in Columns 1, 2, and 3, respectively.

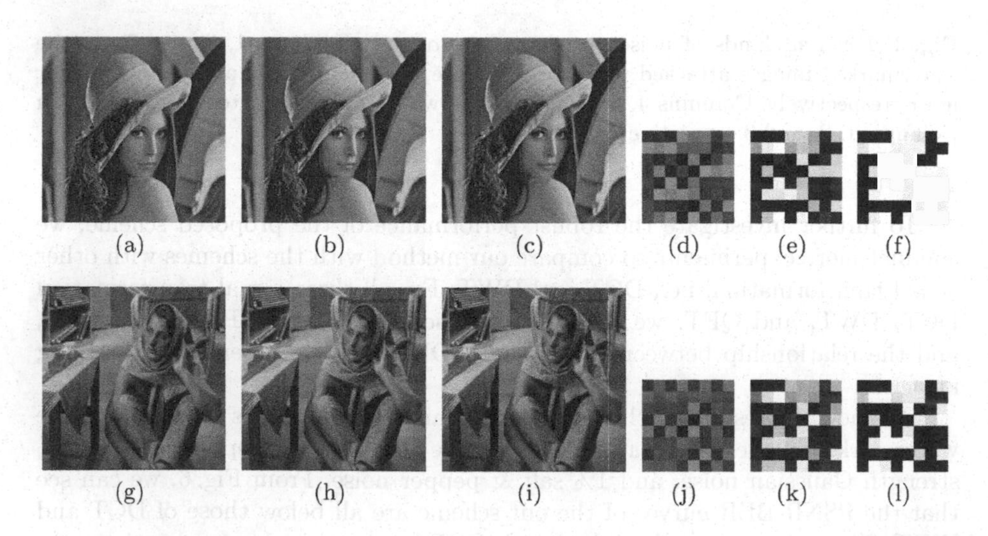

Fig. 5. JPEG-2000 compression attacks with different compression factors. Columns 1, 2, and 3 are the watermarked images attacked with JPEG 2000 compression using the compression factor of 20, 10, and 5, respectively. Columns 4, 5, and 6 are the watermarks extracted from images in Columns 1, 2, and 3, respectively.

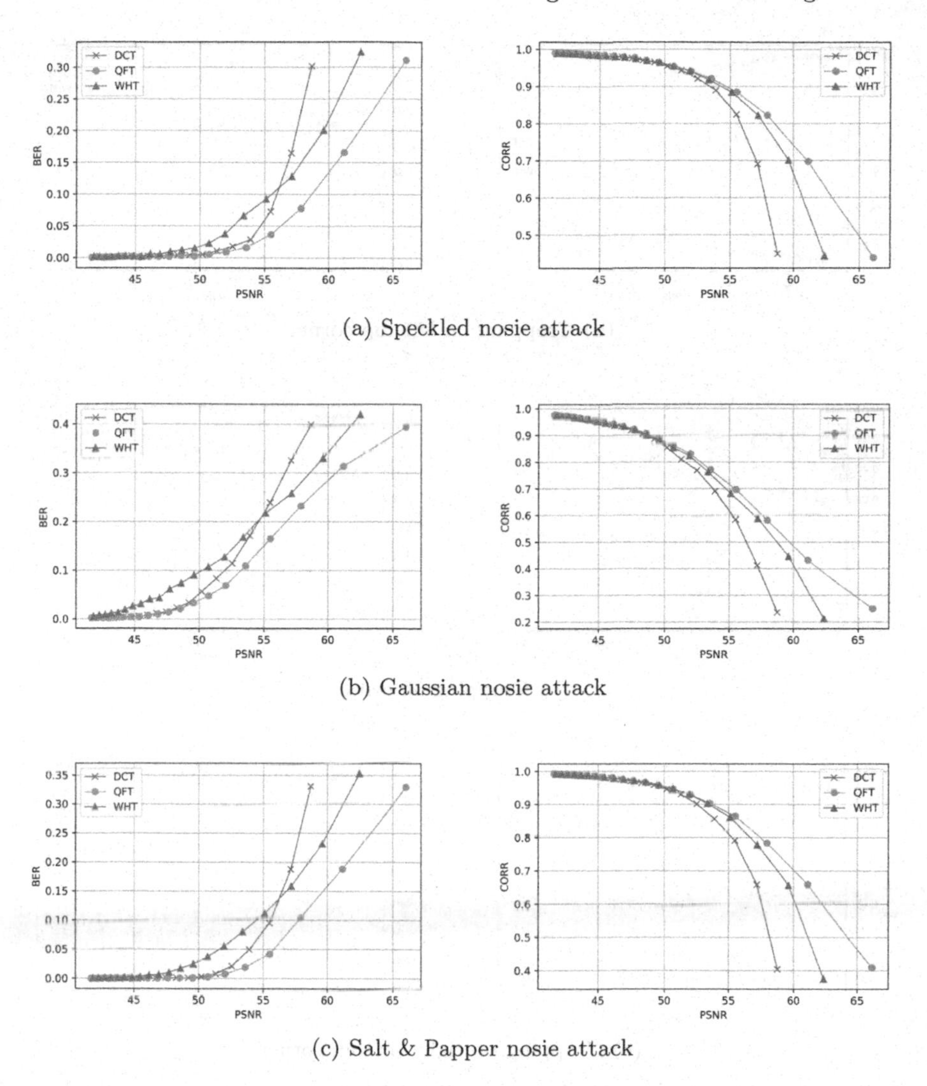

Fig. 6. The results of noise attacks.

The results on the robustness against cropping attacks are shown in Fig. 7. We can see that for When the cropping occurs at the center or the right bottom corner, the PSNR-BER curves of our scheme are all under the curves of DCT and WHT, which indicates that our scheme has a higher robust performance than the compared schemes in this case.

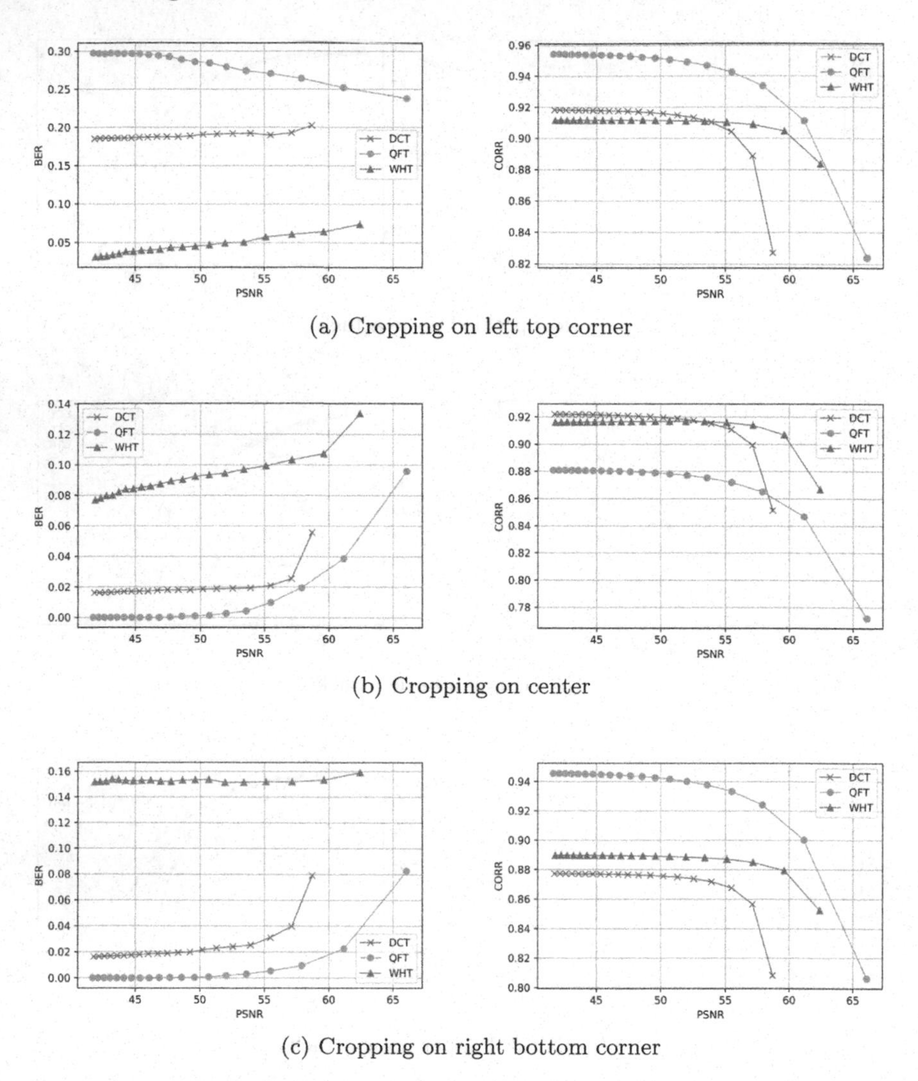

(a) Cropping on left top corner

(b) Cropping on center

(c) Cropping on right bottom corner

Fig. 7. The results of cropping attacks.

In Fig. 8, we show the results on the robustness against JPEG 2000 lossy compression. Our scheme also outperform the scheme with WHT, where the extracted watermarks are badly destroyed under the JPEG 2000 compression attacks. According to our experimental results on the robustness of noise attack, cropping attack, and JPEG 2000 compression attack, we can see that the proposed watermarking scheme with QFT has satisfactory robust performance, and is more robust resisting to most of the test watermarking attacks than the schemes with DCT and WHT.

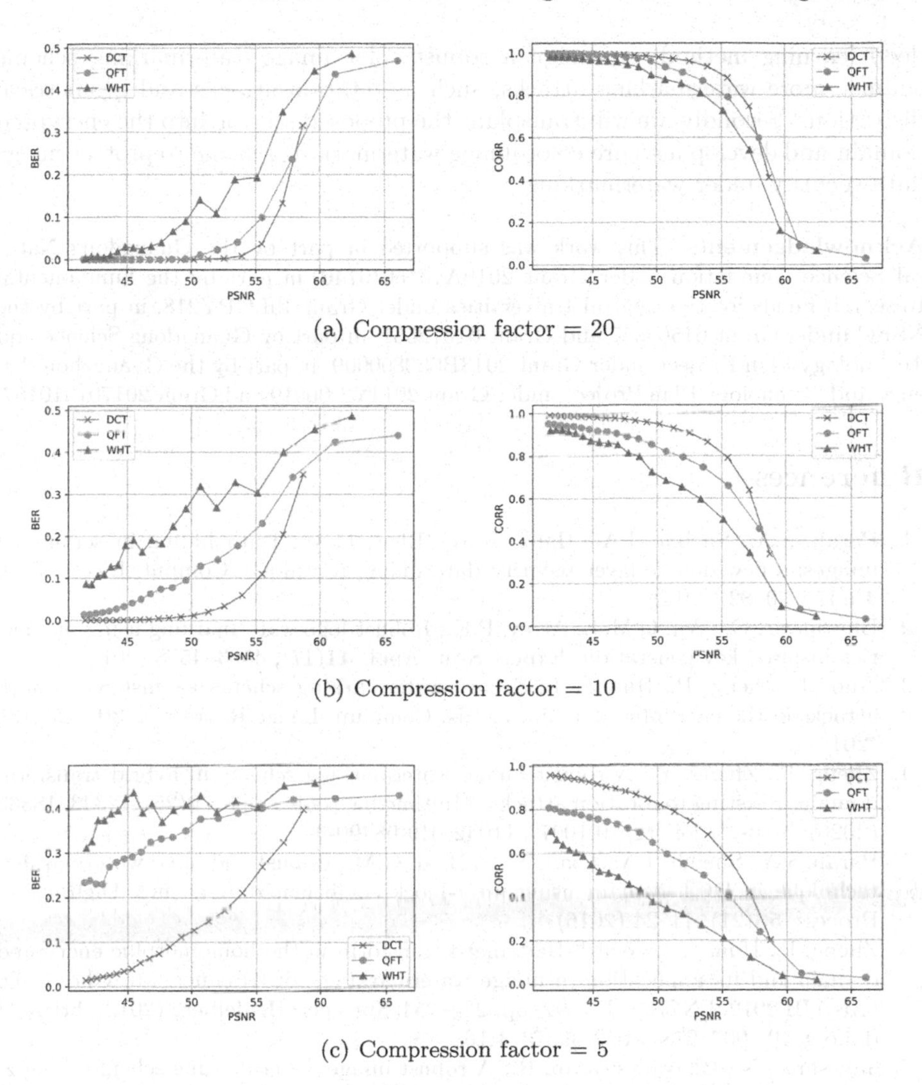

Fig. 8. The results of JPEG 2000 compression attacks.

5 Conclusion

We have proposed a blind robust color image watermarking scheme based on QFT. We spread the watermark message into all the four channels of the QFT to improve the perceptual quality of the watermarked images. Our experimental results demonstrate that the proposed scheme has a good robust performance against noise attack, cropping attack, and JPEG 2000 compression attack. The proposed color watermarking scheme can be applied in nature image copyright protection, source tracking, etc. In the future, our research will focus on the following two aspects. Firstly, we will combine QFT with DWT, SVD and even deep learning method, to design a robust color image watermarking scheme against more watermarking attacks, such as rotation attacks and geometrical distortions. Secondly, we will transplant the proposed scheme into the encrypted domain and develop a secure color image watermarking scheme to protect image data security under watermarking.

Acknowledgements. This work was supported in part by the Guangdong Natural Science Foundation under Grant 2019A1515010746, in part by the Fundamental Research Funds for the Central Universities under Grant 19LGPY218, in part by the NSFC under Grant 61502547 and Grant 61672551, in part by Guangdong Science and Technology Plan Project under Grant 2013B090800009. in part by the Guangzhou Science and Technology Plan Project under Grant 2014Y2-00019 and Grant 201707010167.

References

1. Parah, S.A., Sheikh, J.A., Hafiz, A.M., Bhat, G.M.: Data hiding in scrambled images: a new double layer security data hiding technique. Comput. Electr. Eng. **40**(15), 70–82 (2014)
2. Bhatnagar, G., Wu, Q.M.J., Atrey, P.K.: Robust logo watermarking using biometrics inspired key generation. Expert Syst. Appl. **41**(17), 4563–4578 (2014)
3. Guo, J., Zheng, P., Huang, J.: Secure watermarking scheme against watermark attacks in the encrypted domain. J. Vis. Commun. Image Represent. **30**, 125–135 (2015)
4. Zheng, P., Zhang, Y.: A robust image watermarking scheme in hybrid transform domains resisting to rotation attacks. Multimedia Tools Appl. **79**(25), 18343–18365 (2020). https://doi.org/10.1007/s11042-019-08490-4
5. Parah, S.A., Sheikh, J.A., Loan, N.A., Bhat, G.M.: Robust and blind watermarking technique in DCT domain using inter-block coefficient differencing. Digital Sig. Process. **53**(21), 11–24 (2016)
6. Zheng, P., Huang, J.: Walsh-Hadamard transform in the homomorphic encrypted domain and its application in image watermarking. In: Kirchner, M., Ghosal, D. (eds.) IH 2012. LNCS, vol. 7692, pp. 240–254. Springer, Heidelberg (2013). https://doi.org/10.1007/978-3-642-36373-3_16
7. Jayashree, N., Bhuvaneswaran, R.: A robust image watermarking scheme using z-transform, discrete wavelet transform and bidiagonal singular value decomposition. Comput. Mater. Continua **58**(1), 263–285 (2019)
8. Liu, J., et al.: A robust zero-watermarking based on SIFT-DCT for medical images in the encrypted domain (2019)
9. Liu, J., et al.: A novel robust watermarking algorithm for encrypted medical image based on DTCWT-DCT and chaotic map (2019)
10. Grigoryan, A.M., Jenkinson, J., Agaian, S.S.: Quaternion Fourier transform based alpha-rooting method for color image measurement and enhancement. Sig. Process. **109**, 269–289 (2015)
11. Wang, X., Wang, C., Yang, H., Niu, P.: A robust blind color image watermarking in quaternion Fourier transform domain. J. Syst. Softw. **86**(2), 255–277 (2013)
12. Chen, B., Coatrieux, G., Gang, C., Sun, X., Coatrieux, J.L., Shu, H.: Full 4-D quaternion discrete Fourier transform based watermarking for color images. Digit. Sig. Process. **28**(25), 106–119 (2014)

13. Bas, P., Bihan, N.L., Chassery, J.: Color image watermarking using quaternion Fourier transform, vol. 3, pp. 521–524 (2003)
14. Ell, T.A., Sangwine, S.J.: Decomposition of 2D hypercomplex Fourier transforms into pairs of complex Fourier transforms. In: 2000 European Signal Processing Conference, pp. 1–4 (2000)
15. Lin, T.-Y., et al.: Microsoft COCO: common objects in context. In: Fleet, D., Pajdla, T., Schiele, B., Tuytelaars, T. (eds.) ECCV 2014. LNCS, vol. 8693, pp. 740–755. Springer, Cham (2014). https://doi.org/10.1007/978-3-319-10602-1_48

A Decentralized Multi-authority ABE Scheme in Cooperative Medical Care System

Jinrun Guo, Xiehua Li(✉), and Jie Jiang

College of Computer Science and Electronic Engineering, Hunan University, Changsha 410082, China

beverly@hnu.edu.cn

Abstract. In this paper, we propose an attribute-based encryption (ABE) scheme that can be used in cooperative medical care systems with multiple distrusted principles. Unlike prior multi-authority ABEs, this scheme distributes secret key generation to different medical care centers, which enables distributed authorization and access control. By separating the key generation process among medical care centers and data owners (DOs), our scheme is resilient to collusion between malicious their parties and users. This new Cooperative Medical Care System (CMCS) distinguishes between DO principal and medical care centers: DOs own the data but allows medical care centers to arbitrate access by providing attribute labels to users. The data is protected by access policy encryption over these attributes. Unlike prior systems, attributes generated by medical centers are not user-specific, and neither is the system susceptible to collusion between users who try to escalate their access by sharing keys. We prove our CMCS scheme correct under the Decisional Bilinear Diffie-Hellman (DBDH) assumption; we also include a complete end-to-end implementation that demonstrates the practical efficacy of our technique.

Keywords: Multi-authority ABE · Data sharing · Access control · Cloud storage

1 Introduction

Cooperative medical care is a rapid growth application in health care market. Different medical care centers, pharmaceutical factories, even PSB (Public Safety Bureau) need to share information and massive health care data. For example, for fighting with disease like Ebola, hospitals and PSB need to collaborate. In this scenario, the data being shared is developed not only by the hospitals, but can originate from the PSB. The data originated from these parties may maintain sensitive data about patient health record and should not be accessible by unauthorized user or authority. These types of scenarios are increasingly common, and require the ability to provide secure data access with fine-grained control. It is possible to provision such access using a trusted third party (TTP), such as a central authority (CA) that generates secret keys for specific data; however, TTP-based solutions are particularly problematic in multi-authority scenarios since they require distrusted principals to trust a single party.

X. Sun et al. (Eds.): ICAIS 2020, LNCS 12240, pp. 642–652, 2020.
https://doi.org/10.1007/978-3-030-57881-7_56

In this paper, we introduce a decentralized fine-grained data sharing scheme to achieve access control among users of distrusted medical centers and different DOs, without resorting to TTPs. The main contributions of this paper are as follows:

1) We propose a decentralized scheme that can achieve secure health record sharing among distrusted authorities, users and DOs. Our scheme is resilient to authorities and users collusion attack.
2) We propose a key generation algorithm to distribute secret key generation among all authorized DOs and authorities.
3) We implement our scheme based on cpabe library and measure its efficiency.

The remaining of this paper is organized as follows: We introduces a discussion of related work in Sect. 2. Then, we propose the new scheme and security model in Sect. 3. Next, we analyze the performance of our scheme. We finally conclude this paper in Sect. 5.

2 Related Work

ABE is a promising technique that enables fine-grained access control to encrypted data [1–4], and is widely used in various applications [5, 7]. In many multi-authority ABE-based schemes [6, 8, 17, 18], a trusted central authority (CA) is involved for key management. Most of these schemes use CA as the TTP to generate public key and user secret keys. However, one drawback of the CA-based ABE scheme is that once the CA gets compromised there would be no privacy of the stored data.

Chase firstly proposed a multi-authority ABE method [6] that introduced a global identifier (GID) for CA to generate user secret keys. Lewko and Waters *et al.* propose a multi-authority CP-ABE [9, 10] that uses CA only in the initialization phase. CA distributes the public parameters and verifies AAs according to the user's request. Yang *et al.* propose a DAC-MACS algorithm [11], CA is responsible for the generation of the global public key and private key and distribute a unique identity for the user and all the AAs. Later, they propose two multi-authority ABE access control schemes [12, 13] in which CA is responsible for authenticating all the AAs and users. In addition, CA also assigns GID for users and AID to each AA. In order to eliminate the security risk introduced by CA and protect user's identity, Taeho J *et al.* proposed anonymous privilege control schemes [14, 15]. This scheme generalizes an access tree to a privilege tree. Several trees are required in every encrypted file to verify user's identity and to grant him a privilege accordingly.

3 Preliminaries and Definitions

We will first give the cryptographic background information of bilinear map and security model. Then, we will describe the access structure of our DMA scheme and give the security model.

3.1 Preliminaries

Let G_0 and G_1 be two multiplicative cyclic groups for prime order p and g be a generator of G_0. The bilinear map e is defined as, e: $G_0 \times G_0 \rightarrow G_1$. The bilinear map e has the following properties: 1) Bilinearity: $\forall u,v \in G_0$, and a, $b \in \mathbb{Z}_P$, then $e(u^a, v^b) = e(u, v)^{ab}$. 2) Non-degeneracy: $e(g, g) \neq 1$. 3) Symmetry: $e(g^a, g^b) = e(g, g)^{ab} = e(g^b, g^a)$.

Definition: The Decisional Bilinear Diffie-Hellman(DBDH) assumption in a multiplicative cyclic group G_0 of prime order p with generator g is stated as: given g^a, g^b and g^c for uniformly and independently chosen $a, b, c \in \mathbb{Z}_P$, the following two distributions are computationally indistinguishable:

- G_0, g, g^a, g^b, g^{ab}
- G_0, g, g^a, g^b, g^c

The security of our scheme is based on the DBDH assumption which is widely used in security proof of various ABE schemes. The assumption is reasonable because discrete logarithm problems in large number field are widely considered to be intractable.

3.2 Access Tree Definition in CMCS

Let Γ be a tree representing an access structure. Every non-leaf node of the tree represents a threshold gate that is described by its children and a threshold value. If num_x is the number of children of a node x and k_x is its threshold value, then $0 < k_x \leq num_x$. If $k_x = 1$, the threshold gate is an 'OR' gate. If $k_x = num_x$, it is an 'AND' gate. Every leaf node x of the tree is described by an attribute and a threshold value $k_x = 1$. Attributes contained in the access tree can be issued by different authorities. The access tree in our scheme is defined as follows:

1) Access tree is defined by DO. Leaf nodes represent attributes that are distributed by AAs.
2) Each leaf node of the access tree is described as an attribute which is issued by its authority.
3) If a user's attribute set S satisfies the access tree Γ, he/she can decrypt the data that are encrypted with this access tree.

4 System Architecture and Key Distribution

4.1 System Architecture

There are 4 entities involved in our CMCS system: DO (Data Owner), AA (Attributes Authority), Cloud storage provider (CSP), and user. The system architecture of CMCS is shown in Fig. 1. The entities are described as follows:

1. DO (Data Owner). Data owner is responsible for calculating public key, defining access policy and encrypting data under the access policy. Furthermore, the data

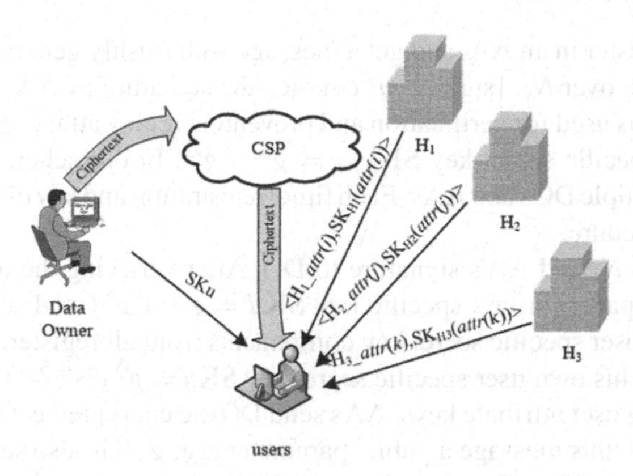

Fig. 1. Access control in HR system

owner needs to upload the encrypted data to the remote cloud storage server. DO keeps the attribute update list (AUL) and user list (UL) to identify the authorized user. For security purpose, DO computes part of users' private keys called user specific key **SK**d and sends it directly to the particular users via secure channels. The reason why DO generates and distributes **SK**d is for preventing authorities collusion attack.

2. AA (Attribute Authority). AA can be regarded as different medical centers or hospitals. It plays the role of attributes distribution and user authorization. It computes users' attributes based on the public parameters and distributes them to DO and users for access policy definition. Every AA can manage multiple attributes and has full control over those attributes. Moreover, AA computes attribute secret keys $\mathbf{SK}_{\mathrm{AA}i}(attr(i))$ and issues them to users via secure channel. AA_i denotes the ith AA in our scheme, $attr(i)$ denotes an attribute issued by a an AA.
3. Cloud Server Provider (CSP). CSP is considered as a semi-trusted storage media that stores data. It is also responsible for updating the ciphertext when attribute revocation occurs. The CSP does not have the secret keys, so it can't decrypt the ciphertext. Based on the semi-trusted assumption, CSP can implement the algorithm honestly, but it will decrypt the ciphertext once it gets the key.
4. User. Ciphertext on the cloud server can be accessed freely by users. But only when the user's attributes satisfy the access policy that defined in the ciphertext, can he/she decrypt the ciphertext. User's attributes are distributed by a number of authorities according to the user privileges so that it can achieve cross-domain access control. In addition, DO prevents the collusion attacks between users by embedding a random number in the private property.

4.2 Key Distribution Algorithm

In our scheme, we refactor the original CP-ABE and extend the key generation algorithm to a multiple authority scenario. In this system, user needs to first register in a set of AAs and get his secret key components from these AAs and DO. The registration and key distribution procedure involves two stages: 1) user registration and user specific secret key distribution stage. 2) attribute key distribution stage.

1. Users first register in an AA, and get a message with freshly generated nonce N_i and AA's signature over N_i. [$\text{sig}_{\text{AA}i}$; m] denotes the signature of AA_i over message m. This message is used for verification and preventing replay attack. AA also generates partial user specific secret key $\mathbf{SK}_{\text{AAi}} = g^{\alpha_i} * g^{\lambda_i}$. In our scheme, user is able to register to multiple DOs and AAs. Each time, registration and key distribution follows the same procedure.
2. User forwards N_i and AA's signature to DO. After verifying the user and AA, DO will generate part of user's specific key $\mathbf{SK}d = g^{\alpha_d} * g^{\lambda_d}$ and send it to the user. User collects user specific secret key components from all registered AAs and DOs, and computes his own user specific secret key $\mathbf{SK}u = g^{\sum \alpha_d + \sum \alpha_i} * g^{\sum \lambda_d + \sum \lambda_i}$.
3. For computing user attribute keys, AAs send DO the encrypted g^{λ_i} with DO's public key, along with this message a public parameter $e(g, g)^{\alpha_i}$ is also sent to DO for data encryption. Each DO collects g^{λ_i} from all AAs and computes $g^{\sum \lambda_i}$ for attribute key calculation. Then, DO sends user a blinded parameter $\mathbf{P}u$ as a partial attribute key.

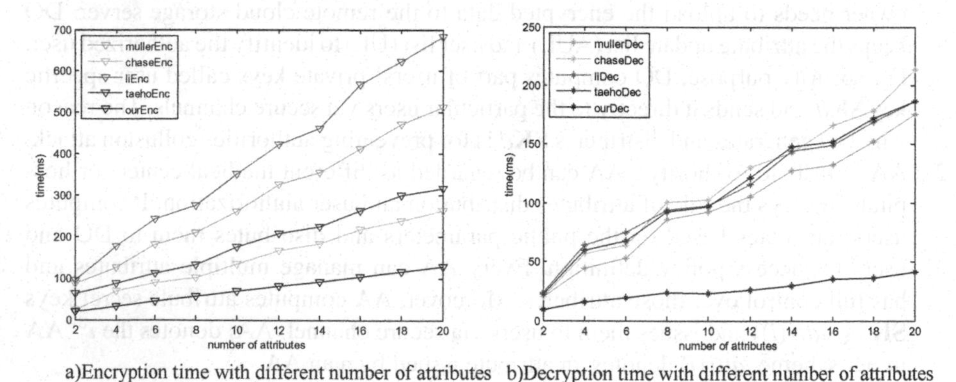

a)Encryption time with different number of attributes b)Decryption time with different number of attributes

Fig. 2 Encryption and decryption time with different number of attributes

$$\mathbf{P}u = A * g^{\sum \lambda_i} \tag{1}$$

Where A is the blinder, such as g^x, $x \in \mathbb{Z}_p$ is a random number chosen by DO when generating user specific secret key. Here we can see that each user has a set of attribute keys that are in accordance with his/her user specific secret key. The reason to blind the partial attribute key $g^{\sum \lambda_i}$ is because whoever gets $g^{\sum \lambda_i}$ can generates whatever attribute keys he/she wants. We will describe this in details in the next section. $\mathbf{P}u$ is encrypted with AA's public key, [$\text{enc}_{\text{K\_AA}i}$; m] denotes encryption of message m over AAi's public key $\text{K}_{\text{AA}i}$.

4. User keeps his user specific secret key and forwards the encrypted $\mathbf{P}u$ to AA. AA decrypts $\mathbf{P}u$. According to user's attributes, AA calculates blinded attribute keys and sends them along with user's attributes and AA's signature back to the user.
5. User forwards this message to DO. DO verifies AA, user's attributes and give back unblinded attribute keys back to the user. In addition, DO raises the unblinded

attribute key from $g^{\sum \lambda_i} H(attr(i))^{r_i}$ to $g^{\sum \lambda_i} H(attr(i))^{r_i + r_s}$, so that user is not able to guess $g^{\sum \lambda_i}$ from **P**u. Since r^i is a random number, the change of r^i won't affect decryption process.

4.3 Access Control Scheme Construction

A. **Setup(PK, MK).** The setup procedure is used for generating public parameters. Any DO can run this procedure and broadcast **PK**. Other DOs and AAs will use this **PK** to generate keys in the future. Let g be the generator of a bilinear group of G, and the prime order of G is p, $e : G \times G \rightarrow G_T$ as the bilinear map. Let $H : \{0, 1\}^* \rightarrow G$ be the hash function which maps attributes to G. DO selects two random exponents $\alpha_d, \eta \in \mathbb{Z}_p$, and computes public key and master key as follows:

$$\mathbf{PK} = \left(g, G, g^{\eta}, g^{1/\eta}, e(g, g)^{\alpha_d}\right) \tag{2}$$

$$\mathbf{MSK} = \left(g^{\alpha_d}, \eta\right) \tag{3}$$

DO publishes **PK** all other principals in this system.

B. **KeyGen I.** In the f-DMA scheme, the KeyGen is divided into two phrases, KeyGen I and KeyGen II.
KeyGen I(MK) is run by all principals in this system. After registration, DO generates and issues $\mathbf{SK}d = g^{\alpha_d} * g^{\lambda_d}$ to user. AA generates and issues $\mathbf{SK}_{\mathrm{AAi}} = g^{\alpha_i} * g^{\lambda_i}$ to user. User collects and combines these key components together to get his user specific secret key **SK**u.

$$\alpha = \sum \alpha_d + \sum \alpha_i, \; \lambda = \sum \lambda_d + \sum \lambda_i \tag{4}$$

$$\mathbf{SK}u = g^{(\alpha+\lambda)/\eta} \tag{5}$$

Where α_i, λ_d, $\lambda_i \in \mathbb{Z}_p$ are random numbers. λ_d, λ_i are unique for every user to prevent user collusion attack.
KeyGen II ((**PK**, **attr**$(i)_{i \in \{1 \ldots n\}}$)) is run by multiple AAs to calculate attribute keys for users. AAs use the blinded parameter to calculate blinded attribute keys for user.

$$\mathbf{BSK}_i = \left(\forall attr(i) \in S_{\mathrm{AA}i}, V_i = A * g^{\lambda} * H(attr(i))^{r_i}, V_i^{'} = g^{r_i}\right) \tag{6}$$

Where $S_{\mathrm{AA}j}$ denotes the set of attributes that AA$_j$ holds. $r_i \in \mathbb{Z}_p$ is a random number for each $attr(i) \in S_{\mathrm{AA}i}$ chosen by AA$_i$. From Eq. (6) we can see that if g^{λ} is given to AA directly, any AA can forge arbitrary attribute keys since $attr(i)$ is a binary string and r_i is a random number. As shown in Fig. 3, DO will send user the final attribute keys by unblinding **BSK**$_i$. The final attribute keys is described as follows.

$$\mathbf{SK}_i = \left(\forall attr(i) \in S_{\mathrm{AA}i}, V_i = g^{\lambda} * H(attr(i))^{r_i}, V_i^{'} = g^{r_i}\right) \tag{7}$$

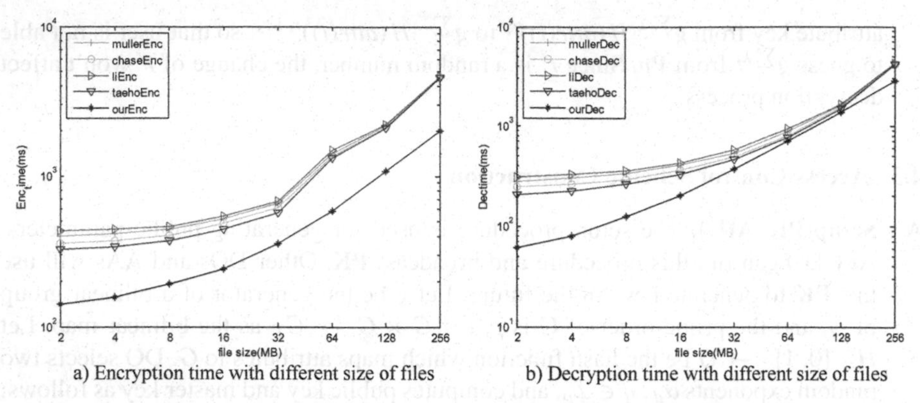

a) Encryption time with different size of files b) Decryption time with different size of files

Fig. 3. Encryption and decryption time with different size of files

C. **Encryption(PK, M, Γ)**. DO encrypts the message M under public key and the access policy Γ. The encryption algorithm chooses a polynomial q_x for each node x in the access tree. For each node x in the tree, set the degree d_x of the polynomial q_x, $d_x = k_x - 1$, where k_x is the threshold value of the chosen node. The algorithm starts at the root node R, it chooses a random exponent $s \in Z_p$ and sets $q_R(0) = s$. Then, for any other node x, the algorithm randomly chooses other coefficients and set $q_{parent(x)}(index(x))$ such that $q_x(0) = q_{parent(x)}(index(x))$, is the index of node x's child nodes, and $parent(x)$ is node x's parent node.
Suppose Y is the set of leaf nodes in Γ, the ciphertext is constructed as follows:

$$CT = \begin{pmatrix} \Gamma, C' = M \cdot e(g, g)^{\alpha s}, C = g^{\eta s} \\ \forall y \in Y, C_y = H(att(y))^{q_y(0)}, C'_y = g^{q_y(0)} \end{pmatrix} \tag{8}$$

D. **Decrypt** (CT, SK). Only when the user's attributes satisfy access policy defined in ciphertext, the user can decrypt the ciphertext. The decryption algorithm takes as inputs secret keys and ciphertext, where $\mathbf{SK} = (\mathbf{SK}_u||\mathbf{SK}_1||\mathbf{SK}_2||\ldots||\mathbf{SK}_n)$. "||" is concatenation operation. We specify our decryption procedure as recursive algorithm. Let $i = attr(y)$, which $attr(y)$ represents the value of leaf node y, if the node x is a leaf node and $x \in S$, then computes

$$\begin{aligned} Decrypt(CT, SK, x) &= \frac{e\left(V_i, C'_x\right)}{e(V'_i, C_x)} = \frac{e(g^{\lambda} \cdot H(i)^{r_i}, g^{q_x(0)})}{e(g^{r_i}, H(i)^{q_x(0)})} \\ &= \frac{e\left(g^{\lambda}, g^{q_x(0)}\right)e\left(H(i)^{r_i}, g^{q_x(0)}\right)}{e\left(g^{r_i}, H(i)^{q_x(0)}\right)} \\ &= e(g, g)^{\lambda q_x(0)} \end{aligned} \tag{9}$$

If $i \notin S$, then we define $Decrypt(CT, SK, x) = \perp$.
When the node x is not a leaf node, for all the nodes z that are the children of x, the outputs is stored as F_z. Let S_x be an arbitrary k_x-sized set of child nodes. If no

such set exists, the function will return $\perp$. The recursive computation is shown as follows:

$$\begin{aligned}F_x &= \prod_{z \in S_x} F_z^{\Delta_{i,S_x'}(0)}, \text{ where} \begin{cases} i = index(z) \\ S_x' = (index(z) : z \in S_x) \end{cases} \\ &= \prod_{z \in S_x} \left(e(g,g)^{\lambda q_z(0)}\right)^{\Delta_{i,S_x'}(0)} \\ &= \prod_{z \in S_x} \left(e(g,g)^{\lambda q_{parent(z)}(index(0))}\right)^{\Delta_{i,S_x'}(0)} \\ &= \prod_{z \in S_x} \left(e(g,g)^{\lambda q_x(i)}\right)^{\Delta_{i,S_x'}(0)} = e(g,g)^{\lambda q_x(0)} \end{aligned} \tag{10}$$

The algorithm recalls the Lagrange polynomial interpolation to decrypt the ciphertext. If the set of attributes satisfy the access tree, we define $A = e(g,g)^{\lambda q_R(0)} = e(g,g)^{\lambda s}$.Then computes:

$$CT/(e(\mathbf{SK}u, C)/A) = CT/(e(g^{(\alpha+\lambda)/\eta}, g^{\eta s})/e(g,g)^{\lambda s}) = M \tag{11}$$

5 Security and Performance Analysis

5.1 Security Analysis

The Decisional Bilinear Diffie-Hellman (DBDH) problem in group G of prime order p with generator g is defined as follows:

on input $g, g^a, g^b, g^c \in G$ and $e(g,g)^{abc} = e(g,g)^z \subset G_T$, where $z = abc$, decide whether $z = abc$ or z is a random element.

Theorem 1 : If Decisional Bilinear Diffie-Hellman assumption holds in group (G, G_T), then our scheme is chosen-plaintext secure in standard model.

Proof: Suppose there exists a probabilistic polynomial time adversary *Adv* can attack our scheme in the security model above with advantage ε. We prove that the following DBDH game can be solved with advantage $\varepsilon/2$. Let $e : G \times G \to G_T$ be a bilinear map, where G is a multiplicative cyclic group of prime order p and generator g. First the DBDH challenger flips a binary coin $\mu = \{0, 1\}$, if $\mu = 1$, he sets $(g, A, B, C, Z) = (g, g^a, g^b, g^c, e(g,g)^{abc})$; otherwise he sets $(g, A, B, C, Z) = (g, g^a, g^b, g^c, e(g,g)^z)$, where $a, b, c \in Z_p$ are randomly selected. The challenger then gives the simulator $(g, A, B, C, Z) = (g, g^a, g^b, g^c, Z)$.The simulator *Sim* then plays the role of a challenger in the following DBDH game.

Init. The adversary *Adv* creates an access tree Γ^* which he wants to be challenged (Nodes inside the tree should be defined by *Adv*).

Setup. *Sim* sets the parameter $Y := e(A, B) = e(g,g)^{ab}$. For all $i \in S$, it will choose a random $d_i \in \mathbb{Z}_p$ and set $H(attr(i)) = g^{d_i}$. Otherwise it chooses a random number

$\beta_i \in \mathbb{Z}_p$ and sets $H(attr(i)) = g^{b\beta_i} = B^{\beta_i}$.Then it will give this public parameters to *Adv*.

Phase 1. *Adv* queries for as many private keys, which correspond to attribute sets $S_1, S_2, \ldots S_q$, where none of them satisfy the Γ^*. After receiving the key queries, *Sim* computes the private key components to respond *Adv*'s requests. *Sim* first defines a polynomial q_x for each node x of Γ^*. For each node x of Γ^*, we know q_x completely if x can be satisfied; if x is not satisfied, then at least $g^{q_x(0)}$ is known. *Sim* sets $q_R(0) = a$, for each node x of access tree, *Sim* defines the final polynomial $Q_x(\cdot) = bq_x(\cdot)$ and let $s = Q_R(0) = ab$. For all $i \in S_k$, he randomly picks $i \in S_k$, and compute $D = g^{(c+\lambda)/\eta}$, $V_i = g^{d_i r_i}$, $V_i^{'} = g^{r_i}$, otherwise $V_i = g \cdot g^{b\beta_i r_i} = g \cdot B^{\beta_i r_i}$. Then, *sim* returns the created private key to *Adv*.

Challenge. The adversary *Adv* submits two equal length challenge messages M_0 and M_1 to the challenger. The challenger flips a binary coin γ, and returns the following ciphertext to *Adv*.

$$CT = \begin{pmatrix} \Gamma^*, C^{'} = M_\gamma \cdot Z, C = g^{\eta s} \\ \forall y \in Y, C_y = B^{d_y q_y(0)}, C_y^{'} = B^{q_y(0)} \end{pmatrix} \tag{12}$$

If $\mu = 1$, $Z = e(g, g)^{abc}$. Let $\alpha = ab$, $s = c$, Then $Z = e(g, g)^{abc} = e(g, g)^{\alpha s}$. Therefore, *CT* is a valid ciphertext of the message m_γ. Otherwise, if $\mu = 0, Z = e(g, g)^z$, $C^{'} = M_b e(g, g)^z$. Since $z \in Z_p$ is a random element, $C^{'} \in G_T$ is a random element, therefore *CT* contains no information about m_γ.

5.2 Performance Evaluation

We present the performance evaluation based on our DMA implementation prototype. Our experiment is implemented on a Linux Ubuntu with Inter (R) Core (TM) i5-6500 @ 3.2 GHz and 2 GB RAM. The code is modified on original CP-ABE library. The code uses the Pairing-Based Cryptography (PBC) library version 0.5.12 to implement the access control scheme. We will compare the computation efficiency of both encryption and decryption in two criteria: the number of authorities and the number of attributes per authority. In order to show the efficiency of our scheme, we implement other similar schemes that are propose in [6, 16–18] for comparison purpose.

Figure 2 shows the comparison of encryption and decryption time with different number of attributes. The encryption and decryption time are measured under file size 100 KB, attributes number varies from 2 to 20.

Figure 3 shows the comparison of encryption and decryption time with different file size. The number of attributes is set to be 20, the file size varies from 2 MB to 256 MB.

From the experiment results we find out that our scheme makes much improvement in key generation, encryption and decryption under different system set. The reason why our scheme is very efficient is that we refactor the original CP-ABE and extend it into multi-authority scenario.

6 Conclusion

Our scheme presents a fully decentralized Multi-Authority ABE that allows multiple distrusted authorities achieve secure data sharing. We introduce a novel key generation and distribution protocol to prevent key forging and authority collusion attack. Theoretical analysis and experiment demonstrate that our scheme can not only achieve ciphertext access control, preventing collision attacks between users, DOs and authorities; but also improve the efficiency of ciphertext encryption and decryption. Therefore, the proposed method can be applied to multi-authority scenario in data sharing systems for efficient data encryption and decryption.

Acknowledgements. This work is supported by the National Natural Science Foundation of China under grant 61402160 and 61872134. Hunan Provincial Natural Science Foundation under grant 2016JJ3043. Open Funding for Universities in Hunan Province under grant 14K023.

References

1. Shamir, A.: Identity-based cryptosystems and signature schemes. In: Blakley, G.R., Chaum, D. (eds.) CRYPTO 1984. LNCS, vol. 196, pp. 47–53. Springer, Heidelberg (1985). https://doi.org/10.1007/3-540-39568-7_5
2. Bethencourt, J., Sahai, A., Waters, B.: Ciphertext-policy attribute-based encryption. In: Proceedings of IEEE Symposium Security and Privacy. Berkeley, CA, pp. 321–334 (2007)
3. Waters, B.: Ciphertext-policy attribute-based encryption: an expressive, efficient, and provably secure realization. In: Catalano, D., Fazio, N., Gennaro, R., Nicolosi, A. (eds.) PKC 2011. LNCS, vol. 6571, pp. 53–70. Springer, Heidelberg (2011). https://doi.org/10.1007/978-3-642-19379-8_4
4. Wang, S., Zhou, J., Liu, J.K., et al.: An efficient file hierarchy attribute-based encryption scheme in cloud computing. IEEE Trans. Inf. Forensics Secur. **11**(6), 1265–1277 (2016)
5. Kwon, H., Kim, D., Hahn, C., et al.: Security authentication using ciphertext policy attribute-based encryption in mobile multi-hop networks. Multimedia Tools Appl. **75**, 1–15 (2016). https://doi.org/10.1007/s11042-015-3187-z
6. Chase, M.: Multi-authority attribute based encryption. In: Vadhan, Salil P. (ed.) TCC 2007. LNCS, vol. 4392, pp. 515–534. Springer, Heidelberg (2007). https://doi.org/10.1007/978-3-540-70936-7_28
7. Liu, Y., Peng, H., Wang, J.: Verifiable diversity ranking search over encrypted outsourced data. Computers, Materials & Continua **55**(1), 037–057 (2018)
8. Chase, M., Chow, S.S.M.: Improving privacy and security in multi-authority attribute-based encryption. In: Proceedings of the 16th ACM Conference on Computer and Communications Security (CCS 2009), pp. 121–130 (2009)
9. Lewko, A., Okamoto, T., Sahai, A., Takashima, K., Waters, B.: Fully secure functional encryption: attribute-based encryption and (Hierarchical) inner product encryption. In: Gilbert, H. (ed.) EUROCRYPT 2010. LNCS, vol. 6110, pp. 62–91. Springer, Heidelberg (2010). https://doi.org/10.1007/978-3-642-13190-5_4
10. Lewko, A., Waters, B.: Decentralizing attribute-based encryption. In: Paterson, Kenneth G. (ed.) EUROCRYPT 2011. LNCS, vol. 6632, pp. 568–588. Springer, Heidelberg (2011). https://doi.org/10.1007/978-3-642-20465-4_31
11. Yang, K., Jia, X., Ren, K.: DAC-MACS: effective date access control for multi-authority cloud storage systems. IEEE Trans. Inf. Forensics Secur. **8**(11), 1790–1801 (2013)
12. Yang, K., Jia, X.: Attribute-based access control for multi-authority system in cloud storage. In: Proceedings of International Conference on Distributed Computing Systems (ICDCS), pp. 536–545 (2012)

13. Yang, K., Jia, X.: Expressive, efficient and revocable data access control for multi-authority cloud storage. IEEE Trans. Parallel Distrib. Syst. **25**(7), 1735–1744 (2013)
14. Taeho, J., Li, X., Wan, Z., et al.: Privacy preserving cloud data access with multi-authorities. In: Proceedings of IEEE INFOCOM, pp. 2625–2633 (2013)
15. Jung, T., Li, X., Wan, Z., et al.: Control cloud data access privilege and anonymity with fully anonymous attribute-based encryption. IEEE Trans. Inf. Forensics Secur. **10**(1), 190–199 (2015)
16. Gentry, C., Silverberg, A.: Hierarchical ID-based cryptography. In: Zheng, Y. (ed.) ASIACRYPT 2002. LNCS, vol. 2501, pp. 548–566. Springer, Heidelberg (2002). https://doi.org/10.1007/3-540-36178-2_34
17. Muller, S., Katzenbeisser, S., Eckert, C.: On multi-authority ciphertext-policy attribute-based encryption. Bull. Korean Math. Soc. **46**(4), 803–819 (2009)
18. Li, J., Huang, Q., Chen, X., Chow, S.S., Wong, D.S., Xie, D.: Multiauthority ciphertext-policy attribute-based encryption with accountability. In: Proceedings of ACM Symposium on Information (ASIACCS), pp. 386–390 (2011)

Big Data and Cloud Computing

Point-to-Point Offline Authentication Consensus Algorithm in the Internet of Things

Xiao-Feng Du[1,3], Yue-Ming Lu[1,3(✉)], and Dao-Qi Han[1,2]

[1] School of Cyberspace Security, BUPT, Beijing, China
ymlu@bupt.edu.cn
[2] School of Information and Communication Engineering, BUPT, Beijing, China
[3] Key Laboratory of Trustworthy Distributed Computing and Service (BUPT), Ministry of Education, Beijing, China

Abstract. In the era of Internet of Things, identity and information authentication have drawn more and more attention. Existing identification mechanisms for identity information generally require networks, biometrics, etc., which exist security vulnerabilities such as security holes, privacy leaks, and vulnerability to attacks. A non-interactive concise offline authentication scheme based on block-chain technology is proposed which no need for complex services such as networking, and hardware implementation is simple. We research anonymity based on block-chain, workload proof and privacy protection technologies, design system models. Integrate device information, data, and confidentiality. The cloud platform function such as key management completes the hardware verification of the offline token authentication algorithm. The experiment results show that the scheme has the characteristics of extremely low recognition rate with a certain difficulty, high performance, high security, low cost, etc. Compared with the existing authentication methods, ours has more advantages.

Keywords: Internet of Things · Offline authentication · Block-chain · Consensus mechanism

1 Introduction

In existing distributed systems and IoT architecture systems, identity authentication generally requires network for authentication. The CA (Certificate Authority) authentication system is widely used, the third-party authentication needs to participate in the authentication process to complete the authentication process. This solution has certain security risks. The attacker may steal the CA certificate for forgery and obtain the authentication authority.

Aiming at the consistency and security of point-to-point information authentication in the existing Internet of Things, we propose a completely different authentication method based on the consensus mechanism. The authentication is based on mathematical calculation and cryptography, without third-party authentication, with high point-to-point authentication. Decentralized authentication structure to ensure security, no need

X. Sun et al. (Eds.): ICAIS 2020, LNCS 12240, pp. 655–663, 2020.
https://doi.org/10.1007/978-3-030-57881-7_57

for real-time networking and collection features, use dynamic address protection device privacy, and use electronic tokens and signatures generated by dynamic private keys as identification elements, simplifying and simplifying. With low power consumption, the certification time is short, one security and one security.

To study the related mechanism design of block-chain, an improved algorithm based on workload proof in block-chain is proposed. In the distributed and multi-participating scenarios, various consensus mechanisms are implemented to achieve consistency and privacy protection, and hashing. Proof, cryptographic algorithms such as elliptic curve cryptography. Design a model that integrates these mechanisms to discuss performance and security.

2 Related Work

The problem of ensuring the consistency of information transmission in traditional distributed systems is mainly to solve the problem of General Byzantine [1]. Internet of Things as a distributed system, information authentication in the network still has information consistency and validity. Lamport [2, 3] first proposed the Paxos algorithm to solve this problem. In 1999, Castro [4] and others proposed the PBFT (Practical Byzantine fault tolerance) algorithm to solve the Byzantine general problem. In 2014, Ongaro [5] and others proposed the Raft algorithm, as opposed to the difficult Paxos algorithm, the Raft algorithm is easier to understand and engineered for distributed systems. In fact, Michael J. Fischer et al. [6] proved that in an asynchronous distributed system, if there is a dishonest node, then Message consistency cannot be solved, so our discussion and discussion are all in synchronous communication systems (point-to-point authentication is a typical synchronous communication system).

Traditional distributed consistency algorithms have a good solution to the consistency problem in centralized distributed systems, but for decentralized distributed systems, such as networks with dishonest nodes (sending false messages), both Without effective implementation consistency, Nakamoto gave a better solution. The consensus mechanism of Proof of Work was first proposed in a 1998 article on B-money [7]. In 2008, Satoshi Nakamoto proposed in the paper "Bitcoin: A Peer-to-Peer Electronic Cash System" [8] to apply the PoW mechanism to bitcoin transactions, and to ensure decentralization by increasing the cost of sending messages and reducing the rate of message transmission. In the distributed system, there is only one or very few nodes at the same time point for information broadcasting, and the broadcast information is accompanied by its own signature. Satoshi Nakamoto finally proposed a decentralized electronic trading system, not only in electronic money trading. To achieve higher security, the related mechanism design is also widely used in distributed computing and peer-to-peer information transmission.

Christidis [9] and others have proposed combining block-chain with the Internet of Things and analyzing the advantages of the new distributed application of the two. For small and medium-sized IoT systems, network nodes already have a certain mutual trust base, and there is no need for transactions in the network, so coin-based consensus algorithms like PoS and DPoS are not suitable for small-scale distributed systems. We propose a new consensus algorithm based on PoW, which does not require real-time communication and does not require a large amount of Computation, based on certain known information, only need to verify the consensus verification algorithm when both parties communicate.

3 Offline Consensus Certification

3.1 Consensus Authentication Algorithm

We denote that the Verifier device is *V* and the Prover device is *P*.

In order to achieve anonymous storage of data, the block-chain's address generation algorithm is used to generate the addresses of the two parties, and multiple pairs of keys are generated, which are randomly selected and used to hide the user's public key and account.

1. Generate a device address, and generate the algorithm as follows:

   First, generate the address public key *PubKey* for *V* and *P*: use the *RIPEMD*160(*SHA*256(*PubKey*)) hash algorithm, take the address public key and hash it twice to get the *PubKeyHash*. Then add the address generation algorithm's *Version* to the prefix of hash.

   For the result generated in step 1, use *SHA*256(*SHA*256(*PubKeyHash*)) to perform two more hashes, and calculate the *Checksum*. The checksum is the first four bytes of the result of this hash.

   Append the checksum to the combination of *Version* + *PubKeyHash* and encode the *Version* + *PubKeyHash* + *Checksum* combination using the Base58 encoding algorithm to get the device address.

2. Generate the required key pair: We define the *KP* (Key Pool), which contains m sets of elliptic curve key pairs for generating passwords:

$$KP = \{[Kpuk_1, Kprk_1], [Kpuk_2, Kprk_2], \ldots [Kpuk_m, Kprk_m]\}$$

   Also generate the *DKP* (Device Key Pool):

$$DKP = \{[Dpuk_1, Dprk_1], [Dpuk_2, Dprk_2], \ldots [Dpuk_m, Dprk_m]\}$$

   Hex code it and store it in Verifier and Prover respectively.

   Before the verification, Prover announced the public key of the equipment used this time $Dpuk_i, i \in [1, m]$, so that Verifier obtained the $Dpuk_i$.

   Define the following variables: the authentication party's information string is *Vs*, the time series is *T*, the behavior information string is *As*, and the device information string is *Ds*.

   The formula for calculating the behavioral string CBS characterizing the appointment activity is:

$$CBS = Vs + T + As + Ds \tag{1}$$

The secp256k1 standard asymmetric encryption algorithm is used to complete the cryptographic signature as *HCBS*, and the SHA256 algorithm is used to hide the information.

$$HCBS = SHA256(CBS) \tag{2}$$

Hash the behavior string before authentication to get *HCBS*. Use the device public key *Dpuk* and the verification public key *Kpuk* to encrypt the HCBS respectively to generate a token *Token* and a signature *Sign*.

$$Token = Kpuk_j(HCBS),\ j \in [1, m] \tag{3}$$

$$Sign = Dpuk_i(HCBS) \tag{4}$$

In order to increase the randomness and the difficulty of cracking, based on: device public key + behavior string hash data *HCBS*, looking for 8 digits random numbers (between 0 and 9), use the special equipment and a certain workload difficulty *dif* to calculate the consensus string *HPC* of PoW mechanism (such as the first 21bit is zero, so $dif = 21$), we call it RPoW (Random Proof of Work).

$$HPC = Dpuk_i + HCBS + \sum_{k=1}^{8} nonce_k \tag{5}$$

Among them, $nonce = \{n_1, n_2, n_3, n_4, n_5, n_6, n_7, n_8\},\ n_k \in [0, 9]$.

Transfer the Token, *dif*, half of the random number $halfnonce(\{n_1, n_2, n_3, n_4\})$, and $Kprk_j$, and the generated consensus string *HPC* to the Prover.

Use the generated strings of the above security processing: device public key + signature data + 4 digits random number, Prover pays a certain amount of work to calculate consensus string HPC under the difficulty requirement *dif*, and verify the consensus string HPC.

3.2 Consensus Verification Algorithm

First, the data sent by Verifier is obtained through offline communication.

Then enter the offline verification phase:

Consistency Verification 1:

After the authentication device obtains the data, the device key pair $[Dpuk_i, Dprk_i]$ is used by the Prover and the verification private key $Kprk_j$ sent by Verifier are used to obtain the hash of the behavior string *HCBS*. Which guarantees that the only Prover can decrypt the obtained *HCBS*.

$$Kprk_j(Token) = HCBS = Dprk_i(Sign) \tag{6}$$

If the two results are inconsistent, the verification is failed.

Consistency Verification 2:

The HCBS obtained by 1 is spliced with the device public key and half of the random number, and the RPoW consensus algorithm is used for verification and look for the other four random numbers as $R = \{r_1, r_2, r_3, r_4\}$:

$$HPC' = Dpuk_i + HCBS + \sum_{k=1}^{4} halfnonce_k + find\left(\sum_{l=1}^{4} r_l\right) \tag{7}$$

If there exists R can satisfy $HPC' = HPC$ under the specified difficulty dif, the verification is successful.

3.3 Algorithm Flow Chart

Based on the above two algorithms, our overall algorithm flow chart is as follows (Fig. 1):

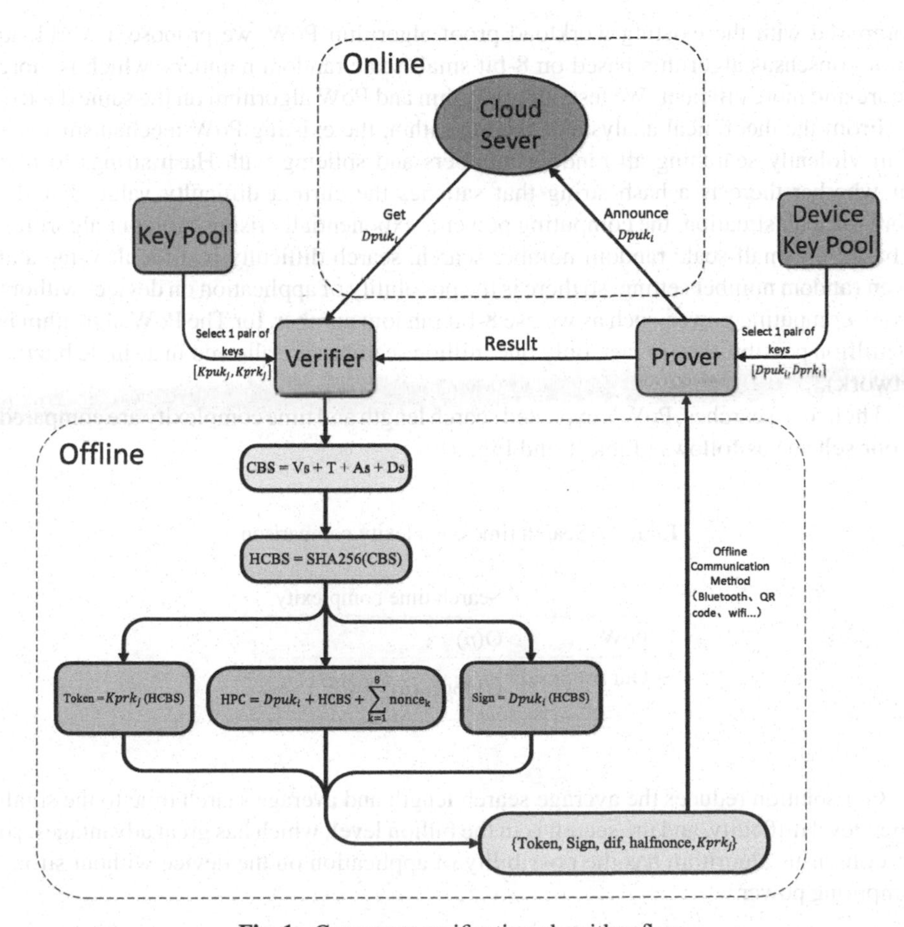

Fig. 1. Consensus verification algorithm flow

It can be seen that the authentication process is completely offline. The network is needed while the Verifier obtains the public key issued by Prover for this verification. The DPoW algorithm process and the verification process are completely offline.

4 Experiments and Tests

4.1 Experiments' Design and Environment

Our experiment is to compare the time performance difference between PoW algorithm and our RPoW algorithm. The test environment is golang runtime environment under macOS. The basic hardware information is: CPU: i7-7700hq (2.8 GHz), RAM: 16 GB, GPU: Radeon Pro 555 (2G). These are unified experimental environment.

4.2 Average Search Length and Speed

Compared with the existing workload proof algorithm PoW, we propose a workload proof consensus algorithm based on 8-bit small-scale random numbers, which is more secure and more efficient. We test our algorithm and PoW algorithm on the same device.

From the theoretical analysis of the algorithm, the existing PoW mechanism starts from violently searching all random numbers and splicing with Hash strings to find out whether there is a hash string that satisfies the current difficulty value. For the more difficult situation, the computing power is exponentially rising. And our algorithm is based on small-scale random number search, search difficulty is difficult value and given random number setting, so there is the possibility of application on devices without strong computing power, such as we use 8-bit random number, for The PoW algorithm is 10 million possibilities (in fact, only one million searches are allowed in a single bitcoin network).

Then for n searches, PoW's expected search length and time complexity are compared to our scheme as follows (Table 1 and Fig. 2):

Table 1. Search time complexity comparison

	Search time complexity
PoW	$O(n)$
Our proposal	$O\left(\log_{10}(n)^{\log_{10}(n)}\right)$

Our solution reduces the average search length and average search time to the small index level difficulty, and the search is in the billion level, which has great advantages, so the consensus algorithm has the possibility of application on the device without strong computing power.

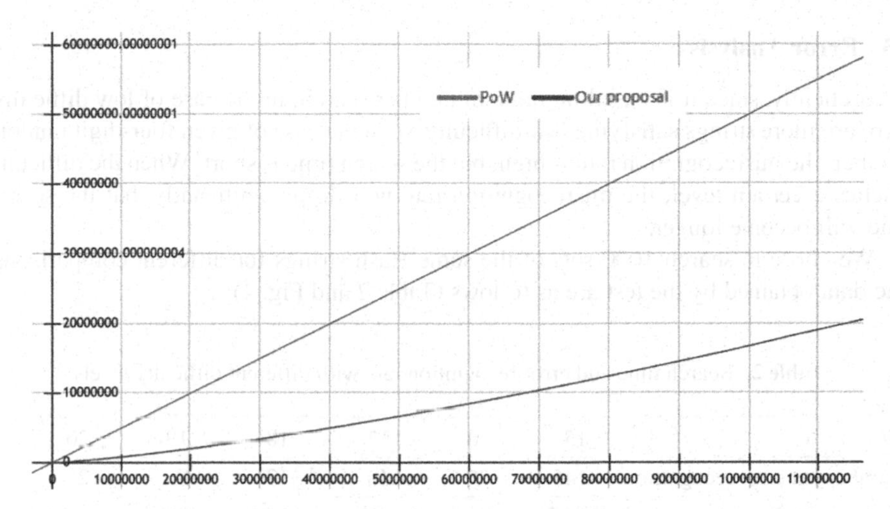

Fig. 2. Algorithm time complexity

Moreover, because it is a verification consensus algorithm, our algorithm half-random number needs to be verified by Prover, and the consensus algorithm is used creatively to complete point-to-point verification. The PoW algorithm is a competitive consensus algorithm, and the first calculation in the block is satisfied. Difficult hash strings are considered valid and broadcast, with no validated designs.

For the same data, we use the same test machine to perform 10 tests under the same difficulty *dif*, and compare the search time. The test results are as follows (Fig. 3):

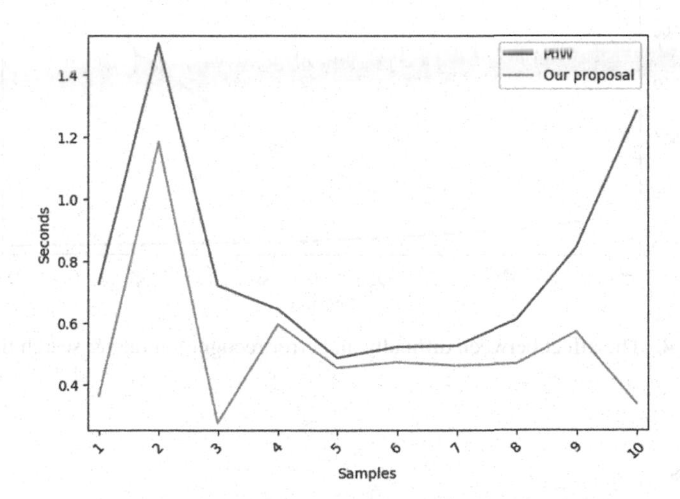

Fig. 3. Comparison of search time on the same difficulty

It can be seen that in general our solution has better performance.

4.3 Error Analysis

Theoretically, since it is based on random number search, in the case of low difficulty, there are more strings satisfying low difficulty, so in the case of given four-digit random number, the misrecognition rate is high, but the search time is short. When the difficulty reaches a certain level, the misrecognition rate will drop significantly, but the search time will become longer.

We chose to search 1000 sets of the same Hash strings for different comparisons. The data obtained by the test are as follows (Table 2 and Fig. 4):

Table 2. Search time and error recognition rate with different difficulty levels

dif	15	16	17	18	19	20
Numbers of error recognition	145	75	40	17	8	2
Error recognition rate	14.5%	7.5%	4%	1.7%	0.8%	0.2%
Total search time (s)	67.89	124.92	245.15	498.42	885.60	1990.60

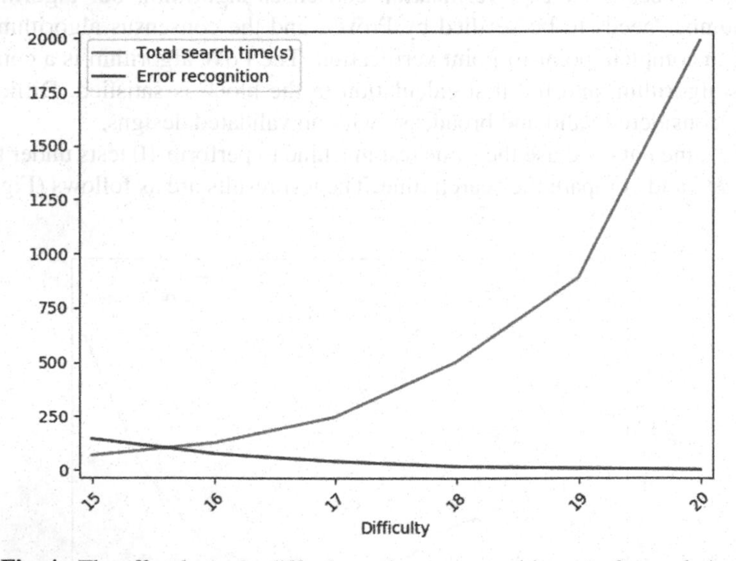

Fig. 4. The affect between difficulty and error recognition rate & search time

4.4 Discuss

We analyze the above experimental results, it is not difficult to conclude that our algorithm has better performance in the original workload proof algorithm within a certain number of digits, and has better performance in the same experimental environment.

And under certain circumstances, the misrecognition rate is also reduced. This is because the difficulty increases, and the hash value that satisfies the condition in the search interval is also small, so the misrecognition rate drops significantly, but at the same time the price of the search time increases. Therefore, this is also worth studying to make a trade-off between time and security.

5 Conclusion

Based on the existing token authentication mechanism, we propose an offline authentication mechanism based on the block-chain core technology and theory, which is more suitable for small scale Internet of Things system. The authentication device only needs to provide relevant information about this verification (such as time, device information, behavior information, etc.), the network only obtains the key used for system computing encryption to obtain locally, and the verification device can perform operations according to the encapsulated information string, match the token, verify the signature, and reach a consensus. In the right mechanism, protect privacy, strengthen authentication security, avoid network interaction. Compared with PoW consensus algorithm in evaluation performance, misrecognition rate, crack difficulty and other indicators. This solution does not need to access centralized authentication service, and eliminate the concurrent authentication. To the best of our knowledge, this is the first attempt to this mathematical non-switched peer-to-peer verification. This mechanism can also be applied to localized verification scenarios like offline, peer-to-peer authentication without revealing privacy.

References

1. Lamport, L., Shostak, R., Pease, M.: The Byzantine generals problem. ACM Trans. Program. Lang. Syst. **4**(3), 382–401 (1982)
2. Lamport, L.: Paxos made simple. ACM SIGACT News **32**(4), 18–25 (2001)
3. Lamport, L.: The part-time parliament. ACM Trans. Comput. Syst. (TOCS) **16**(2), 133–169 (1998)
4. Castro, M., Liskov, B.: Practical Byzantine fault tolerance. In: OSDI 1999, vol. 99, pp. 173–186 (1999)
5. Ongaro, D., Ousterhout, J.: In search of an understandable consensus algorithm (extended version). In: USENIX Annual Technical Conference, pp. 305–320 (2014)
6. Fischer, M.J., Lynch, N.A., Paterson, M.S.: Impossibility of distributed consensus with one faulty process. Massachusetts Inst of Tech Cambridge Lab for Computer Science (1982)
7. Dai, W.: B-money, a scheme for a group of untraceable digital pseudonyms to pay each other with money and to enforce contracts amongst themselves without outside help (1998). http://www.weidai.com/bmoney.txt
8. Nakamoto, S.: Bitcoin: a peer-to-peer electronic cash system (2008). https://bitcoin.org/bitcoin.pdf
9. Christidis, K., Devetsikiotis, M.: Blockchains and smart contracts for the Internet of Things. IEEE Access **4**, 2292–2303 (2016)
10. Koblitz, N., Menezes, A., Vanstone, S.: The state of elliptic curve cryptography. Des. Codes Crypt. **19**(2–3), 173–193 (2000)
11. Hui, H., Zhou, C., Xu, S., Lin, F.: A novel secure data transmission scheme in industrial Internet of Things. China Commun. **17**(1), 73–88 (2020)
12. Su, J., Lin, F., Zhou, X., Lu, X.: Steiner tree based optimal resource caching scheme in fog computing. China Commun. **12**(8), 161–168 (2015)

Classification of Tourism English Talents Based on Relevant Features Mining and Information Fusion

Qin Miao[1(✉)], Li Wu[2], and Jun Yang[3]

[1] Foreign Language School, Aba Teachers University, Wenchuan 623002, Sichuan, China
gufucan6577kbw@163.com

[2] School of Mathematics and Computer Science, Aba Teachers University, Wenchuan 623002, Sichuan, China

[3] Chongqing Vocational College of Transportation, Chongqing 402247, China

Abstract. This paper studied the application of computer aided design in tourism planning. In this paper, the recommendation system of hybrid travel recommendation based on hybrid collaborative filtering was a personalized recommendation system designed for tourism market. Combining the two basic algorithms, through constant analysis and experiment, the algorithm was improved by optimizing the recommendation degree to improve the coverage. The interest of users and items was balanced, adding heat and Raiders city heat parameters. This paper calculated the recommended degrees based on the travel users, travel strategies and tourist cities, and then balanced the three degrees of recommendation to get the final degree of recommendation, which greatly increased the coverage of the recommended strategy and solved the "long-tail effect" recommendation problem.

Keywords: Computer aided · Tourism planning · Collaborative filtering algorithm

1 Introduction

In today's era, the word network is no longer strange, and it can even be said that everybody is dealing with the Internet every day. The Internet has infiltrated all aspects of our lives. We watch movies and TV shows on Ikea, Mango TV and other video websites, read the news on Sina Weibo, Sohu News and other portals, and select a wide range of products from shopping websites such as Taobao and Jingdong Mall. At the same time, it can also enjoy the sounds of nature in music sites such as QQ music and Baidu music. The birth of Internet has brought great convenience to people [1]. But at the same time huge amounts of information have been flocking to us, dazzling us and finding ourselves at a loss, and people have entered the era of the Big Bang [2]. Users spend a lot of energy searching for the information they really need, but useless information is everywhere, really too much, how can they choose the best information from the complex information and how the information is presented effectively In front of users. This undoubtedly brings new challenges to computer workers. To solve this

X. Sun et al. (Eds.): ICAIS 2020, LNCS 12240, pp. 664–673, 2020.
https://doi.org/10.1007/978-3-030-57881-7_58

problem, search engines came into being [3]. The search bar can be found on every website, and can be searched accurately if the user specifically wants to find the product, however, most users do not have a clear search target in most cases [4]. Search engines cannot present information that is not described by the user. For example, a favorite movie and television show is not easy to describe in simple text. Moreover, the search engine search information does not have personalized, when different users search the same keyword, there will be the same page, which means that the information provided is the same, failed to achieve their best and unable to meet the user Personalized needs [5].

2 Related Work

The domestic research on personalized recommendation service started a little later, the related recommendation algorithm has a lower level of research, and has a big gap with the level of international research. Nevertheless, it does not affect its popularity in the country; the vast majority of Internet companies are actively studying it [6]. 2009 Percentage of Companies Established in Beijing, the Company is China's first company to research and recommend technologies. In order to let more people study personalized recommendations and promote their promotion in China, the company shares many documents on recommended technologies [7]. The service for the majority of users provides the appropriate personalized solutions to promote the development of domestic personalized recommendation [8]. In 2011, Baidu CEO Robin Li at the Baidu World Congress proposed in the future for all users will be "tailored" to them one by one personalized recommendation and said the focus of future information services will be the recommendation system, search engine and cloud computing. In order to improve the domestic recommended service level and mobilize the development of the industry, major companies have organized computer competitions one after another [9]. Baidu company organized the national recommended system innovation competition with the intention of improving the quality of recommendation. It not only received the active participation of the majority of university teachers and students, but also achieved better results [10]. Subsequently, in March 2014, the famous Internet company Alibaba held a big data contest, each team needs to be the actual behavior of Lynx users as a data set on the test data set to recommend their own systems to further promote the recommendation system development of [11].

3 Methodology

3.1 Item-Based Collaborative Filtering Recommendation Algorithm

With the development of big data era, people gradually entered the era of information overload from the era of lack of information, and the recommender system came into being [12–15]. It is an important way to solve the problem of information overload and, compared with search engines, it can be tailor-made for each one to meet the unique needs of each user. Nowadays, we can find that the most commonly used method in all fields is the article-based collaborative filtering recommendation algorithm, such as Amazon

Web site, You Tube and so on, based on this algorithm. The User-based collaborative filtering recommendation algorithm has been applied to many websites, but it has some shortcomings [16]. First, the huge user base has led to the difficulty in calculating the user-friend preference similarity. The increase of the number of users and the increase of the time complexity and the space complexity are similar to the quadratic relationship. Second, it is difficult to explain the results it recommends. Therefore, the item-based collaborative filtering recommendation algorithm came in, and it is proposed by Amazon [17].

The item-based collaborative filtering recommendation algorithm is to recommend to the target user items that are similar to the item they were interested in before [18, 19]. For example, user A purchased a "Introduction to Computers" on their website and the algorithm would recommend "Computer Components." However, instead of calculating the similarity between items through the content attributes of the items, it is necessary to analyze the user behavior records and then calculate the similarity between the items. This algorithm is this, like a user a lot of users like the object b, and then it will think that a, b have a great similarity between the two items. Item-based collaborative filtering recommendation algorithm can be summarized as two steps [20, 21]. First the similarity between items is calculated, and then based on the user's historical behavior and has calculated the similarity between objects to generate the user's recommended list. Amazon website recommended articles title is "Customers who browsed this product also browse," starting from the sentence definition. The similarity of the article can be defined by the following formula [22]:

$$w_{ij} = \frac{|N(i) \cap N(j)|}{|N(i)|} \tag{1}$$

The above formula seems plausible, but there are still problems. For example, if item J is quite popular and most people are interested, then the closer the heat is to 1. Therefore, the use of this formula is very likely to such a situation. It is a great similarity between any item and popular items. This is obviously not a good feature for recommended systems [23]. In order to reduce the situation of recommending only popular products to users, the following formula is available [24]:

$$w_{ij} = \frac{|N(i) \cap N(j)|}{\sqrt{|N(i)||N(j)|}} \tag{2}$$

It can be seen from the above definition that in the recommendation algorithm, because many users are interested in these two items, the similarity between the two items will be generated, that is, any user's historical behavior list may contribute to the similarity of items [25]. There is an assumption contained here, each user's preferences are limited to a few aspects, so if there are two items in the same user's list of interests, then they may belong to a limited number of areas. But if two items are not in the same user's list of interests, but belong to the list of interests of many users, so that they may belong to the same area, so there is a great similarity. This part of the algorithm and user-based collaborative filtering recommendation algorithm is similar, when used, you must first create an object-user inverted list to calculate the similarity between items. After that, for each user, they need to add t in the co-occurrence matrix C between the two items in their item list (Fig. 1).

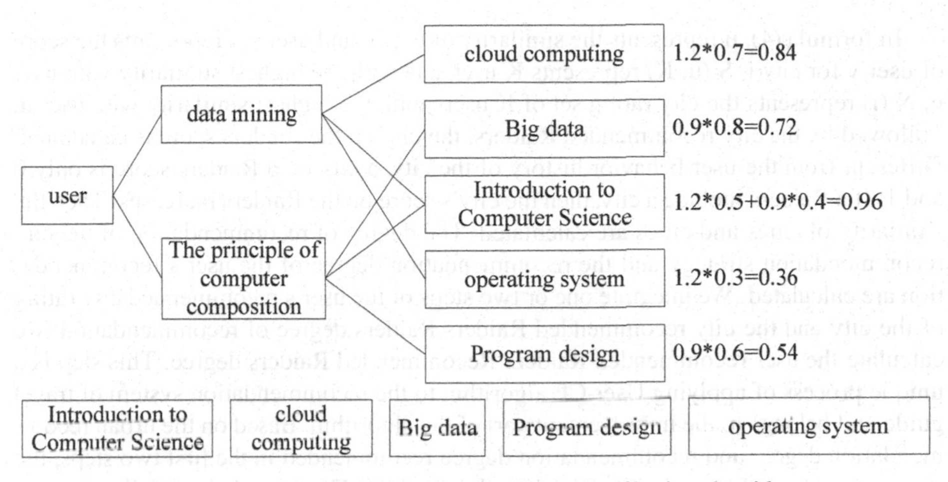

Fig. 1. Example of an item based collaborative filtering algorithm

Figure is a simple example of the recommended algorithm: Data Mining and Computer Principles are two items that the target user specifically likes. The algorithm first finds the three most similar books for each of these two books, and then calculates the target user's preference for each book based on the formula. For example, this algorithm recommends the book Computer Introduction to users because it is similar to Data Mining and has a similarity of 0.5, which is also similar to Computer Theory of Composition with a similarity of 0.4. The degree of user interest in "Data Mining" book is 1.2 and the degree of interest in the book "Principles of Computer Organization" is 0.9, then the user's interest in Computer Introduction is $1.2 * 0.5 + 0.9 * 0.4 = 0.96$.

3.2 User-Based Collaborative Filtering Travel Raiders Recommended Algorithm

In this paper, we call the algorithm User-CF algorithm, the main steps of the algorithm are as follows: According to the user recommended city, according to the user's behavior history to calculate the user to the city score. Each user on the Raiders has different behavior history. This behavior can be praise, browse or create, each Raider has a corresponding tourist city, and different behavior corresponds to the user on the city of different ratings. Therefore, the algorithm for calculating a user's rating on a city is to take the user's rating of the behavior of all the strategies that contain the city. Calculate the similarity between user and user based on the user's rating of city, the formula is as follows:

$$w_{uv} = \frac{|N(u) \cap N(v)|}{\sqrt{|N(u)||N(v)|}} \tag{3}$$

In the formula, N (u) represents the set of cities and scores contained in user u, and N (v) represents the set of cities and scores contained in user v. Calculate the recommended city rating of the city to the user, the formula is as follows:

$$p(u, i) = \sum_{v \in S(u,K) \cap N(i)} w_{uv} r_{vi} \tag{4}$$

In formula (4), u represents the similarity of user u and user v, r represents the score of user v for city i, S (u, K) represents K user sets with the highest similarity with user u, N (i) represents the city rating set of K users with the highest similarity with user u. Followed by the city recommended Raiders, the city on the Raiders score is calculated. Different from the user behavior history of the city, a city on a Raiders score is only 0 and 1. If a Raiders include a city, then the city's score on the Raiders increases. Then the similarity of cities and cities are calculated. The degree of recommendation of the city recommendation strategy and the recommendation degree of the user's recommendation are calculated. We integrate one or two steps of the user's recommended city rating of the city and the city recommended Raiders Raiders degree of recommendation, we calculate the user recommended Raiders Recommended Raiders degree. This step is a unique process of applying User-CF algorithm to the recommendation system of travel guide and belongs to the improvement part of the algorithm. Based on the urban recommendation degree and recommendation degree recommended in the first two steps, the recommendation degree to the user is finally obtained. The formula is as follows:

$$p(u, b) = \sum_{c \in Rc, b \in Rb} r_{uc} * r_{cb} \tag{5}$$

p (u, b) indicates the recommended degree of recommending user b to user u. The Rc below the sums for the equal sign in Eq. 4.3 represents the set of cities recommended to user u. At the same time below the Rb represents the city c recommended Raiders set. The rNc before the multiplication sign on the right side of the equal sign indicates the degree of urban recommendation for the user u recommending the city c. Rcb after the multiplication sign indicates that the city c recommended Raiders b Recommended Raiders. Finally, calculate the recommended coverage of travel strategy, using the traditional method of calculating coverage and the formula is as follows:

$$Coverage = \frac{|U_{u \in U} R(u)|}{|I|} \tag{6}$$

3.3 Data Source

Because improvements are recommended algorithms for travel tips, experiments require the analysis of large amounts of data on travel sites, the need to collect user history records and content of high quality travel tips. System-managed data includes input data, model data, and output data. The input data includes user information, travel guide information, user rating information and the like. The user information is obtained from the personal information that is filled in when the user logs in to the system, including user identity, gender, age, occupation, address, email address and the like. Travel guide information includes travel location, travel city, listing date, type of play, travel items and so on. User rating information refers to the user's evaluation of travel strategy, there are many forms, which may be a direct score, may be to write reviews and may be fuzzy evaluation. This data is based on the user rating of travel Raiders. Model data includes model input data and model output data. Different algorithms correspond to different input data. Therefore, in the "data preprocessing" stage to deal with the input data processing input data for the model. It includes user data, Raiders data, rating data. The model structure data and the user classification data are two parts in the model output data, and the classification result of the original user and the classification result are the user classification data. The output data includes user predictive rating data, forecasting new user data and predicting new user rating data. This experiment dataset is collected from the real users, real behaviors and real data of travel websites. There are totally 42,523 records of user history, 1452 cities and 10,389 high-quality strategy. Experiments use offline experimental methods, carried out on the data set.

4 Result Analysis and Discussion

For the improved algorithm proposed in this paper, a total of three experiments are designed to evaluate, divided into Experiment A, Experiment B and Experiment C. Experiment A is to evaluate User-CF algorithm. User-CF algorithm and User-CF algorithm are used to test the dataset. The proportion of recommended travel strategy to the total number of Raiders is obtained. A slightly improved User-CF Algorithm and the degree of recommendation will be improved after the User-CF algorithm coverage. Experiment B is to evaluate Item-CF algorithm, using Item-CF algorithm and Item-CF algorithm to test on dataset, and get the proportion of recommended travel strategy to the total number of Raiders. Compare Item-CF Algorithm and the recommendation will also be improved Item-CF algorithm coverage. Experiment C is an improved mixed recommendation algorithm, which is tested on a data set using a hybrid algorithm. The proportion of the recommended travel strategy to the total number of Raiders is obtained, and then compared with the coverage of two basic algorithms and two improved algorithms.

Experiment A is as follows. The number of cities recommended to users by User-CF algorithm is 265, and the coverage is 8.25%. The number of recommended cities to cities is 6,445, and the coverage is 62.04%. The number is 774 with a coverage of 7.45%. Using the improved User-CF algorithm in this experiment, the number of cities recommended to users is 623, with a coverage of 42.91%. The number of recommended cities to cities is 6109, and the coverage is 58.8%. The number is 1932 and the coverage is 18.6%. It can be seen that the coverage of the improved User-CF-1 algorithm is greater than the coverage of the underlying User-CF algorithm. The coverage table and the comparison chart are shown in the following table and the figure (Table 1 and Fig. 2):

Table 1. Improved coverage table based on the user's collaborative filtering algorithm

	User-city coverage	City-book coverage	User-book coverage
User-CF	18.25%	62.04%	7.45%
User-CF-1	42.91%	58.8%	18.6%

Experiment B is as follows. The number of cities recommended to users by Item-CF algorithm is 696 with a coverage rate of 7.93%. The number of recommended cities to cities is 6201, and the coverage is 59.69%. Raiders to the user recommended the number

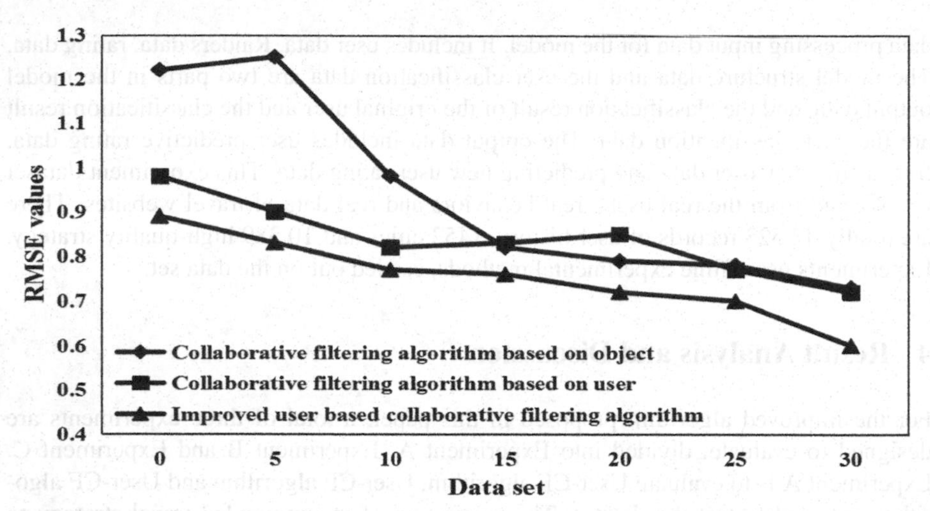

Fig. 2. Based on a user based collaborative filtering recommendation algorithm for improved coverage contrast

is 2279, coverage is 21.94%. Using the improved Item-CF algorithm in this experiment, the number of cities recommended to users is 996, with a coverage of 68.6%. The number of recommended Raiders to the city is 6201, the coverage is 59.69%, to the user recommended number of Raiders is 2762, the coverage is 26.59%. It can be seen that the coverage of the improved Item-CF algorithm is greater than the coverage of the underlying Item-CF algorithm. The coverage table and the comparison chart are shown in the following table and the figure (Table 2 and Fig. 3):

Table 2. An improved coverage table for an improved collaborative filtering algorithm based on items

	User-city coverage	City-book coverage	User-book coverage
User-CF	47.93%	59.69%	21.94%
User-CF-1	68.6%	59.69%	26.59%

Experiment C is as follows. The number of recommended strategies for users using User-CF algorithm is 774, with a coverage rate of 7.45%. The number of recommendations to users based on the improved User-CF based algorithm of the user is 1,932, with coverage of 18.6%. Using Item-CF algorithm to get the number of user-recommended Raiders is 2279, the coverage is 21.94%. Using Item-Based Improved Item-CF Algorithm, the number of recommended users to the user is 2762, with coverage of 26.59%. The number of recommended strategies for users using the hybrid recommendation final algorithm is 4291, with coverage of 41.3%. It can be seen that the recommended result of the hybrid recommendation algorithm is excellent (Fig. 4).

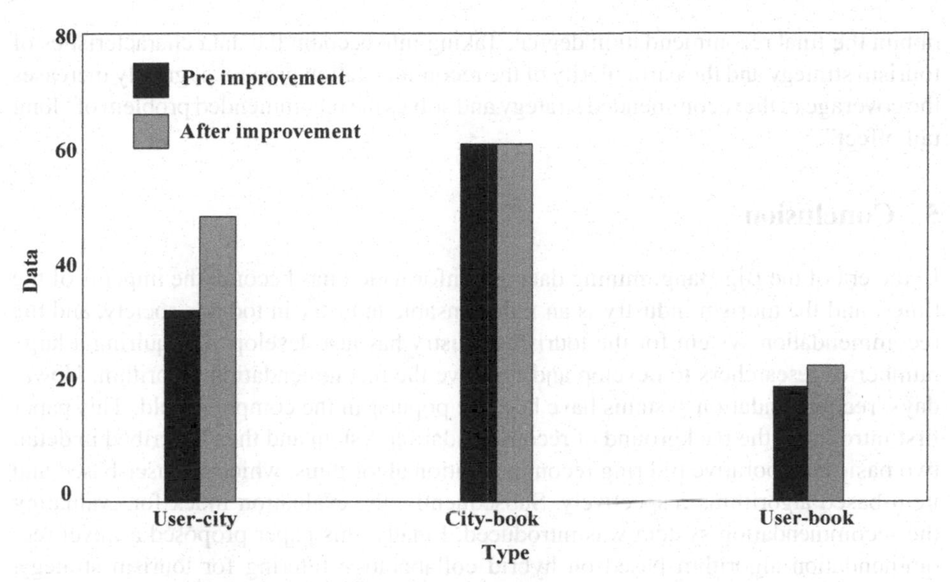

Fig. 3. An improved coverage contrast diagram of an item based collaborative filtering algorithm before and after improvement

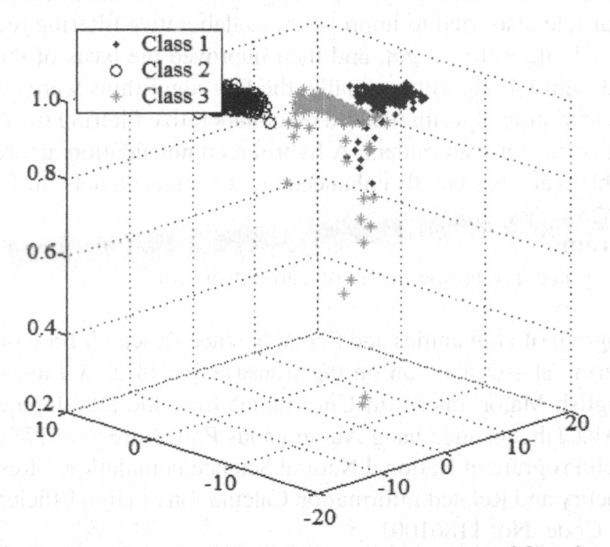

Fig. 4. Results of cluster analysis the coverage contrast of five algorithms

Through the above three groups of experiments and analysis results, we can see that the improvement of each algorithm is based on the original to improve the coverage of a certain degree, and the final hybrid recommendation algorithm compared with the basic algorithm is increased by nearly 35% coverage. The relative effect is better. In short, this paper calculates the recommendation degree based on the travel users, the travel strategy and the tourism city respectively, and then balances the three recommendation degrees to obtain the final recommendation degree. Taking into account the data characteristics of tourism strategy and the particularity of the recommendation process, it greatly increases the coverage of the recommended strategy and solves the recommended problem of "long tail effect".

5 Conclusion

In the era of the Big Bang, mining data and information has become the impetus of the times, and the tourism industry is an indispensable industry in today's society, and the recommendation system for the tourism industry has just developed, requiring a large number of researchers to develop and improve the recommendation algorithm. Nowadays, recommendation systems have become popular in the computer field. This paper first introduced the background of recommendation system and then described in detail two basic collaborative filtering recommendation algorithms, which are user-based and item-based algorithms respectively. Subsequently, the evaluation index for evaluating the recommendation system was introduced. Finally, this paper proposed a travel recommendation algorithm based on hybrid collaborative filtering for tourism strategy, namely the Final algorithm. In this paper, based on User-collaborative filtering recommendation algorithm, the User-CF algorithm was slightly improved, and the improved User-CF algorithm was obtained by improving the recommendation formula. At the same time, the article also tried to improve the collaborative filtering recommendation algorithm. Item-CF algorithm is got, and then improved the basis of this algorithm to get an improved Item-CF algorithm; finally, the four algorithms were combined to get a hybrid recommendation algorithm based on collaborative filtering travel guide, which was recommended for tourism raiders. A hybrid recommendation algorithm can solve the long tail problem and has greatly enhanced the coverage, namely the Final algorithm.

Research Program
This paper is the phased outcome of the research program:

1. General Program of Humanities and Social Science Research Base of Sichuan Education Department—Discussion on the Construction of New Education Model of Tourism English Major Talents in Universities from the Development of Tourism Culture in Aba Tibetan and Qiang Autonomous Prefecture, No: 17SB0004.
2. The Research Program of National Natural Science Foundation—Research on Construction Theory and Related Information Calculation of High Efficiency High Rank Space Time Code, No: 11861001
3. The general program of Aba Teachers University—Applied Research on Factor Analysis in Multivariate Statistics, No: ASB18-05

References

1. Pjerotić, L., Rađenović, M., Tripković-Marković, A.: Stakeholder colaboration in tourism destination planning-the case of Montenegro. Economics **4**(1), 119–136 (2016)
2. Majewska, I.: Strategic planning as a key factor of tourism development in the region. Annales UMCS Geographia Geologia Mineralogia et Petrographia **64**(1), 248–261 (2009)

3. Moscardo, G.: Exploring social representations of tourism planning: issues for governance. J. Sustain. Tour. **19**(4–5), 423–436 (2011)
4. Paül, V., Trillo-Santamaría, J.M., Pérez-Costas, P.: Action research for tourism planning in rural areas? Examining an experience from the Couto Mixto (Galicia, Spain). Geogr. Res. **54**(2), 153–164 (2016)
5. Wan, P.Y.K., Pinheiro, F.V.: Macau's tourism planning approach and its shortcomings: a case study. Int. J. Hosp. Tour. Adm. **15**(1), 78–102 (2014)
6. Zhang, S.L., Jiang, K.F., Chen, Q.: Construction of regional tourism planning system based on meta-synthesis. J. Chongqing Normal Univ. **28**(1), 76–81 (2011)
7. Zhang, S.L., Song, Z.W.: The conceptual planning study of the basic connotation of tourism. J. Chongqing Normal Univ. **27**(5), 71–76 (2010)
8. Wen, T.Z., Xu, A.-Q., Cheng, G.: Multi-fault diagnosis method based on improved ENN2 clustering algorithm. Control Decis. **30**(06), 1021–1026 (2015)
9. Kumar, A., Pooja, R., Singh, G.K.: Design and performance of closed form method for cosine modulated filter bank using different windows functions. Int. J. Speech Technol. **17**(4), 427–441 (2014). https://doi.org/10.1007/s10772-014-9242-8
10. Jiang, Y.Z., Chung, F.L., Wang, S.T., et al.: Collaborative fuzzy clustering from multiple weighted views. IEEE Trans. Cybern. **45**(4), 688–701 (2015)
11. Bi, A., Wang, S.: Transfer affinity propagation clustering algorithm based on Kullback-Leiber distance. JEIT **38**(8), 2076–2084 (2016)
12. Alahmad, B.N., Gopalakrishnan, S.: Energy efficient task partitioning and real-time scheduling on heterogeneous multiprocessor platforms with QoS requirements. Sustain. Comput.: Inform. Syst. **1**(4), 314–328 (2011)
13. Wang, H., Jin, H., Wang, J., Jiang, W.: Optimization approach for multi-scale segmentation of remotely sensed imagery under k-means clustering guidance. Acta Geodaetica Cartogr. Sin. **44**(5), 526–532 (2015)
14. Ou, S.-F., Gao, Y., Zhao, X.-H.: Adaptive combination algorithm and its modified scheme for blind source separation. J. Electron. Inf. Technol. **33**(5), 1243–1247 (2011)
15. Tian, G., He, K.-E., Wang, J., et al.: Domain-oriented and tag-aided web service clustering method. Chin. J. Electron. **43**(7), 1266–1274 (2015)
16. Wu, T., Chen, L.F., Guo, G.D.: High-dimensional data clustering algorithm with subspace optimization. J. Comput. Appl. **34**(8), 2279–2284 (2014)
17. Lei, Y., Yu, X., Yue, S., et al.: Research on PSO-based intuitionistic fuzzy kernel clustering algorithm. J. Commun. **5**, 2015099 (2015)
18. Bo, Z., Jie, H., Gang, M., et al.: Mixture of probabilistic canonical correlation analysis. J. Comput. Res. Dev. **52**(7), 1463–1476 (2015)
19. Sun, C., Yang, C., Fan, S., et al.: Energy efficient distributed clustering consensus filtering algorithm for wireless sensor networks. Inf. Control **44**(3), 379–384 (2015)
20. Guo, X.-Y.: Simulation and analysis on uncertain attenuation property of underwater acoustic signal for oil field pipe. Comput. Simul. **31**(3), 118–121 (2014)
21. Zhang, W., Chen, Q.: Network intrusion detection algorithm based on HHT with shift hierarchical control. Comput. Sci. **41**(12), 107–111 (2014)
22. Luo, H., Qu, Y., Yu, Y.: Oscillation criteria of second order neutral delay Emden-Fowler equations with positive and negative coefficients. Acta Mathematicae Applicatae Sinica **40**(5), 667–675 (2017)
23. Dou, Z., Gang, X., Chen, X., Yuan, K.: Rational non-hierarchical quantum state sharing protocol. Comput. Mater. Continua **58**(2), 335–347 (2019)
24. He, Q., et al.: A weighted threshold secret sharing scheme for remote sensing images based on chinese remainder theorem. Comput. Mater. Continua **58**(2), 349–361 (2019)
25. Guo, F., et al.: Research on the relationship between garlic and young garlic shoot based on big data. Comput. Mater. Continua **58**(2), 363–378 (2019)

Research on Application of Big Data Combined with Probability Statistics in Training Applied Talents

Li Wu[1(✉)] and Jun Yang[2]

[1] School of Mathematics and Computer Science, Aba Teachers University, Wenchuan 623002, Sichuan, China
zhengchun196910@163.com
[2] Chongqing Vocational College of Transportation, Chongqing 402247, China

Abstract. In order to improve the training effect of applied talents, quantitative evaluation and big data analysis are used to evaluate the training of applied talents. A quantitative analysis model of applied talents training based on big data and probability statistics is proposed. The statistical mathematical analysis model of applied talents training is constructed, and the significance of applied talents training is analyzed by using T statistical test analysis method, and the benefit distribution model of applied talents training is established. The cumulative average analysis method is used to evaluate the bilateral reliability of applied talents training, and the big data mining and feature extraction methods are used to analyze the characteristics of applied talents training. The big data robust mining model for the cultivation of applied talents is constructed, and the descriptive statistical analysis method of single variable is taken, the statistical analysis of probability theory and big data analysis method are used to realize the evaluation of benefit index for the cultivation of applied talents. The results of empirical analysis show that the model has good accuracy and high level of confidence in the quantitative evaluation of applied talents training.

Keywords: Big data · Probability statistics · Applied talents · T statistical test

1 Introduction

Paying attention to practical teaching and cultivating applied talents has been paid more attention to in the field of higher education both at home and abroad. The cultivation of applied talents is an important work related to the international people's livelihood and social stability. In recent years, the employment difficulties in colleges and universities in China have become a topic of concern and urgent need to be solved by people from all walks of life. The aim of talent cultivation in colleges and universities must be adapted to the needs of social talents, which has become the primary task of higher education in our country. The independent college aims at training applied talents and becomes a bright spot in cultivating applied talents to meet the needs of society [1–3]. However, how to improve the quality of talents training through the improvement of applied talents training

X. Sun et al. (Eds.): ICAIS 2020, LNCS 12240, pp. 674–685, 2020.
https://doi.org/10.1007/978-3-030-57881-7_59

mode has also become a crucial issue in independent colleges. Different from ordinary colleges and universities to cultivate academic and research talents, independent colleges mainly train applied talents. The so-called applied talents refer to a kind of specialized talents who can apply their professional knowledge and skills to the professional social practice. They are professional talents who master the basic knowledge and basic skills of social production or social activities skillfully. In order to improve the ability of quantitative analysis in the cultivation of applied talents, big data and statistical analysis method of probability theory is used to analyze quantitatively the cultivation of applied talents, and a scientific, operable and characteristic curriculum system is constructed, so as to promote the cultivation and construction of applied talents [4–6].

In order to solve the structural contradiction between talent training and demand, it is necessary to strengthen the training of applied talents and construct a new system of cultivating applied talents. The so-called applied talents refer to a type of specialized talents who can apply their professional knowledge and skills to the professional social practice in which they are engaged [7–9]. It is the basic knowledge and basic skills of mastering social production or first-line social activities. Mainly engaged in front-line production of technical or professional personnel. For the talents trained in applied undergraduate colleges, on the one hand, they should have the basic theoretical knowledge of the social production front, and the basic theoretical basis should be solid; on the other hand, they should be proficient in the basic skills of production and first-line social activities. Strong hands-on ability, after entering the employer can quickly change roles, independently undertake production and social activities of the first line of work. Therefore, the requirements for talent training are different from those for academic and research talents, as well as for skilled personnel in higher vocational and technical colleges. At present, the students' practical ability is not strong, so it is generally reflected by the employing unit. However, the theoretical foundation of talents cultivated in higher vocational colleges is not solid, and the lack of stamina is a common problem reflected by employers [10–12]. The establishment of applied undergraduate colleges and universities can effectively alleviate this contradiction, which also puts forward higher requirements for the mode of talents training in applied undergraduate colleges and universities, as well as a severe challenge to the traditional teaching methods of applied courses.

2 Applied Talent Training Model

2.1 Clear Training Objectives

With the development of the big data era, colleges and universities have launched a series of discussions on how to train applied talents, and put forward different training strategies from different angles [13–15]. Some colleges and universities think that the training of applied talents should be carried out from the aspects of teaching innovation, training orientation, way and mode innovation. The training program should be determined. It should take the combination of production, research and research as the main approach, practice oriented, application ability oriented, practice ability and training, from teaching mode, curriculum system and teacher [16–18]. Construction of the capital team and other aspects. In order to make the school a characteristic and high quality application oriented college, we should focus on the application and strengthen the service as the core and firmly take the way of local, applied and open school. It can be seen that colleges and universities have begun to transform the original elite education mode to adapt to the era of big data reform and innovation and applied talents training. Therefore, colleges and universities should deeply understand the connotation of the training of Applied Talents in the background of the era of big data, and establish the training target of their own applied talents according to the different levels and local development of the school [19, 20].

2.2 Constructing a Reasonable Curriculum System

Building a scientific, operable and characteristic curriculum system is the foundation for training applied talents. According to the characteristics of professional application, a professional curriculum system is established. The professional curriculum system can follow the idea of "Introduction - Basic - Professional - Practice - study", strengthen the existing related courses according to the society and the needs of the times, and make up the vacancy by setting up related courses according to the curriculum system [22]. We should attach importance to the teaching of basic courses: to strengthen the teaching of basic skills courses, to set up the subject and professional courses rationally, and to combine closely with practice in order to meet the needs of the industry and the society [23]. Focusing on key courses, promoting professional team building. According to the characteristics of the economic development of the school districts, we should build characteristic schools. The curriculum group for the local economy attaches importance to the setting up of the quality development platform, and provides different elective courses and quality development courses according to the principles. Teach students in accordance with their aptitude and teach students in accordance with their aptitude. In the course of teaching, teachers should not only understand the existing large data concepts, development, application and other fields to expand their own professional knowledge for further study and further study, but also to study their needs and influence on today's society to teach students related technology, improve students' interest in learning, and make them adapt to social development. In the course assessment, we can improve the way of assessment, the way of assessment and the environment of curriculum development, and make it develop into diversification and diversification, evaluate the professional settings and courses in an all-round way, provide more analysis data and help teachers improve the quality of teaching. The impact of new things and technology, new knowledge and new courses brought by the era of big data makes students not only understand and broaden their knowledge, but also help to improve their interest in learning and understand the social needs. It is helpful to develop self-learning by making large data analysis and making their own learning plans [24].

2.3 Practical Talent Training Implementation Plan and Process

According to the established personnel training objectives, and in combination with the characteristics of the statistics profession and the school itself, an implementation plan for the cultivation of applied talents in statistics is developed, including specific training contents, training methods, management systems, and evaluation systems. The whole process of cultivation covers every aspect of the entrance and exit of students, teachers, personnel selection, talents training, and talent transfer. It truly achieves "one-stop production, production, and sales" operations and avoids the impact on the quality of personnel training due to students and employment. Carrying out practical practice work and strengthening cooperation between schools and enterprises.

In view of the problem that the effect is not ideal in the concrete practice work, this paper puts the practice into the students' examination, and constructs the "two classroom" practice mode. The first is to make students contact and understand the importance of practice to skill development from early grade through teaching methods such as classroom simulation and simulation laboratory [21]. At the same time, to a certain extent, the enterprise cannot fully meet the deficiencies of all students practice. Second, the second classroom practice, that is, the practice activities completed outside the classroom. Examples include student associations, youth volunteering, corporate and social support approaches to practice.

2.4 Cultivating Excellent Teachers with Strong Professional Skills

In the course of basic teaching, teachers should combine theory with practice, transition from previous theory teaching to practical operation to later professional teaching, practice teaching with theory and pay attention to interaction and communication between teachers and students. In the course of theory teaching, teachers should let the students understand the specific concepts, principles, basic methods and application fields of the contents of the course, and combine examples and case teaching to verify the relevant theories, so that students can really grasp the theoretical knowledge and memorizing them from the examples, and actively participate in the whole teaching quality. Students transform from passive acceptance to active learning, and actively encourage students to take the initiative to make use of their knowledge and skills to solve problems. In the course of practical teaching, teachers combine with the practice of professional courses to set up themes for students, explain the matters of attention and key techniques in the process of practice, actively guide and enlighten the students to think actively, carefully review the experimental reports and listen to solutions, and avoid the phenomenon of plagiarism and laziness.

2.5 Determining the Quality Standard System of Applied Talents

In order to improve the quality of applied talents training, we must strengthen the training and management of applied talents, and the basis of quality management is quality standards. All applied talents training institutions should set up their own quality standard system. According to the specifications and characteristics of the training of applied talents, the basic requirements of higher education should be met in the teaching of basic theory, and what is "what is" in the field of professional basic theory and professional theory teaching. For example, the origin of knowledge, the content of knowledge itself, the conditions for the establishment of knowledge itself, the problems that should be paid attention to in the process of applying knowledge, etc., do not have to pay much attention to "why", such as the reasoning process, induction process, deduction process and related principles of knowledge itself. In basic experiments, it mainly trains applied students' basic experimental skills and the most basic comprehensive application ability. In the field of professional foundation and professional practice teaching, we should focus on cultivating the independent experiment ability and comprehensive application skills of applied students, and improve the students' operational skills through practice and practice, so as to accumulate relevant practical experience. In the process of training applied talents, the quality standards for different knowledge modules are also different, as shown in Table 1.

Table 1. The quality standard system for applied talents

Serial number	Knowledge module	Basic content	Quality standard
1	General knowledge	The main body of public basic courses is theoretical teaching, and a bit of experimental content	Meet the basic requirements stipulated by the Ministry of education, and raise no requirement in principle of improve part
2	Professional knowledge	Professional basic knowledge and professional practice content	To apply knowledge to practice, to build a bridge between knowledge and object, to master the skill of operation, and to take the certificate of vocational qualification certificate as the requirement of skill training
3	High quality and innovation ability	Innovating education and expanding education	Cultivate students' awareness and ability of autonomous learning; cultivate innovation and innovation consciousness; stimulate students' desire for reality and imagination

3 Quantitative Analysis Model of Applied Talent Training

3.1 Big Data Analysis Model of Applied Talent Training

The statistical mathematical analysis model of applied talent training is constructed, and the significance of applied talent training is analyzed by using T statistical test analysis method, and the Cauchy kernel optimal boundary is constructed. The big data analysis model of applied talent training is a given multi-objective optimization problem, and its Cauchy kernel optimal boundary is defined as:

$$PF = \{f(X) = (f_1(X), f_2(X), \ldots, f_r(X)) | X \in \{X^*\}\} \quad (1)$$

By using the method of univariate regression analysis, it is obtained that the set $f(x)$ of all periodic points of the statistical characteristic quantity f of applied talents training is a continuous self-mapping on closed interval I. it is expressed as $f: I \to I$, and the benefit index is introduced to carry out quantitative regression analysis. Combined with one-sided test analysis method, the constraint conditions of probability and statistical analysis of applied talents training are obtained, as follows:

1) $\lim_{n\to\infty} \sup|f^n(x) - f^n(y)| > 0, \forall x, y \in S, x \neq y;$
2) $\lim_{n\to\infty} \inf|f^n(x) - f^n(y)| = 0, \forall x, y \in S;$
3) $\lim_{n\to\infty} \sup|f^n(x) - f^n(y)| > 0, \forall x \in S, \forall y \in P(f)$

On the basis of satisfying the above constraint conditions, the sample observation value $x_{n+1} = \mu x_n(1 - x_n)$ is a nonlinear dynamic system with two accumulative and time-varying characteristics. A target lever model for the cultivation of applied talents is established, and the singular decomposition matrix results are obtained by using differential functional submatrix:

$$\begin{aligned}
&\max(H_{ac})\boldsymbol{A}^{(\alpha_1,\cdots,\alpha_m)}(\boldsymbol{A}^{-1})^{(\alpha_1^{-1},\cdots,\alpha_m^{-1})^T} \\
&= \begin{pmatrix} \alpha_1 a_{1,1} & \cdots & \alpha_m a_{1,m} \\ \vdots & \ddots & \vdots \\ \alpha_1 a_{m,1} & \cdots & \alpha_m a_{m,m} \end{pmatrix} \begin{pmatrix} \alpha_1^{-1} t_{1,1} & \cdots & \alpha_1^{-1} t_{1,m} \\ \vdots & \ddots & \vdots \\ \alpha_m^{-1} t_{m,1} & \cdots & \alpha_m^{-1} t_{m,m} \end{pmatrix} \\
&(x_1, \cdots, x_m)^T \\
&= \boldsymbol{EI}(a_i))
\end{aligned} \tag{2}$$

Under the condition of satisfying the boundary condition, the residual statistic of applied talents training satisfies $\boldsymbol{x} = (x_1, \cdots, x_m)^T \in \boldsymbol{GF}(2^n)^m$, the power coefficient and constant of $D_{0+}^{\beta}u(s), G(t, s)f(s, u(s))$ are zero, and the random measure and Brownian motion are independent. The method of continuous function analysis is adopted. In order to evaluate the benefit of applied talent training, the output stable solution of talent training and the statistical average value of $P_0(x_1^0, x_2^0)$ at equilibrium point are obtained by using the method of association feature mining. The series X and $P_0(x_1^0, x_2^0)$ are analyzed in $f(x_1, x_2)$ point, and the applied type is obtained. The approximate linear equations of the linear programming for personnel training are expressed as follows:

$$\begin{cases} \frac{dx_1(t)}{dt} = f_{x_1}(x_1^0, x_2^0)(x_1 - x_1^0) + f_{x_2}(x_1^0, x_2^0)(x_2 - x_2^0) \\ \frac{dx_2(t)}{dt} = g_{x_1}(x_1^0, x_2^0)(x_1 - x_1^0) + g_{x_2}(x_1^0, x_2^0)(x_2 - x_2^0) \end{cases} \tag{3}$$

The significance of applied talents training is analyzed by using T statistical test analysis method, and the benefit distribution model of applied talents training is established. Thus, the stable solution is solved. The big data analysis of multi-applied talents training in Cauchy core is carried out. The stability coefficient matrix of the initial solution is described as:

$$A = \begin{bmatrix} f_{x_1} & f_{x_2} \\ g_{x_1} & g_{x_2} \end{bmatrix} |_{P_0(x_1^0, x_2^0)} \tag{4}$$

The characteristic equation coefficients of the statistical average initial value stability of the big data analysis of applied talents training are expressed as follows:

$$p = -(f_{x_1} + g_{x_2})|_{P_0(x_1^0, x_2^0)} \tag{5}$$

$$q = \det A \tag{6}$$

The stability criterion of big data mining is obtained at the equilibrium point $P_0(x_1^0, x_2^0)$ by combining the multiplex collinear analysis method.

$$G(x) = \begin{cases} \partial f(x), x \in Levf \\ a, x \in Levf \\ \partial C(x) \end{cases} \tag{7}$$

The big data robust mining model of applied talent training is constructed, and the cross-balance point of applied talent training is obtained as:

$$\begin{cases} a(H_{ac}) = 1 - \frac{H_{ac}}{\max(H_{ac}) + l} \\ \max(H_{ac}) = \log_2 k \end{cases} \tag{8}$$

The stability and convergence of initial values of differential equations with multiple complex variables in Cauchy kernel are proved. It is concluded that the quantitative analysis process of applied talent training designed in this paper is stable and convergent.

3.2 Statistical Analysis Modeling of Applied Talent Training Based on Probability Theory

Combining with the descriptive statistical analysis method of single variable, using probability theory statistical analysis and big data analysis method to realize the evaluation of benefit index of applied talent training, the characteristic value of big data probability statistical decomposition of applied talent training is obtained as:

$$f(x_1, x_2, i) - g(y_1, y_2, i) + \int (h(x_1, x_2, i) - g(y_1, y_2, i))xdu < x(|x-y|^2 + x|x-y|^2) \tag{9}$$

Where, $\forall x_1, x_2, y_1, y_2 \in R$, let x^* be a limit point in the optimal solution set $\{x_k\}$ of the statistical big data for the cultivation of applied talents, because the oscillatory amplitude of the statistical big data time series of the training of applied talents is satisfied:

$$C_o(x^*) < 0 \tag{10}$$

The positive semidefinite minimum characteristic of a given stiffness matrix is:

$$\Delta E = -\eta \left[\left(\frac{\partial E}{\partial \omega} \right)^2 + \left(\frac{\partial E}{\partial b} \right)^2 \right] \tag{11}$$

If the probability distribution of applied talent training statistics $C_o(x^*) = 0$, then:

$$Y(P, Q, \beta) = Y\left[red(P, Q, \beta), Q, \beta\right] \tag{12}$$

Using the adaptive beam frequency modulation method, the optimal solution of the statistical big data matrix of applied talents training is obtained as follows:

$$\dot{x}(t) = Ax(t) + BKx(t - d_s(t) - d_a(t)) \tag{13}$$

The minimum information entropy is used for functional weighting and the piecewise linear test method is used for pattern identification:

$$\begin{aligned}\frac{dS_k(t)}{dt} &= O_k(t)\left(P_{os1_k} + P_{os2_k}\right) + T_k(t)P_{tsk} - S_k(t)P_{spk} \\ &- S_k(t)\beta_{ak} - S_k(t)\left(\beta_{u_k(t)} + \beta_{l_k(t)}\right) \\ &- S_k(t)\left(P_{so1_k} + P_{so2_k}\right)\end{aligned} \tag{14}$$

By using linear regression estimation, the control coefficient of statistical big data mining of applied talents training can be obtained. The quantitative analysis results of applied talents training are obtained by using the method of correlation regression analysis:

$$\Psi_1(d_1(t)) = \Psi + d_1(t)K(Z_1 + Z_2 + Z_3)^{-1}K^T + (h_1 - d_1(t))\left[WZ_1^{-1}W^T + L(Z_2 + Z_3)^{-1}L^T\right] \tag{15}$$

$$\Psi_2(d_2(t)) = \Psi + (h_2 - d_2(t))L(Z_2 + Z_3)^{-1}L^T + d_2(t)M^T(Z_2 + Z_3)^{-1}M^T \tag{16}$$

When $\Psi(d_1(t), d_2(t)) < 0$, then:

$$\dot{V}(t) \le \xi^T(t)\Psi(d_1(t), d_2(t))\xi(t) < 0 \tag{17}$$

The cumulative average analysis method is used to evaluate the bilateral reliability of applied talents training, and the big data mining and feature extraction methods are used to analyze the characteristics of applied talents training. The solution vectors are obtained as follows:

$$\begin{aligned}\frac{dS_k(t)}{dt} &= O_k(t)\left(P_{os1_k} + P_{os2_k}\right) + T_k(t)P_{tsk} - S_k(t)P_{spk} - S_k(t)\beta_{ak} - S_k(t)\left(\beta_{u_k(t)} + \beta_{l_k(t)}\right) \\ &- S_k(t)\left(P_{so1_k} + P_{so2_k}\right)\end{aligned} \tag{18}$$

$$\frac{dE_k(t)}{dt} = S_k(t)\left(\beta_{u_k(t)} + \beta_{l_k(t)}\right) + X_k(t)P_{xe_k} - E_k(t)P_{ex_k} - E_k(t)P_{et_k} - E_k(t)\left(P_{ei1_k} + P_{ei2_k}\right) \tag{19}$$

$$\frac{dP_k(t)}{dt} = T_k(t)P_{tp_k} + S_k(t)P_{sp_k} \tag{20}$$

According to whether the Banach fixed point of the autoregressive linear initial value solution of multiple complex differential equations in Cauchy kernel can satisfy the continuous characteristic constraint condition, the convergence is determined. The statistical analysis of probability theory and big data analysis are used to evaluate the benefit of applied talents training, and to improve the ability of quantitative analysis and evaluation of applied talents training.

4 Empirical Analysis and Testing

In order to test the application performance of this method in the realization of quantitative analysis of applied talents training, the experiment adopts SPSS14.5 software and Matlab 7 simulation test, and the maximum iteration of the collection of benefit index parameters of talent training is obtained. The number of times is 100, the discrete sampling rate is $f_s = 10 * f_0$ Hz $= 10$ KHz, the sample length of big data distribution is 1024, the correlation parameter are set as $W_{\min} = 0.4$, $W_{\max} = 0.9$, $C_{\min} = 1.5$, $C_{\max} = 2.0$, according to the above experimental scenario, 50 experiments are carried out on each set of data sets, and the statistics of cultivating applied talents are big. The result of statistical analysis of the rate of return of talent training is shown in Fig. 1.

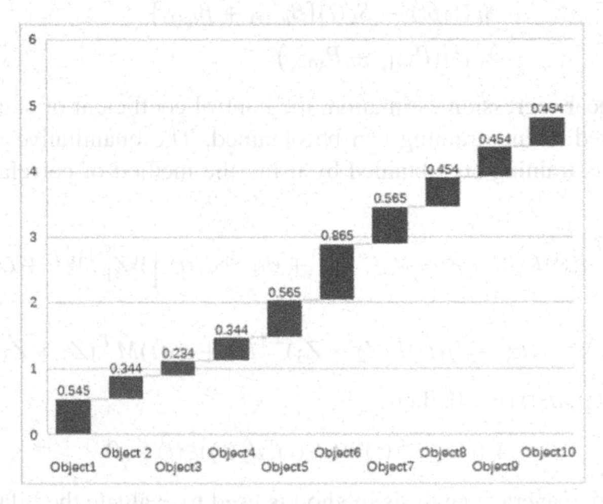

Fig. 1. Rate of return on the benefits of talent training

Figure 1 shows that the probability statistics of talent training by using this method, the rate of return on talent training is increasing gradually, and the results of robust regression analysis of applied talents training are further tested, as shown in Fig. 2.

The simulation results of Fig. 2 show that the quantitative evaluation of applied talents training with this design model has good accuracy, high robustness and high level of confidence, which improves the ability of quantitative regression analysis for the cultivation of applied talents.

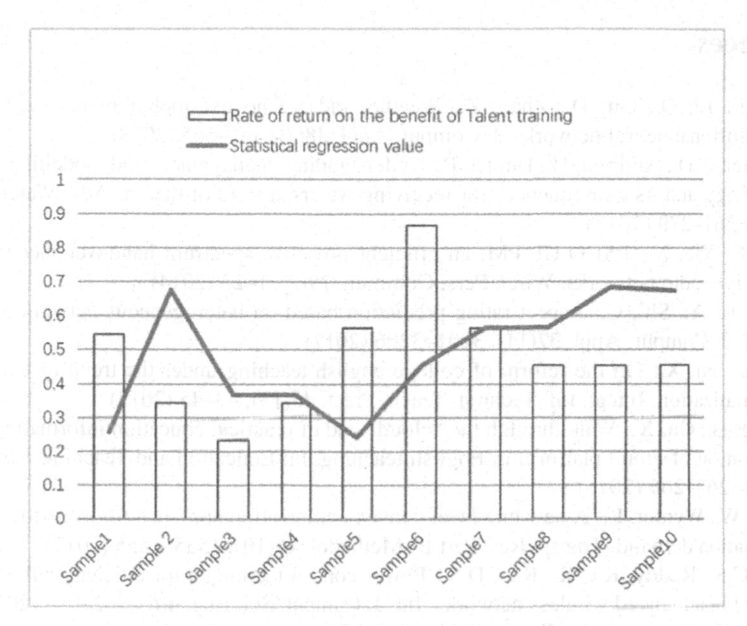

Fig. 2. Steady regression analysis results of applied talent training

5 Conclusions

In this paper, quantitative evaluation and big data analysis are used to evaluate the training of applied talents. A quantitative analysis model of applied talents training is proposed based on big data and probability statistics. The statistical mathematical analysis model of applied talents training is constructed, and the significance of applied talents training is analyzed by using T statistical test analysis method, and the benefit distribution model of applied talents training is established. The cumulative average analysis method is used to evaluate the bilateral reliability of applied talents training, and the big data mining and feature extraction methods are used to analyze the characteristics of applied talents training. The big data robust mining model for the cultivation of applied talents is constructed, and the descriptive statistical analysis method of single variable is taken, the statistical analysis of probability theory and big data analysis method are used to realize the evaluation of benefit index for the cultivation of applied talents. The results of empirical analysis show that the model has good accuracy and high level of confidence in the quantitative evaluation of applied talents training. This method has good application value in quantitative evaluation of applied talents.

Research Programs:

1. The Research Program of National Natural Science Foundation—Research on Construction Theory and Related Information Calculation of High Efficiency High Rank Space Time Code, No: 11861001
2. The general program of Aba Teachers University—Applied Research on Factor Analysis in Multivariate Statistics, No: ASB18-05

References

1. Shi, E., Li, Q., Gu, D., Zhao, Z.: Weather radar echo extrapolation method based on convolutional neural networks. J. Comput. Appl. **38**(3), 661–665 (2018)
2. Fletcher, T.D., Andrieu, H., Hamel, P.: Understanding, management and modelling of urban hydrology and its consequences for receiving waters, a state of the art. Adv. Water Resour. **51**(1), 261–279 (2013)
3. Ma, B., Xie, X.: PSHO-HF-PM: an efficient proactive spectrum handover mechanism in cognitive radio networks. Wirel. Pers. Commun. **79**(3), 1–23 (2014)
4. Ji, Y., Li, Y., Shi, C.: Aspect rating prediction based on heterogeneous network and topic model. J. Comput. Appl. **37**(11), 3201–3206 (2017)
5. Xue, J., Ni, X.: On the reform of college english teaching under the trend of educational informatization. Integr. Inf. Technol. Teach. Pract. **45**(12), 43–45 (2015)
6. Zheng, B., Gu, X.: Walk through the "cloud" end of practical education information–on the application of cloud platform in English teaching. In: Education and Teaching Forum, vol. 03, pp. 263–265 (2016)
7. Shen, W., Wynter, L.: A new one-level convex optimization approach for estimating origin–destination demand. Transp. Res. Part B: Methodol. **46**(10), 1535–1555 (2012)
8. Rao, C.S., Reddy, K.C.K., Rao, D.S.: Power control technique for efficient call admission control in advanced wireless networks. Int. J. Comput. Sci. Eng. **4**(6), 962–973 (2012)
9. Shi, H.-Y., Wang, W.-L., Kwok, N.M., et al.: Game theory for wireless sensor networks: a survey. Sensors **12**(7), 9055–9097 (2012)
10. Zhang, G.-P., Liu, P., Ding, E.-J.: Energy efficient resource allocation in non-cooperative multi-cell OFDMA systems. J. Syst. Eng. Electron. **22**(1), 175–182 (2011)
11. Xiong, X., Yang, L., Ma, Y., Zhuang, Z.: Alerting algorithm of low-level wind shear based on fuzzy C-means. J. Comput. Appl. **38**(3), 655–660 (2018)
12. Zheng, J.F., Zhang, J., Zhu, K.Y., et al.: Gust front statistical characteristics and automatic identification algorithm for CINRAD. Acta Meteorologica Sinica **28**(4), 607–623 (2014). https://doi.org/10.1007/s13351-014-3240-2
13. Hwang, Y., Yu, T.Y., Lakshmanan, V., et al.: Neuro-fuzzy gust front detection algorithm with S-band polarimetric radar. IEEE Trans. Geosci. Remote Sens. **55**(3), 1618–1628 (2017)
14. Sun, H., Zhang, H., Wu, J.: Correlated scale-free network with community: modeling and transportation dynamics. Nonlinear Dyn. **69**(4), 2097–2104 (2012)
15. Killip, R., Visan, M.: The defocusing energy-supercritical nonlinear wave equation in three space dimensions. Trans. Am. Math. Soc. **363**(7), 3893–3934 (2011)
16. Sun, C.L., Qian, M.M.: The research on applied talenttraining mechanism of undergraduate based on the combination of professional programmatic accreditation and double certificates. In: International Computer Conference on Wavelet Active Media Technology and Information Processing, vol. 34, no. 8, 316–319 (2014)
17. Nuo, L.I.: The training scheme of applied talents on combination of enterprises and universities research in local universities. J. Jiaying Univ. **42**(15), 321–327 (2011)
18. Yang, L.Y., Jiang, J.L.: Investigation and research on applied talents training of educational technology specialty in local undergraduate universities—with Jiaying University as example. J. Jiaying Univ. **41**(9), 241–249 (2013)
19. Sun, L., Shen, Q., Zhao, L.: Construction of applied talents training mode for the statistics majors based on the Big Data technology. J. Jilin Inst. Chem. Technol. **34**(6), 35–40 (2017)
20. Yu, Y.: Study on probability and mathematical statistics teaching in the training pattern of applied professionals. J. Langfang Teachers Univ. (Nat. Sci. Ed.) **16**(4), 118–119 (2016)
21. Peng, T., Sun, L., Liu, C.: Research on applied IT talents cultivation based on the OBE model—big data approach. Softw. Eng. **20**(8), 56–58 (2017)

22. Zhang, C., Yang, C., Wu, S., Zhang, X., Nie, W.: A straightforward direct traction boundary integral method for two-dimensional crack problems simulation of linear elastic materials. Comput. Mater. Continua **58**(3), 761–775 (2019)
23. Yuan, F., Zhang, Q., Xia, X.: Effect of reinforcement corrosion sediment distribution characteristics on concrete damage behavior. Comput. Mater. Continua **58**(3), 777–793 (2019)
24. Guo, F., et al.: Research on the law of garlic price based on big data. Comput. Mater. Continua **58**(3), 795–808 (2019)

Region Proposal for Line Insulator Based on the Improved Selective Search Algorithm

Shuqiang Guo, Baohai Yue, Qianlong Bai, Huanqiang Lin, and Xinxin Zhou(✉)

NorthEast Electric Power University, Jilin 132012, China
guoshuqiang@gmail.com, zxx51@qq.com

Abstract. The region proposal algorithm for line insulators is a critical step in line insulator detection. The selective search algorithm provides a rich candidate area for line insulators. However, the selective search algorithm uses SIFT descriptor to extract texture features. It causes a large time overhead. In response to this problem, this paper proposes an improved algorithm. The algorithm use the Harr-Like algorithm to form a feature value map. Then, the formed feature value map is mapped to a fixed-size texture vector HBSN (Harr-Like Based on SPP-Net) by using Spatial Pyramid Pooling Network. In the process of the formation of feature maps and the construction of feature vectors, the integral image is used for acceleration. The integral image can only traverses the pixels in the original image once, forming an integral image of the feature values corresponding to the original image. When calculating the texture vector of different regions in the original image, it only needs to index the corresponding position of the eigenvalue integral image, which overcomes the problem of repeated calculation of texture features in the overlapping region by selective search and achieves fast calculation of texture similarity in different regions.

Experiments show that the improved selective search algorithm for line insulator has a calculation speed increase of 8.21% compared with the selective search.

Keywords: The improved selective search algorithm · Features of Harr-like · SPP-Net · Integral image

1 Introduction

The position of the insulator in the picture is the key to insulator detection. A commonly used method for extracting candidate regions where object may exist is the sliding window method [1]. It scans images of different sizes by using different size windows. This exhaustive algorithm that may have object windows creates a significant amount of time overhead. To this end, Uijlings [2] proposed a selective search algorithm to improve. It uses image segmentation algorithm to divide the image into mutually independent color blocks by taking into account factors such as texture, color, size and overlap. Finally, the divided color blocks are combined to form a new one by combining multiple similarities. The algorithm has a high recall rate and can provide a reasonable target area.

X. Sun et al. (Eds.): ICAIS 2020, LNCS 12240, pp. 686–696, 2020.
https://doi.org/10.1007/978-3-030-57881-7_60

It solves the problem that the sliding window has a large time overhead caused by the exhaustive target area. The algorithm is widely used in Fast R-CNN [3], Faster R-CNN [4]. Li [5] applies the selective search algorithm to vehicle face component detection. Firstly, the detection algorithm proposes to use the Histogram of Oriented Gridients (HOG) descriptor [6–8] instead of the Scale-invariant feature transform (SIFT) [9, 10] descriptor to form texture features to improve the selective search algorithm. Then, the improved algorithm is used to segment different parts of the car face and achieves good segmentation results on the car face data set. Wu [11] proposed using the perceptual hash algorithm and the Hamming distance algorithm to improve the similarity calculation in the selective search algorithm. Experiments show that the improved algorithm on the LFW data set has obvious advantages over the sliding window. These algorithms are mainly used to improve the similarity calculation in the selective search algorithm, and use the new feature extraction method instead of SIFT algorithm to achieve the purpose of fast calculation. However, when calculating the texture similarity of different regions, these overlapping regions cause a problem of double counting due to the overlap of different regions.

In view of the above problems, this paper improves the selective search. The selective search algorithm and the improved selective search are applied to the insulator data separately. The effectiveness of the improved selective search algorithm for insulator regions proposal is examined.

2 The Improved Selection Search Algorithm

Selective search algorithm [2] uses image segmentation algorithm to divide the image into independent image regions and uses a variety of similarity rules to merge the segmented image region to form new image regions. The specific process is as follows.

1. Input a colorful image A.
2. The algorithm proposed by Felzenszwalb [12] is used to segment the image A into different regions. The segmented image regions is recorded as set R, and $R = \{r_1, r_2, r_3, \ldots, r_i, \ldots, r_n\}$.
3. Initialize the similarity set S to be empty, calculate the similarity $s(r_i, r_j)$ for any two elements r_i and r_j in R, and save the result $s(r_i, r_j)$ into the set S, that is $S = S \cup s(r_i, r_j)$.
4. If set S is not an empty set, traversal set S to takes out the maximum of $s(r_k, r_l)$ and merge the region r_k and r_l to form a new image region r_t, that is $r_t = r_k \cup r_l$. Remove the similarity $s(r_k, r*)$ with r_k and the similarity $s(r*, r_l)$ with r_l from set S. Compute the similarity set S_t corresponding to r_t. Merge S_t into S, that is $S = S \cup S_t$. Merge image region r_t into set R, that is $R = R \cup r_t$.
5. If S is empty set, output all image regions in R.

2.1 HBSN Feature

The selective search algorithm uses the Scale-Invariant Feature Transform (SIFT) to obtain the texture feature histogram corresponding to the three channels R, G and B of the image. The SIFT algorithm uses a Gaussian differential of $\sigma = 1$. It causes a large time consumption. In response to this problem, this paper raises the features of HBS N(Harr-Like Based SPP-Net). Firstly, the Harr-Like descriptor is used to extract the feature image to form the feature value map. Then the SPP-Net [13] algorithm is used to map the feature value map to a fixed length texture vector. Finally, the above two processes are accelerated using an integral image.

2.1.1 The Map of Feature Values

The Harr-Like feature descriptor [14] forms a description feature by comparing the gray scale differences of different parts. It has a strong expression of texture feature, which is used as the texture feature of the car in paper [15].

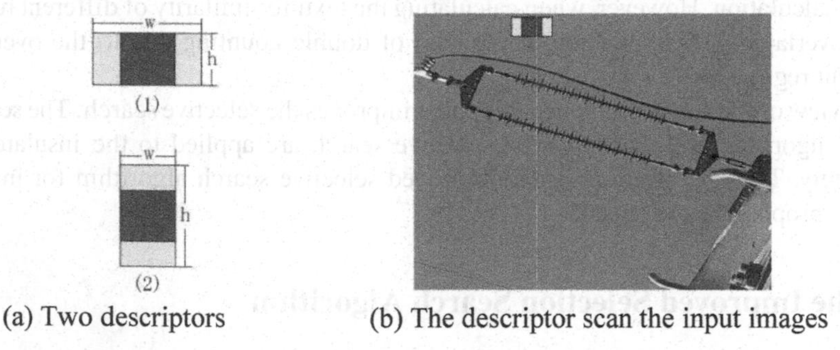

(a) Two descriptors (b) The descriptor scan the input images

Fig. 1. The feature extraction of Harr-like (Color figure online)

Harr-like's descriptors with many different shapes [14], two Harr-Like descriptors selected in this paper, as shown in Fig. 1(a). The picture of the input shown in Fig. 2(b). The process which the Harr-Like descriptor extracts picture features is as follows.

1. As depicted in Fig. 1(b), the Harr-Like descriptor moves from the leftmost end of the input image to the rightmost end of the input pattern in steps of one pixel along the line. Calculating the feature value V using Eq. 1 for the area covered by the Harr-Like descriptor.

$$V = \sum_{p(x,y)\in Y} pixel(x,y) - \sum_{p(x,y)\in R} pixel(x,y) \tag{1}$$

   where $p(x, y)$ represents the position of each pixel and $pixel(x, y)$ represents the corresponding pixel value of each pixel Y and R respectively represents the yellow and red regions corresponding to Fig. 1(a).
2. When the right end of the Harr-Like descriptor reaches the end of the image, the Harr-Like descriptor moves down one pixel.
3. Repeat 1and 2 until the Harr-Like descriptor reaches the lower right corner of the image to be processed. The feature value map corresponding to the input picture is obtained. The size calculation of the feature value map is as shown in Eq. 2.

$$W_1 = \frac{W-w}{s_x} + 1, H_1 = \frac{H-h}{s_y} + 1 \quad (2)$$

where W and W_1 respectively represents the width of the input image and the width of the map of feature values. H and h respectively represents the width of the input image and the width of the map of feature values. w and h respectively represents the width and height of Harr-Like descriptor. s_x and s_y respectively represents the stride of Harr-Like on the x-axis and y-axis of the input image.

2.1.2 HBSN Vector

He [13] proposed the SPP-Net algorithm which maps different size pictures to fixed-length vectors. The algorithm shows excellent performance in neural networks.

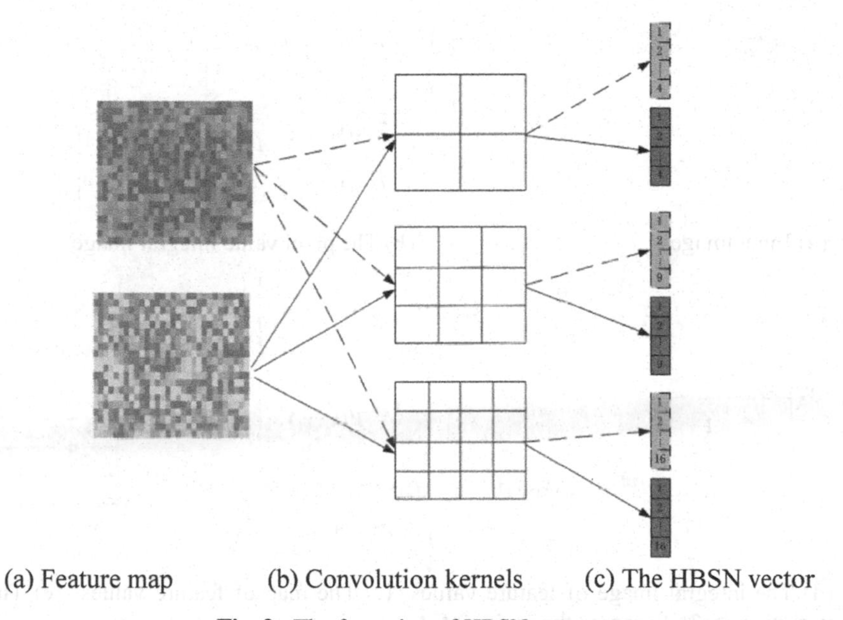

(a) Feature map (b) Convolution kernels (c) The HBSN vector

Fig. 2. The formation of HBSN vectors

In Fig. 2, each patch on the feature map represents one feature value. The length and width of the feature value map are respectively divided into 4 equal parts, 3 equal parts, and 2 equal parts by using three convolution kernels of 4 × 4, 3 × 3 and 2 × 2. Thus, a feature value map is divided into 16, 9 and 4 different block regions. Each block which is the corresponding region in the feature value map is averaged by using Eq. 3 to obtain its corresponding feature value Vj. Thus, a feature value map is mapped into a 29-dimensional feature vector. The 58-dimensional vector formed by the two feature maps is used as the feature vector of a channel of the input image.

$$V_j = \frac{1}{N_j} \sum_{p(x,y) \in Block_j} pixel(x, y) \quad (3)$$

where $Block_j$ represents the jth Block region, and $j \in [1, 2, \ldots, 29]$. N_j represents the total number of pixels in the jth Block region. $p(x, y)$ represents the pixel in $Block_j$ and $pixel(x, y)$ represents the corresponding pixel value of $p(x, y)$ pixel.

2.1.3 Integral Image Acceleration

Accelerating the process of forming map of feature values by integral image algorithm. As shown in Fig. 3(a), the pixel values in the input image are traversed. The Formula 4 is used to calculate the value $T(x, y)$ corresponding to any point (x, y) on the pixel value integral image and the formed pixel value integral image is shown in Fig. 3(b).

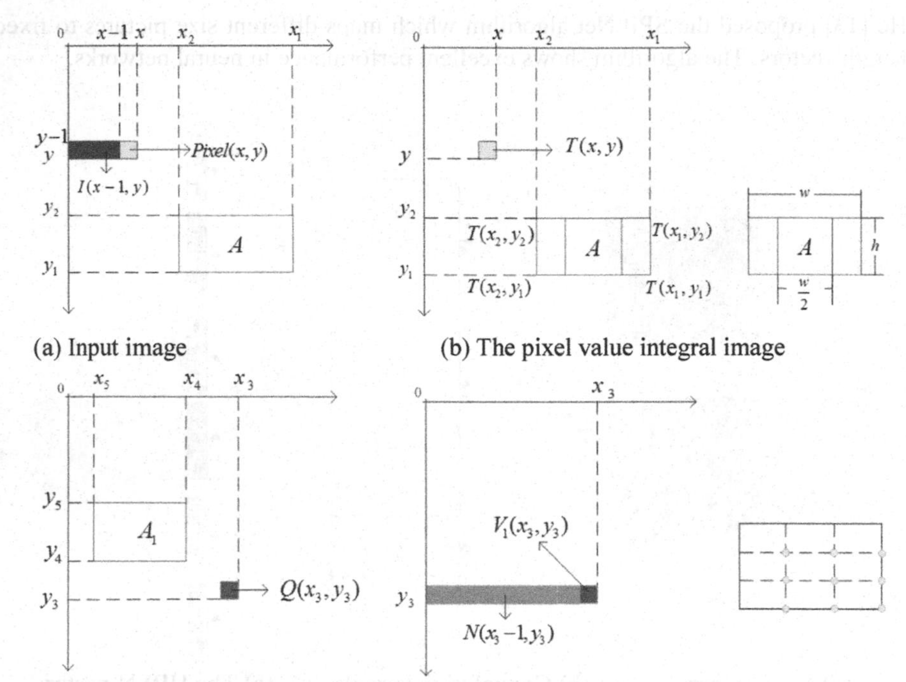

(d) The integral image of feature values (c) The map of feature values (e) The convolution of 3×3 process the region of A_1

Fig. 3. Acceleration of HBSN vector

In Fig. 3(a), the sum of the pixel values in rectangular area A of any size is calculated, which is equivalent to using formula 5 to calculate T_R at the corresponding position of A in Fig. 3(b).

$$I(x, y) = I(x, y - 1) + pixel(x, y)$$
$$T(x, y) = T(x - 1, y) + I(x, y) \tag{4}$$

where $I(x, y)$ denotes the sum of pixel values from the point (0, y) along the x-axis to the point (x, y). $T(x, y)$ denotes the sum of pixels values in the rectangular region from the point (0, 0) to the point (x, y), where $I(-1, y) = 0$, $T(x, -1) = 0$.

$$T_R = T(x_1, y_1) - T(x_2, y_1) - T(x_1, y_2) + T(x_2, y_2) \tag{5}$$

where (x_1, y_1), (x_2, y_1), (x_1, y_2), (x_2, y_2) respectively represent the four vertices of the rectangular region. $x_1 > x_2$ and $y_1 > y_2$.

When the rectangular box A composed of x_1, x_2, y_1 and y_2 is Harr-Like descriptor which is shown in Fig. 3(a), x_1, x_2, x_3, y_1, y_2 and y_3 are satisfy formula 6 in the process of using the descriptor to form the feature value $V_1(x_3, y_3)$ in Fig. 3(c). When the Harr-Like descriptor is the type (1) in Fig. 1(a), the rectangular region composed of x_1, x_2, y_1 and y_2 is divided into three rectangles. The coordinates of three rectangles corresponding to eight points are (x_1, y_1), $(x_1\text{-}w/4, y_1)$, $(x_1 - 3*w/4, y_1)$, (x_2, y_1), (x_1, y_2), (x_2, y_2), $(x_1 - w/4, y_2)$ and $(x_1 - 3*w/4, y_2)$. Using formula 5 and Formula 1, the formula for calculating eigenvalue $V_1(x_3, y_3)$ is obtained as shown in formula 7. Similarly, When the Harr-Like descriptor is the type (2) in Fig. 1(a), the formula for calculating V_2 (x_3, y_3) is shown in Formula 8. Finally, the map of feature values corresponding to different Harr-Like descriptors are obtained, as shown in Fig. 3(c).

$$\begin{aligned} x_1 &= s_x(x_3 - 1) + w, \ x_2 = s_x(x_3 - 1) \\ y_1 &= s_y(y_3 - 1) + h, \ y_2 = s_y(y_3 - 1) \end{aligned} \tag{6}$$

where w and h respectively represent the width and height of Harr-Like descriptor. s_x *and* s_y respectively represent the stride of Harr-Like descriptor on the x-axis and y-axis of the image.

$$\begin{aligned} V_1(x_3, y_3) = {} & T(x_1, y_1) + 2T(x_1 - w/4, y_2) + 2T(x_1 - 3w/4, y_1) + T(x_2, y_2) \\ & - T(x_1, y_2) - 2T(x_1 - w/4, y_1) - 2T(x_1 - 3w/4, y_2) - T(x_2, y_1) \end{aligned} \tag{7}$$

$$\begin{aligned} V_2(x_3, y_3) = {} & T(x_1, y_1) + 2T(x_2, y_1 - h/4) + 2T(x_1, y_1 - 3h/4) + T(x_2, y_2) \\ & - T(x_2, y_1) - 2T(x_1, y_1 - h/4) - 2T(x_2, y_2 - 3h/4) - T(x_1, y_2) \end{aligned} \tag{8}$$

where w and h respectively represent the width and height of the Harr-Like descriptor.

Integral images are used to accelerate the process of mapping the map of feature values for texture vectors. For the feature value $V(x_3, y_3)$ of any point (x_3, y_3) on Fig. 3(c), the value $Q(x_3, y_3)$ corresponding to the (x_3, y_3) is calculated by formula 9 to form the integral image of feature values shown in Fig. 3(d). s_x= 1 and s_y= 1 is used in this article. When the input image is rectangular A of any size in Fig. 3(a), Formula 10 is used to calculate the coordinates of A on the Fig. 3(d) and the area A_1 corresponding area A is obtained. As shown in Fig. 3(e), area A_1 is divided into nine blocks by using convolution kernel which size is 3 × 3 in Fig. 3(e).

$$\begin{aligned} N(x_3, y_3) &= N(x_3, y_3 - 1) + V(x_3, y_3) \\ Q(x_3, y_3) &= Q(x_3 - 1, y_3) + N(x_3, y_3) \end{aligned} \tag{9}$$

where $N(x_3, y_3)$ represents the sum of the feature values from the point (0, y) along the x-axis to the point (x_3, y_3). $Q(x_3, y_3)$ represents the sum of the feature values in the rectangular region from the point (0, 0) to the point (x_3, y_3), where $N(-1, y_3) = 0$, $T(x_3, -1) = 0$.

$$x_4 = x_1 - w+1,\ x_5 = x_2+1$$
$$y_4 = y_1 - h+1,\ y_5 = y_5+1 \tag{10}$$

The process of calculating the mean of each block to form dimension eigenvector value, is equivalent to summing all feature values in each block and dividing them by the total number of feature points in corresponding block. In the above calculation process, the sum of feature values T_R in any rectangle can be calculated by using the four values of the corresponding region on the integral image of feature values. The k = 3 is substituted into Formula 11, and the mean of each block in Fig. 3(e) is obtained, forming a 9-dimensional vector. Similarly, when using convolution kernels of sizes are 4×4 and 2×2, for A, $k = 4$ and $k = 2$ are respectively brought into Formula 11 to form corresponding 16 and 4-dimensional vectors. According to the order of eigenvalues in Fig. 2, two maps of feature values are mapped to a 58-dimensional vector.

$$V(l, m) = \frac{1}{N}[Q(x_5 + al, y_5 + bm) - Q(x_5 + a(l-1), y_5 + bm)$$
$$-Q(x_5 + al, y_5 + b(m-1) + Q(x_5 + a(l-1), y_5 + b(m-1)] \tag{11}$$

where $a = (x_4 - x_5) \div k$, $b = (x_4 - x_5)/k$ and $N = a \times b$. k is constant and $k \in [2, 3, 4]$. $1 \leq l \leq k$ and $1 \leq m \leq k$.

2.2 The Similarity $S(r_i, r_j)$

(1) Computation of texture similarity

Two kinds of Harr-Like descriptors shown in Fig. 2(a) are used to extract features from every color channel of image region r_i. Two kinds of Harr-Like descriptors form a 58-dimensional feature vector on every color channel using Formula 11. The every 58-dimensional vector obtained is normalized by using L_1-norm. The Harr-Like features formed by R, G and B channels are arranged sequentially to form the 174-dimensional vector T_i corresponding to the image region r_i, $T_i = \{t_i^1, t_i^2,..., t_i^k,..., t_i^{174}\}$. Similarly, the texture feature vector of the image region is T_j, $T_j = \{t_j^1, t_j^2,..., t_j^k,..., t_j^{174}\}$. For image region r_i and r_j, formula 12 is used to calculate similarity $S_{text}(r_i,r_j)$.

$$S_{text}(r_i, r_j) = \sum_{k=1}^{174} \min(t_i^k, t_j^k) \tag{12}$$

(2) Computation of color similarity

On the image region r_i, every color channel is divided into 25 groups to form 25 *bins*. Using L_1-norm to regularize the values of the 25 *bins*. A total of 75 *bins* by forming in the R, G and B color channels is denoted C_i,and $C_i = \{c_i^1, c_i^2,..., c_i^k,..., c_i^{75}\}$. Similarly, the color feature vector of the image region r_j is C_j, $C_j = \{c_j^1, c_j^2,..., c_j^k,..., c_j^{75}\}$. The color similarity $S_{color}(r_i,r_j)$ of the image r_i and r_j is calculated by Formula 13.

$$S_{color}(r_i, r_j) = \sum_{k=1}^{75} \min(c_i^k, c_j^k) \tag{13}$$

(3) Computation of size similarity

In order to get as many candidate regions as possible, the smaller regions are merged first. The size similarity $S_{size}(r_i, r_j)$ between the image region r_i and r_j is computed by Formula 14.

$$S_{size}(r_i, r_j) = 1 - \frac{size(r_i) + size(r_j)}{size(im)} \tag{14}$$

where *size*(*im*) denotes the size of the image where r_i and r_j are located in pixel.

(4) Shape compatibility

$S_{fill}(r_i, r_j)$ is defined to indicate the matching degree between the image region r_i and r_j. If r_i is a part of r_j, r_i and r_j are merged. If there is no intersection between r_i and r_j, they are not merged. The formula of $S_{fill}(r_i, r_j)$ is shown in Formula 15.

$$S_{fill}(r_i, r_j) = 1 - \frac{size(BB_{ij}) - size(r_i) - size(r_j)}{size(im)} \tag{15}$$

where $size(BB_{ij})$ denotes the size of the bounding box which is around r_i and r_j.

(5) Computation of similarity $S(r_i, r_j)$

Using the linear combination of the four similarities mentioned above, the calculation formula of $S(r_i, r_j)$ is shown in Formula 16.

$$S(r_i, r_j) = a_1 S_{color}(r_i, r_j) + a_2 S_{texture}(r_i, r_j) + a_3 S_{size}(r_i, r_j) + a_4 S_{fill}(r_i, r_j) \tag{16}$$

where $a_1, a_2, a_3, a_4 \in [0, 1]$.

3 Experiments

3.1 Speed Comparison of Candidate Regions Generation

(a) The original selective search

(b) The improved selective search

Fig. 4. Comparison of candidate regions.

The image shown in Fig. 4(a) is an effect diagram after processing using the original algorithm. The image shown in Fig. 4(b) is an effect diagram after processing using an improved algorithm. 200 images containing insulators were randomly divided into four groups and each group contained 50 insulator images. The selective search algorithm and the improved selective search algorithm are respectively used for each picture to test the average number of generation regions and the corresponding average time consumption. The data obtained is shown in Table 1.

Table 1. Comparison of time and regions

Group	Oriented selective search		Improved selective search	
	Regions	Time (s)	Regions	Time (s)
1	287	1.276	335	1.054
2	324	1.547	389	1.367
3	293	1.335	358	1.246
4	308	1.362	371	1.289

In the Table 1, it can be seen that the improved selective search algorithm produces significantly more candidate regions on each set of insulator pictures than the original algorithm and its corresponding time consumption is lower than the original algorithm. The data analysis shows that the improved algorithm generates a candidate area faster than the original algorithm by 8.21%.

3.2 Comparison of Overlap

This paper uses Average Best Overlap (ABO) [2] as an evaluation index of algorithm performance. Divided 200 images containing insulators into four groups. The position of the insulator is calibrated using the manual calibration for the sample i in the k-th group and the calibrated region is denoted by g_i^k. Sample i is respectively processed using the improved selective search algorithm and the original algorithm, and the corresponding candidate region sets L_1 and L_0 are obtained. The improved ABO_1 of the selective search algorithm is as shown in Eq. 17, and the ABO_0 corresponding to the original algorithm is shown in Eq. 18. The calculation of Overlap(g_i^k, l_j) is shown in Eq. 3. The ABO_1 and ABO_0 data values corresponding to each set of insulator pictures are shown in Table 2.

$$ABO\_1 = \frac{1}{|G^k|} \sum_{g_i^k \in G^k} \max_{l_j \in L_1}(Overlap(g_i^k, l_j)) \tag{17}$$

$$ABO\_0 = \frac{1}{|G^k|} \sum_{g_i^k \in G^k} \max_{l_j \in L_0}(Overlap(g_i^k, l_j) \tag{18}$$

where G^k represents all samples in the k-th group insulator image, and l_j represents the jth candidate region in the set of candidate regions.

$$Overlap(g_i^c, l_j) = \frac{area(g_i^c) \cap area(l_j)}{area(g_i^c) \cup area(l_j)} \quad (19)$$

where area(g_i^k) denotes the area corresponding to g_i^k, and area l_j denotes the area corresponding to l_j in pixels.

Table 2. Comparison of ABO

Group	Oriented selective search	Improved selective search
1	67.32	68.52
2	64.92	66.43
3	63.84	64.24
4	66.82	69.13

In Table 2, data analysis shows that the improved selective search algorithm is better than the original algorithm in ABO performance and can provide candidate regions for insulator pictures.

4 Conclusion

Aiming at the problem that the selective search algorithm uses SIFT algorithm to extract texture features with large time overhead, this paper proposes the HBSN feature for improvement. Meanwhile, the integral image is introduced to accelerate the calculation of the HBSN feature. In the extraction of insulator candidate regions, the improved algorithm is 9.6% faster than the original algorithm. The coincidence contrast of the candidate regions generated by the improved algorithm is higher than the original algorithm. The algorithm overcomes the problem that the original algorithm is slow in extracting the candidate region of the insulator, and can improve the candidate region of the insulator image.

Acknowledgement. This research is partially supported by:

1. Research Foundation of Education Bureau of Jilin Province (JJKN20190710KJ).
2. Science and Technology Innovation Development Plan Project of Jilin city (20190302202).

References

1. Huang, K., Ren, W.: Overview of image object classification and detection algorithms. J. Comput. Sci. 1225–1240 (2014)

2. Uijlings, J.R.R., et al.: Selective search for object recognition. Int. J. Comput. Vis. **1**(1), 154–171 (2013)
3. Girshick, R.: Fast R-CNN. In: 2015 IEEE International Conference on Computer Vision (ICCV), pp. 1440–1448. IEEE, Santiago (2015)
4. Ren, S., He, K., Girshick, R., et al.: Faster R-CNN: towards real-time object detection with region proposal networks. IEEE Trans. Pattern Anal. Mach. Intell. **39**(6), 1137–1149 (2015)
5. Li, X., Zhou, Z., Lu, S.: Face component detection based on selective search algorithms. Comput. Eng. Sci. **34**(09), 1829–1836 (2018)
6. Naiel, M.A., Ahmad, M.O., Swamy, M.N.S.: A vehicle detection scheme based on two-dimensional HOG features in the DFT and DCT domains. Multidimensional Syst. Signal Process. **30**(4), 1697–1729 (2018). https://doi.org/10.1007/s11045-018-0621-1
7. Song, D., Zhao, B.-J.: The object recognition algorithm based on affine histogram of oriented gradient. J. Electron. Inf. Technol. **35**(6), 1428–1434 (2013)
8. Chen, Z., Yao, S., Liu, C., et al.: Sketch face recognition: P-HOG multi-features fusion. Int. J. Pattern Recogn. Artif. Intell. **33**(04), 1–17 (2019)
9. Zhou, H., Yuan, Y., Shi, C.: Object tracking using SIFT features and mean shift. Comput. Vis. Image Understand. **113**(3), 345–352 (2009)
10. Lv, G., Teng, S.W., Lu, G.: Enhancing SIFT-based image registration performance by building and selecting highly discriminating descriptors. Pattern Recogn. Lett. **84**(34), 156–162 (2016)
11. Wu, S., Zhan, Y.: Face detection based on selective search and convolutional neural network. Comput. Appl. Res. **34**(09), 2854–2876 (2017)
12. Felzenszwalb, P.F., Huttenlocher, D.P.: Efficient graph-based image segmentation. Int. J. Comput. Vis. **59**(2), 167–181 (2004)
13. He, K., et al.: Spatial pyramid pooling in deep convolutional networks for visual recognition. IEEE Trans. Pattern Anal. Mach. Intell. **37**(09), 1904–1916 (2015)
14. Yan, Z., Yang, F., Wang, J., Shi, Y., Li, C., Sun, M.: Face orientation detection in video stream based on Harr-like feature and LQV classifier for civil video surveillance. In: IET International Conference on Smart and Sustainable City 2013 (ICSSC 2013), pp. 161–165. IEEE, Shanghai (2013)
15. Wei, Y., et al.: Multi-vehicle detection algorithm through combining Harr and HOG features. Math. Comput. Simul. **155**(12), 130–145 (2019)

A Collaborative Filtering Algorithm Based on the User Characteristics and Time Windows

Dun Li, Cui Wang, Lun Li, and Zhiyun Zheng(✉)

Zhengzhou University, Zhenghou 450001, HN, China
iezyzheng@zzu.edu.cn

Abstract. Collaborative filtering algorithm is a widely used recommendation algorithm. In the traditional collaborative filtering algorithm, a single user similarity calculation method is usually considered, and the user's own attribute characteristics are not used as the basis of neighbor user selection. At the same time, in the process of recommendation, user's interest is considered to be static and given the same weight in different time periods, without thinking the dynamic changes of user's interest. For above problems, this paper proposes a collaborative filtering algorithm based on the user characteristics and time windows. Firstly, a collaborative filtering algorithm based on item rating and user's own attribute characteristics is proposed in the process of calculating similarity. Secondly, the dynamic time windows are divided according to the Ebbinghaus forgetting curve to reflect the user's short-term interests in the recommendation process, the concept of time function is added to assign different time weights to user interests in different periods in the process of interest fusion. Finally, through experimental analysis, the recommended effect of the algorithm is significantly improved compared with the traditional collaborative filtering recommendation algorithm.

Keywords: Collaborative filtering · User characteristics · Time windows · Similarity · Time weight

1 Introduction

In recent years, recommendation system has gradually become an important tool for people to quickly find useful information from massive data. With the development of the Internet, netizens, online enterprises and Internet companies are faced with the explosive growth of information and data. A large amount of data and information makes it impossible for users to obtain the required information quickly and accurately [1], and it takes a lot of time to search, which increases the cost of acquiring effective information, and that is a problem of information overload. In solving this problem, recommendation system shows high efficiency and strong expansibility, and gradually becomes one of the effective methods adopted by e-commerce.

Recommendation system can filter effective information from massive data according to users' interest and recommend it to users. As we all know, our interests and hobbies are constantly changing. Traditional collaborative filtering algorithm ignores the time

X. Sun et al. (Eds.): ICAIS 2020, LNCS 12240, pp. 697–709, 2020.
https://doi.org/10.1007/978-3-030-57881-7_61

factor that information changes [2]. As time goes on, it is a major practical challenge to gain user interests accurately in the recommendation system.

The research on the change of user interest over time has attracted much attention in the recommendation system. Koren [3] used the time sequence diagram to distinguish the long-term interest and short-term interest of users to get more accurate current interest of users. Haitao, C. et al. [4] integrated user interest similarity into user similarity, and measured user interest and its dynamic changes according to the user's proportion matrix and time of purchase of various items. Haiyan, Z. et al. [5] described user interests through tag popularity, obtained user interest similarity by using user's time behavior information, and generated recommendations for users by combining user similarity and time decay. The above studies considered this problem from different aspects, but they all ignored the influence of the specific time period when users visit resources on the prediction.

To solve these problems, this paper uses a hybrid recommendation fusion algorithm based on user characteristics and time windows to improve the above recommendation algorithms. Considering that the interests of users in different periods are different, so different weights are assigned to interests in different periods during interest fusion. Our algorithm assigns weights according to the time function, interests in the recent period can better reflect the interests of users and assign more weights, and the interests of the longer period assign a smaller weight. Therefore, our algorithm can well reflect the characteristics of user interests changing with time.

2 Related Work

Collaborative filtering recommendation is one of the first proposed and most widely used methods in the recommendation system. The concept of collaborative filtering was proposed by Goldberg et al. in 1992 [6] and applied to the Tapestry system. Traditional collaborative filtering recommendations are divided into two categories: memory-based recommendations and model-based recommendations [7].

The memory-based collaborative filtering recommendation algorithm is based on user-item rating matrix for recommendations [8]. It include user-based and item-based collaborative filtering algorithms. The basic idea is that if they are classified according to users, users belonging to the same category will be interested in the same item; if classified by items, items belonging to the same category will be interested by the same user. The model-based collaborative filtering algorithm is very different from the memory-based collaborative filtering algorithm. It mainly describes the user's behavior patterns with statistical methods and machine learning techniques, and builds a prediction model for recommendation with the user-item rating matrix.

Collaborative filtering algorithm is one of the effective methods to provide personalized recommendation for users. However, the traditional collaborative filtering algorithm considers the single problem and has low recommendation efficiency. Researchers have proposed some effective methods to update the traditional recommendation algorithm to improve the recommendation efficiency. In order to solve the problem of data sparsity and cold start, Zelong, L. et al. [9] put forward a collaborative filtering algorithm of calculating similarity based on item rating and attributes. Chenglung, H. et al. [10] introduced time-related similarity calculation and applied user interests to item attributes to recommend new items. Shan, L. et al. [11] improved the correlation coefficient formula by adding news hot parameters when calculating user similarity, and used the hybrid recommendation algorithm to predict user ratings.

Although the above algorithms improved the user similarity calculation to increase the recommendation efficiency, but do not consider the impact of time factors on the recommendation, ignoring the time attribute is also an important feature to reflect the user interest rules. Other researchers have developed new researches, Xiang, L. et al. [12] proposed the Session-based Temporal Graph algorithm and the Injected Interest Fusion recommendation algorithm in order to the interests of users changing over time. Yan, Y. et al. [13] introduced time attributes in the prediction process, and proposed a time-division collaborative filtering algorithm, which was processed by different strategies, which improved the recommendation quality of the system effectively. But they still ignore the role of time factors in the change of user interests, and give them the same weight, failing to reflect the user's interest changes with time. Huaizhen, L. et al. [14] put forward that the enhancement algorithm assigns each score a time weight that gradually decreases over time, and used the weighted score to search for the nearest neighbors. Jiguang, Z. et al. [15] divided the rating history of users into several time periods and analyzed the interest distribution of users in these periods. Changqiong, S. et al. [16] gave the time attribute to the item rating, and thought that the rating closer to the current moment has more reference effect. They believed that the user's interests are not static, but conform to a certain time rule. As time goes, user interests are given corresponding weights.

The above references separately considered the impact of user similarity and time attributes on recommendations, and failed to merge them. Therefore, this paper proposes a hybrid recommendation algorithm based on user characteristics and time windows: 1) In the process of calculating similarity, a collaborative filtering algorithm based on item rating and user characteristics is proposed; 2) In the recommendation process, the time windows are divided to reflect the user's short-term interests; 3) In the final interest fusion process, the concept of time function is added to assign different time weights to user interests in different periods, and then the final recommendation is completed.

3 User Characteristics and Time Windows Based Collaborative Filtering Algorithm

In order to reflect user interests precisely and improve recommendation accuracy, this paper proposes an interest recommendation based on user characteristics and time windows. Firstly, we obtain the user-item rating matrix of the user within a certain time window; Secondly, we calculate separately the similarity based on user characteristics and user ratings, and fuse them to get the final similarity of the user; Thirdly, we select Top-K results to get the nearest neighbor set based on user similarity; Finally, according to the time function, the recommended sets obtained in time windows are given corresponding weights to gain the final recommendation set.

3.1 Similarity Calculation

For the calculation of similarity, traditional methods include Cosine similarity, Pearson correlation coefficient and Euclidean distance. They are all based on vectors, that is, calculating the distance between two vectors. The closer the distance, the greater the similarity. In the recommendation process of this paper, the similarity between users is calculated by using the user characteristic similarity and the user's ratings on all items in the user-item rating matrix.

User Similarity Calculation. In this paper, the similarity is calculated based on the combination of user characteristic similarity and user rating similarity.

Similarity Calculation Based on User Characteristics. The chosen characteristics are:

- User age: People of different ages have different life and experiences, which will lead to different interests. According to the user age distribution of Fig. 1, we divide the ages into 8 spans: 0–9, 10–19, 20–29, 30–39, 40–49, 50–59, 60–69, 70–79, labeled 1, 2, 3, …, 8 respectively.

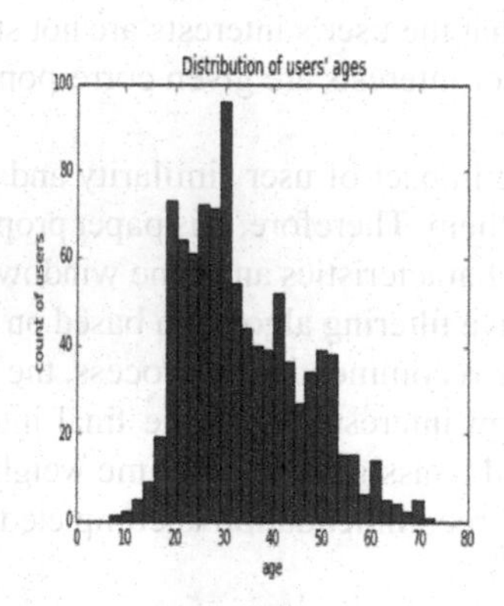

Fig. 1. Age distribution of users

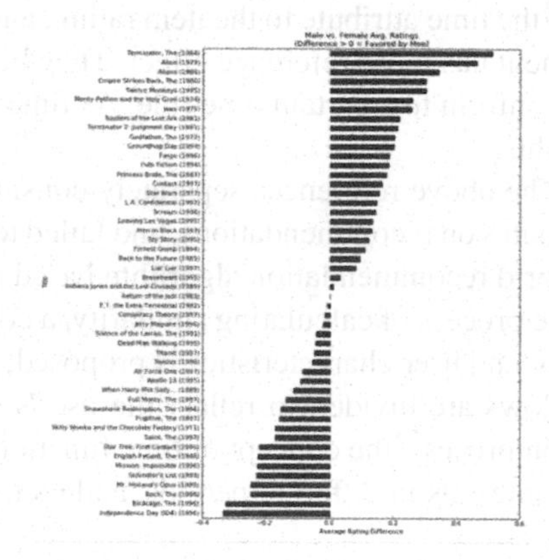

Fig. 2. Distribution of users' interests of different genders

- User gender: Different gender users have different interests in Fig. 2. The user's gender (F, M) is marked with (0, 1).
- User occupation: different occupations lead to different ideas. People with the same occupation are more likely to show similar interest. Similarly, the user's occupation is marked as 1, 2, 3, …, *t* (*t* is the total number of occupations).

According to those user characteristics, the user characteristic similarity is calculated by Formula (1):

$$sim1(i,j) = \frac{\sum_{k=1}^{n} mk}{n} \tag{1}$$

We select 3 user characteristics, so $n = 3$.

$$m_k = \begin{cases} 0, R_{ik} \neq R_{jk} \\ 1, R_{ik} = R_{jk} \end{cases} (i \neq j, k = 1, 2, 3) \tag{2}$$

Where m_k represents the similarity of the *k-th* characteristic of user u_i and user u_j; R_{ik}, R_{jk} represents the *k-th* characteristic value of user u_i and user u_j $(u_i \neq u_j)$.

```
Algorithm 1: Calculating User Characteristic Similarity Algorithm
Input: Characteristics of user u_i and u_j
Output: Characteristic similarity between user u_i and u_j
For u_i, u_j in u:
   m=similarity =0
   if u_i ≠ u_j:
      if u_i.gender== u_j.gender:
         m=m+1
      if u_i.age== u_j.age:
         m=m+1
      if u_i.work== u_j.work:
         m=m+1
      similarity=round(float(m) / 3, 2)    //Normalize the characteristic similarity
EndFor
Return similarity
```

The above algorithm calculates the similarity of user characteristics by traversing the gender, age and occupation of two different users, it fully considers the influence of the user's own factors on the recommendation, takes the user's characteristics as the important basis for the selection of the neighbor user, and improves the accuracy of the neighbor user's selection.

Similarity Calculation Based on User Rating. According to user history behavior, the following user-item rating matrix is obtained to mine user interests more precisely.

$$RAT_{m \times n} = \begin{bmatrix} rat_{1,1} & rat_{1,2} & \dots & rat_{1,n} \\ rat_{2,1} & rat_{2,1} & \dots & rat_{2,n} \\ \dots & \dots & \dots & \dots \\ rat_{m,1} & rat_{m,2} & \dots & rat_{m,n} \end{bmatrix} \tag{3}$$

Where m represents the number of users in the data set, n represents the total number of rated items in the data set, and $rat_{i,j}$ represents the rating of user u_i on item j.

Each row in the matrix represents the distribution of each user's interest, and can be represented by a vector, for example, $RAT_i = \{rat_{i,1}, rat_{i,2}, \dots, rat_{i,n}\}$ represents the distribution of interest of user u_i. Then, the similarity calculation formula based on user rating between two users is as follows:

$$sim2(i, j) = \frac{\overrightarrow{RATi} \cdot \overrightarrow{RATj}}{|RATi||RATj|} = \frac{\sum_{k=1}^{n} rat_{i,k} \times rat_{j,k}}{\sqrt{\sum_{k=1}^{n} rat_{i,k}^2}\sqrt{\sum_{k=1}^{n} rat_{j,k}^2}} \tag{4}$$

The Nearest Neighbor Set. The user characteristic similarity and the user rating similarity are combined linearly to gain the final similarity of the user. When selecting the combination parameter *r*, different data needs to be tested in the experiment to get the optimal parameters.

$$sim(i,j) = (1 - r) \times sim1(i,j) + r \times sim2(i,j) \tag{5}$$

Where *sim (i, j)* represents the similarity between user u_i and user u_j. *r* is a combined parameter in the range of (0, 1), which is used to measure the proportion of characteristic similarity and rating similarity. The higher *r* is, the lower the proportion of characteristic similarity is, and the higher the proportion of rating similarity is. The K users with the highest similarity to the user u_i are selected as the nearest neighbor set *Neig(i)* of the user u_i.

Rating Prediction. According to the user's item-rating matrix, only the item data that the user has rated can be acquired, and the user's interest in the unrated items can't be obtained. Therefore, we need to make predictions on unrated items to get the user's interest which is related to the neighbor users and his rated item records.

$$keys = \frac{N(j)}{Num(j)} \tag{6}$$

Where *N(j)* is the number of rated items of the neighbor user u_j, *Num(j,i)* is the number of items in the rated item records of the neighbor user u_j that the user u_i has not rated. *keys* represents the degree of interest similarity between user u_i and neighbor user u_j. The smaller *Num(j,i)* is, the higher the *keys* value is, and the more similar the interest of user u_i and user u_j is.

$$Score(i, item) = \sum\nolimits_{j \in Neig(i)} sim(i,j) \times keys \tag{7}$$

Where *Score(i,item)* represents the degree of interest of user u_i on the *item.*

3.2 Short-Term Recommendation Within the Time Window

This paper puts forward time windows based collaborative filtering algorithm. Firstly, we get the start time and end time of user's rating, and split them into the time windows $T = \{t_1, t_2, t_3, \ldots, t_k\}$; Then, the user is given a short-term recommendation within the time window based on user rating records.

Algorithm 2: User's short-term recommendation algorithm
Input: Ratings vector of user u_i within the time window t_i
Start time and end time of user u_i in time window t_i
Output: Short-term recommendation set of user u_i
Combining the user's characteristic similarity and rating similarity, and obtaining the final similarity according to Formula (5);
Obtaining the nearest neighbor set based on the similarity;
For u_j, *rating* in *Neig (i)*:
for *item* in *item*(u_j) :
if *item* in *item* (u_i):
Continue
rank[*item*] += *rating* * *keys*
//The value of *keys* is obtained according to Formula (6), and it is the degree of similarity between the user u_i and the neighbor user u_j.
sorted(rank.items(), key=itemgetter(1), reverse=True)[0:N]
//Select the top N items of in terest as a short-term recommendation set
EndFor
Return short-term recommendation set of user u_i

The algorithm traverses the sets of neighbors and rated items of the users' neighbors based on the time window, which reduces the complexity of recommendation time. Users' interest will change with time, so short-term recommendation can reflect users' interest in a certain period and the dynamic change process of users' interest.

3.3 Short-Term Interest Fusion Based on Time Weight Function

The Ebbinghaus Forgetting Curve proposed by Ebbinghaus reveals the law of human forgetting, that is, the phenomenon of forgetting after acquiring new things. Researches have shown that forgetting begins immediately after learning, and the process of forgetting is not uniform. As time goes on, the change of human memory is regular [17], the initial forgetting speed is very fast, and then gradually slows down. After a certain period of time, the memory of knowledge no longer changes. Through analysis, the user's historical behavior forgetting also conforms to this rule [18]. In the recommendation process, the Ebbinghaus Forgetting Curve can be used to describe a person's hobbies change over time. Therefore, the change of user interest can be simulated by human forgetting law.

Over time, when the user's interest tends to stabilize, the time window T is obtained through the interest stabilization time, and then time segments are divided to obtain short-term recommendations. Different users have different time segments $T = \{T_1, T_2, \ldots, T_m\}$, m represents the number of users, and the time windows of each user is divided into $T_i = \{t_{i1}, t_{i2}, \ldots, t_{ik}\}$ according to the start time and the end time of the rating. k is a variable, indicating the number of time segments divided by the user u_i.

$$time(tij) = \frac{ETi - Etij}{ETi} \tag{8}$$

time(t_{ij}) is a time factor function. The larger the value is, the closer the time window is to the current time, and the more the current user's interests and hobbies are reflected; E_{Ti} represents the final end time of the rating of user u_i, and Et_{ij} represents the end time of the rating of the user u_i in the time window t_{ij}. Where, the value of $E_{Ti}-Et_{ij}$ is equal to $cT(c = 0, 1, 2\ldots)$. According to the fitting of forgetting curve, the time weight function can be described as:

$$f(i, tij) = e^{-time(tij)} \tag{9}$$

$f(i,t_{ij})$ is the time weight value of user u_i in time window t_{ij}. Users change their interests under the influence of the outside events, so it is very important to accurately capture their interests. In the process of interest fusion, different weights should be assigned to different periods of interests. Since the interests of the near period reflect the user's interest more, the weight is set larger; and the user's interests for the longer period may not be interested or forgotten, therefore, a lower weight is set. The short-term recommendation sets in time windows obtained in Sect. 3.2 are given different weights:

$$R_i = r_{i1} \times P_{T1} + r_{i2} \times P_{T2} + r_{i3} \times P_{T3} + \ldots + r_{ik} \times P_{Tk} \tag{10}$$

P_{Tj} means the weight value corresponding to the time weight function $f(i, t_{ij})$.

3.4 User Personalized Recommendation

When making personalized recommendation for users, first of all, the user's historical rating record needs to be obtained to understand the user's past interests; then, it is necessary to divide the time window according to the user's rating record time to get the user's short-term interests; due to the decay of user's interests with time [19], so the user's short-term interests in different time windows are set different time weights to recommend finally.

Algorithm 3: User Personalized Recommendation Algorithm
Input: User-item rating matrix $RAT_{m\times n}$,the target user u_i
Output: User personalized recommendation set
Get user u_i's rating start time and end time;
Split time window based on rating start time and end time;
For each time window:
  Obtain user-item rating matrix of user u_i in a time window;
  Combine the user's characteristic similarity and rating similarity, and get the end user similarity according to Formula (5);
  Select Top-K to gain the nearest neighbor set based on user similarity;
  For each item rated by users in the nearest neighbor set:
    Record the degree of interest of the target user u_i for each item using Formula (7);
  Select the Top-N items of interest as a short-term recommendation set;
  Set the corresponding weights for short-term recommendation set within the time window using Formula (9); the interest of the more recent period is set to a larger weight, and the interest of the longer period gives lower weight;
Sort the interest the recommended items in all time windows, and select Top-N as the final recommended set of the target user u_i ;
EndFor
Return user personalized recommendation set

The above algorithm assigns the corresponding weight to the short-term recommendation set in different periods based on the time function, and selects the first N items that users are most interested in as the final recommendation set. It not only takes into account the influence of the user's previous interests on the current interests, but also knows the user's long-term and short-term interests from the process of interest change, thus improving the precision of the recommendation algorithm.

4 Experiment and Analysis

The essence of recommendation algorithm is to predict the target user's interests in the item, and recommend the most interested items to the target user. The public data set MovieLens is selected to test the similarity parameter and time window value involved in our algorithms, and compared with other classical algorithms.

4.1 Data Sets

This paper chose the data from MovieLens recommendation system (https://grouplens.org/datasets/movielens/). MovieLens is the most widely used general data set in collaborative filtering researches. MovieLens has more than 1 million ratings for 1,682 movies by 943 users. In the data set, each user rated at least 50 movies. At the same time, the data set contains some information, such as movie type, age, and so on. The user's rating in the data set lasts approximately 150 days. With the duration of user rating, the rating history of each user is divided into several periods. Because the original data have too much noises, we filtered users with rating time less than 50 days and movies with less than 50 ratings, and obtained the data set of 360489 pieces of data.

4.2 Evaluation Standards

(1) Mean Absolute Error (*MAE*), used to assess the ability of a system to predict. The MAE is applied to measure the deviation between the predicted value and the actual value. *MAE* is defined as:

$$MAE = \frac{\sum_{i=1}^{N} |pi - qi|}{N} \tag{11}$$

(2) The accuracy of the recommendation results (*Precision*). That is, the proportion of hit data in the Top-N set after sorting and selection. *Precision* is defined as:

$$\Pr ecision = \frac{\sum_{u \in U} |R(u) \cap T(u)|}{\sum_{u \in U} |R(u)|} \tag{12}$$

(3) The recall rate of the recommended results (*Recall*). That is, the ratio of the hit data in the Top-N set after sorting and selection to all data. *Recall* is defined as:

$$\text{Re}call = \frac{\sum_{u \in U} |R(u) \cap T(u)|}{\sum_{u \in U} |T(u)|} \tag{13}$$

4.3 Parameter Setting and Experiment Comparison

Experimental Scheme. In order to verify the recommendation efficiency of our algorithm, we set different parameters and compared it with classic algorithms to evaluate the comprehensive performance. The experimental scheme is as follows:

(1) Compare the effect of different similarity parameter r on recommendation;
(2) Adjust the value of time window T to optimize the recommendation effect;
(3) Compare the recommendation effect of our algorithm with other algorithms.

Selection of User Similarity Parameter. The r value of similarity parameter affects the effect of recommendation fusion. In the experiments, the same user data were selected as experiment data set, and different r values were set in each experiment to evaluate the recommendation results which are shown in Fig. 3 and Fig. 4.

In comparison experiments of the similarity parameter r value, the influence of different parameter r values on the precision of the recommendation algorithm and the average absolute error are also different. The experiments show that when the parameter r is 0.95, the *MAE* of the algorithm reaches the minimum value and the *Precision* reaches the maximum value, that is, the algorithm achieves better effect. When the parameter r is 1, it is the traditional collaborative filtering recommendation algorithm and the worse result shows that our fusion algorithm is effective

Selection of Time Window *T*. In order to verify the efficiency of our algorithm using the time window, we changed the value of the time window T, compared and analyzed the influence of different T values on the accuracy. Results are shown in Fig. 5, when the time window T is less than 9 days, the recommended accuracy will be lower than the

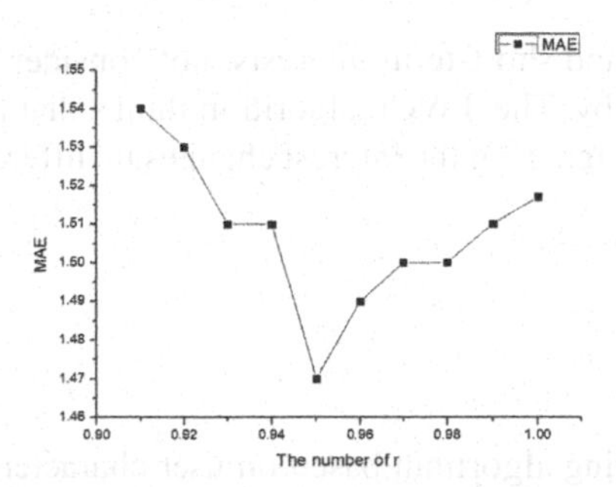

Fig. 3. MAE with different *r* values

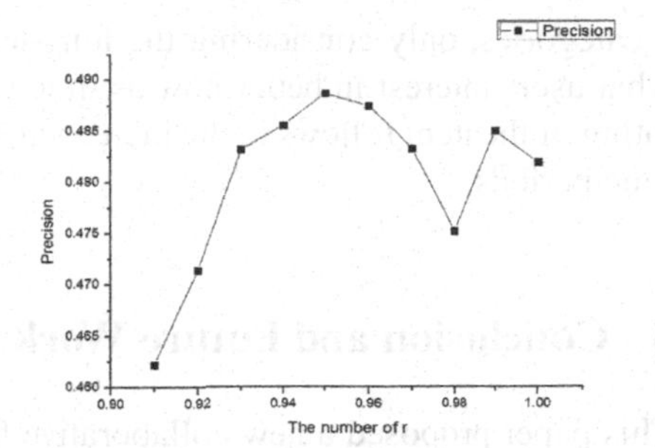

Fig. 4. Precision with different *r* values

traditional collaborative filtering algorithm. Because when the *T* value of time window is small, the same preference of users in a certain period will be divided into different time windows, resulting in scattered user data in the time window, which can't reflect the user's preference well and has a low precision. At the same time, when the time window T is 31 days, the memory capacity of Ebbinghaus forgetting curve tends to be stable, which can more accurately reflect the change process of user interests, at this time, it has better recommendation results.

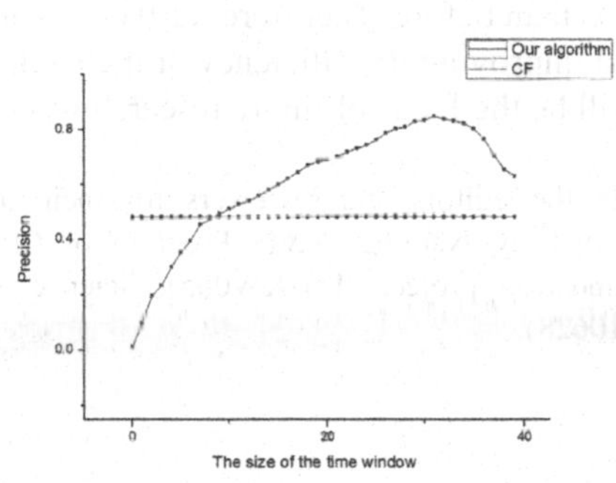

Fig. 5. Precision with different time windows

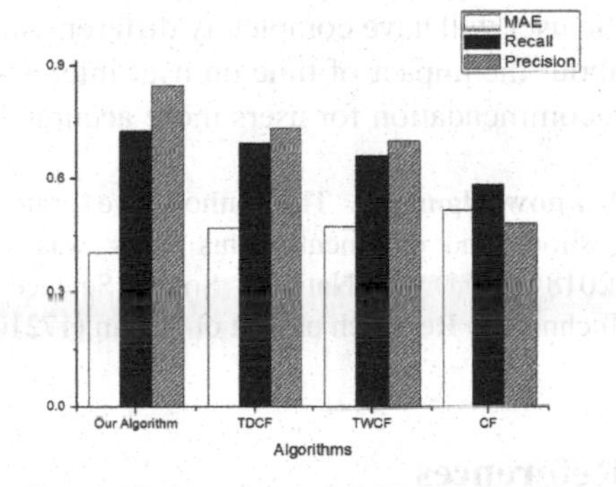

Fig. 6. Comparison of different algorithms

Comparison of Different Recommendation Algorithms. In this section, we compare different recommendation algorithms to test the efficiency of our recommendation algorithm. The TDCF algorithm [15], the TWCF algorithm [20] and the traditional collaborative filtering recommendation algorithm are chosen, and compared with our algorithm in terms of *Precision, Recall,* and *Mean Absolute Error*. The recommendation effect of different algorithms is shown in Fig. 6. It can be seen from the figure that our algorithm is higher than other algorithms in *Precision, Recall,* and *Mean Absolute Error*. Therefore, our algorithm can better reflect the user's interest change process and has a better recommendation effect. By analyzing the reasons, the TDCF algorithm divides the user's rating history into several time periods, and analyzes the user's interest distribution within these time periods. According to the user's interest, the items are divided into

n categories, only considering the long-term and short-term interests, not considering what users interest in better now as time goes by. The TWCF algorithm thinks that the rating of the item follows to the increasing law, ignoring the interest changes in different time periods.

5 Conclusion and Future Work

This paper proposed a new collaborative filtering algorithm based on user characteristics and time windows. In the process of similarity calculation, the user similarity is obtained by combining the item rating and the user's characteristics; in the process of recommendation, the time windows are divided to reflect the user's short-term interests according to Ebbinghaus Forgetting Curve; in the process of interest fusion, the user's interests in different periods are given different time weights, and the user obtain the satisfied personal recommendation. In this paper, we can quantify the short-term interests of users in time windows, but also obtain the personalized recommendation parameters of users' interests in different periods. Experiments show that our algorithm can improve the prediction accuracy compared with the traditional algorithms, especially when the user's rating lasts for a long time. However, this paper understands the user's interests by the user's past rating data to generates recommendations, but if a special event occurs, the user will have completely different interests than before. Therefore, further research about the impact of time on user interests and improving the efficiency of the method, recommendation for users more accurately will be the focus of future research work.

Acknowledgments. The authors are grateful to the editors and reviewers for their suggestions and comments. This work was supported by National K&D Program of China (2018********01),National Social Science Foundation project (17BXW065), Science and Technology Research project of Henan (172102310628).

References

1. Aziguli, W., Yingshuai, W., Dezheng, Z., et al.: A recommendation system based on Fusing Boosting Model and DNN Model. Comput. Mater. Continua **60**(3), 1003–1013 (2019)
2. Fangfang, C.: Research on collaborative filtering recommendation algorithm based on time weight. Dalian University of Technology (2015)
3. Koren, Y.: Collaborative filtering with temporal dynamics. Commun. ACM **53**(4), 89 (2010)
4. Haitao, C., Shanshan, S., Tongqiang, L.: Improved user based collaborative filtering recommendation algorithm. Inf. Stud. Theory Appl. **38**(9), 100–103 (2015)
5. Haiyan, Z., Jingde, H., Qingkui, C.: Collaborative filtering recommendation algorithm combining time weight and trust relationship. Appl. Res. Comput. **32**(12), 3565–3568 (2015)
6. Goldberg, D., Nichols, D., Oki, B.M., et al.: Using collaborative filtering to weave an information tapestry. Commun. ACM. **35**(12), 61–70 (1992)
7. Shuhui, J., Xueming, Q., Jialie, S., et al.: Author topic model-based collaborative filtering for personalized POI recommendations. IEEE Trans. Multimedia **17**(6), 907–918 (2015)
8. Sheng, B., Gengxin, S., Ning, C., et al.: Collaborative filtering recommendation algorithm based on multi-relationship social network. Comput. Mater. Continua **60**(2), 659–674 (2019)

9. Zelong, L., Mengxing, H., Yu, Z.: A collaborative filtering algorithm of calculating similarity based on item rating and attributes. In: Web Information Systems and Applications Conference (WISA), Liuzhou, China, pp. 215–218. IEEE (2017)
10. Chenglung, H., Pohan, Y., Chengwei, L., et al.: Utilizing user tag-based interests in recommender systems for social resource sharing websites. Knowl. Based Syst. **56**, 86–96 (2014)
11. Shan, L., Yao, D., Jianping, C.: Research of personalized news recommendation system based on hybrid collaborative filtering algorithm. In: IEEE International Conference on Computer and Communications (ICCC), Chengdu, China. IEEE (2016)
12. Xiang, L., Quan, Y., Shiwan, Z.: Temporal recommendation on graphs via long-and short-term preference fusion. In: ACM SIGKDD International Conference on Knowledge Discovery & Data Mining, New York, USA, pp. 723–731. ACM Press (2010)
13. Yan, Y., Long, Y.: Notice of retraction collaborative filtering based on time division. In: IEEE International Conference on Computer Science and Information Technology, Chengdu, China, pp. 312–316. IEEE (2010)
14. Huaizhen, Y., Lei L.: An enhanced collaborative filtering algorithm based on time weight. In: International Symposium on Information Engineering & Electronic Commerce, Ternopil, Ukraine, pp. 262–265. IEEE (2009)
15. Jiguang, Z., Xueli, Y., Jingyu, S.: TDCF: time distribution collaborative filtering algorithm. In: International Symposium on Information Science & Engineering, Shanghai, China, pp. 98–101. IEEE (2008)
16. Changqiong, S., Guangwei, X., Jingping, L., et al.: Time weight increasing-based collaborative filtering algorithm. J. Chin. Comput. Syst. **39**(2), 255–261 (2018)
17. Mengzhi, D.: Personalized POI recommendation based on user preferences. Xiamen University (2019)
18. Yunchong, L.: Collaborative filtering algorithm based on improved time and user impat. Anhui University of Science and Technology (2019)
19. Weijin, J., Jiahui, C., Yirong, J., et al.: A new time-aware collaborative filtering intelligent recommendation system. Comput. Mater. Continua **61**(2), 849–859 (2019)
20. Yi, D., Xue, L.: Time weight collaborative filtering. In: International Conference on Information and Knowledge Management, Bremen, Germany, pp. 485–492. ACMCIKM (2005)

MS-SAE: A General Model of Sentiment Analysis Based on Multimode Semantic Extraction and Sentiment Attention Enhancement Mechanism

Kai Yang[1], Zhaowei Qu[1(✉)], Xiaoru Wang[1], Fu Li[2], Yueli Li[1], and Dongbai Jia[1]

[1] Beijing Key Laboratory of Network System and Network Culture, Beijing University of Posts and Telecommunications, Beijing, China
{kaiyang,zwqu,wxr,liyueli,jdb_2017140889}@bupt.edu.cn

[2] Department of Electrical and Computer Engineering, Portland States University, Portland, OR 97207-0751, USA
lif@pdx.edu

Abstract. Recently, there is a lot of research on sentiment analysis. When existing models extract text features, semantic information can't be fully obtained, because they ignore context connection between historical texts and current texts. Meanwhile, models cannot self-optimize extraction algorithms, making key sentiment semantics be neglected, because they can't track and feedback analysis results. Furthermore, models extract insufficient sentiment features, resulting in imperfect semantic extraction and unsatisfactory analysis results, because they only extract POS features. To solve the above problems, we propose a general model of sentiment analysis based on multimode semantic extraction and sentiment attention enhancement mechanism (MS-SAE). The model includes a multimode semantic extraction processing (a multimode semantic extraction module and a sentiment attention enhancement mechanism), an extended dictionary (ExWordNet) and a sentiment analysis module. The multimode semantic extraction module extracts semantic features from multiple perspectives and pays close attention to extracted features, which solves the problem of insufficient semantic extraction. We propose a sentiment attention enhancement mechanism to solve the problem that sentiment semantics is neglected. We construct a general extended dictionary to support MS-SAE in semantic extraction processing. The LSTM-based sentiment analysis module ensures the accuracy of sentiment analysis. We evaluate MS-SAE on SST-2, MR and Subj datasets. Extensive experiments have been conducted and the results demonstrate that MS-SAE could achieve better sentiment analysis performance than the state-of-the-art algorithms in accuracy. It solves the problems including poor understanding of text semantics and errors in analysis results.

Supported by the National Natural Science Foundation of China (No. 61672108, No. 61976025).

X. Sun et al. (Eds.): ICAIS 2020, LNCS 12240, pp. 710–721, 2020.
https://doi.org/10.1007/978-3-030-57881-7_62

Keywords: Multimode · Attention enhancement · Sentiment analysis

1 Introduction

Sentiment analysis is a hot research issue in natural language processing. Its purpose is to judge sentiment categories of texts by mining and analyzing subjective information such as standpoints, opinions, and sentiments in texts. It is widely used in harmful information filtering and social opinion analysis [1].

In general, sentiment analysis methods are divided into sentiment knowledge-based methods and deep learning-based methods. Sentiment knowledge-based methods mainly classify texts by constructing or directly using dictionaries and formulating judgment rules [2–5]. These methods have problems including insufficient accuracy and time-consuming, because of a large number of manual labeling actions. With the development of deep learning, some scholars propose deep learning-based methods [6,7]. These methods use deep learning networks to learn semantic features from text representations. However, such methods only extract POS features and do not take existing sentiment resources and features into account, resulting in insufficient feature extraction. Furthermore, because of the lack of tracking and feedback on the analysis results, these methods can't optimize by themselves, resulting in inaccurate results.

Kim et al. proposed a semantic extraction method based on word2vec, which solved the problem that manual annotating took a long time and the accuracy was not high [2–5]. However, word2vec only paid attention to POS features such as nouns and verbs, it was not sensitive to other semantics, which made semantic extraction incomplete.

Socher et al. proposed a recursive automatic encoder based on a semi-supervised method, which solved the problem that pre-defined dictionaries and analysis rules led to deviation [6]; Irsoy et al. proposed a method based on recurrent neural network, which solved the problem that pre-constructing opinion dictionaries led to time-consuming and inaccurate [7]. However, these models ignored the context connection between historical texts and current texts, which led to insufficient comprehension of texts. Meanwhile, these models lacked tracking and feeding back of analysis results, led them unable to self-optimize extraction algorithms and made important sentiment semantics neglected.

The main contributions of this paper are listed in detail as follows:

1. To extract complete semantics, we propose a multimode semantic extraction module. The module effectively combines current text environment information and extended dictionary to fully extract multimode semantics of texts, improving the accuracy of semantic description.
2. To pay attention to sentiment semantics, we design a sentiment attention enhancement mechanism. It optimizes extraction algorithms and enhances the understanding of sentiment semantics by tracking and feeding back analysis results.

3. To solve the problem of a single semantic extraction perspective, we build a general extended dictionary (ExWordNet) to support the multimode semantic extraction module.

Based on the above, we propose a general model of sentiment analysis based on multimode semantic extraction and sentiment attention enhancement mechanism (MS-SAE). The experiment results show that MS-SAE has improved sentiment analysis effect significantly.

2 Related Work

Sentiment analysis was adopted to social opinion analysis and other fields extensively. Scholars proposed sentiment knowledge-based methods and deep learning-based methods.

Traditional sentiment knowledge-based methods mainly classified texts by using dictionaries and judgment rules [8]. Turney et al. used "excellent" and "poor" as reference words, and determined the polarity of other words by calculating mutual information with points of these reference words [9]. Hu et al. used WordNet to build an adjective dictionary and judged sentiment categories through simple rules [10]. However, these methods had several problems that the analysis of normative texts was insufficient, semantic features were not utilized effectively, and a large amount of manual annotation data was needed [11].

Currently, with the development of artificial intelligence, many scholars used deep learning models to solve sentiment analysis problems [12]. These models used deep learning networks to learn semantic features from representations of texts automatically. These methods were divided into recursive neural network-based methods (RecursiveNN), recurrent neural network-based methods (RecurrentNN) [13], and convolutional neural network-based methods (CNN) [14]. Socher et al. used a semi-supervised encoder to predict sentiment distribution of texts effectively [6]. Then, Socher et al. proposed the tree structure of the RecursiveNN model, which achieved a good analysis effect on fine-grained datasets [15]. Dong et al. proposed an adaptive RecursiveNN to construct representation vectors by using multiple functions of semantic synthesis adaptively [16]. Irsoy et al. used RecurrentNN to extract sequence features for sentiment analysis [7]. However, these methods used deep learning-based models to extract semantic features of texts, then performed sentiment analysis according to these features [17–20]. They only extracted text features from a single perspective and used the same model for different texts without optimization. It was hard to obtain feature representations, when text features were complex, resulting in inaccurate feature descriptions and poor sentiment analysis results.

To solve the above problems, we propose MS-SAE, which fully extracts and feedbacks multimode semantic features, and then performs sentiment analysis. Our method has proved to be very competitive in comparison with other state-of-the-art algorithms.

3 Methodology

3.1 Overview of MS-SAE

As shown in Fig. 1, MS-SAE consists of a multimode semantic extraction processing, an extended dictionary (ExWordNet) and a sentiment analysis module. In multimode semantic extraction processing, we propose a multimode semantic extraction module and a sentiment attention enhancement mechanism, which are used to extract multimode text sentiment feature vectors and enhance modules' attention to sentiment semantics. Besides, we build an extended dictionary to support the multimode semantic extraction module to extract text features. So that the module focuses on multi-dimensional features and extracts complete text semantics. In the sentiment analysis module, we introduce LSTM to ensure the accuracy of sentiment analysis.

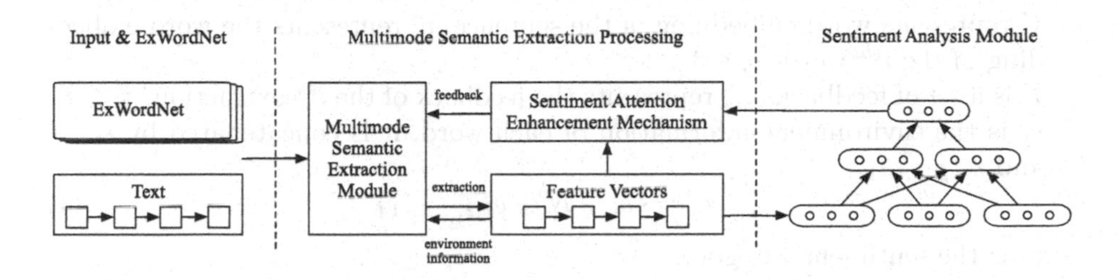

Fig. 1. The architecture of MS-SAE.

3.2 Extended Dictionary

We build a general extended dictionary (ExWordNet) to support the multimode semantic extraction module. ExWordNet consists of words and word embedding vectors. Traditional dictionaries only contain common words such as nouns and adjectives, resulting in insufficient attention to sentiment features.

ExWordNet contains common words, sentiment words, degree adverbs, negative words, and punctuation marks.

We select positive and negative sentiment words and their sentiment scores from SentiWordNet. The principle of choosing a sentiment word is that the word can represent positive or negative sentiment (positive: 1, negative: −1). We select 170-degree adverbs in 6 levels from HowNet. Degree adverbs can strengthen or weaken the sentiment tendency when modifying sentiment words. For example, the sentiment of 'I quite agree with you' is stronger than the sentiment of 'I agree with you' (according to the intensity, degree adverbs are defined as 0, 0.05, 0.1, 0.15, 0.2, 0.25, 0.3). We collect common negative words, which can change the sentiment categories (affirmative words: 0.5, negative words: 0). Punctuation marks are important for expressing the sentiment, we collect punctuation marks to improve the accuracy of semantic feature extraction (exclamation marks: 0.01, question marks: 0.01, otherwise: 0).

3.3 Multimode Semantic Extraction Processing

Multimode Semantic Extraction Module. The multimode semantic extraction module is the key part of MS-SAE. It optimizes the parameters of its algorithm according to the feedback of the sentiment attention enhancement mechanism. It combines the current text environment information to output the most reasonable extraction results. Meanwhile, the multimode semantic extraction module generates complete word embeddings to the sentiment analysis module by using ExWordNet.

Formally, we define the following terms:

- S is a sentence, s_i represents the ith word in the sentence, $s_i \in S$.
- N is the ExWordNet, it consists of words and word embeddings n_j, $n_j \in N$, j represents the number of words.
- o_i represents the extraction option of the ith word, $o_i \in \{0, 1\}$.
- V represents word embedding of the sentence, v_i represents the word embedding of the ith word, $v_i \in V$.
- F is a set of feedbacks, f_i represents the feedback of the ith extraction, $f_i \in F$.
- e_i is the environment information of each word, it is concatenated by s_i, N and $v_{1,2,\ldots,i-1}$.

$$e_i = \{s_i \oplus N \oplus v_{1,2,\ldots,i-1}\} \tag{1}$$

- C is the sentiment category.

As shown in Table 1, the multimode semantic extraction module assumes the following generation process.

Table 1. The generation process of word embeddings.

Process
For each word s_i in a sentence S:
1. Search the word s_i in N.
if $s_i \notin N$: search next word s_{i+1}.
if $s_i \in N$: record all word embeddings n_j that have appeared, and go to step 2.
2. Select the most appropriate word embedding n_j based on the option o_i, given by the multimode semantic extraction module.
where $o_i \in \{0, 1\}$. n_j will not be selected when $o_i = 0$.
Similarly, n_j will be selected when $o_i = 1$.
3. n_j assigned to the word embedding vector v_i of the current word.
4. Repeat 1 - 3, and get the word embedding V of the sentence S.

As shown in Eq. 2, we define the logistic function $\pi_\Theta(o_i, e_i)$ as the probability of feature representations. $P_\Theta(o_i|e_i)$ indicates the probability of taking option o_i in the condition that the environment information is e_i. To obtain accurate probability results, we calculate the delay feedback $P_{\Theta}(o_i|e_i)$ based on word embeddings of the entire sentence.

$$\pi_{\Theta}(o_i, e_i) = P_{\Theta}(o_i|e_i) \tag{2}$$

where $\Theta = \{W, b\}$.

In the training process, the option o_i is determined by Eq. 2. In the testing process, the option o_i' is determined by Eq. 3 to obtain the optimal option.

$$o_i' = \arg\max_{o} \pi_{\Theta}(o_i, e_i) \tag{3}$$

Sentiment Attention Enhancement Mechanism. The accuracy of classification effectively reflects the strength of sentiment attention. As shown in Eq. 4, the sentiment attention enhancement mechanism optimizes the extraction module by calculating feedback $P(C|S)$. The probability $P(C|S)$ represents that sentence S is classified into category C (correct classification). Because sentiment analysis is applied to entire text, $P(C|S)$ is calculated for each sentence which passes through the sentiment analysis module.

$$R = \log P(C|S) \tag{4}$$

This mechanism evaluates the effectiveness of all options generated by the extraction module. It supervises options of the extraction module to maximize the average likelihood of extracted features, which makes the objective function of the extraction module consistent with the sentiment analysis module.

Parameters Optimization. To maximize feedback, we define an objective function as follows:

$$\begin{aligned} J(\Theta) &= E_{o_1,e_1,\ldots,o_n,e_n} f(o_1e_1 \ldots o_ne_n) \\ &= \sum_{o_1e_1\ldots o_ne_n} \prod_i F\pi_{\Theta}(o_i, e_i) \end{aligned} \tag{5}$$

where $o_i \sim \pi_{\Theta}(o_i, e_i)$, $e_{i+1} \sim P(e_{i+1}|e_i, o_i)$. Since environment information e_{i+1} is determined by o_i and e_i, the transition function $P(e_{i+1}|e_i, o_i) = 1$.

The model obtains feedbacks $f(o_1e_1 \ldots o_ne_n)$ through extraction options of the multimode semantic extraction module. We use the reinforce method and policy gradient method to calculate gradient and update parameters of the extraction module. The method is as follows:

$$\Theta + \alpha \sum_{i=1}^{n} F\nabla_{\Theta} \log \pi_{\Theta}(o_i, e_i) \rightarrow \Theta \tag{6}$$

3.4 Sentiment Analysis Module

As shown in Fig. 2, we use LSTM to get sentiment analysis results.

$$C = LSTM(V) \tag{7}$$

where V is the word embedding, which is generated by the extraction module. C is the sentiment category generated by Eq. 7.

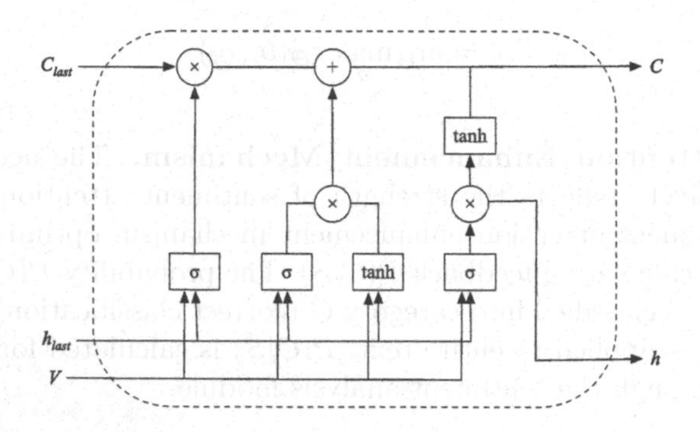

Fig. 2. The structure of LSTM.

The representation of the final layer is sent to a softmax classifier to predict sentiment category:

$$P(C|S) = softmax(W_s h_n^P + b_s) \tag{8}$$

where the parameter $W_s \in \mathbb{R}^{d \times K}$, $b_s \in \mathbb{R}^K$, K is the number of sentiment categories, and d is the dimension of the hidden state. We use cross-entropy as a loss function:

$$Loss = -\sum_{S \in X} \sum_{C=1}^{K} p(C,S) \log P(C|S) \tag{9}$$

where $p(C,S)$ represents the gold probability distribution and X represents training text.

3.5 Model Training

The multimode semantic extraction module and sentiment analysis module are optimized iteratively. We pre-train the extraction module and sentiment analysis module as follows:

- Pre-train LSTM-based sentiment analysis module.
- Calculate feedbacks by using the LSTM-based sentiment analysis module.
- Pre-train multimode semantic extraction module while keeping other parameters unchanged.
- Train all modules together.

In the training process, we use the gradient descent method to minimize objective function and optimize LSTM. And we use the MontoCarlo-based policy gradient method to optimize the multimode semantic extraction module, which makes options with the highest feedbacks are always adopted by the extraction module.

4 Experiments and Evaluation

We propose MS-SAE. The multimode semantic extraction module extracts multimode semantics from texts. The sentiment attention enhancement mechanism tracks and feedbacks analysis results, and pays special attention to contents that affect sentiment categories. The ExWordNet provides support for multi mode semantic extraction module. The LSTM-based sentiment analysis module ensures the accuracy of sentiment analysis results. We evaluate the performance of MS-SAE through the following four experiments:

Experiment 1: Performance evaluation of the multimode semantic extraction module.
Experiment 2: Performance evaluation of the sentiment attention enhancement mechanism.
Experiment 3: Performance evaluation of the ExWordNet.
Experiment 4: Performance evaluation of the MS-SAE.

4.1 Datasets and Metrics

We use three standard datasets: SST-2, MR and Subj.

SST-2. SST-2 is a common dataset of English film comments. It is usually used to divide comments into positive and negative categories.

MR. MR is a dataset of English film reviews with the same classification target as SST-2.

Table 2. Summary statistics for the datasets after tokenization. c: Number of target classes. l: Average sentence length. N: Dataset size. $|V|$: Vocabulary size. $|Vpre|$: Number of words present in the set of pre-trained word vectors. $Test$: Test set size (10-fold CV means there was no standard train/test split and thus 10-fold CV was used).

Data	c	l	N	$\lvert V\rvert$	$\lvert V_{pre}\rvert$	*Test*
SST-2	2	19	9613	16185	14838	1821
MR	2	20	10662	18765	16448	10-fold CV
Subj	2	23	10000	21323	17913	10-fold CV

Subj. Subj is a dataset of subjective and objective sentences. It is used to divide sentences into subjective sentences or objective sentences (Table 2).

We use accuracy to evaluate the performance of sentiment analysis algorithms.

4.2 Settings

Because MR and Subj are not divided into training sets and test sets, experiments use the method of 10-fold cross-validation on two datasets.

In the multimode semantic extraction module, the dimension of word embeddings is 300. Texts are preprocessed by word2vec which is trained by Google on 100 billion words. The α of Eq. 6 is 0.1.

In the sentiment analysis module, the dimension of word embeddings is also 300. To train the module better, we use a dropout rate of 0.5. The learning rate is 0.001. We train the module by using batch gradient descent with a batch size of 32.

The comparison experiments use the same parameter settings as the corresponding papers [2–7].

4.3 Experiment Results and Analysis

Performance Evaluation of the Multimode Semantic Extraction Module. To verify the impact of the multimode semantic extraction module on MS-SAE, we compare the two models with and without this extraction module. To control variables, both experiments use ExWordNet and LSTM-based sentiment analysis module, but do not use sentiment attention enhancement mechanism.

As can be seen from Table 3, the extraction module improves the accuracy of sentiment analysis significantly. The multimode semantic extraction module can focus on previous word embeddings and current environment information, and make the most appropriate feature extraction option each time. When the above text affects the sentiment category of the current text, the module can also get accurate semantic features. Also, the module can extract multimode semantics besides POS.

Table 3. Accuracy evaluation of the multimode semantic extraction module.

Models	SST-2	MR	Subj
Without the extraction module	88.5	82.3	93.9
With the extraction module	**89.2**	**83.5**	**95.1**

Performance Evaluation of the Sentiment Attention Enhancement Mechanism. To verify the effect of the sentiment attention enhancement mechanism on MS-SAE, we compare the two models with and without mechanism. To control variables, experiments use LSTM-based sentiment analysis module and multimode semantic extraction module, without the ExWordNet.

The experiment results in Table 4 show that the mechanism can improve the accuracy of sentiment analysis compared with the model without mechanism. Because the sentiment attention enhancement mechanism can provide feedbacks to the multimode semantic extraction module according to the classification results of the sentiment classification module, and continuously guide the optimization extraction module to adapt to semantic extraction tasks of current texts. The model can fully extract sentiment semantics, and deliver word embeddings that best match current texts, which improves the accuracy of sentiment analysis.

Table 4. Accuracy evaluation of the sentiment attention enhancement mechanism.

Models	SST-2	MR	Subj
Without the mechanism	88.9	82.9	94.2
With the mechanism	**89.7**	**84.2**	**95.7**

Performance Evaluation of the ExWordNet. To verify the effect of the ExWordNet on MS-SAE, we compare the model using ExWordNet with the model without ExWordNet [2]. To control variables, we use the LSTM-based sentiment analysis module, without using the multimode semantic extraction module and sentiment attention enhancement mechanism.

From Table 5, the ExWordNet affects feature extraction. The model without ExWordNet only extracts POS features when pre-training texts by using word2vec. Sometimes, POS features cannot represent sentiment categories accurately. In contrast, the ExWordNet can focus on multimode features, especially for sentiment semantics that affect text sentiment categories.

Table 5. Accuracy evaluation of the ExWordNet.

Models	SST-2	MR	Subj
Without the ExWordNet	84.5	80.3	92.3
With the ExWordNet	**88.5**	**82.3**	**93.9**

Performance Evaluation of the MS-SAE. This experiment is compared with state-of-the-art algorithms (LSTM [2], CNN [3], biLSTM [4], self-attentive [5], RAE [6], DRNN [7]), and experiment results are shown in Table 6.

For semantic information extraction, compared with state-of-the-art algorithms, MS-SAE improves accuracy significantly. LSTM, CNN, self-attentive use word2vec pre-training texts [2,3,5], biLSTM pre-trains tests by Glove/word2vec [4], and RAE, DRNN do not pre-train texts [6,7]. There is the problem of a single perspective in semantic extraction processing. MS-SAE uses the multimode semantic extraction module and ExWordNet. The module can extract semantic features from multiple perspectives in semantic extraction processing. ExWordNet makes the extraction processing more focused on sentiment

Table 6. Accuracy evaluation of the MS-SAE. Results marked with * are re-printed from corresponding papers. The rest are obtained by our implementation.

Models	SST-2	MR	Subj
RAE	83.5	77.7*	91.3
DRNN	82.5*	79.0*	89.8
CNN	88.1*	81.5*	93.4*
LSTM	84.5	80.3*	92.3
biLSTM	85.2*	79.7*	92.8
Self-attentive	83.2	80.1	92.5
Our method	**90.3**	**85.7**	**96.6**

words, negative words, punctuation marks and other contents that affect sentiment categories. It provides support for the extraction module.

For sentiment analysis, MS-SAE also performs better than state-of-the-art algorithms in accuracy. LSTM, CNN, biLSTM, self-attentive, RAE, DRNN can't self-optimize. On the contrary, the sentiment attention enhancement mechanism optimizes extraction algorithms by tracking and feeding back analysis results. It pays higher attention to sentiment semantics.

In summary, MS-SAE improves the accuracy of text comprehension and sentiment analysis effectively. For semantic feature extraction, MS-SAE is more sensitive to contents that affect sentiment categories. It extracts multimode features and has better performance.

5 Conclusion and Future Work

In this paper, we propose a general model of sentiment analysis based on multimode semantic extraction and sentiment attention enhancement mechanism (MS-SAE), to solve sentiment analysis problems. It consists of a multimode semantic extraction processing (a multimode semantic extraction module and a sentiment attention enhancement mechanism), an extended dictionary and a sentiment analysis module. The multimode semantic extraction module extracts multimode features of texts. The sentiment attention enhancement mechanism tracks and feedbacks analysis results effectively, which optimizes the algorithms of semantic extraction. It improves the performance of text descriptions. The general extended dictionary (ExWordNet) pays higher attention to sentiment semantics, it can be used for any feature extraction task. The LSTM-based sentiment analysis module is used to ensure the sentiment analysis effect. A large number of comparative experiments on multiple datasets demonstrate the best performance of MS-SAE.

In the future, we will construct a new sentiment analysis module instead of using LSTM and then exploit multimode semantic features. We'll also apply MS-SAE to other NLP tasks.

References

1. Wang, J., et al.: Dimensional sentiment analysis using a regional CNN-LSTM model. In: ACL (2). The Association for Computer Linguistics (2016)
2. Wang, X., et al.: Predicting polarities of tweets by composing word embeddings with long short-term memory. In: ACL (1), pp. 1343–1353 The Association for Computer Linguistics (2015)
3. Kim, Y.: Convolutional neural networks for sentence classification. arXiv preprint arXiv:1408.5882 (2014)
4. Ma, X., Hovy, E.H.: End-to-end sequence labeling via bi-directional LSTM-CNNs-CRF. CoRR. abs/1603.01354, (2016)
5. Lin, Z. et al.: A structured self-attentive sentence embedding. In: ICLR. OpenReview.net (2017)
6. Socher, R., et al.: Semi-supervised recursive autoencoders for predicting sentiment distributions. In: EMNLP, pp. 151–161. ACL (2011)
7. Irsoy, O., Cardie, C.: Opinion mining with deep recurrent neural networks. In: Moschitti, A. et al. (eds.) EMNLP, pp. 720–728. ACL (2014)
8. Ding, X., et al.: A holistic lexicon-based approach to opinion mining. In: Proceedings of the Conference on Web Search and Web Data Mining (WSDM) (2008)
9. Turney, P.D., Littman, M.L.: Measuring praise and criticism: inference of semantic orientation from association. ACM Trans. Inf. Syst. **21**(4), 315–346 (2003)
10. Hu, M., Liu, B.: Mining and summarizing customer reviews. In: Proceedings of the Tenth ACM SIGKDD International Conference on Knowledge Discovery and Data Mining, Seattle, WA USA, pp. 168–177. ACM (2004)
11. Kim, S.-M., Hovy, E.: Automatic detection of opinion bearing words and sentences. In: Companion Volume to the Proceedings of the International Joint Conference on Natural Language Processing (IJCNLP) (2005)
12. Majumder, N., et al.: Deep learning-based document modeling for personality detection from text. IEEE Intell. Syst. **32**(2), 74–79 (2017)
13. Lai, S., et al.: Recurrent convolutional neural networks for text classification. In: Bonet, B., Koenig, S. (eds.) AAAI, pp. 2267–2273 (2015)
14. Zhang, M., et al.: Gated neural networks for targeted sentiment analysis. In: Schuurmans, D., Wellman, M.P. (eds.) AAAI, pp. 3087–3093. AAAI Press (2016)
15. Socher, R., et al.: Recursive deep models for semantic compositionality over a sentiment treebank. In: Proceedings of the 2013 Conference on Empirical Methods in Natural Language Processing, pp. 1631–1642 (2013)
16. Dong, L., et al.: Adaptive multi-compositionality for recursive neural network models. IEEE/ACM Trans. Audio Speech Lang. Process. **24**(3), 422–431 (2016)
17. Vosoughi, S. et al.: Tweet2Vec: learning tweet embeddings using character-level CNN-LSTM encoder-decoder. CoRR.abs/1607.07514 (2016)
18. Zhang, Y.H., Wang, Q.Q., Li, Y.L., et al.: Sentiment classification based on piecewise pooling convolutional neural network. Comput. Mater. Continua. **56**(2), 285–297 (2018)
19. Xu, F., Zhang, X.F., Xin, Z.H., et al.: Investigation on the Chinese text sentiment analysis based on convolutional neural networks in deep learning. Comput. Mater. Continua **58**(3), 697–709 (2019)
20. Yan, X.D., Song, W., Zhao, X.B., et al.: Tibetan sentiment classification method based on semi-supervised recursive autoencoders. Comput. Mater. Continua **60**(2), 707–719 (2019)

Information Processing

Research on Psychological Counseling and Personality Analysis Algorithm Based on Speech Emotion

Zhaojin Hong, Chenyang Wei, Yuan Zhuang, Ying Wang, Yiting Wang, and Li Zhao(✉)

School of Information Science and Engineering, Southeast University, Nanjing, China
zhaoli@seu.edu.cn

Abstract. The main work of this paper is based on the research of psychological counseling and personality analysis algorithms of speech emotions. First, the speech emotion recognition and related work are briefly introduced. Starting from the aspect of deep learning, after researching the relevant neural network architecture, a hybrid model of convolutional neural network and long-term and short-term memory network is constructed, which realizes the recognition of speech emotion. We used the CASIA Chinese data set, which contains 7,200 speeches to train and test the model. The recognition rate of the model on this data set reached 0.8365, showing a good recognition function. At the same time, the natural speech data set collected by the experiment was tested in the experiment, which proved that the model has certain recognition ability. In addition, through comparative experiments, we found that the hybrid neural network model is significantly better than that of CNN or LSTM network alone. Finally, according to the Five-Factor Model, we conducted a personality analysis.

Keywords: Speech emotion recognition · CNN · LSTM · MFCC · Five-Factor model

1 Introduction

The speech signal is a collection of many kinds of information. In addition to the most basic semantic information, the speech signal also contains all kinds of complex emotions of human beings, and the emotion plays an important role in human communication and life. The importance of speech emotion recognition in the field of artificial intelligence is self-evident. With the development of science and technology, artificial intelligence is becoming more and more mature. However, the self-cognition and self-consciousness of artificial intelligence are developing slowly, which hinders the natural interaction between human and machine. An important way to solve the self-cognition and self-awareness of artificial intelligence is to let the machine understand the emotional state of human beings and make the intelligent machine have emotional ability.

Speech emotion recognition is to analyze the human voice and the changes contained in it, use the computer's super-computing ability to extract the relevant parameters in

X. Sun et al. (Eds.): ICAIS 2020, LNCS 12240, pp. 725–737, 2020.
https://doi.org/10.1007/978-3-030-57881-7_63

the speech, and then classify the parameters by different levels through the neutral network to predict the speaker's possible emotions. Emotional analysis mainly analyzes the emotional polarity of texts, sentences, phrases, etc. The analysis results can be positive, negative or neutral, or the strength of emotional polarity.The traditional sentiment analysis related research methods can be roughly divided into two types: dictionary-based methods and machine learning-based methods. An emotional dictionary-based approach requires the use of an artificially annotated sentiment dictionary. This method relies heavily on the quality of the sentiment dictionary, and the maintenance of the dictionary requires a lot of manpower and material resources. With the continuous emergence of new words, it can no longer meet the application requirements, and needs to be improved and optimized. Although machine learning-based methods have achieved good results, both features and templates require manual design, and the features of the design often do not have good adaptability, and need to be redesigned when the field changes. In order to avoid too many artificial design features, researchers begin to use deep learning methods.

In recent years, deep learning algorithms have achieved excellent results in the field of natural language, whose classic result is Recurrent Neural Networks (RNN). RNN was first used by Socher in syntactic parsing. Irsoy [1] and the others combined RNN into a deep structure and became a typical three-layer deep learning model. RNN has been proved to be effective in solving serialization problems and utilize context information, however, RNN has gradient explosion and disappearance problems in the solution process, and the processing effects on long text are not good. The Long Short-Term Memory (LSTM) proposed in the later stage effectively solves this problem. The Long Short-Term Memory controls the cell state by adding three kinds of gates in the hidden layer, which is suitable for solving long sequence problems.

Speech emotion recognition technology has enormous application prospects in many fields such as information inquiry system, computer-aided teaching, e-commerce field, virtual character dialogue and medical assistant. For example, it can be used for emotional tracking of patients with depression, emergency level sorting of call center users and driver fatigue testing [2–4].

2 Related Work

In recent years, the research on speech emotion recognition has made great progress in various fields such as the introduction of emotion description models, the construction of emotional speech databases and the analysis of emotional characteristics [5, 6]. At present, there are two types of mainstream emotion description models, namely discrete emotion description model and dimension emotion description model. Correspondingly, according to the different ways of emotion description, the emotional corpus can be divided into discrete emotional corpus and dimensional emotional corpus. The former describes emotions as forms of discrete, adjective label, such as happiness, anger, etc., which are comprehensively used in human daily communication process, and are also commonly used in early emotional correlated research. Abundant language tags describe a large number of emotional states, and it is generally acknowledged that who can across different human culture, and even the emotional category shared by human and social mammals is the basic emotion. The latter describes a emotional state as a point in the multidimensional emotional space. The emotional space here is actually a Cartesian space, and each dimension of the space corresponds to a psychological property of the emotion. For example, the valence attribute which indicates the intensity of emotion and the positive and negative degree of emotion. Although in theory, the emotional description ability of emotional space in the dimensional model can cover all emotional states, but the dimensional model has to face the problem of how to convert between qualitative emotional states and quantitative spatial coordinates, which is more difficult. However, the discrete description model, although the single emotion category leads to the limited ability of emotion description, it cannot satisfy the description of spontaneous emotions, but it is simple and easy to understand, which is conducive to the initiation and development of research work.

At the present stage, the analysis of both Chinese speech emotion and Chinese text sentiment is relatively insufficient [7]. Therefore, this paper used the CASIA Chinese Emotional Corpus, which is commonly used for speech emotion recognition as the discrete emotional corpus. In the meanwhile, in order to make the emotional recognition model more realistic, we collected a natural speech database containing neutral, fear, anger, joy, sadness and surprise as the testing set of the model.

Extracting the emotional features in the speech signal accurately is the premise and basis for speech emotion recognition. At present, the emotional features extracted generally include rhythmic features, sound quality features and spectral features [8]. Pitch, energy, fundamental frequency and duration are the most commonly used rhythmic features; Luengo [9] et al., analyzed and obtained the six-dimensional characteristics of the fundamental frequency mean, energy mean, fundamental frequency log, the dynamic range of the fundamental frequency logarithm and the dynamic range of the energy logarithm, which have the best ability to distinguish emotions. In addition, scholars have found that the prosody feature area has limited ability to distinguish emotions. Fundamental frequency features of anger, fear, happiness and surprise have similar performance. Sound quality features mainly include breathing sound, brightness characteristics, formants, etc.; studies by Gobl, Johnstone, Pereira et al. have proved the sound quality of speech signals can not only express valence dimension information of the three-dimensional emotion model including Arousal dimension-Valence dimension-Power dimension, but also can reflect the power dimension [10–12] information in the three-dimension, which can distinguish emotion better.

In the spectral features, the Mel-Frequency Cepstral Coefficients (MFCC) and the Linear Prediction Cepstral Coefficients (LPCC) are most concerned. This paper extracted the MFCC feature of the speech signal.

In recent years, the research of the Convolutional Neural Network (CNN) [13] has made a major breakthrough. A typical network structure consists of an input layer, a convolutional layer, a pooling layer (downsampling layer), a fully connected layer and an output layer.

The pooling layer is a feature mapping layer, which samples the features obtained after the convolution layer to get a local optimum value. The CNN structure is shown in Fig. 1.

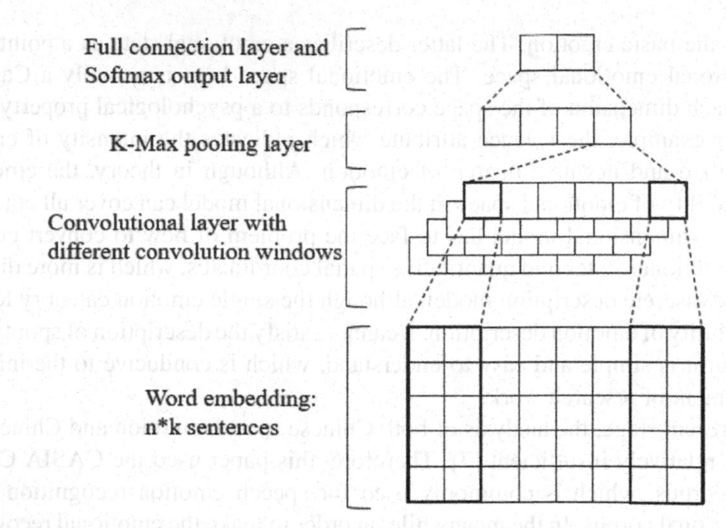

Fig. 1. Convolutional neural network structure.

Similar to Back Propagation (BP), forward propagation is used to calculate the output value, and backpropagation adjusts the weight and offset. The biggest difference between CNN and BP is that the neural unit between adjacent layers in CNN is not fully connected, but part Connection, that is, the sensing area of a certain neural unit comes from some of the upper neural units, rather than all the neural units like BP.

Long Short-Term Memory (LSTM) network structure has many forms, but they are all similar, mainly including input gates, output gates, and forgetting gates, as shown in Fig. 2.

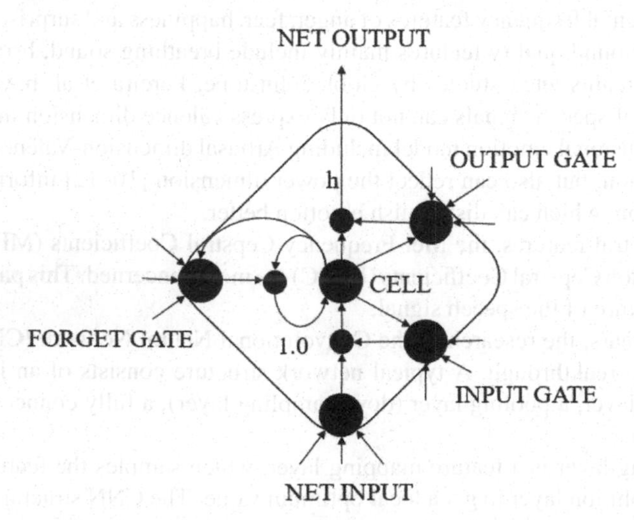

Fig. 2. Long Short-Term Memory network structure.

CNN, LSTM [14] and other deep network models have achieved success on image classification and recognition [15], speech recognition, natural language processing, etc., bringing new ideas to speech emotion recognition. For instance, Ossama Abdel-Hamid et al. added CNN to the traditional DNN and conducted experiments in the TIMIT speech library. Compared with the method using only traditional DNN, the recognition error rate was reduced by more than 10% [16]. Sak et al. [17] proved through experiments that LSTM is more suitable for acoustic modeling of large-scale speech, which can effectively extract time domain information of speech signals. Based on the AlexNet network model, Badshah et al. [18] designed a convolutional neural network with three convolutional layers and two fully connected layers for sentiment analysis of speech signals.

The experiment in this paper built a hybrid model of CNN and LSTM. Through comparison experiments, we found that the accuracy of the hybrid neural network model was significantly better than that of CNN or LSTM alone, which achieved more accurate speech emotion recognition. Finally, we conducted personality analysis based on the Five-Factor Model theory.

3 Model Introduction

3.1 Feature Introduction

The speech feature used in the experiment is the Mel Frequency Cepstral Coefficient (MFCC) [19, 20]. Studies have shown that due to the special structure of the human ear, whose filtering effect is linear scale below 1000 Hz, and the logarithmic scale above 1000 Hz, which makes the human ear more sensitive to low frequency signals. The human ear can't distinguish all the frequency components. Only when the two frequency components differ by a certain bandwidth, can humans distinguish them, otherwise people will listen to the two tones as the same tone. This is called the shielding effect, and the bandwidth is called the critical bandwidth. The MFCC is based on this auditory characteristic of the human ear, which uses a nonlinear frequency unit (Mel frequency) to simulate the human auditory system.

The relationship between MFCC and frequency is similar to the following:

$$Mel(f) = 2595 \times lg\left(1 + \frac{f}{700}\right) \tag{1}$$

In the above formula, f is the frequency and the unit is Hz. The relationship between Mel frequency and linear frequency is shown in the Fig. 3.

3.2 Model Introduction

The neural network constructed in this experiment is 2-layer CNN plus linear layer, LSTM layer and full connection layer, as shown in Fig. 4.

The neural network was specifically built by the Keras library, using the TensorFlow library as the back end.

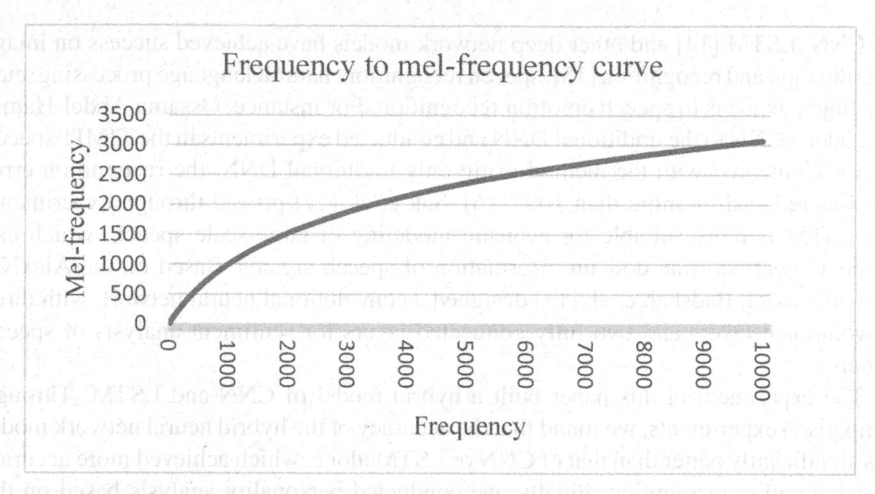

Fig. 3. The relationship between Mel frequency and linear frequency.

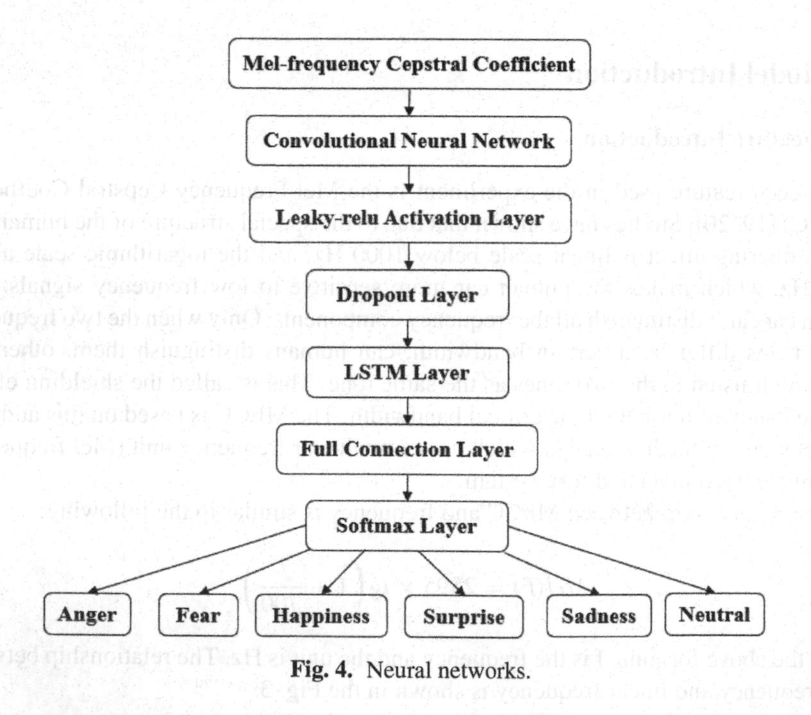

Fig. 4. Neural networks.

The network used the MFCC features of the speech signal and its first-order, second-order differential features as inputs. Here we used the python-speech-feature built-in function to extract these features.

In the experiment, we firstly used CNN, which can use the invariance of convolution to overcome the diversity of the speech signal itself, with an output dimension of 128.

The dropout layer can make a certain proportion of neurons in a silent state when updating weights, effectively preventing overfitting.

Leaky-relu is an activation function:

$$f(x) = max(0.01x, x) \tag{2}$$

Compared with the function relu, it can avoid the phenomenon that learning cannot be performed when x less than zero.

In the experiment, we used the LSTM layer for long-term domain modeling, the output dimension is 128. Then we joined the fully connected layer and the SoftMax layer to complete the prediction of six kinds of sentiment classification.

In addition to the LCNN model proposed above, the paper also includes the following comparative experiment:

First, we set up a two-layer CNN network with the same specifications as the original model without the LSTM layer, then compare the experimental results.

Second, we set up a LSTM network with the same specifications as the original model without CNN, then compare the experimental results.

3.3 Experiment Procedure

First, we traversed all the wav files of the CASIA Chinese database and saved the path of all voices. Then we used the interface of the convenient wav format file to read and write provided by the python package, and wrote a function to read the wav file, which was used to return some information of the file, including an array of N * 1 representing the voice, the sample rate, etc. For the path traversal of all the voices that have been obtained, we used the path truncation to add a label to each file, and obtained the information of the wav file by calling the function that has been written to read the wav file. Then the information was passed to the function of extracting the MFCC and its first-order second-order features, and the data frame length threshold is set to 400, thus obtaining a feature array of the voice signal size of 400 * 40 * 3. The pre-emphasis process of the speech signal was encapsulated inside the function. The pre-emphasis factor was 0.97.

The formula is:

$$H(z) = 1 - az^{-1} \tag{3}$$

That is, a = 0.97.

Finally, we can get a 7200 * 400 * 40 * 3 size large array containing all the voice MFCC features (7200 is the number of voice data.), and randomly divide it to get the training set, test set and verification set that can be fed into the neural network.

The input of the convolution layer in the first layer CNN was set to 400 * 40 * 3, corresponding to the feature array of the speech signal. The convolution kernel was 3 * 3, the step size was 1 and the number of filters was 128. Through the normalization layer, pooling layer and dropout layer, the data entered the second layer of the CNN, which is similar to the first l, but the number of filters of the convolution layer was set to 256. The next LSTM layer had an input time step of 200 and a depth of 768. Therefore, it was necessary to flatten and change the shape before the data was set. The number of LSTM units was set to 128, and then the two layers were fully connected and classified by

SoftMax. The other two layers of CNN and LSTM needed to set the activation function. We used the advanced activation function Leaky-Relu in the experiment.

When compiling the network model., we decided to use the Adam optimizer. The loss function we selected was categorical_crossentropy (multi-class loss function), because we have used one-hot encoding to vectorize the labels, and trained the built model with training and validation sets. We set the number of training rounds to 20 rounds and obtained the curve of the loss value and accuracy as a function of the number of rounds. The experimental results showed that after 10 rounds of training, the change in loss value and accuracy was very slow. In addition to testing the test set, the experiment also tested the natural voice data set collected by the experiment, and the recognition effect was better.

4 Experimental Results and Analysis

4.1 Experimental Environment

The running machine configuration of this experiment: the computer runs memory size was 8 GB, the CPU was Intel Core i5-8250 1.60 GHz, the IDE used Win10 (64-bit) on Anaconda integration, Python 3.6.8 and TensorFlow 1.12.0 and Keras 2.2.4.

4.2 Database Introduction

The speech library adopted in this experiment was the CASIA Chinese Emotional Corpus which was recorded by the Institute of Automation of the Chinese Academy of Sciences. It contains voice data from 4 professional speakers covering 6 emotional categories of neutral, happiness, sadness, anger, surprise and fear. Each emotion contains 1200 voices, a total of 7200 recordings.

4.3 Data Processing

In this experiment, the labels of the speech samples were vectorized and converted into one-hot codes. One-hot encoding is essentially a binary vector, with only one index being one and the rest being zero. The labels used in this experiment and their corresponding relationships are shown in Table 1.

In addition, the extracted samples were randomly scrambled and divided into test set, training set and verification set according to a ratio of 1:8:1.

Since the length difference of the recording time of the voice samples is difficult to be equal, samples shorter than the threshold must be filled with zero value, and samples above a certain threshold must be truncated.

4.4 Experimental Results

Result analysis:

As shown in Fig. 5 and Fig. 6, during the training process, the training set loss value is continuously decreased and the accuracy is improved. The accuracy and loss values

Table 1. Emotional classification label.

English label	Digital label	One-hot code
Anger	0	[1 0 0 0 0 0]
Fear	1	[0 1 0 0 0 0]
Happiness	2	[0 0 1 0 0 0]
Neutral	3	[0 0 0 1 0 0]
Sadness	4	[0 0 0 0 1 0]
Surprise	5	[0 0 0 0 0 1]

of the validation set generally rise or fall during the training process. After about 10 rounds, the convergence speed is obviously slowed down, and the accuracy at this time is 83.91%. After 18 rounds of training, the accuracy of the model decreased slightly due to overfitting, and the generalization ability of the model continued to decrease. As shown in Table 2, it can be seen from the comparison of experimental results that the LCNN model combined with CNN and LSTM can achieve an accuracy of 83% after the same number of rounds of training, and the prediction accuracy is obviously better than using CNN network alone. Slightly better than using the LSTM network alone. Furthermore, among the three models, LCNN can train the least parameters, while LSTM can train the most parameters. Therefore, it can be proved that the proposed model has validity in the classification of speech emotions.

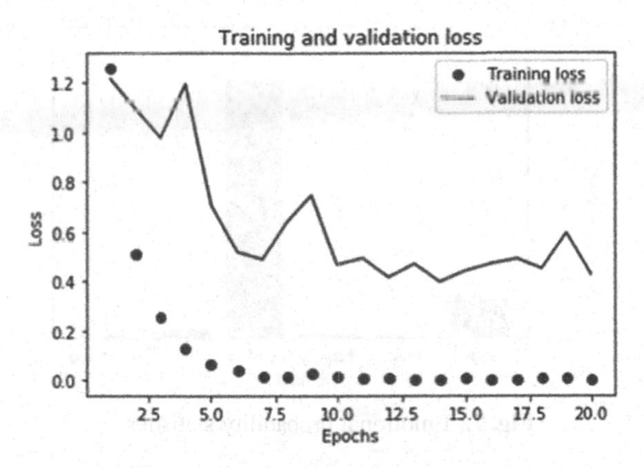

Fig. 5. Number of rounds and loss value.

In addition to testing the extracted model on a given CASIA test set, probabilistic predictions are also made on the actual recorded speech data set, as shown in Fig. 7. The results show that the model can distinguish the happiness, neutral and fear emotions more accurately, but the recognition rate of anger, surprise and sadness is lower.

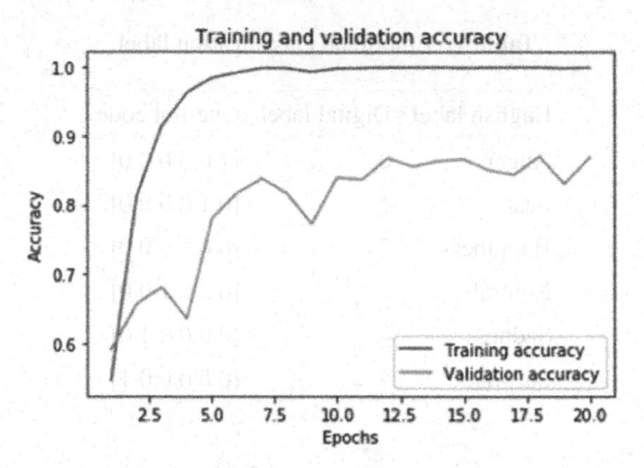

Fig. 6. Number of rounds and accuracy.

Table 2. Model and best accuracy.

Experiment	Precision	Loss value
LCNN	0.8365	0.4943
CNN	0.4464	4.6838
LSTM	0.8244	0.5083

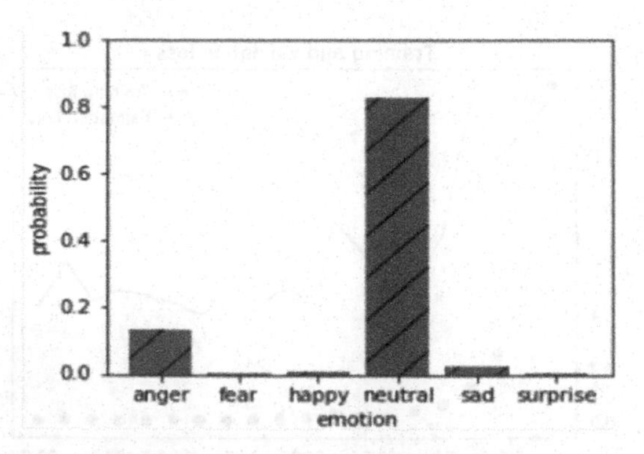

Fig. 7. Emotional probability statistics.

4.5 Personality Model Analysis

The Five-Factor Model proposed by psychologist Goldberg is adopted [21]. According to the theory, personality factors are composed of the big five personality factors: extroversion, agreeableness, conscientiousness, neuroticism, and openness. In the study, personality factors are closely related to the emotions that people often express in words.

For example, people with high emotional stability are more likely to express angry emotions, while those with lower scores tend to show neutral emotions. There is a certain connection between personality traits and emotional expression [18]:

$$\begin{bmatrix} P \\ A \\ D \end{bmatrix} = \begin{bmatrix} 0.21 & 0.59 & 0.19 & 0 & 0 \\ 0 & 0.3 & -0.26 & 0.15 & 0 \\ 0.6 & -0.32 & 0 & 0.25 & 0.17 \end{bmatrix} \begin{bmatrix} \Delta E_x \\ \Delta A_g \\ \Delta S_t \\ \Delta S_o \\ \Delta C_O \end{bmatrix} \tag{4}$$

$(E_x, A_g, S_t, S_o, C_O)$ are five factors.

In the discrete emotion model:

$$\begin{bmatrix} anger \\ fear \\ happy \\ neutral \\ sad \\ surprise \end{bmatrix} = \begin{bmatrix} -2.04 & 2.36 & 1.00 \\ -0.95 & 0.32 & -0.63 \\ 2.77 & 1.21 & 1.42 \\ 1.57 & -0.79 & 0.38 \\ -1.60 & -0.80 & -2.00 \\ 1.72 & 1.71 & 0.22 \end{bmatrix} \begin{bmatrix} P \\ A \\ D \end{bmatrix} \tag{5}$$

5 Conclusion

In this paper, we have detailed the emotional theory model, thc Five-Factor Model, and the commonly used speech emotional characteristics, especially the theory of MFCC, and applied deep learning to the classification of phonetic emotion, and proposed a classification model of speech emotion combining CNN and LSTM. Through testing on the CASIA dataset, the model exhibited a strong recognition function that effectively identifies the type of emotion in the voice. Furthermore, the natural speech database collected by ourselves was tested. The experiment proved that the model had certain recognition ability. At the same time, compared with traditional CNN and LSTM in terms of accuracy and loss values, we found that LCNN had a higher accuracy, which was very suitable for use in speech emotion recognition and related scenes.

Although this paper has gained certain achievements in emotional classification, there are still many improvements in the research due to the limitations of our abilities.

In this experiment, although we tested the natural speech database we collected, however, due to the small amount of data, the persuasiveness for testing is not enough, and the database is even more difficult to train the neural network. Therefore, the data set applied to train the neural network was the CASIA data set, the voice source of which was relatively single. All voices were recorded by professional actors, so it was a performance-type data set, which was obviously less natural. Compared with the actual speech, the background environment and accent of the speaker are ignored. Therefore, it is necessary to build a natural voice data set with more extensive sources and richer contents.

Actually, in the real life, human emotions are difficult to be reflected by a single voice. Voice emotions are constantly changing when people are speaking. The implementation of this model was mainly based on the Keras learning framework in Python, and the recognition speed was relatively slow (about 7–10 s), which made it difficult to apply directly to real life, especially real-time speech emotion recognition scenes. To accurately identify the speaker's emotions, it is essential to rely on more complex real-world models and real-time analysis techniques.

References

1. Irsoy, O., Cardie, C.: Deep recursive neural networks for compositionality in language. In: Advances in Neural Information Processing Systems, pp. 2096–2104 (2014)
2. France, D.J., Shiavi, R.G., Silverman, S., Silverman, M., Wilkes, M.: Acoustical properties of speech as indicators of depression and suicidal risk. IEEE Trans. Biomed. Eng. **47**(7), 829–837 (2000)
3. Lee, C.M., Narayanan, S.S.: Toward detecting emotions in spoken dialogs. IEEE Trans. Speech Audio Process. **13**(2), 293–303 (2005)
4. Boril, H., Omid Sadjadi, S., Kleinschmidt, T., Hansen, J.H.: Analysis and detection of cognitive load and frustration in drivers' speech. In: Proceedings of INTERSPEECH 2010, China, pp. 502–505 (2010)
5. Shengting, W., Liu, Y., Wang, J., Li, Q.: Sentiment analysis method based on K means and online transfer learning. Comput. Mater. Continua **60**(3), 1207–1222 (2019)
6. Wenjing Han, H.L., Ruan, H.: Review of research progress in speech emotion recognition. J. Softw. **25**(01), 37–50 (2014)
7. Feng, X., Zhang, X., Xin, Z., Yang, A.: Investigation on the Chinese text sentiment analysis based on convolutional neural networks in deep learning. Comput. Mater. Continua **58**(3), 697–709 (2019)
8. Ayadi, M.E., Kamel, M.S., Karray, F.: Survey on speech emotion recognition: features, classification schemes, and databases. Pattern Recogn. **44**(3), 572–587 (2011)
9. Luengo, I., Navas, E., Hernaez, I., Sanchez, J.: Automatic emotion recognition using prosodic parameters. In: Proceedings of the ISCA-INTERSPEECH, Lisbon, pp. 493–496 (2005)
10. Gobl, C., Ni Chasaide, A.: The role of voice quality in communicating emotion, mood, and attitude. Speech Commun. **40**(1), 189–212 (2003)
11. Johnstone, T., Scherer, K.R.: The effects of emotions on voice quality. In: Presented at the XIVth International Congress of Phonetic Science, San Francisco, pp. 2029–2032 (1999)
12. Pereira, C.: Dimensions of emotional meaning in speech. In: Proceedings of the ISCA-Workshop on Speech and Emotion, Belfast, pp. 25-28 (2000)
13. Hubel, D.H., Wiesel, T.N.: Receptive fields and functional architecture of monkey striate cortex. J. Physiol. **195**(1), 215–243 (1968)
14. Hochreiter, S., Schmidhuber, J.: Long short-term memory. Neural Comput. **9**(8), 1735–1780 (1997)
15. Fang, W., Zhang, F., Sheng, V.S., Ding, Y.: A method for improving CNN-Based image recognition using DCGAN. Comput. Mater. Continua **57**(1), 167–178 (2018)
16. Abdel-Hamid, O., Mohamed, A., Jiang, H., Penn, G.: Applying convolutional neural networks concepts to hybrid NNHMM model for speech recognition. In: Proceedings of the ICSAAP, Kyoto, Japan, pp. 4277–4280 (2012)
17. Sak, H., Senior, A., Beaufays, F.: Long short-term memory recurrent neural network architectures for large scale acoustic modeling. In: Proceedings of the of the ISCA (2014)
18. Badshah, A.M., Ahmad, J., Rahim, N., et al.: Speech emotion recognition from spectrograms with deep convolutional neural network. In: Proceedings of the PlatCon, pp. 1–5 (2017)

19. Song, Z.: Application of MATLAB in Speech Signal Analysis and Synthesis. Beijing University of Aeronautics and Astronautics Press, pp. 25–27 (2013)
20. Kari, B., Muthulakshmi, S.: Real time implementation of speaker recognition system with MFCC and neural networks on FPGA. Indian J. Sci. Tech. **8**(19), 0974–5645 (2015)
21. Guo, G., Tang, W.: Comparison of Mai's personality model and big five personality model. Psychol. Sci. **26**(03), 487–490 (2003)

Underwater Image Enhancement Based on Color Balance and Edge Sharpening

Yan Zhou[1,2](✉), Yibin Tang[1], Guanying Huo[1,2], and Dabing Yu[1]

[1] College of Internet of Things Engineering, Hohai University, Changzhou 213022, China
yanzhou@hhu.edu.cn

[2] Changzhou Key Laboratory of Sensor Networks and Environmental Sensing, Changzhou 213022, China

Abstract. Underwater optical images often suffer from color cast, edge-blurring and low contrast due to the medium absorption and scattering in water. To solve these problems, we propose an effective technique to improve underwater image quality. First, we introduce an effective color balance strategy based on affine transform to address the color distortion. Then we convert the underwater image from RGB color space to CIE-Lab color space for contrast improvement. In 'L' component's nonsubsampled contourlet transform (NSCT) domain, global contrast adjustment and multi-scale edge sharpening are conducted respectively for lowpass and bandpass direction subbands. Finally, a color-corrected and contrast-enhanced output image can be generated by inverse NSCT and conversion back to RGB color space. The propose method is a single image approach that does not require prior knowledge about the underwater imaging conditions. Experimental results show that our method outperforms state-of-the-art methods both in qualitative and quantitative evaluation. It generally results in good perceptual quality, with significant enhancement of the global contrast, the color, and the image structure details.

Keywords: Underwater image enhancement · Color correction · Contrast adjustment · Multi-scale edge sharpening

1 Introduction

In recent years, underwater imaging is more and more widely used in practical applications such as water conservancy project, marine biology research, underwater archaeology and underwater robot vision [1]. Acquiring clear images from underwater environments is one of primary issues in these scientific missions [2]. However, underwater images often suffer from poor visibility, due to the absorption and scattering of light as it propagates through the water. The attenuation of light depends both on the light's wavelength and the distance it travels. The attenuation of light reduces the light energy, and the wavelength dependent attenuation causes color distortions that increase with an object's distance. In addition, the scattering caused by suspended particles in water results in contrast degradation and hazy appearance, making edge details lost. Often, an

X. Sun et al. (Eds.): ICAIS 2020, LNCS 12240, pp. 738–747, 2020.
https://doi.org/10.1007/978-3-030-57881-7_64

artificial light source is added to the imaging device to try to increase the visibility range in the scene, which leads to uneven illumination. Thus, restoring images degraded by an underwater environment is challenging and improving the visual quality of underwater images is of great importance to further image processing, such as edge detection, image segmentation [3], target recognition [4] and so on.

Generally, underwater images can be restored and enhanced by two categories of algorithms and techniques: physics-based methods and image-based methods [5]. The physics-based methods enhance underwater images by considering the basic physics model of light propagation in water. Dark Channel Prior (DCP) dehazing method [6] for outdoor foggy scenes is widely used in underwater image restoration [7–12], due to the similar effect of light scattering. The algorithm based on wavelength compensation and image dehazing (WCID) [7] segments the foreground and the background regions based on DCP, and uses this information to remove the haze and color distortion by compensating attenuated light along the propagation path. Galdran *et al.* [11] introduce the Red Channel prior (RCP) which is interpreted as a variant of the DCP to recover colors associated with short wavelengths in underwater. Drews *et al.* [12] present the Underwater Dark Channel Prior (UDCP), which basically considers that the blue and green color channels are the underwater visual information source.

Since physics-based methods are sensitive to modeling assumptions and parameters, these methods may fail in extremely variable underwater environment. Image-based methods are usually simpler and more efficient than physics-based methods without considering any physical model for the image formation. Li *et al.* [13] propose an underwater image enhancement method which includes underwater image dehazing based on a minimum information loss principle and contrast enhancement based on histogram distribution prior. Fu *et al.* [14] introduce a two-step enhancing strategy to address color shift and contrast degradation. The approach proposed by Ancuti *et al.* [15] builds on the fusion of two images that are directly derived from the color correction version of original degraded image. For shallow-water images, Huang *et al.* [16] present relative global histogram stretching (RGHS) method based on adaptive parameter acquisition. In these algorithms, global contrast, edges sharpness and color naturalness are improved to generate a high quality image.

In this paper, we propose a novel enhancement method for single underwater image. First, we introduce an effective color balance algorithm to correct the color distortion caused by the selective absorption of light with different wavelength along scene depth. After removing the undesired color castings, the underwater image in RGB color space is converted into CIE-Lab color space. In CIE-Lab color space, 'L' component, which is equivalent to the image lightness, is used to enhance image contrast and details, while 'a' and 'b' components are preserved. Then, we adopt global contrast adjustment strategy for lowpass subband and multi-scale edge sharpening for bandpass direction subbands in 'L' component's nonsubsampled contourlet transform (NSCT) domain. The final enhanced image is obtained by inverse NSCT and color space conversion to RGB. In this way, image contrast and edges are both enhanced while colors are balanced. By qualitative and quantitative comparisons, experiment results show the superiority of our method to existing methods. Moreover, our method can be effectively applied in underwater machine vision systems.

The rest of this paper is organized as follows: Sect. 2 describes the proposed underwater image enhancement method. The experimental results and performance evaluation are given in Sect. 3. The conclusions are drawn finally in Sect. 4.

2 The Proposed Approach

In our approach, color balance aims at compensating for the color cast caused by the selective absorption of colors with depth, while image enhancement is considered to enhance the edges and details of the scene, to mitigate the loss of contrast resulting from backscattering. Figure 1 shows the entire processing of our proposed method. Firstly, we apply an affine transform based on cumulative histogram statistics to each channel in RGB color space for color correction. In CIE-Lab color space, 'L' component is corresponding to the image brightness, while 'a' and 'b' components are corresponding to the image chrominance. Considering the independence of luminance and chromaticity in CIE-Lab color space, we convert the underwater image from RGB color space to CIE-Lab color space for contrast enhancement. Then, we adopt multiscale NSCT transform tool to obtain the lowpass subband coefficients and bandpass directional subband coefficients of 'L' component image. Since each pixel of the transform subbands corresponds to that of the original image in the same spatial location, we can gather the geometrical information pixel by pixel from the NSCT coefficients. By using global contrast enhancement strategy in lowpass subband, shadows and uneven illumination is eliminated and the global dynamic range is well stretched. In addition, adaptive multi-scale nonlinear mapping is conducted in bandpass direction subbands to sharp edges and remove noise. Finally, a color-corrected and contrast-enhanced output image can be generated by inverse NSCT and conversion back to RGB color space.

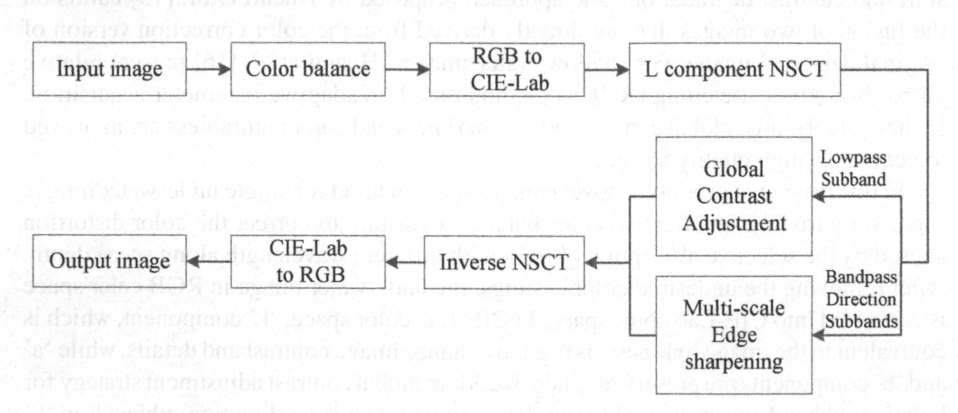

Fig. 1. Flowchart of the proposed underwater image enhancement method.

2.1 Underwater Color Balance

Light attenuation underwater leads to different degrees of color change, depending on the wavelength of each light beam. As the depth increases, red light with a longer wavelength declines faster than green and blue light, resulting in a characteristic bluish-greenish tone in underwater images. To correct the color cast, we apply an affine transform based on cumulative histogram distribution to each channel in RGB color space.

The assumption [17] underlying this algorithm is that the highest values of R, G, B observed in the image must correspond to white, and the lowest values to obscurity. The values of the three channels R, G, B, can be fully stretched to occupy the maximal possible range [0, 255] by using an affine transform to each channel. Considering the possible outliers in the histogram, we clip a small percentage of the pixels at the beginning and end of the histogram, and saturate the pixels to the boundary values of the remaining pixels range, before applying the affine transform. When performing a color balance on the N pixels of image in each channel, we saturate a percentage c_1 of the pixels on the left side of the histogram to $V_{\min}$, and a percentage c_2 of pixels on the right side to $V_{\max}$. $V_{\min}$ and $V_{\max}$, the saturation extrema, can be found by the cumulative histogram of the pixel values. The cumulative histogram bucket labeled i contains the number of pixels with value lower or equal to i. $V_{\min}$ is the lowest histogram label with a value larger than $N \times c_1$, and $V_{\max}$ is the label immediately following the highest histogram label with a value lower than or equal to $N \times (1 - c_2)$. Finally, the pixel interval $[V_{\min}, V_{\max}]$ is mapped to [0, 255] by affine transformation as follows:

$$f(x) = \frac{(x - V_{\min})}{(V_{\max} - V_{\min})} \times 255 \tag{1}$$

where x denotes the input pixel value, and $f(x)$ denotes the output pixel value. The percentage c_1 and c_2 are chosen according to:

$$c_1 = 0.005 \times ratio(\lambda) \tag{2}$$

$$c_2 = 0.005 \times ratio(\lambda) \tag{3}$$

where $ratio(\lambda)$ is a adjustment factor for $\lambda \in \{R, G, B\}$ channel, which is defined as:

$$ratio(\lambda) = \frac{\max\limits_{\lambda \in \{R,G,B\}} (\mathrm{mean}(I_\lambda))}{\mathrm{mean}(I_\lambda)} \tag{4}$$

where I_λ represents $\lambda \in \{R, G, B\}$ color channel of the image I.

Despite color balance is crucial to recover the color, using this correction step is not sufficient since the edges and details of the scene have been affected by the scattering. Therefore, we propose an effective enhancement approach to deal with the hazy nature of the color corrected image in the next section.

2.2 Adaptive Underwater Image Enhancement in CIE-Lab Color Space

In CIE-Lab color space, ‘L’ component is corresponding to the image luminance, while ‘a’ and ‘b’ components are corresponding to the image chrominance. After removing the undesired color castings, the underwater image in RGB color space is converted into CIE-Lab color space for contrast enhancement. Since NSCT [18] can effectively capture geometry and directional information of images in different scales and directions, we adopt multiscale NSCT-based enhancement method to improve ‘L’ component image, while ‘a’ and ‘b’ components are no change. After conducting NSCT on ‘L’ component image, we obtain the lowpass subband which contains overall contrast information and is almost noiseless, and a series of the bandpass directional subbands which contain not only edges but also noise. Then, we handle the lowpass subband and the bandpass directional subbands respectively.

Global Contrast Adjustment for Lowpass Subband. The NSCT lowpass subband represents overall contrast of the image. Taking into account of the presence of non-uniform illumination in the underwater image, uneven brightness tends to appear in the NSCT lowpass subband. Thus, we use the multiscale retinex (MSR) approach [19, 20] to remove the inhomogeneous illumination. Then, an adaptive global contrast adjustment function is proposed to manipulate the lowpass subband coefficients. In this stage, the global dynamic range of the image can be sufficiently and accurately adjusted to improve contrast, eliminate shadows and uneven illumination. The adaptive global contrast adjustment function is defined as follows:

$$\hat{C}_{(1,0)} = C_{\max(1,0)} \cdot sign\left(C_{(1,0)}\right) \cdot \left[\sin\left(\frac{\pi}{2}\frac{\left|C_{(1,0)}\right|}{C_{\max(1,0)}}\right)\right]^{\alpha} \tag{5}$$

where

$$\alpha = \frac{\log\left(\frac{\bar{C}_{(1,0)}}{C_{\max(1,0)}}\right)}{\log\left[\sin\left(\frac{\pi}{2}\cdot\frac{\bar{C}_{(1,0)}}{C_{\max(1,0)}}\right)\right]} \tag{6}$$

$C_{(1,0)}$ is the NSCT lowpass subband coefficient at the first scale, i.e., the coarsest scale. $\hat{C}_{(1,0)}$ is the processed NSCT lowpass subband coefficient at the first scale.

$$C_{\max(1,0)} = \max\left(abs\left(C_{(1,0)}\right)\right) \tag{7}$$

$C_{\max(1,0)}$ denotes the maximum absolute coefficient amplitude in the lowpass subband.

$$\bar{C}_{(1,0)} = mean\left(abs\left(C_{(1,0)}\right)\right) \tag{8}$$

$\bar{C}_{(1,0)}$ denotes the mean value of absolute coefficient amplitude in the lowpass subband.

Multi-scale Edge Sharpening for Bandpass Direction Subbands. The NSCT bandpass directional subbands contain edges of the image at different scales and directions with noise. Because edges tend to correspond to the large NSCT coefficients and noise corresponds to the small NSCT coefficients in bandpass directional subbands, noise can be effectively suppressed by thresholding. The hard-thresholding rule is used for estimating the unknown noiseless NSCT coefficients. In this paper, the thresholds for each bandpass directional subband can be chosen according to:

$$T_{s,d} = k\sigma\sqrt{\tilde{\sigma}_{s,d}} \tag{9}$$

We set $k = 4$ for the finest scale and $k = 3$ for the remaining ones. The noise standard deviation σ and individual variances $\tilde{\sigma}^2_{s,d}$ are calculated by using Monte-Carlo simulations [21, 22].

Subsequently, we present a multi-scale nonlinear mapping function to modify the NSCT bandpass direction subbands coefficients at each scale and direction independently and automatically so as to achieve multiscale edge sharpening. The proposed multi-scale nonlinear mapping function is given by:

$$\hat{C}_{(s,d)} = C\max_{(s,d)} \cdot sign\left(C_{(s,d)}\right) \cdot \left[\sin\left(\frac{\pi}{2}\frac{\left|C_{(s,d)}\right|}{C\max_{(s,d)}}\right)\right]^{\sqrt{\beta}} \tag{10}$$

where

$$\beta = \frac{\log\left(\frac{\bar{C}_{(s,d)}}{C\max_{(s,d)}}\right)}{\log\left[\sin\left(\frac{\pi}{2}\cdot\frac{\bar{C}_{(s,d)}}{C\max_{(s,d)}}\right)\right]} \tag{11}$$

$C_{(s,d)}$ is the NSCT bandpass direction subband coefficient in the subband indexed by scale s and direction d. $\hat{C}_{(s,d)}$ is the processed NSCT bandpass direction subband coefficient.

$$C\max_{(s,d)} = \max\left(abs\left(C_{(s,d)}\right)\right) \tag{12}$$

$C\max_{(s,d)}$ denotes the maximum absolute coefficient amplitude in the subband indexed by scale s and direction d.

$$\bar{C}_{(s,d)} = mean\left(abs\left(C_{(s,d)}\right)\right) \tag{13}$$

$\bar{C}_{(s,d)}$ denotes the mean value of absolute coefficient amplitude in the subband indexed by scale s and direction d.

The final output enhanced image can be obtained by inverse NSCT and conversion back to RGB color space.

3 Experimental Results and Analysis

In this section, the effectiveness of the proposed algorithm is validated through computer simulation. We compare our method with the existing specialized underwater restoration/enhancement methods (*i.e.*, RCP [11], UDCP [12], HDP [13], TSA [14]) both qualitatively and quantitatively. All the implementations are implemented under MATLAB R2018a environment on a PC platform with a 3.6 GHz Intel Core i7 CPU, 16 GB RAM and 64-bits OS.

3.1 Qualitative Evaluation

We test many images in our experiments, and some typical test images and the results of different methods are shown in Fig. 2. The test images contain different color distortion,

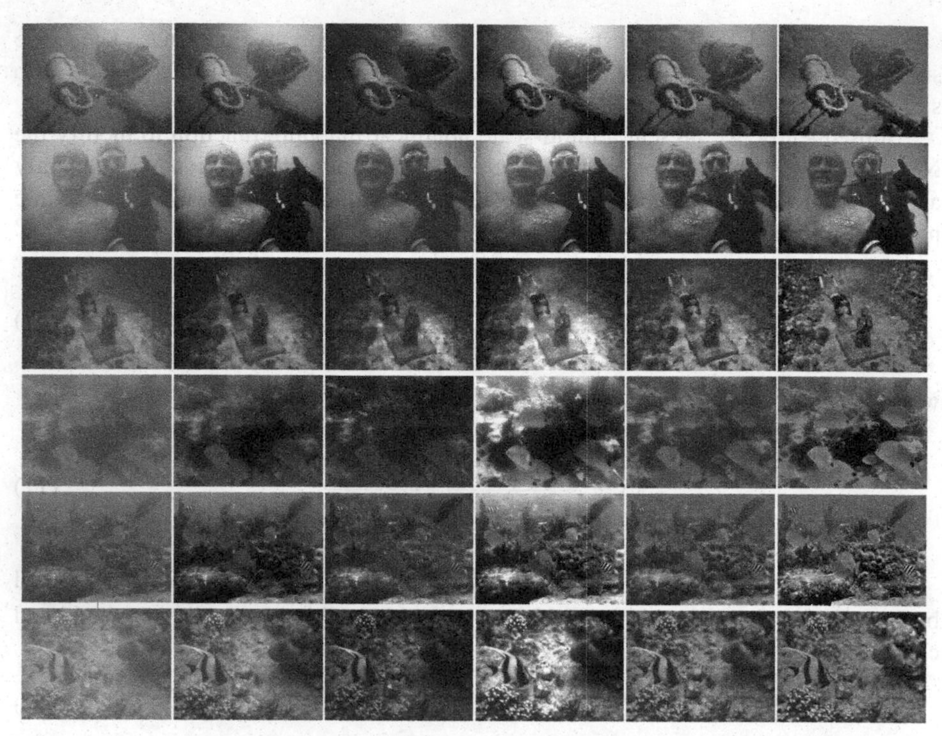

Fig. 2. Comparison on different underwater scenes. From left to right are raw underwater images, and the results of RCP [11], UDCP [12], HDP [13], TSA [14] and the proposed method.

such as bluish and greenish color, and also contain various distance from objects to camera.

As shown in Fig. 2, RCP [11] shows high robustness in recovering the visibility of the considered scenes, but the contrast and detail sharpness of the restored image are not enough. UDCP [12] brings in obvious color deviation and tends to produce artifacts on restored results in some cases. HDP [13] significantly increases the brightness and contrast of underwater images, but tends to generate over-enhancement and obscure the details due to uneven illumination. TSA [14] can well deal with the color cast and increase the contrast of underwater images, but edge sharpness is less than our proposed method. In summary, the proposed method is the only approach which can simultaneously enhance the contrast, sharpen the edges and eliminate the non-uniform illumination while balancing colors, and has relatively good performance on a variety of underwater images.

3.2 Quantitative Evaluation

To quantitatively evaluate the performance of different methods, we employ three metrics (*i.e.*, image information entropy [23], average gradient [24] and UCIQE [25]). The image information entropy measures the information level of an image. The higher the entropy value is, the more information contains. The average gradient [26] measures the sharpness of an image, and the larger the average gradient is, the clearer the image is. The UCIQE metric is designed specifically to quantify the non-uniform color cast, blurring and low-contrast that characterize underwater images. A higher UCIQE score indicates the result has better balance among the chroma, saturation and contrast. The average scores are shown in Table 1 with the optimal values in bold.

Table 1. Comparisons of average value of quantitative evaluation metrics.

Method	RCP [11]	UDCP [12]	HDP [13]	TSA [14]	Proposed
Entropy	6.8580	6.2683	6.3925	7.1673	**7.3745**
Average gradient	5.1340	4.0998	6.6495	5.3648	**8.4558**
UCIQE	0.4551	0.5723	0.4950	0.3019	**0.5830**

As shown in Table 1, our proposed method has the highest average value among all these metrics, in accordance with the lowest distortion and the highest quality of enhanced results in general. Both qualitative and quantitative assessments proof that our method is more effective to enhance the visibility, improve details, correct colors and eliminate noise in underwater images.

4 Conclusions

In this paper, we propose a new underwater image enhancement method based on color balance and edge sharpening. The proposed method includes underwater image color balance algorithm and a contrast enhancement algorithm. Considering the color distortion caused by different wavelength attenuation, linear stretching based on cumulative histogram statistics in RGB color space is firstly introduced to compensate for the loss of the colors. Then, by converting to CIE-Lab color space, we deal with 'L' component in multiscale NSCT domain for contrast and edge enhancement. In the NSCT lowpass subband of 'L' component, MSR algorithm is applied to remove the inhomogeneous illumination, and a new nonlinear transform is presented to stretch the global dynamic range. At the same time, the multiscale nonlinear mapping is designed to enhance the edges and suppress noise in each NSCT bandpass direction subband. Therefore, the proposed algorithm allows for underwater image denoising, color correction, edge sharpening and contrast enhancement to be achieved automatically and simultaneously. Experimental results show that our method can generate promising results while comparing with other methods.

Acknowledgments. This work was supported by the National Natural Science Foundation of China (No. 41706103) and the Natural Science Foundation of Jiangsu Province (No. BK20170306), and the Fundamental Research Funds for the Central Universities (No. 2017B17714).

References

1. Kocak, D.M., Dalgleish, F.R., Caimi, F.M., et al.: A focus on recent developments and trends in underwater imaging. Marine Technol. Soc. J. **42**(1), 52–67 (2008)
2. Sahu, P., Gupta, N., Sharma, N.: A survey on underwater image enhancement techniques. Int. J. Comput. Appl. **87**(13), 19–23 (2014)
3. Liu, Z., Xiang, B., Song, Y., et al.: An improved unsupervised image segmentation method based on multi-objective particle swarm optimization clustering algorithm. Comput. Mater. Continua **58**(2), 451–461 (2019)
4. Wang, N., He, M., Sun, J., et al.: ia-PNCC: noise processing method for underwater target recognition convolutional neural network. Comput. Mater. Continua **58**(1), 169–181 (2019)
5. Schettini, R., Corchs, S.: Underwater image processing: state of the art of restoration and image enhancement methods. EURASIP J. Adv. Signal Process. **2010**, 1–14 (2010)
6. He, K., Sun, J., Tang, X.: Single image haze removal using dark channel prior. IEEE Trans. Pattern Anal. Mach. Intell. **33**(12), 2341–2353 (2011)
7. Chiang, J.Y., Chen, Y.C.: Underwater image enhancement by wavelength compensation and dehazing. IEEE Trans. Image Process. **21**(4), 1756–1769 (2012)
8. Wen, H., Tian, Y., Huang, T., et al.: Single underwater image enhancement with a new optical model. In: IEEE International Symposium on Circuits and Systems, pp. 753–756 (2013)
9. Drews, P., Nascimento, E., Moraes, F., Botelho, S., Campos, M.: Transmission estimation in underwater single images. In: Proceedings of the IEEE International Conference on Computer Vision Workshops, pp. 825–830 (2013)
10. Sathya, R., Bharathi, M., Dhivyasri, G.: Underwater image enhancement by dark channel prior. In: International Conference on Electronics and Communication Systems, pp. 1119–1123 (2015)
11. Galdran, A., Pardo, D., Picon, A., et al.: Automatic Red-Channel underwater image restoration. J. Vis. Commun. Image Represent. **26**(1), 132–145 (2015)
12. Drews, P.L.J., Nascimento, E.R., Botelho, S.S.C., et al.: Underwater depth estimation and image restoration based on single images. IEEE Comput. Graphics Appl. **36**(2), 24–35 (2016)
13. Li, C.Y., Guo, J.C., Cong, R.M., et al.: Underwater image enhancement by dehazing with minimum information loss and histogram distribution prior. IEEE Trans. Image Process. **25**(12), 5664–5677 (2016)
14. Fu, X., Fan, Z., Ling, M., et al.: Two-step approach for single underwater image enhancement. In: 2017 International Symposium on Intelligent Signal Processing and Communication Systems (ISPACS), pp. 789–794 (2017)
15. Ancuti, C.O., Ancuti, C., Vleeschouwer, C.D., Bekaert, P.: Color balance and fusion for underwater image enhancement. IEEE Trans. Image Process. **27**(1), 379–393 (2018)
16. Huang, D., Wang, Y., Song, W., Sequeira, J., Mavromatis, S.: Shallow-water image enhancement using relative global histogram stretching based on adaptive parameter acquisition. In: Schoeffmann, K., Chalidabhongse, T.H., Ngo, C.W., Aramvith, S., O'Connor, N.E., Ho, Y.-S., Gabbouj, M., Elgammal, A. (eds.) MMM 2018. LNCS, vol. 10704, pp. 453–465. Springer, Cham (2018). https://doi.org/10.1007/978-3-319-73603-7_37
17. Limare, N., Lisani, J.L., Morel, J.M., et al.: Simplest color balance. Image Process. Line **1**, 297–315 (2011)
18. Cunha, A.L., Zhou, J.P., Do, M.N.: The nonsubsampled contourlet transform: theory, design, and applications. IEEE Trans. Image Process. **15**(10), 3089–3101 (2006)
19. Jobson, D.J., Rahman, Z., Woodell, G.A.: A multiscale retinex for bridging the gap between color images and the human observation of scenes. IEEE Trans. Image Process. **6**(7), 965–976 (1997)

20. Jobson, D.J., Rahman, Z., Woodell, G.A.: Properties and performance of a center/surround retinex. IEEE Trans. Image Process. **6**(3), 451–462 (1997)
21. Donoho, D.L., Johnstone, J.M.: Ideal spatial adaptation via wavelet shrinkage. Biometrika **81**(3), 425–455 (1994)
22. Po, D.D.Y., Do, M.N.: Directional multiscale modeling of images using the contourlet transform. IEEE Trans. Image Process. **15**(6), 1610–1620 (2006)
23. Xie, H.L., Peng, G.H., Wang, F., et al.: Underwater image restoration based on background light estimation and dark channel prior. Acta Optica Sinica **38**(01), 18–27 (2018)
24. Jiang, Z.X., Pu, Y.: Underwater image color compensation based on electromagnetic theory. Laser Optoelectron. Progress **55**(08), 237–242 (2018)
25. Yang, M., Sowmya, A.: An underwater color image quality evaluation metric. IEEE Trans. Image Process. **24**(12), 6062–6071 (2015)
26. Jin, M., Wang, T., Ji, Z., et al.: Perceptual gradient similarity deviation for full reference image quality assessment. Comput. Mater. Continua **56**(3), 501–515 (2018)

Stereo Matching Using Discriminative Feature-Oriented and Gradient-Constrained Dictionary Learning

Jiale Zhang[1], Yan Zhou[1,2(✉)], Qingwu Li[1,2], Huixing Sheng[1,2], Dabing Yu[1], and Xinyue Chang[1]

[1] College of Internet of Things Engineering, Hohai University, Changzhou 213022, China
yanzhou@hhu.edu.cn

[2] Changzhou Key Laboratory of Sensor Networks and Environmental Sensing, Changzhou 213022, China

Abstract. Sparse coding has been widely utilized in the field of image processing and proved to have great accuracy. The unsupervised learning ability of sparse coding makes it possible to find a group of over-complete basic vectors to represent samples more efficiently. This method is also applied to the field of stereo matching. However, most stereo matching method based on sparse coding prefer to consider each pixel of the image equally instead of segmenting the images with similar features thus ignoring the optimal sparse domain. We propose a stereo matching algorithm based on dictionary learning using discriminative feature and gradient constrain in this paper. First, we introduce a dictionary learning method based on super-pixel patches. We select matching pairs on super-pixel patches for the dictionary learning. Simultaneously, we embed the gradient information of images into the training data to further improve the discriminating ability of the dictionary. Then, the matching cost of matching points is measured by its sparse representation over the learning dictionary. Semi-global cost aggregation and a winner-takes-all strategy are adopted to calculate the initial disparity maps. Finally, a disparity filling method is proposed based on the distance and color information between the mismatched points and its surrounding points. The superiority of the proposed method is demonstrated by comparison with other methods on the Middlebury datasets.

Keywords: Stereo matching · Sparse representation · Dictionary learning · Disparity optimization

1 Introduction

Nowadays, binocular stereo vision is an important research direction in computer vision and has been applied to different visual tasks, such as 3D modeling, obstacle detection and so on [1, 16]. Stereo matching is a critical problem in this field [2]. The purpose of stereo matching is to find matching points from different images and then calculate the disparity. Scharstein and Szeliski summarize the four steps of stereo matching [3],

X. Sun et al. (Eds.): ICAIS 2020, LNCS 12240, pp. 748–759, 2020.
https://doi.org/10.1007/978-3-030-57881-7_65

including matching cost calculation, cost aggregation, disparity calculation and disparity optimization. Several current classical matching cost calculation methods, such as absolute difference (AD) and non-parametric transformation (rank and census) [4], have poor performance in weak texture regions and unstable exposure environments. In the case of surface discontinuities, two or more matching cost calculation methods are combined to improve the robustness, such as the AD-census joint matching cost calculation proposed by Xing et al. [5]. In addition, Hamzah et al. [6] utilizes a combination of AD, gradient difference and census transform to fix the size of the window where the pixel is located, and then matching cost can be calculated with higher accuracy. Siti et al. [7] uses the sum of absolute differences (SAD) and threshold adjustment to determine the window size. In the recent researches of stereo matching, using convolutional neural networks to obtain image feature information has been proven to be a high-precision method [8, 17]. However, the weights in the network need to be trained by many samples.

We propose a stereo matching method based on sparse representation in this paper. A group of over-completed vectors are found as basic vectors in training data and then the basic vectors are used to construct an over-complete dictionary. Through this over-complete dictionary, each image can be represented precisely by fewer basis vectors [23]. Compared with neural networks, building an over-complete dictionary not only saves the weights' training time, but also extracts the image features well.

The main contributions of this paper can be summarized as:

1. In order to solve the problem that existing dictionary learning methods do not consider regional features in each training image, we segment each training image to form super-pixel patches and perform matching pairs selection on those patches for the dictionary learning. We also utilize the essential gradient information of training dataset to present a discriminative feature-oriented and gradient-constrained dictionary learning algorithm.
2. Aimed at the problem that the mismatched points which are at the intersection of multi-layer background overlap area may result in failing to make effective surface decision, a disparity filling method based on color and distance information of mismatch points is presented.

The rest of the paper is organized as follows. Section 2 covers the proposed methods, including dictionary learning and stereo matching. Section 3 demonstrates the superiority of the proposed method in extensive experimental results. Finally, the whole work is discussed and concluded in the Sect. 4.

2 Stereo Matching Based on a Discriminative Feature-Oriented and Gradient-Constrained Dictionary

This section introduces the matching pairs selection method and the dictionary learning model. After that, we continue to describe how to apply the learning dictionary to the stereo matching task. We elaborate on each important step and give diagrams and tables if necessary.

2.1 Matching Pairs Selection

In order to consider the regional features, we use the simple linear iterative clustering (SLIC) super-pixel segmentation algorithm to segment image into super-pixel patches $\boldsymbol{Y} = \{\boldsymbol{Y}_1, \boldsymbol{Y}_2, \ldots, \boldsymbol{Y}_n\}$. Compared with other algorithms such as Mean-shift or Quick-shift, SLIC can control the size and number of patches, as shown in Fig. 1:

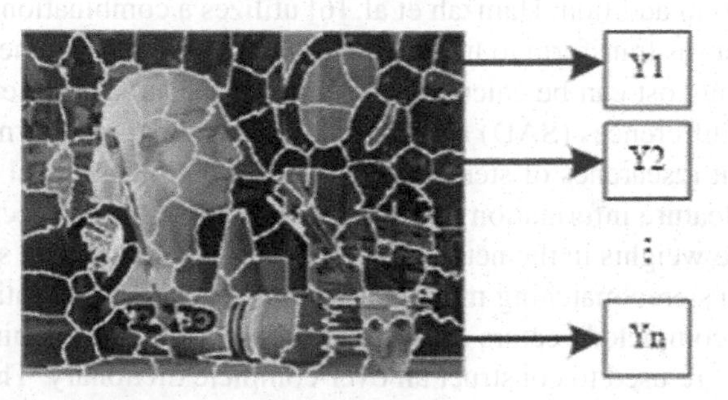

Fig. 1. Image segmentation based on SLIC

Before segmentation, we perform an occlusion region removal process on each image. The specific training data selecting process is as follows: We select pixel matching pairs $\boldsymbol{X}^{\mathrm{p}} = [\boldsymbol{X}_{\mathrm{lp}}, \boldsymbol{X}_{\mathrm{rp}}]$ and gradient matching pairs $\boldsymbol{X}^{\mathrm{g}} = [\boldsymbol{X}_{\mathrm{lg}}, \boldsymbol{X}_{\mathrm{rg}}]$ on the super-pixel patches, where $\boldsymbol{X}_{\mathrm{lp}}$ and $\boldsymbol{X}_{\mathrm{rp}}$ represent image pairs of size $n \times n$ from left and right super-pixel patches, and $\boldsymbol{X}_{\mathrm{lg}}$ and $\boldsymbol{X}_{\mathrm{rg}}$ represent image gradient pairs corresponding to $\boldsymbol{X}_{\mathrm{lp}}$ and $\boldsymbol{X}_{\mathrm{rp}}$. Then, all the matching pairs on each super-pixel patches are obtained by a $n \times n$ sliding window. Before learning, we normalize the training pairs and select the top S variances as the final training data $\boldsymbol{X} = [\boldsymbol{X}_{\mathrm{l}}, \boldsymbol{X}_{\mathrm{r}}]$, where $\boldsymbol{X}_{\mathrm{l}} = [\boldsymbol{X}_{\mathrm{lp}}, \boldsymbol{X}_{\mathrm{lg}}]$, $\boldsymbol{X}_{\mathrm{r}} = [\boldsymbol{X}_{\mathrm{rp}}, \boldsymbol{X}_{\mathrm{rg}}]$.

2.2 Dictionary Learning Model

Sparse Representation. For image pixel matrix $\boldsymbol{Y} = \{\boldsymbol{Y}_1, \boldsymbol{Y}_2, \ldots, \boldsymbol{Y}_{\mathrm{n}}\}$, $\boldsymbol{Y}_{\mathrm{n}} \in \boldsymbol{R}^{\mathrm{n}}$, the basic assumption of sparse theory is that $\boldsymbol{Y}$ can be approximated as a linear representation of a group of basic atoms $\{d_{\mathrm{i}}\}_{\mathrm{i}=1}^{\mathrm{m}}$ in an over-complete dictionary $\boldsymbol{D} \in \boldsymbol{R}^{\mathrm{n} \times \mathrm{m}(\mathrm{n} \geq \mathrm{m})}$ [24], as follows:

$$\boldsymbol{Y} = \sum_{i=1}^{m} \boldsymbol{d}_i \boldsymbol{\alpha}_i = \boldsymbol{D}\boldsymbol{\alpha} \tag{1}$$

Where $\boldsymbol{\alpha} = (\boldsymbol{\alpha}_1, \boldsymbol{\alpha}_2, \ldots, \boldsymbol{\alpha}_{\mathrm{n}})$ is an unknown sparse coefficient vector. Given the image $\boldsymbol{Y}$ and the dictionary $\boldsymbol{D}$, the process of finding the sparse coefficient $\boldsymbol{\alpha}$ is called sparse representation. Basing this principle, we can learn an over-complete dictionary in advance, and then sparsely represent image through the dictionary, that is, map the image to the sparse space. Finally, the matching cost is calculated by the norm of each matched pixels under the sparse representation.

Dictionary Initialization. Mairal J et al. [9] proposed a method of initializing a dictionary, as follows:

$$\min_{D_0 \in C} \frac{1}{N} \sum_{i=1}^{N} \min_{\alpha_i \in R^k} \left(\frac{1}{2} \|x_i - D_0 \alpha_i\|_2^2 + \lambda_1 \|\alpha_i\|_1 \right) \tag{2}$$

Where K is the basic atomic number of dictionary $\mathbf{D}_0$, N is the number of training datas, x_i is the i-th training data, and α_i is the sparse representation of x_i. The 2-norm makes the dictionary more sparsely represent the training data, while the 1-norm makes the dictionary avoid overfitting.. This dictionary initialization is publicly available, and most of existing algorithms [18] directly initialize the dictionary D_0 for the unclassified and unmarked training set by the formula (2). In order to obtain the optimal sparse solution, we propose a dictionary learning method based on the super-pixel patches combined with the image gradient information. We redefine a new function:

$$\begin{aligned} \min_{\{D_0, \alpha_{lp}, \alpha_{rp}, \alpha_{lg}, \alpha_{rg}\}} & \|X_{lp} - D_0 \alpha_{lp}\|_F^2 + \|X_{rp} - D_0 \alpha_{rp}\|_F^2 + \|X_{lg} - D_0 \alpha_{lg}\|_F^2 + \|X_{rg} - D_0 \alpha_{rg}\|_F^2 \\ & + \lambda_1 \|\alpha_{lp}\|_1 + \lambda_1 \|\alpha_{rp}\|_1 + \lambda_1 \|\alpha_{lg}\|_1 + \lambda_1 \|\alpha_{rg}\|_1 + \Gamma(\alpha_{lp}, \alpha_{rp}, \alpha_{lg}, \alpha_{rg}) \end{aligned} \tag{3}$$

Where α_{lp}, α_{rp} are the sparse representations of pixel matching pairs in left and right images respectively, α_{lg} and α_{rg} are the sparse representations of gradient matching pairs in left and right images respectively. Compared with formula (2), we define a penalty term Γ which constrains the dictionary learning by the differences between matching pairs. The definition of Γ is as follows:

$$\Gamma(\alpha_{lp}, \alpha_{rp}, \alpha_{lg}, \alpha_{rg}) = \left\| v^T (\alpha_l - \alpha_r) \right\|_2^2 \tag{4}$$

Where $\alpha_l = (\alpha_{lp}, \alpha_{lg})$, $\alpha_r = (\alpha_{rp}, \alpha_{rg})$, $v \in \mathbf{R}^{K \times 1}$ is the projection vector associated with α_l and α_r.

The existing stereo matching algorithms are mostly based on the assumption that the binocular image is subjected to the same intensity of illuminations and exposures, but it is difficult to guarantee this precondition in the actual situation [19]. Therefore, we selectively set the weighting coefficient to weaken the influence of different illuminations and exposures on the binocular image. The model is further optimized as:

$$\min_{\{D_0, \alpha\}} \frac{1}{2} \|X - D_0 \alpha\|_F^2 + \lambda_1 \|diag(w) \alpha\|_1 + \Gamma(\alpha_{lp}, \alpha_{rp}, \alpha_{lg}, \alpha_{rg}) \tag{5}$$

Where $\alpha = (\alpha_{lp}, \alpha_{rp}, \alpha_{lg}, \alpha_{rg})$, $w \in R^{K \times 1}$ is the weight vector of α and $diag(\cdot)$ generates a diagonal matrix from w. When left and right image have an exposure difference and cause a large change in the amplitude of the sparse representation, we can infer that the larger element in α must be multiplied by the relatively small value in $diag(w)$ to minimize formula (5). According to formula (4) and (5), we determine the inverse relationship between w and v and define w by formula (6):

$$w_k = \frac{1}{|v_k| \times \exp(|v_k|) + \varepsilon_1} \tag{6}$$

Where ε_1 is a constant close to zero.

Dictionary Learning. It should be noted that $\boldsymbol{w}$ and Γ in formula (5) are only used in the dictionary learning. In the dictionary initialization, we set $\boldsymbol{w}$ to the unit matrix and the Γ to 0. In order to demonstrate the superiority of the proposed algorithm better, we save the initial dictionary $\boldsymbol{D}_0$ and make a comparison in Sect. 3.2.

When dictionary $\boldsymbol{D}_0$ and $\boldsymbol{v}$ are fixed, the sparse representation of $\boldsymbol{\alpha}$ is converted into a Lasso regression problem, as follows:

$$\min_{\boldsymbol{\alpha}} \frac{1}{2}\|\boldsymbol{X}-\boldsymbol{D}_0\boldsymbol{\alpha}\|_F^2+\lambda_1\|diag(\boldsymbol{w})\boldsymbol{\alpha}\|_1 \tag{7}$$

The optimization of formula (7) can be properly solved by using proximal gradient descent method. Since $\boldsymbol{v}$ is the factor that reflects the difference of matching pairs, we define the update method for $\boldsymbol{v}$ as follows [20]:

$$\boldsymbol{Av} = \lambda \boldsymbol{Bv} \tag{8}$$

Where $\boldsymbol{A} = (\boldsymbol{\alpha}_{\text{lp}} - \boldsymbol{\alpha}_{\text{rp}})^{\text{T}}(\boldsymbol{\alpha}_{\text{lp}} - \boldsymbol{\alpha}_{\text{rp}})$, $\boldsymbol{B} = (\boldsymbol{\alpha}_{\text{lg}} - \boldsymbol{\alpha}_{\text{rg}})^{\text{T}}(\boldsymbol{\alpha}_{\text{lg}} - \boldsymbol{\alpha}_{\text{rg}})$, λ is the generalized eigenvalues. After $\boldsymbol{\alpha}$ is known, the learning dictionary is updated by formula (9):

$$\min_{\boldsymbol{D}_0} \frac{1}{2}\|\boldsymbol{X}-\boldsymbol{D}_0\boldsymbol{\alpha}\|_F^2 \tag{9}$$

With the above learning algorithm, dictionary $\boldsymbol{D}_0$ is determined by the termination condition after each round of learning:

$$\frac{|f_t - f_{t-1}|}{f_{t-1}} \leq \varepsilon_2 \tag{10}$$

Where ε_2 is the termination constant and f_t is the value of formula (5) at the t-th learning state. The dictionary learning will stop and save current dictionary as the final learning dictionary $\boldsymbol{D}$ for the stereo matching task when the termination condition is met.

So far, we elaborate the dictionary initialization and learning. As shown in Fig. 2, the left and right images of the training dataset are segmented into super-pixel patches, and the matching pairs for dictionary learning are selected on each super-pixel patches. Then the learning dictionary is obtained through the dictionary initialization and learning methods proposed in Sect. 2.2. Each test image can be sparsely expressed in the dictionary and used for the next stereo matching task.

2.3 Stereo Matching

For the left and right test images, the size $n \times n$ sliding window is directly used to divide the image into image pairs. Then each column $\boldsymbol{I}_{\text{c}}$ of a pair is sparsely represented as $\boldsymbol{\beta}$ based on the learning dictionary $\boldsymbol{D}$ by formula (11):

$$\min_{\boldsymbol{\beta}} \frac{1}{2}\|\boldsymbol{I}_{\text{c}} - \boldsymbol{D}\boldsymbol{\beta}\|_F^2 + \lambda_1\|\boldsymbol{\beta}\|_1 \tag{11}$$

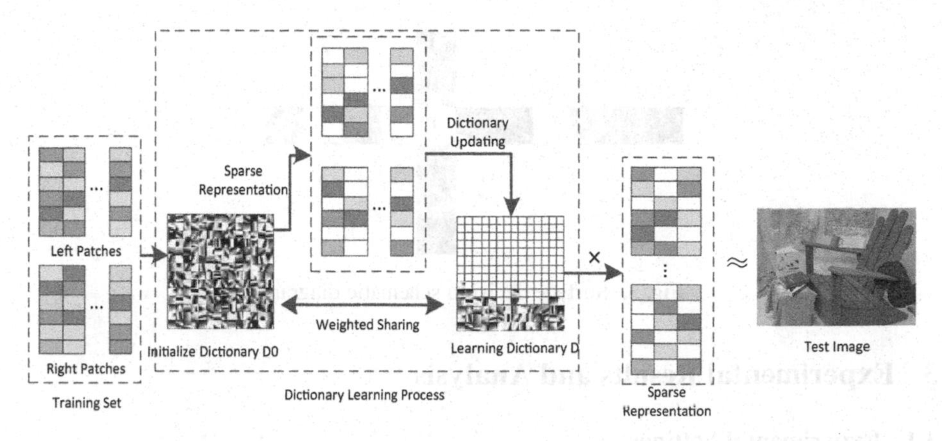

Fig. 2. Dictionary learning and sparse representation structure

After sparse representation, each image can be mapped into the sparse space, the matching cost at pixel $\boldsymbol{p} = (x, y)$ is calculated as:

$$C_{\mathrm{ad}}(\boldsymbol{p}, d) = \left|\boldsymbol{\beta}_{\mathrm{L}}(\boldsymbol{p}) - \boldsymbol{\beta}_{\mathrm{R}}(\boldsymbol{p} - d)\right| \tag{12}$$

Where $C_{\mathrm{ad}}(\boldsymbol{p}, d)$ is the matching cost, $\boldsymbol{\beta}_{\mathrm{L}}$ and $\boldsymbol{\beta}_{\mathrm{R}}$ are the sparse representations of point in left and right image, d is the disparity range.

The initial matching cost of a single pixel is defined by formula (12), but it is easily susceptible by mismatched points and other noise. Therefore it is necessary to aggregate the matching cost of surrounding pixels [10]. The matching cost is aggregated along the up, down, left and right directions to obtain the cumulative cost $S(\boldsymbol{p}, d)$. Then the winner-takes-all strategy is used to select the disparity with the lowest cumulative cost of each pixel as the initial disparity d_0.

When the initial disparity is obtained, the left and right consistency check will be applied to find mismatched points. However, when the mismatched points are in the multi-layer background overlap area, it can not make an effective surface decision to distinguish which background they belong to.

To solve this problem, we propose a disparity filling method based on color and distance information of matching points. As shown in Fig. 3, $\boldsymbol{P}$ is the point that needs to be filled with a reliable disparity. Red points are already matched and the black are mismatched points. Points $\boldsymbol{U}$, $\boldsymbol{D}$, $\boldsymbol{L}$, and $\boldsymbol{R}$ are the first matched points in the four directions of $\boldsymbol{P}$. The color and distance information in four directions are calculated by the following formula to determine which direction the point $\boldsymbol{P}$ belongs to.

$$F(\boldsymbol{p}, \boldsymbol{q}) = \frac{1}{L}\exp\left(-\frac{pixel_{\boldsymbol{p},\boldsymbol{q}}}{\sigma^2}\right) \tag{13}$$

Where $\boldsymbol{p}$ is the mismatched point, $\boldsymbol{q}$ is the first matched points in the four directions of $\boldsymbol{p}$, L is the pixel distance between $\boldsymbol{p}$ and $\boldsymbol{q}$. $pixel_{\boldsymbol{p},\boldsymbol{q}}$ is the difference between the colors of $\boldsymbol{p}$ and $\boldsymbol{q}$. We choose to fill $\boldsymbol{p}$ by taking the disparity in a direction with a higher color uniformity, in other words, by choosing the point which maximizes $F(\boldsymbol{p}, \boldsymbol{q})$.

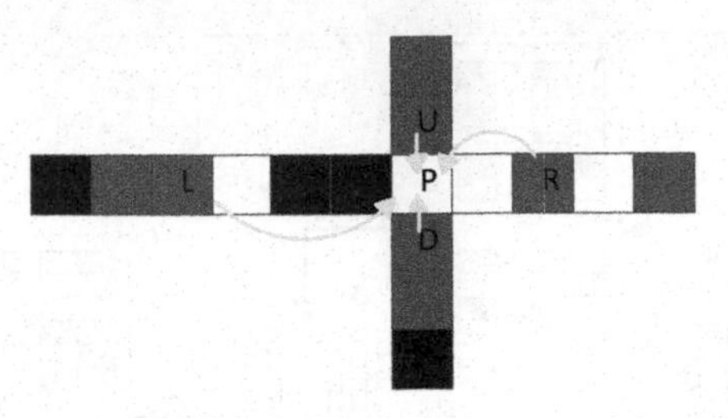

Fig. 3. Surface decision schematic diagram

3 Experimental Results and Analysis

3.1 Experimental Settings

The algorithm is implemented by MATLAB and C code on the PC with Intel i5 CPU and 16 GB RAM. The parameter settings involved are outlined in Table 1.

Table 1. Parameter settings

Parameter	λ_1	ε_1	ε_2	σ	n	S
Value	0.24	10^{-4}	10^{-3}	4.9×10^{-4}	5	12000

In the process of initializing the dictionary, the size of the matching pairs is similar to that of the window in the local stereo matching method, which both satisfies the calculation efficiency and ensures that the window involves enough feature information [21]. In this paper, we test the case of $n = 3, 5, 7, 9$. The average matching error rate is depicted in Fig. 4. Considering the accuracy, $n = 5$ is chosen to be the most suitable case for our algorithm.

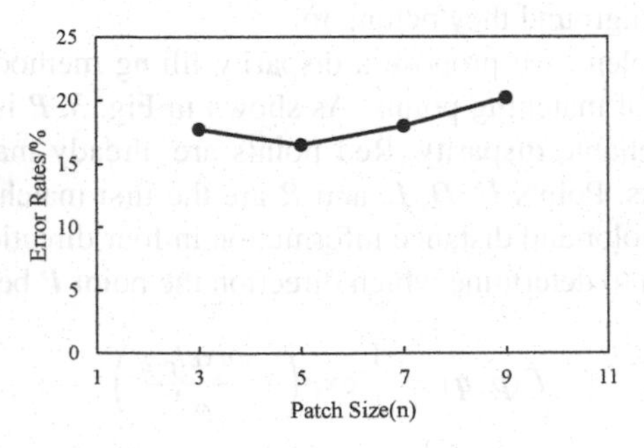

Fig. 4. Average matching error rate

3.2 Evaluation of Stereo Matching Methods

The performance of the proposed method is tested on the dataset provided by the middlebury platform [22], the true disparity and the estimated disparity differed by more than 1 pixel in the non-occlusion region are considered to be mismatched. The newest datasets from Middlebury benchmark v3 include training datasets and test datasets. We use the training datasets as our training data. Then we test the proposed algorithm on the benchmark v2 and v3. All the test images selected in the experiment are images with a quarter resolution.

We compared the dictionary $\boldsymbol{D}_0$ and the learning dictionary $\boldsymbol{D}$ on the training datasets from Middlebury benchmark v3 for stereo matching. Table 2 displays the mismatch rate between our algorithm and other stereo matching algorithms, including IGF [11], ISM [6] and SM-AWP [7]. It can be seen that the matching accuracy based on the sparse representation is higher than the other sterero matching algorithms. What is more, the dictionary $\boldsymbol{D}_0$ is lower in matching accuracy than the $\boldsymbol{D}$ after the dictionary learning and the average mismatch rate is decreased from 17.7% to 16.4%. It can be concluded that the discriminative feature-oriented and gradient-constrained dictionary for stereo matching method is effective and superior.

Table 2. Mismatch rate of 15 groups images from Middlebury benchmark v3

Datasets	IGF	ISM	SM-AWP	$\boldsymbol{D}_0$	$\boldsymbol{D}$
Adirondack	14.9	15.5	14.9	8.1	6.7
ArtL	15.9	16.5	24	13.2	13.1
Jadeplant	24.2	25.1	30.6	17.9	17.4
Motorcycle	9.63	10	10.1	10.8	8.3
MotorcycleE	10.5	11	11.9	8.9	7.4
Piano	24.1	23.5	23.4	16.8	16.6
PianoL	32.1	32.6	34.7	29.8	27.6
Pipes	15.1	15.5	16.7	12.8	11.1
Playroom	21.5	21.6	22.8	18.1	18.0
Playtable	39.6	39.6	19.9	30.6	25.6
PlaytableP	20.9	20.9	18.7	12.0	12.2
Recycle	14.4	15	13.5	10.1	9.5
Shelves	36.0	34.4	35.3	41.6	38.4
Teddy	7.0	7.2	11.4	7.2	6.4
Vintage	34.6	34.4	37.5	27.8	28.7
Ave error/%	21.4	21.5	21.7	17.7	16.4

We select the results of Adirondack, Jadeplant and PlaytableP, which are compared with the groundtruth, to better illustrate the experiment in Fig. 5.

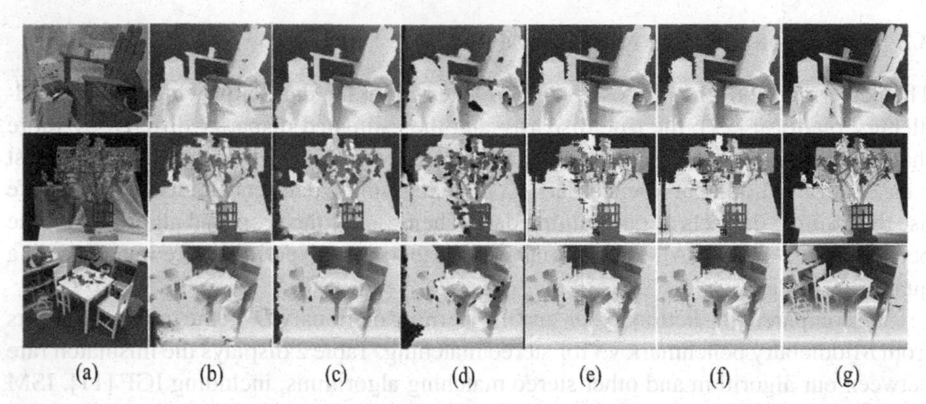

Fig. 5. Adirondack, Jadeplant, and PlaytableP disparity maps obtained by using (b) IGF; (c) ISM; (d) SM-AWP; (e) $\boldsymbol{D}_0$; (f) $\boldsymbol{D}$ and (a) is the original image; (g) is the groundtruth.

It can be seen from Fig. 5 that the proposed algorithm performs better for the weak texture region at the bottom of Jadeplant. We effectively utilize the gradient information in dictionary learning and combine the disparity optimization method to solve the weak texture region.

We also compared our results with other algorithms in Middlebury benchmark v2. Table 3 shows the average mismatch rate of the 4 groups of classic images from Middlebury benchmark v2. We compare the proposed algorithm with AdaptAggrDP [12], LCVB-DEM [13], GC-OCC [14] and RINCensus [15] algorithm. The proposed algorithm outperforms other algorithms with overall error rate of 4.13%.

Table 3. Mismatch rate of 4 groups of classic images from Middlebury benchmark v2

Datasets	SMPF	AdaptAggrDP	LCVB-DEM	GC-OCC	RINCensus	Our method
Tsukuba	0.98	1.57	4.49	1.19	4.78	3.37
Venus	0.25	1.53	1.32	1.64	1.11	0.96
Teddy	9.93	6.79	9.99	11.18	9.76	7.01
Cones	6.51	5.53	6.56	5.36	8.09	5.18
Ave error/%	4.41	3.86	5.59	4.48	5.97	4.13

The corresponding disparity pseudo color map are given in Fig. 6. By comparing results, the proposed method achieves better edge accuracy on Teddy and Venus and has better disparity smoothing effect in non-occluded areas. For Cones, our algorithm is superior to other algorithms for the processing of object edges as well.

Finally, we verify the disparity optimization method proposed in this paper. As shown in Table 4, 'No optimization' means that the initial disparity map is directly processed with scanline filling after the left and right consistency detection. We compare our method on 15 groups images and the average mismatch rate is decreased by about 1%, which proves that our disparity filling method can correct the error effectively.

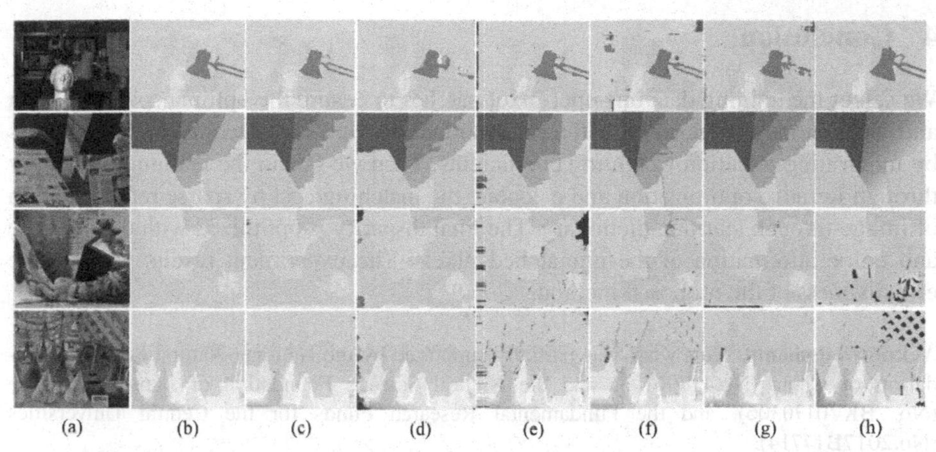

Fig. 6. Disparity maps of Tsukuba, Venus, Teddy, and Cones generated by using: (b) SMPF; (c) AdaptAggrDP; (d) LCVB-DEM; (e) GC-OCC; (f) RINCensus; (g) Our method and Column (a) is the original image; Column (h) is the groundtruth.

Table 4. Disparity filling based on color and distance information

Datasets	No optimization	Disparity optimization
Adirondack	8.1	6.7
ArtL	13.2	13.1
Jadeplant	17.9	17.4
Motorcycle	10.8	8.3
MotorcycleE	8.9	7.4
Piano	16.8	16.6
PianoL	29.8	27.6
Pipes	12.8	11.1
Playroom	18.1	18.0
Playtable	30.6	25.6
PlaytableP	12.0	12.2
Recycle	10.1	9.5
Shelves	41.6	38.4
Teddy	7.2	6.4
Vintage	27.8	28.7
Ave error/%	17.7	16.4

4 Conclusion

We select the training data on super-pixel patches to ensure the optimal sparse domain and weaken the difference of different illuminations and exposures of matching pairs by integrating radiation weighting coefficients. Then we obtain the learning dictionary through iterative optimization and calculate the matching cost by sparse representation of image through learning dictionary. The final disparity is optimized with the distance and color information of the unmatched pixels. The experiment results validate the effectiveness of the proposed method.

Acknowledgments. This work was partially supported by the National Natural Science Foundation of China (No.41706103) and the Natural Science Foundation of Jiangsu Province (No. BK20170306), and the Fundamental Research Funds for the Central Universities (No.2017B17714).

References

1. Li, Q.W., Ni, J.Y., Ma, Y.P.: Stereo matching using census cost over cross window and segmentation-based disparity refinement. J. Electron. Imaging **27**(2), 1–13 (2018)
2. Sui, J., Jing, W.Q.: The realization and development of binocular stereo vision technology. Electron. Technol. Appl. **10**, 3–7 (2014)
3. Scharstein, D., Szeliski, R.: A taxonomy and evaluation of dense two-frame stereo correspondence algorithms. Int. J. Comput. Vis. **47**(1–3), 7–42 (2002)
4. Zabih R.: Non-parametric local transforms for computing visual correspondence. In: European Conference on Computer Vision 1994, pp. 151–158 (1994)
5. Xing, M., Xun, S., Zhou, M.: On building an accurate stereo matching system on graphics hardware. In: IEEE International Conference on Computer Vision 2011, pp. 467–474 (2011)
6. Hamzah, R.A., Kadmin, A.F., Hamid, M.S.: Improvement of stereo matching algorithm for 3D surface reconstruction. Sig. Process. Image Commun. **65**, 165–172 (2018)
7. Siti, S.A.R., Mohd, A.O., Ahmad, F.K.: The effect of adaptive weighted bilateral filter on stereo matching algorithm. Int. J. Eng. Adv. Technol. **8**(3), 284–287 (2019)
8. Liu, Z., Wang, X., Kuntao, L., David, S.: Automatic arrhythmia detection based on convolutional neural networks. Comput. Mater. Continua **60**(2), 497–509 (2019)
9. Mairal, J., Bach, F., Ponce, J.: Online learning for matrix factorization and sparse coding. J. Mach. Learn. Res. **11**(1), 19–60 (2009)
10. He, K., Sun, J., Tang, X.: Guided image filtering. In: European Conference on Computer Vision 2010, pp. 1397–1409 (2010)
11. Hamzah, R.A., Ibrahim, H., Abu Hassan, A.H.: Stereo matching algorithm based on per pixel difference adjustment, iterative guided filter and graph segmentation. J. Vis. Commun. Image Represent. **42**, 145–160 (2017)
12. Wang, L., Yang, R., Gong, M.: Real-time stereo using approximated joint bilateral filtering and dynamic programming. J. Real-Time Image Process. **9**(3), 447–461 (2014)
13. Martins, J.A., Rodrigues, J.M.F., Hans, D.B.: Luminance, colour, viewpoint and border enhanced disparity energy model. PLoS ONE **10**(6), 1–24 (2015)
14. Zabih, Z.: Computing visual correspondence with occlusions using graph cuts. In: Eighth IEEE International Conference on Computer Vision 2001, vol. 2, pp. 508–515 (2001)
15. Ma, L., Li, J.J., Ma, J.: Modified census transform based on the related information of neighborhood for stereo matching algorithm. Comput. Eng. Appl. **50**(24), 16–20 (2014)

16. Ni, J., Li, Q., Liu, Y.: Second-order semi-global stereo matching algorithm based on slanted plane iterative optimization. IEEE Access **6**, 61735–61747 (2018)
17. Mayer, N., Ilg, E., Häusser, P.: A large dataset to train convolutional networks for disparity, optical flow, and scene flow estimation. In: Proceedings of IEEE Conference on Computer Vision and Pattern Recognition 2016, pp. 4040–4048 (2016)
18. Mairal, J., Bach, F., Ponce, J.: Online learning for matrix factorization and sparse coding. J. Mach. Learn. Res. **11**(Jan), 19–60 (2010)
19. Mouats, T., Aouf, N., Richardson, M.A.: A novel image representation via local frequency analysis for illumination invariant stereo matching. IEEE Trans. Image Process. **24**(9), 2685–2700 (2015)
20. Brown, M., Hua, G., Winder, S.: Discriminative learning of local image descriptors. IEEE Trans. Pattern Anal. Mach. Intell. **33**(1), 43–57 (2011)
21. Kehtarnavaz, N.: Stereo matching via selective multiple windows. J. Electron. Imaging **16**(1), 013012 (2007)
22. Scharstein, D.: High-resolution stereo datasets with subpixel-accurate ground truth. In: German Conference on Pattern Recognition 2014, pp. 31–42 (2014)
23. Liu, J., Sun, N., Li, X., Han, G., Yang, H., Sun, Q.: Rare bird sparse recognition via part-based gist feature fusion and regularized intraclass dictionary learning. Comput. Mater. Continua **55**(3), 435–446 (2018)
24. Wang, R., Li, Y., Gomes, S.: Multi-task joint sparse representation classification based on fisher discrimination dictionary learning. Comput. Mater. Continua **57**(1), 25–48 (2018)

Acoustic Emission Recognition Based on Spectrogram and Acoustic Features

Wei Wang[1], Weidong Liu[1](✉), and Jinming Liu[2]

[1] School of Information and Control Engineering, Xuzhou 221000, China
lwdcumt@163.com

[2] School of Information Science and Engineering, Southeast University, Nanjing, China

Abstract. In order to improve the level and efficiency of fault diagnosis, an acoustic emission recognition method based on spectrum and acoustic features is proposed. The proposed system is composed of CNN and BiLSTM networks. Firstly, the amplitude spectrum and group delay phase spectrum of AE signals are extracted, and the amplitude-phase spectrum composed of the two extracted spectrum is input into CNN network to obtain the global features of AE signals. Secondly, the acoustic features such as short-term energy, zero crossing rate and kurtosis of AE signals are extracted to obtain the features of AE signals Finally, the features extracted from CNN network and BiLSTM network are fused to get the fused features, which are classified and recognized by softmax, so as to realize acoustic emission recognition. Simulation results show that the performance of the proposed system is improved by more than 17% compared with the other algorithm, and the effectiveness of the feature fusion model is verified by experiments.

Keywords: Convolutional neural network · Acoustic emission · Fault detection · BiLSTM

1 Introduction

Effectively identifying the rub-impact transmission signal of the rotor is of great significance to the early diagnosis of mechanical faults, the analysis of rub-impact status and the early warning of the trend of fault development. There are many studies on the occurrence of rotor rub-impact. In view of the important influence of features on rub-impact recognition, how to extract effective and robust rub-impact signal features has attracted many scholars' attention. For example, in 2010, Deng Aidong and others [1] proposed a method of recognition of Acoustic Emission Signal Based on the Algorithms of TDNN and GMM, and achieve better results. In recent years, with the development of machine learning technology, deep learning has attracted wide attention.

Convolutional Neural Network (CNN), proposed by Yann LeCun [2] of New York University, is a neural network mainly used to deal with high-dimensional grid data. In 2012, the AlexNet [3] network designed by Hinton and his student Alex Krizhevsky won the ImageNet Classification Competition by an overwhelming advantage in 2012,

X. Sun et al. (Eds.): ICAIS 2020, LNCS 12240, pp. 760–768, 2020.
https://doi.org/10.1007/978-3-030-57881-7_66

demonstrating the specific potential of deep neural networks. At present, CNNs have been widely used in image segmentation [4], noise processing [5], text detection [6], image classification [7–10] and other scenarios [11]. Meanwhile, CNN has been preliminarily applied in fault diagnosis, and the level and efficiency of fault diagnosis have been improved.

LSTM network [12–16], as a kind of RNN network, is widely used in the modeling of time series data. BiLSTM is a network composed of forward LSTM and backward LSTM. Compared with LSTM network, BiLSTM network can encode information from backward to forward, which attracts wide attention.

In this paper, an acoustic emission recognition method based on spectral and acoustic features is proposed. Firstly, the amplitude spectrum and group delay phase spectrum of AE signal are extracted, and the amplitude phase spectrum composed of the two extracted spectrum is input into CNN network to obtain the global features of AE signal; secondly, the acoustic features such as short-term energy, zero crossing rate and kurtosis of AE signal are extracted, and the extracted features are input into BiLSTM network to improve the acoustic features. Finally, the features extracted from CNN network and BiLSTM network are fused to get the fused features, which are classified and recognized by softmax to realize acoustic emission recognition. The simulation results show the effectiveness of the proposed method.

This contribution includes the following two points:

(1) the group delay phase spectrum is introduced into the acoustic emission signal recognition, and more dynamic information is introduced into the acoustic emission signal recognition;
(2) an acoustic emission recognition method based on spectrum and acoustic features is proposed. The designed CNN network is used to extract the global features, and the BiLSTM network is used to process the acoustic features, and finally the fusion features for acoustic emission recognition are obtained.

2 Main Principle

2.1 Input Information

2.1.1 Amplitude Spectrum

AE signal of rotating machinery rotor has similar acoustic characteristics with natural language. Therefore, AE signal can be analyzed and recognized by reference to the method of language recognition. We assume that the rub-impact AE signal is short-sighted stable. For any time n, the spectrum of small-scale signal near n time can be analyzed to obtain two-dimensional spectrum image of AE signal. Considering the strong learning ability of CNN and the dynamic characteristics of AE signal, besides the original spectrum image, the first-order difference spectrum along the time axis and the first-order difference spectrum along the frequency axis of the amplitude spectrum are extracted simultaneously. Three graphs are composed of three 3D spectrograms and input into CNN network.

A long signal is divided into frames, windowed, and FFT (Fast Fourier Transform) is performed for each frame. Finally, the results of each frame are stacked along another dimension to obtain the STFT amplitude spectrum. The STFT amplitude is calculated as follows:

The acoustic emission signal s is divided into frames, and the sequence $s_u(n), (u = 1, \cdots, U)$ is obtained, where U represents the number of frames; according to formula (1), the STFT coefficient of the speech signal $s_u(n)$ at the u frame is calculated, and the STFT coefficient $S_u(k)$ of $s_u(n)$ is obtained, where $\omega(n_0 - n)$ is the window function.

$$S_u(k) = \sum\nolimits_{n=0}^{N-1} x_u(n)\omega(n_0 - n)\, e^{-\frac{2\pi j}{N}kn}, (0 \le k \le N-1) \quad (1)$$

After that, we let the STFT coefficient matrix of sequence $x_u(u = 1, \cdots, U)$, is $S = \left[S_1 \cdots S_U \right] \in R^{N \times U}$, N for the interval length of STFT, and N for the number of frames.

2.1.2 Group Delay Phase Spectrum

Group delay [17] is the rate of change of phase information relative to frequency of a signal at a certain frequency, which contains abundant dynamic information of speech. Therefore, the modified group delay phase spectrum is selected as another way to generate the spectrum, in which the modified group experiment function can be defined as:

$$\beta_m(\omega) = \left(\frac{\beta(\omega)}{|\beta(\omega)|} \right) (|\beta(\omega)|)^{\alpha} \quad (2)$$

Where $\beta(\omega) = \frac{X_R(\omega)Y_R(\omega) + X_I(\omega)Y_I(\omega)}{R(\omega)^{2\gamma}}$. $R(\omega)$ is smooth version. The parameters α and γ are designed to reduce spikes and the range of modified group delay functions.

The group delay phase spectrum of AE signal is extracted, and the first order differential spectrum along the time axis and the first order differential spectrum along the frequency axis are extracted to form a 3D spectrum.

The two spectra are combined to form an amplitude-phase spectrogram.

2.2 The Network Structure of CNN

The basic structure of convolutional neural network is two layers. The first layer is the feature extraction layer. The input of neurons is connected with the local receiving domain of the previous layer to extract local features. The second layer is the feature mapping layer. Each feature mapping layer is a plane, and the weights of neurons on the plane are equal, which is used to reduce the number of free parameters of the network. Convolution neural network generally consists of input layer, convolution layer, activation function, pooling layer and full connection layer.

The input layer is the input image data.

Convolution operation of convolution layer realizes feature extraction, convolution operation is shown in formula (3):

$$x_i^k = \sum\nolimits_{j \in N_i} W_{ij}^k * x_j^{k-1} + b_i^k \quad (3)$$

In the formula, x_i^k represents the k-th neuron of the k-th layer, N_i represents the set of input eigengraphs of the previous layer network, * is the convolution operator, W_{ij}^k represents the convolution kernel parameter connecting the i-th neuron of the k-th layer and the j-th neuron of the $(k-1)$-th layer, b_i^k and represents the bias term.

The activation function makes the deep neural network have the ability of nonlinear mapping learning. The activation function is usually realized by some nonlinear functions.

The main function of the pooling layer is to reduce the calculation load, reduce the use of memory, and prevent over fitting. The pooling layer has no weight value, which is mainly achieved by calculating the local maximum or average value.

The full connection layer connects the neurons in the output layer and each neuron in the input layer.

As shown in Fig. 1, the network structure of CNN used in this paper is shown. The input data of the network is a three-channel image of the amplitude-phase spectrum of rub-impact AE signal with a frame overlap rate of 50%.

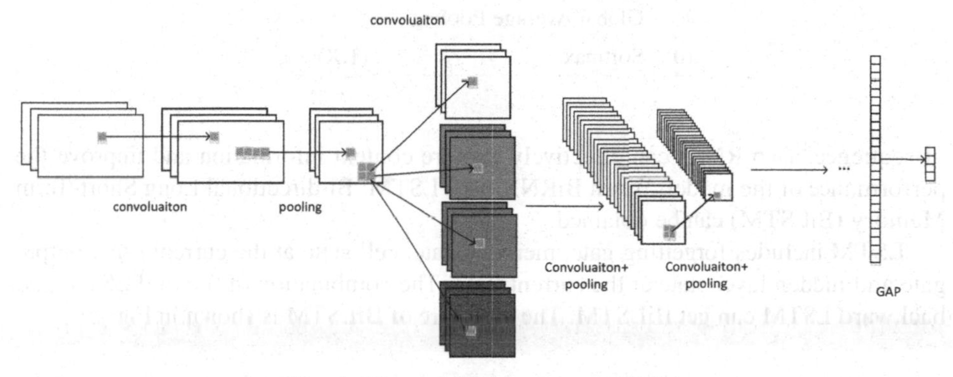

Fig. 1. The network structure of CNN

Firstly, the convolution layer with 1*1 convolution kernel is used to transform the input data nonlinearly, and then the rectangular graph of the previous layer is organized into a 1:1 feature graph by using the pooling operation.

Inception structure is introduced to extract the characteristics of each level of the network.

Finally, the mean pooling is used instead of the full connection layer in the traditional convolutional neural network to alleviate the over-fitting of the network.

Table 1 shows the network structure of CNN

2.3 BiLSTM

The state of traditional RNNs at t-Time can only capture information from the input sequence of the past time and the input of the current time. In 1997, Bi-directional Recurrent Neural Networks, proposed by Schuster, effectively solved this problem. BiRNN includes RNN moving from the beginning of sequence and RNN moving from the end

Table 1. The network structure of CNN

No.	Layer	Kernel size
1	Convolution	(1,1)
2	Pooling	(4,1)
3-1	Convolution	(1,1)
3-2	Convolution	(1,1), (3,3)
3-3	Convolution	(1,1), (3,3)
3-4	Max Pooling–Convolution	(4,4), (1,1)
4	Concatenate	–
5	Convolution	(2,2)
6	Pooling	(2,2)
7	Convolution	(,2)
8	Pooling	(3,3)
9	Global Average Pooling	–
10	Softmax	(1,X)

of sequence. Two RNNs can effectively capture context information and improve the performance of the model. When BiRNN uses LSTM, Bi-directional Long Short-Term Memory (BiLSTM) can be obtained.

LSTM includes forgetting gate, memory gate, cell state at the current time, output gate and hidden layer state at the current time. The combination of forward LSTM and backward LSTM can get BiLSTM. The structure of BiLSTM is shown in Fig. 2.

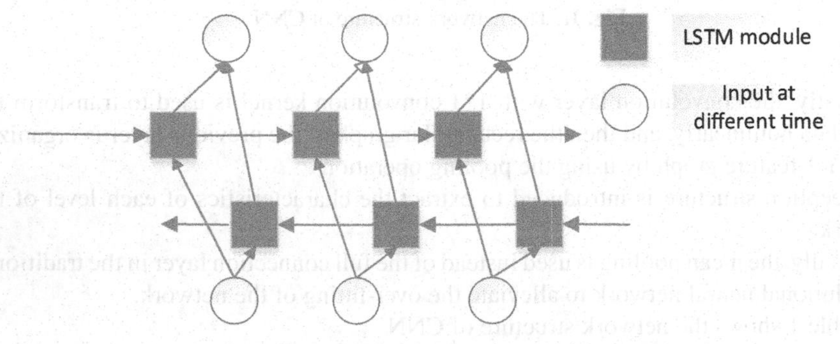

Fig. 2. The network structure of BiLSTM

2.4 Input Features

Each frame of three types of AE signals is extracted. The extracted frame level features include short-time energy, zero crossing rate, kurtosis, spectrum flux, etc. Features are shown in Table 2.

Table 2. Features list

Feature name	Feature number
Short-time Energy	1
Zero-crossing Rate	2
Kurtosis	3
Spectral Contrast Feature	4–10
MFCC Feature	11–22
Spectrum flux	23
Spectral Centroid	24
RMS Energy	25
LPC Coefficient	26–38

3 Algorithmic Architecture

We construct a recognition model of AE signal based on feature fusion of BiLSTM and convolution neural network. The overall architecture of the algorithm is shown in the following Fig. 3.

The extracted acoustic features are input into the BiLSTM network, and the AE signals are converted into amplitude-phase spectrogram and input into CNN network. Finally, the features obtained after the training of two networks are combined, and the softmax classifier is used for classification and recognition.

4 Experiments

4.1 Setting of Experimental Parameters

In this paper, the rub-impact AE signal with speed of 600 rad/s is used for experiment. Hanning window is used to add windows and frame discrete AE signals. The choice of frame length mainly depends on the validity of FFT point representation. CNN network adopts ReLU activation function, where the ReLU is shown in (4). This experiment uses Adam optimization algorithm to train. For the pooling operation, the padding method is adopted.

$$\mathrm{Re}LU(x) = \max(0, x) \tag{4}$$

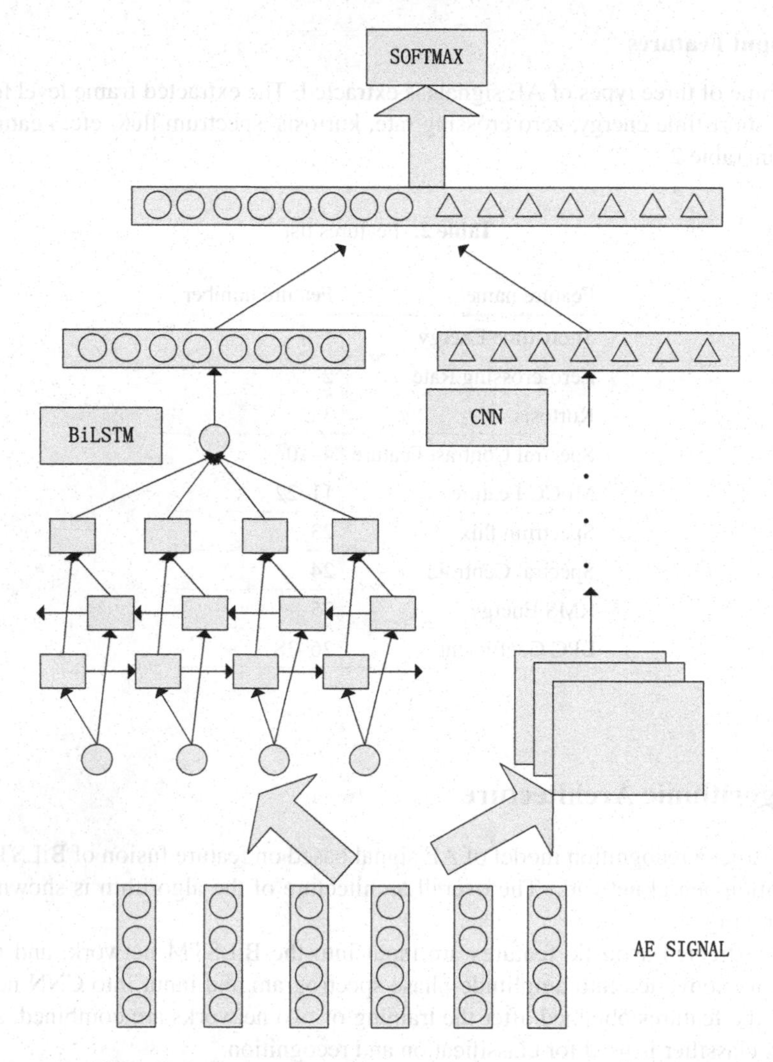

Fig. 3. Recognition model of AE signal based on feature fusion of BiLSTM and convolution neural network CNN

4.2 Analysis of Experimental Results

4.2.1 Display of Recognition Effect

Table 3 shows the confusion matrix of the proposed algorithm for classification of 600 rad/s rotating rub-impact AE signals. From the confusion matrix, it can be seen that the proposed recognition method has a recognition rate of more than 85% for three types of AE signals: normal, bearing crack and spindle rubbing, and the classification effect is excellent.

Table 3. Analysis of Confusion Matrix for proposed method at 600 rad/s Speed (%)

		Predicted value		
		Normal	Bearing crack	Spindle rubbing
True value	Normal	85.87	6.14	7.99
	Bearing crack	3.41	92.13	4.46
	Spindle rubbing	7.96	5.50	86.54

4.2.2 Model Contrast Experiments

In order to further explore the validity of the model proposed in this paper, the traditional recognition methods of acoustic emission signals are compared. Table 4 shows the performance of different classifiers in the classification of rub-impact AE signals, the features used by the comparison classifier are shown in Table 4.

Table 4. Recognition rates of three types of AE signals by different models at 600 rad/s speed (%)

Models	AE			Average recognition rate
	Normal	Bearing crack	Spindle rubbing	
KNN	59.31	56.27	52.32	55.97
DNN	68.14	72.12	66.34	68.87
SVM	66.21	75.35	69.14	70.23
Proposed method	85.87	92.13	86.54	88.18

It can be seen from the table that the recognition accuracy of the proposed method is the highest, reaching 85.87%, 92.13% and 86.54% respectively, whether for normal, bearing cracks or AE signals of spindle rubbing, and the overall average recognition rate is over 17% higher than that of SVM. The performance of SVM in traditional AE signal recognition method is the best, reaching 70.23%.

Compared with DNN, which is also used as a neural network, CNN and BILSTM can extract more effective features from spectral and acoustic features, so the proposed methods achieve higher recognition results.

5 Conclusions

In this paper, an acoustic emission recognition method based on spectral and acoustic features is proposed. The global characteristics of acoustic emission signals are obtained by inputting multi-channel spectrum of acoustic emission signals into CNN network; the information of short-term energy, zero-crossing rate and kurtosis is input into BiLSTM network to extract acoustic features of emission signals. Finally, the features extracted from CNN network and BiLSTM network are fused and the signal recognition is realized by soft max. The experimental results show that the proposed recognition method is effective for acoustic emission signals. The recognition accuracy of the classifier is much higher than that of the traditional recognition method.

References

1. Deng, A., Cao, H., Tong, H., Zhao, L., Qin, K., Yan, X.: Recognition of acoustic emission signal based on the algorithms of TDNN and GMM. Appl. Math. Inf. Sci. **8**(2), 907–916 (2014)
2. LeCun, Y., Bottou, L., Bengio, Y., et al.: Gradient-based learning applied to document recognition. Proc. IEEE **86**(11), 2278–2324 (1998)
3. Krizhevsky, A., Sutskever, I., Hinton, G.E.: Imagenet classification with deep convolutional neural networks. In: Advances in Neural Information Processing Systems, pp. 1097–1105 (2012)
4. Mesbah, R., Mccane, B., Mills, S., et al.: Improving spatial context in CNNs for semantic medical image segmentation. In: 2017 4th IAPR Asian Conference on Pattern Recognition (ACPR), pp. 25–30. IEEE Computer Society (2017)
5. Wang, N., et al.: ia-PNCC: noise processing method for underwater target recognition convolutional neural network. Comput. Mater. Continua **58**(1), 169–181 (2019)
6. Xianyu, W., Luo, C., Zhang, Q., Zhou, J., Yang, H., Li, Y.: Text detection and recognition for natural scene images using deep convolutional neural networks. Comput. Mater. Continua **61**(1), 289–300 (2019)
7. Zhao, K., He, T., Wu, S., et al.: Application research of image recognition technology based on CNN in image location of environmental monitoring UAV. EURASIP J. Image Video Process. **2018**(1), 150 (2018)
8. Fu, L., Feng, Y., Elkamil, T., et al.: Image recognition method of multi-cluster kiwifruit in field based on convolutional neural networks. Trans. Chin. Soc. Agric. Eng. **34**(2), 205–211 (2018)
9. Shen, Y., Tao, H., Yang, Q., et al.: CS-CNN: enabling robust and efficient convolutional neural networks inference for internet-of-things applications. IEEE Access **6**, 13439–13448 (2018)
10. Zheng, J., Liang, Z., Wang, Z.J.: Mid-level deep Food Part mining for food image recognition. IET Comput. Vis. **12**(3), 298–304 (2018)
11. Liu, Z., Wang, X., Kuntao, L., David, S.: Automatic arrhythmia detection based on convolutional neural networks. Comput. Mater. Continua **60**(2), 497–509 (2019)
12. Liu, J., Wang, G., Duan, L.Y., et al.: Skeleton based human action recognition with global context-aware attention LSTM networks. IEEE Trans. Image Process. **27**(4), 1586–1599 (2018)
13. Liu, J., Shahroudy, A., Xu, D., et al.: Skeleton-based action recognition using spatio-temporal LSTM network with trust gates. IEEE Trans. Pattern Anal. Mach. Intell. **40**(12), 3007–3021 (2018)
14. Ma, Y., Peng, H., Khan, T., et al.: Sentic LSTM: a hybrid network for targeted aspect-based sentiment analysis. Cogn. Comput. **4**, 1–12 (2018)
15. Ullah, A., Ahmad, J., Muhammad, K., et al.: Action recognition in video sequences using deep bi-directional LSTM with CNN features. IEEE Access **6**(99), 1155–1166 (2018)
16. Xu, B., Shi, X.F., Zhao, Z.H., et al.: Leveraging biomedical resources in bi-LSTM for drug-drug interaction extraction. IEEE Access **6**, 33432–33439 (2018)
17. Deng, J., Xu, X., Zhang, Z., et al.: Exploitation of phase-based features for whispered speech emotion recognition. IEEE Access **4**, 4299–4309 (2017)

Polyhedron Target Structure Information Extraction Method in Single Pixel Imaging System

Jin Zhang, Mengqiong Ge, Xiaoyu Shi, Zhuohao Weng, and Jian Zhang(✉)

College of Communication Engineering, Nanjing Institute of Technology, Nanjing, China
zhangjian@njit.edu.cn

Abstract. In a single-pixel imaging system, the illumination information of the reconstructed target is also different due to the different locations of single pixel detectors. For the object of polyhedra, this effect can make it difficult to extract the edge image representing the structural information. Aiming at this problem, this paper proposes to use multiple detectors for composite detection. By using the complementarity between images formed by multiple detectors, the reconstructed edge image is much more complete as compared to the one obtained by traditional method. The study promoted the development of single-pixel imaging technology.

Keywords: Single-pixel imaging · Edge extraction

1 Introduction

In three-dimensional imaging, since its data dimension is one more dimension than a two-dimensional image, the three-dimensional image needs to be compressed and stored. Among many imaging methods, single-pixel imaging has better adaptability, that is to say, measurement can be performed according to the characteristics of the target or the region of interest.

Single-pixel imaging is a new imaging method, using the correlation between calculate reference pattern and the measurement values of bucket detector. The technology was first proposed by Shapirio of MIT in 2008 [1] and was experimentally proven in 2009 [2]. Subsequently, the technology received wide attention. The research of single-pixel imaging are mainly concentrated in two aspects. First of all, in adaptive imaging, because its spatial measurement range can be controlled in real time, single-pixel imaging can flexibly perform high-resolution measurement of the region of interest according to practical requirement [3]. Second, for three-dimensional imaging, single-pixel imaging technology combined with lidar technology and photometric stereo vision technology respectively. For long-distance targets, single-pixel imaging laser radar for long-distance targets is developed [4–6]. For the object close to the projector, stereo imaging technology using single pixel detectors are derived [7, 8]. In addition, the use of deep learning algorithm has also been applied to single-pixel imaging, improving the efficiency of this system [9–11].

X. Sun et al. (Eds.): ICAIS 2020, LNCS 12240, pp. 769–775, 2020.
https://doi.org/10.1007/978-3-030-57881-7_67

For a polyhedral target, its structural features are reflected in the edge information of its image, so imaging this feature can better characterize its three-dimensional information. However, the difficulty lies in the fact that the edge information formed by different detectors is inconsistent, which makes it difficult to extract the complete structural information of the target. Aiming at this problem, this paper proposes to use multiple detectors for composite detection. By using the complementarity between images of multiple detectors, a more complete structural image is obtained.

2 Methodology

2.1 Setup

The setup of imaging system is shown in Fig. 1. This system consists of a digital light projector (Texas Instruments, DLP4500), to illuminate an object with a set of Hadamard patterns, and three bucket detectors (Thorlabs, DET100A), are located in three different positions to collect the reflected light. The signal of the bucket detector is collected by a data acquisition board (National Instruments, USB-6351). The data acquisition rate is 1 kHz and the exposure time of each bucket detector for each measurement is 0.3 s. The spatially separated bucket detectors are located in a plane 325 mm away from the object. The 2D image of object can be reconstructed

$$G_k(x, y) = \left\langle I_k^{(i)} \cdot P^{(i)}(x, y) \right\rangle - \left\langle I_k^{(i)} \right\rangle \left\langle P^{(i)}(x, y) \right\rangle \quad (1)$$

where $\langle \ldots \rangle$ denotes the assemble average over N patterns, $P^{(i)}$ is the i-th projected pattern, and $I_k^{(i)}$ is the i-th measurement by the bucket detector k (index k = up, left, right). The formed images G_{up}, G_{left}, G_{right} are shown in Fig. 2(b).

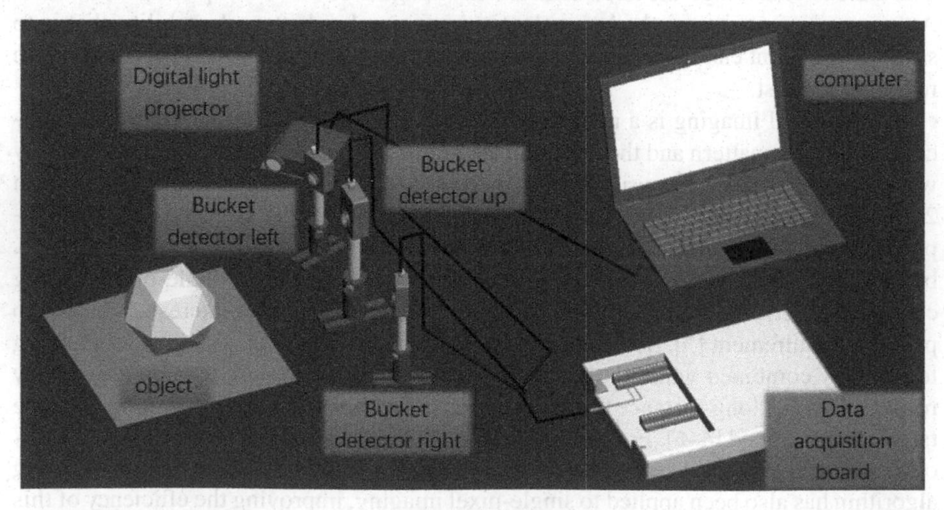

Fig. 1. Setup

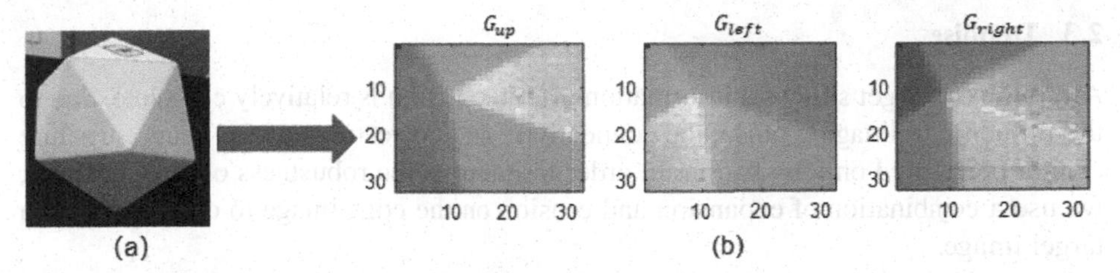

Fig. 2. (a) Object (b) I_1, I_2, I_3 represent G_{up}, G_{left}, G_{right} respectively

2.2 Method of Edge Extraction

To extract the edges of the image, Gaussian convolution core is first used to filter the noise and smooth the image $G_k(k = \text{up, left, right})$. Secondly, the images of gradient magnitude of G_k in horizontal and vertical directions are calculated by the finite difference of the first-order partial derivative (see Eq. (2)), marked by E_{kx} and E_{ky}.

$$\begin{cases} E_{kx}(x, y) = G_k(x, y + 1) - G_k(x, y) \\ E_{ky}(x, y) = G_k(x + 1, y) - G_k(x, y) \end{cases} \tag{2}$$

After threshold filtering, the edge image E_k consists of the corresponding E_{kx} and E_{ky}, as seen in Eq. (3). The three edge images E_{up}, E_{left}, E_{right} are shown in Fig. 3(a), where the structure of polyhedron of any single edge image is not complete.

$$E_k = E_{kx} \cup E_{ky} \tag{3}$$

$$E = E_{up} \cup E_{left} \cup E_{right} \tag{4}$$

In order to obtain a complete structure of object, three edge images are fused as seen in Eq. (4), and the corresponding result is shown in Fig. 3(b). These three images actually complement each other on the edge information, so the structure can be made complete. The above method relies on the placement of the barrel detector. Specifically, this approach works when the three bucket detectors are far enough away to detect the target from three distinct perspectives. On the contrary, if the three bucket detectors are close to each other, the corresponding three edge image information similarities are too high and do not have complementarity.

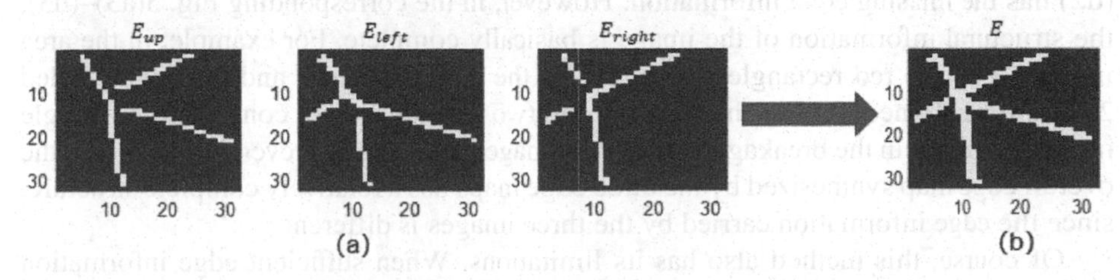

Fig. 3. (a) The three edge images (b) The corresponding result

2.3 Denoise

Although the target structure information after the fusion is relatively complete, due to the influence of imaging noise, the connectivity and correctness of the image structure cannot be ensured only by fusion. In order to improve the robustness of the algorithm, we use a combination of expansion and erosion on the edge image to obtain a smooth target image.

For the edge image E of the target (obtained by Eq. (4), the structural element s is first generated. s is a flat disc-shaped structure with a radius R of 1. Based on s, the E is first expanded and then etched. The result of the expansion is marked as E', and the result after the corrosion is E''. The result after the operation is shown in Fig. 4. In the experimental part, we will prove that this method can effectively fill the holes, ensure structural integrity, and remove redundant information.

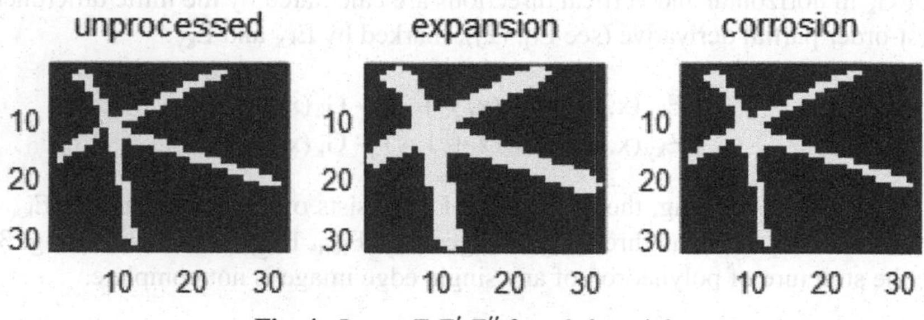

Fig. 4. Image E E' E'' from left to right

3 Results

First, we show the effect of using multiple bucket detectors for edge extraction. The results are shown in Fig. 5. In Fig. 5, the imaging objects are respectively Fig. 5(a1)–(d1), wherein the red straight line marks the structural information of the polyhedron under ideal conditions, that is to say, the position at the edge. Figure 5(a2)–(d2) are images reconstructed for each target by three bucket detectors, respectively, and Fig. 5(a3)–(d3) are the edge images after fusion using Eq. (4).

It can be seen that the edge image acquired by a single detector in Fig. 5(a2)–(d2) has the missing edge information. However, in the corresponding Fig. 5(a3)–(d3), the structural information of the image is basically complete. For example, in the area indicated by the red rectangle in Fig. 5(b2), the edge labeled 1 and the edge labeled 2 should be connected to each other, but the two images are not connected in a single image, resulting in the breakage of the edge image structure. However, in Fig. 5(b3), the overall edge map synthesized by the three edge maps has a relatively complete structure, since the edge information carried by the three images is different.

Of course, this method also has its limitations. When sufficient edge information cannot be provided in E_{up}, E_{left}, and E_{right}, the proposed method will not be able to extract the image completely, as shown in Fig. 5(c1)–(c3). In this experiment, the threshold

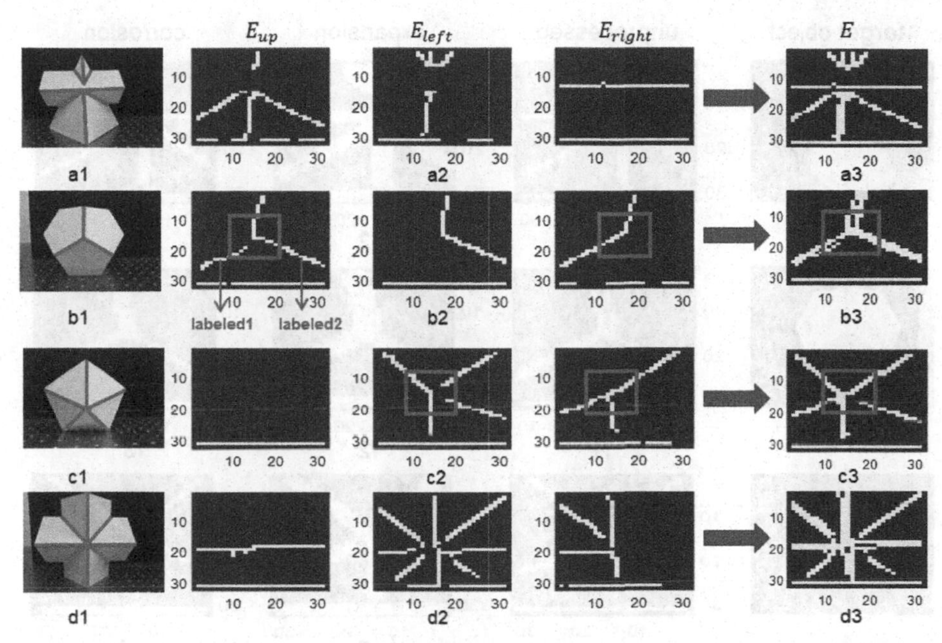

Fig. 5. Edge extraction

which can extract useful information and remove noise cannot be achieved at meantime, thus affecting the integrity of the extracted image.

Secondly, we verify the effectiveness of the proposed denoising method in this paper. The results are shown in Fig. 6. Figure 6(e1)–(h1) are edge images without expansion corrosion, Fig. 6(e2)–(h2) are the images after expansion, and then Fig. 6(e3)–(h3) are the images after corrosion.

It can be seen that the image after the fusion, although the structure is complete, the edge morphology changes drastically and the continuity is poor. After expansion and corrosion, the noise points smaller than the structural elements are eliminated, and the voids in the target are filled. It is worth noting that this method preserves the structural information of the image (this structural information can be understood as the 'skeleton' of the target), and also reduces the redundant image attached to the 'skeleton', thus making the 'skeleton' as narrow as possible.

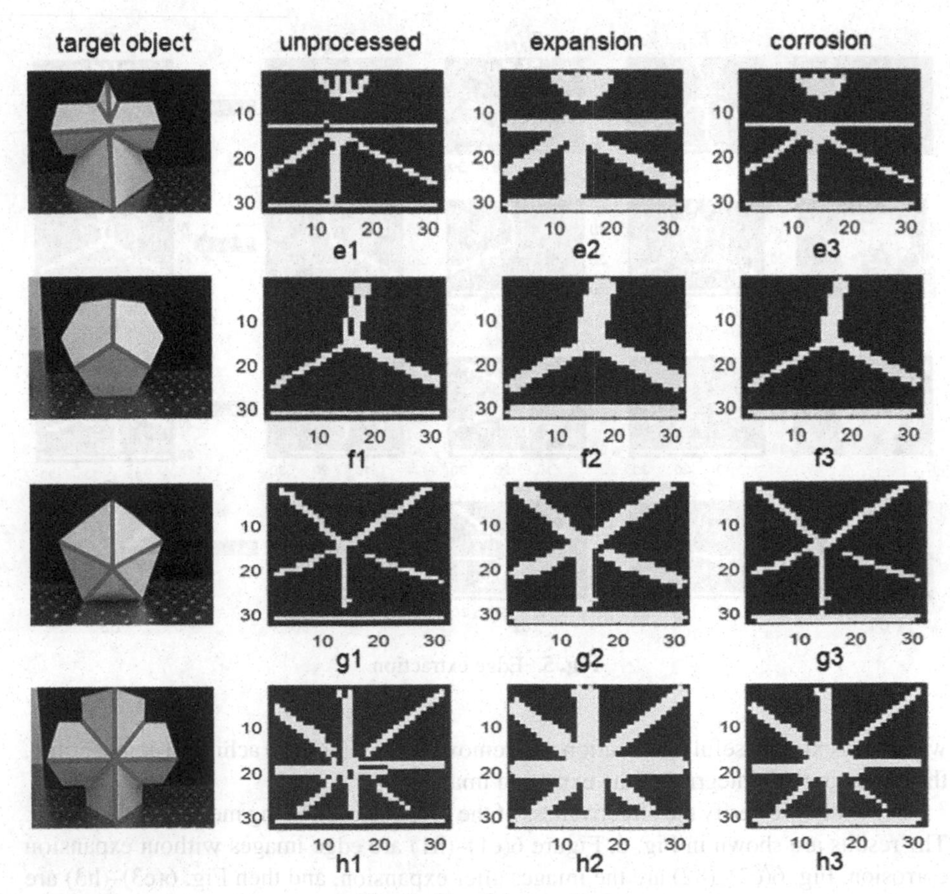

Fig. 6. Denoising method

4 Conclusion

In this paper, for the problem of difficult extraction of polyhedral structure information in single-pixel system, a multi-view edge image fusion method is proposed, and the fused image is denoised, which effectively maintains the structural integrity of the object. It should be emphasized that the algorithm still has room for improvement. For example, for some edge information that should exist, which is not included in the three images and cannot be performed by morphological operators, it cannot be reconstructed by this method. In general, this approach has driven the development of single-pixel imaging technology.

References

1. Shapiro, J.H.: Computational ghost imaging. Phys. Rev. A **78**, 061802 (2008)
2. Bromberg, Y., Katz, O., Silberberg, Y.: Ghost imaging with a single detector. Phys. Rev. A **79**, 1744–1747 (2008)
3. Phillips, D.B., et al.: Adaptive foveated single-pixel imaging with dynamic supersampling. Sci. Adv. **3** (2017)

4. Zhao, C., Gong, W., Chen, M., Li, E., Wang, H., Xu, W., Han, S.: Ghost imaging lidar via sparsity constraints. Appl. Phys. Lett. **101**(14), 141123 (2012)
5. Li, E., Bo, Z., Chen, M., Gong, W., Han, S.: Ghost imaging of a moving target with an unknown constant speed. Appl. Phys. Lett. **104**(25), 251120 (2014)
6. Sun, M.J., et al.: Single-pixelthree-dimensional imaging with time-based depth resolution. Nat. Commun. **7**, 12010 (2016)
7. Sun, B., Edgar, M.P., Bowman, R., Vittert, L.E., Welsh, S., Bowman, A., Padgett, M.J.: 3D computational imaging with single-pixel detectors. Science **340**(6134), 844–847 (2013)
8. Zhang, Y., Edgar, M.P., Sun, B., Radwell, N., Gibson, G.M., Padgett, M.J.: 3D single-pixel video. J. Opt. **18**, 035203 (2016)
9. Meng, L., et al.: Deep-learning-based ghost imaging. Sci. Rep. **7**, 1–6 (2017)
10. Ota, S., et al.: Ghost cytometry. Science **360**, 1246–1251 (2018)
11. Higham, C.F., Higham, R., Padgett, M.J., Edgar, M.P.: Deep learning for real-time single-pixel video. Sci. Rep. **8**(1), 2369 (2018)

Design and Implementation of Four-Meter Reading Sharing System Based on Blockchain

Baoyu Xiang[1(✉)], Zhuo Yu[2], Ke Xie[3], Shaoyong Guo[1], Meiling Dai[1], and Sujie Shao[1]

[1] Beijing University of Posts and Telecommunications, Beijing, China
473429384@qq.com, {syguo,mldai,buptssj}@bupt.edu.cn
[2] Beijing Zhongdian Puhua Information Technology Co., Ltd, Beijing, China
yuzhuo@sgitg.sgcc.com.cn
[3] State Grid ICT Industry Group Research Institute, Beijing, China
xieke@sgitg.sgcc.com.cn

Abstract. The four-meter reading refers to the unified collection and treatment of the energy data of water, electricity, gas and heat, and is a part of the national energy strategy. Although the state has already proposed four-meter reading, it has developed slowly and has little effect. The main reason is the problem of the reliability of the transmission of data in the channel and the collaboration problems caused by the industrial barriers of various energy companies. In this paper, we explore the problem of using the blockchain technology to solve the problem of the application of the four-meter reading application scenario. The research content of this paper is based on the fabric alliance chain, design data format and data management model, guarantee data security isolation, and realize a set of four-meter reading sharing system based on blockchain. For the research content of the four-meter reading scenario, firstly we investigate the knowledge of the four-meter reading application scenario, and then learn the blockchain technology and the fabric blockchain open source project. After that, the system is analyzed and designed, including requirements analysis, data isolation design, and system design. Next, the implementation of the system, including the construction of the fabric blockchain network, the preparation of smart contracts, the development of the client interface. Finally, the system is tested to verify that the system is functioning properly.

Keywords: Four-meter reading · Blockchain · Hyperledger fabric · Data isolation

1 Introduction

In the field of energy Internet, the meter record industry is an indispensable part of the daily energy consumption of residents [1]. Although in the field of the industry, the country has spent a lot of manpower and financial resources, combined with many advanced science and technology, but in the course of its development, there are still many problems still unresolved, one of which is the unified collection and treatment of the four energy data of water, electricity, gas and heat [2].

X. Sun et al. (Eds.): ICAIS 2020, LNCS 12240, pp. 776–785, 2020.
https://doi.org/10.1007/978-3-030-57881-7_68

At the national strategic level, the State Grid Corporation of China has been working hard to promote the development of the four-meter reading. At the same time, the relevant departments of the state have repeatedly emphasized in various conferences and publications that the infrastructures of electricity, gas, water and heat should be coordinated to achieve resource sharing and avoid redundant construction [3].

Although the country has been promoting the four-meter reading in various places over the years, because each of the four companies has its own management style, even if the four energy sources are very similar in meter reading, it has not been able to reach a consensus on cooperation and form a resource sharing. So that a lot of resources are wasted on the data collected by the meter reading, and it also increases the trouble for the daily life of the user.

The blockchain, known as the "trust cooperation puzzle solution", is naturally the best solution to this problem [4].

At the term of the perspective of the technical characteristics of blockchain, blockchain technology has five typical features such as decentralization, trust mechanism, open transparency, time series tamperability and traceability, which makes blockchain technology not only successfully applied in the field of digital currency such as Bitcoin, but also applicable to other scenes in real life [5]. Therefore, the blockchain and smart contract technology can be used to realize the process of data transmission, storage and verification in the energy field, and finally realize the open and transparent data value, avoid artificial malicious interference, and make the information untamperable [6].

Therefore, the application of blockchain technology can explore and promote the application of "four-meter reading" and solve the problem of trust cooperation that hinders promotion for many years, and at the same time, the combination of the blockchain technology and the "four-meter reading" scene is another exploration of the actual landing of blockchain technology [7].

The knowledge system involved in this paper includes blockchain technology, knowledge of fabric platform development, and application scenarios of four-meter reading. The main research contents include the following:

(1) For the application environment of the four-meter reading, use the blockchain technology to design the service bearer mode of the untrusted channel
(2) Design data format and data management model to ensure data security isolation
(3) Combine the Hyperledger Fabric for platform development and testing.

2 Related Work

2.1 Four-Meter Reading

Introduction to the Application Scenario. The four-meter reading refers to process the meters' data of electricity, water, gas and heat together. The four-meter reading is proposed to solve the problem of China's current energy construction development. At present, the four energy is divided into different company management.

Four-meter reading is based on the improvement of smart meter collection, combining smart water meter, intelligent heat meter and intelligent gas meter for unified meter reading, and transmitting the collected data to a unified management platform through channels.

The Problem Facing. At present, there are two main problems encountered in the process of promotion of the four-meter reading: reliable transmission of collected data in the channel and collaboration problems arising from industry barriers.

2.2 Blockchain Technology

Blockchain technology uses blockchain data structures to validate and store data, use distributed node consensus algorithms to generate and update data, use cryptography to ensure data transfer and access security, and utilize smart contracts made up of automated script code to program and manipulate data [8].

In terms of technology implementation, the blockchain has the characteristics of security, privacy, transparency, traceability, non-tampering and trusted sharing of information, so it can solve the problem of reliable transmission and sharing of data in various fields [9]. At the same time, the blockchain can perform the functions of self-written smart contracts to realize business logic. Therefore, blockchain technology will bring huge development to the entire Internet and even all walks of life in the future [10].

Throughout the development of blockchain technology and its future vision, the blockchain is divided into three phases based on the functions that the blockchain has achieved and the functions that may be realized in the future [11] (Fig. 1).

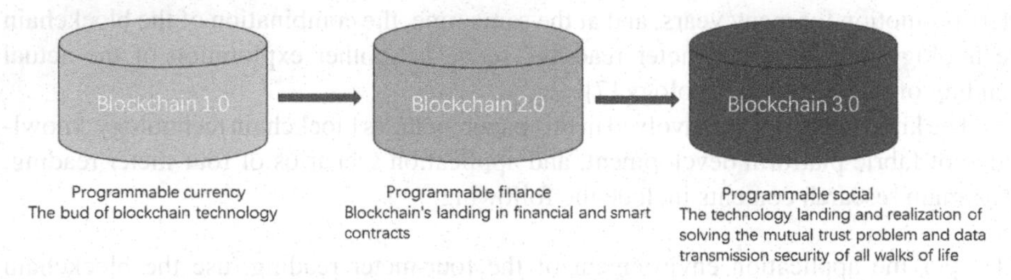

Fig. 1. Blockchain evolution trend

2.3 Hyperledger Fabric

Fabric is a sub-project under the Hyperledger project. It is an open source project initiated by the Linux Foundation for the enterprise blockchain project and belongs to the alliance chain. One of the biggest features of Fabric is its high degree of modular design, and its modules are pluggable, easy to personalize according to different business needs. It is also the versatility of Fabric that makes it attractive in all areas [12].

Figure 2 shows the system logic architecture of the fabric design.

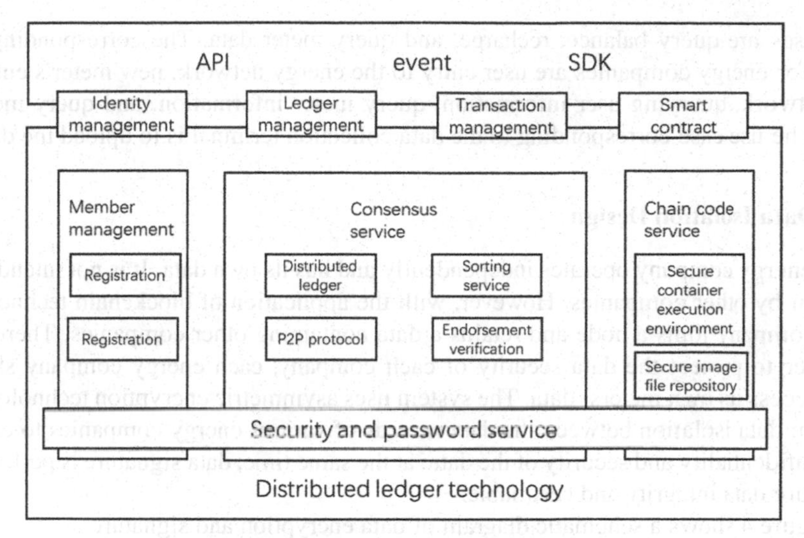

Fig. 2. Fabric system architecture diagram

3 Demand Analysis and Design

3.1 Demand Analysis

The entities that interact with the system mainly include users, energy companies, and data collection terminal.

As shown in the use case diagram in Fig. 3.

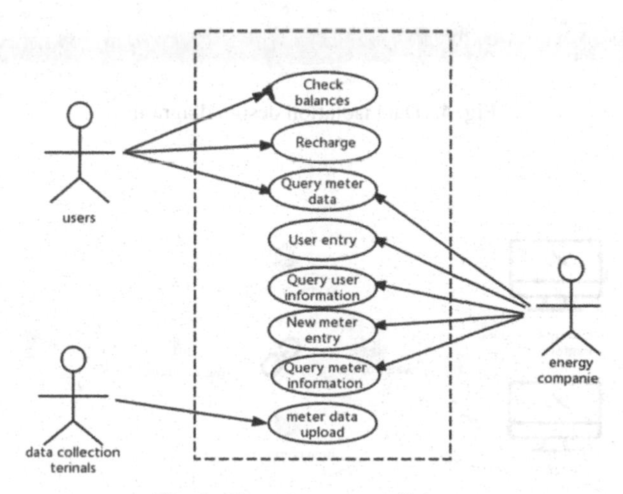

Fig. 3. System use case diagram

The roles that interact with the system are primarily energy-consuming users, companies that provide energy, and data collection terminal that upload data. The user-specific use cases are query balance, recharge, and query meter data. The corresponding use cases for energy companies are user entry to the energy network, new meter's entry to the network, querying user information, query meter information, and query meters' data. The use case corresponding to the data collection terminal is to upload the data.

3.2 Data Isolation Design

Each energy company operates independently and has its own data. It is not intended to be seen by other companies. However, with the application of blockchain technology, each company joins a node and retains a data containing other companies. Therefore, in order to protect the data security of each company, each energy company should only access its own meters' data. The system uses asymmetric encryption technology to achieve data isolation between the data records of various energy companies to ensure the confidentiality and security of the data; at the same time, data signature is performed to ensure data integrity and certifiable.

Figure 4 shows a schematic diagram of data encryption and signature.

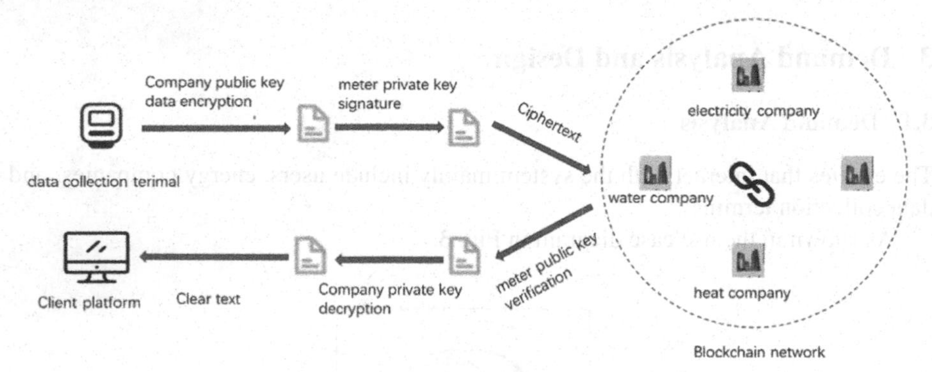

Fig. 4. Data isolation design diagram

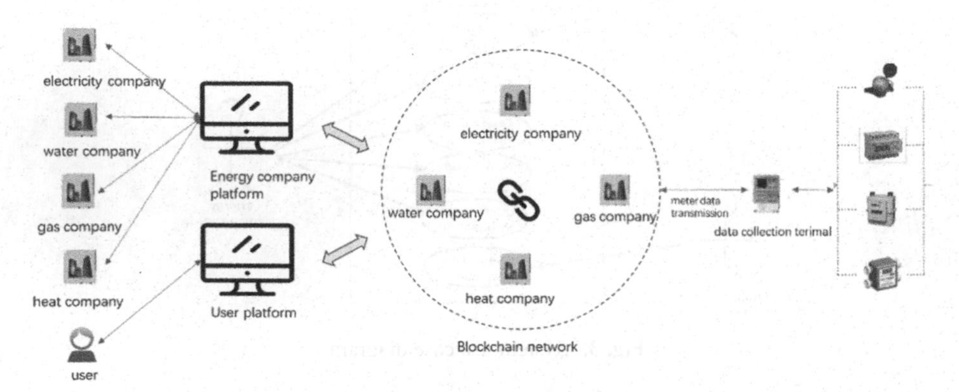

Fig. 5. System scene architecture diagram

3.3 System Design

The system consists of three parts: the client, the blockchain network, and the data collection terminal. Figure 5 shows the system scenario.

As shown in the above figure, the scene architecture of the four-meter reading system is mainly the data collection terminal, process the data recorded by the four energy companies' meters, and then transmitted to the blockchain platform for storage. At the same time, each energy company interacts with the user through the system client and the blockchain platform to realize the query of the data and the operation of related information.

Blockchain Network. Based on the Fabric project, the four-meter reading system builds a blockchain network to achieve reliable storage of data. At the same time, Fabric can not only provide blockchain basic services such as consensus services, ledger services, transaction services, and ledger services, but also support smart contract custom business logic. Therefore, by writing a smart contract, the business logic of the four-meter reading can be directly deployed on the blockchain, which can alleviate the burden on the client business logic and make full use of the characteristics of the fabric blockchain network platform.

Client Platform. The client of the system is a visual web interface for the user to interact with the energy company and the blockchain system. It provides the user with the function of querying the balance, recharging, and querying the data. At the same time, it provides the company user with access to the network, new meter access, and query form. The function of information, querying user information, and querying data.

The user operates the client, inputs data, and the client receives the user request, and then sends the request to the business logic layer, that is, the background processing interface, where the business logic is written in the chain code and deployed in the blockchain network.

The client is developed in web form, that is, the user can use any browser to access, only need to access a specific URL, and then enter their own username and password for identity authentication, then enter the system, operate related functions, and interact with the blockchain network.

Data Collection Terminal. The data collection end is a hub for the blockchain network to interact with each meter in reality, and the energy meter timing collection data of each company is transmitted to the data collection end. The data collection end processes the relevant meter data and transmits it to the blockchain network platform and stores it on the blockchain.

4 System Implementation

4.1 Fabric Blockchain Platform Construction

The system server is Tencent's cloud server, and its configuration is as follows (Table 1):

Table 1. Server configuration

Configuration	Parameter
Core	1
Member	1 GB
Bandwidth	1 Mbps
System	Ubuntu 16.04

The specific software environment installation is shown in Table 2 below.

Table 2. Software installation environment

Software environment	Installation version
Core	1
Member	1 GB
Bandwidth	1 Mbps
System	Ubuntu 16.04

The source code of the fabric is cloned from the git repository. Since the fabric is constantly being updated, the system is developed based on version 1.0, and the source code provides a tool for quickly building a blockchain network.

4.2 Implementation of Business Smart Contracts

The system functions mainly include user function modules, company function modules and table data uploading three blocks. The smart contract is written in the go language and communicates with the endorsement node by implementing and calling the SDK of the chain code (i.e. the shim provided by the fabric official).

Three structures are defined in the system's smart contract, representing the user, meter, and meter data. The three data structures are described in turn below.

User. User information is the main body of energy consumed in the energy chain, representing an individual in the real world.

```
type User struct {
    UserId string `json:"UserId"`
    Money string `json:"Money"`
    Meter []string `json:"Meter"`
}
```

Meter. The meter information is divided into four meters of electricity, gas, water and heat according to the identification of the meter, and the data is collected.

```
type Meter struct {
   MeterId string `json:"MeterId"`
   RegisterTime string `json:"RegisterTime"`
   Owner string `json:"Owner"`
}
```

EnergeCost. The table data is transmitted by the table collection through the data collection end and stored in the blockchain network.

```
type EnergeCost struct {
   MeterId string `json:"MeterId"`
   Timestamp string `json:"Timestamp"`
   CostOfenerge string `json:"CostOfenerge"`
}
```

The function functions finally realized by the system are summarized, as shown in Table 3.

Table 3. Smart contract function table

Name	Function
Check balances	{"Function":"getMoney", "Args":["userid"]}
Recharge	{"Function":"addMoney", "Args":["userid","moneycount"]}
Query bill	{"Function":"queryForUser", "Args":["userid","meterid"]}
User entry	{"Function":"registerUser", "Args":["userid"]}
Meter entry	{"Function":"registerMeter","Args":["meterid","registertime","userid"]}
Query user information	{"Function":"getUser", "Args":["userid"]}
Query meter information	{"Function":"getMeter", "Args":["meterid"]}
Query meter data	{"Function":"queryHistory", "Args":["company","meterid"]}
Meter data upload	{"Function":"save", "Args":["meterid","value"]}

4.3 Client Development

The client is developed using the vue framework. According to different functions, axios is used to call the back-end http interface to implement front-end connection.

Vue is a progressive framework for building user interfaces that focuses only on the user view layer. Its features are simple project construction, lightweight architecture, faster page rendering, and support for two-way binding of instructions and data. On the other hand, vue is developed and maintained by Chinese. The ecological environment in China is relatively better. There are many Chinese reference materials and tools to facilitate learning and development.

The system uses the element-ui component developed by the company "eleme" to build the front-end client of the system. The following is the main client of the system.

5 System Testing

The performance test of the system is stress tested by controlling the number of concurrent requests and the number of requests executed. In the actual test process, use the ab test tool in the httpd-tools toolkit to test.

The table below gives the performance test results of this system (Tables 4 and 5).

Table 4. System write performance test

Concurrent	Requests	Tps	Ms/Req
10	100	8.7	114.934
10	1000	4.42	226.373
100	100	8.89	112.491
100	1000	7.48	133.759

Table 5. System read performance test

Concurrent	Requests	Tps	Ms/Req
10	100	32.84	30.454
10	1000	34.99	28.580
100	100	36.36	27.505
100	1000	36.40	27.469
500	1000	38.70	25.841

During the system write performance test, when the number of concurrent calls is 300, the system crashes. The final test result is a system concurrency of approximately 100 and a tps of approximately 9. During the system read performance test, when the number of concurrency is 1000, the system crashes. The final test result is that the system concurrency is approximately 500 and the tps is approximately 39.

6 Summary

The problem of four-meter reading is an important part of promoting the development of social infrastructure and benefiting people's lives. The emergence of blockchain technology and its unique technical characteristics have brought new directions to the solution of this problem.

This paper designs and implements a four-meter reading and sharing system based on blockchain for the trust cooperation problems of various energy companies in the process of four-meter reading promotion. Although the system has been initially completed, it is still not perfect, there are many problems, some designs and functions have not been realized or just a simple practice.

With the advancement of technology, the arrival of 5G and the combination of blockchain technology will eventually solve the problem of four-meter reading.

Acknowledgement. This work is supported by National Key R&D Program of China (2018YFB1402704) and the National Natural Science Foundation of China (61702048).

References

1. Tang, C., Zhang, F.: Combined electricity-gas-heat energy internet scheduling with power-to-gas and renewable power uncertainty. In: 2018 2nd IEEE Conference on Energy Internet and Energy System Integration (EI2), Beijing, pp. 1–5 (2018)
2. Zhu, J., Xie, P., Xuan, P., Zou, J., Yu, P.: Renewable energy consumption technology under energy internet environment. In: 2017 IEEE Conference on Energy Internet and Energy System Integration (EI2), Beijing, pp. 1–5 (2017)
3. Kalyanaraman, S.: Back to the future: lessons for internet of energy networks. In: IEEE Internet Computing, vol. 20, no. 1, pp. 60–65, January-February 2016
4. Wessling, F., Ehmke, C., Hesenius, M., Gruhn, V.: How much blockchain do you need? Towards a concept for building hybrid dapp architectures. In: 2018 IEEE/ACM 1st International Workshop on Emerging Trends in Software Engineering for Blockchain (WETSEB), Gothenburg, Sweden, pp. 44–47 (2018)
5. Mechkaroska, D., Dimitrova, V., Popovska-Mitrovikj, A.: Analysis of the possibilities for improvement of blockchain technology. In: 2018 26th Telecommunications Forum (TELFOR), Belgrade, pp. 1–4 (2018)
6. Jiang, X., Liu, M., Yang, C., Liu, Y., Wang, R.: A blockchain-based authentication protocol for WLAN mesh security access. Comput. Mater. Continua **58**(1), 45–59 (2019)
7. Deng, Z., Ren, Y., Liu, Y., Yin, X., Shen, Z., Kim, H.-J.: Blockchain-based trusted electronic records preservation in cloud storage. Comput. Mater. Continua **58**(1), 135–151 (2019)
8. Sun, G., et al.: Research on public opinion propagation model in social network based on blockchain. Comput. Mater. Continua **60**(3), 1015–1027 (2019)
9. Lee, H., Sung, K., Lee, K., Lee, J., Min, S.: Economic analysis of blockchain technology on digital platform market. In: 2018 IEEE 23rd Pacific Rim International Symposium on Dependable Computing (PRDC), Taipei, Taiwan, pp. 94–103 (2018)
10. Kodali, R.K., Yerroju, S., Yogi, B.Y.K.: Blockchain based energy trading. In: TENCON 2018–2018 IEEE Region 10 Conference, Jeju, Korea (South), pp. 1778-1783 (2018)
11. Thakkar, P., Nathan, S., Viswanathan, B.: Performance benchmarking and optimizing hyperledger fabric blockchain platform. In: 2018 IEEE 26th International Symposium on Modeling, Analysis, and Simulation of Computer and Telecommunication Systems (MASCOTS), Milwaukee, WI, pp. 264–276 (2018)

Edge-Feedback ICN Cooperative Caching Strategy Based on Relative Popularity

Huansong Li[1(✉)], Zhuo Yu[2], Ke Xie[3], Xuesong Qiu[1], and Shaoyong Guo[1]

[1] Beijing University of Posts and Telecommunications, Beijing, China
lhspeppa@gmail.com
[2] Beijing Zhongdian Puhua Information Technology Co., Ltd, Beijing, China
[3] State Grid ICT Industry Group Research Institute, Beijing, China

Abstract. Information-centric networking (ICN) improve content-oriented services by enabling in-network caching and providing the best request forwarding path. The On-path cooperative caching in in-network caching is an effective way to reduce the service delay. It provides higher performance by coordinating between nodes on the request forwarding path. However, the current strategy is to push popular content to the edge of the network, resulting in a large amount of cache redundancy at the edge of the network. This will reduce the cache utilization of the network, and cause additional communication overhead between nodes on the forwarding path. This paper proposes an edge-feedback cooperative caching strategy, which advances the caching decision to the request forwarding phase, from the edge of the network to the core network. The node on the forwarding path recalculates the relative popularity of the content based on the feedback of the downstream node, and determines the content items that need to be cached in the network according to the calculation result.

Keywords: Information-Centric network · Cooperative caching · Edge feedback · Relative popularity

1 Introduction

In the research of ICN networks, in-network caching is one of the important research directions. The research mainly consider two aspects: from the perspective of users, improve the service quality of users; from the perspective of the network itself, improve the utilization of network resources and reduce the communication overhead between nodes.

However, the two aspects cannot be satisfied at the same time. In order to improve the user's service quality and reduce the user service delay, the popular content needs to be cached in a node closer to the user, which causes a large number of copies of the same content in the network, resulting in cache redundancy; On the other hand, in order to improve the network resource utilization and enrich the resource diversity, it is necessary to reduce the number of content copies. The extreme method is to uniquely limit the same content within the network, which will sacrifice the user experience. And

X. Sun et al. (Eds.): ICAIS 2020, LNCS 12240, pp. 786–797, 2020.
https://doi.org/10.1007/978-3-030-57881-7_69

because of the increase in transmission distance, the communication cost caused by the extra transmission is increased. Due to the cache capacity limitation of ICN routers, the advantages of ICN cache can be maximized only by placing more popular content closer to the user. In addition, the request and content in the ICN are forwarded by using a symmetric path. When the downstream node in the request forwarding process caches the content, the probability of requesting the content by the upstream node will decrease, and the gain of the content cache in the upstream node will decrease. Therefore, the decision of the cache location is also an important factor in determining the cache gain.

2 Related Work

It has been proved that service performance is poor when using single-node cache decisions. Current research on ICN in-network caching focuses on collaborative caching decisions. Collaborative caching effectively utilizes cache resources in the network through cache coordination between nodes.

In the existing caching strategy, Laoutaris et al. [6] proposed several meta-algorithms for hierarchical web caching. When the packet is forwarded to the downstream node, Leave Copy Down (LCD), Move Copy Down (MCD) is directly applied to the forwarding path, which can slightly improve cache efficiency and service delay. Psaras et al. [7] proposed a ProbCache method, which uses the cache capacity and the distance from the node to the source server as the standard for measuring the cache probability. However, this method does not consider that content popularity is an important factor to reduce the traffic transmission within the network. Suksomboon et al. [8] proposed a PopCache method, which obtained a lower service delay than ProbCache by calculating the content popularity. However, each node in PopCache needs to collect global access information to calculate the cache probability, which will affect the network service performance.

Based on the above research, this paper proposes an Edge-Feedback ICN cooperative caching strategy based on relative popularity. According to the feature that the request forwarding and content propagation in the ICN network are forwarded by a symmetric path, in the request forwarding phase, each node on the forwarding path recalculates the relative popularity of the content based on the cache decision of the downstream node to determine the set of content cached by nodes within the network. Estimation of relative popularity at the node serves as an important basis for content caching. By adjusting the weight of the content in the cache in real time, the transmission delay of the active content can be effectively shortened, and the edge cache space can be fully utilized.

3 Edge-Feedback Strategy

3.1 Problem Model

This paper describes the Edge-Feedback cooperative caching Strategy (EFS) based on the NDN network model. The symbols used in this article are shown in Table 1.

Table 1. Symbol list

Symbol	Annotation
N	The set of nodes in the edge network
E	The set of node paths in the edge network
C	The total contents in the edge network
h_n^c	Cache hit ratio for content c at node n
$\sigma_n, \overline{\sigma}_n$	Request satisfied/forwarding number at node n
D^c	Request delay for content c
θ^c	Popularity correction parameter in content request
CL^c	List of cache nodes in the content request
λ_n^c	Average request arrival rate for content c at node n
p_c	Request probability of content c
RP_n^c	The relative popularity of content c at node n
S_n^c	Binary variable identifies whether content c is cached at node n
T^c	The default forwarding path for content c
P^c	The set of nodes on the default forwarding path of content c
Q^c	The edge set on the default forwarding path for content c

In the system model, undirected graph G = $< N, E >$ represents the network topology, where **N** is the set of nodes in the network, **E** is the set of transport links between the routes,$e_{<i,j>} \in$ E represents that the content can be transferred from node n_i to n_j, C_n is the buffer capacity of node n. For the content c $\in$ C in each request, it will eventually point to a content server **S** closest to the requesting node when forwarding, and forward it along the shortest path to S. $P^c(V^c, E^c)$ is the default forwarding path of content c, $v_i \in V^c$ denotes that node v_i is on the forwarding path of content c, and $e_{i,j} \in E^c$ denotes that edge $< i, j >$ is an edge on the default forwarding path. To simplify the problem model, make the following assumptions:

- **Assume I:** The content blocks are of the same size and can be carried in a single transmission message, so there is no content fragmentation.
- **Assume II:** The content can have multiple copies in the network, but there is only one content source. The request forwarding is based on the NDN's own routing protocol (such as OSPFN). This article does not consider how to transmit packets.

The cache allocation optimization model can be derived as follows:

$$\min \sum\nolimits_{c \in C} \sum\nolimits_{n \in N} \lambda_n p_c d_n^c (1 - S_n^c) \tag{1}$$

$$\text{s.t.} \begin{cases} \sum_{c \in C} S_n^c \le C_n & \forall n \in N \\ \sum_{c \in C} p_c = 1 & \\ S_n^c \in \{0, 1\}, & \forall n \in N, \forall c \in C \end{cases} \tag{2}$$

Where d_n^c represents the distance from the node n requesting the content c to the node providing the content; λ_n is the request arrival rate at each node; p_c is the proportion of content c in all requests. Equation (1) indicates that for all content requests, the ultimate goal is to minimize the request forwarding distance, i.e. maximize the cache gain. S_n^c indicates whether the content c is cached at the node n, which is a boolean value. For all content requests, the sum probability is 1, and the amount of cached content per node does not exceed its cache size.

3.2 Strategy Design

This section proposes a cache optimization strategy for the ICN edge network—the Edge-Feedback caching Strategy (EFS). The EFS strategy prioritizes the service performance of the network edge, advances the cache decision to the request forwarding section, and allocates the cache space reasonably through local dynamic coordination between nodes on the forwarding path.

Specifically, in the request forwarding section, the forwarding node first calculates the current content Relative-Popularity (**RP**) according to the cache decision situation from the downstream node. The forwarding node decides whether to cache the content according to the RP. The downstream node then feeds the decision back to the upstream node, which assists the upstream node in modifying its RP. By introducing RP as a measure of whether to cache content, content popularity can be updated in real time to achieve better network performance.

In order to satisfy the transmission of information between nodes on the path, two fields are newly added based on the transmission package of the original NDN model. Add Cache-List (**CL**) and Popularity-Correction-Parameter (**PCP**) to the interest packet representing the content request. When the cache hits, add the CL to the returned data packet. The CL is used to mark the cache decision of the downstream node, and when the node on the forwarding path decides to cache the content in the request, the node tag is added to the CL. The PCP is used by the downstream node to deliver the content popularity change to the upstream node, and each node on the forwarding path recalculates the relative popularity of the content according to the PCP received in the interest packet.

Next, we illustrate the effectiveness of the EFS strategy with an example, and explain the dynamic adjustment process of relative popularity in the decision process. In Fig. 1, there are five in-network nodes ($N_1 \sim N_5$) and one content server **S** in the network topology. We consider that there are only two types of content X and Y, and the request probabilities for X and Y are 0.6 and 0.4. If using absolute content popularity as the cache decision indicator, the content X will be cached in the node N_1 and N_2. but when we consider the request forwarding to the upstream node due to request missing, caching X at N_1, and caching Y at N_2 is a better caching scheme.

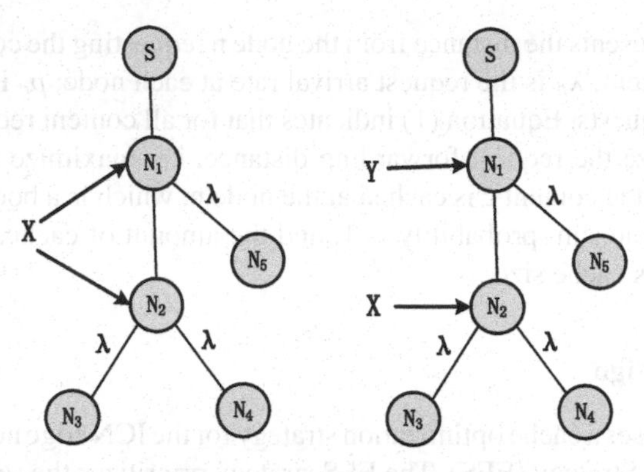

Fig. 1. Algorithm example

According to previous discussion, the calculation of relative popularity is the core of the EFS strategy. There are three main influencing factors in the calculation process: (1) **Time Factor**, the relative popularity of content decreases with time; (2) **Request Arrival Frequency**, with the number of requests increases, the relative popularity of the content increases; (3) **Downstream Node Cache Decision**. When the downstream node has decided to cache the content, the relative popularity is recalculated according to the popularity correction parameter in the content request packet.

Next, we introduce the cache decision in the request forwarding process and the replacement decision in the content packet propagation process.

Cache Decision. At node n, the average request arrival rate of content c is r_n^c; $S_n^c \in \{0, 1\}$ indicates whether content c is cached at node n, and when $S_n^c = 1$ indicates that node n caches content c; RP_n^c indicates the Relative-Popularity (RP) of content c; $T^c = \langle P^c, Q^c \rangle$ represents the node and edge on the default forwarding path of content C, where P^c is a subset of node set N, Q^c is a subset of edge set E, for i, j $\in P^c$, $\langle i, j \rangle \in Q^c$ indicates that upstream node j is the unique next hop of node i. All requests for content c initiated by node n are forwarded along the nodes and edges on T^c, and the content is also returned along this path.

σ_n^c represents the case where the content at node n hits. Instead, $\overline{\sigma}_n^c$ represents the number of content requests forwarded to the upstream node. Considering that in the actual network topology, the missed content request of the downstream node will be forwarded to the upstream node and become part of the upstream node's incoming request. Therefore, the impact of cache hits of adjacent nodes needs to be considered in the calculation. The number of content requests that can be satisfied/forwarded at node n is calculated as Eqs. 3 and 4.

$$\sigma_n^c = \left(r_n^c + \sum\nolimits_{n' \in Q^c_{<n',n>}} \overline{\sigma}_{n'}^c \right) \times S_n^c \tag{3}$$

$$\overline{\sigma}_n = \left(r_n^c + \sum\nolimits_{n' \in Q^c_{<n',n>}} \overline{\sigma}^c_{n'} \right) \times \left(1 - S_n^c\right)$$
$$= r_n^c + \sum\nolimits_{n' \in Q^c_{<n',n>}} \overline{\sigma}^c_{n'} - \sigma_n^c \tag{4}$$

θ^c represents the Popularity-Correction-Parameter(PCP) in the content request packet. For each content request, the updated relative popularity $RP_n^{c,new}$ is equal to the pre-update relative popularity $RP_n^{c,old}$ plus the PCP from the downstream node request and the current node's request arrival rate r_n^c. The expression for updating the relative popularity of the content at node n is as shown in Eq. 5.

$$RP_n^{c,new} = RP_n^{c,old} + \sum\nolimits_{n' \in Q^c_{<n',n>}} \theta^c_{n'} + r_n^c \tag{5}$$

If $S_n^c = 0$, indicating that the local cache cannot satisfy the content request, the request needs to be forwarded up. The Cache-List (CL) and Popularity-Correction-Parameters(PCP) in the request package need to be updated before forwarding. $CL^c = \{n_1, n_2, \ldots, n_m | \mathrm{m} \leq |N|\}$ denotes a list of downstream nodes that decide to de-cache content on the forwarding path. After each hop node updates the relative popularity, it is determined whether the current node is added to the Cache-List of the request packet, calculated as Eq. 6.

$$CL^c = \begin{cases} CL^c \cup \{n\}, & RP_n^{c,new} > \min\left(\{RP_n^m | m \in C_n\}\right) \\ CL^c, & otherwise \end{cases} \tag{6}$$

C_n represents the cached content list at node n. When the relative popularity of the requested content c is greater than the minimum relative popularity of all the content maintained by the node, the node will cache the content and replace the content that has the lowest relative popularity. Finally, the node is added to the Cache-List of the request packet.

At node n, the popular content in a certain period of time may not be requested in the subsequent time, so for the time decay factor $\tau \in [0, 1]$, each node periodically recalculates the relative popularity of the content according to τ degree, the frequency of replacement of popular content is controlled by adjusting τ during the network process.

$$RP_n^{c,new} = \tau \times RP_n^{c,old} \tag{7}$$

The cache decision process is as follows:

Algorithm 1 The cache decision process in request forwarding

(1) node **n** receive the interest packet.

(2) get the **cache list** H^c and **popularity parameter** M^c from the interest packet.

(3) $RP_n^{c,old} = RP_n^c$

(4) **if** $M^c \neq \emptyset$

(5) $\quad RP_n^c = RP_n^{c,old} + M^c$

(6) **if** $S_n^c = 1$

(7) $\quad$ generate data packet, and put H^c in it.

(8) $\quad$ **return** data packet

(9) **else**

(10) $\quad$ **if** $(\sum_{i \in C} S_n^i < B_n) \,||\, (RP_n^c > \min\{RP_n^i | S_n^i = 1\})$

(11) $\quad\quad H^c = H^c + n$

(12) $\quad\quad M^c = -RP_n^{c,old}$

(13) $\quad$ **else**

(14) $\quad\quad M^c = RP_n^c$

(15) **forward request packet to the next hop**

Replacement Decision. The process of EFS cache replacement decision is: when the node receives the data packet, if the current node cache space has remaining, the content is directly cached; otherwise, it checks whether the Cache-List in the data packet contains this node. If the node is included, the cache is replaced according to the Relative-Popularity of the cached content; otherwise, the content packet is forwarded to the next node.

When the data in the cache space needs to be replaced, the relative popularity of the cached content is sorted from low to high, and the one with the lowest popularity is replaced. According to Eq. (5, 7), the Relative-Popularity is:

$$RP_n^{c,new} = \tau RP_n^{c,old} + (1-\tau)\left(\sum\nolimits_{n' \in Q^c_{<n',n>}} \theta_{n'}^c + r_n^c\right) \quad (8)$$

The replacement decision process is as follows:

Algorithm 2 The replacement decision process in content distribution process

(1) node **n** receive the data packet.

(2) **if** $\sum_{i \in C} S_n^i < B_n$

(3) $S_n^c = 1$

(4) **else**

(5) get the **cache list** H^c from the data packet

(6) **if** n $\in H^c$

(7) $t = \{i | \{\min(RP_n^i) | S_n^i = 1\}\}$

(8) $S_n^t = 0$

(9) $S_n^c = 1$

(10) **forward data packet to the next hop**

Cache Example. An example is given to illustrate the working effect of the above decision process. Consistent with the assume, assuming that each node's cache space is one unit, the request arrival rate is λ, and the content includes X and Y.

As shown in Fig. 2a, in the first section, the probability ratio of X and Y in the request flowing from N_4 and N_5 is 6:4. The relative popularity at N_2 is as shown in the table, so N_2 decide to cache X.

According to EFS, the request for Y flowing into N_2 cannot be satisfied and continues to be forwarded to the upstream node N_1, and the Popularity-Correction-Parameter is 0.4. Then N_1 recalculates the relative popularity based on the information acquired from the downstream node and determines to cache Y.

As shown in Fig. 2b, in the second section, N_4 and N_5 no longer receive new requests, and in the N_3 inflow request, the ratio of X and Y is 2:8. When the request for Y finds that the Relative-Popularity is greater than the RP of the currently cached content, the current node (N_2) is added to the cache node list in the request forwarded to the upstream node (N_1), indicating that cached at N_2 when the content packet arrives. At the same time, the PCP −0.4 is added to the request packet, indicating the impact on the RP of upstream nodes after the current node is cached. N_1 modifies the RP of the local content according to the PCP in N_2 incoming interest request, and determines that N_1 should cache the content X. The above is the process of adjusting the cache according to EFS.

4 Performance Analysis

In order to verify the performance improvement of EFS strategy, this section uses the ndnSIM tool to simulate the EFS strategy and several other caching strategies. The effectiveness of the EFS strategy is verified by comparing the experimental results.

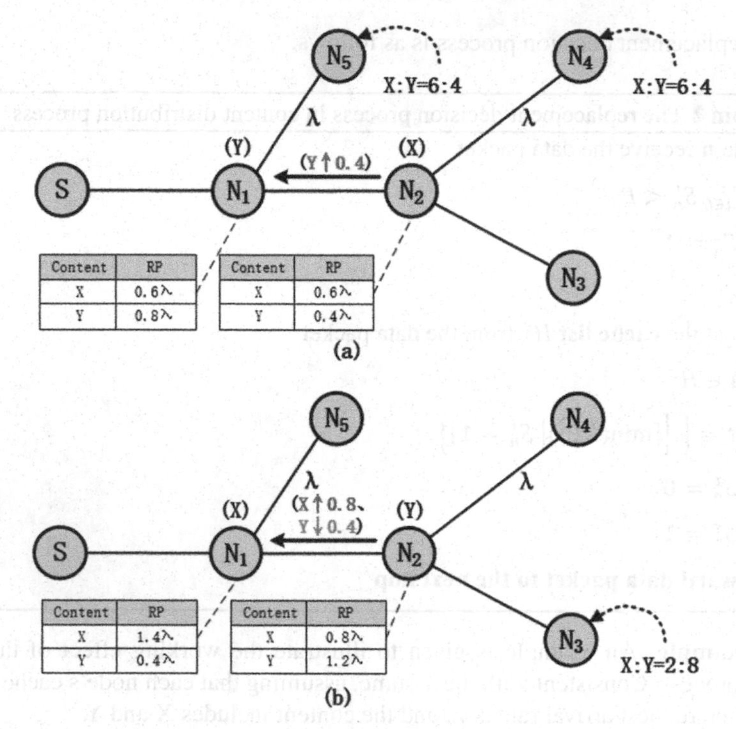

Fig. 2. EFS cache adjustment example

4.1 Simulation Environment

Under the Linux (Ubuntu 16.04) OS, simulations were performed using an NS3 based ndnSIM emulator. By adding the required fields in the transmission packet and modifying the cache decision logic, the EFS strategy proposed in this paper is realized and compared with other cache strategies. The parameters used are as follows:

The simulation topology is assumed to be a tree structure, which is divided into seven layers. Each layer has 0–2 child nodes. There are 65 nodes in the network, including 33 leaf nodes, which are the nodes that generate requests for users; 31 intermediate nodes; a content source node, that is the root node of the tree.

The content source provides 1000 different content, each content is 1 MB, all content popularity meets the Zipf distribution, and the Zipf parameter ranges from 0.5 to 1.5. The arrival of the request meets the Poisson distribution, and the arrival frequency is 10req/s.

The total capacity of the CS in the network is 10–50, which is 1% ~ 5% of the total content, and the link delay between the nodes is 10 ms. In the EFS strategy, the time decay factor $\tau = 0.98$, the update period is 1 s, and the simulation time is 60 s. The first 10 s are the initialization phase, and the last 50 s are used for the final statistical analysis.

The simulation parameters are as shown in Table 2.

Table 2. Simulation parameters

Parameter	Value
Content request frequency	10 req/s
Cache capacity	10 ~ 50
Content size	1 MB
Popularity parameter	Zipf(q = 0.7)
Time decay factor	0.98

Choose several common caching strategies for comparison, including:

- LCE (Leave Copy Everywhere): The most primitive Cache Policy for NDN.
- LCD (Leave Copy Down): The content is only cached at the next hop node of the hit node.
- Probability: Content is cached with a certain probability (0.7).
- ProbCache: The content cache probability is proportional to cache capacity and cache revenue.
- Betw: The content is placed at a node with a high median of the node. Median is used to measure the importance of the node in the topology.

4.2 Performance Parameter

The performance evaluation indicators selected during the simulation process include:

1. In-network Hit Rate (IHR)
   The proportion of requests that are satisfied by the in-network cache in all requests. Reflects the source node service load reduced by cache allocation. The higher the intranet hit rate, the higher the cache performance. R_{total} is the total number of content requests, R_{hit} is the number of hits by other nodes except the source node, then:

$$\mathrm{IHR} = \frac{R_{hit}}{R_{total}} \tag{9}$$

2. Average Access Delay (AAR)
   The average time the user has sent a content request until the corresponding content is received. Reflects the efficiency of content retrieval. The larger the value, the worse the user experience.
3. Average Node Hit Rate (ANHR)
   Cache hit ratio at node n:

$$NHR_n = \frac{Hit\_request_n}{Incoming\_request_n} \tag{10}$$

   $Hit\_request_n$ is total number of requests that can be satisfied at node n, $Incoming\_request_n$ is number of requests received by n, then ANHR is:

$$\mathrm{ANHR} = \frac{\sum_{n=1}^{N} NHR_n}{N} \tag{11}$$

4.3 Simulation Result

The performance comparison of the six cache strategies as the cache capacity changes is as follows.

Figure 3 shows the change in the hit rate of the user's request in the network when the cache capacity changes from 10 to 60. It can be seen that as the cache capacity increases, the hit rate in the network increases. The EFS strategy has a higher intranet hit rate. This is because the strategy updates and replaces the cached content in real time, so that the distribution of the cache in the network is more consistent with the popularity of the content, thereby providing a higher intranet hit rate.

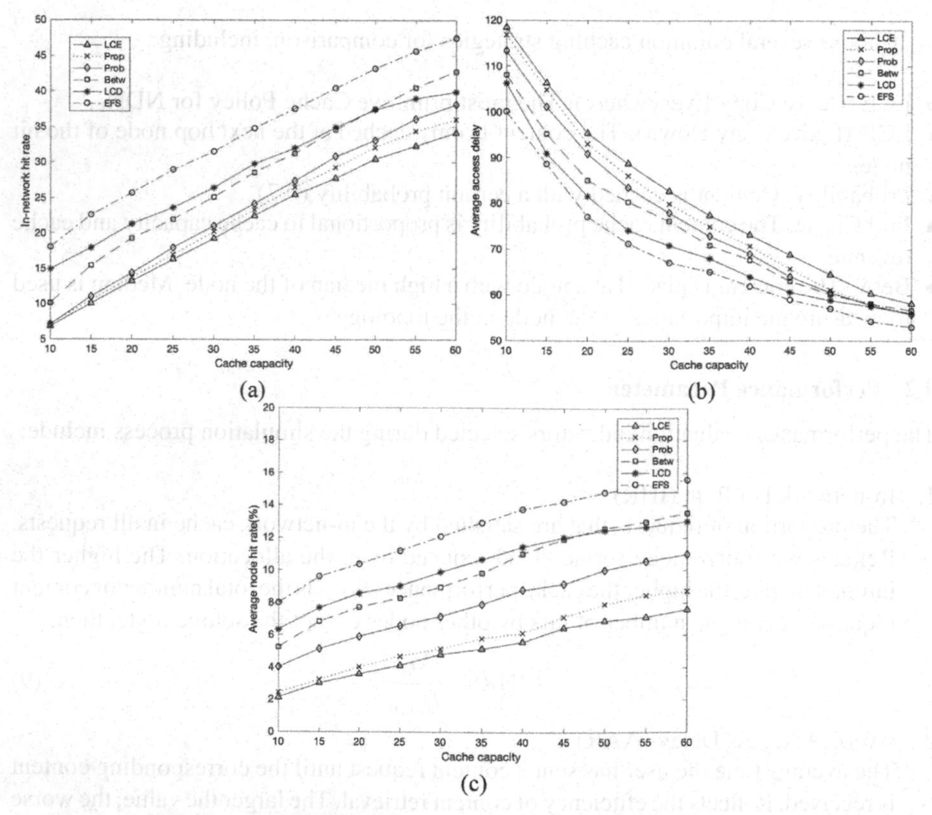

Fig. 3. (a) IHR (b) AAR (c) ANHR changes with cache capacity

Figure 4 shows the average delay variation of different strategies when the intranetwork cache increases. It can be seen that the EFS strategy has a lower average delay than other strategies. Because the cache decision of the EFS policy is from the edge to the core of the network, the cache is placed closer to the user edge, so that the user request can be satisfied more quickly, so that the average delay of the requested content is lower.

Figure 5 shows the average node hit rate as a function of cache capacity. As the cache capacity increases, the average node hit rate increases. EFS strategy through the coordination between the nodes on the path, the average node hit rate is significantly improved than several other strategies.

5 Summary

This paper firstly analyzes the existing caching strategies in ICN networks, and proposes an edge-following strategy based on relative popularity. This strategy prioritizes the service performance of the edge side, advances the caching decision to the request forwarding stage, and forwards the nodes on the path. Between the content feedback, the network cache is allocated reasonably. The simulation results show that this strategy increases the hit rate of the data in the network and reduces the request delay of the user, thus making full use of the limited cache space in the edge network and improving the overall performance of the network. There are still many shortcomings in the work done in this paper. The next step will be based on the research of this paper, combined with the intra-regional caching mechanism, using the domain cache information to design a more efficient hybrid caching strategy.

Acknowledgement. This work is supported by National Key R&D Program of China (2018YFB1800704) and the National Natural Science Foundation of China (61702048).

References

1. Deng, M., Liu, F., Zhao, M., Chen, Z., Xiao, N.: GFCache: a greedy failure cache considering failure recency and failure frequency for an erasure-coded storage system. Comput. Mater. Continua **58**(1), 153–167 (2019)
2. Wang, S., Yanpiao Zhang, L., Zhang, N.C., Pang, C.: An improved memory cache management study based on spark. Comput. Mater. Continua **56**(3), 415–431 (2018)
3. Wei, Y., Wang, Z., Guo, D., Yu, F.R.: Deep Q-learning based computation offloading strategy for mobile edge computing, Comput. Mater. Continua **59**(1), 89–104 (2019)
4. Suksomboon, K., et al.: PopCache: cache more or less based on content popularity for information-centric networking. In: 38th Annual IEEE Conference on Local Computer Networks, Sydney, NSW, pp. 236–243 (2013)
5. Cho, K., Lee, M., Park, K., Kwon, T.T., Choi, Y., Pack, S.: WAVE: popularity-based and collaborative in-network caching for content-oriented networks. In: 2012 Proceedings IEEE INFOCOM Workshops, Orlando, FL, pp. 316–321 (2012)
6. Laoutaris, N., Syntila, S., Stavrakakis, I.: Meta algorithms for hierarchical Web caches. In: IEEE International Conference on Performance, Computing, and Communications, 2004, Phoenix, AZ, USA, pp. 445–452 (2004)
7. Psaras, I., Chai, W.K., Pavlou, G.: In-network cache management and resource allocation for information-centric networks. IEEE Trans. Parallel Distrib. Syst. **25**(11), 2920–2931 (2014)
8. Rath, H.K., Panigrahi, B., Simha, A.: On cooperative on-path and off-path caching policy for information centric networks (ICN). In: 2016 IEEE 30th International Conference on Advanced Information Networking and Applications (AINA), Crans-Montana, pp. 842–849 (2016)

Reliability Improvement Algorithm of Power Communication Network Based on Network Fault Characteristics

Ruide Li(✉), Feng Wang, XinXin Zhang, Jiajun Chen, and Jie Tong

Jiangmen Power Supply Bureau of Guangdong Power Grid Co., Ltd, Jiangmen 529000, Guangdong, China
1125893346@qq.com

Abstract. The reliability optimization of power communication networks has become an important research in power communication. However, the research has not considered the value of historical data of network operation on the improvement of network reliability. As a result, amount of accumulated data in the network operation process has not played a role in improving network reliability. In order to solve this problem, based on the historical operational fault data, we analyze the characteristics of network faults from three aspects: historical data characteristics of network, equipment remaining service life and external factors of municipal construction. And we propose reliability improvement algorithm of power communication network based on network fault characteristics. In the simulation experiment, it is verified that the proposed algorithm has achieved good results in the reliability and communication efficiency improvement.

Keywords: Power communication network · Power service · Fault · Reliability

1 Introduction

With the rapid construction and application of smart grids, the scale of power communication networks and network equipment are both increasing, which brings great challenges to the normal operation of the network. The reliability optimization has become an important research direction in power communication networks [1, 2]. According to different objects of improving the reliability of power communication networks, the reliability optimization research of power communication networks can be divided into two research fields: network reliability improvement and power equipment reliability improvement. In terms of network reliability improvement, typical research methods include routing strategy optimization based on intelligent algorithm [3], optimization of network anti-attack capability [4], optimization of network system and transmission efficiency [5], etc. In terms of reliability improvement of power equipment, typical research methods include optimization of distribution system based on improved hybrid cloud model [6], network reliability improvement based on frequency control [7], etc. In order to further improve the reliability of the power communication network, new technologies such as SDN [8] and dynamics theory [9] have been gradually applied, and have achieved notable research results.

X. Sun et al. (Eds.): ICAIS 2020, LNCS 12240, pp. 798–804, 2020.
https://doi.org/10.1007/978-3-030-57881-7_70

However, existing research does not consider the value of historical data of network operations on improving reliability. As a result, massive accumulated data during network operation does not play a role in improving network reliability. To solve this problem, we analyze the characteristics of network faults from three aspects: the historical data characteristics, the remaining service life of network equipment, and the characteristics of external factors in municipal construction. In simulation part, it is verified that the algorithm in this paper has achieved good results in terms of reliability and communication efficiency improvement of power communication networks.

2 Problem Description

The power communication network includes two resources, network nodes and links, denoted by $G = (N, E)$, where N denotes a set of network nodes $n_i \in N$, and E denotes a set of network links $e_j \in E$. When a network node or network link fails, it will affect the connectivity of the network and cause some power services to be unavailable. In order to reduce the occurrence of fails, we released the value the three aspects in Introduction, so as to improve the reliability of the network.

In terms of fault history data, those nodes ever failed show higher probability to fail again. We can discover faulty nodes and back them up based on historical data. In terms of equipment service life, Devices remaining life closer to end are easier to malfunction. We can find the target device by sorting the remaining service life. In terms of municipal construction, due to the development of the city, the government will repair and demolish roads and houses, which will easily lead to network failure. We should backup the nodes which are easier influenced by construction.

3 Characteristics of Network Failure

3.1 Historical Data Characteristics of the Network

In order to mine the historical data characteristics of the network, we build a fault feature matrix, a symptom feature matrix, and a fault symptom feature matrix. Use M_k^N to represent the fault feature matrix, the dimension is equal to the number of network nodes n, and $a_{ii} \in M_k^N$ is used to indicate the number of failures of the network node n_i in time period k. The M_k^L is used to represent the symptom feature matrix, and $b_{ij} \in M_k^L$ is used to indicate the number of power communication services between the network node n_i and the network node n_j that are not available in the time period k.

Based on above two matrixes, the fault symptom feature matrix M_k^{NL} can be obtained. Because the number of rows and columns of the fault feature matrix and the symptom feature matrix are both the number n, the fault symptom matrix can be obtained by summing the fault feature matrix and the symptom feature matrix.

3.2 Remaining Service Life of Network Equipment

After the network equipment is officially launched, it runs all day. As the usage time is too long, the aging of the main components such as the built-in chip and motherboard card will affect the performance of the network device. In general, after one year of use, network device performance will be reduced by about 5% to 10%. The performance of the network device is represented by $R(n_i)$, and the (1) gives the calculation of $R(n_i)$, wherein $L_{n_i} - U_{n_i}$ represents the remaining life of the network device, L_{n_i} represents the total life of the product, and U_{n_i} represents the service life.

$$R(n_i) = \frac{L_{n_i} - U_{n_i}}{L_{n_i}} \quad (1)$$

3.3 External Factors of Municipal Construction

Network failure caused by demolition and road construction is inevitable So, a backup mechanism is needed to avoid its impact. By analyzing the fault point and realizing the backup of the initial node and the end node of the fault area, the impact of municipal construction on the network can be effectively avoided. For example, Fig. 1. contains three regions C1, C2, and C3. Assuming that C2 is an area affected by the municipal construction, in order to ensure that C1 and C3 can communicate normally, n5 and the e2, e4 of C2 need to be migrated or backed up.

Because identifying the fault node starting node and the end node is not the focus of this paper, this paper uses the network connectivity identification method to make the fault zone to externally connected nodes and links as key resources for redundancy.

4 Algorithm

The reliability improvement algorithm of power communication network based on network fault feature proposed in this paper includes the following three steps. (1) Analyze key nodes and links based on historical data and perform backup; (2) Calculate the service life of nodes to sort and back them up; (3) Calculate network resources affected by external factors of municipal construction and perform backup.

In step 1, first, the fault history data matrix and the symptom history data matrix are constructed by collecting historical data (process a of step 1); secondly, the symptom and fault history data matrix in the time period k is calculated to obtain a historical data

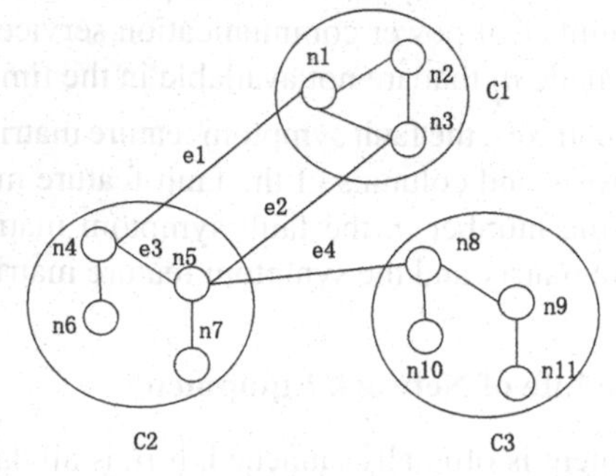

Fig. 1. Power communication network affected by municipal construction factors

matrix (Process b) of step 1; Finally, redundancy backup is performed on network nodes and network links with a large number of historical failures (process c of step 1). In step 2, the lifetime performance of the network device is first calculated (process a of step 2); secondly, the network resources with poor lifetime performance are backed up (process b of step 2). In step 3, first construct a network resource matrix affected by external factors of municipal construction (process a of step 3); secondly, identify key nodes and links (process b of step 3); and finally, back up perform key network devices therein. I (Process c of Step 3) (Table 1).

Table 1. Algorithm Description

Algorithm: Reliability Improvement Algorithm for Power Communication Network Based on Network Fault Characteristics
1. Analyze key nodes and links based on historical data and perform backups. (a) collecting historical data, constructing a fault history data matrix M_k^N, a symptom history data matrix M_k^L (b) Calculating the symptom and fault history data matrix in the time period k to obtain a historical data matrix M_k^{NL}; (c) constituting a set R_H^n of the largest H network nodes in the diagonal element $S_{ii} \in M_k^{NL}$, and the largest L network links in the non-diagonal element $S_{ij} \in M_k^{NL}$ constitute a set R_H^e, and backing up R_H^n and R_H^e; *2. Sort the life of the nodes and back them up.* (a) Calculate the service life performance $R(n_i)$ of the network equipment using (1); (b) Forming a set R_H^R for D network devices with the smallest $R(n_i)$ value, and backing up the network devices that do not belong to R_H^n and R_H^e; *3. Calculate network resources affected by external factors of municipal construction and make backups.* (a) Building a matrix of network resources affected by external factors of municipal construction; (b) Identifying key nodes and links to form a set of critical network device resources R_H^K; (c) backing up of network devices that do not belong to R_H^n, R_H^e, and R_H^R in R_H^K.

5 Simulation

5.1 Environment

In order to simulate the network environment, the BRITE tool was used in the experiment to generate the network topology [10]. In order to verify the performance of the algorithm under different network scales, the number of network nodes in the experiment increased from 100 to 600 with a step size of 50. When simulating the power communication service, 10% of the network nodes are used as the starting node of the power service, and any unused nodes among the remaining network nodes are used as the target nodes. In the fault simulation of network nodes, the LLRD 1 model is used to generate network faults [11]. The reliability of the network is evaluated by simulating network failures and analyzing the availability of power communication services. In order to verify the performance of the network algorithm, the network performance of the power communication network of the algorithm RIAoNFC and the traditional algorithm ENIoNI [12] is evaluated from two aspects of network reliability and communication efficiency.

In terms of network reliability, $R(G)$ is used to represent network reliability, and calculation is performed using (2), where $S_u(G)$ is the total number of services available, and $S_o(G)$ is the total number of services.

$$R(G) = \frac{S_u(G)}{S_o(G)} \tag{2}$$

In terms of communication efficiency, $E(G)$ is used to represent the average of the communication efficiency between any two nodes in the whole network, and is calculated using (3). Where N represents the number of network nodes and d_{ij} represents the shortest distance between network node n_i and n_j.

$$E(G) = \frac{1}{N(N-1)} \sum_{i,j \in G, i \neq j} \frac{1}{d_{ij}} \tag{3}$$

It can be known from (3) that when the network resource fails, the shortest distance between the network node n_i and n_j may be blocked, resulting in a longer routing distance between the network node n_i and n_j, and a lower communication efficiency of the network. In order to unify the characteristics of network communication efficiency, $E_o(G)$ is used to indicate communication efficiency when the network is not faulty. The normalized communication efficiency is calculated using (4).

$$E_f(G) = \frac{E(G)}{E_o(G)} \tag{4}$$

5.2 Result Analysis

Network Reliability Evaluation. The results of network reliability comparison are shown in Fig. 2, where the x-axis represents the network node size and the y-axis represents the network reliability value. As can be seen from the figure, the network reliability of the traditional algorithm ENIoNI is maintained at around 0.75. The network reliability of the algorithm RIAoNFC is maintained at 0.85. Therefore, the algorithm RIAoNFC improves the network reliability.

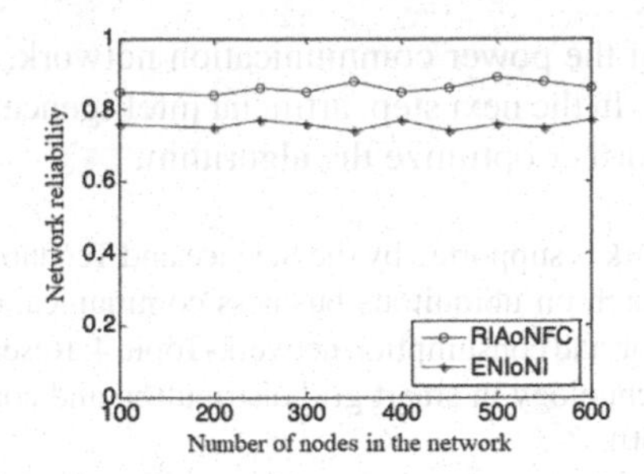

Fig. 2. Network reliability comparison

Communication Efficiency. The result of network communication efficiency comparison is shown in Fig. 3, where the x-axis represents the network node size and the y-axis represents the network communication efficiency value. As can be seen from the figure, the network reliability of the traditional algorithm ENIoNI is maintained at about 0.61. The network reliability of the algorithm RIAoNFC is maintained at 0.77. Therefore, the algorithm RIAoNFC improves the efficiency of network communication.

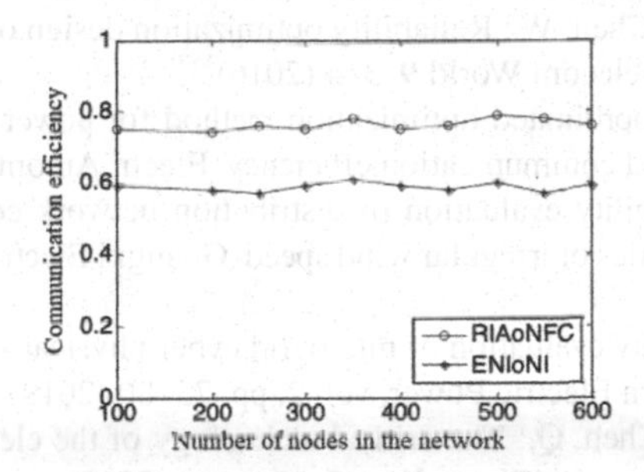

Fig. 3. Network communication efficiency comparison

6 Conclusion

With the rapid construction and application of smart grids, the scale of power communication networks is increasing and network equipment is increasing, which brings great challenges to the normal operation of the network. The reliability optimization of power communication networks has become an important research content. In order to improve the reliability of the network, based on the characteristics of historical operational fault data, this paper analyzes the characteristics of network faults from three aspects: historical data characteristics of the network, remaining service life of network equipment and external factors of municipal construction, and proposes characteristics based on network faults. Power communication network reliability improvement algorithm. In the simulation experiment, it is verified that the proposed algorithm has achieved good results in the reliability and communication efficiency improvement of the power communication network. Although the algorithm based on the network fault feature effectively improves the reliability of the power communication network. However, the algorithm is less efficient to execute. In the next step, artificial intelligence algorithms such as deep learning will be used to further optimize the algorithm.

Acknowledgment. This work is supported by the Science and Technology Project of Guangdong Power Grid Co., Ltd: Research on ubiquitous business communication technology and service mode in smart grid distribution and consumption network-Topic 4: Research on smart maintenance, management and control technology in smart grid distribution and consumption communication network(GDKJXM20172950).

References

1. Yang, L., Sun, S.: Evaluation algorithm of node importance of power communication network. Smart Grid **7**, 656–659 (2017)
2. Li, L.: Service-based node identification of power communication network. In: Electric Power Information and Communication Technology, vol. 6, pp. 62–66 (2018)
3. Sun, F., et al.: Research on routing optimization of power communication network based on improved genetic algorithm. Autom. Instrum. **6**, 25–28 (2018)
4. Peng, L., Yaoqun, W., Chen, W.: Reliability optimization design of computer network based on genetic algorithm. Telecom World **9**, 3–4 (2016)
5. Zeng, W., Yuan, Y.: Coordinated optimization method for power communication networks balancing reliability and communication efficiency. Electr. Autom. **4**, 102–104 (2018)
6. Chen, S., et al.: Reliability evaluation of distribution network considering the probability distribution characteristics of irregular wind speed. Guangxi Electric Power, vol. 2, pp. 17–21 (2019)
7. Guo, J., et al.: Reliability evaluation of microgrid cyber physical system considering control function failure. Modern Electric Power, vol. 2, pp. 73–80 (2019)
8. Luming, W., Qiu, Y., Chen, Q.: The critical technology of the electric power telecommunication based on SDN. Jiangsu Electr. Eng. **3**, 134–144 (2018)
9. Wang, Y., Zhang, Y., Wang, C.: A method of vulnerability assessment for power communication network based on mechanical model. Telecommun. Sci. **1**, 54–61 (2019)
10. Brite. http://www.cs.bu.edu/brite/
11. Padmanabhan, V.N., Qiu, L., Wang, H.J.: Server-based inference of Internet link lossiness. In: IEEE INFOCOM 2003 Twenty-second Annual Joint Conference of the IEEE Computer and Communications Societies (IEEE Cat. No. 03CH37428), San Francisco, CA, vol. 1, pp. 145–155 (2003)
12. Zhou, M., Rui, L., Qiu, X., Xia, Z., Li, B.: Evaluation of the node importance in power grid communication network and analysis of node risk. In: NOMS 2018–2018 IEEE/IFIP Network Operations and Management Symposium, Taipei, pp. 1–5 (2018)

Link Prediction Based on Modified Preferential Attachment for Weighted and Temporal Networks

Xuehan Zhang[1(✉)], Xiaojuan Wang[1], and Lianping Zhang[2]

[1] School of Electronic Engineering, Beijing University of Posts and Telecommunications, No.10, Xitucheng road, Haidian district, Beijing, China
xuehanz@126.com
[2] Alibaba Cloud Computing, Beijing, People's Republic of China

Abstract. Due to the huge amount of data, network in real world is expanding rapidly. The structure of network is so complicate that a handy tool for analyzing is in desperate need. Link prediction provides forecast about entity interactions. A large range of areas can gain useful insights applying the prediction result. It can also help us understand the evolutionary mechanism of complex networks theoretically. This paper is aimed at promoting the performance of link prediction. We worked on two aspects to meet this end. Firstly, the weight and timestamp information were included in the data we referred to. We brought up the assumption of time attenuation effect. Secondly, we modified the PA index by dividing the links around targeted node pairs into different types. Instead of just focusing on the degree of two targeted nodes, links of various types weighs differently on the final prediction result. We conducted experiments on four real world datasets. The AUC using PA index considering link types was improved indeed.

Keywords: Link prediction · Temporal network · Preferential attachment

1 Introduction

Link prediction task aims at forecasting links that may occur in a future time or mining the possible links missing from the current network. From social network to power grid, almost everything interacting with each other can be abstracted to a network. Although the elements forming a network seem simple to understand, just nodes and edges connecting them, the tons of information they carry with is of great use. We can detect terror activities from the terrorist communication network before any damage was made [1,2]. In bioinformatics, unknown connections are found in genes causing disease and protein interactions [3]. According to our previous browsing and clicking records, more similar commodities or websites will be recommended to us [4]. In social network, your interactions with

X. Sun et al. (Eds.): ICAIS 2020, LNCS 12240, pp. 805–814, 2020.
https://doi.org/10.1007/978-3-030-57881-7_71

others can be used to predict who you may know in real life or who you probably are interested in [5]. Besides, when constructing a network for analysing or visualization, there may exist some inaccurate information that reflects the real world in a wrong way. We can identify spurious links through link prediction [6]. In addition to the practical use, it also has important theoretical significance. We can propose a network model using link prediction to evaluate the evolving mechanism of real network [7]. It can be used as an auxiliary mean to learn about the structure and evolving pattern of networks.

In the past twenty years, network science has drawn much attention from various areas. Therefore, a large number of researches were conducted around link prediction which working as a powerful tool for network analysing and data mining. In computer science area, Markov chain [8] and machine learning [9] methods were utilized in prediction task. Sarukkai [10] built a model using Markov chain for analysing network path. Instead of using topological information, Pavlov et al [11]. Estimated similarity between nodes by including many aspects of node attributes. Although the prediction result can benefit a lot using this kind of extra information, the attributes are always privacy-sensitive and hard to get.

To solve the challenge of inadequate information, more and more scientists focused on the topological information of network. Generally, there are similarity-based algorithm and learning-based method. The former one gives the possibility on the connectivity of a node pair by calculating their similarity. It extracts specific topological characteristics of node pairs and gives each potential pair a score according to certain rules. The higher they score, the more likely they are going to be linked. A bunch of similarity indexes were brought up during the exploring process. Newman et al. [12] put forward the most basic one common neighbors. Sorensen Index (SI) [13], Jaccard's Coefficient (JC) [14], Local Path (LP) [15], Local Random Walk (LRW) [16] and so forth were used for grading node pairs. The later takes link prediction as a binary classification problem. As thus, many machine learning methods such as SVM [17], BP neural network [18], Bayesian method [19] can be used for prediction. Instead of assign a score for node pairs, they build models on known information and learn through multidimensional features.

Along with the basic methods for prediction, researches on directed, weighted, temporal network attract more attention. *Bütün* et al. [20] proposed a pattern-based prediction method for directed complex network. They divided triad units into nine categories according to their link directions. Munasinghe et al. [21] came up with a new feature called time score for prediction in temporal network. It captures the important aspects of time stamps of interactions and the temporality of the link strengths. *Özcan* et al. [22] introduced time series forecasting models such as AR, MA, VAR to process time information.

The rest of the paper is organized as follows. Section 2 introduces the baseline methods for prediction and our modified PA measure targeting at weighted temporal network. Experiment details and datasets used in this paper are referred in Sect. 3. We also give a clear demonstration on experimental results in Sect. 3. In Sect. 4, a brief conclusion is made about this paper.

2 Method

2.1 Baseline Methods

Adamic-Adar Coefficient (AA): AA measure assigns different weight to the common neighbors of two specific nodes. The common neighbor with a smaller degree will be weighted more heavily. In other words, two nodes are more likely to be connected if they have a common neighbor with a lower degree. Empirically, two users who both follow a less popular person indeed have a higher possibility to know each other.

Common Neighbor (CN): CN index presumes that the probability of two nodes connecting to each other is proportional to the number of their common neighbors.

Jaccard's Coefficient (JC): JC index defines the probability as the percentage of common neighbors in the total neighbors of two nodes.

Preferential Attachment (PA): In this measure, the probability of a new link connecting two nodes is proportional to the product of the degrees of the two nodes.

2.2 Weighted and Temporal Extensions

All the baseline methods mentioned above have not taken weight and appearing time of edges into account. However, weight captures the importance of edges and appearing time shows the growing and changing process of the whole network. In this part, we attached weight and timestamp information to all of the edges in networks.

Table 1. Traditional prediction indexes and their Weighted and temporal extensions

Baseline definition	Weighted and temporal extensions
$AA_{xy} = \sum_{z \in \Gamma(x) \cap \Gamma(y)} \frac{1}{\lg k(z)}$	$AA\_WT_{xy} = \sum_{z \in \Gamma(x) \cap \Gamma(y)} \frac{wtS_{xz} + wtS_{yz}}{2 \lg(1 + \sum_{c \in \Gamma(z)} wtS_{zc})}$
$CN_{xy} = \lvert\Gamma(x) \cap \Gamma(y)\rvert$	$CN\_WT_{xy} = \sum_{z \in \Gamma(x) \cap \Gamma(y)} \frac{wtS_{xz} + wtS_{yz}}{2}$
$JC_{xy} = \frac{\lvert\Gamma(x) \cap \Gamma(y)\rvert}{\lvert\Gamma(x) \cup \Gamma(y)\rvert}$	$JC\_WT_{xy} = \sum_{z \in \Gamma(x) \cap \Gamma(y)} \frac{wtS_{xz} + wtS_{yz}}{2(\sum_{a \in \Gamma(x)} wtS_{xa} + \sum_{b \in \Gamma(y)} wtS_{yb})}$
$PA_{xy} = k(x) \times k(y)$	$PA\_WT_{xy} = \sum_{a \in \Gamma(x)} wtS_{xa} \times \sum_{b \in \Gamma(y)} wtS_{yb}$

First of all, we can assume reasonably that more recently occurred edges have a larger impact on the network during following time stages. Besides, the attenuation rate should be gradually smooth as the occurring time of edges more and more ancient. We can take this factor into consideration by adding a time-related weight to edges which defined in (1) where p > 0.

$$tS_{xy} = \ln(t_{xy} - t_{start} + p) \tag{1}$$

The time attenuation effect has been taken care of. Then the weighted and temporal extensions can be described as (2). In Table 1, you can find the baseline methods and the corresponding weighted and temporal extensions.

$$wtS_{xy} = w_{xy} * tS_{xy} \tag{2}$$

2.3 Link Prediction Method Based on PA Index

From the definition of baseline methods, it is obvious that PA index failed to take advantage of different types of links around two targeted nodes. PA index just focuses on the degree of two targeted nodes, while other indexes at least bring up the concept of common neighbors. Say the least of it, there are three types edges around two nodes. It is very likely that they will have diverse effect on link prediction tasks.

Taking node x and node y as two targeted nodes, Fig. 1 introduces three types of edges around them.

1) e_1: the edge between node x and node y
2) e_2: the edges between node x/y and their common neighbors
3) e_3: the edges between node x/y and other neighbors

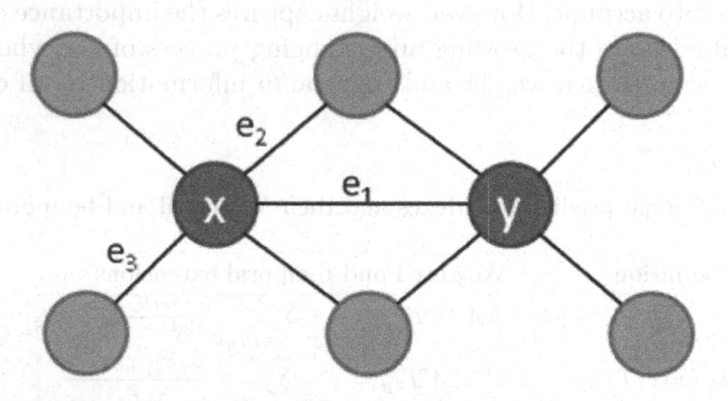

Fig. 1. Different types of link around targeted node pairs

In order to find out whether the three types of links have different even inverse impact on network evolving, we rewrote the PA index for node x and node y as (3), where $q = (\lambda, \mu, \nu)$ represent the importance of the three types of links respectively.

$$\begin{aligned} PA_{xy} &= \lambda e_{x1}e_{y1} + \mu e_{x2}e_{y2} + \nu e_{x3}e_{y3} \\ &= q \begin{pmatrix} score_{xy1} \\ score_{xy2} \\ score_{xy3} \end{pmatrix} \end{aligned} \tag{3}$$

where e_{x1}, e_{x2}, e_{x3} can be written as (4)

$$\begin{aligned} e_{x1} &= wtS_{xy} \\ e_{x2} &= \sum_{i \in \Gamma(x) \cap \Gamma(y)} wtS_{xi} \\ e_{x3} &= \sum_{j \in [\Gamma(x) - \Gamma(x) \cap \Gamma(y) - y]} wtS_{xj} \end{aligned} \quad (4)$$

2.4 The Methods Used to Deal with Imbalanced Datasets

In link prediction problem, we always consider node pairs with direct connection as positive samples while others as negative. Due to the large scale of networks, node pairs having direct connection belong to the minority class. We can solve the problem caused by imbalanced datasets from two perspectives. One is changing the data distribution to balance the size of data of different class. The other way is to improve the classification algorithms so that they can weigh the minority class more heavily. In this work, we solved this problem by resampling the data.

We combined under-sampling and over-sampling to achieve a better result. The under-sampling process was carried out through Tomek links method and random sampling. We used SMOTE to oversample the data of minority class. Suppose that two sample x_i and x_j belong to different classes, and d $(x_i,\ x_j)$ represents the distance between them. If there is no other sample x_l making $d(x_l, x_i) < d(x_i, x_j)$ or $d(x_l, x_j) < d(x_i, x_j)$ true. We will call $(x_i,\ x_j)$ a Tomek link pair. Firstly, we removed the one from majority class in all Tomek link pairs. Then, we performed random sampling to complete the under-sampling process. As for the over-sampling process, we used SMOTE to balance the classes by generating new samples near some data of minority classes. Since it is not simply copying the minority classes, overfitting can be avoided to some extent.

3 Experiments

3.1 Problem Statement

We focus on the link prediction problem of weighted and temporal network. Thus the datasets are divided into two parts in chronological order. Take the graph M constructed with the links from the former part of datasets as our experimental subject. Then 80% of the links in M chosen randomly work as training set. The testing set consists of the rest links in M.

The latter part of datasets will provide class labels for all the node pairs in graph M. If two nodes connect directly in the latter part, they are called a positive instance, while others not directly linked are negative instances. Classification algorithm will learn from the specific features about all the data in training set. Combined with their class labels, a model for classification task will be built. We then use it to classify data from testing set. Compare to their actual class labels, we can find out whether our method is good enough.

3.2 Evaluation Metric

It is generally assumed that the sample size of different class is close to each other in common classification algorithms. For imbalanced datasets, most of samples tend to be classified into the majority class to attain a higher precision by classification methods. For example, suppose we want to tell whether the traffic data is normal or abnormal. The proportion of normal traffic data in the dataset is always extremely high such as 99%. The precision can be as high as 99% if the algorithm just put all the samples into the majority class. But obviously, this kind of indicators are meaningless.

To compare the performance of various link prediction algorithm effectively, we choose the area under the receiver operating characteristic curve (AUC) as our evaluation indicator.

The calculation method of AUC is introduced as follows. Pick a link in testing set and a non-existent link between nodes in testing set randomly. Compare the possibility they will be classified as positive by the classification model. Repeat this process for n times. There are n' times the possibility of existing links considered positive is higher than that of non-existent links and n' times they are equally likely to be classified as positive. The final expression of AUC can be written as (5).

$$AUC = \frac{n' + 0.5n''}{n} \tag{5}$$

3.3 Datasets

To testify whether the result of our method is good enough, we choose four real-world networks as datasets. Due to the specific of our method, all the edges in these networks are considered undirected, weighted and timestamped. We also excluded edges connecting a node with itself.

fb-messages [23]. The Facebook-like Social Network originate from an online community for students at University of California, Irvine. The dataset includes the users that sent or received at least one message.

fb-forum [23]. The Facebook-like Forum Network was attained from the same online community as the online social network; however, the focus in this network is not on the private messages exchanged among users, but on users' activity in the forum.

DNC-emails [24]. This is the directed network of emails in the 2016 Democratic National Committee email leak. Nodes in the network correspond to persons in the dataset. A directed edge in the dataset denotes that a person has sent an email to another person. Since an email can have any number of recipients, a single email is mapped to multiple edges in this dataset, resulting in the number of edges in this network being about twice the number of emails in the dump.

email-Eu [25]. The network was generated using email data from a large European research institution. The emails only represent communication between institution members (the core), and the dataset does not contain incoming messages from or outgoing messages to the rest of the world. A directed edge (u, v, t) means that person u sent an email to person v at time t. A separate edge is created for each recipient of the email. The details of four networks mentioned above are listed in Table 2.

Table 2. Details about real world datasets

Datasets	Number of nodes	Number of edges	Number of unique edges	Start time	End time
fb-messages	1,899	61,734	13,838	2004.03.24	2004.10.26
fb-forum	899	33,720	7,036	2004.05.15	2004.10.26
DNC-emails	2,029	39,264	4,384	2013.09.16	2016.05.25
email-Eu	1,005	332,334	16,064	1970.01.01	1972.03.15

4 Results

4.1 The Impact of Links Belonging to Different Types

To verify whether the various types of links have different even inverse impact on predicting result, we performed the following experiment on four datasets. The links were divided according to their occurring time as mentioned before. We added link weight and timestamp information them and set the parameter p value to 10. We modified PA methods by taking three types of links into consideration. This method was written in short as PA_LT. After using random forest to perform the link prediction task, we got the importance score for these three features as shown in Fig. 2. It can be seen that the most distinctive feature is not always the same depending on datasets. But the links between x, y and those between x/y and their common neighbors are of more importance overall.

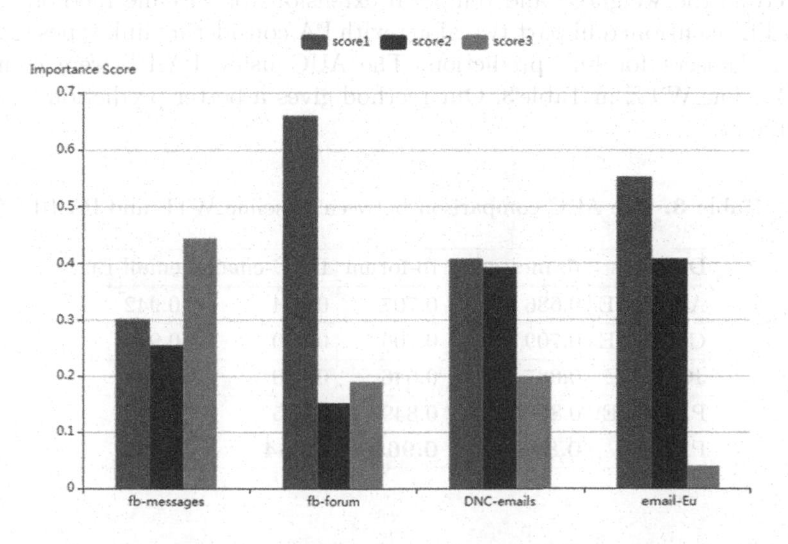

Fig. 2. The importance score for links of different types

4.2 The Data Balancing Process

On the training set of email-Eu dataset, there were two classes detected. The original amount of positive and negative samples is 3,830 and 372,918 respectively. To clear the boundary of two classes, we removed all the samples belonging to the majority class in the 1542 Tomek links found. Then we randomly picked 188,375 negative samples from the remaining majority data and used SMOTE algorithm to generate 184544 new positive samples to balance data from different classes. The sample distribution in this process can be seen in Fig. 3.

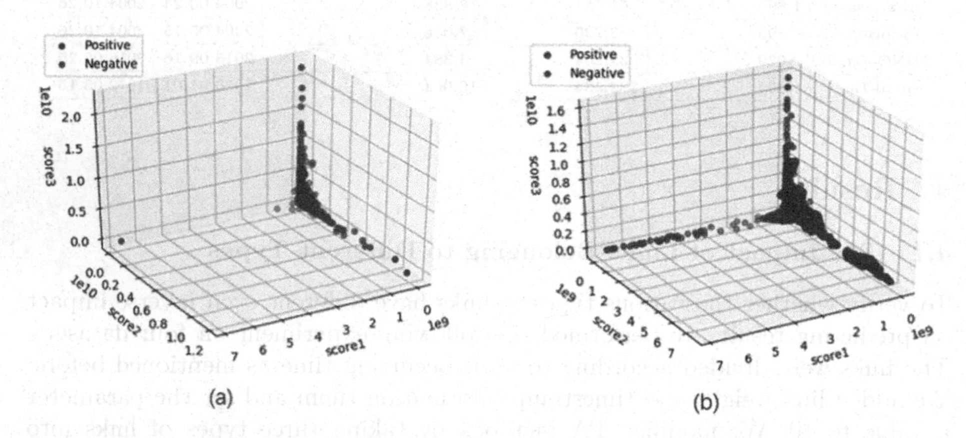

Fig. 3. Positive and negative sample distribution on email-Eu dataset, (a) before data balancing, (b) after data balancing

4.3 The Link Prediction Result for Different Methods

We used all the weighted and temporal extensions of baseline methods (Baseline_WTE) mentioned in part two along with PA considering link types (PA_LT) on four datasets for link prediction. The AUC using PA_LT were compared with Baseline_WTE in Table 3. Our method gives a better prediction tested by experiments.

Table 3. The AUC comparison between Baseline_WTE and PA_LT

Datasets	fb-messages	fb-forum	DNC-emails	email-Eu
AA_WTE	0.686	0.707	0.874	0.942
CN_WTE	0.709	0.706	0.880	0.939
JC_WTE	0.687	0.646	0.890	0.941
PA_WTE	0.816	0.849	0.955	0.810
PA_LT	**0.898**	**0.965**	**0.984**	**0.972**

5 Conclusion

This paper started at taking the weight and timestamp information into consideration. The temporal information can be of positive use for prediction saying from experience because the network is evolving constantly through a long range of time. We introduced the time attenuation effect which means more recently occurred edges have a larger impact on the prediction problem. Then, we divided the links around targeted node pairs into three types in order to improve the performance of PA index. The links of different type may have a diverse impact on prediction. They contributed diversely in the classification process. The datasets used in this work are actually directed networks. We just ignored the direction information of links for problem reduction. The link prediction problem on directed networks has been left for further research.

Acknowledgement. This work is supported by National Natural Science Foundation of China (Grant No. 61871046).

References

1. Anil, A., et al.: Link prediction using social network analysis over heterogeneous terrorist network. In: 2015 IEEE International Conference on Smart City/SocialCom/SustainCom (SmartCity), pp. 267–272. IEEE (2015)
2. Fang, L., Fang, H., Tian, Y., Yang, T., Zhao, J.: The alliance relationship analysis of international terrorist organizations with link prediction. Phys. A **482**, 573–584 (2017)
3. Yu, H., et al.: High-quality binary protein interaction map of the yeast interactome network. Science **322**(5898), 104–110 (2008)
4. Xie, F., Chen, Z., Shang, J., Feng, X., Li, J.: A link prediction approach for item recommendation with complex number. Knowl. Based Syst. **81**, 148–158 (2015)
5. Ma, C., Zhou, T., Zhang, H.F.: Playing the role of weak clique property in link prediction: a friend recommendation model. Sci. Rep. **6**, 30098 (2016)
6. Zhang, P., Qiu, D., Zeng, A., Xiao, J.: A comprehensive comparison of network similarities for link prediction and spurious link elimination. Phys. A **500**, 97–105 (2018)
7. Kumar, R., Novak, J., Tomkins, A.: Structure and evolution of online social networks. In: Yu, P., Han, J., Faloutsos, C. (eds.) Link Mining: Models, Algorithms, and Applications, pp. 337–357. Springer, New York (2010). https://doi.org/10.1007/978-1-4419-6515-8_13
8. Zhu, J., Hong, J., Hughes, J.G.: Using Markov models for web site link prediction. In: Proceedings of the uhirteenth ACM Conference on Hypertext and Hypermedia, pp. 169–170. ACM (2002)
9. Soundarajan, S., Hopcroft, J.E.: Use of supervised learning to predict directionality of links in a network. In: Zhou, S., Zhang, S., Karypis, G. (eds.) ADMA 2012. LNCS (LNAI), vol. 7713, pp. 395–406. Springer, Heidelberg (2012). https://doi.org/10.1007/978-3-642-35527-1_33
10. Sarukkai, R.R.: Link prediction and path analysis using Markov chains. Comput. Netw. **33**(1–6), 377–386 (2000)

11. Pavlov, M., Ichise, R.: Finding experts by link prediction in co-authorship networks. FEWS **290**, 42–55 (2007)
12. Newman, M.E.: Clustering and preferential attachment in growing networks. Phys. Rev. E **64**(2), 025102 (2001)
13. Sørensen, T.J.: A method of establishing groups of equal amplitude in plant sociology based on similarity of species content and its application to analyses of the vegetation on Danish commons. Munksgaard, I kommission hos E (1948)
14. Jaccard, P.: Étude comparative de la distribution florale dans une portion des alpes et des jura. Bull. Soc. Vaudoise Sci. Nat. **37**, 547–579 (1901)
15. Lü, L., Jin, C.H., Zhou, T.: Similarity index based on local paths for link prediction of complex networks. Phys. Rev. E **80**(4), 046122 (2009)
16. Liu, W., Lü, L.: Link prediction based on local random walk. EPL (Europhysics Letters) **89**(5), 58007 (2010)
17. Fernandez, P., et al.: Improving the vector auto regression technique for time-series link prediction by using support vector machine. In: MATEC Web of Conferences. vol. 56, p. 01008. EDP Sciences (2016)
18. Li, J.C., Zhao, D.L., Ge, B.F., Yang, K.W., Chen, Y.W.: A link prediction method for heterogeneous networks based on BP neural network. Phys. A **495**, 1–17 (2018)
19. Xiao, Y., Li, X., Wang, H., Xu, M., Liu, Y.: 3-HBP: a three-level hidden bayesian link prediction model in social networks. IEEE Trans. Comput. Soc. Syst. **5**(2), 430–443 (2018)
20. Bütün, E., Kaya, M.: A pattern based supervised link prediction in directed complex networks. Phys. A **525**, 1136–1145 (2019)
21. Munasinghe, L., Ichise, R.: Time score: a new feature for link prediction in social networks. IEICE Trans. Inf. Syst. **95**(3), 821–828 (2012)
22. Özcan, A., Öğüdücü, Ş.G.: Multivariate temporal link prediction in evolving social networks. In: 2015 IEEE/ACIS 14th International Conference on Computer and Information Science (ICIS), pp. 185–190. IEEE (2015)
23. Rossi, R.A., Ahmed, N.K.: The network data repository with interactive graph analytics and visualization. In: AAAI (2015). http://networkrepository.com
24. Kunegis, J.: KONECT - The Koblenz Network Collection. In: Proceedings of the International Conference on World Wide Web Companion, pp. 1343–1350 (2013). http://userpages.uni-koblenz.de/kunegis/paper/kunegis-koblenz-network-collection.pdf
25. Paranjape, A., Benson, A.R., Leskovec, J.: Motifs in temporal networks. In: Proceedings of the Tenth ACM International Conference on Web Search and Data Mining, pp. 601–610. ACM (2017)

Author Index

Abbas, Ghulam I-173
Abbas, Ziaul Haq I-173
An, Xing-Shuo I-49

Bai, Qianlong II-686
Bai, Xiwei I-510
Bombie, Ninfaakanga Christopher I-586
Bu, Rongjing II-143

Cai, Zengyu II-417
Cai, Zhenhua I-95
Cao, Dandan I-350
Cao, Ning I-719
Chang, Sheng I-646
Chang, Xinyue II-748
Chang, Xu II-109
Chang, Yan II-246, II-280, II-289
Chen, Fang I-575
Chen, Haipeng I-729
Chen, Hanwu I-522
Chen, Jiajun II-798
Chen, Jiansen II-557
Chen, Ming I-362
Chen, Minghan II-191
Chen, Ning II-463
Chen, Tie II-143
Chen, Xi I-350
Chen, Xiu-Bo I-219
Chen, Zhiyu I-291
Cheng, Jiajia I-232
Cheng, Jieren II-191
Cheng, Wenqing II-393
Chi, Laixin II-131

Dai, Dong I-635
Dai, Fangyan II-191
Dai, Jialing II-258
Dai, Jinqiao II-246
Dai, Juan I-142
Dai, Meiling II-776
Dai, Qianning II-166
Dai, Qin-yun I-219
Dai, Xiaodong I-810
Dai, Yongyong II-225
Deng, Huiyun I-232
Deng, Tao I-151
Deng, Zhenrong I-670
Ding, Fei I-775
Dong, Huiwen I-86
Dong, PingPing I-786
Dong, Xu I-62
Dong, Zangqiang I-532
Dou, Bennian I-107
Du, Chunlai II-353
Du, Jiawei II-55
Du, Ming I-350
Du, Shouxu I-719
Du, Wei II-341
Du, Xiao-Feng II-655
Du, Xiaohu I-422
Du, Xiaoni II-55, II-65
Duan, Ec II-55

Fan, Shurui I-350
Fan, Xiaoping II-475
Fang, Yanmei II-629
Fang, Yixiang II-593
Feng, Guorui II-109
Fu, Xiaohui I-489
Fu, Zhangjie I-133

Gan, Jianchao II-307
Gao, Jilong I-729
Gao, Ran I-763
Gao, Yingjian II-215
Ge, Jinhui I-501
Ge, Mengqiong II-769
Gu, Zhihao II-557
Guo, Chao I-564, I-575
Guo, Jianwei I-291
Guo, Jinrun II-642
Guo, Juan I-564, I-575
Guo, Lantu I-317
Guo, Peng I-207, I-232

Guo, Qun I-267
Guo, Shaoyong II-440, II-776, II-786
Guo, Shuqiang II-686
Guo, Wenzhong II-405
Guo, Yanhui II-353
Guo, Zichuan I-244

Han, Dao-Qi II-655
Han, Jin II-30, II-42
Han, Weihong II-380
He, Jia I-184
He, Lin II-475
He, Peilin II-18, II-258
He, Xianfeng II-557
He, Yong II-499
He, Zhentao I-729
Hei, Xinhong II-318
Hong, Zhaojin II-725
Hu, Aiqun I-658
Hu, Fei II-86
Hu, Guangmin I-751
Hu, Hangyu I-751
Hu, Jiajia I-142
Hu, Jinxia II-65
Hu, Kun II-523
Hu, Qiyan I-375
Hu, Yue II-570
Huang, Hai I-327
Huang, He I-219
Huang, Qiuyang I-798
Huang, Xinru I-257
Huang, Yuanyuan II-18, II-258
Hui, Zhao I-739
Huo, Guanying II-738
Huo, Wei I-719
Huo, ZhanXiong II-523

Ji, Meixia I-719
Ji, Xuan I-798
Jia, Dongbai II-710
Jia, Ke-bin I-279
Jiang, Die I-162
Jiang, Hongyi I-707
Jiang, Jie II-642
Jiang, Qing I-162
Jiang, YingXiang II-523
Jiang, Yu I-658
Jiao, Jia I-384
Jiao, Lingling II-617
Jin, Qiao I-455
Jin, Tong II-353
Jin, Yong I-455

Kai, Lorxayxang I-810
Kang, Xiangui II-98
Kang, Yan II-143

Lan, Rushi I-670
Lei, Hang I-586
Li, Chaochao I-244
Li, Dun II-697
Li, Duohui II-475
Li, Fu II-710
Li, Hao II-143
Li, Huansong II-786
Li, Hui II-166, II-179
Li, Huiling II-203
Li, Jie I-207, I-564
Li, Jingbing II-166, II-179
Li, Lun II-697
Li, Qi II-545
Li, Qingwu II-748
Li, Ruide II-798
Li, Shasha I-422
Li, Shudong II-380
Li, Shuxun I-646
Li, Tao II-143
Li, Weihai II-86
Li, Weijun I-95
Li, Weiwei II-604
Li, Wencui II-440
Li, Wenxin I-317
Li, Xiaojun I-707
Li, Xiaoyu I-586
Li, Xiehua II-642
Li, Xueyang II-246
Li, Xue-Yang II-280
Li, Yan II-74
Li, Yang II-581
Li, Yi I-467
Li, Yiting II-42
Li, Yuanshu I-327
Li, Yueli II-710
Li, Zhiqiang I-142
Li, Zirui I-350
Liang, Guangjun I-695
Liang, Jiahao I-479
Liang, Renjie II-629

Liao, Junguo II-510
Lin, Huanqiang II-686
Lin, Ruijie I-543, I-554
Liu, Alex X. I-532
Liu, Baolu I-411
Liu, Chang I-279
Liu, Dianfeng II-604
Liu, Fang I-95
Liu, Feng I-327
Liu, Fenlin II-534
Liu, Gang I-291
Liu, Gui II-225
Liu, Jing II-166, II-179
Liu, Jinming II-760
Liu, Lihua II-109
Liu, Min II-417
Liu, Mingda II-236
Liu, Peng-yu I-279
Liu, Shenghui II-353
Liu, Tong I-479
Liu, Wei I-195, II-74, II-557, II-581
Liu, Weidong II-760
Liu, Wenjie I-532
Liu, Xiaoliang I-501
Liu, Yan II-534
Liu, Yanan I-317
Liu, Yang I-305
Liu, Yangchun I-305
Liu, Yanlin II-179
Liu, Yizhuo II-30
Liu, Yufeng II-428
Liu, Yujiao II-475
Liu, Yuliang I-646
Liu, Yu-liang I-621
Liu, Zhenghui II-393
Liu, Zhigui II-156
Liu, Zhihao I-522
Long, Zhe I-74
Lou, Sijia I-775
Lu, Changlong II-593
Lu, Kai II-215
Lu, Mingming I-133
Lu, Mingxin II-30, II-42
Lu, Xuhong II-557
Lu, Yue-Ming II-655
Luo, Da II-393
Luo, Huan II-405
Luo, Qun I-244
Luo, Weiqi II-393
Luo, Xiangyang II-534
Luo, Xiaozhao II-475
Lv, Xinyue I-62, II-451

Ma, Bin II-545
Ma, Jun I-422
Ma, Liangyu I-195
Ma, Ming I-810
Ma, Yi II-203
Ma, Yuntao I-151
Ma, Zihao I-62, II-451
Mai, Yubo I-607
Mao, Han-de I-646
Mao, Ha-de I-621
Mao, Kaiyin I-184
Mao, Li II-341
Meng, Lingxiao II-380
Mensah, Martey Ezekiel I-586
Miao, Qin II-664
Miao, Yi I-38

Nawaz, Saqib Ali II-166, II-179
Ni, Qian I-719
Ni, Xueli I-695
Ning, Zhiyan II-215
Niu, Jiqiang II-604

Obed, Appiah I-586
Ou, Yikun I-362

Pan, Yupeng I-339
Pei, Qingqi II-617
Peng, Anjie II-98
Peng, Guo II-475
Peng, Han II-341
Peng, Jiansheng II-487
Peng, Jinbo I-434
Peng, Xiaoyu I-479
Peng, Xizi II-328
Peng, Yi II-593
Peng, Yufei II-581

Qi, Feng II-440
Qi, Guilin I-607
Qi, Hong I-327
Qi, Jin I-151
Qian, Cheng I-621
Qian, Junyan I-682
Qin, Fengming II-179
Qin, Qinzhi I-670

Qiu, Xiaojun I-362
Qiu, Xuesong II-786
Qu, Meixuan I-597
Qu, Zhaowei II-710
Qu, Zhiguo I-532

Rehman, Sadaqat Ur I-173
Ren, Jiayi I-207

Shao, Sujie II-776
She, Wei II-557, II-570, II-581
Shen, Gongxin II-267
Shen, Jian II-166
Shen, Jianjing II-74
Shen, Jing II-440
Shen, Xiajiong I-607
Shen, Yatian I-607
Sheng, Feng I-564
Sheng, Huixing II-748
Sheng, Zhiwei II-307
Shi, Bowen I-729
Shi, Jiaxing II-463
Shi, Jin II-30, II-42
Shi, Xiaoyu II-769
Shi, Yijuan II-236
Shi, Yunqing II-545
Shou, Zeng II-215
Si, Lei II-353
Song, Chenming II-143
Song, Tingting II-629
Song, Yanyan I-62, II-451
Song, Yu II-156
Su, Yihong I-95
Sun, Guang I-232, II-475
Sun, Xing I-707
Sun, Yanbin II-365

Tan, Jiangxia I-244
Tan, Jie I-510
Tan, Li I-62, II-451
Tan, Wenpan II-499
Tan, Yusong I-422
Tang, WenSheng I-786
Tang, Xiaoxiao II-440
Tang, Yibin II-738
Tang, Yuliang I-195
Tao, Lan I-14
Tian, Ying I-86
Tian, Zhao II-557, II-570, II-581
Tong, En I-775
Tong, Jie II-798
Tong, Luyan I-339
Tong, Mingyang I-670
Tu, Kai II-593
Tu, Shanshan I-173

Wan, Guogen II-18
Wan, Pin I-467
Wang, Bo II-3
Wang, Cheng II-405
Wang, Chunpeng II-545
Wang, Cui II-697
Wang, Dan II-215
Wang, Feng II-798
Wang, Hong I-362
Wang, Hua II-604
Wang, Jing I-207
Wang, Jinying I-38
Wang, Junxiang II-593
Wang, Kai I-327, I-339
Wang, Li II-225, II-463
Wang, Libing I-86
Wang, Mingda I-751
Wang, Qun I-695
Wang, Tianxin II-55
Wang, Wei II-760
Wang, Weipeng II-405
Wang, Xiaojuan II-805
Wang, Xiaoru II-710
Wang, Xiaoyu II-545
Wang, Yabo I-3
Wang, Yichuan II-318
Wang, Ying II-725
Wang, Yiting II-725
Wang, Yong I-14
Wang, Yonghua I-467
Wang, Yunfei I-597
Waqas, Muhammad I-173
Wei, Chenyang II-725
Wei, Fan I-120
Wei, Fenglin I-327, I-339
Weibo, Wei I-739
Wen, Peizhi I-670
Weng, Zhuohao II-769
Wu, Jian II-109
Wu, Jie I-444
Wu, Ji-Zhong II-298
Wu, Kai II-593
Wu, Li II-119, II-664, II-674

Wu, Qinbo I-422
Wu, Qing-Qing I-49
Wu, Shuixiu I-14
Wu, Xi I-142
Wu, Xiaobo II-380
Wu, Xiuting I-658
Wu, Yang I-810
Wulamu, Aziguli II-463

Xia, Lingling I-695
Xiang, Baoyu II-776
Xiang, Ke I-682
Xiang, Wang II-617
Xiao, He II-156
Xiao, Yanru I-133
Xiao, Yao II-510
Xie, Dingbang I-564, I-575
Xie, JingYun I-786
Xie, Ke II-776, II-786
Xie, Liangbo I-162, I-489
Xie, Qianyu II-131
Xin, Jianfang I-695
Xiong, Zijia I-532
Xu, Bin I-151
Xu, Chungen I-107
Xu, Gang I-219
Xu, Haijiang II-225
Xu, Haitao I-543, I-554
Xu, Hanwen I-244
Xu, Huibo II-570
Xu, Jianbo I-411
Xu, Jie I-695
Xu, Li II-365, II-570
Xu, Lu I-798
Xu, Mengke I-522
Xu, Rui II-534
Xu, Shuning I-3
Xu, Yinsong I-532
Xu, Yuanbo I-798
Xu, Zhipeng I-646
Xu, Zi-Xiao I-219
Xue, Chen I-184
Xue, Xianwei I-257

Yan, Bingjie II-191
Yan, Hong-zhe I-621
Yan, Lili II-289, II-298
Yang, Bo I-786
Yang, Canji II-617
Yang, Chen II-18
Yang, Donghan I-142
Yang, Funing I-798
Yang, Haixia I-635
Yang, Jun II-119, II-664, II-674
Yang, Kai II-710
Yang, Kehua II-428
Yang, Lijia II-463
Yang, Shanshan I-399
Yang, Tongfeng II-109
Yang, Xiaoyu II-557
Yang, Xusheng I-455
Yang, Xutao II-131
Yang, Yi-Xian I-219
Yao, Yuanzhi II-86
Ye, Ziyi II-203
Yi, Wenjia II-203
Yi, Zibo I-422
Yin, Mingxi I-173
Yin, Xinyue II-318
You, Fan I-479
Yu, Dabing II-738, II-748
Yu, Jie I-422
Yu, Jinliang I-173
Yu, Nenghai II-86
Yu, Wenbin I-532
Yu, Xiaoling I-107
Yu, Zhezhou I-798
Yu, Zhuo II-776, II-786
Yuan, Fuxiang II-534
Yuan, Huaqiang II-393
Yuan, Song I-434
Yuan, Xin I-434
Yuan, Zhongyun I-257
Yue, Baohai II-686

Zeng, Hui II-98
Zeng, Wu II-523
Zhai, Xuemeng I-751
Zhai, Zhongyi I-682
Zhang, Aobo I-554
Zhang, Baili I-479
Zhang, Fujun II-65
Zhang, Hanxiao II-487
Zhang, Jiale II-748
Zhang, Jian II-769
Zhang, Jianjun I-207, I-232
Zhang, Jianming I-305
Zhang, Jin II-769
Zhang, Jinquan II-328

Zhang, Jixian II-131
Zhang, Liang I-26, I-375
Zhang, Lianping II-805
Zhang, Qi II-215
Zhang, Qikun II-417
Zhang, Ran II-417
Zhang, Shi-Bin II-280, II-289
Zhang, Shibin II-18, II-246, II-258, II-307, II-328
Zhang, Shunchao I-467
Zhang, Tian I-729
Zhang, Tong II-341
Zhang, Wei I-142, II-428
Zhang, Weigong I-635
Zhang, Xiaopeng I-543
Zhang, Xiaoqing II-570
Zhang, Ximing I-151
Zhang, XinXin II-798
Zhang, Xuehan II-805
Zhang, Xuejie II-131
Zhang, Yachuan II-143
Zhang, Yan II-42
Zhang, Yan-yan I-49
Zhang, Yongwei I-467
Zhang, Yuzhu I-291
Zhang, Zhenyuan I-95
Zhang, Zhibin II-3
Zhang, Zhijun II-215
Zhang, Zuping I-74
Zhao, Changming I-184
Zhao, Li II-725
Zhao, Lingzhong I-682
Zhao, Ming I-384, I-399
Zhao, Xiaobing I-26, I-375
Zhao, Yanchao I-763
Zhao, Yang II-307
Zhao, Yingnan I-444
Zhao, Yunkai I-151
Zheng, Peijia II-629
Zheng, Quan II-405
Zheng, Sihao I-522
Zheng, Tao II-246, II-289
Zheng, Zhiyun II-697
Zhong, Hua I-786
Zhong, Maosheng I-14
Zhong, Tie I-729
Zhou, Hangjun I-207
Zhou, Jingjun II-166
Zhou, Lina II-451
Zhou, Mu I-489
Zhou, Naqin II-341
Zhou, Qingfeng I-635
Zhou, Ti I-38
Zhou, Wei II-225
Zhou, Xinxin II-686
Zhou, Yan II-738, II-748
Zhou, Yinian I-479
Zhu, Bing II-499
Zhu, He II-318
Zhu, Lei II-428
Zhu, Lina II-341
Zhu, Ruofei I-267
Zhu, Zhengzhou I-267
Zhuang, Jiawei I-467
Zhuang, Wei I-635
Zhuang, Yuan II-725
Zhuzhu, Gao I-739
Zong, Yuzhuo I-575
Zou, Wei II-74
Zou, Yayi II-203

Printed in the United States
By Bookmasters

Printed in the United States
By Bookmasters